Switchgear and Control Handbook

OTHER McGRAW-HILL HANDBOOKS OF INTEREST

Switchgear and Control Handbook

ROBERT W. SMEATON Editor

Editorial Consultant, "Allis-Chalmers Engineering Review"
Editor, "Motor Application and Maintenance Handbook"
Registered Professional Engineer, Wisconsin
Member, WSPE, IEEE, and ESM

McGRAW-HILL BOOK COMPANY

New York   St. Louis   San Francisco   Auckland   Bogotá   Düsseldorf
Johannesburg   London   Madrid   Mexico   Montreal
New Delhi   Panama   Paris   São Paulo   Singapore
Sydney   Tokyo   Toronto

Library of Congress Cataloging in Publication Data
Main entry under title:

Switchgear and control handbook.

Includes index.
1. Electrical switchgear—Handbooks, manuals, etc.
2. Electric controllers—Handbooks, manuals, etc.
3. Automatic control—Handbooks, manuals, etc.
I. Smeaton, Robert W.
TK2821.S88   621.31′7   76-17925
ISBN 0-07-058439-7

*The editors for this book were Harold B. Crawford, Ross J. Kepler,
and Betty Gatewood, the designer was Naomi Auerbach, and the
production supervisor was George E. Oechsner. It was set in Caledonia
by Quinn & Boden Company, Inc.*

It was printed and bound by The Kingsport Press.

Contents

Contributors

ANNIS, R. R., P.E. *Assistant Vice President, Engineering, Oster Corporation, Milwaukee, Wis.* (SECTION 29)

BELLINGER, T. F. *Consulting Engineer, Controls Division, Allis-Chalmers Corporation, Milwaukee, Wis.* (SECTION 22)

BLAKEY, R. C. *Manager, Repair Division (Retired); Westinghouse Electric Corporation, Phoenix, Ariz.* (SECTION 31)

BLOODWORTH, T. H. *Chief Systems Engineer, Process Electrical Systems (Retired), Allis-Chalmers Corporation, Milwaukee, Wis.* (SECTION 24 — PART 2)

BRANDT, T. F. *Engineering Manager, Switchgear Division, Distribution and Controls Group, I-T-E Imperial Corporation, Spring House, Pa.* (SECTION 19 — PART 1)

CATALDO, J. B. *Vice President, Engineering, Distribution and Controls Group, I-T-E Imperial Corporation, Spring House, Pa.* (SECTION 19 — PART 2)

CHAMPNEY, G. W. *Senior Design Engineer, Low Voltage Switchgear (Retired), Westinghouse Electric Corporation, Pittsburgh, Pa.* (SECTION 18)

DOWNEY, R. A. *Supervisor, Application Engineering, Square D Company, Milwaukee, Wis.* (SECTION 3)

ECKENSTALER, G. F. *Chief Engineer, Dc Drives, Electronics Operation, Allis-Chalmers Corporation, New Orleans, La.* (SECTION 6 — PART 2)

EVERETT, M. *Senior Engineer (Retired), Switchgear Division, Allis-Chalmers Corporation, Milwaukee, Wis.* (SECTION 17)

EVERSON, H. K. *Senior Application Engineer (Retired), Motor and Generator Division, Allis-Chalmers Corporation, Norwood, Ohio* (SECTION 7)

EWY, A. *Manager of Projects and International Services (Retired), Transformer Division, Allis-Chalmers Corporation, Milwaukee, Wis.* (SECTION 1)

FUNKE, L. A. *Instrument Product Planner, Relay — Instrument Division, Westinghouse Electric Corporation, Newark, N.J.* (SECTION 14)

GERG, R. A. *Electrical Project Engineer, Reduction Systems Division, Allis-Chalmers Corporation, Milwaukee, Wis.* (SECTION 23)

HALTER, A. C., P.E. *Supervisor, Technical Services, Eastern European Operations, Processing Group, (Formerly Senior Staff Engineer, Electrical Systems Projects) Allis-Chalmers Corporation, Milwaukee, Wis.* (SECTION 28)

HEDIN, R. H. *Manager, Engineering Analysis and Simulation, Power Systems Technology, Allis-Chalmers Corporation, Milwaukee, Wis.* (SECTION 27)

IBACH, W. R., P.E. *W. R. Ibach and Associates, Electrical Engineers (Formerly Chief Substation Design Administrator, Wisconsin Electric Power Company), Milwaukee, Wis.* (SECTIONS 4 and 5)

JOUNG, K. S., Ph.D, P.E. *Associate Director, Planning and Control (Formerly Manager, Engineering Analysis and Computer Applications, Mechanical Systems Technology), Advanced Technology Center, Allis-Chalmers Corporation, Milwaukee, Wis.* (SECTION 9)

KAVANAUGH, T. H. *Senior Project Engineer, Square D Company, Milwaukee, Wis.* (SECTION 2)

KUSSY, F. W., D.Eng., P.E. *Director of Engineering, Control and Instrumentation Division, I-T-E Imperial Corporation, Westminster, Md.* (SECTION 13)

LEET, C. H., P.E. *Market Manager, Power Systems, Exide Lightguard Division, ESB Incorporated, Philadelphia, Pa.* (SECTION 6 – PART 3)

LEFFERTS, W. G. *Senior Engineer (Retired), Allis-Chalmers Corporation, Milwaukee, Wis.* (SECTION 10)

McDOWELL, S. E., P.E. *Senior Staff Engineer, Switchgear Division, Allis-Chalmers Corporation, Milwaukee, Wis.* (SECTION 20)

MOKRYTSKI, B., P.E. *Manager, Ac Drives Engineering, Robicon Corporation, Pittsburgh, Pa.* (SECTION 25)

MOORE, R. C., P.E. *Adjunct Professor, Milwaukee School of Engineering, Milwaukee, Wis.* (SECTION 11)

MYLES, A. H. *Manager, Product Planning Systems (Retired), Square D Company, Cleveland, Ohio* (SECTION 24 – PART 1)

NICKELS, L. E., P.E. *L. E. Nickels and Associates – Converter Engineering (Formerly Chief Development Engineer of Semiconductor Products, Allis-Chalmers Corporation), New Berlin, Wis.* (SECTION 26)

PAAPE, K. L., P.E. *Director, Industrial Control Development, Allen-Bradley Company, Milwaukee, Wis.* (SECTIONS 8 and 16)

PERKINS, W. B. *Manager, Commercial and Industrial Product Marketing Section, Distribution Equipment Division, Square D Company, Lexington, Ky.* (SECTION 21)

SCHOOF, R. F., P.E. *Corporate Director, Safety and Security, Jos. Schlitz Brewing Company, Milwaukee, Wis.* (SECTION 30)

SELLERS, J. F. *Consulting Engineer, Dc Machines (Formerly Chief Engineer, Dc Machines), Allis-Chalmers Corporation, Milwaukee, Wis.* (SECTION 6 – PART 1)

STOUPPE, R. L., Jr., P.E. *Senior Engineer, Systems Relaying, Detroit Edison Company, Detroit, Mich.* (SECTION 12)

THOMAS, R. C., P.E. *President, Acutran Instrument and Specialty Transformers, Inc., Donora, Pa.* (SECTION 15)

UBERT, W. H. *Staff Engineer, Electrical, Miller Brewing Company, Milwaukee, Wis.* (SECTION 2)

Preface

This Handbook was written to provide practical engineering data for assisting engineers in selecting and specifying switchgear and control. It also provides installation and maintenance information for plant and maintenance engineers and for technicians and electricians who install and care for this equipment.

Through this Handbook, eminently qualified engineers give the reader the benefit of their many years of practical experience. The thirty-four contributors have become widely known through their published books, technical articles, engineering papers, and through committee activities in their engineering societies.

Sections 1, 2, and 3 cover fundamental information on switchgear, its classifications and arrangements, control standards, general purpose motor starters, and control circuit devices used with both switchgear and control.

System considerations affecting switchgear and control selection are given in Sections 4 through 7.

Sections 8 and 9 cover installation and service requirements, including earthquake considerations, that must be understood when specifying switchgear and control.

An important background on load and motor characteristics affecting control selection is given in Sections 10 and 11.

Sections 12 through 15 provide an excellent guide to the coordination of system protection and instrumentation.

Plant engineers and engineering consultants are given information in Sections 16 through 21 for planning and specifying distribution and utilization systems which may include switchgear, control, substation, bus duct, distribution panelboards and switchboards.

Process design and application engineers are provided a wide variety of control system information in Sections 22 through 29, enabling them to specify controls from standardized polyphase motor control and control centers to the very complex automated control systems.

Section 30 provides plant engineers and management with comprehensive safety planning information.

Maintenance engineers and technicians are given valuable information on the care of power distribution equipment in Section 31.

SAFETY

Since switchgear and control govern and protect electrical systems and equipment operating on voltages which can be dangerous to personnel, safety is a prime consideration in their application and maintenance. A thorough understanding of safe procedures in installation and servicing of electrical equipment by those who design, specify, apply, and work with this equipment will greatly reduce accidents. Some facet of safe practice is covered in almost every section of this book. Section 30, which is concerned with plant safety in general, should be carefully studied by all who work with electrical equipment, their supervision and management.

STANDARDS

Because modern electric machines are precisely designed to exact ratings, their control and protection require careful adherence to up-to-date standards of the electrical industry.

The IEEE, ANSI, and NEMA standards, the National Electrical Codes, and Underwriters Laboratory Codes are continually improved and updated by the addition of new requirements or by changes in existing requirements. It is, therefore, recommended that those specifying, applying, and maintaining electrical equipment avail themselves of the latest applicable standards and codes and note the changes as they are made. This Handbook will help the reader to understand and apply these standards and codes when using or specifying switchgear and control.

Acknowledgments

The editor wishes to express his most sincere appreciation to the many engineers who gave freely of their advice and encouragement during the planning and organization of this book, to Margaret H. Barbo, who typed the manuscript and prepared the index, and to David L. Hendrickson, who prepared the illustrations and graphs for reproduction.

Robert W. Smeaton

Switchgear and Control Handbook

1

Functions of
Industrial Switchgear

A. EWY *

FOREWORD

Industrial switchgear consists of assemblies of switchgear devices such as circuit breakers, switches, protective equipment, metering, instrumentation, control and associated equipment for power generation, transmission, and distribution. This section outlines the basic application information and terminology involving this equipment.

PURPOSE AND TYPES

Switchgear provides protection for such equipment as generators, transformers, motors, and transmission lines. It also performs certain switching functions for the proper operation of generating and industrial plants.

Switchgear equipment comes in various forms and ratings depending on the particular functions it is to perform (Table 1). The components used in this equipment are

---

* Manager of Projects and International Services, Transformer Division; Allis-Chalmers Corporation (Retired); Registered Professional Engineer (Wis.); Member, IEEE.

TABLE 1. Switchgear Assemblies

Metal-enclosed power switchgear	Metal-enclosed bus	Switchboards	
		Power	Control
Metal-clad	Nonsegregated	Enclosed	Enclosed
Low-voltage power circuit breaker	Segregated phase	Dead front	Dual
Metal-enclosed interrupter	Isolated phase	Live front	Duplex
Station-type cubicle			Control desk
			Benchboard
			Dual benchboard
			Duplex benchboard
			Vertical panel

covered in other sections of this handbook. More details on switchgear selection are found in Sec. 17.

1. Metal-clad switchgear is metal-enclosed and has the following features:

1. The circuit breaker is removable and can be moved into the operating position, test position, and disconnect position.

2. The primary and secondary contacts are self-aligning and self-coupling.

3. The primary circuit such as the circuit breaker, buses, and potential transformers are enclosed in metal grounded compartments.

4. Insulating materials cover the primary bus conductors and connections (Table 2).

TABLE 2. Rated Voltages and Insulation Levels, Ac Switchgear Assemblies

Rated voltages, rms		Insulation levels, kV	
Rated nominal voltage	Rated max voltage	60-Hz withstand 1 min, rms *	Impulse withstand
Metal-enclosed Low-voltage Power-Circuit-Breaker Switchgear			
240 volts	250 volts	2.2	
480	500	2.2	
600	630	2.2	
Metal-clad Switchgear			
4.16 kV	4.76 kV	19	60
7.2	8.25	36	95
13.8	15.0	36	95
34.5	38.0	80	150
Interrupter Switchgear			
4.16 kV	4.76 kV	19	60
7.2	8.25	26	75
13.8	15.0	36	95
14.4	15.5	50	110
23.0	25.8	60	125
34.5	38.0	80	150
Station-Type Cubicle Switchgear			
14.4 kV	15.5 kV	50	110
34.5	38.0	80	150
69.0	72.5	160	350

* When switchgear is installed in the field and a 60-Hz, 1-min withstand test is applied, 75% of these values should be used.

5. When the circuit breaker is removed, automatic shutters close off and prevent exposure of the primary conductors.

6. Interlocks are provided for safe removal and insertion of the circuit breaker into the structure.

7. The door or panel through which the circuit breaker enters the structure may be used to mount instruments and panels. The instruments, relays, and their wiring are isolated from high voltage by grounded metal barriers.

Some typical circuits and arrangements of metal-clad switchgear are given in Figs. 1 through 11. Device numbers used in the figures are given in Table 3. Metal-clad-switchgear ratings are given in Table 4.

2. Metal-enclosed low-voltage switchgear is an assembly of equipment rated 600 V ac. It is completely enclosed on all sides and top with sheet metal (Table 5). The assembly contains primary power circuit breakers, switching or interrupting devices,

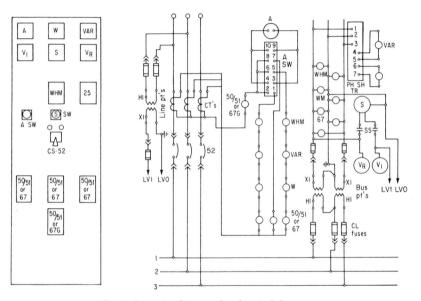

Fig. 1   Incoming-line panel and typical diagram.

with buses and connections. It may contain control, measuring, or protective devices (Fig. 12).

Metal-enclosed low-voltage switchgear includes the following equipment and features:

1. Low-voltage power circuit breakers that are mounted stationary or removable and contained in individual grounded metal compartments.

2. Bus bars and connections.

3. Instrument and control transformers.

4. Instruments, meters, and protective relays.

5. Control wiring and accessory devices.

6. Circuit breakers may be controlled at the switchgear or from a remote point.

7. When the circuit breakers are removable, interlocks are provided so circuit breakers may be removed or inserted safely.

3. Metal-enclosed interrupter switchgear is metal-enclosed power switchgear which may include the following equipment:

1. Interrupter switches

2. Power fuses

3. Bare buses and connections

4. Instrument transformers

5. Control wiring and accessory devices

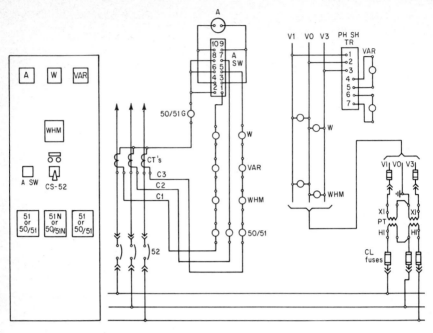

Fig. 2   Feeder panel and typical diagram.

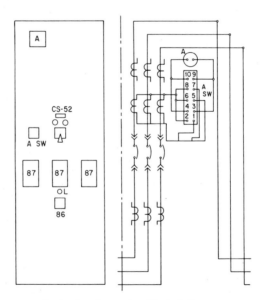

Fig. 3   Bus-tie panel and typical diagram.

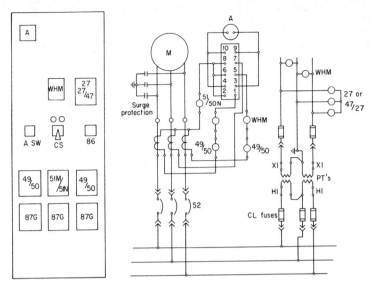

Fig. 4   Induction-motor control panel for 3-phase full-voltage start with typical diagram.

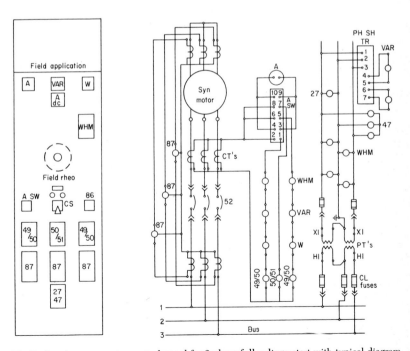

Fig. 5   Synchronous-motor control panel for 3-phase full-voltage start with typical diagram.

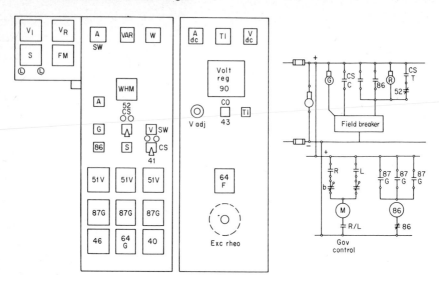

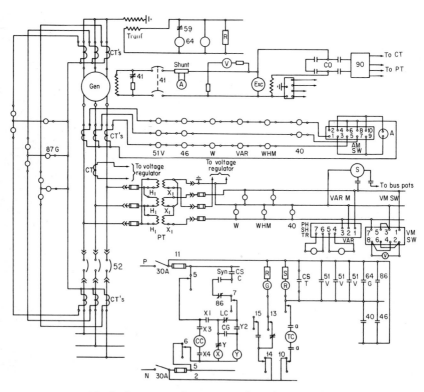

Fig. 6 Generator and exciter control panel with diagram.

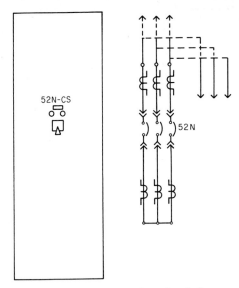

Fig. 7 Generator neutral control panel with diagram.

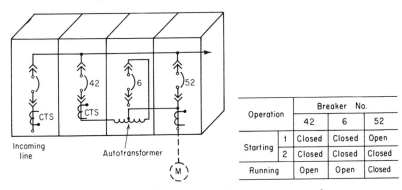

Operation		Breaker No.		
		42	6	52
Starting	1	Closed	Closed	Open
	2	Closed	Closed	Closed
Running		Open	Open	Closed

Fig. 8 Closed-circuit motor-starting scheme using autotransformer.

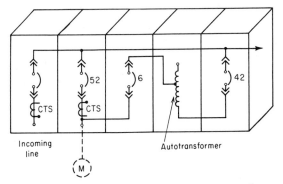

Operation		Breaker No.		
		52	6	42
Starting	1	Open	Closed	Closed
	2	Open	Open	Open
Running		Closed	Open	Open

Fig. 9 Open-circuit motor-starting scheme using autotransformer.

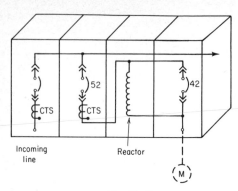

Operation	Breaker No.	
	52	42
Starting	Closed	Open
Running	Closed	Closed

Fig. 10   Reactor-type motor-starting scheme.

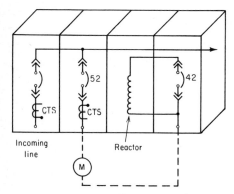

Operation	Breaker No.	
	52	42
Starting	Closed	Open
Running	Closed	Open

Fig. 11   Neutral-reactor-starting scheme.

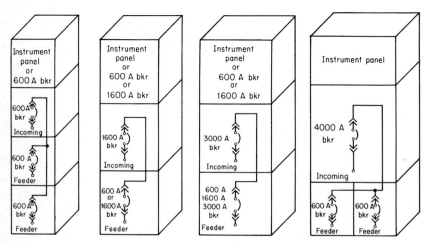

Fig. 12   Metal-enclosed low-voltage switchgear arrangements.

TABLE 3. Standard Device-Function Numbers (ANSIC37.2 – 1962)

1. Master element
2. Time-delay starting or closing relay
3. Checking or interlocking relay
4. Master contactor
5. Stopping device
6. Starting circuit breaker
7. Anode circuit breaker
8. Control power-disconnecting device
9. Reversing device
10. Unit sequence switch
11. Reserved for future application
12. Overspeed device
13. Synchronous-speed device
14. Underspeed device
15. Speed- or frequency-matching device
16. Reserved for future application
17. Shunting or discharge switch
18. Accelerating or decelerating device
19. Starting-to-running transition contactor
20. Electrically operated valve
21. Distance relay
22. Equalizer circuit breaker
23. Temperature-control device
24. Reserved for future application
25. Synchronizing or synchronism-check device
26. Apparatus thermal device
27. Undervoltage relay
28. Flame detector
29. Isolating contactor
30. Annunciator relay
31. Separate excitation device
32. Directional power relay
33. Position switch
34. Master-sequence device
35. Brush-operating or slip-ring short-circuiting device
36. Polarity device
37. Undercurrent or underpower relay
38. Bearing protective device
39. Mechanical-condition monitor
40. Field relay
41. Field circuit breaker
42. Running circuit breaker
43. Manual transfer or selector device
44. Unit-sequence starting relay
45. Atmospheric-condition monitor
46. Reverse-phase or phase-balance current relay
47. Phase-sequence voltage relay
48. Incomplete-sequence relay
49. Machine or transformer thermal relay

50. Instantaneous overcurrent or rate-of-rise relay
51. Ac time overcurrent relay
52. Ac circuit breaker
53. Exciter or dc generator relay
54. Reserved for future application
55. Power-factor relay
56. Field-application relay
57. Short-circuiting or grounding device
58. Rectification-failure relay
59. Overvoltage relay
60. Voltage- or current-balance relay
61. Reserved for future application
62. Time-delay stopping or opening relay
63. Liquid- or gas-pressure or vacuum relay
64. Ground-protective relay
65. Governor
66. Notching or logging device
67. Ac directional overcurrent relay
68. Blocking relay
69. Permissive control device
70. Electrically operated rheostat
71. Liquid- or gas-level relay
72. Dc circuit breaker
73. Load-resistor contactor
74. Alarm relay
75. Position-changing mechanism
76. Dc overcurrent relay
77. Pulse transmitter
78. Phase-angle measuring or out-of-step protective relay
79. Ac reclosing relay
80. Liquid- or gas-flow relay
81. Frequency relay
82. Dc reclosing relay
83. Automatic selective control or transfer relay
84. Operating mechanism
85. Carrier or pilot-wire receiver relay
86. Locking-out relay
87. Differential protective relay
88. Auxiliary motor or motor generator
89. Line switch
90. Regulating device
91. Voltage directional relay
92. Voltage and power directional relay
93. Field-changing contactor
94. Tripping or trip-free relay
95 to 99. Used only on specific applications on individual installations where none of the assigned numbered functions from 1 to 94 are suitable

When the type of device is incidental to the function, a relay, contactor, circuit breaker, switch, or device may be substituted as applicable.

A similar series of numbers, starting with 201 instead of 1, will be used for those device-function numbers in equipment controlled directly from a supervisory system.

Suffix letters X, Y, or Z are added to the appropriate device-function numbers to denote separate auxiliary relays.   Example: 27X, 52Y, 79Z.

If two or more devices with the same function number and suffix letter (if used) are present in the same equipment, they may be distinguished by numbered suffixes.   Example: 27-1, 27-2, 27X-1, 27X-2.

TABLE 4. Metal-clad-Switchgear Preferred Ratings

Nominal voltage	4.16 kV	4.16 kV	4.16 kV	7.2 kV	13.8 kV	13.8 kV	13.8 kV
3-phase interrupting capacity, MVA........	75	250	350	500	500	750	1000
Continuous-current rating, A..................	1200	1200–2000	1200–2000	1200–2000	1200–2000	1200–2000	1200–3000
Max ampere interrupting capability, kA...	12	36	49	41	23	36	48
Closing and latching capability, kA	19	58	78	66	37	58	77
Momentary current capability, kA	20	60	80	70	40	60	80

The interrupter switches and power fuses may be removable or stationary. When they are the removable type, mechanical interlocks are provided for safe operation.

4. Station-type cubicle switchgear is metal-enclosed power switchgear and consists of the following:

1. Stationary power circuit breakers.
2. The primary equipment for each phase segregated and enclosed by metal.
3. Group-operated disconnect switches are provided to isolate the circuit breaker. For safe operation, the disconnect switches are interlocked with the circuit breakers.
4. Bare buses and connections.
5. Instrument transformers.
6. Control wiring and accessory devices.

5. Metal-enclosed bus consists of an assembly of primary conductors with associated connections, joints, and insulated supports within a grounded metal enclosure (Sec. 19).

The following are types of metal-enclosed bus:

1. Nonsegregated-phase bus has all phases of a primary circuit conductor in a common metal enclosure without barriers between phases.
2. Segregated-phase bus has all phases of a primary circuit in a common metal enclosure, but each phase is segregated by metal barriers.
3. Isolated-phase bus has each phase of primary conductor enclosed by an individual metal enclosure and separated from adjacent conductors by an air space.

6. Switchboards and benchboards are assemblies which consist of one or more panels with a framework on which electrical devices are mounted. They are considered a type of switchgear assembly (Fig. 13). Switchboards and benchboards may be described as follows:

1. Power switchboard includes primary power circuit breakers and associated electrical devices.
2. Live-front switchboard has exposed live parts on the front of the panel.
3. Dead-front switchboard has no exposed live parts on the front of the panel.

TABLE 5. Metal-enclosed Low-Voltage Switchgear Preferred Ratings

Rated voltage, V	Rated max voltage, V	Short-circuit rating, symmetrical, A	Frame size, A	Continuous-current and trip rating, A
600	630	14 000	225	15–225
600	630	22 000	600	40–600
600	630	42 000	1600	200–1600
600	630	65 000	3000	2000–3000
600	630	85 000	4000	4000
480	500	22 000	225	20–225
480	500	30 000	600	100–600
480	500	50 000	1600	400–1600
480	500	65 000	3000	2000–3000
480	500	85 000	4000	4000
240	250	25 000	225	30–225
240	250	42 000	600	150–600
240	250	65 000	1600	600–1600
240	250	85 000	3000	2000–3000
240	250	130 000	4000	4000

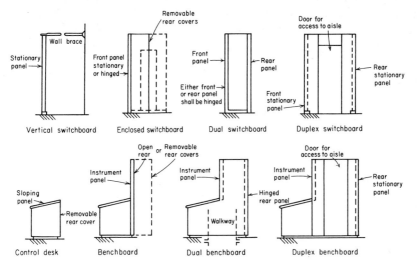

Fig. 13 Various switchboard and benchboard configurations.

4. Control switchboard is an assembly that includes instrumentation, metering, protective relays, and control for controlling remote equipment.

5. Vertical switchboards consist of only vertical panels. No rear enclosure is included.

6. Dual switchboard has front and rear panels on which electrical devices are mounted. Both ends and the top are enclosed. On at least one side the panels are hinged to provide access to panel wiring.

7. Duplex switchboards consist of panels, front and rear, forming a common aisle which is enclosed at the ends and top. Entry doors are provided at each end for access to the common aisle between the panels.

8. Benchboard is a combination of a control desk and a vertical or enclosed switchboard in a common assembly.

9. Dual benchboard is a combination assembly of benchboards and a vertical hinged panel switchboard placed back to back and enclosed at the top and ends.

10. Duplex benchboard is a combination of a benchboard and a vertical control switchboard placed back to back to form a common aisle. The aisle is enclosed at the ends and top. Entry doors are provided to the aisle between the benchboard and the vertical control switchboard.

7. A circuit breaker is a device for making and interrupting an electrical circuit. A high-voltage power circuit breaker is an ac circuit breaker which has a rating of 1500 V or greater. These circuit breakers are panel- or frame-mounted (Table 6).

8. Circuit-Breaker Ratings.   When a power circuit breaker is applied to an electrical power system, the following ratings should be considered:

1. Ac circuit breakers are rated for 60 Hz. Application of a circuit breaker on a higher frequency will reduce the continuous rating ability. The interrupting capability will also be affected.

2. Circuit breakers are rated at normal rated voltage and maximum operating voltage. This maximum operating voltage should not be exceeded by the power system to which the circuit breaker is applied.

3. The circuit-breaker rated current is the continuous current it can carry without exceeding the standard temperature rise. There are no short-time continuous rating values on circuit breakers because of the short time it requires to reach the maximum temperature rise.

4. The temperature of the air surrounding a circuit breaker shall be between 10 and 40°C, and the circuit breaker shall be located so the air is free to circulate around it to permit adequate cooling.

TABLE 6. Preferred Ratings for Outdoor Oil Circuit Breakers

						Current values		
Nominal voltage class, kV, rms	Nominal 3-phase MVA class	Rated max voltage, kV, rms	Low frequency, kV, rms	Impulse, kV, crest	Rated continuous current at 60 Hz, A, rms	Max symmetrical interrupting capacity, kA, rms	Rated short-circuit current (at rated max kV), kA, rms	Closing and latching capability, kA, rms
14.4	250	15.5	50	110	600	24	8.9	38
14.4	500	15.5	50	110	1200	23	18	37
23	500	25.8	60	150	1200	24	11	38
34.5	1500	38	80	200	1200	36	22	58
46	1500	48.3	105	250	1200	21	17	33
69	2500	72.5	160	350	1200	23	19	37
115	*	121	260	550	1200	20	20	32
115	*	121	260	550	1600	40	40	64
115	*	121	260	550	2000	40	40	64
115	*	121	260	550	2000	63	63	101
115	*	121	260	550	3000	40	40	64
115	*	121	260	550	3000	63	63	101
138	*	145	310	650	1200	20	20	32
138	*	145	310	650	1600	40	40	64
138	*	145	310	650	2000	40	40	64
138	*	145	310	650	2000	63	63	101
138	*	145	310	650	2000	80	80	128
138	*	145	310	650	3000	40	40	64
138	*	145	310	650	3000	63	63	101
138	*	145	310	650	3000	80	80	128
161	*	169	365	750	1200	16	16	26
161	*	169	365	750	1600	31.5	31.5	50
161	*	169	365	750	2000	40	40	64
161	*	169	365	750	2000	50	50	80
230	*	242	425	900	1600	31.5	31.5	50
230	*	242	425	900	2000	31.5	31.5	50
230	*	242	425	900	3000	31.5	31.5	50
230	*	242	425	900	2000	40	40	64
230	*	242	425	900	3000	40	40	64
230	*	242	425	900	3000	63	63	101
345	*	362	555	1300	2000	40	40	64
345	*	362	555	1300	3000	40	40	64

* Not applicable.

5. Short-circuit rating is the 3-phase MVA rating of a circuit breaker and should not be exceeded by the power system under fault conditions.

6. Maximum short-circuit interrupting current rating is the maximum current the circuit breaker can interrupt at reduced voltage.

7. Short-time current rating of a circuit breaker is the maximum rms current including the dc components it can carry under fault conditions in the power system.

8. Momentary current rating of a circuit breaker is the maximum rms current that it can withstand under fault conditions in the power system. This value includes the dc component.

9. Rated interrupting rating of a circuit breaker is the highest MVA it must interrupt at specified operating voltage under the operating duty specified.

10. Reclosing capability of a circuit breaker applied on a system for reclosing duty requires that it be derated in accordance with the standards which apply.

Control Standards and General-Purpose Starters

W. H. UBERT

T. H. KAVANAUGH

FOREWORD

A controller is defined as "a device, or group of devices, which serve to govern in some pre-determined manner, the electric power delivered to the apparatus to which it is connected" (National Electrical Code). This definition, which was originally issued by the National Electrical Manufacturers Association in 1916, takes into account any switch, relay contactor, or overload-disconnecting means, or any combination of these devices used to control the power to a system or to a single machine. This section covers in a very general sense the functions that these controls perform, the methods used to accomplish this control, the components that make up a controller, and the proper application of these devices. This section was written to aid in applying the information given in greater detail in other sections of this book.

Part 1

Standards, Components, and Application

W. H. UBERT *

STANDARDS FOR CONTROLS

1. **The National Electrical Manufacturers Association.** Most of the standards for industrial controls in existence today are the result of the activities of the National Electrical Manufacturers Association (NEMA). This organization, and its predecessors, have been engaged in a drive to standardize the electrical industry since the turn of the century. Their efforts have led to the development of a group of publications that govern standard requirements of control devices in use throughout industry. These publications, known as NEMA Standards, also provide practical information concerning construction, test, performance, and the manufacture of industrial control equipment NEMA Standards are constantly being updated as requirements and capabilities change. The NEMA Standard ratings include ratings for volts, amperes, frequency horsepower, watts, etc. They also spell out standard ambient conditions for temperature and altitude.

2. **The National Electrical Code.** Another influential organization that concerns itself with standards for industrial controls is the National Fire Protection Association (NFPA). This organization sponsors the National Electrical Code (NEC), under the auspices of the American National Standards Institute (ANSI). The NEC was originally developed in 1897 as a result of the joint efforts of insurance, electrical, architectural, and allied interests. The NEC deals with all types of electrical apparatus from large-scale industrial motors to residential wiring. It serves as a guideline from which all other standards, codes, and specifications are written. It has been sponsored by the NFPA since 1911. A revised, up-to-date version of this code is published every 3 years.

3. **Other Standardizing Organizations.** Besides NEMA Standards and the NEC, other organizations that are involved in standardization include:

   a. *Institute of Electrical and Electronic Engineers (IEEE).* This organization is composed of the former American Institute of Electrical Engineers (AIEE) and the Institute of Radio Engineers (IRE). It publishes standards for such factors as temperature rise, rating methods, classification of insulating materials, and test codes.

   b. *Underwriters' Laboratories, Inc. (UL).* This independent organization concerns itself with testing devices, systems, and material, with particular reference to their effect on the safety of personnel coming in contact with these apparatuses. The Underwriters' Laboratories also develop standards for mass-produced controls of all types for hazardous locations through cooperation with manufacturers. The UL seal of approval is one of the more familiar symbols in the area of electric devices appearing on the

---

* Staff Engineer, Electrical, Miller Brewing Company.

market. Because of the extensive testing necessary, it is not practical to supply custom-built control devices with UL approval.

c. *American National Standards Institute (ANSI)*. This organization represents manufacturers, distributors, consumers, and other groups. They are responsible for sponsoring the ANSI Standards, which cover such things as dimensions, specifications of materials, test methods, and definition of terms. The NEC is an example of an ANSI standard.

d. *Occupational Safety and Health Act (OSHA)*. This law, which was enacted by Congress in 1970, establishes agencies at the federal, and eventually state, level of government to enforce the standards which they adopt. OSHA administration is primarily political and is under the direction of the Secretary of Labor.

Besides these agencies, many other organizations are involved in the field of standardization. Different industries have their own committees charged with the responsibility of protecting the interests of their members. Since industrial controls are in use in literally every industry throughout the country, it is obvious that the problems of the industrial-control designer regarding standardization are complex. Specific industry standards as well as local zoning and code regulations must therefore be consulted whenever a custom control is being considered. Overseas applications will generally come under the jurisdiction of the International Electro-Technical Commission (IEC) as well as the national agency within the country involved.

4. Standard Ambient Conditions. Controls are designed to regulate the amount of power delivered at nameplate ratings under standard ambient conditions. Standard ambient conditions are 40°C (104°F) at an altitude up to 6000 ft. When the ambient temperature in the vicinity in which the control device is to be located exceeds this value, it is essential to derate the device. The manufacturer normally has this derating information available to users who request it. When controls must operate at altitudes greater than 6000 ft, a special design is often necessary to accommodate this condition. The controller might also be located in a hostile atmosphere, in which high humidity, water sprays, flammable or explosive particles, or gases are present (Sec. 8). To protect the device, and personnel in the area of such a location, a special enclosure is required (Sec. 16).

5. Standard Voltage and Frequency Ratings. Standard voltages and frequencies, in use in the electrical industry throughout the world, vary from country to country. There are two distinct types of voltage ratings: the utilization rating and the nominal system rating. Utilization voltages are the voltages present at the point of control, whereas nominal system voltages refer to the voltage level at the supply terminals. Standard utilization voltages for 3-phase 60-Hz systems are 115, 200, 230, 460, and 575 V. The corresponding nominal system-voltage ratings are 120, 208, 240, 480, and 600 V. Single-phase 60-Hz ac systems have utilization-voltage ratings of 115 and 230 V and nominal system ratings of 120 and 240 V.

Most ac controls are rated for both 50- and 60-Hz operation. This dual rating is useful, since many European countries that are large purchasers of industrial-control apparatus use 50-Hz power. Coils, transformers, etc., that have only 60-Hz ratings must be voltage-derated when they are used at 50 Hz. Often components are marked with voltage ratings for each frequency.

6. Standard Electrical Symbols. There are two basic types of diagrams in wide use in the electrical industry—the elementary or schematic diagram and the connection or wiring diagram.

The schematic diagram describes the electrical functioning of the controllers. Various components are laid out so as to facilitate visualization of the sequence of electrical events which take place during operation of the control. This type of diagram ignores the physical aspects of the controller with regard to location of components, wires, or cables.

The connection or wiring diagram illustrates the physical layout of the components as well as the electrical connections and wiring which make up a controller. The electric circuit can be understood and traced with this type of diagram for some of the very basic controllers. However, a schematic diagram should always be supplied in conjunction with the connection diagram for controllers which involve any degree of complexity.

Figure 1 shows some of the more common symbols used in diagrams for controllers.

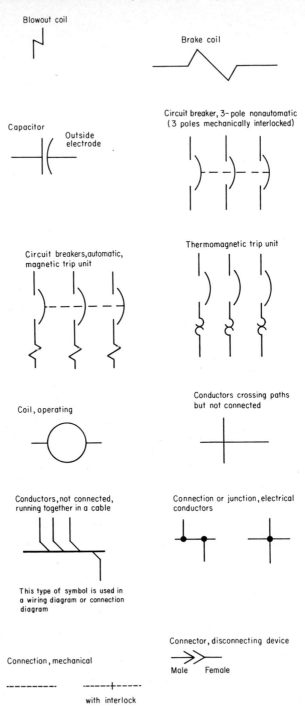

Fig. 1 Schematic diagram symbols for electrical components.

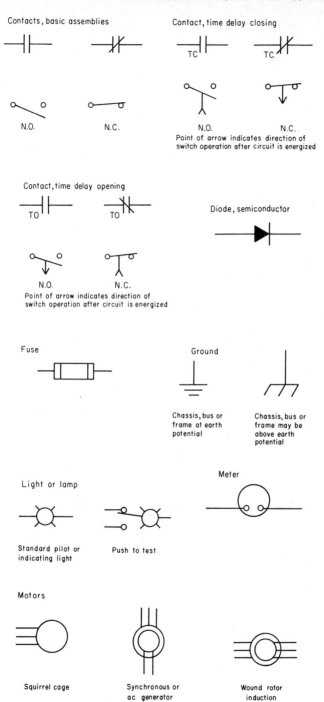

Contacts, basic assemblies

N.O.   N.C.

Contact, time delay closing

TC   TC

N.O.   N.C.
Point of arrow indicates direction of
switch operation after circuit is energized

Contact, time delay opening

TO   TO

Diode, semiconductor

N.O.   N.C.
Point of arrow indicates direction of
switch operation after circuit is energized

Fuse

Ground

Chassis, bus or
frame at earth
potential

Chassis, bus or
frame may be
above earth
potential

Light or lamp

Meter

Standard pilot or
indicating light

Push to test

Motors

Squirrel cage

Synchronous or
ac generator

Wound rotor
induction

Fig. 1 Schematic diagram symbols for electrical components (*cont.*)

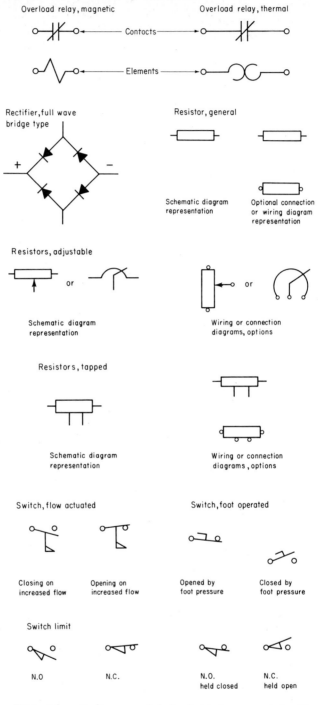

Fig. 1 Schematic diagram symbols for electrical components (*cont.*)

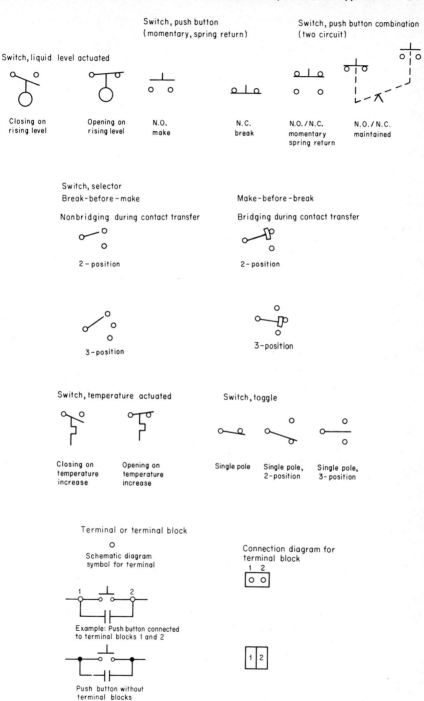

Fig. 1 Schematic diagram symbols for electrical components (*cont.*)

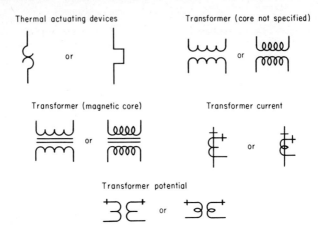

Thermal actuating devices

Transformer (core not specified)

or

or

Transformer (magnetic core)

Transformer current

or

or

Transformer potential

or

Fig. 1 Schematic diagram symbols for electrical components (*cont.*)

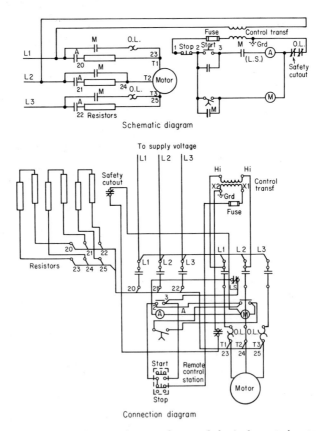

Schematic diagram

Connection diagram

Fig. 2 Schematic diagram (above) and wiring diagram (below) of a typical resistance-type reduced voltage motor starter.

Unless otherwise stated, the symbols show the schematic representation only. For the most part, the only difference between the symbol shown in schematic diagrams and those shown in the connection or wiring diagrams is that the latter always show electrical-connection terminals for each individual component.

Figure 2 shows an example of a schematic and wiring diagram combined for a primary resistance reduced-voltage starter. This type of combined diagram should accompany a control device of any degree of complexity to aid the technician who may have to troubleshoot the circuit. It can be noted that the schematic portion of the diagram depicts how the circuit functions with regard to the electrical sequence of events, while the wiring diagram shows where the various components are physically located within the starter.

COMPONENTS OF MOTOR CONTROLLERS

A basic combination controller contains three components. One is a disconnecting device such as a circuit breaker or fusible disconnect switch to open the circuit between the incoming power lines and the controller, in the event of a high-current fault. It also contains a contactor to open and close the circuit between the energy source and the motor during normal operation, and an overload-protective device to prevent long-term overcurrents from reducing the operating life of the motor.

Each of these components will be covered in depth in other sections of this handbook. The following paragraphs, however, will briefly describe, and list, some of the ratings for several types of these basic components.

7. Circuit Breakers. A circuit breaker is a device capable of opening and closing an electric circuit, either by nonautomatic means or automatically. It opens a circuit at some predetermined value of current without injury to itself so that it can be manually closed and again function properly. This component serves the same function as a line fuse with the exception that it is reusable after it opens the circuit. Circuit breakers are located wherever branch-circuit protection is required in the electrical layout of the plant. Where motors are involved, a single breaker is sometimes employed to protect one or more machines under certain conditions (Art. 430–53 of the NEC). While some consider this arrangement poor practice, it is often employed where low-horsepower motors are installed (Sec. 13).

The primary consideration in selecting a circuit breaker is that it can be set to trip at the proper value of fault current, and that it possesses a high enough interrupting rating to perform its function without being damaged. The NEC lists the maximum ratings or settings for ac-motor branch protection as being 700% of full-load current for magnetic circuit breakers, 250% for dc motors up to 50 hp, and 150% for dc motors above 50 hp. The code, however, follows this rule with an exception, which allows up to 1300% where it is necessary to start the motor. For time-limit breakers, all motors should be protected by circuit breakers that are rated between 200 and 250% of full-load motor current with the exception of Code Letter A motors and wound-rotor and dc motors, which should be supplied with breakers rated at a maximum of 150% of the full-load motor current.

Some varying viewpoints exist on the function that a circuit breaker performs in the control scheme. Two types of circuit breakers are commercially available, the thermal-magnetic type and the magnetic-trip type. The thermal-magnetic-type circuit breaker has two separate parts. It has a thermal-trip unit, which will open the circuit if an overload condition exists for a period of time. It also has a magnetic-trip unit, which will open the circuit instantly at a much higher value of current. Regardless of the application, the thermal-trip point should be such that it will protect the power lines or power bus from drawing excessive currents over a period of time, which will cause abnormal heating of these conductors. The magnetic portion of the circuit breaker should provide immediate removal from the line during short-circuit or ground-fault conditions.

Magnetic-trip or instantaneous circuit breakers are generally used wherever circuit protection against short circuit and ground faults is the only consideration. For motor-control circuits, the instantaneous breaker, when used, should be selected such that it will allow the inrush current to pass without tripping but will open the circuit at some value in excess of this current.

Although a circuit breaker is said to be capable of opening a circuit without damaging itself, many feel that they should be subject to routine maintenance checks after operating to make certain that they are still in proper adjustment.

8. Power-Line Fuses.   Line fuses are sometimes preferred over circuit breakers for the following reasons:

1. Fuses have no moving parts, are maintenance-free, and require no periodic checking. Once properly selected and installed, fuses can be relied upon to protect a circuit for an indefinite period of time.

2. Fuses are generally considered the most reliable and accurate means of circuit protection.

3. A blown fuse provides more incentive for an electrician to correct the cause of the failure than a tripped circuit breaker. It also provides management personnel with a means of keeping records of circuit malfunctions.

It should be remembered that fuses like circuit breakers are thermal devices and must be selected and applied accordingly. Ambient temperature in the location of the fuses can be a factor in the selection. Other factors of consequence are the type of fuse clips used and the amount of heating at the junction between the fuse clip and the fuse.

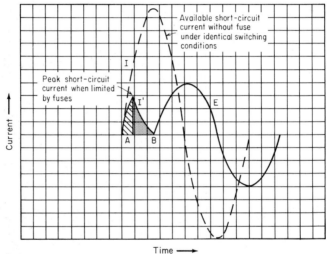

Fig. 3 Typical characteristic curve for current-limiting fuse from actual test. A = meting time, B = arcing time and A + B = total clearing time.

The three major types of fuses in use throughout industry are the dual-element (time-delay) type, the non-time-delay (single-element) type, and the current-limiting type.

a. The dual-element fuse has a built-in time delay, which allows large values of current in excess of its rating to pass for a short period of time without opening. The fuse can therefore be selected with a rating closer to the normal operating current without making allowances for the initial current inrush.

b. Single-element fuses have one inherent drawback which decreases their usefulness and appeal to the industrial-control designer. It is necessary to select a single-element fuse rated much higher than the normal operating conditions to allow for initial surge conditions. Single-element fuses are also considered by many to be prone to nuisance tripping.

c. Current-limiting fuses are designed to open the circuit at a specific point on the first cycle of an overcurrent fault. This clearing time is normally less than 0.25 cycle for standard 60-Hz power circuits (Fig. 3).

Similar types of fuses have been manufactured in various ways and of different materials by different fuse manufacturers. The characteristics of these fuses will therefore differ regardless of the fact that the type, ratings, and physical size are identical. For this reason, it is wise to consult the manufacturers' specifications in all cases where time-current relationships are critical.

9. Low-Voltage Contactors for Motor Control. The contactor is the central device of all control systems. It is the component in the controller which opens and closes the circuit from the energy source to the apparatus under control. Low-voltage contactors operate at 600 V maximum. They are supplied in 11 basic sizes depending on the voltage, horsepower, or current and the duty cycle of the machinery under control. Table 1 shows the sizes of contactors to be used for across-the-line full-voltage motor controllers, for normal starting duty. The ratings shown assume continuous duty without exceeding any of the established limitations.

TABLE 1. Low-Voltage-Contactor Ratings

| NEMA size | Continuous-current rating, A | | Horsepower rating | | | | | | | |
| | | | 3-phase 60-Hz | | | | | Single-phase 60-Hz | | 3-phase 50-Hz |
	Open	Enclosed	115 V	200 V	230 V	460 V	575 V	115 V	230 V	380 V
00	10	9	³/₄	1¹/₂	1¹/₂	2	2	¹/₃	1	1¹/₂
0	20	18	2	3	3	5	5	1	2	5
1	30	27	3	7¹/₂	7¹/₂	10	10	2	3	10
2	50	45	7¹/₂	10	15	25	25	3	7¹/₂	25
3	100	90	15	25	30	50	50	7¹/₂	15	50
4	150	135	25	40	50	100	100	–	–	75
5	300	270	–	75	100	200	200	–	–	150
6	600	540	–	150	200	400	400	–	–	300
7	900	810	–	–	300	600	600			
8	1350	1215	–	–	450	900	900			
9	2500	2250	–	–	800	1600	1600			

Adapted from NEMA Standards 2-321B-1 and 2-321B-2.

Motors draw high inrush currents during starting which range to approximately 600% of the full-load motor current (FLMC). The contactor must therefore be capable of closing on and carrying these higher currents without experiencing abnormal degradation. It must also be capable of interrupting these same magnitudes of current. NEMA Standards provide for these factors by establishing maximum operating conditions of six times FLMC for ac motors and four times FLMC for dc motors used in conjunction with reduced-voltage starters and ten times FLMC for dc across-the-line full-voltage starters.

Normally, when a controller is to be used for plugging or inching, the contactor must be derated. This precaution is necessary to accommodate the additional heating which results from these frequent inrush-current surges. Table 2 lists some appropriate derating factors recommended for plugging and inching duty.

10. High-Voltage Contactors for Motor Control. The use of high-voltage control for larger motors, generally above 250 hp, is becoming increasingly popular. The long-range economic advantages are often being found to overcome the increased initial costs of installing high-voltage equipment.

NEMA Class E general-purpose magnetic controllers are rated at 2500 V maximum and 5000 V maximum, 3-phase, 50- and 60-Hz with air break and oil-immersed contactors.

Tables 3 and 4 list the standard ratings of contactors used for high-voltage applications.

11. Motor-Overload Protection. The purpose of overload protection is to afford protection for the motor during the normal course of its operation and to prevent the motor from operating beyond its full-load capability, during both starting and running.

TABLE 2. Plugging or Inching Service

Volts	Horsepower rating for NEMA sizes							
	Size 0	Size 1	Size 2	Size 3	Size 4	Size 5	Size 6	Size 7
3-phase:								
115	1	2						
200–230	1½	3	10	20	30	75	150	225
460–575	2	5	15	30	60	150	300	450
Single-phase:								
115	½	1	2					
230	1	2	5					

Adapted from Table 3, 2-321B-3 and 2-321B-4, NEMA Standards.

These horsepower ratings are recommended when contactors are required to open the stalled motor repeatedly, for example, plug-stop or jogging (inching repeatedly), which requires continuous operation with more than five openings per minute. Size 00 contactors are not recommended for plugging or inching service.

Whenever a motor is subjected to such conditions, its normal life is generally reduced.

Overload-protective devices are designed to protect the motor from overheating because of low-voltage conditions, heavy-duty load cycles, excessive loads, extended periods of locked rotor, high-inertia loads, and sometimes single phasing.

The basic criterion in the selection of an overload device is that it responds to current increases in the same manner as the motor itself. It must take the motor off the line before the current produces enough heat to cause damage to the motor.

CONTROLLERS FOR NONMOTOR LOADS

12. Ratings and Requirements. Controllers for low-voltage nonmotor loads are identical to motor controllers with the exception that they are rated differently and they do not include thermal-overload protection. The most common nonmotor loads include tungsten-filament-lamp loads, fluorescent-lamp loads, electric furnace or oven loads, capacitor banks, sign flashers, etc. Some of these loads have extremely high ratios of inrush to normal operating current. Tungsten-filament lamps, as an example, may have inrush currents from 15 to 17 times the normal operating current. The contactor must be capable of withstanding continued current peaks of this nature. An industrial-lamp bank is normally operated only twice during a 24-h period, once to illuminate the facility and once to turn the lights off. The contacts are never required to take the circuit off the line at currents above normal operating current.

TABLE 3. Synchronous-Motor Ratings

Controller size	Continuous-current rating, A	Voltage rating	Power factor	Horsepower
H2	180	2200–2400	100	900
			80	700
		4000–4600	100	1500
			80	1250
H3	360	2200–2400	100	1750
			80	1500
		4000–4600	100	3000
			80	2500

Adapted from Table 2-324-1, NEMA Standards.

TABLE 4. Induction-Motor Ratings

Controller size	Continuous-current rating, A	Voltage	Horsepower
H2	180	2200–2400	700
		4000–4600	1250
H3	360	2200–2400	1500
		4000–4600	2500

Adapted from Table 2-324-1 of NEMA Standards.

This type of low-voltage contactor is available in 11 sizes, which are normally listed in accordance with the maximum continuous ampere rating of the device. The available sizes for these contactors are identical to the sizes for motor contactors. They range from size 00, which is rated at 5 A continuous for tungsten and 9 A for all other uses including fluorescent and mercury-vapor-lamp loads, to size 9, which is capable of switching loads up to 2000 A for tungsten and 2250 A for general-use loads. Commercial manufacturers of nonmotor contactors do not follow the NEMA listed ratings, particularly the tungsten-lamp loads for sizes 1 through 4.

In some applications of the nonmotor controller, such as hospital lighting, it may be desirable to have a completely noise-free magnetic contactor. This type of device may be designed with permanent-magnet latching contacts. When the latch circuit closes the contact, the coil is automatically deenergized and the contactor is held closed by a permanent magnet. The "stop" or "off" position of the pilot device energizes an unlatch circuit, which momentarily demagnetizes the permanent magnet, allowing gravity (or springs) to separate the magnetic yoke and the armature a safe distance to where the armature cannot move to close the circuit until the latch circuit provides the momentary magnetic kick necessary. Silicon diodes mounted in a module adjacent to the coil provide correct latch and unlatch polarities. When using this type of contactor, it should be remembered that when power fails and then is restored, the device remains in the energized position.

Other types of loads which utilize this type of controller, such as resistive elements, transformer primary switching, and capacitor switching, present unique problems which must be recognized and accounted for in selecting the rating necessary for the job. For resistive loads, derating is necessary if the switching frequency is in excess of the manufacturers' recommended duty cycle. Ratings for switching transformer primaries are generally supplied; however, the nature of the transformer secondary load may necessitate additional derating or an adjustment in the duty cycle.

SELECTING MOTOR CONTROLLERS

13. Application Data. There are a number of factors which govern the selection of a controller for motor applications. When this selection is made, consideration must be given not only to the ratings but also to such factors as power-company regulations, accelerating time, torque requirements, and economy. An application checklist should be developed to aid in selecting the controller which will best suit a particular situation. The following are items which should be included on such a list:

1. *Type of Motor.* Squirrel-cage, wound-rotor, synchronous, and dc machines all have different types of controllers, which are designed to be compatible with the specific requirements of these different types of motors.

2. *Ratings and Duty Cycle.* The controller must be capable of operating under the specified conditions without overheating or nuisance tripping of the overload-protective relays.

3. *Ambient Conditions and Location.* The temperature, altitude, and atmosphere surrounding the controller can govern the type of enclosure, the electrical layout of the components, and the type and rating of the overload protection used.

4. *Power-Company Regulations.* Power companies have restrictions on the amount of energy a customer can take from the line for a given period of time. These restrictions will often determine whether full-voltage starting or reduced-voltage starting may be used. These regulations take into account such factors as distance between the motor and the nearest substation, whether the location is in an industrial, commercial, or residential neighborhood, and the total power consumed by the entire facility.

5. *Local Electrical Safety and Fire Codes.* Some locations have regulations which are unique, and specific modifications to a standard controller are therefore necessary. These may involve external reset provision, pilot or warning lights or bells, push-button stations, etc.

6. *Torque Requirements during Starting.* When a controller employing reduced-voltage starting is used, the controller must be so chosen that it will allow enough power to be supplied to the motors to accommodate the torque requirements of the load during the accelerating period. Where long accelerating periods are required, overload-protective devices must be set to allow for these conditions.

7. *Motor Load.* Knowledge of the type of load a motor is driving is of great use to the engineer designing the controller. If he knows the specific application of the motor, he has more insight and a better overall picture, which may be of value in deciding which way to go on some of the parameters that are "borderline."

14. **Nameplates.** Each controller is supplied with a nameplate that lists its electrical ratings. The following minimum information should be spelled out on the controller nameplate:

1. Manufacturer's name and the city and state of the corporate or division headquarters

2. The name of the country where the device was manufactured (this is general practice)

3. A descriptive catalog number

4. A serial number if the controller is not a standard catalog device

5. The NEMA size and/or electrical rating of the controller

6. All electrical ratings of the control circuit that differ from the ratings of the power circuit

Components that have several different settings, such as circuit breakers and some overload-protective devices, should carry separate nameplates which list these various settings or ratings.

The nameplate should be constructed and marked such that it will be legible throughout the normal lifetime of the device. Most nameplates meet these requirements by using stamped metal or engraved phenolic.

Part 2

Motor Starters

T. H. KAVANAUGH *

FULL-VOLTAGE GENERAL-PURPOSE STARTERS

The electrical power service and the motors that receive it are important elements in the electric-motor system, but neither is as significant, in terms of its responsibility, as the motor controller. The motor controller receives the power service and directs it accordingly to the appropriate motors to perform useful work efficiently. Typical functions performed by a motor controller include starting, accelerating, stopping, reversing, and protecting motors.

* Senior Project Engineer, Square D Company.

Most ac motors, 600 V or less, are started directly across the supply lines. However, a motor must have windings capable of taking a sudden mechanical stress, accelerating fast enough to prevent overheating, and holding the inrush current within limits established by the power company. There are two basic across-the-line 600-V motor controllers, the simple manual starter and the magnetic starter.

15. Magnetic Starters. A magnetic starter is basically an electromagnetic on-off switch which starts a motor when voltage is applied to its magnet coil and stops the motor when voltage to the coil is disconnected. It also provides motor-overload protection.

Occasionally it is desirable to locate the overload protection separately. In some applications it is not needed at all, although an electromagnetic switch is still required. In these cases, a basic starter without overload protection is used. This device is commonly called a magnetic contactor. It is merely an electromagnetic on-off switch.

Magnetic starters control motors of all kinds. They are used to start, accelerate, stop, protect, and in some cases, to reverse motors that operate machine tools, compressors, saws, pumps, conveyors, grinders, presses—practically any type of machine.

Since motors are rated in voltage, horsepower, and current, starters and contactors are selected in accordance with these ratings. Most motors run at voltages less than 600 V; so the majority of starters and contactors controlling them are designed to operate up to the 600-V level. They are usually rated at 600 V maximum.

Also, starters and contactors vary in size to control the wide variety of motors in use, some of which are rated as high as 1600 hp. Lower-horsepower motors draw little current whereas higher-capacity motors require much more electrical energy. As they vary in size, commonly used motors draw as little as 1 to over 1200 A.

Motor controllers are therefore rated to carry current continuously in a specific range from 9 to 1215 A. This range covers almost all motor requirements. NEMA has established standards to identify the maximum current that individual starters and contactors must be capable of carrying. NEMA standard sizes and their corresponding continuous-current ratings are show in Table 1.

These maximum continuous-current values are related to a specific controller, designated by its NEMA size. For example, a NEMA size 00 controller must be able to carry 9 A continuously, a NEMA 0 device 18 A, a NEMA 1 device 27 A, a NEMA 2 device 45 A, etc. The current-carrying capacity increases with the discrete NEMA size. Generally, as the NEMA size of the device increases, so does the controller's physical size. This increased size is usually the result of the need for large contacts to carry high values of current and a strong, reliable mechanism to open and close them.

Selection of a starter or contactor is based on many factors related to the particular motor that is being controlled, such as voltage, horsepower, and current. The choice of a controller should be based on all these factors and care must be taken to accumulate all available data when selecting a motor controller for a specific application.

16. Magnetic vs. Manual Starters. A magnetic starter is similar in many ways to a manual starter. Both devices provide an on-off operation and overload protection for motors. However, the magnetic starter is much more versatile.

Manual starters are usually capable of fewer switching operations than magnetic starters. This characteristic results from the common requirements of a magnetic starter to turn circuits on and off frequently and quickly. Manual starters are rarely called upon to perform in such severe applications because of the usual nature of their operation—to control simple machines that run continuously.

Magnetic starters are often selected because they can be mounted either directly on the machine or remotely. The pilot device that operates the magnetic starter can be mounted on the machine that is being controlled, while the starter is located at a distance from the machine where mounting space and accessibility are often more convenient. Since the start-stop operation is an integral part of a manual starter, remote control is not possible. In addition, multiple start-stop functions can be used with magnetic starters for safe and convenient control. This is not possible with manual starters, which can utilize only a single start-stop operator per device.

Whenever the voltage in the system is reduced to a value below its normal operating level, the current increases and damage to the motor and other circuit components can occur.

One of the most important reasons for using magnetic starters is that, when wired appropriately, they provide low-voltage protection. If a power failure occurs, the magnetic starter opens the motor-circuit power contacts and stops the motor. If the failure is corrected and power is once again restored, the magnetic starter, if properly wired, will not pick up until its controlling pilot device is again operated.

This safety feature is particularly important because without it, personal injury and equipment damage can occur. Factory personnel, for example, could erroneously assume that a motor controller has been shut off when actually there has been a power failure. When electricity in the factory is restored, a magnetic starter that is wired properly will not allow the machine it controls to start unexpectedly until it is deliberately reenergized.

In the case of a manual starter, its power contacts remain closed until it has been manually turned off. If the manual starter is left in the "on" position, the machine it controls starts up immediately when power in the factory is restored. This condition could lead to serious personal injury and costly damage to the machine.

A summary of the comparative features of the manual and magnetic starters is shown in Table 5.

TABLE 5. Comparison of Magnetic and Manual Starters

	Starters	
	Magnetic	Manual
Hp ratings	Broad range	Limited to small sizes
Operation	Continuous and frequent switching	Continuous
Location	Direct and remote	Within sight
Low-voltage release	Yes	No
Low-voltage protection	Yes	No

17. Standard Wiring for Magnetic Starters and Contactors. Whenever voltage is applied to a contactor or starter coil, a set of contacts closes which completes current paths to the motor, making it run. These contacts are usually referred to as the power contacts and are wired in the starter power circuit (Fig. 4).

A magnetic contactor has power contacts only, while a magnetic starter utilizes power contacts and overload protection in its power circuit. In a magnetic starter, current flows from a power source through the closed power contacts, through the overload-relay thermal units, to the motor.

The portion of the wiring that includes the coil of the starter is called the control circuit (Fig. 5) and is usually connected to lines 1 and 2 in the power circuit. In addition to the starter coil, this circuit contains the overload contact, which opens when excess current is sensed by the thermal units in the starter power circuit.

Overload current drawn by the motor is detected by the thermal units in the power circuit and causes the normally closed overload-relay contact in the control circuit to open (Fig. 6). This operation interrupts voltage to the coil and opens the power contacts, stopping the motor.

To energize the starter or contactor coil conveniently, a control-circuit device, sometimes referred to as a pilot device, is wired in the control circuit. When voltage at L_1

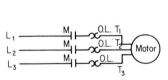

Fig. 4   Starter power circuit.

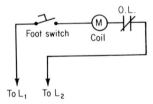

Fig. 5   Starter control circuit.

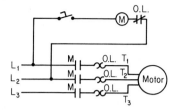

Fig. 6   Starter power and control circuit.

and L_2 in the power circuit is applied and the control-circuit device contacts are closed, the coil in the control circuit is energized.   This action closes the starter power contacts to start the motor.

The control-circuit device can be one of many devices such as a push button, foot switch, limit switch, or relay contact (Sec. 3).   Control-circuit devices are selected according to the type of mechanical or physical actuation that can be provided.

18. Two- and Three-Wire Control.   Two basic wiring schemes for controlling a starter or contactor are two-wire control and three-wire control.   Two-wire control devices rely primarily on maintained-contact control-circuit devices.   That is, the contacts of the device are physically or mechanically held closed to keep the starter or contactor coil energized.

If a common two-wire control scheme (Fig. 6) uses a foot switch in the control circuit, the starter or contactor remains picked up as long as the foot pedal is depressed and held.   Allowing the pedal to return to its initial position interrupts the control circuit and deenergizes the starter coil, opening the motor circuit.

Note that if the pilot device were located remotely from the starter, there would be two wires running between it and the starter.   Caution must be used when applying a magnetic controller wired with two-wire control.   When this scheme is used, low-voltage protection is not possible.

When the control-circuit-device contacts are held closed, there is always a complete current path in the starter-coil circuit.   If a voltage failure occurs, the power circuit opens, but when electricity is restored, the starter picks up immediately and starts the motor.   This arrangement may be dangerous, especially if there is a possibility that there are people in the area who assume that the machine has merely been shut off when suddenly power resumes.

Three-wire control schemes (Fig. 7) provide low-voltage protection by utilizing momentary-contact control-circuit devices and an auxiliary contact, called a holding-circuit interlock.   This contact, which is physically located on the starter, operates simultaneously with the power contacts and is wired in parallel with the normally open start button in the coil circuit.

When the momentary-start button is depressed, current flows from L_1 through the normally closed stop button, the start button being held closed, the coil and the overload relay contact, and back to L_2 to complete the control circuit.   The starter electromagnet operates and closes all its contacts, including the holding-circuit interlock.

Releasing the start button will not deenergize the starter, since there is still a closed current path in the control circuit from L_1 through the stop button, through the closed

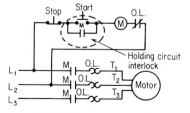

Fig. 7   Three-wire control.

holding-circuit interlock, through the coil, the overload-relay contact, and back to L_2. There is a holding circuit even though the momentary-start button has been released.

The coil circuit can be deenergized by depressing the stop button, by a low- or no-voltage condition, or if an overload condition occurs to open the overload-relay contact. In these instances, the control circuit is disconnected and remains open because the holding-circuit contact opens whenever the coil is deenergized. It can be reclosed only by pushing the start button to reactivate the coil.

Three-wire control schemes are safe because they provide low-voltage protection. Under normal conditions, if the starter is energized, the power contacts close and the motor runs.

If a power failure occurs, the coil will be deenergized and the armature falls away from the magnet. This action releases the control circuit and opens the power contacts to shut off the motor.

When power resumes, the starter or contactor coil in a three-wire control scheme will not immediately be energized because there are no closed current paths to the coil. The momentary-start button is in the open position and the holding-circuit interlock has opened with the release of the coil. There cannot be a closed current path to the starter coil until the start button is again depressed.

There are three wires that are required to connect a remotely located start-stop push button properly to the starter, which includes some factory wiring. Because three external wires are all that are needed for starter energization with a holding circuit, this scheme is commonly called three-wire control.

Three-wire control prevents accidental start-up by providing low-voltage protection. It aids operator safety and eliminates possible equipment damage.

19. Common Control. There are many variations in wiring the control circuit depending on the type of power available and individual safety requirements. The schematic diagrams of Figs. 5 to 7 show the control circuit wired directly to lines 1 and 2 in the power circuit.

Since voltage is applied to L_1, L_2, and L_3 for current flow in each line of a polyphase motor, the control circuit is energized because of its direct connection to the power circuits. This is called "common" control. Both the control circuit and power circuit are tied in together and powered from a single common source.

The voltage in the control circuit is the same as the voltage between L_1 and L_2 in the power circuit. According to electrical standards, the control circuit is wired to L_1 and L_2, although any two of the three power lines of a 3-phase system would produce the same single-phase voltage required for the control circuit.

20. Separate Control. Another type of control scheme for wiring the power and control circuit is called separate control. Here, the control circuit (Fig. 8) is wired to a voltage source totally separate from the power circuit. This arrangement is used when the power circuit is being operated at a high voltage such as 460 V as required by the motor. However, high voltage is usually not desirable in the control circuit for reasons of operator safety.

In a separate control scheme, the high voltage is connected only to the power circuit to operate the motor. A lower voltage from another separately located source is wired to the control circuit, which is the only circuit the operator would probably come into physical contact with. When the control circuit is at a lower voltage, there is less chance of serious injury if the operator accidentally touches a faulty pilot device or a live control-circuit wire.

Operation of a separate control system is as follows: Assume there is voltage at each

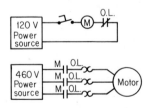

Fig. 8 Separate control.

source. When the pilot device closes, 120 V, for example, is supplied to the coil circuit of the starter. This voltage causes all the starter contacts to close. The higher voltage, 460 V, is connected across L_1, L_2, and L_3 and is applied directly to the motor through the now closed power contacts. The motor runs at 460 V and is operated by a starter control circuit at 120 V.

Separate control allows simple operation of high-voltage motors and yet provides the safety of low-voltage control.

21. Control-Circuit Transformer. Another way to obtain low voltage at the control station without wiring the control circuit to a separate source is to install a fused transformer, as shown in Fig. 9.

The transformer is wired between the control and power circuit. It takes the high voltage in the power circuit and reduces it to a safer level in the control circuit. The fuse protects the control circuit against damage from short-circuit currents.

The voltage at L_1 and L_2 is connected to the primary side of the transformer, which results in a low voltage on the secondary side of the transformer. The low voltage operates the coil circuit of a starter, which controls the motor at a higher voltage with its power contacts.

A fused transformer provides the safety offered by a low-voltage control circuit with the convenience of only one power source.

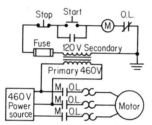

Fig. 9 Control-circuit transformer.

22. Versatility of Magnetic Controllers. A major advantage of magnetic controls is that they are very versatile. Modification of contactors and starters is often necessary to accommodate the particular needs of each motor system. Standard kits are available to add start-stop, selector-switch, and pilot-light functions conveniently for particular starter energization and indication requirements.

In addition to pilot devices, electrical interlocks, timers, and special control modules can be combined with contactors and starters to meet further requirements for complex motor control systems.

a. Electrical Interlocks. Electrical interlocks provide simultaneous contact actuation of control circuits whenever the starter or contactor they are mounted on is energized. The interlocks, either normally open or normally closed, switch control circuit loads. These loads draw little current, usually less than 6 A.

Electrical interlocks are often wired to obtain a remote indication of starter energization with a pilot light or to actuate the coil circuit of other magnetic-control devices — such as relays, timers, or other starters and contactors. The interlock is mechanically linked to the controller so that its contacts change position simultaneously with the power contacts.

If a normally open electrical-interlock contact is wired to a pilot light in the control circuit of a starter (Fig. 10), its power contacts and its normally open interlock closes when the starter coil is energized. The motor runs and the pilot light is illuminated. If the starter is deenergized, the motor stops and the light goes out.

The pilot light can be physically located on the starter or at a distance to indicate when a motor is running. The light location is chosen depending on the particular installation.

Similarly, a relay, timer, or any magnetic-control device can be wired in series with other electrical interlocks which are attached to the controller (Fig. 11). When the controller picks up, its electrical interlocks are actuated to energize the coil circuits of

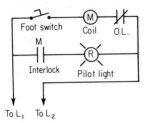

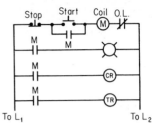

Fig. 10 Electrical interlock and pilot light.

Fig. 11 Electrical interlocks in series with control devices.

these devices. One signal to a magnetic starter or contactor can be used to initiate the simultaneous operation of many control devices.

b. Timer Attachment. A timer attachment can also be added conveniently to magnetic controllers. It is a device containing normally open and normally closed auxiliary contacts whose contacts have a time delay. The time delay can occur upon energization or deenergization of the starter or contactor.

After a starter with a time-delay-upon-energization timer (Fig. 12) coil is energized, the power contacts close and after a set time period elapses, the timed contact on the timer will be actuated. This contact, like an electrical interlock, is wired to activate the control circuits of other magnetically operated devices.

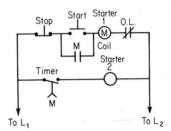

Fig. 12 Control circuit with timer.

A typical application for a starter timer attachment is a drill press. Lubricant is often required to flow over the workpiece to drill at high speed and to keep the drill bit from burning up. If an on-delay timer is attached to the starter that controls the lubricant pump, and its normally open timed contact is wired in series with another starter controlling the rotation of the spindle motor, the drill spindle will not be allowed to rotate until lubricant flows for the set time period.

Pushing the start button picks up the lubricant-pump starter immediately. After the selected time period elapses, the normally open contact on the timer closes to energize the second starter controlling the drill spindle. When this starter is energized, the spindle rotates to drill the well-lubricated workpiece, thereby preserving the drill bit and reducing drilling time.

23. Functions Compared. In summary, it can be said that the basic functions of a magnetic motor starter are to start, stop, and protect a motor. The major advantages (Table 7) of magnetic starters over manual include:

1. They are available in a wider variety of sizes with capacities ranging up to 1600 hp.
2. They are capable of frequent switching and have a long mechanical and electrical life.
3. They can be mounted directly on the machine or remotely located from it.
4. They have low-voltage release.
5. If wired appropriately, they offer increased personnel safety because of their use in a low-voltage protection scheme.
6. Magnetic starters are, overall, the most versatile motor-control devices available because of their easy modification and the many variations possible to meet practically any motor application.

3

Control-Circuit Devices

R. A. DOWNEY *

* Supervisor, Application Engineering, Square D Company.

FOREWORD

Control-circuit devices are available in an endless variety of types and arrangements. Because they govern all control functions, they are the basis of control circuitry, and their application and operation determine the characteristics and reliability of control and switchgear systems. This section describes the most common types and forms of control-circuit devices and provides an important background on their application. Control-circuit devices are often referred to as pilot devices.

GENERAL

A control system is made up of input devices, logic or decision-making devices, and output devices. The output devices, such as motor starters and contactors, handle power loads where current levels are relatively high and where horsepower ratings are required for motor loads. Input and logic devices are used in control circuits to switch nonmotor loads where current levels are relatively low. In general, power devices control currents of approximately 15 A or higher, and control-circuit devices control nonmotor loads of approximately 15 A or lower.

1. Ratings. Most control applications today involve nonmotor inductive loads such as starter, contactor and relay coils, solenoids, and clutches. To apply a device correctly, the inductive or control-circuit duty rating must be known. Ratings are generally listed in three ways:

Resistive or noninductive
Inductive or control-circuit duty
Continuous

a. Resistive or Noninductive Ratings. This rating indicates the resistive load only that the contacts can make or break. Resistive ratings are generally based on a 75% PF for ac.

TABLE 1. Control-Circuit Device Ratings

Contact rating designation			Max ac voltage, 50 or 60 Hz	Amperes		Continuous carrying current	Voltamperes	
				Make	Break		Make	Break
A600	A300	A150	120	60	6	10	7200	720
A600	A300	......	240	30	3	10	7200	720
A600	......	......	480	15	1.5	10	7200	720
A600	......	......	600	12	1.2	10	7200	720
B600	B300	B150	120	30	3	5	3600	360
B600	B300	......	240	15	1.5	5	3600	360
B600	......	......	480	7.5	0.75	5	3600	360
B600	......	......	600	6	0.6	5	3600	360
C600	C300	C150	120	15.5	1.50	2.5	1800	180
C600	C300	......	240	7.5	0.75	2.5	1800	180
C600	......	......	480	3.75	0.375	2.5	1800	180
C600	......	......	600	3.00	0.30	2.5	1800	180

Max contact ratings per pole (spanning header over Max ac voltage, Amperes, Continuous carrying current, Voltamperes)

b. Inductive or Control-Circuit Duty Ratings. The control-circuit duty rating indicates the nonmotor inductive load such as contactors, relays, and other remote-controlled devices that the contacts can make or break. These ratings are usually based on a 35% PF for ac.

c. Continuous Rating. The continuous rating indicates the load that the contacts can carry continuously without making or breaking the circuit.

The control-circuit duty rating is always less than the resistive or continuous rating. When the contacts break an inductive circuit, the inductance in the load tends to keep the current flowing in the same direction. The result is an arc across the contacts which causes heating and burning of the contacts. Because of the extra heat generated, the allowable inductive current must be less than resistive current for equal contact life.

Control-circuit devices are often referred to incorrectly by their resistive rating. For example, a relay may be called a "10-A relay" without reference to the fact that this rating is a resistive rating. This procedure can lead to misapplication because the device may be capable of breaking only 6 A or less inductive current.

The user should check the device ratings closely and determine all the ratings, inductive, resistive, and continuous.

Quick-make and quick-break or snap-action contacts reduce the arcing time and allow higher control-circuit duty ratings than with slow make-and-break contacts.

Inductive-coil loads have a momentary inrush current of approximately ten times the sealed current (Art. 58). Contacts must be able to make the inrush current and break the sealed current for a reasonable number of operations. Contacts must also be able to break or interrupt the inrush current in an emergency. NEMA has established ratings for control-circuit devices. Some of the more common ac ratings listed in the NEMA Standards ICS 2-125 are shown in Table 1.

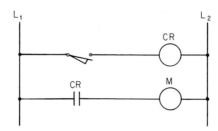

Fig. 1   Relay or contactor can be used where load is too heavy for control-circuit device.

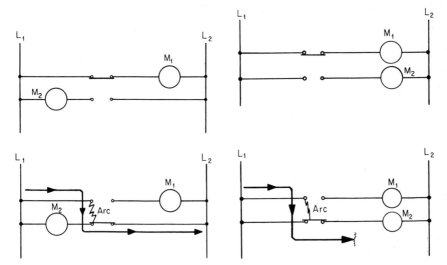

Fig. 2 Contacts connected to opposite polarities. Line-to-line short (bold line) can occur through arc drawn when contacts operate.

Fig. 3 Contacts connected to same polarity. Line-to-line short cannot occur when contacts operate.

d. Excessive Load. If the load exceeds the control-circuit-device ratings, a relay or contactor with the required ratings can be used as in Fig. 1.

The dc interrupting ratings of a device can be increased by placing contacts in series. This arrangement effectively increases the contact gap, allowing higher current rating. In general, the following ratings can be applied:

Two contacts in series — dc rating × 2.5

Three contacts in series — dc rating × 5

The current should never exceed the maximum continuous-current rating of the device.

e. Opposite Polarities. Single-pole, double-throw contacts of a control-circuit device should not be used on opposite polarities. When one contact of a switch is connected to the L_1 side of a load and the other contact is connected to the L_2 side of a load (Fig. 2), a line-to-line short can occur through the arc which may be drawn as the contacts operate. When the contacts are connected to the same polarity (Fig. 3), this line-to-line short cannot occur.

f. Power from Different Sources. Single-pole, double-throw contacts of a control-circuit device should not be connected to two different power sources (Fig. 4). If two

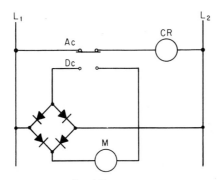

Fig. 4 Contacts connected to different power sources — wrong method.

different power sources must be controlled by the same device, two isolated poles should be used as shown in Fig. 5.

CONTACT RESISTANCE

2. Resistance of Control-Circuit Duty Contacts and Snap Switches. Contact or snap-switch resistance is the total electrical resistance that a device adds to the circuit and consists of:

a. Conducting Path. Includes all terminals, inserts, stationary contacts, material resistance, movable-blade assembly, and any other parts in the conducting circuit.

b. Constriction resistance of all joints, discontinuities, or interfaces. This resistance is caused by limited mating surfaces through which the load must pass. If the movable and stationary contact tips are viewed through a microscope, it can be seen that they touch at only a very few points. Thus, increased resistance is presented to the current.

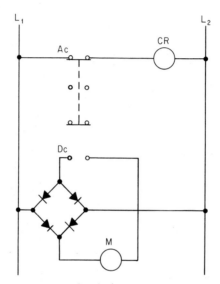

Fig. 5   Contacts connected to different power sources—right method.

If the current is high enough, the points of constriction are softened and enlarged through thermal effects, and the resistance decreases.

c. Film Resistance. It is well known that silver accumulates a surface resistance because of chemical reaction with its environment. The most common reaction is with sulfur and oxygen, which creates a sulfide and an oxide of silver. Such a surface is known to have some resistance which can be read with a low-voltage instrument such as an ohmmeter. These surfaces, however, have the characteristic of being self-cleaning. When current is passed through such a surface, it creates heat, which in turn reduces the compound to pure silver again and restores the contact to a low ohmic value. This characteristic has made silver a good selection for contact material.

d. Particle Resistance. Contamination in the form of foreign material can also produce resistance. Carefully controlled production processes are used to prevent contamination during assembly.

3. Contact Voltage. Control-circuit devices are designed for use with industrial-control equipment where voltages are relatively high (normally 120 V or higher) and current levels are high (normally 0.25 A or higher). An increase in contact resistance may appear to be more critical in high-energy circuits because it represents a greater percentage of total circuit impedance; however, the arcing produced breaks down or burns away the contaminants, reducing the actual resistance seen by the load.

Contact resistance in dry-circuit (low-voltage, low-current) applications is more of a problem, but the resistance must be quite high to affect operation significantly. For example, a solid-state circuit operating at 20 V and 1 mA has a circuit resistance of $20/(1 \times 10^{-3})$ or 20,000 Ω. A contact with 10-Ω resistance would have no effect. The problem arises with film or particle resistance, which can be quite high or even present an open circuit. The use of relay contacts, snap switches, etc., with dry circuits is generally not recommended, because these voltages or currents are not high enough to reduce the silver sulfide or silver oxide to pure silver or to burn away other contaminants present.

An ohmmeter test on contacts or snap switches is unreliable because the most common voltage source in an ohmmeter is a 1.5-V battery, and to the contacts being tested this is just another dry circuit. Several ohmmeter readings on the same contacts may vary from a few hundredths of an ohm to several ohms; yet the contacts will work perfectly with a nominal coil load.

4. Recommendation for Application and Testing of Control-Circuit Duty Contacts. The following points are important in selecting control-circuit devices:

1. Control-circuit duty contacts are normally recommended for use with industrial-control devices, not dry circuits. If lower voltages are present, the current drawn through the contacts should be at least 0.25 A to maintain proper continuity.

2. To test continuity in the field, a 6-V, 0.25-A pilot light is recommended.

3. *Do not use an ohmmeter to test continuity.* An ohmmeter is a reliable test only if the contacts are to be used with a dry circuit, and as started in item 1 above, this practice is normally not recommended.

4. For improved circuit reliability, contacts can be wired in parallel. Even the best of contacts may not be completely reliable for every operation, since foreign material from adverse environmental conditions can come between the contacts at some time. Placing two or more contacts in parallel provides an additional factor of reliability.

5. If standard contacts must be used as input to solid-state circuitry, for nominal reliability, a dummy load can be used to increase current through the contacts to a more desirable level.

For better reliability, the contacts can be used at 120 V ac or dc and converted to the logic-system voltage. For example, standard contacts operating at 120 V through a signal converter into logic-level systems have worked satisfactorily.

For best reliability using standard components, 2-pole devices with contacts wired in parallel and a signal converter can be used.

Standard contacts can be gold-plated to reduce contact resistance but may not provide sufficient increased reliability to warrant the added cost. For example, if the contact material underneath is exposed at any point owing to excessive current or normal mechanical wear, a contaminating silver sulfide film may develop and will eventually creep over the plated portion. Also, particles of dirt could still cause continuity problems. Long-term reliability is then no better than with standard contacts.

6. Probably the most reliable contacts used with solid-state circuitry are the sealed type such as reed contacts. Many reed devices such as relays and limit switches are available as logic-input devices.

TYPES OF CONTROL-CIRCUIT DEVICES

Input control-circuit devices convert an initiating force or movement into an electrical signal and include devices such as push buttons, selector switches, limit switches, pressure switches, and float and temperature switches. The electrical signal from the input devices is received by the logic devices, sometimes called the "brain" elements, which in turn make the control-circuit decisions and provide the proper signal paths called for by the various machine-operating sequences. The logic pilot devices include electrically held relays, mechanically held relays, and time-delay relays.

Only industrial-control components are covered in this section. Logic systems, either electromechanical or solid-state, will not be covered.

5. Push Buttons. Two types of pilot devices are available, manually operated and automatically operated. Push buttons are manually operated devices. The term push button is often used as a general term to cover not only push buttons but selector switches, pilot lights, and various other types of operators.

The push button is a button operator which is mechanically linked to one or more contacts and which makes or breaks the contacts when manually operated. When the push button is pressed, the contacts operate. When a momentary-contact-type push button is released, the contacts reset by spring action. When a maintained-contact-type push button is released, the contacts remain in the operated position until manually reset.

The most common application of push buttons is three-wire control of magnetic starters. Three-wire control or low-voltage protection is a control scheme using momentary-contact push 'buttons or similar pilot devices to energize the starter coil. This scheme is used to prevent the unexpected starting of motors which could result in possible injury to machine operators or damage to driven machinery.

In Fig. 6 the starter is energized by pressing the start button. An auxiliary holding-circuit interlock on the starter forms a parallel circuit around the start-button contacts, holding the starter in after the button is released. If a power failure occurs, the starter will drop out and will open the holding-circuit interlock. Upon resumption of power, the start button must be operated again before the motor will restart. The term three-wire control arises from the fact that in the basic circuit at least three wires are required to connect the control-circuit devices to the starter.

When a motor must be started and stopped from more than one location, any number of start and stop push buttons may be wired together as required. The start buttons are wired in parallel and the stop buttons are wired in series (Fig. 7). It is also possible to use only one start-stop station and have several stop buttons at different locations to serve as emergency stops.

6. Selector Switches. A selector switch is an operator which is mechanically linked to one or more contacts and which makes or breaks the contacts when rotated to two or more positions. When a maintained-contact-type selector-switch operator is rotated to a particular position, the operator remains in that position until manually rotated to another position. When a momentary-contact-type selector-switch operator is rotated to a particular position, the operator will spring return to the initial position when released. The most common application is two-wire control of magnetic starters using a hand-off-automatic maintained-contact selector switch where it is desirable to operate the starter manually as well as automatically. In Fig. 8 the starter coil is energized manually when the switch is turned to the hand position, and is energized automatically by the control-circuit device when the switch is in the automatic position.

If a power failure occurs while the contacts of the control-circuit device are closed, the starter will drop out. When power is restored, the starter will pick up automatically through the closed contacts. The term two-wire control arises from the fact that in the basic circuit, only two wires are required to connect the control-circuit device to the starter. This is also known as low-voltage release. The control-circuit device may be a limit switch, pressure switch, temperature switch, etc.

The target table shown in Fig. 8 shows the condition of the contacts in the various positions of the selector switch. X indicates a closed contact in that position. The arrow indicates the position of the selector switch shown in the diagram. In the hand

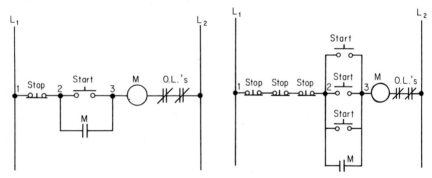

Fig. 6   Three-wire control.

Fig. 7   Three-wire control, multiple-push-button station.

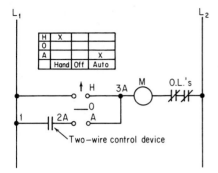

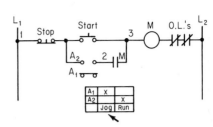

Fig. 8   Two-wire control with hand-off-automatic selector switch.

Fig. 9   Jogging using a selector switch, jog with start button.

position, contact H is closed.   In the automatic position, contact A is closed.   In the off position, contacts A and H are open.

An example of a two-position selector switch is shown in Fig. 9.   Jogging or inching is defined as a momentary operation of a motor from rest for the purpose of accomplishing small movements of the driven machine.   The selector switch disconnects the holding-circuit interlock, and jogging may be accomplished by pressing the start button.   With the selector switch in the run position, normal three-wire control with the start button is accomplished.

7. Selector Push Buttons.   A selector push button is a combination of a selector switch and push button in one unit.   The outer guard of the push button rotates to two or more positions and provides different contact actions when the push button is free and depressed in these positions.   Common use of a selector push button is to obtain jogging (Fig. 10).   In the run position, the selector push button gives normal three-wire control.   In the jog position, the holding circuit is broken and jogging is accomplished by depressing the button.   This arrangement provides the same function as the circuit shown in Fig. 9, but it is accomplished with one device instead of two.

8. Pilot Lights.   Pilot lights fall into the push-button family of devices.   The pilot light is an indicating means by which it can be visually determined if a control circuit or power circuit is energized.   The pilot light is wired in parallel with the load to indicate an "on" condition.   When wired in series with a normally closed interlock on the load device, the pilot light indicates an "off" condition.   A typical application for an on pilot light is shown in Fig. 11.   The pilot light is wired in parallel with the starter coil to indicate when the starter is energized and thus shows that the motor is running.

A typical application of an off pilot light is shown in Fig. 12.   If a pilot light is required to indicate when the motor is stopped, this function can be accomplished by wiring a normally closed auxiliary contact on the starter in series with the pilot light

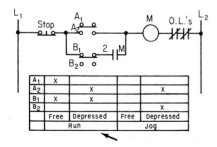

Fig. 10   Jogging using a selector push button.

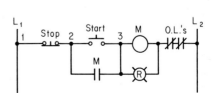

Fig. 11   Pilot light indicates when motor is running.

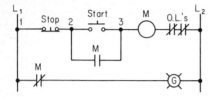

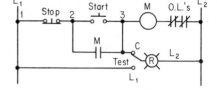

Fig. 12 Pilot light indicates when motor is stopped.

Fig. 13 Push-to-test pilot light.

as shown. When the starter is deenergized, the pilot light is on. When the starter picks up, the auxiliary contact opens, turning off the light.

9. Types of Pilot Lights. The full-voltage type has the lamp placed directly across the line. It has the advantage of lower cost, but the disadvantage of shorter lamp life because the higher-voltage filaments are susceptible to the effects of shock and vibration.

The transformer type steps the line voltage down to a lower voltage, generally 6 V, and a lower-voltage lamp which is less susceptible to shock and vibration is used.

The resistor type has a resistor placed in series with the lamp which allows a lamp of less than full voltage to be used. Resistance to shock and vibration is better than when a full-voltage lamp is used but not as good as the lower-voltage lamp used in the transformer type.

10. Push-to-Test Pilot Light. In a standard pilot light, when the lamp is not lighted, there may be doubt as to whether the circuit is open or the lamp is burned out. The push-to-test pilot light allows testing of the lamp simply by pushing on the color cap which connects the lamp directly across the line (Fig. 13).

11. Illuminated-Push-Button Type. Here the functions of a push button and pilot light are combined into one unit. Pushing on the lens closes the contacts and also lights a lamp to indicate an "on" condition (Fig. 14). Space is saved by requiring only a two-unit push-button station instead of three.

CONTROL STATIONS

Push buttons, selector switches, and pilot lights can be mounted in the cover of a magnetic-starter enclosure, in a control panel, or in a separate enclosure (control station) for remote control of a magnetic device.

12. Advantages of Control Stations. The small control station can be mounted in a convenient location and the more bulky starter can be placed out of the way. Lower voltages can be used in the control circuit to increase operator safety and smaller-sized wire can be used from the control station to the starter, since the current flow is relatively small.

13. Construction. Control stations today fall into three basic categories: standard-duty, heavy-duty, and heavy-duty oiltight.

a. Standard-Duty Type. Standard-duty control stations are used for light- and medium-duty applications and have a maximum rating of 3 A at 110 to 120 V. These units are generally suitable for making and breaking magnetic contactor and starter coils through NEMA size 4 where the coil inrush and sealed currents do not exceed the

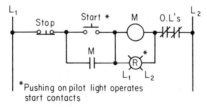

*Pushing on pilot light operates start contacts

Fig. 14 Illuminated push button.

push-button rating. Standard-duty control stations are available in surface or flush-mounted enclosures and are the least costly of the control stations offered.

b. Heavy-Duty Type. Heavy-duty control stations are used for heavy- or severe-duty applications that require frequent operation. They have twice the current rating of the standard-duty type, 6 A maximum at 110 to 120 V, and will make and break magnetic contactor and starter coils NEMA size 5 and larger. Heavy-duty control stations are medium-cost—more than the standard duty, but less than the heavy-duty oiltight type.

c. Heavy-Duty Oiltight Type. These control stations are specially constructed to prevent the entrance of oil, coolant, and dust into live parts. They have a NEMA 13 rating and are used for heavy-duty applications where oil, coolant, and dust are present, such as with machine tools. The heavy-duty oiltight control station is the most flexible of the three types, with a wide selection of push buttons, selector switches, and various other operators available for factory or field assembly. They are also the most attractive and the most costly of the control-station types.

14. Other Control-Station Types. *a. Pendant Types.* These control stations are designed as portable stations which allow the operator to move about. They are generally suspended from a power cord and are used to control hoists, monorails, and other overhead material-handling equipment.

b. Corrosion-Resistant (NEMA 4X) Type. These control stations are designed to provide protection against the corrosive liquids and atmospheres encountered in sewage-disposal plants, chemical and fertilizing plants, and tanneries. They are also suitable for breweries, dairies, and meat- and food-processing plants where cleanliness is essential and equipment is hosed down frequently. Because of the possibility of frequent hose-down applications, these control stations must meet NEMA 4 watertight requirements. Since most oiltight operators are not watertight, protective watertight boots must generally be used with these control stations.

LIMIT SWITCHES

Limit switches are used to convert a mechanical motion into an electrical control signal. The mechanical motion is usually in the form of a cam, a machine component, or an object moving toward a predetermined position. The cam engages the limit-switch lever or plunger and in turn makes or breaks an electrical contact inside the switch. This electrical control signal is then used to limit, position, or reverse machine travel or initiate another operating sequence. It can also be used for counting, sorting, or as a safety device.

Typical limit-switch applications are in the control circuits of solenoids, control relays, and motor starters which control the motion of machine tools, presses, conveyors, hoists, elevators, and practically every type of motor-driven machine.

Experience has shown that most limit-switch failures are the fault of the installation. In some cases, an improper installation cannot be avoided, but in the majority of cases, proper application of the limit switch would have prevented failure.

15. Definition of Limit-Switch Terms. Many terms that are common to limit switches are not used with other control devices. Before proceeding, definitions of the commonly used terms should be understood, as these terms will be used throughout this section.

Limit Switch. A device that converts a mechanical motion into an electrical control signal (Fig. 15).

Cam. A machine part or component that applies force to the switch actuator, causing it to move as intended. It is also known as a "dog."

Actuator. A mechanism of the limit switch that operates the contacts, i.e., lever arm, plunger, wobble stick.

Pretravel. The distance the limit-switch actuator must move to trip the contacts (Fig. 16).

Overtravel. The distance the limit-switch actuator moves beyond the trip point.

Differential. The distance the limit-switch actuator moves from the trip point back to the reset point of the contacts.

Operating Force. The force required to move the limit-switch actuator to trip the contacts.

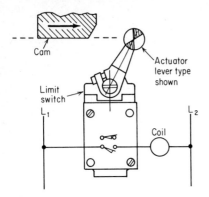

Fig. 15 Typical limit switch with cam and actuator lever arm.

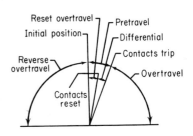

Fig. 16 Travel-angle terms.

Spring Return. A means of returning contacts to their original position when the actuating force is removed.

Maintained Contact. A device in which contacts remain in the tripped position until the return travel of the cam moves the switch actuator back and resets the contacts.

Cam-Track Dimension. The distance from the switch-mounting surface to some point on the roller or actuator (Fig. 17).

Linear Cam. A cam that moves in a straight line in either the forward or reverse direction.

Rotary Cam. A cam that moves in circular motion and may travel in one direction only or in both forward and reverse directions.

16. Types of Limit Switches. There are three basic types of limit switches—mechanical, proximity, and interrupted-beam.

a. Mechanical. A lever arm or push rod on the switch housing is mechanically connected to the electrical switching element inside. The moving object comes into direct contact with the limit switch. Types include lever-operated, push-rod, wobble-stick, and rotary-cam.

b. Proximity. The presence of an object close to the sensor disturbs an electrical or magnetic circuit and triggers a switch. There is no physical contact with the object being detected. Types include ac magnetic field, permanent-magnet field, and capacitance or electrical conductivity.

c. Interrupted-Beam. A beam of light, sound, or fluid is aimed across the path of the object to be detected. Interruption of the beam indicates the presence of an object. The object does not have direct contact with the limit switch. Types include photoelectric, infrared, ultrasonic, and air and fluid jet.

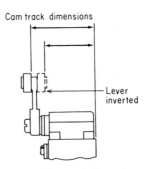

Fig. 17 Cam-track dimensions.

17. Selection of Limit-Switch Actuator.   The first step in the proper application of limit switches is the selection of a suitable actuator.   Selection depends on the shape, speed, direction, and total travel of the cam or part being used to trip the limit switch and the accuracy desired.   The following general rules should aid in the initial selection of the limit switch.   In some cases a trial-and-error method may be required to select the final limit-switch type used.

a. Lever-Arm Type.   The lever-arm type will handle approximately 90% of all limit-switch applications.   It is easiest to apply, has more safety overtravel, infinite adjustment of the lever arm, a wide selection of lever-arm types and lengths, and is generally the most rugged mechanism available.   This type is suitable for actuation by any cam or dog whose motion is perpendicular to the shaft axis about which the lever rotates.   Selection of lever-arm length should be based on the relative distance between the switch mounting and cam.   If lateral shifting of the cam is possible, a wider roller should be used to assure continued contact with the cam.   When in doubt about which limit-switch type to use, the lever-arm type is the most logical first choice.

Although readily available, other limit-switch types are really specials.   If the application is such that the lever-type limit switch cannot be used, it falls into a "special" category.   Other types of limit switches should be chosen only when the lever type cannot be used for some specific reason.

b. Top-Push-Rod-Plunger Types.   These are suitable for use where short, controlled machine movements are to be detected and where mounting space prohibits a lever-arm type.   Actuation must be in line with the plunger axis.   Adjustable-plunger types have an adjusting nut which allows a small adjustment of plunger length, typically $1/4$ in or less.   These switches are used where the initial plunger position is critical.

c. Side-Push-Rod-Plunger Types.   These switches are suitable for the same applications as the top-push-rod-plunger types but should be used where the allowable mounting space permits operation from the side only and not from the top of the switch.

d. Top-Roller-Plunger Types.   These switches can be used instead of roller-lever types where space or mounting is limited or where lever-arm snapback cannot be eliminated.   Plunger-type switches inherently have no recycling or telegraphing of contacts due to rapid-dropoff cams.   Actuation can be in line or perpendicular to the plunger axis.

e. Side-Roller-Plunger Types.   These switches can be used in the same application as the top-roller-plunger type but should be used where the allowable mounting space permits operation from the side only and not from the top of the switch.   Because two plunger bearings are generally used, the side-roller-plunger types are more rugged than top-roller-plunger types in applications where actuation is perpendicular to the plunger axis.   The roller position usually can be changed in the field from vertical to horizontal.

f. Side-Push-Rod-Plunger, Maintained-Contact Types.   These switches can be used as a memory device to signal movement in one direction even though the cam has been removed.   The switch can be reset manually or by another cam pushing on the reset plunger.   A typical application is detecting an improper movement of a machine component where manual reset is required to alert personnel that this condition has occurred.

g. Palm-Operated Types.   These devices are compact, oiltight push-button stations generally used as emergency stop buttons.   The mounting of the palm-operated type is the same as for other limit switches on the machine and is used to avoid the different drilling and tapping of mounting holes which would be required if a standard one-unit push-button station were used.

h. Wobble-Stick and Cat-Whisker Types.   Wobble-stick or cat-whisker operators extend from the top of the turrethead-type limit switch and can be deflected in any direction to operate the switch.   Typical applications include conveyors, where the product or a box rather than a cam is used to operate the switch, or as a safety device where the machine operator can deflect the switch by hand from any direction to shut off the machine in an emergency.   The wire-extension wobble-stick type uses a small-diameter steel wire which can be bent to any desired shape.   Where the steel wire might scratch the product being detected, a nonmetallic extension can be used.   Where extremely light parts are to be detected, such as empty shoe boxes, the cat-whisker type should be used.   Where greater flexibility of the operator is required to prevent damage to the operator, the coil-spring extension should be used.

i. Air-operated Type.   Increasing air pressure applied to the limit switch causes the contacts to trip.   Decreasing pressure allows the contacts to reset.   The air-operated limit switch is suitable for use where the switch must be remotely mounted.   The limit switch is connected into a machine's air line and can be used to detect the position of a machine part or workpiece.   Guide holes in the air line allow air to escape until the part to be detected is in the correct position.   When this happens, the guide holes are covered and allow the air pressure to build up, tripping the switch.   The air-operated type should not be used where the accuracy and versatility of a standard pressure switch are required.   It is generally used where a simple pressure device is required in addition to other limit switches and to avoid the different drilling and tapping of mounting holes which would be required if a standard pressure switch were used.

j. Remote-Cable Types.   Limit switches with remote-cable operators are used where limited space prevents mounting or maintaining a standard limit switch or where unfavorable environmental conditions such as excessive coolant or flooding are present. The remote operator and cable assembly are generally oiltight, dusttight, and watertight similar to the standard limit switch.   The cable assembly should have a swivel action to allow it to be installed without twisting.   Twisting the cable can shift the center member and cause a shift in the operating point of the limit switch.   Since remote-cable types require considerable care in mounting and installation, the manufacturer's instructions should be followed closely.   Remote-cable types should not be used in place of standard limit switches unless the application requires this type of limit switch.

k. Proximity Type.   Like standard mechanical limit switches, proximity limit switches convert mechanical motion or presence of an object into an electrical control signal.   The difference lies in the methods of actuation, since a proximity switch requires no physical contact with the actuating object.   Proximity switches are used in those applications where limit-switch operation is necessary but environmental or operational requirements do not permit the use of standard mechanical limit switches. Proximity limit switches are usually selected to detect parts or materials with small mass that would be unable to operate a standard limit switch.

18. Selection of Limit-Switch Enclosure.   The selection of a suitable enclosure depends on the environmental conditions present and the type of mounting required.

a. Surface-Mounting Type.   If possible, a surface-mounting limit switch should be used because it is the most common and has the simplest type of mounting.   Generally, only two holes need be drilled and tapped for switch mounting.   Two basic types of surface-mounted switches are available, plug-in and non-plug-in.

Non-Plug-in.   Wiring is made directly to snap-switch terminals.   When the switch needs replacing, wiring must be removed, conduit must be removed, and the complete limit switch must be replaced.

Plug-in.   Wiring is made to a plug-in receptacle which becomes a permanent part of the machine.   The plug-in assembly includes the actuator and the contact mechanism. When the switch needs replacing, only the cover screws need loosening and the plug-in portion is replaced—wiring is not disturbed.   Plug-in switches reduce user downtime and expense to a minimum.

Surface-mounting enclosures are generally oil-, water-, and dust-tight (NEMA 13, 4, and 2) and can be used on most machine-type applications where oil, coolant, dust, chips, etc., are present.   The oiltight limit switch, as it is generally called, is also a good choice for general-duty applications, as it will give maximum trouble-free life.   Cost may be a factor in selecting a less expensive switch for general-duty applications.

b. Multiple-Unit Types.   When several surface-mounting limit switches are mounted in a row and actuated by one or more cams, it may be more economical to install a multiple-unit limit switch.   Only one enclosure need be installed and only one conduit need be run.   Also, any interwiring between switches can be done internally. Lever arms are available to provide several different cam tracks if required.

c. Manifold-Mounting Type.   Occasionally it may be more convenient to run wiring into a wiring trough, panel, or raceway and bring the wire into the limit switch through the underside instead of through the standard conduit hole.   Manifold-mounting switches are provided with a gasket and hole in the underside, and the conduit hole is plugged.   The gasket prevents leakage into the manifold hole.   This type of mounting eliminates external conduit runs, and the wiring to the limit switch comes from within the machine cavity itself.

d. Flush-Mounting Types. These switches are used where a machine cavity is provided and a standard limit-switch box is not required. A gasket on the flush plate prevents entrance of contaminants. Wiring is brought in through the machine cavity.

e. Hazardous-Location Types. Explosive dusts or gases may be present in the atmosphere and require an explosionproof enclosure. NEMA 7 to 9 limit-switch enclosures are available to meet National Electrical Code requirements for Class 1, Groups B, C, and D gases, and Class 2, Groups E, F, and G dusts.

APPLICATION OF LIMIT SWITCHES

Once a limit-switch type has been selected, it must be applied properly to provide maximum operating life of the device and prevent unnecessary and premature failure.

19. Lever Type. As noted previously, the lever-type limit switch can handle approximately 90% or more of all applications. It is the easiest to apply because pretravel is generally not as critical as in plunger types. Usually the cam and/or lever arm can be adjusted to provide the desired trip point. More overtravel is available than on plunger types, and lever types now provide repeat accuracy as good as or better than plunger types. There are, however, many lever-type limit switches. Most manufacturers list a standard pretravel type which should usually be the first choice. The other types should be used only when necessary to obtain special characteristics.

a. Low-Differential Types. These are suitable for use where the differential (the distance between trip point and reset point) must be short. In general, these switches should not be selected to provide shorter trip angles, as either the lever arm or cam can be adjusted on standard pretravel switches to provide the required trip point. Low-differential types may also be selected for the better repeat accuracy available.

b. Light-Operating-Torque Types. Spring-return types are suitable for use where the object to be detected is lightweight and the operating force required for standard switches would prevent proper actuation. Lever arms with heavy rollers that must reset against gravity should be avoided. Gravity-return types are suitable for use where extremely light operating torques are required and where the actuator can be gravity-return. The lever arm should be mounted so it is straight down in the limit-switch initial position. Rod-type lever arms are recommended. These switches generally have no shaft seal and are not oiltight or watertight.

c. Maintained-Contact Types. These switches are suitable for use where the limit switch must remember that a cam or component has passed a certain point. It is a memory device and eliminates the need for a latching relay or a hold-down cam. The switch is reset by operating the lever arm in the opposite direction. Forked lever arms are recommended with maintained-contact switches. When the lever arm reaches the trip point, typically 45 to 50°, it will mechanically flip the remaining distance. This action eliminates contact recycling due to high impact or snapback of the lever arm.

20. Selection and Application of Lever Arms. *a. Length of Lever Arm.* Length of arm is determined by the distance between the cam and the limit switch. Length of arm is defined as the distance from the center of the roller to the center of the shaft. Where permissible, the recommended length of lever arm is 1½ in, because it is a standard length used by most manufacturers.

b. Rollers. The diameter and width of the roller are determined by the size and shape of the cam used. If the cam is free to shift, a larger diameter or width should be used. In general, large heavy rollers are to be avoided to minimize resetting and snapback problems. For general applications where the cam surface may be rough, a metal roller is recommended. Where the actuating surface may be scratched by a metal roller or where an explosionproof switch is used, nylon rollers are recommended. Also certain nylon-to-metal combinations may have wear characteristics superior to those of metal-to-metal. The recommended roller size is ¾ in diameter, ¼ in wide where possible.

c. Special Lever Arms. The standard lever arm listed by most manufacturers is a straight lever 1½ in long with a ¾-in-diameter, ¼-in-wide roller. These lever arms can be used in the majority of applications, but occasionally a special lever arm is required.

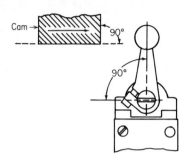

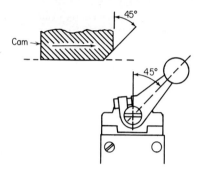

Fig. 18 Satisfactory for low cam speeds
—50 ft/min or less—nonoverriding cams.

Fig. 19 Recommended for moderate cam
speeds up to 90 ft/min, nonoverriding cams.

The following arms may be used:

1. Offset type. Used to obtain different cam-track dimensions.
2. Forked type. Used with maintained-contact lever-arm-type switches. With rollers on the same side, one cam trips and resets the switch. With rollers on opposite sides, one cam will trip the switch and a second cam will reset the switch.
3. Adjustable length. Used where the length of arm required is not known when devices are ordered or where length requirements vary.
4. Angular adjustable. Used where arm adjustment is critical. Arms generally can be adjusted through a 360° arc.
5. Ball-bearing-roller type. Used where abrasive dust, such as in cement mills, would cause undue wear of standard rollers and roller pins. Also used with high-speed cams.
6. One-way-roller type. Used with reversible cams where operation in one direction only is required.
7. Rod type. Used generally on conveyor systems or where the rod must be bent into a desired shape.
8. Spring-rod type. Used on conveyor or similar systems where a jam-up may occur. The material backing up against a standard rod arm could cause it to move in a non-operating direction, resulting in damage to the arm and/or the switch. The flexible spring rod allows movement in any direction and eliminates damage.

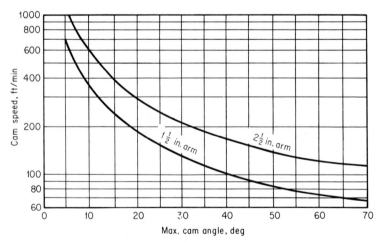

Fig. 20 Maximum cam angle for high-speed cams (Fig. 21).

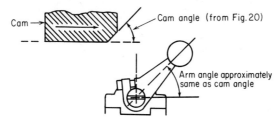

Fig. 21   For high-speed cams 90 ft/min or higher. For a given cam speed, find maximum cam angle from Fig. 20.

21. Cam Design—Lever Type.   Cams and limit switches can be applied with considerable accuracy. However, for the average application, a few simple design rules will suffice.

1. Adjust the lever arm parallel to the leading edge of the cam (Fig. 18). For maximum limit-switch life, the force applied to the lever arm by a cam should be perpendicular to the lever arm. Thus, the cam angle and lever-arm angle should be the same. The arrangement shown is satisfactory only at cam speeds below 50 ft/min. At higher speeds, the impact due to high lever acceleration causes excess roller bounce. Rapid deceleration occurs at the lower cam edge.

A good rule of thumb for moderate cam speeds (up to 90 ft/min) is to make the cam leading edge and lever-arm angle 45° (Fig. 19). Here lever-arm acceleration is less and deceleration is also less at the lower cam edge. Recommended cam and lever angles are shown in Figs. 20 and 21 for cam speeds greater than 90 ft/min.

2. Do not allow the lever arm to snap back. The cam trailing edge on overriding cams must be considered for maximum switch life (Fig. 22). Lever-arm snapback causes shock loads which reduce switch life. Also, with reversing cams, the trailing edge becomes a leading edge on the return stroke and a 60° maximum angle between lever arm and cam edge is recommended.

3. Overtravel should not be exceeded. For normal applications, overtravel of 5 to 15° beyond trip point should be used to allow for variations in machine or actuating cam. Additional overtravel available should preferably be used only for setup and emergencies, since extreme overtravel tends to shorten overall switch life.

22. Determining Cam Travel and Cam Rise.   Regardless of the cam angle and lever angle selected, total cam travel and cam rise can be calculated from Eqs. (1) and (2) and Fig. 23.

$$\text{Cam rise } B = L[\cos\theta - \cos(\theta + \phi)]$$
$$\text{Cam travel } A = C + D + E$$

(1)

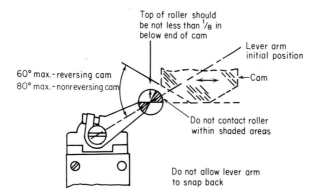

Fig. 22   For overriding cams up to 90 ft/min.

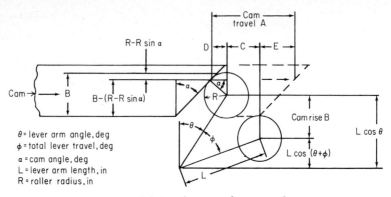

Fig. 23   Relation of cam speed to cam angle.

where $C = L[\sin (\theta + \phi) - \sin \theta]$
$\quad D = R \cos \alpha$
$\quad E = \tan \alpha[B - (R - R \sin \alpha)]$

$\quad$ Cam travel $A = L[\sin (\theta + \phi) - \sin \theta] + R \cos \alpha + \tan \alpha[B - (R - R \sin \alpha)] \qquad (2)$

$\quad$ Note that Eq. (2) applies only when the lower cam edge travels to the vertical center-line of the roller as shown.   If the cam does not travel this far, refer to Eq. (5).   If the cam travels beyond this point, add the additional cam travel to Eq. (2).

$\quad$ If the lower cam edge is rounded, smoother operation will result.   The cam arc radius (Fig. 24) can be determined from Eq. (3).

$$R_1 = \frac{B - (R - R \sin \alpha)}{\cos \alpha}$$

See Eq. (1) for B.

$$R_2 = \frac{R_1}{\sin [(90 - \alpha)/2]}$$

$$R_3 = R_2 \cos \left(\frac{90 - \alpha}{2}\right) \qquad (3)$$

$$R_3 = \frac{B - (R - R \cos \alpha)}{\cos \alpha \tan [(90 - \alpha)/2]}$$

$\quad$ When cam-rise and cam-travel data are required for small increments of lever-arm movement before the cam corner travels beyond the point of initial cam contact with the roller (distance $C_1 + E_1$ in Fig. 25), Eqs. (4) and (5) can be used.   For example,

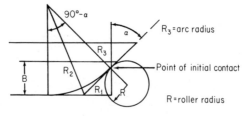

Fig. 24   Factors to consider for smooth limit-switch operation.

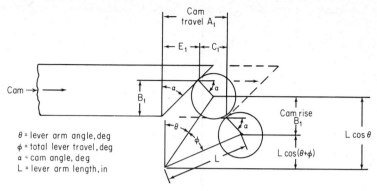

θ = lever arm angle, deg
ϕ = total lever travel, deg
a = cam angle, deg
L = lever arm length, in

Fig. 25 Cam-travel considerations.

when $\theta = 0°$, $\alpha = 0°$, $L = 1.5$ in, and $\theta = 45°$, $\alpha = 45°$, $L = 1.5$ in, computed data are given in Tables 2 and 3, respectively.

$$\text{Cam rise } B_1 = L[\cos\theta - \cos(\theta + \phi)]$$

$$\text{Cam travel } A_1 = C_1 + E_1 \tag{4}$$

where $C_1 = L[\sin(\theta + \phi) - \sin\theta]$
$E_1 = B_1 \tan\alpha$

$$\text{Cam travel } A_1 = L[\sin(\theta + \phi) - \sin\theta] + B_1 \tan\alpha \tag{5}$$

Note that Eq. (5) applies only when the lower cam edge shown does not travel beyond the point of initial cam contact with the roller as shown.

TABLE 2. Data Relating Cam Rise (A_1) and Cam Travel (B_1) to Total Lever Angle of Travel (ϕ) (Fig. 25) when $L = 1.5$ in, $\theta = 0°$, and $\alpha = 0°$

ϕ	B_1	A_1	ϕ	B_1	A_1	ϕ	B_1	A_1
1	0.0002	0.026	21	0.100	0.538	41	0.368	0.984
2	0.0009	0.052	22	0.109	0.562	42	0.385	1.000
3	0.002	0.078	23	0.119	0.586	43	0.403	1.023
4	0.003	0.105	24	0.130	0.610	44	0.421	1.042
5	0.005	0.131	25	0.140	0.634	45	0.439	1.061
6	0.008	0.157	26	0.151	0.658	46	0.458	1.079
7	0.011	0.183	27	0.163	0.681	47	0.477	1.097
8	0.015	0.209	28	0.176	0.704	48	0.496	1.115
9	0.018	0.235	29	0.188	0.727	49	0.516	1.132
10	0.023	0.260	30	0.201	0.750	50	0.536	1.149
11	0.028	0.286	31	0.214	0.773	51	0.556	1.166
12	0.033	0.312	32	0.228	0.795	52	0.577	1.182
13	0.038	0.337	33	0.242	0.817	53	0.597	1.198
14	0.045	0.363	34	0.256	0.839	54	0.618	1.214
15	0.051	0.388	35	0.271	0.860	55	0.640	1.229
16	0.058	0.413	36	0.286	0.882	56	0.661	1.244
17	0.066	0.439	37	0.302	0.903	57	0.683	1.258
18	0.073	0.464	38	0.318	0.923	58	0.705	1.272
19	0.082	0.488	39	0.334	0.944	59	0.727	1.286
20	0.090	0.513	40	0.351	0.964	60	0.750	1.299

Note. If arm length used is not 1.5 in, multiply A_1 and B_1 above by arm length/1.5.

TABLE 3. Data Relating Cam Rise (A_1) and Cam Travel (B_1) to Total Lever Angle of Travel (ϕ) (Fig. 25) when $L = 1.5$ in, $\theta = 45°$, and $\alpha = 45°$

ϕ	B_1	A_1	ϕ	B_1	A_1	ϕ	B_1	A_1
0	0.000	0.000	16	0.334	0.585	31	0.698	1.093
1	0.019	0.037	17	0.357	0.620	32	0.724	1.124
2	0.038	0.074	18	0.380	0.656	33	0.749	1.155
3	0.057	0.111	19	0.403	0.691	34	0.775	1.186
4	0.075	0.148	20	0.427	0.726	35	0.801	1.217
5	0.097	0.185	21	0.451	0.760	36	0.826	1.247
6	0.117	0.222	22	0.475	0.795	37	0.852	1.277
7	0.136	0.258	23	0.499	0.829	38	0.878	1.306
8	0.157	0.295	24	0.523	0.863	39	0.904	1.355
9	0.179	0.332	25	0.558	0.897	40	0.920	1.364
10	0.200	0.368	26	0.573	0.930	41	0.956	1.392
11	0.232	0.405	27	0.597	0.963	42	0.982	1.419
12	0.254	0.441	28	0.622	0.996	43	1.009	1.447
13	0.266	0.477	29	0.658	1.028	44	1.035	1.474
14	0.288	0.513	30	0.673	1.061	45	1.061	1.500
15	0.310	0.549						

Note. If arm length used is not 1.5 in, multiply A_1 and B_1 above by arm length/1.5.

23. Cam Design—Plunger Type. *a. Top- and Side-Push-Rod-Plunger Type.* The cam must actuate the push rod in line with the rod axis. Care must be taken to avoid exceeding the overtravel. A mechanical stop should be used where the possibility of excessive overtravel exists.

b. Top-Roller-Plunger Type. Because the bearing length is relatively short on top-plunger types, the maximum leading edge recommended is 30° (Fig. 26). Otherwise, the side thrust is too great and the plunger may be bent. The recommended leading edge is 10 to 15°. The leading and trailing edges of high-speed cams must have a very gradual rise. The lever type is shown in Fig. 21.

c. Side-Roller-Plunger Type. Side-plunger types generally have two bearings to support the plunger, and the maximum recommended leading edge can be 45° (Fig. 27). However, the recommended leading edge is the same as for the top-roller-plunger type, 10 to 15°.

d. Side-Push-Rod Plunger, Maintained-Contact. Operation of one plunger trips the contacts; operation of the opposite plunger resets the contacts. The cam must actuate the push rod in line with the rod axis. Care must be taken to avoid exceeding the overtravel, and a mechanical stop should be used where the possibility of excessive overtravel exists.

24. Cam Design—Wobble-Stick Type. The cam design is not critical for these types. The switches may be actuated by boxes, cams, or bottles, or the operator may be deflected by hand as in an emergency-stop device. Care should be taken to avoid ex-

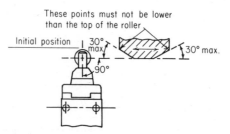

Fig. 26   Cam design for top-roller-plunger types.

cessive snapback of the wobble stick or cat whisker, as this tends to shorten switch life and may cause recycling or telegraphing of the contacts.

25. Application of Proximity Limit Switches. *a. Typical Applications.* A proximity limit switch may be used to obtain the following types of actuation:

1. Actuation by delicate materials or those with small mass. Thin aluminum foils will actuate a proximity switch without scratching, bending, or deflecting the material, since physical contact is not necessary.

2. Actuation through barriers. A proximity switch can sense without difficulty through any nonconductive, nonmagnetic material without disturbing the continuity of the device or the product.

3. Actuation by fast-moving objects. The proximity switch can detect fast-moving objects or can be used where high rates of operation (high duty cycles) are experienced such as in strip-mill automation or can-process lines. They are also ideal for counting, sorting, or inspection, giving long service life and savings in downtime and maintenance cost.

4. Actuation in contaminated atmospheres. Proximity switches can be used in contaminated atmospheres like those found on paint-spray lines and processes involving chemicals, lint, or applications which generate dust and dirt. They are immune to wet or high-humidity conditions or exposure to corrosive fumes or dust.

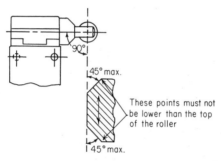

Fig. 27   Cam design for side-roller-plunger types.

b. Sensitivity Factor. Sensitivity is defined as the maximum distance between the transducer sensing surface and the object to be detected which will produce an output response. The following factors should be considered in applying the proximity limit switch:

1. The sensitivity required by the application. The sensitivity range is defined as the perpendicular distance from the center of the sensing coil to the surface of the item being detected.

2. The size of the item to be sensed. The sensitivity range of the proximity sensor is influenced by the area of the item used. If the area is equal to or greater than the sensing-coil diameter, the range will be maximum. When the area is less than the coil diameter, the sensing distance will be decreased proportionately.

3. The material of the item to be detected. Aluminum foil and steel are most easily detected and may be sensed to a greater range than low-resistance metal such as copper or brass.

4. The working area and mounting requirements. Most proximity switches may be mounted in any position. Some devices may be mounted side by side without adverse effects; however, other devices must have a minimum spacing between them.

5. The speed of operation required. Maximum rate of operation of the limit switch must be known.

26. Limit-Switch Contacts. *a. Standard Contacts.* Most limit switches are furnished with single-pole double-throw or two-pole double-throw contacts. Double-throw contacts should not be used on opposite polarities (Figs. 2 and 3).

b. Neutral-Position Contacts. These contacts are similar to single-pole double-

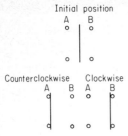

Initial position

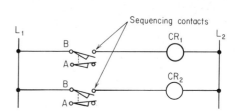

Counterclockwise Clockwise

Fig. 28 Neutral-position contacts.

Fig. 29 Limit-switch sequence contacts.

throw contacts except that they can assume a center off or neutral position (Fig. 28). Operating the limit switch in one direction closes one contact; operating it in the other direction closes the other contact. Thus the neutral-position type is a direction-sensing device.

c. Sequencing and Overlapping Contacts. Many two-pole limit switches can be adjusted so that one snap switch operates after the other snap switch. If the normally open contacts are wired as shown in Fig. 29, relay CR_1 will pick up before CR_2, providing sequencing action. Overlapping contacts can be obtained as shown in Fig. 30. The normally open contact of the upper switch will close before the normally closed contact of the lower switch opens, providing the contact overlap desired.

27. Pilot Lights. One or two neon pilot lights can usually be furnished in a limit switch to indicate the circuit condition at a glance. When connected in parallel with the normally open contact (Fig. 31), the light is normally on and operation of the switch turns the light off. When connected in parallel with the normally closed contact (Fig. 32), the light is normally off and operation of the switch turns the light on. Pilot lights are normally used on machines with many limit switches to make it easier to pinpoint limit-switch failure.

28. Installation of Limit Switches. *a. Mounting and Wiring.* The following general rules should be followed when installing a limit switch:

1. Make sure the electrical load is within the limit-switch contact ratings.
2. Do not connect the double-throw contacts to the opposite polarity.
3. Adjust the lever arm parallel to the leading edge of the cam; 45° is recommended.
4. Overtravel of the limit switch should not be exceeded; 5 to 15° overtravel is recommended.
5. Do not allow the lever arm to snap back. A gradual trailing edge should be used on cams which travel beyond the roller.
6. When possible, avoid mounting limit switches where they will constantly be exposed to coolant, chips, etc. Although designed for such applications, the switches will last longer when not exposed to these contaminants.
7. Make sure cover screws are tightened to assure good oiltight seal.
8. When possible, avoid use of fire-resistant coolants of the phosphate-ester type. Equipment exposed to these coolants requires special seals and gaskets.

b. Installation of Conduit. A limit-switch manufacturer has no control over leakage through the conduit system into a standard limit switch. Oiltightness is dependent upon a good conduit connection and seal. Recommendations for installing conduit to limit switches are as follows:

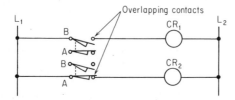

Fig. 30 Limit-switch overlapping contacts.

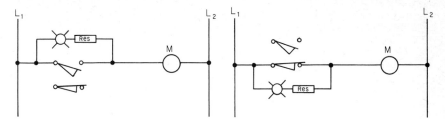

Fig. 31 Pilot light normally on.   **Fig. 32** Pilot light normally off.

1. A thread seal, conduit seal, or sealing bushing must be used around the conduit fitting to ensure an oiltight seal. Contrary to popular belief, a leakage path around the threads may exist, because the top of the conduit thread is flat and may not completely fill the thread grooves in the switch conduit.

2. If conduit leakage is severe, prewired and potted switches should be used.

3. Often a junction box fills with coolant and/or condensation which backs up into the limit switch through the conduit. A simple solution is to drill a hole in the bottom of the junction box to allow the liquid to drain out.

4. Experience has shown that the liquidtight type of flexible conduit does not consistently provide a good oiltight connection to the limit-switch conduit hole. UL Type SO, 600-V flexible-cord type used with a good sealing grip has proved much better and is recommended.

PRESSURE SWITCHES

Pressure switches are control-circuit devices which respond to the varying pressure of mediums such as air, water, oil, gases, and hydraulic fluids. Generally there are three basic types — diaphragm-, bellows-, and piston-actuated.

29. Diaphragm-actuated Switches. These types are designed for the automatic control of heavy-duty electrically driven air compressors and liquid-pumping units. Diaphragm-actuated pressure switches are selected where the medium is air or liquids that are not harmful to the diaphragm material, where maximum pressures do not exceed approximately 500 lb/in², where sensitive operation is required, and where low cost is a factor. Pressures greater than approximately 500 lb/in² would deform or damage the diaphragm and cause poor repeat accuracy or failure. In most cases oil-resistant diaphragms can be obtained for applications requiring this feature.

Two-way and three-way valves can be furnished in applications where the pressure does not exceed approximately 250 lb/in² to exhaust the air trapped between the check valve and the compressor head automatically at the end of a pumping operation. The next pumping operation starts against no load.

30. Bellows-actuated Switches. These types are designed for use in control circuits on machine tools, lubricating systems, and welding equipment. Bellows-actuated types are used where the medium is air, water, oil, and many other liquids and gases and where maximum pressures do not exceed approximately 2500 lb/in². The bellows are generally made of brass, and the pressure medium should not be corrosive to brass. Stainless-steel or monel bellows are available for corrosive applications. Monel is particularly well suited for ammonia service.

To provide maximum life expectancy for the bellows type, the pressure switch should be matched to the application. Bellows life can vary from a few thousand to millions of operations depending upon operating pressure, bellows stroke, frequency of operation, presence of corrosive elements, and pressure surges.

There are two types of operations on the bellows type. In one the switch directly controls the pressure and the pressure varies only between the high and low trip points. Thus the bellows and operating lever travel only a small percentage of the total possible stroke.

In the second type of operation the switch is used as a signal device only and does not control the pressure directly. The pressure can then vary between zero and some point beyond the high trip point, possibly to the maximum allowable pressure. Bellows and operating lever thus operate over a larger percentage of the total possible stroke.

The life of the bellows type in the first application may be several times that in the second application.

31. Selection of Bellows Types.   The following points should be remembered in selecting a bellows-type pressure switch to obtain maximum life:

1. Select a switch which will meet the requirements of the application with the highest maximum allowable pressure.

2. Select a switch where the actual working pressure is the smallest possible percentage of the maximum allowable pressure consistent with acceptable cost. The less flexing there is of the bellows, the longer will be the switch life. For example, if the working pressure is 100 to 185 lb/in², a switch with 600 lb/in² maximum allowable pressure will have longer life than one rated 255 lb/in². The cost of these two devices is approximately the same. However, a device rated 2000 lb/in² maximum will have even longer life, but the cost is approximately double that of a 255 or 600 lb/in² maximum device.

3. Make sure the pressure medium is not corrosive to the bellows material.

32. Piston-actuated Switches.   These types are designed for use in control circuits of machine tools and high-pressure lubricating systems. Piston-actuated types are selected where the pressure medium is oil or hydraulic fluids and where maximum pressures may run as high as approximately 25 000 lb/in². Other pressure mediums should not be used because deposits (such as mineral deposits from water) between the piston and mating surfaces may cause erratic operation or jamming. Most piston types have damping mechanisms to eliminate false or premature operation due to surges and pulsations in the system. A timing relay may also be used to eliminate these effects.

Slight oil leakage past the piston is normal on devices that have no piston seal. A tapped drain hole in the cylinder wall on the low-pressure side of the piston permits piping of the leakage oil back to the reservoir. This hole should never be plugged nor should oil-return lines be connected to a high-volume discharge system because backpressure on the drain side can damage the diaphragm. Devices with piston seals have no leakage, and although an oil-return line is not needed, the drain hole should never be plugged.

33. Operating Characteristics.   To select and apply a pressure switch properly, an understanding of the operating characteristics is necessary.

a. Maximum Allowable Pressure.   Maximum pressure that can be applied without damaging the switch.

b. Differential.   The pressure difference between the trip point and reset point of the switch (or between the high operating point and the low operating point).

c. Range.   Variation of pressure settings to which the pressure switch can be adjusted. The relationship of the differential to the specified range settings varies depending on the type of pressure switch used.

34. Diaphragm-Type Range.   The range for diaphragm-type switches is specified to cover the adjustment of the cutout pounds per square inch. In a pump application, the pressure-switch contacts close when the pressure drops to a certain point, causing the pump to "cut-in." When the pressure rises to a certain point, the contacts open and cause the pump to "cutout." The differential adjustment specified is subtracted from the cutout pounds per square inch to obtain the cut-in pounds per square inch. For example, a switch has a range adjustment for cutout pounds per square inch of 20 to 180 lb/in² and a differential adjustment of 10 to 40 lb/in². If the switch is adjusted for 150 lb/in² cutout, the cut-in can be adjusted for 110 minimum (150 minus 40) or 140 maximum (150 minus 10).

35. Bellows-Type Range.   The range of bellows-type switches is specified to cover the adjustment of an operating point on falling pressure. In a machine-tool application, contacts must operate if the system pressure drops to a certain level. The differential specified is added to range pounds per square inch setting to obtain the high operating point. For example, a switch has a range pounds per square inch adjustment of 10 to 300 lb/in², a differential adjustment of 25 to 125 lb/in², and a maximum allowable pressure of 600 lb/in². If the range adjustment is set to 300 lb/in², the contacts will operate when pressure falls to 300 lb/in². The high operating point can then be adjusted to 325 lb/in² minimum (300 plus 25) or 425 lb/in² maximum (300 plus 125). If the pressure switch does not control the pressure directly, pressure can increase to 600 lb/in² maximum without damaging the switch.

36. Piston-Type Range. The range for these high-pressure hydraulic switches is specified to cover the adjustment of the operating point on rising pressure. The differential specified is subtracted from the range pounds per square inch setting to obtain the low operating point. For example, a switch has a range pounds per square inch adjustment of 400 to 3000 lb/in², a differential adjustment of 100 to 400 lb/in² and a maximum allowable pressure of 10 000 lb/in². If the range adjustment is set to 2000 lb/in², the low operating point can be adjusted to 1600 lb/in² minimum (2000 minus 400) or 1900 lb/in² maximum (2000 minus 100).

37. Other Applications. When it is necessary to have control over the difference between two pressures, a differential-pressure switch may be used. Two diaphragms, bellows, or pistons are used in these types, and the contacts operate at a predetermined difference in the two pressures applied.

Air-hydraulic-ratio pressure switches are similar in operating principle to the differential switches except that a diaphragm is used on the air side and a piston on the hydraulic side. The controls automatically adjust to changes in the supply pressure and operate on a ratio of air pressure to hydraulic pressure. For example, any given pressure applied on the low-pressure-diaphragm side of the unit must be balanced out by a pressure approximately 25 times as great on the hydraulic side before the switch will operate. This principle is particularly useful on certain press installations in which air pressure is used to drive a hydraulic cylinder to develop clamping or working pressure. Use of the ratio switch ensures that the hydraulic pressure is proportional to the air-supply pressure before the work can be performed. A variety of ratios are available. The ratio switch can also be used for remote control of the hydraulic pressure.

Certain applications require dual functions or sequencing. For example, as pressure reaches a certain level, an alarm sounds, but if pressure increases further, the system is shut down. One dual-stage switch can be used on these "alarm-shutdown" applications in place of two separate pressure switches. Here two sets of contacts are used, with one operating at a slightly lower pressure than the second.

TEMPERATURE SWITCHES

Industrial temperature switches are designed for the automatic control of temperature on diesel engines and large compressors, oil-temperature control for various machine tools, and a variety of liquid-bath applications.

They are rugged, reliable, industrial-type controls and should not be confused with light-duty residential or commercial heating controls or refrigeration controls.

There are two general categories of temperature switches, non-cross-ambient and cross-ambient. The non-cross-ambient is further broken down into heating-type and cooling-type.

38. Non-Cross-Ambient. The heating type of control should be used on applications where the temperature at the sensing element or bulb will be at a higher temperature than the ambient around the bellows housing (Fig. 33) when the switch is operating between the trip and reset points. Conversely, the cooling type should be used on applications where the bulb will be at a cooler temperature than the ambient around the bellows housing (Fig. 34) when the switch is operating between the trip and reset points. It is only under these conditions that control is maintained.

39. Cross-Ambient. The cross-ambient type is a combination of the heating and cooling types and should be used in applications where control must be maintained whether the temperature in the bulb is hotter or cooler than the ambient temperature around the bellows housing (Fig. 35). One example of an application for a cross-ambient switch is a diesel locomotive operating in or through desert country where the ambient temperature during the day may be higher than the jacket-water temperature but drops considerably below the jacket-water temperature at night.

40. Switch Mechanism. The basic mechanism of the temperature switch is a pressure-switch mechanism which has the pressure-sensing element containing a measured liquid fill. The temperature switch is then simply a pressure switch which responds to changes in temperature due to the vapor pressure developing in the sensing element. Vaporization is taking place at the same rate as condensation, and the system is in balance. The surface of the fill liquid is called the liquid-vapor interface and is the control point. The control point must remain in the bulb for proper control.

The range specified is referenced to the operating point on falling temperature. The range adjustment should be made first to set the low operating point.

The differential is the degrees of temperature between operating points on the switch. The differential adjustment is the second adjustment to be made and affects the operating point on rising temperature only.

41. Heating-Type Operation. In a typical heating-type temperature switch, control is maintained only when the temperature of the bulb is higher than the ambient around the bellows housing (Fig. 33). If at any time the temperature in the bulb becomes cooler than the temperature of the bellows housing, condensation of the liquid fill will take place in the bulb. Since the bulb is not large enough to contain all the fluid, vaporization will take place in the bellows housing and the capillary transferring the liquid-vapor interface (the control point) from the bulb to the bellows or capillary. The

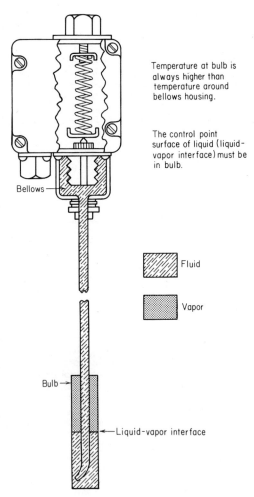

Fig. 33 Non-cross-ambient, heating-type temperature switch.

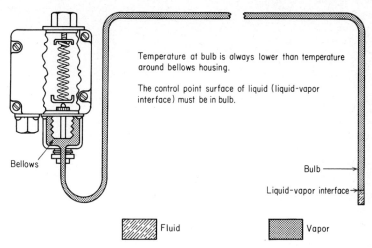

Temperature at bulb is always lower than temperature around bellows housing.

The control point surface of liquid (liquid-vapor interface) must be in bulb.

Bellows

Bulb ⟶

Liquid-vapor interface ⟶

Fluid   Vapor

Fig. 34   Non-cross-ambient, cooling-type temperature switch.

switch will no longer properly control the temperature in the bath because it will be responding to the temperatures surrounding the bellows and capillary where the liquid-vapor interface is now located.

To see how a typical heating-type temperature switch operates, assume a switch setting of 150 to 180°F.   The bulb is immersed in a bath, and when the temperature of the bath is 180°F, the vapor pressure in the bulb is 4 lb/in². This pressure is felt by the

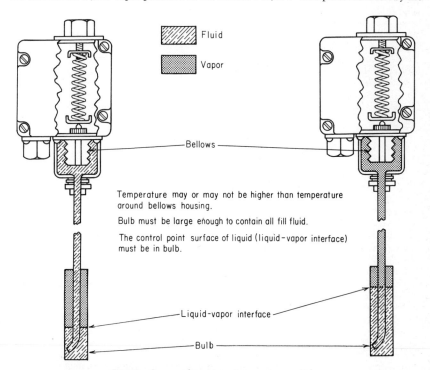

Fluid

Vapor

Bellows

Temperature may or may not be higher than temperature around bellows housing.

Bulb must be large enough to contain all fill fluid.

The control point surface of liquid (liquid-vapor interface) must be in bulb.

Liquid-vapor interface

Bulb

Fig. 35   Cross-ambient-type temperature switch.

bellows of the switch through hydraulic action. As the temperature increases, the vapor pressure increases. When the temperature reaches 180°F, the vapor pressure is 21 lb/in², enough to trip the switch. On decreasing temperature, the vapor pressure also decreases. When the temperature drops to 150°F, which is equal to a vapor pressure of 4 lb/in², the switch resets.

42. Cooling-Type Operation.   The basic principle of operation of a typical cooling-type temperature switch is the same as the heating type. To maintain control, however, the bulb must always be cooler than the bellows temperature (Fig. 34). Only a small amount of fluid is placed in the element so that the liquid-vapor interface, or control point, remains in the bulb and limits the amount of pressure developed at room temperature. Under normal operating conditions, the vapor is in direct contact with the bellows, since the capillary and bellows is the hotter portion of the element.

43. Cross-Ambient-Type Operation.   The cross-ambient-type temperature switch is used where it is desired to maintain control whether the ambient temperature is higher or lower than the bulb-controlling temperature. Notice that the bulb on the cross-ambient switch on the right in Fig. 35 is large enough to contain all the liquid fill. This size is necessary so that when a crossover occurs and the temperature around the bulb becomes cooler than the temperature around the bellows, the liquid-vapor interface remains in the bulb and control is maintained. For this reason the longer the capillary, the larger the bulb must be.

44. Effects of Barometric Change.   The vapor-pressure type of temperature switch is affected by barometric change. Temperature switches are generally set for the barometric reading at sea level. If the switch is used at altitudes higher than sea level, it will result in a downward shift of the temperature setting. This downward shift varies depending on type of fill, altitude, and temperature settings. Normally, the shift is not enough to make a big difference in the setting. At extremely high altitudes, however, the error is appreciable. Using the previous heating-type-switch example with a setting of 150 to 180°F at an altitude of approximately 2300 ft above sea level, the operating points are 146.8 to 177.9°F. At Denver, a mile high, the operating points on the same device are 142.8 to 175.3°F. It is evident that in critical applications, correction should be made. These data can usually be obtained from the temperature-switch manufacturer. Solid-metal expansion-type pressure switches are not affected by barometric changes as is the vapor-pressure type, but these are not as sensitive and are limited to direct-connected bulbs (see Art. 45).

45. Mountings.   Temperature switches are available with two types of mountings. The direct-connected type may be screwed into a tank with no accessories required. However, wells are available if the user wishes to avoid draining the tank when the switch is removed. The capillary and bulb type usually comes with 6 ft or more of capillary tubing and allows remote operation.

TYPICAL APPLICATIONS OF TEMPERATURE SWITCHES

46. Diesel Locomotives.   On diesel locomotives, temperature switches have been used to control fans for cooling the water jacket. Several switches are used to control fans, and another is used for safety shutdown.

When the water temperature of the water jacket reaches a predetermined temperature, the first fan starts drawing in outside air to cool the water jacket. Should the water temperature continue to rise, the second fan is cut in, then the third, etc. A safety shutdown follows if the water temperature continues to rise.

Switches for this type of service are usually cross-ambient types.

47. Air Compressors.   Temperature switches are used on large air compressors for sensing the temperature in both air and water used for cooling. The switch will shut down the compressor if the air or water becomes too hot.

A temperature switch on a compressor that takes air in from the outside is used to control heaters. When the outside air becomes too cold, it may start to frost up the intake. The temperature switch then cuts in the heaters to heat up the air in the intake.

48. Liquid Baths.   When it is desirable to maintain a liquid bath at a certain temperature above the ambient, a temperature switch may be used to control heater elements.

When the desired temperature is below the normal ambient, the switch controls cooling elements.

FLOAT SWITCHES

Float switches are used in two basic applications:
1. Automatic control of open-tank or sump-pump equipment
2. Automatic control of closed-tank condensate pumps where it is necessary to regulate the liquid level

49. Open-Tank Switches. The float-rod operating lever may be designed for use with a free-moving vertical rod attached to a float with a tapped hole at the top. Stop collars attached to the rod on either side of the lever provide the operating force when actuated by the float. These types are recommended where short lengths of rod and small liquid-level changes are required.

In some applications the liquid level may vary over a wide range and require a very long float rod. If there is not much overhead room, a free-moving rod can hit the top before tripping the float switch. In these applications a fixed rod and a center-hole float which can move freely are recommended. Stop collars above and below the float provide the operating force. Liquid-level changes are adjusted by moving the stop collars. When long rods are used to control relatively large liquid-level changes, the weight of the rod and the stops may be more than the switch can support. In this case, compensating springs may be required.

50. Closed-Tank Switches. Closed-tank float switches are furnished with a watertight flange and mount on the side or the top of the tank.

The angular-movement type mounts on the side of the tank. Movement of the float rod is transferred through a bellows seal to the outside switch. The float is rigidly attached to the float rod.

The vertical-movement type mounts at the top of the tank and is used in applications involving restricted tank space where large liquid-level changes are required. The center-hole float on a fixed rod is generally used. Adjustment of the float travel is made by moving stop collars on either side of the float.

MECHANICAL ALTERNATORS

Mechanical alternators are designed to provide a simple positive means of mechanically alternating the operation of two pumps installed in a duplex system with a common tank. The function of these controls is to provide alternate cycles of operation to ensure even wear of the pumping units. They thus eliminate the problems of long standby for one of the two pumps. These alternators further provide the additional function of starting the second pump in cases where extra capacity under peak conditions is required.

FOOT SWITCHES

The foot switch is a manually operated control-circuit device whose contact action can be controlled by the operation of a foot pedal. Foot switches are used where a machine operating cycle must be manually started and the operator's hands are both engaged in loading or handling the materials involved. Typical applications include resistance-welding machines, punch presses, industrial sewing machines, and small machine tools.

51. Standard-Duty. Standard-duty foot switches are used for general-purpose, light- to medium-duty applications such as stapling machines, banding machines, and conveyors. They are the most economical foot-switch types. Standard-duty foot switches generally have NEMA B600 or B300 ratings and can control magnetic starters through NEMA size 4.

52. Heavy-Duty. These foot switches are used for heavy-duty or severe-duty applications requiring frequent operation. They are generally rated NEMA A600 and can generally control magnetic starters NEMA size 5 or larger. Heavy-duty foot switches are usually available in a wider variety of enclosures than the standard-duty types. General-purpose, watertight, dusttight, and hazardous-duty enclosures are common.

Pedal locations can be on the right or left or on both sides of the switch to handle a variety of applications.

53. Heavy-Duty Oiltight.   These foot switches are especially constructed to prevent entrance of oil, coolant, and dust into the enclosure and have a NEMA 13 rating. They are used for heavy-duty service or applications in which industrial dusts, cutting oils, etc., are encountered.

Heavy-duty oiltight foot switches are generally rated NEMA A600 and are the most flexible of the three types.   Some of the features generally available are single-pole double-throw or two-pole double-throw contacts, two-stage or sequencing contacts, ratchet type, maintained contacts, spring-return or mechanical-latch types, and pedal guards with or without side shields.   The ratchet-type contacts provide operation on alternate strokes of the foot pedal similar to the light-dimmer switch in an automobile. Pedal guards and side shields are both used to prevent accidental operation of the foot switch.

CONTROL RELAYS

Relays are magnetically operated control-circuit devices that make and break the control circuit of magnetic starters, contactors, other relays, and control-circuit-duty solenoids.   Relays make the control-circuit decisions and provide the proper signals and signal paths called for by various machine operating sequences.   Sometimes called logic elements, relays are the "brains" of a control system.   Relays are really small contactors that are designed for use only in control circuits rather than power circuits.   Typical relay loads as compared with contactor loads are as follows:

Relays, control-circuit-duty loads	*Contactors, power loads*
Relay coils	Motors
Timer coils	Heavy-duty solenoids
Contactor coils	Lighting
Starter coils	Resistor banks
Control-circuit-duty solenoids	Capacitor banks

TYPES OF RELAYS

Industrial-control relays can be classified into the following basic types:
1. Electrically held relays
2. Latching or mechanically held relays
3. Timing relays

54. Electrically Held Relays.   Industrial-control relays are generally classified as general-purpose or machine-tool-type relays.

a. General-Purpose Relays.   A general-purpose relay is designed for light-duty applications requiring one or two poles and infrequent operation.   They are the least costly of all relay types and are smaller physically.   Many have tube-socket-type mounting.   Although higher voltage ratings are available, 150 V maximum is the most common rating.

b. Machine-Tool Relays.   These relays are designed for use in complex machine-tool panels, material-handling equipment, and automatic processing systems where operation is frequent and long, reliable life is required.   A choice of 600-V relays or the more compact 300-V relays is available.   Relays can be furnished with fixed contacts or wherever field modification is required, with convertible contacts.   Whereas general-purpose relays usually have 1 to 3 poles and power contactors 1 to 5 poles, industrial-control relays commonly have up to 8, 10, or 12 poles.

55. Operation of Electrically Held Relays.   An electrically held relay must have power applied to the coil to remain picked up.   If power is removed, the relay will drop out.

The heart of a relay is an electromagnet which consists of a coil of wire placed on an iron core (Fig. 36).   Briefly reviewing operation, when current flows through the coil, the iron of the magnet becomes magnetized, attracting the iron bar, called the armature. The electromagnet can be compared with the permanent magnet (Fig. 37), as both would attract the iron bar.

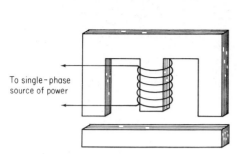

Fig. 36 Electromagnet.

Fig. 37 Permanent magnet.

To single-phase
source of power

The field of the permanent magnet, however, will hold the armature against the pole faces of the magnet indefinitely until it is physically pulled away. In the electromagnet, interrupting the current flow through the coil causes the armature to drop out.

The magnetic circuit of a relay consists of a magnet assembly, a coil, and an armature. The magnet assembly is the stationary part of the magnetic circuit. The coil is supported by and surrounds part of the magnet assembly to induce magnetic flux into the magnetic circuit. The armature is the moving part of the magnetic circuit. When it has been attracted to its sealed-in position, it completes the magnetic circuit.

In the construction of a relay, the armature is mechanically connected to a set of contacts so that when the armature moves to its closed position, the contacts also close (Fig. 38). To provide maximum pull to close the contacts and to help ensure quietness, the faces of the armature and magnet assembly are ground to a very close tolerance.

56. Air Gap. When a relay's armature has sealed in, it is held closely against the magnet assembly; however, a small gap is always deliberately left in the iron circuit. When the coil is deenergized, some magnetic flux (residual magnetism) always remains, and if it were not for the air gap in the iron circuit, the residual magnetism might be sufficient to hold the armature in the sealed position.

57. Shading Coil. A shading coil is a single turn of conducting material, generally copper or aluminum, mounted in the face of the magnet assembly (Fig. 39). The alternating main magnetic flux induces currents in the shading coil (similar to the secondary of a transformer), and these currents set up an auxiliary magnetic flux which is out of

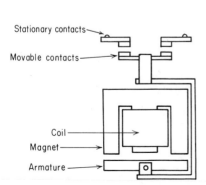

Stationary contacts

Movable contacts

Coil
Magnet
Armature

Fig. 38 Basic relay construction.

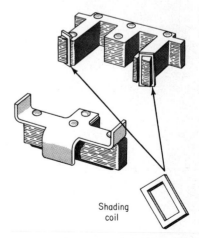

Shading
coil

Fig. 39 Magnet assembly with shading coils.

phase with the main flux (Fig. 40). The auxiliary flux produces a magnetic pull out of phase from the pull due to the main flux, and this pull keeps the armature sealed in when the main flux falls to zero, which occurs 120 times per second with 60-Hz control voltage. Without the shading coils, the armature will tend to open each time the main flux goes through zero. Excessive noise, wear on the magnet faces, and heat will result.

58. Magnet-coil Inrush and Sealed Currents. When the relay is deenergized, there is a large air gap (not to be confused with the built-in air gap in the magnetic circuit) since the armature is at its farthest distance from the magnet. The impedance of the coil is relatively low because of the air gap and is mainly the resistance of the coil winding. When the coil is initially energized, it draws a relatively high current. As the armature moves closer to the magnet assembly, the air gap is reduced. The reduction in air gap increases the impedance in the coil, and the coil current decreases. When the armature has sealed in, the final current is referred to as the sealed current. The inrush current, at the moment the coil is energized, is approximately six to ten times the sealed current. Thus, control-circuit-duty devices require contact make ratings ten times greater than the break ratings.

Ac magnet coils should never be connected in series. If one device seals in ahead of the other, the increased circuit impedance will reduce the coil current so that the slow device will not pick up or, having picked up, will not seal. Ac coils should be connected in parallel.

Magnet-coil data are usually given in voltamperes (VA). For example, given a relay whose coils are rated 120 VA inrush and 12 VA sealed, the inrush current of a 120-V coil is 120/120 or 1 A and the sealed current is 12/120 or 0.1 A.

59. Coil Voltages.

a. Pickup Voltage. The minimum control voltage which will cause the armature to start to move is called the pickup voltage.

b. Seal-in Voltage. The seal-in voltage is the minimum control voltage required to cause the armature to seat against the pole faces of the magnet.

c. Dropout Voltage. If the control voltage is reduced sufficiently, the relay will drop out. The voltage at which this happens, called the dropout voltage, is somewhat lower than the seal-in voltage.

60. Magnet-Coil-Voltage Variations. NEMA standards require that a magnetic device operate properly at varying control voltages from a high of 110% to a low of 85% of rated coil voltage. This range, established by design, ensures that the coil will withstand a specified temperature rise at voltages up to 10% higher than rated voltage and that the armature will pick up and seal in, even though the voltage may drop to 15% below the nominal rating.

If the voltage applied to the coil is too high, the coil will draw more than its designed current. Excessive heat will be produced and will cause early failure in the coil insulation. The magnetic pull will be too high and will cause the armature to slam home with excessive force. The magnet faces will wear rapidly, leading to shortened life for the relay. In addition, contact bounce may be excessive, resulting in reduced contact life.

Low control voltage produces low coil currents and reduced magnetic pull. If the voltage is greater than the pickup voltage but less than the seal-in voltage, the relay may pick up but will not seal. With this condition the coil current will not fall to the sealed value. Since the coil is not designed to carry continuously a current greater

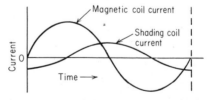

Fig. 40   Magnet-coil and shading-coil current displacement.

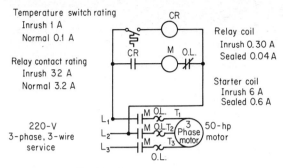

Fig. 41   Current amplification with a relay.

than its sealed current, excessive heat will be produced and it will burn out. The armature will also chatter, and in addition to the noise produced, excessive wear on the magnet-pole faces results. If the armature does not seal, the contacts will not close with adequate pressure. Arcing and possible welding of the contacts will occur as the relay attempts to carry current with insufficient contact pressure.

61. Application of Electrically Held Relays. Relays are generally used to amplify the contact capability or multiply the switching functions of a control-circuit device. Figure 41 represents a current amplification. The ampere rating of the temperature switch is too low to handle current drawn by the starter coil M. A relay is interposed between the temperature switch and starter coil. The current drawn by the relay coil CR is within the rating of the temperature switch, and the relay contact CR has a rating adequate for the current drawn by the starter coil.

Figure 42 represents a voltage application. Here the voltage rating of the temperature switch is too low to permit its direct use in a starter control circuit operating at some higher voltage. In this application, the coil of the relay and control-circuit device are wired to a low-voltage source of power compatible with the rating of the control-circuit device. The relay contact, with its higher voltage rating, is then used to control the operation of the starter.

Figure 43 represents another use of relays which is to multiply the switch functions of a control-circuit device with a single or limited number of contacts. In the circuit shown, a single-pole push-button contact can, through the use of an interposing 6-pole relay, control the operation of a number of different loads such as a pilot light, starter, contactor, solenoid, and timing relay.

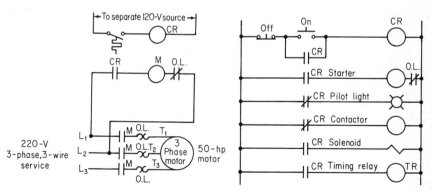

Fig. 42   Voltage amplification with a relay.

Fig. 43   Multiplying control device functions with a relay.

SELECTION OF ELECTRICALLY HELD RELAYS

62. Voltage. In selecting a relay for a particular application, one of the first steps is to determine the control voltage at which the relay will operate. Relays are available in 150-, 300-, and 600-V types. Because most control circuits today use 120 V, 150-V relays will theoretically do the job. However, relays are generally chosen for other reasons such as ease of converting contacts, physical size, ease of wiring, ease of maintenance, number of contacts required, and probably the most important factor, reliability.

The 150-V relays are usually general-purpose tube-socket types, although industrial 150-V relays are beginning to make their appearance. The choice then is generally between a 300- and 600-V industrial relay.

Most 300-V relays are selected based on small size, convertible contacts, availability of attachments which allow conversion to latching or timing relays, or the addition of 4 or 8 poles to the basic 4-pole relay.

The 600-V relays are generally larger physically than the 300-V types and are selected where the higher voltage rating is actually required and where the larger size makes the relay easier to work on. Convertible or fixed contacts and the usual attachments are also available. Many 300- and 600-V relays are the same basic device and are referred to as 300/600-V relays.

63. Contacts. Contacts may be fixed either normally open or normally closed or convertible from one to the other. Fixed contacts are selected where contact arrangements are established or where it is desirable that no one change the contact arrangement either during assembly of the panel or in the field. Convertible contacts are selected either so that any last-minute specification changes can be easily accomplished or for simplified stocking purposes. Most manufacturers list normally open contacts only with normally closed versions available at higher prices. Specifying only normally open contacts simplifies inventory, and the required contact arrangements can be set up during the assembly period.

64. Overlapping Contacts. Occasionally relay contacts are required where one contact must operate slightly before the other. Generally it is a normally open contact closing before a normally closed contact opens and is used to provide a positive signal during the overlapping period. Overlapping contacts may be used to shift a control signal from a momentary input LS_2 to another circuit without interrupting the signal to the coil (Fig. 44). Normally open contact CR_1 must close before the normally closed contact CR_1 opens to ensure that the relay coil CR_1 will pick up and seal in. If the normally closed contact CR_1 opens too soon, the relay will drop back out. Two contacts, a normally open and a normally closed, are always involved in each pair of overlapping contacts.

65. Attachments. Most relays are basic 4-pole or 6-pole devices. Additional poles can be added to the top or side of the relay. Latching attachments can be added to convert from electrically held to latching relays. Timer attachments can also be added to provide either on-delay or off-delay functions. The various attachments provide the basic elements for a complete logic control system and, similar to convertible contacts, simplify stocking and reduce inventory.

Other relay types are built with the above functions as an integral part of the relay and no attachments are furnished. Selection of the latter type might be based on eliminating a possible need for on-the-job adjustment of the attachments after installation and the problems that might follow misadjustment.

66. Specifications. The selection of relays is one of the most complicated of all

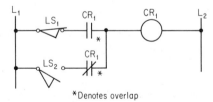

*Denotes overlap

Fig. 44 Overlapping contacts.

control-circuit devices because of the many types available. A logical selection should be based on the following specifications:

Coil Voltage
☐ _____ Volts ac _____ Hz
☐ _____ Volts dc
☐ + _____ % of nominal
☐ − _____ % of nominal

Temperature Range
☐ _____ °C to _____ °C
☐ _____ °F to _____ °F

Contacts
☐ Convertible   ☐ Isolated   ☐ Visible
                 (single-throw)   (for inspection)
☐ Fixed   ☐ Nonisolated   ☐ Enclosed
                 (double-throw)
☐ _____ N.O.
☐ _____ N.C.

Contact Load Rating
☐ _____ Volts ac   ☐ Resistive
☐ _____ Volts dc   ☐ Inductive
☐ _____ Amperes make   ☐ Duty cycle: _____
☐ _____ Amperes break   Operations per
☐ _____ Amperes continuous   _____

Attachments   *Coil*
☐ _____ Latching   ☐ Replaceable
☐ _____ Timer   ☐ Not replaceable
☐ _____ Additional poles

Size   *Mounting*
☐ _____ Height   ☐ Panel   ☐ Plug-in
☐ _____ Width
☐ _____ Depth

Termination
☐ Pressure wire   ☐ Slip-on connectors
   connectors   ☐ Solder
☐ Pan-head screws   ☐ Other _____
   (for ring or
   spade lugs)
Wire Size AWG
☐ 12   ☐ 14   ☐ 16   ☐ 18   ☐ Other _____

67. Maintenance of Relays. The maintenance required on modern relays is minimal. The use of molded coils has virtually eliminated burned-out coils. Properly applied contacts may never need replacing unless the relay is controlling a load near its maximum contact rating or is operated at a relatively high rate. Contacts should be inspected at regular intervals to ensure proper operation. Contacts should not be filed, as this only wastes contact material. Pitting and discoloration of contacts is normal and, excluding dry-circuit applications, will have no effect on operation.

68. Troubleshooting Relay Problems. Table 4 lists the common relay problems and the recommended remedies.

LATCHING RELAYS

A latching relay is similar to a control relay, but with a latch mechanism that holds in the contacts even when power is shut off to the latch coil. It serves as a memory device in case of power failure and can also be used where ac hum would be objectionable. An unlatch coil must be energized to release the contacts. An attachment is often furnished to convert a basic control relay to a mechanically held relay. Although commonly called mechanically held relays, the term latching relay is more correct, as the latching means may be either mechanical or magnetic.

69. Latching-Relay Operation. As shown in Fig. 45, pressing the "on" button energizes the latch coil *CR* (*L*), causing the relay to pick up and latch in. The relay will remain latched in even though the "on" button is released. To unlatch the relay,

TABLE 4. Troubleshooting Common Relay Problems

Trouble	Cause	Remedy
Contact chatter	1. Broken shading coil	1. Replace shading coil
	2. Poor contact in control circuit	2. Improve contact or use holding-circuit interlock (3-wire control)
	3. Low voltage	3. Correct voltage condition. Check momentary voltage dip during starting. Check control transformer regulation
Contact welding or freezing	1. Abnormal inrush current	1. Use higher-rated device or check for grounds, shorts, or excessive load current
	2. Insufficient contact-tip pressure	2. Replace contact springs; check contact carrier for deformation or damage
	3. Low voltage preventing magnet from sealing	3. Correct voltage condition; check momentary voltage dip during starting
	4. Foreign matter preventing contacts from closing	4. Clean contacts with Freon or other approved cleaner
	5. Short circuit	5. Remove short fault and check to be sure fuse or breaker size is correct
Short contact-tip life	1. Filing or dressing	1. Do not file contact tips. Rough spots or discoloration will not harm tips or impair their efficiency
	2. Interrupting excessive high currents	2. Install higher-rated device or check for grounds, shorts, or excessive currents
	3. Weak tip pressure	3. Replace contact springs; check contact carrier for deformation or damage
	4. Dirt or foreign matter on contact surface	4. Clean contacts with Freon or other approved cleaner
Noisy magnet	1. Broken shading coil	1. Replace shading coil
	2. Magnet faces not mating	2. Replace magnet assembly or realign
	3. Dirt or rust on magnet faces	3. Clean and realign
	4. Low voltage	4. Check system voltage and voltage dips during starting
Burned-out coil	1. Overvoltage or high ambient temperature	1. Check application and circuit
	2. Incorrect coil	2. Check coil rating and if incorrect replace with proper coil
	3. Shorted turns caused by mechanical damage	3. Replace coil; use molded coil if possible
	4. Low voltage, failure of magnet to seal in	4. Correct system voltage
	5. Dirt or rust on pole faces increasing air gap	5. Clean pole faces
Failure to pick up and seal	1. Low voltage	1. Check system voltage and voltage dips during starting
	2. Coil open or shorted	2. Replace coil
	3. Wrong coil	3. Check coil number and if necessary replace with correct coil
	4. Mechanical obstruction	4. With power off, check for free movement of contact and armature assembly
Failure to drop out	1. Gummy substance on pole faces	1. Clean pole faces
	2. Voltage not removed	2. Check coil circuit
	3. Worn or rusted parts causing binding	3. Replace part
	4. Residual magnetism due to lack of air gap in magnet path	4. Replace worn magnet and armature parts

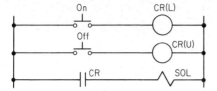

Fig. 45   Latching relay.

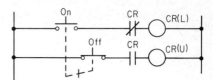

Fig. 46   Coil-clearing contacts.

the "off" button must be pressed.   The unlatch coil CR (U) is energized to release the latching mechanism, and the relay will return to its initial condition.

70. Coil-clearing Contacts.   When a maintained-contact initiating device is used, the latch coil will remain energized after the relay has latched in and could produce an annoying hum or consume extra power.   A normally closed coil-clearing contact can be used to deenergize the coil (Fig. 46).   The contact must open after the armature has sealed and the relay is latched in.   If it opens too soon, the armature will drop out and the relay will continue to attempt to pick up and drop out, resulting in considerable noise and a probable coil burnout.   Usually two coil-clearing contacts are required, one for the latch coil and one for the unlatch coil.   These are in addition to the poles normally required for the application.   With momentary-contact initiating devices, coil-clearing contacts are not required.

OTHER INDUSTRIAL-RELAY TYPES

Certain applications require other than conventional relay types.   Two common types are solid-state relays and reed relays.

71. Solid-State Relays.   Solid-state relays are used where electromechanical relays are not suitable.   For example, a small number of relays in a large system may be subjected to a high rate of operation and electromechanical relays would have to be replaced quite frequently.   Because of the extremely long life of a solid-state relay, its substitution in the high-duty-cycle application would greatly improve the reliability of the whole system at moderate cost compared with the much higher cost of a complete solid-state relay system or a solid-state logic system.   Solid-state relays are also suitable in applications involving severe shock and vibration, temperature variations, and the presence of dust, dirt, moisture, and corrosive vapors or fumes.   Unaffected by gravity, they are also indifferent to mounting positions.   There is no contact bounce or arcing, and since the load is always turned off at current zero, the problem of rf interference is minimized.   The output element in solid-state relays is usually a triac.

The triac is a member of the thyristor SCR family.   The SCR is a unidirectional device and passes current in one direction only (dc) when a positive gate signal is applied.   Because the SCR is self-latching, the load is energized until the dc supply is interrupted.

The triac is a bidirectional device and passes current in either direction (ac) when triggered by either a positive or negative gate signal.   Once turned on, the triac cannot be turned off until the load current falls near zero.   With an ac load, current passes through zero twice per cycle.   With the gate signal present, the triac will automatically turn on and off at the beginning and end of each half cycle, and virtually all the available power is applied to the load.   Since the triac performs zero current switching, no transients or RFI are produced by this output device.   Protection from external line transients can be provided in the form of an RC filter and/or thyristor surge suppressor across the triac.   The load can be isolated from the triac input by a transformer, a light-dependent resistor (LDR), and a pilot light or a reed relay in the gate circuit.

72. Reed Relays.   Because reed relays have hermetically sealed contacts, they are used in industrial-control applications where conventional open-contact devices are unsuitable.   When a relay is required as an interface between conventional input devices and solid-state logic systems, the reed relay provides reliable switching for the dry-circuit (low voltage, low current) loads present.   Reed relays are also suitable for applications where atmospheric contaminants may cause continuity problems with conventional contacts and where high switching rates and long life are required.

The basic reed switch consists of a set of contacts on two long, flat reeds of ferromagnetic material fused into the opposite ends of a glass tube and hermetically sealed in an inert gas. The contacts are generally plated or diffused with gold to provide a high degree of reliability and low contact resistance.

In the absence of a magnetic field, the normally open contact remains open. Application of a magnetic field (Fig. 47) causes an attraction between the two reeds, and the contact closes. Instead of a permanent magnet, a magnet coil may be used (Fig. 48). With no current flowing through the coil, the contact remains open. With sufficient current through the coil, the contact closes and remains closed while the coil is energized. Upon removal of the magnetic field, the contact returns to its initial position.

A normally closed contact can be produced using a permanent magnet as a biasing magnet to hold the contact closed. Energizing the coil produces a magnetic field that opposes the bias magnet and causes the contact to open.

A latching relay can be furnished by using a permanent-magnet bias that is not strong enough to open the contact but strong enough to hold it in the open position once it has been operated and coil power is removed. An unlatching coil winding is used to return the contact to its initial position.

A disadvantage of the reed relay is the low contact pressure and the resulting possi-

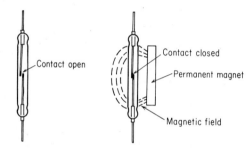

Fig. 47   Reed switch operated by permanent magnet.

bility of contacts welding when inductive loads with high current are used. Recent developments have produced reed relays capable of handling loads of several amperes. The basic application, however, is with dry-circuit loads. Also, unless the reed switch is shielded, stray magnetic fields may cause false operation. Most reed relays provide the necessary shielding.

73. Application of Reed Relays.   Conventional relays are still recommended for the majority of relay applications, but the reed relay can complement other relays where dry-circuit or logic-level switching is required, where atmospheric contaminants such as dust, dirt, humidity, corrosive fumes, etc., are present, and where high-speed switching and long life are required.

TIMING RELAYS

In general, a timing relay is a device which will sense an input and, after a specified time delay has elapsed, will produce an output. A timing relay is also known as a *timer*.

74. Terminology.   A failure by industry to develop standard terminology has resulted in much confusion in specifying timing relays. Several different terms are used to describe one mode or characteristic, and conversely, one term may have several meanings. The following terminology is recommended to describe the more common operating characteristics. Other terminology is found below under Modes of Operation.

a. Repeat accuracy (also known as repeatability, tolerance, and accuracy) is the maximum variation in the time delay of successive operations. Repeat accuracy in percent is determined from the equation

$$\frac{T_{max} - T_{min}}{T_{max} + T_{min}} \times 100$$

where T_{max} equals maximum time delay and T_{min} equals minimum time delay recorded for successive operations. Repeat accuracy may be specified:

1. At constant voltage and temperature and within specified reset times
2. Over specified voltage and temperature range and within specified reset times

b. *Reset time* (also known as recycle time and recovery time) is the length of time the control voltage must be interrupted (or applied):

1. After a timing period has been completed to produce a subsequent time delay within the repeat accuracy specified
2. During a timing period to produce a subsequent time delay within the specified repeat-accuracy percentage of the previous full time-delay period
3. To return the output switch to its original state

The three times above may or may not be the same, depending on the type of timer. The longest of the three will determine the device reset time.

Reset time is important on applications requiring successive timing cycles. All timing relays require some finite time to return to their original state. If the proper reset time is not provided, the subsequent timing period will deviate considerably from

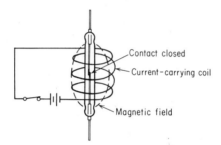

Fig. 48   Reed switch operated by magnet coil.

previous timing periods. For pneumatic timers, reset time is governed by the pickup or dropout time of the magnet, armature assembly, and timed contacts. For solid-state timers, minimum reset time is generally governed by the discharge time of the timing capacitor and maximum allowable reset time that will provide the specified repeat accuracy is governed by the dielectric polarization of the capacitor.

c. *Fixed Time Delay.* Time delay is preset at the factory to a specified value.

d. *Tolerance* is the maximum variation of any timing period from the specified fixed time delay at nominal voltage and 25°C.

It is not possible to manufacture a timing relay to the exact specified time delay because of manufacturing "tolerances" of component parts, assembly, adjustment, etc. When fixed-time-delay relays are required, the tolerance must be taken into consideration.

e. *Dial-setting accuracy* (also known as dial accuracy, setting accuracy, and tolerance) is the maximum variation of any timing period from that indicated by the dial at nominal voltage and 25°C.

f. *Transient Protection.* Protection of devices against damage caused by external transient input voltages.

g. *Polarity protection* is the protection of dc devices against reversal of input-voltage polarity.

h. *False operation* is the momentary transfer of the output switch when control power is removed before the end of the timing period.

i. *Burden* is the power required by the device during and after timing.

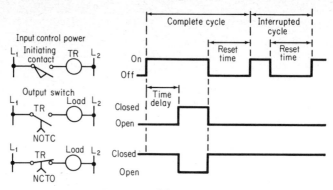

Fig. 49 On-delay timing logic.

MODES OF OPERATION

75. On Delay. On-delay operation is also called time delay after energization (TDE), delay on energization, delay on pull-in (DOPI), and delay on make. On-delay timing logic is shown in Fig. 49.

When the initiating contact closes, a time-delay period begins. After a preselected time delay, the output switch operates and remains in that state. When the initiating contact is opened, the output switch resets or returns to its initial state immediately. If the initiating contact opens before the end of the timing period (cycle interrupted), the output switch does not operate and the timer resets to its initial state. Solid-state devices may require continuous control power to operate and require an isolated initiating contact. Otherwise, timing logic is identical.

76. Off Delay. Off-delay operation is also called time delay after deenergization (TDD), delay on drop-out (DODO), delay on break, delay on release, pulse timer, and time latch. Off-delay timing logic is shown in Fig. 50.

When the initiating contact closes, application of power causes the output switch to reset immediately and remain in that state. When the initiating contact opens, a time-delay period begins. After a preselected time delay, the output switch operates. When the initiating contact recloses, the output switch immediately resets. If the initiating contact closes before the end of the timing period (interrupted cycle), the output switch does not operate and the timer resets. Solid-state devices require continuous control to operate and require an isolated initiating contact. Otherwise, timing logic is identical.

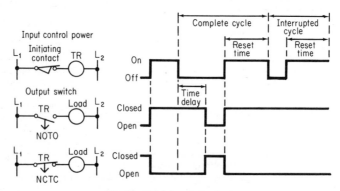

Fig. 50 Off-delay timing logic.

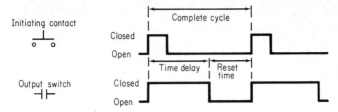

Fig. 51   Interval, momentary-start timing logic.

77. Interval, Momentary-Start.   These timers are also called interval, momentary-actuation, one-shot, single-pulse, momentary-start, push-button-start, period, and pulse-interval.   Interval, momentary-start timing logic is shown in Fig. 51.

When the initiating contact is closed, the output switch operates and the time-delay period begins.   The time-delay period is not affected by the duration of the initiating contact or by reclosing the initiating contact during the timing period.   After a preselected time delay, the output switch returns to its initial state.   Although these timers are designed for momentary-contact initiation, maintained-contact initiation can also be used.

78. Interval, Maintained-Start.   These timers are also called interval, maintained-actuation, and interval-duration (Fig. 52).

When the initiating contact is closed, the output switch operates and a time-delay period begins.   After a preselected time delay, the output switch returns to its initial state.   The initiating contact must be opened to reset the timing relay.   If the initiating contact opens before the end of the timing period, the output switch immediately returns to its initial state.   Solid-state devices may require continuous control power to operate and require an isolated initiating contact.   Otherwise, the timing logic is identical.

79. Repeat Cycle.   These timers are also called sequencing, on-off timer, cycle timer, flasher, and recycling timer (Fig. 53).

When the initiating contact is closed, a time-delay period begins.   After the time delay T_1, the output switch operates.   After the time delay T_2, the output switch resets. The cycle is repeated continuously while the initiating contact is closed.   The output switch may be started in the open state and sequence the load off-on-off-on or start in the closed state and sequence the load on-off-on-off.   Opening the initiating contact resets the timer.   Solid-state devices require continuous control power to operate and require an isolated initiating contact.   Otherwise the timing logic is identical.

80. Timing Symbols.   The timing symbols are made up of two parts (Fig. 54).   The first part shows the state of contacts, normally open or normally closed.   The second part consists of a tail and arrow with the arrow indicating the direction of timing.   Thus the upper left-hand symbol is a normally open, timed closed contact since the arrow indicates that the direction of timing is toward the closed position.

The use of the timing symbol is shown in the elementary diagram of Fig. 55.

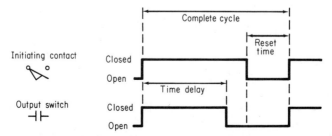

Fig. 52   Interval, maintained-start timing logic.

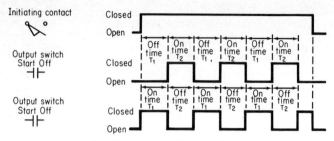

Fig. 53   Repeat-cycle timing logic.

TYPES OF TIMING RELAYS

Many types of timing relays are available, each with various advantages and disadvantages. The most popular types are:

1. Pneumatic
2. Solid-state
3. Motor-driven
4. Dashpot
5. Thermal
6. Mechanical or escapement

81. Pneumatic Timing Relays.   The pneumatic timer works on the principle of transferring air from one chamber to another through an adjustable orifice which controls the rate of airflow and in turn controls the movement of a diaphragm or piston and contact assembly.

A pneumatic timing head is shown in the timed-out position in Fig. 56. To reset the timing head, plunger A is rapidly pushed upward. An operating lever attached to the plunger causes snap switch B to reset. Diaphragm C is attached to the plunger and is also moved upward, compressing the air in chamber D. The air is forced through outlet E and upward against valve seat F. The air pressure on the reset stroke is high enough to force valve seat F upward, allowing the air to escape into chamber G, through filter H and out into the surrounding atmosphere through opening J. Valve seat F then returns to its normal position. All these operations happen in a few milliseconds.

To begin the timing period, plunger A is released. Diaphragm C tries to move downward under spring pressure, but a vacuum is created in chamber D, holding the diaphragm and plunger in position. Air is then drawn in through filter H, through inlet hole K, past the needle valve L and into the lower chamber D. As air enters chamber D, the diaphragm and plunger move downward. The timing period ends when plunger A and the operating lever cause the snap switch to operate. The above reset and timing cycles can then be repeated.

The length of the timing period is adjusted by needle valve L. Adjusting the needle

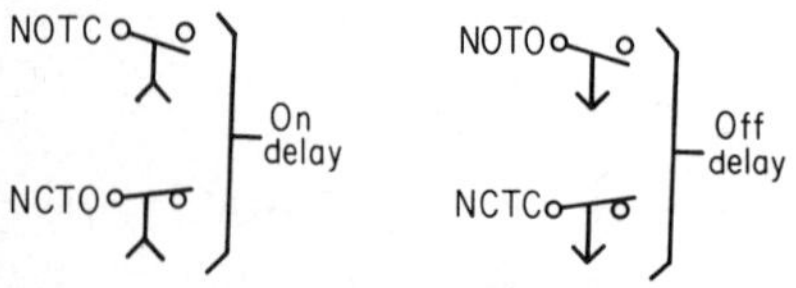

Fig. 54   Timing symbols.

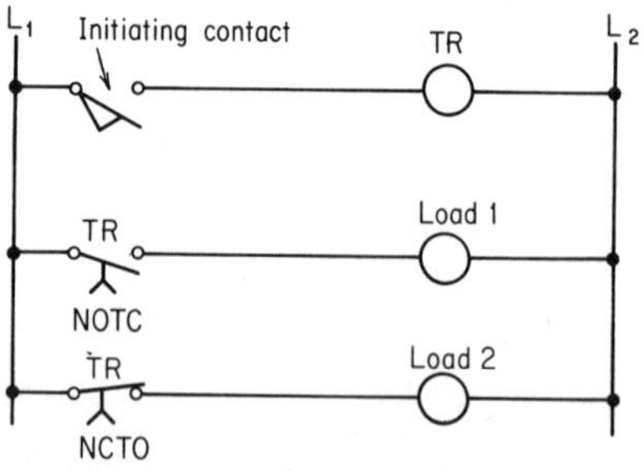

Fig. 55   Application of timing symbols.

valve inward reduces airflow and lengthens the timing period. Adjusting it outward allows more airflow and shortens the timing period.

The diaphragm and valve seat are generally made of silicone rubber, which exhibits excellent operating stability over temperature extremes and retains its flexibility and sealing properties indefinitely.

The timing head operates as an open-air system in which external air is drawn in and expelled through a filter. Closed-air systems recirculate the air internally and do not use external air. Closed-air systems, however, are not truly closed-air, as they depend on a seal to keep out external air and dust. Although closed-air systems are designed to eliminate using external air, a well-designed open-air system can be equally or more reliable.

a. Other Timing-Head Designs. Other types of timing heads use a bellows arrangement instead of a diaphragm. Bellows allow a larger amount of air to be transferred in a smaller-sized chamber but generally wear out faster than diaphragms because of the flexing action. Another variation is the use of a timing disk with a circular groove instead of a needle valve. Rotating the timing disk controls the effective length of the

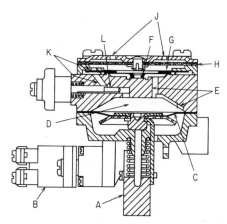

Fig. 56   Cross section of typical pneumatic timing head.

groove and adjusts the timing period. The disk type provides a linear timing adjustment.

b. Converting Mode of Operation. The timing head is operated by a magnet and armature assembly. Energizing the coil causes the armature to move and operate the timing-head plunger. Generally, the magnet and armature assemblies are invertible; that is, they can be rotated 180° so that the timing relay can be made on delay or off delay.

c. Instantaneous Contacts. Instantaneous contacts may be added to certain timing relays, and these contacts will operate at the same time that the magnet assembly operates. The instantaneous contacts thus provide action identical to that of a standard control relay, and the functions of a control relay and timing relay are combined into one device.

d. Advantages of Pneumatic Timing Relays. Pneumatic timing relays have been used for many years and have the following advantages:
1. Good repeat accuracy
2. Operation relatively unaffected by temperature and voltage variations
3. Short reset time
4. Variety of timed and instantaneous contact arrangements
5. Good mechanical and electrical life
6. Low cost

e. Disadvantages of Pneumatic Timing Relays
1. No remote adjustment
2. Repeat accuracy poor at longer timing periods (5 to 10 min and higher)

82. Solid-State Timing Relays.   The solid-state timing relay features:
1. Extremely long life
2. Excellent repeat accuracy, generally ±1 to 2% (±10 to 20 ms for digital timers)
3. Local or remote timing adjustment
4. Immunity to industrial atmospheric contaminants such as dust, dirt, and moisture
5. Resistance to shock and vibration
6. Bounceless switching (solid-state output)

Most solid-state timing relays use the series resistor-capacitor (RC) circuit to obtain a time delay (Fig. 57). When switch S_1 is closed, capacitor C begins to charge through resistor R. As the charge on the capacitor increases, the voltage across the capacitor E_C approaches that of the source voltage E_S. The charging rate of the capacitor is shown in Fig. 58. The time constant of the circuit is defined as the product of $R \times C$ and is useful in determining the voltage across the capacitor as a function of time. After one time constant, the voltage is 63.21% of E_S and for all practical purposes equal to E_S after 5 RC. E_S varies exponentially according to the equation $E_C = E_S(1 - e^{-t/RC})$, where t is time in seconds, R is resistance in megohms, C is capacitance in microfarads, and e equals 2.71828. When $t = RC$, $e^{-t/RC} = e^{-1} = 0.3679$. Thus, $E_C = E_S(1 - e^{-1}) = 0.6321$ after one time constant.

If the value of R or C is increased, the time required to reach a specific E_C is increased provided E_S remains constant. The RC circuit therefore provides an output voltage that varies as a function of time. This change in voltage is used to trigger a switch, such as a unijunction transistor, which in turn operates an output device (Fig. 59). The output may be an electromechanical relay as used in hybrid solid-state timing relays or a triac or silicon controlled rectifier (SCR) in complete solid-state devices. The ac input voltage must be converted to dc to operate the solid-state components. This conversion is made in a dc power supply which includes a rectifier, filter, and often a zener-diode voltage regulator.

Timing in a digital timer is based on line frequency, and accuracy is dependent on the accuracy of the frequency produced by the power company. Integrated circuits are generally used to count the cycles, and repeat accuracy is often in the ±10 to 20 ms range.

a. Reset Time.   The reset time of a solid-state timing relay is governed mainly by the time required to discharge the timing capacitor so that the next timing period can begin. If the capacitor is not allowed to discharge fully, the next timing period may be too short.

If the reset time is too long, that is, if the timing relay is deenergized for a considerable length of time, the first timing period may be greater than the subsequent period. This charge is primarily caused by dielectric polarization of the timing capacitor. The use of tantalum capacitors can minimize or reduce this effect to an insignificant value.

b. False Operation.   In some on-delay solid-state timing-relay circuits, interrupting the control power before the timing relay times out will cause the output device to transfer momentarily. This false operation of the output could cause serious circuit problems. The user should therefore check device specifications regarding false operation before selecting the timing relay.

c. Output Devices.   Hybrid solid-state timing relays use electromechanical relays as output devices. The most common solid-state output device is the triac. Complete solid-state devices use triacs or SCRs as output devices.

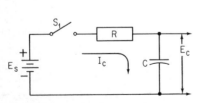

Fig. 57   Basic RC timing circuit.

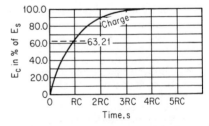

Fig. 58   Voltage-response curve for RC timing circuit.

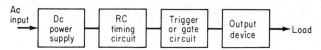

Fig. 59 Block diagram of solid-state timing relay.

The advantages of electromechanical relays are:
1. They are not affected by transients.
2. They provide complete isolation of load and input.
3. Contacts completely break the circuit to the load (no leakage current).
4. They have generally higher inductive ratings.
The disadvantages of electromechanical relays are:
1. Relatively short life.
2. Contact bounce may affect the load circuit.
3. Contacts may be affected by atmospheric contaminants.
4. Contact switching may cause radio-frequency interference (RFI) or transients.
Solid-state output devices have the following advantages:
1. No moving parts.   Life is not affected by high duty cycle or number of operations.
2. No contact bounce.
3. Zero current switching.   The triac always opens at current zero; so it produces no arcing or transients due to stored energy in the inductive load.
4. Not affected by gravity.   They can be mounted in any position.
5. Unaffected by dust, dirt, moisture, and other environmental conditions.
6. Unaffected by shock and vibration.
Solid-state output devices have the following disadvantages:
1. Leakage current.   When the triac is off, a small leakage current on the order of a few milliamperes flows.   When a small load current is being controlled, the leakage current may become an important factor.   If several "contacts" are connected in parallel, the leakage current may be large enough to prevent the load from deenergizing.
2. Series or parallel limitations.   The number of solid-state "contacts" that can be wired in parallel is limited by leakage current.   The number of series connections is limited by the voltage drop across the triac, which may be on the order of 1 or 2 V with high load currents.   Relay specifications should be checked for series or parallel limitations.
3. Minimum load.   A minimum load current is required to ensure firing of the triac.
4. Load isolation.   The load is not isolated from the input in some devices.
5. Transients.   External transients could possibly damage solid-state output devices if not properly protected.
83. Motor-driven Timing Relays.   A synchronous ac motor is used to drive a clutch and gear train which reduce the shaft speed to a usable output.   Cams mounted on the shaft provide the required switching action.   Accuracy is similar to a clock and is approximately as good as the accuracy of power-line frequency.   Motor-driven timing relays feature:
1. Extremely high repeat accuracy, generally less than $\pm1\%$
2. Long time delays, up to 30 h and higher
3. Elapsed-time indication
84. Dashpot Timing Relays.   Operation of the timed contact is governed by the movement of an iron core lifted by the field of a solenoid magnet coil against the retarding force of a piston moving through an oil-filled dashpot.   The timing range is adjusted by opening or closing bypass holes in the bottom of the piston.   Repeat accuracy is not good, and a relatively long reset time must be allowed to let the piston settle down to the bottom of the dashpot.   Dashpot timing relays feature:
1. Low cost
2. Limited timing range, generally 30 s maximum
3. On delay only
4. Poor repeat accuracy
5. Long reset time
6. Good temperature characteristics generally unaffected by temperature variations
85. Thermal Timing Relays.   Thermal timing relays use a bimetallic element which heats on energization and expands or bends, performing the required switching action.

These devices are generally not adjustable, but when they are, the adjustment is small and indicated settings are inaccurate. They cannot be used where short cycles or fast reset times are required. Thermal timing relays are not used frequently today, but the basic principle is used in the bimetallic thermal-overload relay and circuit breaker. Thermal timing relays feature:

1. Low cost
2. Limited timing range
3. Poor repeat accuracy
4. Long reset time

86. Mechanical or Escapement Timing Relays. A mechanical timing relay is similar to an alarm clock. The device is set by winding up the spring. The proper selection of cams, levers, and gears provides operation of the timed contacts at preselected intervals. These timing relays are rarely used in an industrial application because they have to be rewound after each operation.

SELECTION OF A TIMING RELAY

Timing relays, like other control devices, are part of a control system and must be matched to the application. The following is a checklist of factors to be considered:

> *Mode of Operation*
> ☐ On delay—provides a delayed start of an
>   action
> ☐ Off delay—provides a delayed termination
>   of an action
> ☐ Interval (one-shot)—provides the accurate
>   timing of an action after initiation and
>   independent of operator control during
>   the timing period
> ☐ Repeat cycle (sequencing)—provides continuous
>   sequencing or on-off control of two
>   actions
>
> *Timing Range*
> ☐ Adjustable: From _____ to _____
> ☐ Fixed: _____ Nominal
> ☐ Repeat cycle adjustable:
>   On time adjustable from _____ to _____
>   Off time adjustable from _____ to _____
>   Start in (on) (off) mode
> ☐ Repeat cycle fixed: On time _____ Nominal
>   Off time _____ Nominal
>   Start in (on) (off) mode
> ☐ Reset time (relay interrupted during
>   timing period): _____ ms
>   (relay interrupted after
>   timing period): _____ ms
>
> *Repeat Accuracy*
> ☐ ± _____ % over ± voltage and temperature
>   range noted below

> *Input Voltage*
> ☐ _____ Volts ac _____ Hz
> ☐ _____ Volts dc
> ☐ ± _____ % of nominal
>
> *Output*
> ☐ _____ N.O.
> ☐ _____ N.C.
>
> *Output Load Rating*
> ☐ _____ Volts ac
> ☐ _____ Volts dc
> ☐ _____ Amperes make
> ☐ _____ Amperes break
>
> ☐ _____ Amperes continuous
>
> *Mounting*
> ☐ Panel (surface)

> *Temperature Range*
> ☐ _____ °C to _____ °C
> ☐ _____ °F to _____ °F

> ☐ Resistive
> ☐ Inductive
> ☐ Duty cycle: _____
>   Operations
>   per _____

☐ Flush   ☐ Plug-in

87. Which Type of Timer?   After the operating characteristics and mode of operation have been determined, the basic type of timing relay must be selected.   Pneumatic, solid-state, and motor-driven timers are the most commonly used devices.   The major points of difference are shown in Table 5.

In general, pneumatic timing relays are selected to provide:
1. A chunk of time where accuracy is unimportant or ±10% is acceptable
2. Low cost
3. Relatively good reliability
4. Adjustable timing period up to 1 or 3 min

In general, solid-state timing relays are selected to provide:
1. Long life in high-cycle-rate applications
2. Excellent repeat accuracy, ±1 to 2% (±10 to 20 ms for digital timers)
3. Bounceless contacts, zero current switching
4. Immunity to shock and vibration
5. Remote timing adjustment
6. More dependability in unfavorable environments such as dust, moisture, and humidity

In general, motor-driven timing relays are selected to provide:
1. Long timing periods—up to 60 h or longer
2. Elapsed-time indication
3. Excellent repeat accuracy, ±0.5%

APPLICATION OF ON-DELAY TIMING RELAYS

The basic on-delay circuit diagram is shown in Fig. 60.

When the initiating contact closes, the timing relay *TR* is energized and the time-delay period begins.   After a preselected delay, NOTC timed contact *TR* closes and energizes load 1, and NCTO timed contact *TR* opens and deenergizes load 2.

88. Typical On-Delay Applications.   The following are common uses of timing relays with on delay:

1. Conveyor Control (prevent jam-up).   When a product moves down the conveyor, it closes a limit-switch initiating contact and then as it moves past, resets or opens the limit-switch contact.   *TR* is adjusted so that as long as products continue moving, it will not be allowed to time out.   If products jam up, the limit-switch contact is held closed, *TR* times out, and the NOTC contact energizes an alarm (load 1).

2. Parts Feeder On Machine.   Closing the limit switch and initiating contact begins

TABLE 5. Comparison of Timer Types

Pneumatic	Solid-state	Motor-driven
Adjustable timing ranges up to 1 min, 3 min typical; up to 30 min or longer at reduced repeat accuracy	Adjustable or fixed timing up to 5 min typical	Adjustable timing ranges up to 30–60 h typical
±10% repeat accuracy typical	±1–2% repeat accuracy typical ±10 to 20 ms for digital timers	±0.5% or less repeat accuracy typical
Local adjustment only	Local or remote adjustment	Local adjustment only
No elapsed-time indication	No elapsed-time indication (possible on digital type)	Elapsed-time indication
Contact bounce—several ms typical	No contact bounce	Contact bounce—several ms typical
Possibly affected by shock and vibration	Unaffected by shock and vibration	Possibly affected by shock and vibration
Relatively good life, 1–10 million operations typical	Excellent life—independent of number of operations	Relatively good life, 1–10 million operations typical
Complete load isolation with open output contact	Small leakage current present with open output contact	Complete load isolation with open output contact

the machining sequence. Timing relay *TR* allows time for parts feeding before actual operation can take place by the load 1 device.

3. *Sequence or Programming Control.* The initiating contact energizes one or more timing relays to provide a sequence of operations at the preset time-delay intervals. For example, on an injection-molding machine, the timing relays would control solenoid valves feeding material to the machine and would provide timed intervals for filling, curing, etc.

4. *Machine-Tool-Coolant Control.* The operating head (for example, a drill) moves toward part and closes a limit-switch, initiating contact. The initiating contact energizes a coolant-pump motor starter (not shown in Fig. 60) and timer *TR*. After a preset delay, which ensures that coolant is flowing, the drill motor starter (load 1) is energized by the NOTC contact *TR*.

5. *Safety Control.* After a limit-switch initiating contact is closed, if another function does not occur within a specified time to reopen the initiating contact, *TR* will time out and energize an alarm or shut off the machine.

6. *Automatic Filling or Packaging Operation.* With a continuous flow of material, the limit-switch initiating contact will be closed and a fill solenoid (load 2) will be energized by the NCTO contact *TR*. After a preset delay, which corresponds to a filled container, the NCTO contact opens, deenergizes the fill solenoid, and stops the flow of material.

7. *Industrial Lathes.* An on-delay timing relay can be used to sense the end of a cut. When the workpiece moves into position, the limit-switch initiating contact closes and energizes *TR*. After a preset time delay, which is based on the speed of the workpiece and the distance it travels, NCTO contact *TR* opens and stops the workpiece (load 2).

8. *Delayed-Sequence Starting.* An on-delay timing relay can be used to provide delayed sequence starting of a series of motor starters, preventing power-line overloading which would occur if all motors came on at the same time.

9. *Batching Machines.* Several timing relays could be used to proportion materials (cement, sand, and gravel, for example) in a batching process. Each timing relay would control the length of time that load 2 is energized and thus control the quantity of each material that would flow.

10. *Motor-Generator Starting.* After a shutdown, the motor-generator set is started and energizes *TR* through the closed initiating contact. Other control devices (load 1) are prevented from operating by NOTC contact *TR*, which allows the motor-generator set to come up to speed before the control sequences are started.

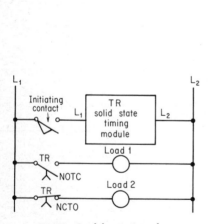

Fig. 60 On-delay timing relay.

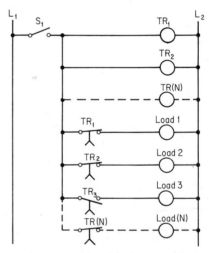

Fig. 61 On-delay timing relay controls processing times.

11. Food Processing. On-delay timing relays can be used to control processing times. In making a cake mix, for example, the timing relays can automatically control the amount of ingredients used. In Fig. 61, S_1 closes and initiates the mixing sequence. Timing relays TR_1, TR_2, etc., through TR (N) are energized. Each is set for the proper time to energize solenoid valves (loads 1 through N) to dispense the correct amount of flour, sugar, shortening, milk, etc., into the mixing vat. Note that the normally open TR_3 timed contact may control the point at which the mixing begins (load 3). TR_4 and subsequent timing relays then control the dispensing of further ingredients during the mixing cycle.

12. Metalworking Processes. Heat-treat and annealing cycles can also be controlled like the food-processing sequence described above.

APPLICATION OF OFF-DELAY TIMING RELAYS

The basic off-delay circuit diagram is shown in Fig. 62.

When the limit-switch initiating contact closes, timing relay TR is energized. NOTO contact TR immediately closes and load 1 is energized; NCTC contact TR immediately opens and load 2 is deenergized. When the initiating contact opens, TR is deenergized,

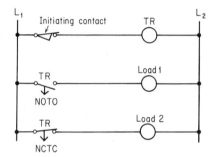

Fig. 62   Off-delay timing relay.

and the timing period begins. After a preset time delay, NOTO timed contact TR opens and deenergizes load 1; NCTC contact TR closes and energizes load 2.

89. Typical Off-Delay Applications.

1. Dwell-Time Control. On a machine tool, for example, a drilling operation, it is desirable for the drill bit to dwell in the hole for a period of time before withdrawal to ensure proper clearance of the hole. The limit-switch initiating contact opens and deenergizes the forward drive-motor control (not shown in Fig. 62) and timing relay TR. After a preset delay, NOTO contact TR opens and turns off the drill motor (load 1). NCTC contact TR then turns on the reverse drive-motor control (load 2) to remove the drill.

2. Conveyor Monitoring. A part moving on a conveyor closes the limit-switch initiating contact and resets timing relay TR. As the part passes, the initiating contact opens and TR begins timing. However, if another part comes along before TR times out, TR will reset and the timed contact will not operate. If parts stop or the interval between parts is too long, the timed contact will operate and sound an alarm or turn off the conveyor.

3. Conveyor Clearing (automatic shut-off). As parts move off the conveyor, the limit-switch initiating contacts open and the timing relay TR begins to time. If another part comes along before TR times out, the limit-switch contact is closed and TR resets. After the last part has passed, the limit-switch contact remains open, TR times out, and NOTO contact TR opens, shutting off conveyor-motor control (load 1). This procedure allows the conveyor to clear and shut off automatically.

4. Escalator Control. When a person steps on a treadle approaching an escalator, a limit-switch initiating contact closes and energizes TR. NOTO contact TR immediately closes and energizes escalator drive-motor control (load 1). When the person steps off

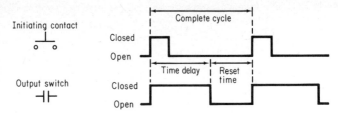

Fig. 63 Interval, momentary-start timing relay.

the treadle onto the escalator, the limit-switch contact opens and *TR* begins to time out. After a preset time, which allows the person to be carried to the top, NOTO contact *TR* opens and stops the escalator. Thus the escalator runs only when there is someone to ride on it.

5. *Equipment Cooling.* When the limit-switch initiating contact is opened, the equipment is turned off (control not shown) and *TR* is deenergized. NOTO contact *TR* remains closed and fan-motor control (load 1) remains energized. After a preset delay, which allows the fan to cool the equipment, NOTO contact *TR* opens and shuts off the fan motor.

6. *Elevator-Door Control.* After the last breaking of the door-safety-photocell initiating contacts, timing relay *TR* times out and NCTC contact *TR* closes and operates door-closing-motor control (load 2).

7. *Hydraulic-Pressure Protection.* An off-delay timing relay can be used to protect against low hydraulic pressure on a machine. When pressure drops, pressure-switch initiating contact opens and deenergizes *TR*. If the pressure is below normal for more than the preset 1 or 2 s, NOTO contact *TR* opens and stops the automatic cycle (load 1). The timing relay thus eliminates nuisance shutdown by inconsequential transient pressure drops.

APPLICATION OF INTERVAL, MOMENTARY- OR MAINTAINED-START TIMING RELAYS

The basic interval, momentary-start timing logic is shown in Fig. 63. Closing the initiating contact causes the output switch to close immediately. After a preset delay, the output switch opens. The initiating contact may be closed momentarily or held closed and produce the same output sequence.

On-delay timing relays may be wired for interval, momentary-start timing (Fig. 64). Closing S_1 energizes *TR*. Instantaneous interlock *TR* closes and along with the left-hand normally closed timed contact *TR* provides a holding circuit around S_1. The right-hand normally closed timed contact *TR* energizes the load immediately. After a preset delay, both timed contacts *TR* open, deenergizing the load. The timing relay coil is also deenergized and ready for the next cycle. If S_1 is held closed, *TR* will remain energized, but the load is deenergized by the opening of timed contacts *TR*. S_1 must be opened to reset the timer for the next cycle.

The interval, maintained-start diagram is shown in Fig. 65. Closing the initiating contact energizes the load immediately through normally closed time contact *TR*. After a preset delay, the timed contact *TR* opens and deenergizes the load. The initiating contact must be opened to reset the timer for the next cycle. Opening the initiating contact during the timing period will immediately reset the timer.

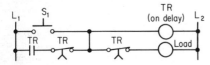

Fig. 64 On-delay timing relay wired for interval, momentary-start timing.

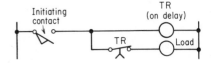

Fig. 65 On-delay timing relay wired for interval, maintained-start timing.

90. Typical Interval-timing Applications.   The following are typical uses for interval timing:

1. Packaging machines. Heat sealing of plastic bags is a critical operation. Too little heat and an improper seal is made—too much heat and the plastic is damaged. To make the time of heat application independent of the operator, an interval timer can be used (Fig. 63). Operating the push button causes the output switch to close immediately, and heat is applied by the load. After a preset time delay, the output switch opens and removes the heat. The operator can momentarily push the button or hold it down and the timing will not be affected.

2. Other applications which require a timed operation independent of operator control are:

a. Photograph printing	*f.* Bottle filling
b. Spot welding	*g.* Food processing
c. Heat treating	*h.* Tube testing
d. Electroplating	*i.* Paint spraying
e. Bag filling	*j.* Mixing machines

APPLICATION OF SEQUENCING OR REPEAT-CYCLE TIMING RELAYS

A typical solid-state timer is shown in Fig. 66. Closing the initiating contact begins the cycle. Without the jumper, the load is off during the first timing period T_1 and then on during the second timing period T_2, and it continues off-on, off-on until the initiating contact is opened. If a jumper is connected to terminals 5 and 6, the load begins on during the first timing period T_1, then is off during T_2 and continues on-off, on-off until the initiating contact is opened.

In Fig. 67, two pneumatic timers may be connected for repeat-cycle timing. Closing the initiating contact begins the cycle. TR_1 is energized through the normally closed contact TR_2. During the first timing period T_1, TR_2 and load 1 are off. At the end of T_1, the normally open timed contact TR_1 closes, turning TR_2 and load 1 on. At the end of T_2, timed contact TR_2 opens and turns off timing relay TR_1. Timed contact TR_1 immediately opens and turns off timing relay TR_2 and load. Timed contact TR_2 immediately closes and turns on timing relay TR_1, starting the cycle all over again. The cycle repeats until the initiating contact opens. If the normally closed timed contact TR_1 is used, load 2 continues to cycle on-off, on-off until the initiating contact opens.

91. Typical Repeat-Cycle Applications.   The following are typical repeat-cycle applications:

1. *Life Testing.*   Life testing of a device, such as a relay or magnetic starter, may require several hundred thousand operations per day. A solid-state repeat-cycle timing relay can reliably switch the test device on and off as required.

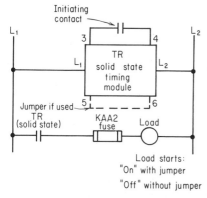

Fig. 66 Repeat-cycle timing relay.

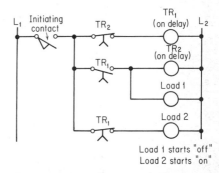

Fig. 67 Repeat-cycle timing using two on-delay timing relays.

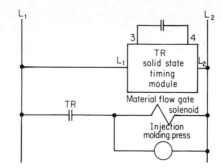

Fig. 68   Repeat-cycle timing relay on injection-molding-press application.

2. *Injection Molding.*   A repeat-cycle timing relay can be used to control the filling and operation of an automatic injection-molding press (Fig. 68).   During T_1, contact *TR* opens the material-flow-gate solenoid and allows material to enter the press.   During T_2, *TR* closes the valve and the press is operated for the desired molding time.

3. *Conveyor Sorting.*   A repeat-cycle timer can be used to control distribution of material on a gated conveyor feeding to manual loading stations (Fig. 69).   The gate first allows the material to flow onto the conveyor and then switches the material onto a second conveyor.   The timer operates the gate (load 1 in Fig. 66 or 67), transferring it at the preselected intervals to equalize the number of packages being handled at the loading station.

4. *Temperature Control.*   A repeat-cycle timer can be used to apply on-off control to the heating elements to provide less overshoot and more accurate control.

5. *Heat Cycling.*   A repeat-cycle timer can be used to control the on-off time in certain heat-treating applications.

INSTRUMENT AND CONTROL SWITCHES

Standard-duty instrument and control switches are widely used in both switchgear and control.   They are designed for panel mounting and are available in either maintained or spring-return types.

These rotary-type switches are designed to provide exceptionally high reliability because of the importance of the functions, systems, or processes that they are used to control.

Mill-duty instrument and control switches are heavy-duty switches for use in steel mills, chemical plants, petroleum refineries, and other industrial applications requiring constant and hard usage.

92. Circuit Arrangements.   These switches are adaptable to an unlimited number of circuit arrangements.   Frequently used momentary-type control switches with pistol grip or oval handles include circuit-breaker control (Fig. 70), voltage control, motor control, voltage control with pullout, and speed control.   Frequently used maintained-type instrument switches with round notched handles include ammeter, voltmeter, power factor, temperature indicator, and synchronizing switches.

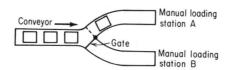

Fig. 69   Repeat-cycle timing relay on conveyor sorting application.

93. Operation.   Switches are made for partial or full 360° rotation with 2 to 12 radial positions.   Spring-return switches are made for three positions but can be furnished for a maximum of seven positions, three from either side of center position.   Maintained-contact switches are available from 2 to 12 positions.   An increase in the number of positions decreases the travel between positions and reduces the interrupting capacity.

Circuit-breaker control switches have a window in the position plate with a red and green flag indicator to show the last operation of the switch.   In the trip position, the signal-lamp circuit can be opened by pulling the handle into the latch position.   The lamp is not lighted when the handle is in this position, indicating that the circuit controlled by the breaker is not in use.   Voltage-control, speed-control, or motor-control switches are not usually furnished with the flag indicator or pullout feature.

The rotating contacts of standard switches are mounted on a square insulated shaft (Fig. 71).   Contacts may or may not be electrically insulated from each other, depending upon the desired circuit.   In operation, circuits are completed across various members of stationary contacts or a continuous circuit is maintained across two or more stationary fingers for a number of switch positions by means of conductive spacers.   The stationary contact and positioning fingers are mounted on finger supports.   These supports in turn are mounted on a base of insulating material.   Contact pressure is provided by a spring.   External connections are made to the stationary contact finger through terminal studs below the base.

The positioning wheel, two roller arms, and springs hold the shaft assembly in each position on maintained-contact switches.   Mounted on the operating shaft, the star-shaped positioning wheel corresponds to the number of switch positions.   Spring action engages the roller arms by the notches in the positioning wheel.   This action holds the switch in the desired position.

Because interrupting capacity depends on current, voltage, and inductance of the

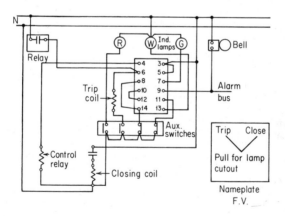

Contacts	Positions				
	Close	Normal after close	Trip	Normal after trip	Pull out in trip
3-4	x				
5-6			x		x
7-8	x	x	x	x	
9-10		x		x	
11-12	x	x			
13-14	x	x			

(Spring return)

Fig. 70   Typical circuit-breaker instrument and control-switch application.

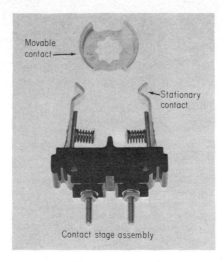

Contact stage assembly

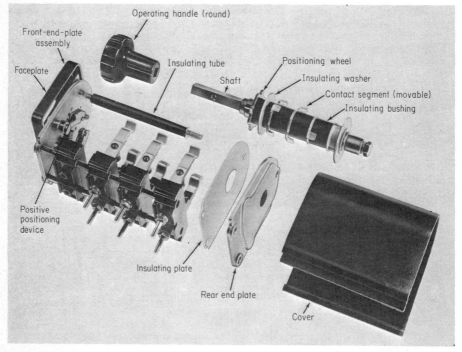

Fig. 71 Parts arrangement for typical instrument and control switch used for voltmeter switch. (*Allis-Chalmers Corp.*)

circuit and speed of contact opening, control relays should be used with control switches on heavy currents. These switches are made in standard construction for up to 14 circuits. In many cases additional circuits may be provided.

Standard switch circuits are usually connected from a stud on one side of the terminal base through a segment of the shaft assembly to a stud on the other side. This constitutes a single contact or stage.

A stage is a single-pole, double-break, single-throw contact. In other switches involving various sequences and connections to the same circuit, the entire switch may serve as a single pole.

Groups of switches can be gang-operated with a single handle. One such an arrangement with six gang-operated control switches can control 84 circuits with three positions from a single pistol-grip handle.

4

System Considerations for Switchgear

WILLIAM R. IBACH *

  * W. R. Ibach and Associates, Electrical Engineers (formerly Chief Substation Design Engineer, Wisconsin Electric Power Company); Registered Professional Engineer (Wis.); Senior Member, IEEE and Engineers and Scientists of Milwaukee.

FOREWORD

This section presents system considerations that have an effect upon the application of switchgear. Knowledge of the factors covered here will aid in selecting the type of switchgear needed in a specific application.

PLANNING

Before a basic system is decided upon, it is necessary to obtain as complete information as possible on the initial and ultimate load and to plan and coordinate the system.

1. Initial planning involves the preparation of a one-line diagram showing the basic components. Although a standard industrial and commercial power system does not exist, certain standard components of equipment and voltages should be used as much as possible to minimize special equipment.

Standardization will reduce initial cost, minimize the cost of future expansion, and also assist in selling the equipment, if necessary. The initial design is important, since it is the nucleus of any plant expansion, and after the basic system is decided upon, any change will be very costly. The system should be flexible so that the service voltage will be ample for the projected increase in load for many years to come. The system should be planned so that additional load centers can be easily and conveniently added.

2. Preparation for Future. The size of future loads is generally underestimated because of the expanding use of electrical energy, improved technology, and increased laborsaving devices, as well as the general growth and expansion of industry in this country. Provisions for equipment capacity should be liberal, since the equipment cost does not double if the capacity is doubled. Experience has shown that it is expensive to replace equipment with larger capacity; and in addition, it is difficult to dispose of or sell the removed equipment. Often the operation of partially loaded equipment will provide better reliability and extend its life.

3. Economy. Considerable savings can be realized by accurate and complete plans, specifications using competitive bidding, standard equipment, and avoidance of special equipment. Fortunes can be saved in a large plant by using more economical materials such as aluminum conductors, thin-wall conduit, interlocked armor cable, and aluminum-sheathed cable where permitted by the code and where applicable. Efforts should be made to use factory-made and -assembled components, distribution voltages as high as possible, three-phase instead of single-phase for lighting distribution, cable and cable racks for multiple runs instead of conduit and wire, and to minimize the length of lighting and power feeders. Care should also be taken to limit the transformer size at load centers so as to reduce the secondary short-circuit capacity.

4. Reliability. Plans should include means of providing continuity of service by either dual lines, loop feed, automatic transfer, uninterruptible power, or spare equipment, with consideration given to isolating the fault and providing protective grounding on the faulted section while work is being done.

5. Maintenance. A companion to reliability is proper periodic testing, inspection, cleaning, and painting, as well as a record system of all motors, equipment, and wiring failures. A program should be adopted for the control and exclusion of rodents, squirrels, birds, roaches, cats, etc. These small animals can often cause serious outages which are difficult to analyze, as the resulting fire usually destroys all evidence. In addition, a complete set of plans should be readily available showing all wiring diagrams, location of all underground cables, size of all cables, locations of transformers, generators, motors, and fuses. In addition, operating instructions should be available for all equipment. This information may be needed quickly in case of trouble or in an emergency.

6. Safety is one of the most important considerations. It requires a periodic review of all equipment and operations by a team of qualified, safety-minded operating and maintenance men (see Sec. 30). In addition, it is necessary that all operating men be trained so that they know, and are familiar with, the equipment which they operate. This training must be periodically reviewed and upgraded with particular attention given to new employees, with their know-how subject to examination.

Whether high-voltage lines or feeders should be grounded after being deenergized so that men can safely work on them depends upon the qualifications of the men doing the grounding, since it is absolutely necessary for the line to be dead before it is grounded. The hazards are great and the consequences are very serious if energized lines are grounded. However, it is recommended that on an extended system where there are qualified personnel, protective grounding be provided after the source has been disconnected and the lines tested to be sure they are deenergized. This applies to circuits 2000 V and higher.

7. Service from Utility or Isolated Plant. The majority of switchgear is connected to private utilities, municipal utilities, or REA or TVA systems. A very small portion of switchgear will be connected to isolated private generating systems. Being connected to any one of the utility systems, however, involves a certain amount of responsibility. This responsibility requires overcurrent and other necessary service protection so that any trouble on the customer's service will not jeopardize the service of other customers supplied by the feeder or system.

a. Power Requirements. Before preparing specifications or ordering equipment it is necessary to contact the utility and supply them with the following information:

1. Initial and estimated future lighting loads.
2. Initial and estimated future power loads.
3. Any extraordinary initial or future loads that may require special considerations such as large motors or motors requiring large inrush currents, arc furnaces, and welders.
4. The location of the load.
5. The length of time or the period of operation of the load.
6. The date at which service is desired.

b. Information to be requested from the utility may include the following:

1. The service rules of the utility.
2. The number of sources (or lines) to supply the load.
3. The location at which service will be supplied.
4. If an automatic transfer scheme is desired, a schematic diagram of the system recommended by the utility or provision for the utility to accept the system proposed by the customer or his engineer.
5. The voltage and characteristics of the service to be supplied, such as whether it is 3- or 4-wire, whether it is effectively or solidly grounded or ungrounded, and if it is grounded, whether it is grounded through a resistor or reactor and if so what is the size of each.
6. Type and characteristics, and number of overload relays required.
7. The initial and ultimate size of service fuses permitted and the time-current characteristics of them if this characteristic is critical.
8. The initial and planned ultimate symmetrical interrupting duty of the service protection and the X/R system ratio at this point.
9. If parallel or synchronizing operation between the customer's generating facilities and the utility's system is used, specifications for and location of the protective relays for this purpose such as reverse-power relays and undervoltage relays.
10. Test facilities for the relays and tripping circuit so that the secondary of the tripping-current transformers can be short-circuited and the effectiveness of the tripping circuit tested.
11. Recommended physical arrangement and wiring diagram of the utility's metering current and potential transformers, fuses, and meters.
12. Specifications for the tripping source such as 48-V (minimum) battery and automatic charger and indication of whether capacitor or reactor tripping or whether instantaneous transformer trip coils are permitted.
13. Specifications for the lighting and power transformers, the voltage rating, and taps recommended.
14. Recommendations for spare equipment such as fuse-refill units, tank lifter (for oil circuit breakers), ground and test device, test racks, switch sticks, if needed, and fuse tongs. A complete set of spare refill units is recommended for each size fuse as well as a fuse holder if fuse holders are used; also a set of spare parts for circuit breakers.
15. If the utility standardizes on potheads, specifications for the potheads so that in case of trouble spare potheads or parts are readily available.

c. Extraordinary loads should be carefully planned and coordinated with the utility engineers.

Arc-furnace load, because of its erratic and wide fluctuations, may require an isolated line, a subtransmission, or a transmission line so that its fluctuation load will not affect other customers on the line. Likewise transformer arc welders, because of their fluctuations, may require special considerations. This may require an isolated transformer or service from a subtransmission line (Art. 70).

X-ray services are such that they cannot tolerate a variation in voltage while the x-ray pictures are being taken. Separate services are often provided so that they can be fed through isolated transformers should this later become necessary and the x-ray picture will not be affected by the starting current of motors or other loads fed from the same transformer.

d. Computer Supply. Some computers operate either from a 208-V, 3-wire, 3-phase, 60-Hz system or from a 230-V, 3-wire, 3-phase, 60-Hz system. The voltage range for satisfactory operation is from +10 to −8% and within a frequency range of 59.5 to 60.5 Hz. Computers cannot satisfactorily operate when subject to surges caused by lightning, switching, short circuits, starting of large motors, welders, etc. If the data handled by the computer are critical and the loss of any of them cannot be tolerated, an uninterruptible power system (UPS) is necessary.

8. Uninterruptible Power Supply for Critical Loads (Ref. 1). The integration of large computer, data-handling, and communications systems into business, manufacturing, and transportation network operations demands high performance and reliability (see Sec. 7). However, improved speed of operation and increased data-storage capacity reduce the input power ride through capabilities of these systems. Therefore it is becoming increasingly important to isolate these systems from the utility power supply by an interface which will suppress transients and provide short-time ride-through capability.

The short-time ride-through requirement may extend from a couple of milliseconds up to 1 h. The lower limit is determined by the electronic system supplied, whereas the upper limit is determined by various factors which include maximum utility-disturbance period, on-site prime-mover response time, importance of maintaining system operation, and orderly system-shutdown period. Unfortunately, the influence of the utility power supply on electronic-system performance is frequently not realized until operation commences. Adapting a power-supply interface later may introduce space problems and delay electronic-system utilization for a significant period of time. Clearly, then, the need for a reliable source of system power should be considered early in the electronic-system design.

Electronic-system manufacturers can advise a user of the effects of transients on their equipment. Utility-service reliability data cannot be easily obtained, since they are somewhat beyond the control of the utility service and vary within their network. They are subject to both natural and man-made disturbances such as lightning strokes, icing, or wind rupture, highway rupture, transient loading, switching surges, and faults. These disturbances frequently occur within utility customers' distribution systems or through the lines to other utilities. One basic fact is that a 100% reliability of electric utility service can never be attainable, no matter how much time, effort, and money are spent.

a. Interface Equipment. A wide variety of interface equipment has been proposed and installed in an attempt to provide better electronic-system operation. It includes alternators with flywheel, prime movers with alternator, and static power-conversion equipment. Each type has a particular range of capabilities. A general rule of thumb is that the performance of this interface power-supply equipment is approximately proportional to its cost, and the cost for the best equipment is usually not more than 2 to 3% of the cost of the electronic system it serves. The best power-supply interface equipment is the static parallel-redundant continuous or "floated-battery" type. This type of equipment requires no switching sequence and likewise no switching time to begin operating in the ride-through mode, since it is connected to the load at all times and utilizes a battery which is also always connected to the inverter dc bus.

Static-thyristor-inverter equipment is different from rotating machines in that the no-load stress levels are a significant percentage of the full-load stress level. Therefore, there is no advantage in operating this equipment in a standby mode. Actually, standby operation is a disadvantage because of the shock-loading effects and the inability to detect continuously a marginal or deteriorated ability to commutate within thyristor inverters. Shock-loading inverter equipment at the time of greatest need is the surest way to get into trouble.

A second characteristic of static-thyristor-inverter equipment is that the no-load losses are a significant portion of the full-load losses. Therefore the system efficiency is not adversely affected by continuous loaded operation.

  b. Static Uninterruptible Power Supply. A continuous or "floated-battery" type uninterruptible power supply (UPS) consists of a controlled rectifier section, a battery, and an inverter section. The basic block diagram is shown in Fig. 1a. The controlled rectifier converts incoming ac power to regulated dc which supplies the inverter sections and maintains the battery at float charge. The inverter section converts the dc into a suitable quality of ac to operate the user's electronic system. The battery provides the necessary stored energy to operate the inverter section during a utility outage. Clearly, then, the critical load is isolated from any disturbance of either voltage or frequency on the utility lines. A wide range of UPS equipment can be fabricated to meet a variety of requirements for reliable power. The battery is rated to supply emergency power to the inverter to support the critical load for the desired time. Battery support times of 5 to 60 min are most common. The basic large, i.e., 250-kVA, nonredundant

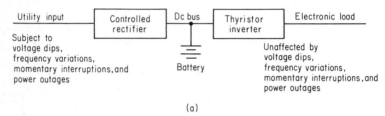

(a)

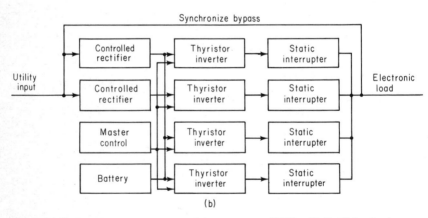

(b)

Fig. 1 (a) Continuous-type uninterrupted power system (UPS). (b) Parallel-redundant continuous-type UPS.

UPS field reliability is much better than that of rotating machinery. However, it can be expected to fail eventually.
  c. Parallel-redundant UPS. Various combinations of equipment can produce a UPS mean time between failures (MTBF) of over 80,000 h. One such configuration includes two fully rated controlled rectifiers, a battery, and four thyristor inverters rated so that full-load peaks can be supplied with one inverter out of service. The thyristor inverters are operated from a common master control, and each has a static interrupter on its output. A one-line diagram is shown in Fig. 1b.
  A simple parallel connection of inverters will not achieve improved reliability, since a failure in one will cause the others to fail because of the fault currents developed. The result of four parallel-connected inverters would be a MTBF time one-fourth that of a single inverter.
  The only way to improve reliability with multiple inverters is to ensure that a single inverter failure does not cause others to fail or cause a significant load-voltage transient. The only successful method of isolation is to use a static interrupter on each inverter output to disconnect the faulted inverter. The static interrupter must be capable

of removing the faulted inverter before it can produce a low impedance on the electronic-system bus. Although design considerations required to assure UPS reliability are very important, equally serious consideration must be given to minimizing the transient-response characteristics. Obviously, the lower the inverter transient impedance the faster the static interrupter must operate.

  d. Controlled Rectifier. The UPS controlled rectifier utilizes a basic 3-phase full-wave thyristor bridge configuration. Utility power is transformed to match the input voltage to the UPS system dc operating voltage level. The thyristor bridge rectifier provides rectification and regulation by the phase-control technique. High-capacity output is achieved by operating a number of these thyristor bridge rectifiers directly in parallel. Output current is filtered by a conventional low-pass network. The controlled rectifier provides a regulated output voltage up to a current limit and then becomes current-regulated. The current limit can be adjusted typically from 50 to 110% of rating. This regulation is required when prime ac power equipment is utilized. The initial output current is retarded for several seconds by means of a current walk-in feature. This feature prevents excessive inrush on the supply whenever the output is connected to a battery.

  The controlled-rectifier rating is determined by electronic-system load, thyristor-inverter efficiency, battery rated voltage, and battery charging current. Added reliability is achieved by operating two controlled rectifiers in parallel.

  e. Battery. The stored-energy source for the UPS system is a lead-calcium-type battery connected to the dc bus. The controlled rectifier normally supplies power to the dc bus which is maintained at the battery-float charge voltage. The battery rating is determined by electronic-system load, inverter efficiency, minimum inverter input-voltage rating, and maximum support time. The lead-calcium-type battery is preferred for UPS systems because of the following characteristics: life (typically 20 years), float-charging ability, float-charging current (about 1 A for large ratings), hydrogen evolution (0.5 ft^3/min for large ratings), and low water consumption. These data are for a 177-cell 2000-Ah battery.

  f. Master Control. A master-control concept is used in the redundant-thyristor-inverter UPS system. The master control ensures that all redundant thyristor inverters operate in phase by controlling all the inverters with one set of control signals. This provision, combined with automatic regulation to the critical ac bus voltage, provides the capability to take individual inverters in and out of service while the system is in operation without having to synchronize to the system frequency or adjust the inverter to the system voltage. These features improve operational reliability. The automatic regulation to the critical ac bus feature is achieved by multiple voltage-feedback signals.

  The master control is designed for high reliability by derating components, the output buffer amplifier, and certain redundant functions. Two oscillators are operated in a parallel-redundant mode so that either circuit failure does not adversely affect the master control signals. Each oscillator includes an automatic alarm circuit. These oscillators operate at 12 times desired frequency to obtain positive phase relationships. Desired phase relationships are established by sequential switching circuitry. The required pulse-width voltage-control signals are derived by solid-state phase-control techniques. The output buffer-amplifier circuitry further increases reliability by isolating the master control from the redundant thyristor inverters.

  g. Thyristor Static Inverter. The thyristor static inverter is a 12-phase circuit with 24 switching elements. The resultant square waves are combined in the output transformer zigzag wye secondary windings. The lowest harmonic voltage developed is therefore the 11th, since all 3rd, 5th, 7th, and multiples of these are canceled. Voltage control is achieved by means of the pulse-width method. A simple low-pass filter is used to reduce the total harmonic distortion to less than 5%. Figure 2 is a block diagram of the redundant 12-phase thyristor inverter. Thyristor gating signals are developed in the pulse amplifier, which is operated from its low-voltage power supply and driven from the master control. A static interrupter is included for redundant operation. An inverter monitor network is utilized to detect inverter malfunction and actuate the static interrupter.

  The inverter switching circuits utilized are the type described as complementary-impulse commutated with limited inverse voltage. This circuit incorporates some of the better features and avoids some of the disadvantages of other inverter switching circuits (Fig. 3*a*). In particular, the *LC* circuit that generates the commutating-current pulse provides a local low-impedance path for almost one half cycle of natural oscillation. This arrangement enables a high percentage of the stored energy in a commutating capacitor to be transferred to its complementary commutating capacitor. Therefore, only the losses have to be made up from the dc supply (Fig. 3*b*). Inverter efficiency is about 90% at normal UPS operating levels.

Operation of this switching circuit is described with an initial reactive load current I_L flow from the positive dc line through thyristor CR_1 through the upper half of the center-tapped winding of L_3 to the ac terminal. Capacitor C_2 is fully charged to the dc voltage level. The commutation interval is initiated at time t_0 by applying a gate signal to thyristor CR_4. The commutating current I_1 impulse is generated by the discharge of capacitor C_2 through the lower half of the center-tapped winding of L_3 through thyristor CR_4 and through reactor L_2. This current impulse is transformed to the upper half of the center-tapped winding of L_3, through thyristor CR_1, in opposition to the load current I_L, and returns through reactor L_1 and capacitor C_1. Simultaneously a current I_0 builds up through thyristor CR_1, reactor L_3, and thyristor CR_4. When the

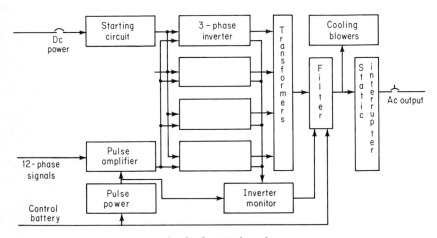

Fig. 2 Typical redundant 12-phase thyristor inverter.

commutating current I_1 attains a magnitude greater than the load current I_L plus current I_0, thyristor CR_1 becomes back-biased. This value is shown at time t_1. Current continues through diode D_1 until the commutating current I_1 decays below the load current I_L plus current I_0, which occurs at time t_2. The time interval t_2 minus t_1 is the available cleanup or turnoff time for thyristor CR_1. The load current I_L is now transferred to the lower half of the center-tapped winding of L_3 and through diode D_4. The secondary winding of L_3 provides a means to return energy to the supply during the subsequent oscillation of the commutating circuit, i.e., capacitor C_2 and reactor L_2.

h. Static Interrupter. The static interrupter is the only known way to isolate a failing redundant thyristor inverter properly in UPS service. Typical thyristor inverters fail in a short-circuit mode across the dc supply which is subsequently removed by fuse action. However, this failure also rapidly creates a short circuit across at least a portion of the output terminals. If the failing thyristor inverter remains connected to the ac bus, fault currents will be drawn from that bus and produce significant voltage transients.

Conventional methods of fault isolation depend upon considerable fault current for actuation and significant time periods for operation. Therefore, conventional methods of fault isolation cannot protect an electronic-system ac bus.

The static interrupter is an independent thyristor switching device that is capable of rapidly interrupting inverter power flow whenever actuated. Impending thyristor-inverter failure can be detected, and circuit isolation can be achieved in less than $1/25$ of a cycle, on a 60-Hz basis. It literally isolates the failing inverter from the output

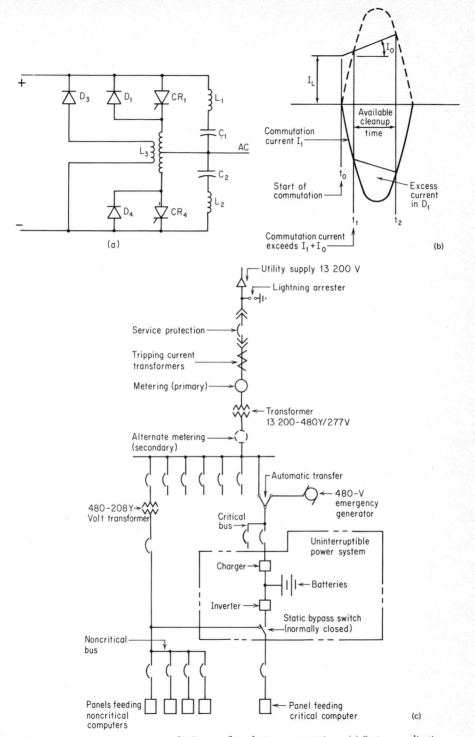

Fig. 3 (*a*) Commutation circuit. (*b*) Current flow during commutation. (*c*) System application of UPS.

ac bus before an ac fault develops. Since it is an independent thyristor switching device, all modes of thyristor-inverter failure are isolated. Specific inverter-component failures isolated by the static interrupter include thyristor, diode, commutating circuit, pulse amplifier, and pulse power supply.

The static interrupter is provided with another mode of operation, i.e., current limit. This limits each thyristor-inverter output current to a predetermined value in the event of an electronic-system bus fault. Fault current is supplied for several seconds to actuate feeder-circuit protective devices. If the fault persists beyond the allotted time, the bus is deenergized. Typical current pickup value is 130%, and current limit is 120% rated output.

The continuous or "floated-battery" parallel-redundant thyristor inverter meets all design specifications and provides adequate isolation from the utility system. Both voltage and frequency disturbances on the utility are buffered from the electronic-system bus. UPS systems rated to 2000 kVA are designed and practical for such service. They offer high efficiency, low maintenance, modular construction, easy installation, fast transient response, extremely stable frequency, and quiet operation as advantages over other approaches to the "buffer" system problem. The rapidly growing technology of solid-state power-conversion devices promises improved performance at decreasing cost in the future for UPS systems (Fig. 3c).

i. UPS Module. In the design of uninterruptible power systems, serious consideration should be given to the module design. Modularity is important to increase reliability through redundant operation, to afford connection or disconnection of the equipment to change its capacity, to provide isolation of components for maintenance while maintaining continuity of service, and to exchange equipment at different locations.

9. Large Motors—Light-Flicker Limitations. The starting currents taken by large motors driving air-conditioning compressors may cause an abnormally large voltage drop which may affect other customers on the line. The utility generally includes the maximum motor-starting currents in their service rules. However, they can tolerate a higher voltage drop if they are started infrequently. Often it is necessary for the customer to control the starting current of the motor by means of reduced-voltage starters, wound-rotor motors, etc.

10. Effect of Motor Starting on Distribution System (Ref. 2). Frequently, in the case of purchased power, there are transformers and/or cables between the starting motor and the generator. Most of the drop in this case is within the distribution equipment. When all the voltage drop is in this equipment, the voltage falls immediately (because it is not influenced by a regulator as in the generator case) and does not recover until the motor approaches full speed, since the transformer is usually the largest single impedance in the distribution system and therefore takes almost all the total drop. Figure 4 has been plotted in terms of motor-starting kVA which is drawn if rated transformer secondary voltage is maintained.

11. Reactor Voltage Drop. The approximate circuit voltage drop introduced by a current-limiting reactor can be obtained from the curves plotted in Fig. 5 when the load power factor and percent reactance are normally given as part of the reactor rating.

12. Light-Flicker Problems. Although a change in voltage changes the output of lamps, slow changes in voltage such as those due to normal load variations generally do not affect the output enough to be noticeable or irritating. The effect of sudden changes in voltage repeated at short intervals is termed "light flicker." It may become an annoying problem. In industrial plants, flicker is caused primarily by the following types of load: resistance welders, arc furnaces, and fluctuating motor loads such as compressors and punch presses. The irritation caused by light flicker is a function of the amount of change in the light output, the frequency of change, the rate of change, the duration, and the acuity of the individual observer. The dip-limit curves of Fig. 6 are a composite of several studies.

Fluorescent and mercury lamps are less subject to flicker during voltage changes than are incandescent lamps, provided that the lower limit of voltage remains above that value at which the fluorescent and mercury lamps will be extinguished.

The system should be designed to eliminate objectionable light flicker so as to adhere to the limits of Fig. 6. Wider limits may be used under conditions without complaint from the personnel occupying the affected area.

Figure 7 has been included because it represents some up-to-date thinking on the part of electric utilities with regard to flicker limits.

One way to reduce light flicker is to separate electrically the flicker-producing load from the lights.

13. Harmonics and Nature of Harmonics.   Harmonic voltages and currents are becoming of increasing importance in industrial power systems, particularly with regard to their effect on fluorescent lighting, communication systems, capacitor installations, and more recently, electronic-process control systems.   They are, for the most part, caused by nonlinear loads, such as electric welders, arc furnaces, and rectifiers, transformer-magnetizing current, and to a lesser extent, synchronous and induction machines.

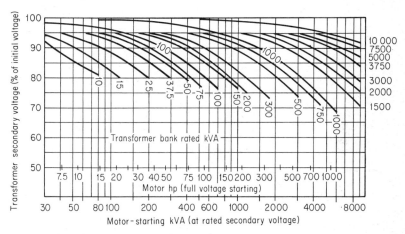

Fig. 4   Voltage drop in transformer due to starting a motor (for estimating purposes only).

Notes:

1. Scale of motor horsepower based on starting current being equal to approximately 5.5 times normal.

2. Short-circuit kVA of primary supply is assumed to be as follows:

Bank kVA	Primary short-circuit kVA
10–300	25 000
500–1000	50 000
1500–3000	100 000
3760–10 000	250 000

3. Transformer impedances are assumed to be as follows:

Bank kVA	Bank impedance, %
10–50	3.0
75–150	4.0
200–500	5.0
750–2000	5.5
3000–10 000	6.0

4. Representative values of primary system voltage drop as a fraction of total drop are as follows, for the assumed conditions:

Bank kVA	System drop/total drop
100	0.09
1000	0.25
10 000	0.44

The harmonic content and magnitude existing in any power system is largely unpredictable and will have a wide variation at different parts of the same system. Consideration of the effects of harmonics in the design of an industrial power system is seldom practicable or necessary, except where the following are included:
1. Mercury-arc or mechanical rectifiers
2. Arc furnaces
3. Large arc welders
4. Large fluorescent or mercury-vapor lighting systems
5. Local generation, particularly if directly connected to an overhead distribution system at generated voltage
6. Voltage stabilizers.

14. Arc Loads. Rectifiers, furnaces, and welders are seldom troublesome except where capacitors are installed or where there is the possibility of inductive coupling to

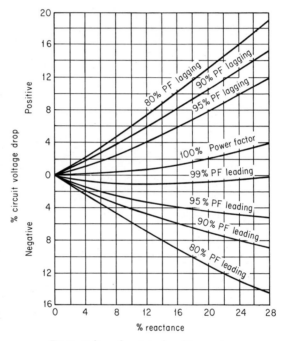

Fig. 5   Voltage drop introduced by reactors.

telephone circuits, either within the plants or, more likely, on the utility system. (Capacitors do not generate harmonics, but they may reduce or increase harmonics, depending upon the particular circumstances.)

Where capacitors are installed for power-factor improvement on a power system with arc-producing loads and the system reactance happens to equal the capacitor reactance at one of the harmonic frequencies generated, the harmonic currents in the capacitors and the interconnecting circuits will be substantially higher than the harmonic currents normally generated and may damage the capacitors through overheating. The resultant harmonic voltages on the distribution system may also cause local telephone interference. Possible remedial measures include the use of tuning inductances or changing the total kVA of connected capacitors.

15. Lighting Systems. The arc discharges of fluorescent or mercury-vapor lamps, combined with their associated capacitors and ballasts, are a source of harmonics, particularly the third. Experience shows that the third-harmonic current may be as high as 30% of the fundamental in the phase conductors, and 90 to 95% in the neutrals.

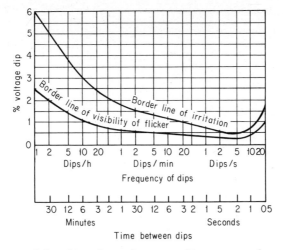

Fig. 6 Flicker of incandescent lamps caused by recurrent voltage dips.

Therefore feeder circuits serving such lighting predominantly should have the neutral conductor rated 100% of the lighting load.

16. Rotating Machines. It is commercially impracticable to build synchronous machines that generate a pure sine wave. Also, unbalanced supply voltages to induction machines will result in the generation of harmonics. While well-designed generators and motors seldom cause harmonic problems in the average industrial power-distribution system, it is well to consider the possibility while specifying generators, particularly where unbalanced loading or possible telephone interference from exposed transmission lines or cables is anticipated.

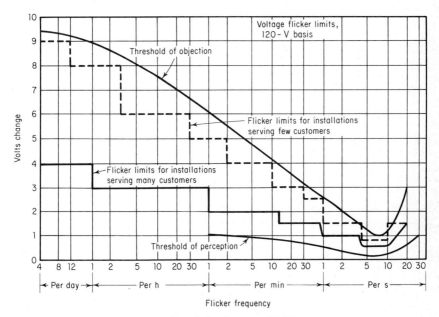

Fig. 7 Voltage flicker limits on 120-V basis.

VOLTAGE CONSIDERATIONS

Service voltage for other than isolated plants is determined by the utility and its policy, service rules, and economics. The location of the proposed load, the availability of the service, as well as future additional load requirements may have an important bearing on the service voltage.

17. Nominal System Voltages. The basic pattern of voltage identification followed in the United States is shown in Table 1. Rated generator voltage, transformer secondary voltage, and motor and control voltages are tabulated for each nominal system voltage. In general, the rated generator and transformer secondary voltages are the same value as the nominal system voltage and are multiples of 120, such as 240 and 480 V. Motor and control rated voltages are lower than the nominal system voltages to compensate for system-voltage drops and are therefore multiples of 115 V such as 230 and 460 V. Similar practice is followed in regard to voltage ratings of other utilization equipment.

TABLE 1. Standard System Voltages

Nominal system voltage (preferred voltages are underlined)						Generator or transformer secondary rated voltage	Motor and control rated voltage
2-wire	3-wire	3-wire wye	3-wire delta	4-wire wye	4-wire delta		
Low-Voltage Systems, Lighting or Combination Lighting and Power Circuits							
120	120/240	480	240 480 600	208Y/120 480Y/277	240/120	120 120/240 240/120 208Y/120 240 480 480Y/277 600	115 115/230 230/115 200Y/115 230 460 460Y/265 575
Power Circuits Only							
		208Y 480	240 480 600			208 240 480 600	200 230 460 575

However, listing a system as a nominal 480-V (3-wire) system or a nominal 480Y/277-V (4-wire) system does not specify the type of grounding. The grounding method used must be stated in addition to the nominal system designation as shown on the diagram, along with the nominal system designation as illustrated in Table 1. For each of the nominal system voltages listed those underlined are recommended voltages for new systems for the purpose of conveniently designating the voltage class.

It is evident that 120/240 V refers to the conventional 3-wire system. However, reversing this to 240/120 V refers to a 3-phase, 4-wire delta system.

18. Selection of Proper Plant Voltage Levels. The selection of system voltages in an industrial plant is, broadly speaking, an economic problem. However, a determination based on strict economic analysis may, in practice, be modified by plant or industry

standardization, availability of equipment, and other factors. The economic analysis will take into consideration:
1. Class of service available from utility
2. Total size of the installation
3. Planning for future growth
4. Characteristics of the equipment supplied
5. Density of the load
6. Safety considerations and qualifications of personnel available for operation and maintenance
7. Whether the installation is an extension or reconstruction of an existing plant, or whether it is a completely new isolated installation

For an extension to an existing plant, the principal decision to be made will be whether in the interest of long-range economy, a different (usually higher) voltage should be selected for the new equipment. If it should, there will be the temporary disadvantage of operating two different systems, either superimposed in the same building or separated in different buildings, the latter being the least objectionable. The superseded system may, in the future, become a small part of the total and ultimately disappear as processes change, as machines are moved about, and as motors are rewound.

Several factors have a tremendous influence on the overall cost and selection of system voltage. They are the feeder circuits, switchgear, system-fault duty, circuit arrangements, total motor horsepower, and motor-horsepower ratings. These items are beyond the scope of this book, since in various voltage classes different factors have the most pronounced effect on voltage selection. The various voltage classes that may be considered are as follows:
1. Selection of voltages of 600 V or less (low)
2. Selection of voltages of 601 to 15 000 V (medium)
3. Selection of voltages above 15 000 V (high)

19. Selection of Voltages of 600 V or Less. In most industrial plants, the majority of loads are integral-horsepower polyphase motors and welders which are most suitable for operation on systems of 600 V or less. The choice of nominal system voltages in this class for serving these loads is 208, 240, 480, or (480Y/277).

20. 480- vs. 600-V Systems. Although 600-V load-center systems cost about 2 to 7% less (Fig. 8) than 480-V load centers, they have decreased in popularity because standard 575-V utilization equipment is not available from manufacturers' and distributors' stocks in the United States.

When ordering machine tools or other utilization equipment with considerable electric control circuits, it is often difficult to obtain 575-V equipment, particularly on short shipment. Pumps and other equipment which are stocked by the manufacturers with motors already mounted are generally stocked with 230- or 460-V ratings and not with 575-V ratings. Availability of utilization equipment is the only major problem when choosing between 600- and 480-V systems. Today 600-V systems are limited primarily to expansion of those plants which already operate at 600 V, or to some textile plants where most motors are of special design and thus are not widely available from manufacturers' stocks.

Another advantage of the 480-V over the 600-V system is the possibility of using 480Y/277-V distribution with 277-V fluorescent lighting.

21. 480- vs. 240-V Systems. Economically speaking, there is seldom any reason for selecting 240 V instead of 480 V. Load-center systems at 240 V cost from 25 to 50% more (Fig. 8) than 480-V load-center systems. Lower-voltage systems cost more because there is more current per kVA to be carried, thereby increasing the size of the circuit breakers and conductors required.

Generally 240-V systems have higher losses and higher percentage voltage drop than 480-V systems. If enough copper or aluminum is used in the 240-V feeders, the losses and percentage voltage drop can be more comparable with those in 480-V systems, but in practice the heavy feeders are seldom used because this practice is expensive.

Some industries in which there is considerable dampness such as dairies and slaughterhouses have selected 240 V because they feel it is safer than 480 V. Operating records show that the biggest factor in safety is to ground all non-current-carrying metal parts properly and securely. When working on circuit conductors while energized, there is a greater chance for injury from electric shock with higher potentials to ground or phase to phase. However, any voltage above 50 can be lethal, and therefore the only safe way to handle these circuits in damp or other locations is to enclose the current-carrying conductors in securely and properly grounded metal enclosures and to work on current-carrying metal parts only when they are deenergized (Sec. 30).

In areas where the load is predominantly electric furnaces, 240 V may be most advantageous because of voltage limitations applied to these furnaces. In general, furnaces are a large spot load and cover only a small portion of the area of the plant. The furnaces may therefore be supplied by a separate load-center substation stepping down to 240 V for the furnace only and the rest of the load may be supplied at 480 V.

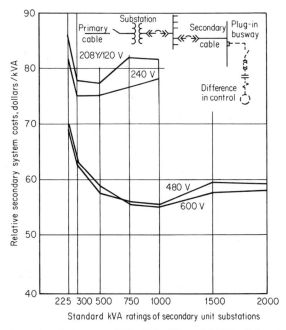

Fig. 8 Approximate installation cost of 208-, 240-, 480-, and 600-V radial secondary systems.

It is seldom economical to use 240 V for general distribution in a plant, even though a sizable portion of the total load may be 240-V furnaces.

22. Where 208Y/120-V Systems Are Applicable. In certain areas 208Y/120-V systems are more economical than 480-V systems because the types of utilization equipment involved must be operated at 120 V. Where such utilization equipment constitutes a major portion of the load (more than 50 to 65% of the total load), 208Y/120-V systems may be more economical than 480-V systems.

Typical of such a load is that of a clothing-manufacturing establishment where practically all the power is utilized by motor-operated hand shears. Other areas in which 208Y/120 V may be desirable are on assembly benches where small components are assembled and small portable tools such as soldering irons, electric hand drills, and electric nut tighteners are used. A typical case would be a small electronic-equipment assembly line. Again the choice of the lower voltage is primarily based on the desire to limit the voltage in the hand tools to 120 V. In these assembly areas, most of the power is utilized at this low voltage.

23. General Practice—Systems of 600 V and Less. In general practice, 480 V is most widely used and recommended by engineers, as a general secondary distribution voltage in industrial plants. Where a lower secondary system voltage is required, 208Y/120 V may be used, since it provides 120 V for incandescent lamps, office machines, hand tools, etc. These two voltages should cover practically all requirements for secondary voltages for offices, commercial establishments, and modern industrial plants except for electric furnaces and some welders.

In locations where at least one-third to one-half the load can operate on 480-V systems and the remainder of the load must operate at a lower voltage or where distances are greater than 200 ft, 480-V main distribution and small step-down transformers to 120 or 240 V may be most economical. Or instead these step-down transformers should be 208Y/120 V, since they provide the economy of 3-phase distribution and 3-phase, 4-wire multiwire branch circuits.

One additional point which should be mentioned in connection with low-voltage distribution is the trend toward the 480Y/277-V system. Such a system with the neutral readily available takes advantage of the savings inherent in the 480-V system and, in addition, provides an economical method for supplying the lighting load. By extending the neutral circuit in the lighting feeders only, 277 V fluorescent or mercury lighting can be distributed between the 3-phase conductors and neutral. The power load would be supplied at 480 V from 3-phase conductor circuits. It is not necessary to extend the neutral in the feeders supplying only power load.

24. Selection of Voltages for Systems of 601 V to 35 kV or Higher. Voltages in this class are used mainly for primary power distribution in industrial plants, public institutions, high-rise buildings, large office buildings, shopping centers, etc. All locations using a primary voltage employ voltages up to 35 kV except some of the very large chemical plants, steel mills, etc. The latter may employ subtransmission voltages above 35 kV.

When the utility voltage is below 15 kV, there is no problem of selecting the primary voltage.

Since the National Electrical Code permits 35 kV or higher inside buildings, without special restrictions, there is generally no reason for transforming voltages of the order of 35 kV to 4160 V or 15 kV for distribution through the building. Some engineers prefer not to exceed 15 kV in plants of moderate capacities not exceeding 20 MVA. The higher voltage can be carried to the load-center substations and there transformed to utilization voltage.

25. High-Voltage Distribution. Section 230-201h of the 1975 NEC states that when service-entrance conductors exceed 15 kV they shall enter either metal-enclosed switchgear or a transformer vault. An exception would be conductors placed under at least 2 in of concrete beneath a building or conductors within a building in conduit or duct and enclosed by concrete or brick not less than 2 in thick which would be considered outside the building. However, this voltage limitation applies to service-entrance conductors only.

After the service and feeders are properly provided with switching facilities and overcurrent protection as required by the code, conductors of any voltage can be extended through the building if they conform to Article 710 of the NEC. Article 710 does not place any limitation on conductor voltage for conductors extended through the building. Likewise askarel-filled or dry-type transformers up to and including 35 kV may be installed inside a building without a vault according to Section 450-21. The extension of higher-voltage conductors through a building is a new tool and can be used to feed load-center substations in large steel mills, chemical plants, etc., having large widely scattered loads. The trend during the past few years has been the selection of higher voltages to obtain greater economy and flexibility for expansion.

Studies have shown that the following voltages can be economically used for distribution (Fig. 9): 4160 V for plants having a supply transformer or generating capacity of 10 MVA or less, and 13.8 kV for plants having a supply transformer or generating capacity of 20 MVA or greater. For the range between 10 and 20 MVA, either 4160 V or 13.8 kV may prove to be more economical. While 4160 V may be slightly less expensive for a 15-MVA plant, if the plant grows, 13.8 kV will be more economical. However, since the capacity of a 4160-V feeder is quite limited, and inasmuch as a 13.8-kV distribution is more flexible and expandable, 13.8 kV is recommended if at all possible. Also there is little difference between the cost of 4160-V and 13.8-kV transformers and switchgear. However, if large motors at the distribution voltage are used, it may be justified to use 4160-V distribution. Obviously if the utility's service voltage is 13.8 kV, the same voltage must be used for distribution. It is very improbable that a 4160-V service of more than 1000 kVA will be provided directly from the utility's 4160-V distribution system.

Distribution from a 34.5-kV class line can be economically used for loads exceeding 20 MVA, since it may eliminate one of the transformations and instead use a 34.5-kV class distribution system and load center. It is predicted that the 34.5-kV class distribution system will be the fastest-growing distribution system.

26. Where 2400 V Is Particularly Applicable. In plants which are served at 2400 V directly from the utility system, of course it would be more economical to use 2400 V directly and place all motors rated 200 hp and above directly on the primary feeders. However, in these cases, most motors rated less than 200 hp should be operated on a 480-V system, stepping down from the 2400-V primary system. In some cases there are existing 2400-V systems in industrial plants, and it is often more economical to extend these at 2400 V rather than change to 4160 V. However, it may be more desirable to extend them at some higher voltage such as 13.8 kV rather than extend the system at 2400 V.

Where the primary voltage is above 5000 V, 2400 V is applicable to concentrated loads which supply motors rated 200 hp and above. In these cases it is preferable to step down from the higher voltage to 2400 V rather than to 4160 V, if the capacity per 2400-V bus is limited to less than approximately 7500 kVA and nearly all this load is utilized at the bus directly by motors rated above 200 hp. For single loads larger than 7500 kVA per bus for motors of 250 hp, it is often necessary to go to 4160 V to enable the switchgear to handle the short-circuit currents available from such large buses. A typical application where 2400 V would be most economical is a pumping station, in which the control may be of the fused combination type to give fast short-circuit protection or of

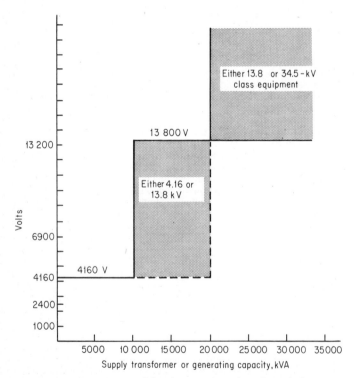

Fig. 9 Most economical distribution voltage for industrial plants.

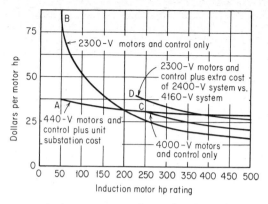

Fig. 10 Approximate cost of induction motors and control, and appropriate system cost as a function of motor horsepower.

the circuit-breaker type where the frequency of starting is not too great. If this preference is followed, it will be necessary to make a transformation for these large motors. The cost of the transformation is substantially the same whether it is made to 2400 or to 4160 V. Thus the power-system cost is not a significant factor as all the load is concentrated in one place. The 2300-V motors and their starters are less expensive than 4000-V motors and their starters in most cases.

The influence of motors on the choice between 2400 and 4160 as the primary system voltage is shown in Fig. 10.

TYPE SYSTEM

27. Grounded or Ungrounded Systems. Whether a system is to be grounded or ungrounded depends upon the purpose and use. Section 250-5 of the NEC requires that ac systems shall be grounded if the voltage to ground does not exceed 150 V or if a neutral is used as a circuit conductor or if a service conductor is uninsulated. Higher-voltage systems of any voltage may be grounded. However, if supplied by a utility providing utilization voltage, it is necessary to accept the utility's service, whether it is grounded or ungrounded.

a. Grounded System. It is a general practice to ground systems, since this provides an automatic clearing of the accidental ground on the ungrounded conductor by the operation of the overcurrent device protecting the system (Ref. 2).

In grounded systems, phase-to-ground faults produce currents of sufficient magnitude to be useful in the operation of overcurrent relays or fuses, which automatically detect the fault, determine which feeder has faulted, and initiate the tripping of the correct circuit breakers to deenergize the faulted portion of the system without interruption of service to unfaulted portions. Moreover, if the system neutral is grounded through a well-chosen impedance, the value of the fault current can be made sufficient for dependable relaying, yet insufficient for extensive damage at the point of the fault.

b. Ungrounded System. To provide continuity of service from a temporary ground, systems are ungrounded. Some engineers prefer a supervised ungrounded system (with ground detectors). Therefore, if one conductor is accidentally grounded, the circuit is not interrupted. This system is used in food-processing plants, etc., where continuity of supply is of the utmost importance and, if interrupted, will cause a large financial loss. It is necessary that this system be continually monitored by means of a ground detector which indicates when an accidental ground occurs. This is also recommended by the NEC. Qualified and reliable maintenance personnel, therefore, must follow and remove the accidental ground during the next shutdown. The disadvantage of this system is that if the first accidental ground is not removed, and the second ground develops, this will often cause a failure of one or more motors. Trouble from this source is quite common and is caused by induced voltages due to the making and breaking of the accidental grounds (Ref. 2).

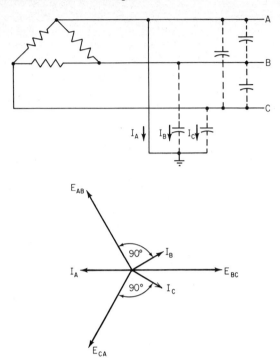

Fig. 11 Source of ground current is distributed capacitance to ground of the unfaulted conductors in an ungrounded system and phase-to-ground fault.

In ungrounded systems, phase-to-ground faults produce relatively insignificant values of fault current. In a small isolated-neutral industrial installation, the ground-fault current may be well under 1 A; while the largest plant, containing miles of cable to provide electrostatic capacitance to ground, may produce not more than 20 A of ground-fault current. These currents are not useful for the operation of overcurrent relaying to locate and remove such faults, not only because of the extreme sensitivity of the relays that would be required, but also because of the complexity of the flow pattern resulting from the fact that the "source" of the ground current is the distributed capacitance to ground of the unfaulted conductors (Fig. 11). It is possible, however, to provide a neutral voltage relay which will operate an alarm on the occurrence of a ground but which cannot provide any indication of its exact location.

The one advantage of an ungrounded system lies in the possibility of maintaining service on the entire system, including the faulted section, until the fault can be located and the equipment shut down for repair. Against this advantage must be balanced such disadvantages as the impossibility of relaying the fault automatically, the difficulty of locating the fault, the long-continued overstressing of the insulation of the unfaulted phases (1.73 times operating voltage in the case of solid grounds and perhaps much more in the case of arcing grounds), and the hazard of multiple ground faults and transient overvoltages.

28. Distortion of Phases during Faults. Balanced 3-phase faults do not cause voltage distortion or current unbalance. The balanced relationships of voltages and currents are shown in Fig. 12. Other types of faults, phase-to-phase, single-phase-to-ground, and two-phase-to-ground, cause distorted voltages and unbalanced currents. The voltage distortion is greatest at the fault and minimum at the generator or source.

Currents and voltages which exist during a fault vary widely for different systems. They vary on a given system depending on type and location of the fault and the degree of system grounding. The vector diagrams of Fig. 13a, b, and c show voltage and current relations which exist for different types of faults on a solidly grounded system in which the currents lag the voltages by 60°. Load currents are not included.

These diagrams are typical of the fault conditions which cause relays to operate. The distortion can be greater or less than that shown, depending on the severity of the fault and its distance from the relay.

29. Practical Limits of Protection. When the industrial power system is in normal operation, all parts should have some form of automatic relay protection; however, some fault possibilities may be deliberately set aside as too improbable to justify the cost of specific protection. Before accepting a risk on this basis, however, the magnitude of the probable damage should be seriously considered; otherwise, too much protection may be provided for troubles which occur frequently but cause only minor difficulties, while rare but serious causes of trouble may be neglected. For example, internal transformer failures rarely occur, but the consequences may be very serious since such faults can cause oil fires and endanger personnel.

Most systems have some flexibility in the manner in which circuits are connected together. The various possible arrangements should be considered in planning the relay system, so that some emergency operating condition is not left without protec-

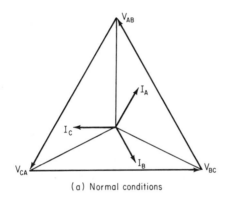

(a) Normal conditions

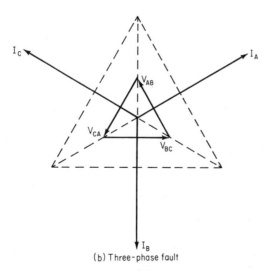

(b) Three-phase fault

Fig. 12 Balanced 3-phase faults do not cause voltage distortion or current unbalance.

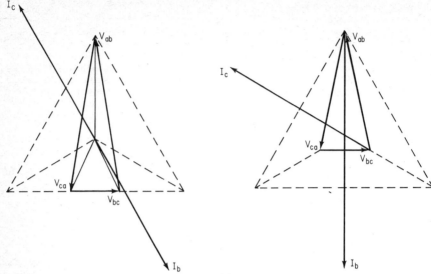

(a) Phase-to-phase fault between phases b and c

(b) Two-phase-to-ground fault between phases b and c and ground

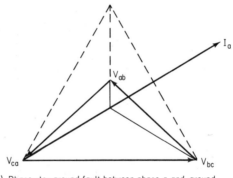

(c) Phase-to-ground fault between phase a and ground

Fig. 13 Voltage and current relation for different types of faults on solidly grounded system with current lagging voltage by 60°.

tion. Some types of systems have so many possible operating combinations that relay protection cannot be applied to operate correctly for all combinations. In such cases, the operating connections for which the protection is inadequate should be avoided if possible.

30. Arcing Ground Fault. The selection of an overcurrent device should be based on its effectiveness for protection of ground faults. Arcing ground faults have been prevalent on the 480Y/277-V and to a lesser extent on the 208Y/120-V system. Arcing faults have occurred on the higher-current-rated circuits 1000 or 1200 A or larger. It has generally been conceded that the trouble is caused by high-resistance grounds with the result that the arc is maintained but the arc current is insufficient to operate the overcurrent device, and the arc continues with disastrous results. Unfortunately, adequate ground tests have not been made after severe fires caused by arcing grounds. The 1975 NEC now requires ground-fault protection for solidly grounded wye electrical services of more than 150 V to ground, but not exceeding 600 V phase-to-phase for each service-disconnecting means rated 1000 A or more.

Many cases have been observed where a very high impedance was introduced in the ground-fault circuit through the use of one grounding-electrode conductor for the service and a different one for the equipment-grounding circuit. This procedure is a violation of the present NEC requirements. Such a connection more than doubles the impedance of the ground-fault circuit. Under such conditions a ground fault may not clear quickly enough to avoid extensive damage when using normal overcurrent devices. With the higher-voltage system (480 V) it is very unlikely that the arc will be self-clearing and thus more damage will be done; consequently, more attention will have to be called to such ground faults.

A large number of distribution systems that do not comply with the present Section 250-50 of the National Electrical Code continue to remain in service. This condition is due to the fact that prior to the 1962 National Electrical Code a common grounding-electrode conductor for service conductors and non-current-carrying metal parts was not mandatory but optional. However, arcing ground faults have resulted in the destruction of entire plants, killing people and causing lengthy interruptions with large monetary losses.

Protection against arcing ground faults can be obtained through the following three methods: First, overcurrent protection of a sensitive nature can be provided which will monitor and interrupt a ground current that could develop into an arcing ground fault. Second, the system can be designed to ensure a low ground impedance so that the overcurrent device will operate quickly enough so that no damage is done. The third method is to prevent the arc from starting by insulating all or most of the live parts and closing all voids so that rodents and foreign objects cannot enter any of the enclosures.

31. Overcurrent Protection to Interrupt Ground Current (Ref. 3). Arcing occurs in almost all fault situations, because a bolted or solid fault is in itself extremely rare. However, since in metal-enclosed equipment using bare busing, the ground path is usually involved, it is the arcing ground faults which cause the most problems. Arcing ground faults can cause fault currents appreciably below expected three-phase and single-phase bolted-fault currents because of the high impedance of the ground return path and the high-voltage drop in the arc itself. Recent studies have shown that minimum arc fault currents can reach somewhat less than 20% of the expected three-phase bolted-fault values. It is not surprising then that normal-phase protective devices are unable to cope with this situation. These devices can clear these fault currents only after they have developed into line-to-line faults where currents are substantial enough to be recognized. Unfortunately, serious equipment damage is involved by that time.

It has become apparent that these lower fault currents can sometimes cause greater equipment damage than heavy fault currents. Circuits of 208 Y/120 V may or may not be self-sustaining, but on 480 Y/277-V circuits, arcing is practically always self-sustaining. This arc releases tremendous amounts of energy that will concentrate at the point of the arcing rather than be dissipated throughout the system as in a bolted-fault situation. Thus, unless a more sensitive method of protection is utilized, almost complete destruction of equipment is possible, with a lengthy shutdown period for replacement or repair.

32. Choice of Methods. Ground-fault protection can be achieved by various methods and devices, with final choice dictated by systems considerations as to the sensitivity desired, coordination necessary, and economics. Two of the more popular methods in use are as follows:

a. Combination Static Tripping Devices on 600-V Air Circuit Breakers (Ref. 4). Several switchgear manufacturers offer optional ground-fault protection by incorporating this extra-sensitive protection within their static trip units. Thus, in addition to the normal time-delayed overcurrent protection and instantaneous fault protection for line-to-line faults, they afford the ability to protect against ground faults below normal full-load currents. In one case, a 20 to 80% adjustable pickup range of minimum trip-coil setting is available with the tripping characteristics. In another case, definite pickup points, 100 through 3000 A, and time-delay settings of 6 to 30 s are available. Separate devices by one maker are used for 3- and 4-wire circuits, while another uses the same device with one extra current transformer in the neutral circuit of the equipment. The basic advantages of this combination phase and ground device lie in its low cost, the compactness of the unit, and the fact that space and mounting problems of additional equipment are not present and an external tripping power supply is not needed since it is "direct-acting." The added cost is 4% for a 3-wire system and 9% for a 4-wire system.

b. Doughnut CT Sensing System.   Doughnut or window-type current transformers (CT) provide an effective way of sensing low-current ground faults, when used in combination with either a conventional electromagnetic overcurrent relay or a static type of overcurrent relay (Fig. 14).   Since these are not conventional ratio-type CTs, the makers refer to them as current sensors or zero-sequence current transformers. Here, the phase conductors and the neutral conductor, if present, are all passed through the CT.   Under normal and abnormal conditions not involving ground, no flux is produced in the CT core.   But when a ground fault occurs, the ground-fault current bypasses the CT, which in turn produces proportional flux and thereby CT current to the relay coil.

With careful matching of CT and relay, ground-fault detection can be made quite sensitive and operate down to 5 A or less. Different relay characteristics are available with fast operating times to limit ground-fault damage.   They are adjustable for current and time to obtain selectivity.   Although the relay portion is usually separate from the CT, one maker also can combine the relay and CT in one compact package.   While this

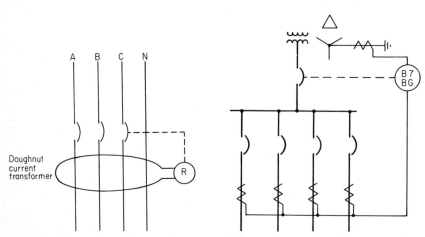

Fig. 14 Doughnut CT ground-sensing system.   **Fig. 15** Ground-differential relaying scheme.

method is quite flexible, it is also quite costly.   It requires mounting and wiring of several components, an external tripping power source must be provided, and the breaker must have a shunt trip device.   The added cost is approximately 17%.

c. Ground-Differential Relaying.   A ground-differential scheme can be effectively used to clear bus ground faults within the equipment itself where tripping of ground faults external to the feeders is not desired.   This uses window CTs encompassing all feeder buses or cables including the neutral and another CT located in the transformer neutral connection to ground (Fig. 15).   Only ground-fault current within the equipment will result in operation of the differential overcurrent relay used here, and it results in sensitive recognition of low ground-fault currents with inherent selectivity. As an example, one user of this scheme clears minimum phase to ground-fault currents of less than 1% of the symmetrical short-circuit ampere capacity of the unit substation in less than 10 cycles.

Here the difficulties encountered in mounting of a number of window CTs in restricted enclosures, space, and cost must be weighed against advantages gained.   The added cost of this arrangement is 19%.

d. Residual Ground Relaying.   This scheme is quite commonly used in 5- and 15-kV distribution systems, and would result in using relays on low-voltage switchgear instead of direct-acting trip devices.   Here the actual ground current is not involved, as CTs are used for two of the 3-phase lines with an overcurrent relay connected in the CT neutral to respond to a current proportional to the ground-fault current (Fig. 16). This method loses its effectiveness in situations where normal neutral currents flow, as settings would have to be above normal neutral current.

Although effective on 3-wire systems, the high cost of separate relaying just about rules out this method on low-voltage systems. The added cost of this arrangement is approximately 50%.

33. Providing Low-Ground Impedance (Ref. 5). The demands for safety are bounded by two opposing factors which must both be given consideration. When a ground fault occurs, it must be cleared as quickly as possible and yet continuity of service must be provided without nuisance outages. Hence, the optimum values of time and current can be arrived at only through a compromise.

A distribution system should be designed so that each circuit will clear on a ground fault within the time limit set up by the ground-fault protective device (GFPD). If it should not, the GFPD will come to the rescue and clear the circuit at the end of the time selected.

The keys to safety are: (1) The distribution center and circuits are to be designed to have a ground-fault circuit of low enough impedance to pass the current required to open the overcurrent (OC) device selected within the time set up as a standard. (2) Keep the possibility of an arcing fault to a minimum through the use of insulation, isolation, and guarding wherever feasible. (3) Use a GFPD to keep all ground faults to limited time.

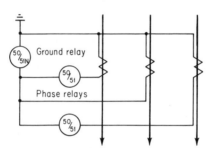

Fig. 16   Residual ground relaying scheme.

It is assumed that the distribution system is "effectively grounded" per Section 250-51 of the National Electrical Code, to keep the impedance of each circuit as low as practical. It is essential to do so if successful operation is to be attained regardless of what time and current settings are used. This plan does not cover a distribution system using more than one GFPD. Where more than one is used, it is possible to design to open any circuit on which a ground fault has developed automatically in a matter of cycles.

34. Ground Fault Defined. A ground fault is one that is created, accidentally or otherwise, between a phase conductor and the grounded equipment or conductor enclosure or other grounded metallic object. It is the result of insulation depreciation or failure, which may be caused by a variety of reasons. Usually it starts through direct contact between the phase conductor and the grounded object.

Ground faults may be divided into three classes:

1. A bolted fault.

2. Faults resulting from insulation depreciation starting as what may be considered a leakage of from 10 to 250 mA. Such a breakdown may be observed in equipment such as motors and transformers.

3. Faults resulting from immediate insulation breakdown that produces an arc.

35. Component Parts of Ground-Fault-Current Circuit. The ground-fault-current circuit can be broken down into its component parts as follows:

E = voltage from phase to grounded conductor

E_1 = arc impedance, which may be expressed as volts drop at the arc

Z_2 = impedance of circuit from point of arc to circuit disconnect

Z_3 = impedance of ground-fault circuit from circuit disconnect to source of supply

The effective voltage that will drive current through the ground-fault circuit will be E minus E_1 (the voltage drop across the arc).

36. Typical Ground-Fault-Current Circuit. The values which may be used in calculating the ground-fault-current flow are shown in a circuit (Fig. 17). They include the following:

1. Overcurrent-device rating and its time-current characteristics (TCC)
2. Ground-fault circuit voltage—phase to grounded conductor E
3. Impedance of arc Z_1 expressed as a voltage drop E_1
4. Impedance of the conduit run Z_2
5. Impedance of the balance of the ground-fault circuit to the source Z_3

There are some variations in the component parts of the ground-fault circuit. Some allowance must thus be made to cover these variations and arrive at a figure which will have practical value. The five values may be broken down as follows:

1 and 2. The OC-device rating and the ground-fault circuit voltage can be determined accurately within a very small tolerance.

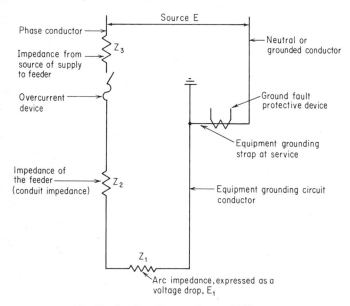

Fig. 17 One-line diagram of ground-fault circuit.

3. The impedance of the arc expressed as a voltage drop can be determined reasonably but not accurately.

4. The impedance of the conduit run may be calculated to within about 25% of accuracy. Such a difference is largely accounted for by normal manufacturing tolerances which affect the dc resistance and thus the impedance.

5. The impedance of the ground-fault circuit from its point of origin to the source varies for each circuit. An average figure is used here as equivalent to 50 ft of additional feeder.

Because of the variables 3, 4, and 5, a factor of safety is introduced to arrive at a ground-fault current flow that will not be likely to be less than that which will actually occur. However, regardless of what values may be used to compute ground-fault current flow as a basis of selecting proper current and time values for a ground-fault protective device, these values are not in accordance with effective grounding per Section 250-51 of the NEC. Considerable experience shows that safety has been neglected.

37. Hypothesis for Setting Current and Time Values Using Single GFPD. The following factors must be weighed:

1. Maximum current sensitivity of the current relay (minimum current setting) and minimum time delay provide maximum protection but can result in nuisance tripping for low-level ground faults. Increasing the current and time settings will improve service continuity and increase possible damage.

2. The object is to use a current setting such that all overcurrent devices in the system, except the main, have been selected to clear in less than the maximum time delay that has been selected.

3. Avoid opening the main but take a calculated risk and design to clear the feeder first, opening the main only as a last resort.

4. Pay special attention to insulation from the source of power to the main distribution center (switchboard) as well as insulation in the main distribution center itself, to reduce the possibility of a ground fault in that part of the system to a minimum.

5. Limit all ground faults anywhere in the system to a fixed time which will be the time value selected for the timing relay of the GFPD. This is, of course, a maximum only. The ground fault may and often can clear in less than the time selected by making full use of the TCC built into the OC devices used in the system if the system is properly designed and effectively grounded.

6. A GFPD may also be depended on to clear a stubborn high-impedance arcing phase-to-phase fault, which usually involves ground in a very short time.

7. The GFPD will provide backup protection for all normal overcurrent devices, which will be the first line of defense.

8. To improve coordination, the GFPD will not function on a ground fault on a circuit 100 A or smaller for services up to and including 2500 A; 125 A or smaller for services up to 3000 A; and 150 A or smaller for services over 3000 A.

9. Select 2 s as the time for the GFPD to operate on any ground fault, to coordinate with item 8, and set up $1\frac{1}{2}$ s as the maximum clearing time for all circuits.

10. The impedance of a circuit in steel conduit is equal to the impedance of the conduit alone.

11. Use an ampere setting of the GFPD in accordance with Table 2.

TABLE 2. Setting of a Single GFPD at Main, Based on Service Size and on 2-s Timing

Service size, A	A setting
1200	1000
1600	1000
2000	1000
2500	1000
3000	1200
4000	1500

12. Introduce an inverse time action in the main relay to decrease time delay as the fault current exceeds the values shown in item 11.

38. Length of Circuit an Important Factor. After wire and conduit size are selected for a circuit, its length is determined to find if the equipment-grounding-circuit (EGC) conductor (conduit or metallic conductor as may be found necessary) is adequate to provide safety.

Consider a 100-A circuit in $1\frac{1}{4}$-in steel conduit with an OC device that requires 900 A to clear in $1\frac{1}{2}$ s. The impedance of 100 ft of the ground-fault circuit is (Tables 3 and 4)

$Z_2 = 0.0222 \ \Omega$ (based on a current density of 1320 A/in^2 in the conduit)

$Z_3 = 0.0111 \ \Omega$ (50% of Z_2)

Total $= 0.0333 \ \Omega$

To pass 900 A through a circuit with a driving voltage of $120 - 50$, or 70 V, would require a circuit impedance of 70/900, or 0.0777 Ω. A factor of safety of 2 would require a circuit impedance of one-half that figure, or 0.0388 Ω. This factor of safety applies to the entire length of the circuit and not to the feeder length alone. The first 100 ft of conduit plus the impedance to the source would have an impedance of 0.0333 Ω. The impedance of the balance of the circuit must thus be 0.0388 minus 0.0333, or 0.0055 Ω. The impedance of 100 ft of conduit is 0.0222 Ω. Hence the feeder length

TABLE 3. Dimensions and Dc Resistance of Rigid Steel and Aluminum Conduit

Size of conduit, in *	Area of wall, in² *	Dc resistance, Ω per 100-ft aluminum conduit	Dc resistance, Ω per 100-ft steel conduit
½	0.254	0.008	0.0320
¾	0.337	0.00605	0.0242
1	0.53	0.00385	0.0154
1¼	0.68	0.003	0.0120
1½	0.79	0.00258	0.0103
2	1.03	0.00198	0.0079
2½	1.71	0.00119	0.00476
3	2.31	0.00088	0.00352
3½	2.70	0.000752	0.00301
4	3.40	0.000598	0.00239
5	4.30	0.000473	0.00189

* From NEMA "Handbook on Rigid Conduit."

can be increased $0.0055/0.0022 \times 100$, or 25 ft, making the maximum safe circuit length 125 ft.

Table 5 gives conduit lengths of 125, 119, and 110 ft, respectively, for clearing times of 1½, 1, and ½ s. Table 6 recommends a length of 120 ft.

In a similar manner, using Tables 3 and 4, the impedance of any steel-conduit run and thus its maximum safe length may be determined. If the required circuit length is greater than the safe length, the conduit must be supplemented with a metallic EGC conductor to provide the lower impedance that must be obtained.

39. Maximum Safe Length of Steel Conduit. Calculating the impedance of all conduit runs provides safe maximum lengths as shown in Table 5. From Table 5, it can be seen that the maximum safe length varies very little with circuits 100 A and larger or with whether the time of clearing is set at 1½ or ½ s. This apparent paradox results from including in the ground-fault circuit a magnetic material (the conduit) whose impedance decreases as the fault current increases. That same fact markedly limits any attempt to add any percentage figure to increasing the conduit lengths recommended. The length remains virtually constant regardless of clearing time within wide limits, as can be seen by comparing the figures in Table 5.

TABLE 4. Factors for Multiplying Dc Resistance of Rigid Steel Conduit to Obtain Impedance at Various Current Densities

Current density, A/in²	Conduit size, in			
	½, ¾	1, 1¼, 1½, 2	2½	3, 3½, 4, 5
200	5.0	5.2	6.1	7.1
250	4.7	5.0	5.6	6.3
300	4.3	4.6	5.1	5.8
400	3.8	4.0	4.4	4.9
500	3.3	3.6	3.8	4.15
600	3.0	3.25	3.4	3.7
700	2.7	2.9	3.1	3.4
800	2.5	2.65	2.75	3.0
900	2.4	2.45	2.6	2.8
1000	2.2	2.3	2.4	2.5
1250	1.8	1.9	2.0	2.2
1500	1.6	1.7	1.75	1.8
1750	1.4	1.45	1.55	1.65
2000	1.3	1.35	1.4	1.45

TABLE 5. Recommended Safe Maximum Length in Feet of Rigid Steel Conduit Which May Safely Be Used to Clear a Ground Fault; Conduit Used as the Equipment-Grounding-Circuit Conductor

		208Y/120 V			480Y/277 V			
		Time to clear fault, s						Size of EGC
OC device rating, A	Size of conduit, in	1½	1	½	1½	1	½	conductor [*]
30	½	150	143	133	500	500	453	10
50	¾	141	132	127	478	452	438	10
60	1	136	137	133	461	469	453	10
100	1¼	125	119	110	432	416	394	8
125	1½	116	112	111	410	396	396	6
150	2	103	113	115	416	400	404	6
200	2½	110	118	112	411	413	397	6
300	3	105	109	107	374	390	388	3
400	4	111	105	106	396	378	379	3
500	3½	100	98	96	360	362	352	1
500	4	107	106	101	382	383	368	1
600	5	100	109	100	372	391	366	1

[*] Table 250-95, 1968 NEC. These sizes must increase depending on the circuit length (see Table 7). Calculations assume (1) 50-V drop at the arc for 120 V to ground and 75 V for 277 V to ground, (2) impedance from feeder to source equal to 50 ft of feeder.

Based on using an OC device with an average K factor of about 10.4 for 1 s, it includes a factor of safety of 0.5 applied to the calculated ground-fault-circuit impedance to provide for the variables. The factor of safety applies to the whole length of the circuit and not to the length of the feeder only. If aluminum conduit is used, these circuit lengths may be increased about three times.

From Table 5, we can derive Table 6, which gives recommended maximum safe lengths for rigid steel conduit. When lengths exceed the values given in Table 6, it will be necessary to add a metallic equipment-grounding-circuit conductor in parallel with the conduit at points not greater than 100 ft apart for 208Y/120-V systems, or 300 ft apart for 480Y/277-V systems. The size of the EGC conductor cannot be obtained from Table 250-95 of NEC, not because the Code is wrong, but because the Code is a minimum standard and thus does not cover lengths such as those involved here.

TABLE 6. Recommended Maximum Safe Length of Circuits in Rigid Steel Conduit Using Conduit as the Equipment-Grounding-Circuit Conductor

OC device rating	Conduit size, in	208Y/120-V circuits	480Y/277-V circuits
30	½	150	485
50	¾	140	460
60	1	135	460
100	1¼	120	415
125	1½	115	400
150	2	110	400
200	2½	110	410
300	3	105	385
400	4	105	385
500	3½	100	360
500	4	105	375
600	5	100	370

If the length of the circuit exceeds the maximum lengths given, a metallic EGC conductor must be included with the phase conductors in the conduit, sized as required; see Table 7.

The EGC conductor must be bonded to the conduit every 100 ft for circuits of 120 V to ground and every 200 ft for circuits of 268 V to ground.

Table 7 illustrates the ineffectiveness of using EGC conductors as specified in the Code in any but relatively short lengths. A comparison of Tables 5 and 7 also shows the value of using the conduit as the EGC conductor supplementing it where conditions demand. Note that Table 7 does not include either a factor of safety or allowance for the added impedance of the balance of the ground-fault circuit from the circuit disconnect to the source. The maximum safe length of the circuits as given in Table 7 would thus be more nearly in the order of about 50% of the figures given.

TABLE 7. Comparison of Length of Circuit in Nonmetallic Conduit Using Different Sizes of Equipment-Grounding-Circuit Conductors

OC device rating, A	Phase conductor, AWG, MCM	EGC conductor, AWG, MCM	K factor for OC device for 1½ s *	Length of circuit, ft †
100	No. 3	No. 8	9.0	90
100	No. 3	No. 3	9.0	180
400	500 MCM	No. 3	10.25	70
400	500 MCM	500 MCM	10.25	340
600	1500 MCM	No. 1	10.0	80
600	1500 MCM	1500 MCM	10.0	350

* OC device rating times K factor equals current to open circuit in 1½ s.
† No factor of safety included, nor is any allowance made for impedance from feeder to source.

The following is a quotation from an early paper (Ref. 6) published on this subject as far back as 1954: "Since the Code does not specify the length of ground circuit and consequently does not limit voltage rise on the ground connection under fault condition, it is the responsibility of the engineer to specify a connection that will be effective."

40. Maximum Safe Length of Aluminum Conduit. In addition to using a metallic EGC conductor with steel conduit when the length of the circuit exceeds the values given in Table 6, the engineer has the further choice of using aluminum conduit. A multiplier of 3 applied to Table 6 will give maximum lengths for aluminum conduit. Using an internal bonding conductor, sized as per Table 250-95 of the NEC, with aluminum conduit has virtually no value. However, it is very important to maintain conductivity as well as continuity by assuring perfect joint connections on all metallic enclosures. For the rare case of an extremely long circuit where even aluminum conduit will not provide the answer, a separate GFPD will have to be used directly on the feeder in question to provide safety and avoid nuisance outage of the main.

41. Maximum Safe Length of Electrical Metallic Tubing (EMT). The impedance of EMT can be determined by using the factors shown in Table 4. For circuits with 120 V to ground the maximum safe length of circuits, rated from 60 to 400 A, in EMT from 1 to 4 in. in size will be from 80 to 101 ft.

42. Maximum Safe Length of Bus Duct. Bus duct may be divided into two general classifications:

1. All-steel enclosure
2. Enclosures having one or more sides of aluminum

If the enclosure is to serve safely as the equipment-grounding-circuit conductor, it must be capable of carrying a minimum current of about ten times the rating of the bus and not exceed 70-V drop for the length of the bus for 120 V to ground and 190-V drop for 277 V to ground. It is important for the engineer to assure himself that the standard 10-ft sections of bus have joints of adequate capacity not to disturb the continuous conductivity of ten times the rating of the bus. Where necessary, a metallic EGC conductor must be used. A rule-of-thumb figure for the EGC conductor is that it should be about 25% of the capacity of the bus or more if the voltage drop, as calculated, for the ground-fault circuit exceeds the figures given above for the respective voltage to ground. On the basis of an EGC conductor sized at 25% of the capacity of the bus, the maximum safe length is 115 ft for 120 V to ground or 317 ft for 277 V to ground. If an EGC conductor smaller than 25% of the main is used, the safe bus-duct length will decrease below the values of 115 or 317 ft, for voltages of 120 or 277 V to ground, respectively.

The size of EGC conductor given above is in conflict with the sizes given in Table 250-95. It is not that the NEC is wrong; the Code does not specify length but rather establishes a minimum requirement.

43. Maintaining Low-Impedance Ground Path. The importance of keeping the path to ground of low enough impedance to facilitate the operation of the overcurrent devices in the circuit, Section 250–51 of the Code, is shown by three examples as follows. All examples cover a 300-A feeder 100 ft long in 3-in rigid conduit.

CASE 1. A current of 3100 A (as per TCC of the overcurrent device) will open the circuit in $1\frac{1}{2}$ s. If the circuit impedance is doubled, the current will reduce to 1550 A and the clearing time will increase from $1\frac{1}{2}$ to 15 s, ten times as much.

CASE 2. A current of 3300 A will open the circuit in 1 s. If the circuit impedance is doubled, the current will reduce to 1650 A and the clearing time will increase from 1 to 13 s, thirteen times as long.

CASE 3. A current of 3600 A will open the circuit in $\frac{1}{2}$ s (30 cycles). If the circuit impedance is doubled, the current will reduce to 1800 A and the clearing time will increase from $\frac{1}{2}$ to 10 s, twenty times as long.

If careful attention is not paid to design as well as installation, it is quite easy to more than double the impedance of the ground-fault circuit. Two glaring examples are (1) using separate grounding-electrode conductors for the grounded conductor and for the equipment, a violation of Section 250–53 of the NEC, and (2) having the circuit length too long, a violation of Section 250–51.

44. Ground Faults Should Clear More Quickly on Higher-Voltage Circuits. Contrary to a sometimes expressed opinion, ground faults should clear more easily on a higher voltage to ground. Many cases have been observed where a very high impedance was introduced in the ground-fault circuit through the use of one grounding-electrode conductor for the service and a different one for the equipment-grounding circuit. Such a connection more than doubles the impedance of the ground-fault circuit. Under such conditions a ground fault may not clear quickly enough to avoid extensive damage when using normal OC devices. With the higher-voltage system, it is unlikely that the arc will be self-clearing, and thus more damage is done. Consequently more attention will be called to such ground faults. To arrive at the conclusion that ground-fault protective devices are more necessary on higher-voltage circuits is to confuse cause and effect.

Many recent cases of ground faults on 480Y/277-V systems that would not clear have been investigated. In every case it was found that the use of one grounding-electrode conductor for the service and another for the equipment was the chief cause by markedly increasing the impedance of the ground-fault circuit and thus the time for the ground fault to clear. A large number of distribution systems that do not comply with the present Section 250–53 of the National Electrical Code are now in service because prior to the 1962 Code a common grounding-electrode conductor for service and equipment was not mandatory but optional.

The timing of 2 s and the current values as given for a GFPD appear to be very practical. Their selection is a matter of judgment. The values suggested attempt to offer a satisfactory compromise between safety and service continuity. It is neither difficult nor costly to design an average distribution system to clear ground-fault currents through its normal OC devices and then add a GFPD to provide "police" protection or as a factor of safety.

Better insulation becomes of prime concern from the power company's service up to and including the main distribution center.

Effective grounding and the use of a single grounding-electrode conductor for a system can play a big part in safety.

All circuit lengths must be established and steps taken to see that the impedance of each circuit is sufficiently low to facilitate the operation of the OC devices in the circuit.

When a magnetic material is used as the conductor enclosure, the maximum safe circuit length is virtually constant regardless of the clearing time within practical limits.

All ground faults must be limited as to time.

High-current ground faults above those listed in Table 2 introduce an inverse time action in the current relay of the GFPD.

Equipment-grounding-circuit conductors, sized as per the NEC, are effective in short lengths only.

Safety can be greatly enhanced by following the above outlined principles.

45. Insulation.   A ground-fault protective device (GFPD) is useless to arcing faults if the fault is located on the source side of the service OC protective device. Therefore it is absolutely necessary that all live parts on the source side of the service OC protective device be insulated, since this area is generally not provided with overcurrent protection, and uncontrolled arcing grounds can easily start at this point.

It is a good policy not to have any bare parts in any switchboard, load center, or panelboard, since most arcing grounds start at bare or uninsulated parts.

46. Isolation.   It is necessary to be sure all voids in switchgear enclosure are closed so as to exclude all vermin, roaches, dirt, etc., which often are the means of starting arcing grounds between exposed live parts and the ground.

47. Guarding.   In order to guard against contacts with exposed live parts, it is often necessary to screen all sides of an outdoor substation to prevent squirrels, birds, cats, etc., from contacting live parts.

POWER-SUPPLY FAULT CAPABILITY

Protection of industrial and commercial power systems against failures in circuits and equipment has become of increased importance with the expanded utilization of electric power in industry. Even the best-designed electric systems will occasionally experience short circuits resulting in abnormally high currents. Overcurrent protective devices must operate to isolate such faults safely with a minimum of damage to circuits and equipment and with a minimum amount of shutdown of plant operation. The circuit breakers and fuses ordinarily used to perform this protection job must be selected to handle and interrupt safely the largest currents to which they may be subjected. Other parts of the system such as cables, bus duct, and disconnect switches must be capable of withstanding the mechanical and thermal stresses resulting from maximum flow of fault current through them (Ref. 2).

The current flow during a fault at any point in a system is limited by the impedance of circuits and equipment from the source or sources to the point of fault and is not directly related to the load on the system. However, additions to the system, made to handle a growing load, while not affecting the normal load being carried in some existing parts of the system, may well cause protective devices in those parts to be subjected to drastically increased fault currents. Expansion of an existing system as well as installation of a new system should include an accurate knowledge of available fault currents for proper application of overcurrent protective devices. The purpose of this section is to present a relatively simple method for calculating fault currents and to furnish typical data which can be used in making such calculations.

The size and complexity of many modern industrial systems may well make long-hand fault calculations so time-consuming as to be impractical. Analog- and digital-computer facilities are available in many locations, and the cost of using these facilities may be justified where a major fault study is planned. Whether or not these facilities are used, a knowledge of the nature of fault currents and calculating procedures is essential to such a study.

48. Sources of Fault Current.   Current which flows during a fault comes from two basic sources, synchronous and induction rotating machines. These may be operating as generators, motors, or synchronous condensers. The current which each of these delivers to a fault on its own terminals is limited by the impedance of the machine. Each of these rotating machines produces fault current which decreases with time after the initiation of a fault. In other words, these sources exhibit a variable reactance to the flow of fault current.

a. Generators.   Fault current from a generator decreases exponentially from a relatively high initial value to a lower steady-state value some time after the initiation of the fault. Since a generator continues to be driven by its prime mover and to have its field energized from its separate exciter, the steady-state value of fault current will persist unless interrupted by some switching means.

For purposes of fault-current calculation, the variable reactance of a generator can be represented by three reactance values:

X''_d = subtransient reactance. Determines current during first cycle after fault occurs. In about 0.1 s this value increases to the value X'_d.

X'_d = transient reactance. This value increases in about $\frac{1}{2}$ to 2 s to the value X_d.

X_d = synchronous reactance. This is the value that détermines the current flow after a steady-state condition is reached.

As most fault-protective devices, such as circuit breakers or fuses, operate before steady-state conditions are reached, generator synchronous reactance is seldom used in calculating fault currents for application of these devices.

b. Synchronous Motors. Synchronous motors supply current to a fault in much the same manner as do synchronous generators. Upon the drop in system voltage due to a fault, the synchronous motor receives less power from the system for rotating its load. At the same time the internal voltage will cause current to flow to the system fault. The inertia of the motor and its load acts as a prime mover, and with field excitation maintained, the motor acts as a generator in supplying fault current. This fault current diminishes as the motor slows down.

The same designation is used to express, in useful terms, the variable reactance of a synchronous motor as previously described for a generator. However, numerical values of the three reactances X''_d, X'_d, and X_d will often be different for motors than for generators.

c. Induction Motors. The fault-current contribution of an induction motor results from generator action produced by inertia driving the motor after the fault occurs. In contrast to the synchronous motor the field flux of the induction motor is produced by induction from the stator rather than from a dc field winding. Since this flux decays on removal of source voltage resulting from a fault, the contribution of an induction motor drops off, disappearing completely after a few cycles. As field excitation is not maintained, there is no steady-state value of fault current as for synchronous machines. As a consequence of these factors, induction motors are assigned only a subtransient value of reactance X''_d. This value will be about equal to the locked-rotor reactance, and hence the fault-current contribution will be about equal to the full-voltage starting current of the machine.

Wound-rotor induction motors normally operate with their rotor rings shorted. In this circumstance they will contribute fault current in the same manner as a squirrel-cage induction motor. Occasionally, however, large wound-rotor motors are operated with some external resistance maintained in their rotor circuits. These machines may then have sufficiently low short-circuit time constants that their fault contribution is not significant. A specific investigation should be made before neglecting the contribution from a wound-rotor motor.

d. Capacitors. The discharge current from power capacitors to a system fault is of a high-frequency nature with a time constant of only 1 or 2 cycles in most cases. Therefore the effect from power capacitors on system-fault currents can usually be neglected. Unnecessary tripping of the capacitor bank breaker on capacitor inrush or discharge currents is avoided by the use of overcurrent relays having a short time delay. Accordingly it is not usual practice to calculate capacitor discharge currents.

49. Fundamentals of Fault-Current Calculation (Ref. 2). Ohm's law, $I = E/Z$, furnishes the relationship used in determining fault current, where I is the desired current, E is normal system voltage at point of fault, and Z is the impedance from source to fault including the impedance of the source. Rigorous calculations generally introduce tedious and time-consuming complications, and experience has shown that simplifying assumptions can be made which detract little from accuracy and much from labor.

50. Types of Fault. In the usual procedures of fault-current calculation it is assumed that the fault is a zero-impedance "bolted" fault with no current-limiting effect due to the fault itself. Such calculations are used to determine the maximum short-circuit current value for the purpose of selecting devices of adequate interrupting rating, momentary rating, and to determine the maximum value of current at which time-current coordination need exist in relay studies. The 3-phase fault condition is frequently the only one considered since, in an industrial system, this type of fault generally results in maximum current.

In medium- and high-voltage systems line-to-line fault currents are approximately 87% and line-to-ground fault currents can range from about 60 to possibly 125% of the 3-phase value. However, line-to-ground fault currents of more than the 3-phase value are rarely encountered in industrial systems. Therefore only the method of calculating 3-phase fault currents is presented here. Calculations are vastly simplified by assuming a 3-phase fault condition because the system including the fault remains symmetrical about the neutral point whether or not the neutral point is grounded and regardless of star or delta transformer connections. The current can be calculated on a single-phase basis using only line-to-neutral voltage and impedance.

It should be recognized, however, that faults which occur in practice generally involve the impedance of an arc, with its varying limitation on the magnitude of fault current. In low-voltage systems there is a trend toward taking this factor into account to calculate the minimum value of fault current, which is of interest when specifying the necessary sensitivity for protective devices (Ref. 4). Approximate minimum values of arcing-fault current per unit of 3-phase voltage value for typical cases equal 0.89 at 480 V, 0.12 at 208 V for 3-phase arc; 0.74 at 480 V, 0.02 at 208 V for line-to-line single-phase arc; 0.19 at 480 V, 0.00 at 208 V for line-to-neutral single-phase arc.

51. Voltage and Impedance. The voltage which serves as a basis for fault-current calculation is derived from the rated nameplate voltage of the generator or transformer supplying the faulted element of the system. As explained previously, this value will be a line-to-neutral voltage or

$$\frac{\text{Rated line-to-line voltage}}{\sqrt{3}}$$

In ac circuits the impedance is the vector sum of resistance and reactance. In major elements such as generators and transformers, the reactance is usually at least five times the resistance. The fault current calculated by neglecting the resistance of the major elements and using only reactance will be in error by only a few percent. This error is on the safe side for calculating interrupting-duty for fault-protective devices. The resistance of generators, transformers, motors, reactors, and large bus work is therefore not considered regardless of system voltage.

In systems above 600 V the resistance of other circuit elements, such as cables, is generally neglected. In systems below 600 V when calculating faults out on branch-feeder circuits, resistance should generally be considered. If conductor resistance is important in calculating faults at any point in the system, it is likely to be in low-voltage branch-feeder circuits where the wire size is small. It is suggested that the following procedure be used for low-voltage-feeder circuits. If the resistance of the feeder circuit is one-fourth or more of the total reactance from source to fault, resistance should be included in the calculations. The resistance of the system up to the feeder should now be included and an approximate total resistance obtained by adding to the resistance of the feeder a resistance equal to one-fourth of the total reactance of the system from source to feeder. Using this value of total equivalent resistance R, and the total equivalent reactance X, the impedance to the fault is found by the expression

$$Z = \sqrt{R^2 + X^2}$$

It is important to consider the reactance of all circuit elements in calculating fault currents on low-voltage systems. The importance of small elements of reactance in limiting total fault current becomes great in systems below 600 V.

52. Symmetry of Fault-Current-Dc Component. In determining the maximum value of fault current which can occur at some point in a system, it must be considered that the fault-current wave is likely not to be symmetrical about the zero-current axis for several cycles after the fault occurs. System voltage and fault current are substantially sine wave in shape and are related in phase angle by the impedance angle of the system to the point of fault. Since the resistance will usually be negligible in comparison with the reactance, the fault current will usually lag the source voltage by nearly 90°. When a fault occurs at or near the peak of the voltage wave, the fault-current wave therefore starts at zero and is symmetrical about the zero axis. When a fault occurs at or near the zero point of the voltage wave, the fault-current wave again starts at zero on the original zero axis. However, at the start of the fault the current magnitude should be at or near peak because the voltage leads by 90° and is at zero magnitude. To meet this requirement, the fault-current wave becomes symmetrical about a new zero axis and is offset or asymmetrical to the zero axis of the voltage wave.

The magnitude of offset for any fault will be between the two extremes described above. When the resistance of the system to the point of fault is not negligible, to produce maximum or minimum asymmetry, the point of the voltage wave at the time the fault occurs will be different from that for the circuit containing negligible resistance. Maximum asymmetry occurs at a time angle equal to 90° + θ (measured in degrees from the zero point of the voltage wave), where tangent θ equals the X/R ratio of the circuit.

For convenience purposes the asymmetrical fault-current wave can be considered as composed of two basic components. With the zero axis of the voltage wave as a reference, the symmetrical ac component of fault current (as determined by E/Z) has superimposed on it a dc component whose magnitude is determined by the point on the voltage wave at which the fault occurs.

The magnitude of the dc component can vary from zero to a maximum equal to the peak of the symmetrical ac component. The initial magnitude of the dc component

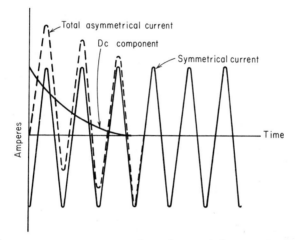

Fig. 18   Components of typical system-fault current.

is equal to the value of the ac symmetrical component at the instant the fault occurs. At any instant after the fault, the magnitude of current is equal to the sum of the ac and dc components (Fig. 18).

In a theoretical system with zero resistance, the dc component of fault current would remain at a constant value. However, in an actual system where resistance is present, the dc component decays to zero as the stored energy it represents is expended in the form of I^2R loss in the resistance of the system. The rate of decay of the dc component is proportional to the ratio of reactance to resistance (X/R ratio) of the system from source to fault. The lower the X/R ratio, the more rapid is the decay of the dc component. The effect of this decay is that the fault current gradually changes from an asymmetrical to a symmetrical current with respect to the zero-voltage axis.

There are thus two factors which result in the initial magnitude of fault current being greater than the steady-state fault current. One of these, the variable reactance of rotating machines, is accounted for in fault-current calculations by using the initial value of machine reactance, that is, the subtransient value. The second factor is the initial asymmetry explained in terms of a decaying dc component. This second factor does not readily lend itself to a mathematical analysis and is most easily accounted for by the use of simple multipliers applied to the calculated symmetrical value of fault current.

In the past these multipliers were applied to determine asymmetrical fault-current

TABLE 8. Multiplying Factors and Machine Reactances to Be Used for Calculating Short-Circuit Currents for Circuit Breaker, Fuse, and Motor-Starter Applications

Classification	Circuit voltage
Power circuit breakers: *	
8-cycle or slower (general case)...	Above 600 V
5-cycle	Above 600 V
General case...	Above 600 V
Less than 4 kV	601–5 kV
Medium-voltage fuses:	
All types, including all current-limiting fuses	Above 600 V
Non-current-limiting types only	601 V to 15 kV
Medium-voltage, fused motor starters:	
All horsepower ratings	2400 and 4160 V
Medium-voltage motor starters:	
Circuit breaker or contactor type...	601 V to 5 kV
Circuit breaker or contactor type...	601 V to 5 kV
Circuit breaker or contactor type...	601 V to 5 kV
Apparatus, 600 V and below:	
Low-voltage power molded-case circuit breakers, of 1-V fuses.................	600 V or less
Low-voltage motor starters (with fuses or molded-case breakers)...............	600 V

* Revisions to ANSI C37.10 have been proposed. These revisions eliminate the use of these multiplying factors in applying power circuit breakers.

values. These fault currents were then used in the selection of equipment which had asymmetrical current-capability ratings. A growing trend now exists toward the elimination of the use of asymmetrical equipment ratings. Table 8 lists the asymmetrical multiplying factors still in use at the time of this writing. Multipliers are applied to the calculated rms symmetrical fault current to obtain the rms asymmetrical value desired for application of protective devices.

53. Detailed Calculation Procedure Using Ohms, Percent Reactance, or Per Unit Reactance. The determination of short-circuit currents has been shown to be dependent principally upon the reactance X from the source (or sources) to the fault. The principal problem of short-circuit-current determination is one of determining

Location in system	Multiplying factor	Machine reactances to use		
		Generators, synchronous converters, synchronous condensers, frequency changers	Synchronous motors	Induction motors
		Interrupting duty		
Any place where symmetrical short-circuit kVA is less than 500 MVA	1.0 † 1.1 †	Subtransient Subtransient	Transient Transient	Neglect Neglect
		Momentary duty		
Near generating station	1.6	Subtransient	Subtransient	Subtransient
Remote from generating station (X/R ratio less than 10)	1.5	Subtransient	Subtransient	Subtransient
		Max rms A interrupting duty		
Anywhere in system	1.6	Subtransient	Subtransient	Subtransient
Remote from generating station (X/R ratio less than 4)	1.2	Subtransient	Subtransient	Subtransient
		Max rms A interrupting duty		
Anywhere in system	1.6	Subtransient	Subtransient	Subtransient
		Interrupting duty		
Anywhere in system	1.0	Subtransient	Transient	Neglect
		Momentary duty		
Anywhere in system	1.6	Subtransient	Subtransient	Subtransient
Remote from generating station (X/R ratio less than 10)	1.5	Subtransient	Subtransient	Subtransient
		Interrupting or momentary duty		
Anywhere in system	1.0 ‡	Subtransient	Subtransient	Subtransient
Anywhere in system	1.25 ‡	Subtransient	Subtransient	Subtransient

† These factors are increased to 1.1 and 1.2, respectively, if the symmetrical fault level is above 500 MVA, and the system is fed predominantly by generators or through current-limiting reactors.
‡ Fuses which operate in under 0.004 s have a multiplying factor of 1.4 to 1.6.

this reactance. To obtain it, the reactance of each significant element in the electric circuit must be determined and the elements combined in series or parallel.

At the outset a decision must be made concerning the system used for expressing the reactance of the elements. Three systems can be used with reactance expressed either in ohms or in percent or per unit on a chosen base value. It is often convenient to use the per unit system in calculations involving a system with several different voltage levels. When reactances are expressed as per unit quantitites on a chosen kVA base, they can be combined directly without regard for transformer turns ratio in systems utilizing more than one voltage level. In this section the per unit system is used.

Conversion equations for the three systems are as follows:

$$\text{Per unit reactance} = \frac{\% \text{ reactance}}{100} \tag{1}$$

$$\text{Per unit reactance (on chosen kVA base)} = \frac{\text{ohms} \times \text{kVA base}}{1000 \times \text{kV}^2} \tag{2}$$

$$\text{Per unit reactance (on chosen MVA base)} = \frac{\text{ohms} \times \text{MVA base}}{\text{kV}^2} \tag{2a}$$

$$\% \text{ reactance (on chosen kVA base)} = \frac{\text{ohms} \times \text{kVA base}}{10 \times \text{kV}^2} \tag{3}$$

$$\% \text{ reactance (on chosen MVA base)} = \frac{\text{ohms} \times \text{MVA base}}{\text{kV}^2} \times 100 \tag{3a}$$

where ohms = line-to-neutral values (single-conductor)
   kVA base = 3-phase base kVA
      kV = line-to-line voltage
   MVA = 1000 kVA

The per unit and percent systems are methods of expressing numbers in a form that allows them to be easily compared.   Actually they are a ratio based on a prechosen base number.   In determining fault currents, a convenient base kVA number must be chosen.   Reactances for buses, cables, lines, current transformers, air circuit breakers, etc., will be given in ohms, and it will be necessary to convert these values into percent or per unit reactances on a chosen kVA base by means of Eqs. (2) and (3).

Elements in the system such as transformers, generators, and motors normally will have their reactance given in percent based on their own kVA rating.   These reactances must be converted to the chosen kVA base as follows:

Per unit reactance on base kVA

$$= \text{per unit reactance on kVA rating} \times \frac{\text{base kVA}}{\text{kVA rating}} \tag{4}$$

For example, assume a value of 1000 kVA has been chosen as a kVA base.   One of the elements of reactance is a 500-kVA transformer with a reactance of 5% on its kVA rating.   This transformer would then be represented by the following per unit reactance on the chosen kVA base:

$$X_{pu} \text{ (on kVA rating)} = \frac{5\%}{100} = 0.05$$

$$X_{pu} \text{ (on base kVA)} = (0.05) \frac{1000}{500} = 0.10$$

54. Preparing System Diagrams.   Preparation of a one-line diagram is the first step in making a short-circuit study.   This diagram should show all the sources of short-circuit current and all the significant circuit elements.   Figure 19 is a one-line diagram of a hypothetical industrial system.

The following tabulation will serve as a guide to proper choice of reactances and impedances to be used for determining fault currents above and below 600 V.

For faults above 600 V:
   Utility-supply reactance
   Main-plant substation-transformer reactance
   Plant generator reactance
   Plant primary-feeder reactances
   Plant secondary-substation reactance
   Plant distribution-circuit reactance
   High-voltage (above 600 V) motor reactances
   Contribution through transformers from low-voltage motors
For faults below 600 V:
   Utility-supply reactance

Main-plant substation-transformer reactance
Plant generator reactances
Plant primary-feeder reactances
Plant secondary-substation reactance
Plant distribution-circuit reactance
Branch-feeder-circuit reactance and resistance (Art. 51)
Motor reactances (both above and below 600 V)
Impedance of any significant low-voltage bus runs
Low-voltage circuit-breaker reactances
Current-transformer reactances

The one-line diagram should be simplified when desirable by making a reactance diagram in which all the significant equipment reactances are indicated. Figure 20 is a reactance diagram for the one-line diagram shown in Fig. 19. In many cases clarity of the one-line diagram can be maintained and it will prove most convenient to add the reactance data to the one-line diagram.

55. Assigning Reactance Values. Reactance values should be assigned to all system elements shown on the reactance diagram (Tables 8 through 20 and Figs. 21 through 26).

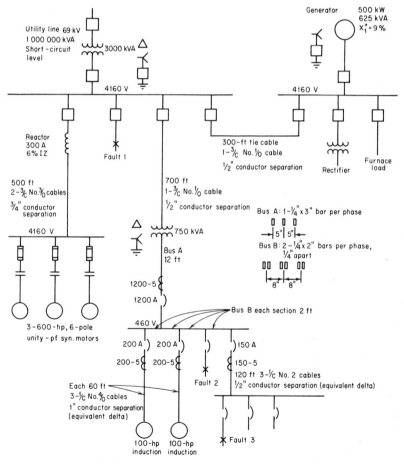

Fig. 19   Hypothetical industrial-system one-line diagram.

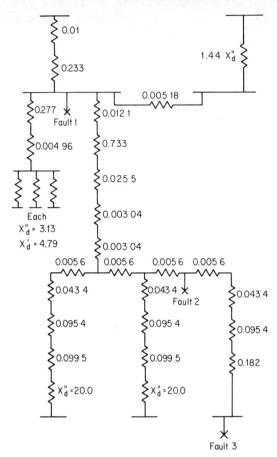

Fig. 20 Reactance diagram of system in Fig. 19.

TABLE 9. Typical Reactance Values for Synchronous Machines per Unit Values on Machine kVA Rating *

	X''_d	X'_d
Turbine generators: †		
2-pole..	0.09	0.15
4-pole..	0.15	0.23
Salient-pole generators with damper windings: †		
12 poles or less..	0.16	0.33
14 poles or more..	0.21	0.33
Synchronous motors:		
6-pole..	0.15	0.23
8- to 14-pole...	0.20	0.30
Synchronous condensers †	0.24	0.37
Synchronous converters: †		
600 V dc..	0.20	
250 V dc..	0.33	

Note. Synchronous-motor kVA bases can be found from motor horsepower ratings as follows:

0.8-PF motor — kVA base = hp rating
1.0-PF motor — kVA base = 0.8 × hp rating

* Use manufacturer's specified values if available.
† X'_d not normally used in short-circuit calculations.

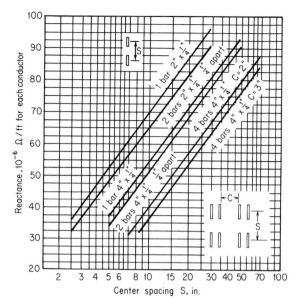

Fig. 21 Curves of reactance at 60 Hz vs. spacing for 2- by ¼-in bars and for 4- by ¼-in bars in edgewise arrangement; ohms per foot of each conductor at a distance S from the return conductor. In a balanced 3-phase circuit, the mean reactance per foot of each conductor is determined approximately for the mean spacing $S = (S_1S_2S_3)^{1/3}$ where S_1 = center spacing between phases 1 and 2, S_2 = center spacing between phases 2 and 3, S_3 = center spacing between phases 1 and 3.

**TABLE 10. Typical Reactances of
Induction Motors per Unit Values
on Machine kVA Base (Horsepower
Rating) ***

	X''_d	X'_d
Above 600 V	0.17	
600 V and below	0.20	

$$\text{Per-unit reactance of induction motors} = \frac{1}{\text{per-unit locked-rotor current † at rated stator voltage}}$$

* Use manufacturer's specified values if available.
† Usually five or six times rated full-load current.

**TABLE 11. Typical Reactances of Transformers per Unit Reactance
on Transformer kVA Rating ***

Primary voltage rating	Bank kVA (3-phase or 3 single-phase)		
	25–100	100–500	Above 500
2400/4160 V	0.015–0.018	0.050	0.055
13.8 kV	0.015–0.025	0.050	0.055
46 kV	. . .	0.060	0.065
69 kV	. . .	0.065	0.070

* Use manufacturer's specified values if available.

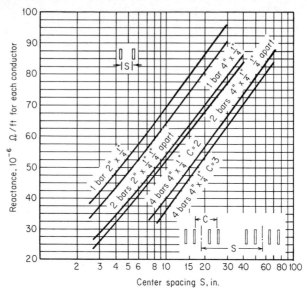

Fig. 22 Curves of reactance at 60 Hz vs. spacing for 2- by ¼-in bars and for 4- by ¼-in bars in facewise arrangement; ohms per foot of each conductor at a distance S from the return conductor. In a balanced 3-phase circuit, the mean reactance per foot of each conductor is determined approximately for the mean spacing $S = (S_1 S_2 S_3)^{1/3}$ where $S_1 =$ center spacing between phases 1 and 2, $S_2 =$ center spacing between phases 2 and 3, $S_3 =$ center spacing between phases 1 and 3.

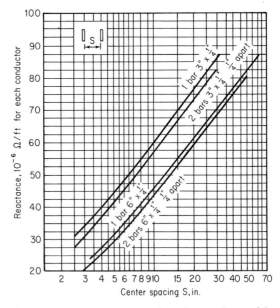

Fig. 23 Curves of reactance at 60 Hz vs. spacing for 3- by ¼-in bars and for 6- by ¼-in bars in facewise arrangement; ohms per foot of each conductor at a distance S from the return conductor. In a balanced 3-phase circuit, the mean reactance per foot of each conductor is determined approximately for the mean spacing $S = (S_1 S_2 S_3)^{1/3}$ where $S_1 =$ center spacing between phases 1 and 2, $S_2 =$ center spacing between phases 2 and 3, $S_3 =$ center spacing between phases 1 and 3.

TABLE 12. Representative Conductor Spacings for Overhead Lines

System nominal voltage	Equivalent delta spacing, in
120	12
230	12
460	18
575	18
2 400	30
4 160	30
6 900	36
13 800	42
23 000	48
34 500	54
69 000	96
115 000	204

TABLE 13. Constants of Copper Conductors for 1-ft Symmetrical Spacing (Use Tables 15 and 16 of Spacing Factors for Other Spacings)

Size of conductor		r_a Resistance, Ω/conductor/1000 ft at 50°C, 60 Hz	x_a Reactance at 1-ft spacing; 60 Hz, Ω/conductor/1000 ft
Cir mils	AWG		
1 000 000		0.0130	0.0758
900 000		0.0142	0.0769
800 000		0.0159	0.0782
750 000		0.0168	0.0790
700 000		0.0179	0.0800
600 000		0.0206	0.0818
500 000		0.0246	0.0839
450 000		0.0273	0.0854
400 000		0.0307	0.0867
350 000		0.0348	0.0883
300 000		0.0407	0.0902
250 000		0.0487	0.0922
211 600	4/0	0.0574	0.0953
167 800	3/0	0.0724	0.0981
133 100	2/0	0.0911	0.101
105 500	1/0	0.115	0.103
83 690	1	0.145	0.106
66 370	2	0.181	0.108
52 630	3	0.227	0.111
41 740	4	0.288	0.113
33 100	5	0.362	0.116
26 250	6	0.453	0.121
20 800	7	0.570	0.123
16 510	8	0.720	0.126

For a 3-phase circuit the total impedance line to neutral is
$$Z = r_a + j(x_a + x_d)$$

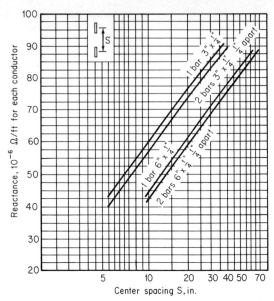

Fig. 24 Curves of reactance at 60 Hz vs. spacing for 3- by ¼-in bars and for 6- by ¼-in bars in edgewise arrangement; ohms per foot of each conductor at a distance S from the return conductor. In a balanced 3-phase circuit, the mean reactance per foot of each conductor is determined approximately for the mean spacing $S = (S_1 S_2 S_3)^{1/3}$ where $S_1 =$ center spacing between phases 1 and 2, $S_2 =$ center spacing between phases 2 and 3, $S_3 =$ center spacing between phases 1 and 3.

TABLE 14. Constant of Aluminum Cable Steel Reinforced for 1-ft Symmetrical Spacing

Size of conductor		r_a Resistance, Ω/conductor/1000 ft at 50°C, 60 Hz	x_a Reactance at 1-ft spacing, 60 Hz, Ω/conductor/1000 ft
Cir mils	AWG		
1 590 000		0.0129	0.0679
1 431 000		0.0144	0.0692
1 272 000		0.0161	0.0704
1 192 500		0.0171	0.0712
1 113 000		0.0183	0.0719
954 000		0.0213	0.0738
795 000		0.0243	0.0744
715 500		0.0273	0.0756
636 000		0.0307	0.0768
556 500		0.0352	0.0786
477 000		0.0371	0.0802
397 500		0.0445	0.0824
336 400		0.0526	0.0843
266 800		0.0662	0.1145
	4/0	0.0835	0.1099
	3/0	0.1052	0.1175
	2/0	0.1330	0.1212
	1/0	0.1674	0.1242
	1	0.2120	0.1259
	2	0.2670	0.1215
	3	0.3370	0.1251
	4	0.4240	0.1240
	5	0.5340	0.1259
	6	0.6740	0.1273

For a 3-phase circuit the total impedance line to neutral is

$$Z = r_a + j(x_a + x_d)$$

TABLE 15. 60-Hz Reactance Spacing Factor X_d, Ω/conductor/1000 ft

					Separation, in						
Ft	0	1	2	3	4	5	6	7	8	9	10	11
0	...	−0.0571	−0.0412	−0.0319	−0.0252	−0.0201	−0.0159	−0.0124	−0.0093	−0.0066	−0.0042	−0.0020
1	0	0.0018	0.0035	0.0051	0.0061	0.0080	0.0093	0.0106	0.0117	0.0129	0.0139	0.0149
2	0.0159	0.0169	0.0178	0.0186	0.0195	0.0203	0.0211	0.0218	0.0255	0.0232	0.0239	0.0246
3	0.0252	0.0259	0.0265	0.0271	0.0277	0.0282	0.0288	0.0293	0.0299	0.0304	0.0309	0.0314
4	0.0319	0.0323	0.0328	0.0333	0.0337	0.0341	0.0346	0.0350	0.0354	0.0358	0.0362	0.0366
5	0.0370	0.0374	0.0377	0.0381	0.0385	0.0388	0.0392	0.0395	0.0399	0.0402	0.0405	0.0409
6	0.0412	0.0415	0.0418	0.0421	0.0424	0.0427	0.0430	0.0433	0.0436	0.0439	0.0442	0.0445
7	0.0447	0.0450	0.0453	0.0455	0.0458	0.0460	0.0463	0.0466	0.0468	0.0471	0.0473	0.0476
8	0.0478											

TABLE 16. 60-Hz Reactance Spacing Factor X_d, Ω/Conductor/1000 ft

	Separation, 0.25 in			
In	0	1/4	2/4	3/4
0	...	...	−0.0729	−0.0636
1	−0.0571	−0.0519	−0.0477	−0.0443
2	−0.0412	−0.0384	−0.0359	−0.0339
3	−0.0319	−0.0301	−0.0282	−0.0267
4	−0.0252	−0.0238	−0.0225	−0.0212
5	−0.0201	−0.01795	−0.01790	−0.01684
6	−0.0159	−0.01494	−0.01399	−0.01323
7	−0.0124	−0.01152	−0.01078	−0.01002
8	−0.0093	−0.00852	−0.00794	−0.00719
9	−0.0066	−0.00605	−0.00529	−0.00474
10	−0.0042			
11	−0.0020			
12	0.00			

TABLE 17. Reactance of Typical Three-Phase Cable Circuits, Ω/1000 ft

	230 V	460 V	575 V	2400 V	4160 V	6900 V	13 800 V
Cable size 4 to 1:							
3 1/C cables in magnetic conduit	0.0521	0.0546	0.0520	0.0620	0.0618		
1 3/C cable in magnetic conduit	0.0381	0.0400	0.0381	0.0384	0.0384	0.0522	0.0526
1 3/C cable in nonmagnetic duct	0.0310	0.0326	0.0310	0.0335	0.0335	0.0453	0.0457
Cable size 1/0 to 4/0:							
3 1/C cables in magnetic conduit	0.0495	0.0515	0.0490	0.0550	0.0550		
1 3/C cable in magnetic conduit	0.0360	0.0380	0.0360	0.0346	0.0346	0.0448	0.0452
1 3/C cable in nonmagnetic duct	0.0291	0.0300	0.0290	0.0300	0.0300	0.0386	0.0390
Cable size 250 to 750 MCM:							
3 1/C cables in magnetic conduit	0.0450	0.0478	0.0450	0.0500	0.0500		
1 3/C cable in magnetic conduit	0.0325	0.0342	0.0325	0.0310	0.0310	0.0378	0.0381
1 3/C cable in nonmagnetic duct	0.0270	0.0284	0.0270	0.0275	0.0275	0.0332	0.0337

Note. Above values may also be used for magnetic and nonmagnetic armored cables.

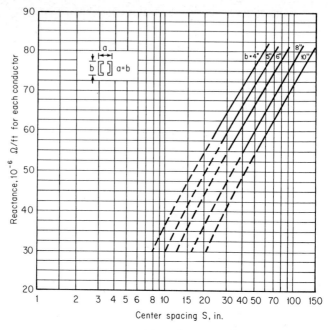

Fig. 25 Reactance values of bus bars in pairs of channels. Curves of reactance at 60 Hz vs. spacing for 4-, 5-, 6-, 8-, and 10-in channels in pairs; phases spaced horizontally or vertically; ohms per foot of each conductor at a distance S from the return conductor. In a balanced 3-phase circuit, the mean reactance per foot of each conductor is determined approximately for the mean spacing $S = (S_1 S_2 S_3)^{1/3}$ where S_1 = center spacing between phases 1 and 2, S_2 = center spacing between phases 2 and 3, and S = center spacing between phases 1 and 3.

TABLE 18. Reactance of Low-Voltage Power Circuit Breakers

Breaker interrupting rating, A	A rating	Reactance, Ω
15 000 and 25 000	15– 35	0.04
	50– 100	0.004
	125– 225	0.001
	250– 600	0.000 2
50 000	200– 800	0.000 2
	1000–1600	0.000 07
75 000	2000–3000	0.000 08
100 000	4000	0.000 08

Note. Owing to the method of rating low-voltage power circuit breakers, the reactance of the breaker which is to interrupt the fault is not included in calculating fault current.

TABLE 19. Reactance of Low-Voltage Disconnect Switches

The reactance of disconnecting switches for low-voltage circuits (600 V and below) is in the order of magnitude of 0.000 08 Ω/pole to 0.000 05 Ω/pole at 60 Hz for switches rated 400–4000 A, respectively.

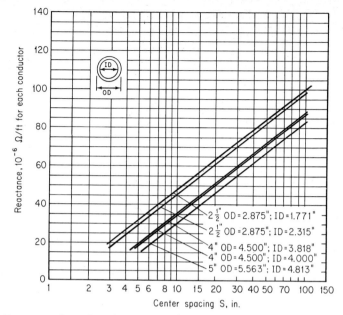

Fig. 26 Reactance values of tubular bus. Curves of reactance at 60 Hz vs. spacing for tubular copper bus; ohms per foot of each conductor at a distance S from the return conductor. In a balanced 3-phase circuit, the mean reactance per foot of each conductor is determined approximately for the mean spacing $S = (S_1 S_2 S_3)^{1/3}$ where S_1 = center spacing between phases 1 and 2, S_2 = center spacing between phases 2 and 3, and S_3 = center spacing between phases 1 and 3.

When conductors are not arranged in a delta, the following formula may be used to determine the equivalent delta:

$$d = \sqrt[3]{A \times B \times C} \tag{5}$$

When the conductors are located in one plane and the outside conductors are equally spaced from the middle conductor, the equivalent spacing is 1.26 times the distance between the middle conductor and an outside conductor; for example,

$$\text{Equivalent delta spacing} = \sqrt[3]{A \times B \times C} = 1.26A$$

56. Combining Reactances. Reactances to the point of fault are combined into a single equivalent reactance.

Most industrial distribution systems are relatively simple and merely require the combination of branches in series or parallel. In some instances, the layout may involve three branches tied together. Wye-delta or delta-wye transformations state that the reactances joining three different points in a system may be optionally represented

TABLE 20. Approximate Reactances of Current Transformers

Primary current ratings, A	Reactance, Ω for various voltage ratings		
	600–5000 V	7500 V	15 000 V
100– 200	0.002 2	0.004 0	0.000 9
250– 400	0.000 5	0.000 8	0.000 2
500– 800	0.000 19	0.000 31	0.000 07
1000–4000 (through type)	0.000 07	0.000 07	0.000 07

by a wye or delta connection. One of these connections may be more suitable for combination with external branches of the circuit than the other. Figure 27 gives equations for transforming and combining system reactances.

57. Determining Fault Current or kVA. Short-circuit current or short-circuit kVA are determined at the point of fault using the equivalent reactance.

After combining the reactances of the reactance diagram to a single equivalent reactance, it is possible to arrive at a value of symmetrical short-circuit kVA.

If the equivalent reactance is X per unit,

$$\text{Symmetrical short-circuit kVA} = \frac{\text{base kVA}}{X \text{ per unit}} \tag{6}$$

then

$$\text{Symmetrical short-circuit current} = \frac{\text{base kVA}}{X \text{ per unit } \sqrt{3} \times \text{kV}} \tag{7}$$

where kV is the line-to-line kV.

Since it is convenient to use a 3-phase base kVA and reactance values are single-phase values, it becomes necessary to use line-to-line voltage in Eq. (7).

To determine the short-circuit current that the protective device will be called upon to interrupt and the momentary short-circuit current, refer to Table 8. Table 8 also indicates types of machine reactances to be used in making calculations.

58. Calculating Utility-System Reactance. The short-circuit kVA of the utility is the maximum 3-phase short-circuit kVA that the utility can produce; therefore, the per unit

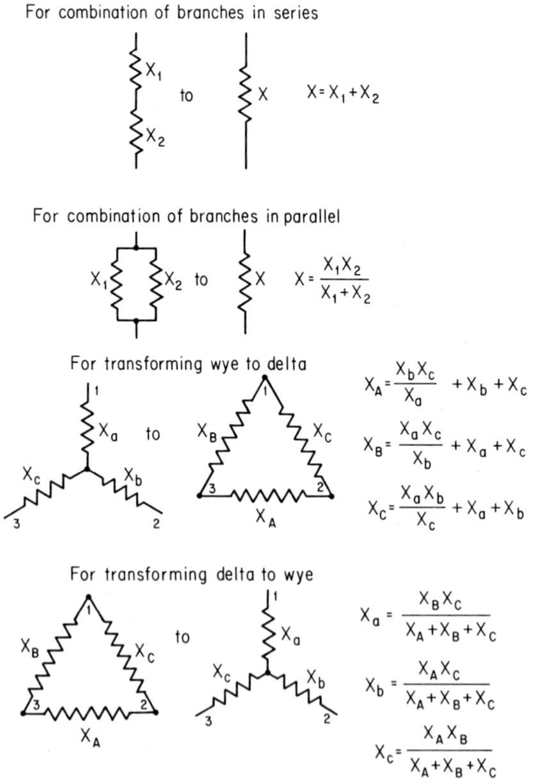

For combination of branches in series

$$X = X_1 + X_2$$

For combination of branches in parallel

$$X = \frac{X_1 X_2}{X_1 + X_2}$$

For transforming wye to delta

$$X_A = \frac{X_b X_c}{X_a} + X_b + X_c$$

$$X_B = \frac{X_a X_c}{X_b} + X_a + X_c$$

$$X_C = \frac{X_a X_b}{X_c} + X_a + X_b$$

For transforming delta to wye

$$X_a = \frac{X_B X_C}{X_A + X_B + X_C}$$

$$X_b = \frac{X_A X_C}{X_A + X_B + X_C}$$

$$X_c = \frac{X_A X_B}{X_A + X_B + X_C}$$

Fig. 27 Equations for transforming and combining system reactances.

reactance of the utility on its own short-circuit-kVA base is 1.0.   It follows from Eq. (4) that the per unit reactance of the utility on the chosen kVA base is therefore equal to

$$\frac{1.0 \times \text{chosen kVA base}}{\text{Utility short-circuit kVA}} \tag{8}$$

59. Example of Ac Fault-Current Calculations.   The procedure to be followed in calculating fault currents is best illustrated by an example.   Figure 19 shows a hypothetical power-distribution system typical of those to be found in industrial plants. The circuit components and arrangement do not necessarily conform to practices outlined in other chapters but have been chosen to illustrate the fundamentals and procedures involved in making fault-current calculations.

A complete fault-current study would involve the calculation of fault currents at all locations in the system.   This example will illustrate the calculations for only a few significant locations.

This example will be worked in the per unit system.   A similar procedure would be followed if it were worked in percent or ohmic values.

The base kVA will be chosen as 10 000 kVA.   This will result in per unit values that are easy to work with, that is, values which are neither too large nor too small.   Equation (2) gives the value by which ohmic values of reactance must be multiplied to convert them to per unit values on the chosen kVA base.

At 4160 V:

$$\text{Per unit reactance} = \frac{(\text{ohms reactance}) \ (10 \ 000\text{-kVA base})}{(4.16 \ \text{kV})^2 \ (1000)}$$

$$= (\text{ohms reactance}) \ (0.576)$$

At 480 V:

$$\text{Per unit reactance} = \frac{(\text{ohms reactance}) \ (10 \ 000\text{-kVA base})}{(0.480 \ \text{kV})^2 \ (1000)}$$

$$= (\text{ohms reactance}) \ (43.4)$$

These multipliers will be used in the example to simplify the conversion from ohmic to per unit values.

The first step is to calculate the per unit reactance value for each significant circuit element that will contribute to or limit the fault current.

UTILITY-SUPPLY EQUIVALENT REACTANCE

From Eq. (8):

$$\text{Per unit reactance} = \frac{(1.0) \ (10 \ 000\text{-kVA base})}{1 \ 000 \ 000 \ \text{kVA}} = 0.01$$

3000-kVA TRANSFORMER

From Table 11:

Per unit reactance = 0.07 on transformer rating kVA base

From Eq. (4):

$$\text{Per unit reactance on 10 000-kVA base} = 0.07 \ \frac{10 \ 000 \ \text{kVA}}{3 \ 000 \ \text{kVA}}$$

$$= 0.233$$

625-kVA GENERATOR

Given $X''_d = 9\%$ reactance. From Eq. (1):

$$\text{Per unit reactance} = \frac{9}{100} = 0.09 \text{ on generator rating kVA base}$$

From Eq. (4):

$$\text{Per unit reactance on 10 000-kVA base} = 0.09 \frac{10\ 000\ \text{kVA}}{625\ \text{kVA}}$$

$$= 1.44$$

300-FT TIE CABLE

From Table 13: $X_a = 0.103\ \Omega/1000\ \text{ft}$
From Table 16: $X_d = -0.073\ \Omega/1000\ \text{ft}$ for $\frac{1}{2}$-in spacing
Total reactance $X_T = X_a + X_d$
    $X_T = 0.103 - 0.073 = 0.030\ \Omega/1000\ \text{ft}$

For 300 ft $X_T = 0.030 \dfrac{300}{1000}\ 0.00900\ \Omega$

Per unit reactance $= (0.00900)(0.576) = 0.005184$

500-FT FEEDER CABLE

From Table 13: $X_a = 0.0981\ \Omega/1000\ \text{ft}$
From Table 16: $X_d = -0.0636\ \Omega/1000\ \text{ft}$ for $\frac{3}{4}$-in spacing
Total reactance $X_T = X_d + X_a$
    $X_T = 0.0981 - 0.0636 = 0.0345\ \Omega/1000\ \text{ft}$

For 500 ft $X_T = 0.0345 \dfrac{500}{1000} = 0.01725\ \Omega$

For two parallel conductors per phase,

$$XT = \frac{1}{2}\ (0.01725) = 0.008625\ \Omega$$

Per unit reactance $= (0.008625)(0.576) = 0.00496$

700-FT FEEDER CABLE

From the calculation for the 300-ft tie cable,

$$X_T = 0.030\ \Omega/1000\ \text{ft}$$

For 700 ft $X_T = 0.30 \dfrac{700}{1000} = 0.21\ \Omega$

Per unit reactance $= (0.021)(0.576) = 0.0121$

BUS A

From Fig. 23, for an equivalent delta spacing of

$$\sqrt[3]{(5)(5)(10)} = 6.3\ \text{in}$$

Reactance $= 0.000049\ \Omega/\text{ft}$
For 12 ft, reactance $= (12)(0.000049) = 0.000588\ \Omega$
Per unit reactance $= (0.000588)(43.4) = 0.0255$

BUS B

From Fig. 22, for an equivalent delta spacing of

$$\sqrt[3]{(8)(8)(16)} = 10.08\ \text{in}$$

    Reactance $= 0.0000645\ \Omega/\text{ft}$
For 2 ft   Reactance $= (2)(0.0000645) = 0.000129\ \Omega$
    Per unit reactance $= (0.000129)(43.4) = 0.0056$

60-FT MOTOR FEEDER CABLES

From Table 13: $X_a = 0.0953\ \Omega/1000$ ft
From Table 16: $X_d = -0.0572\ \Omega/1000$ ft for 1-in spacing
Total reactance $X_T = X_a + X_d$
$X_T = 0.0953 - 0.0572 = 0.0381\ \Omega/1000$ ft

For 60 ft $\quad X_T = 0.0381\ \dfrac{60}{1000} = 0.00229\ \Omega$

Per unit reactance $\quad = (0.00229)(43.4) = 0.0995$

For 700 ft $\quad X_T = 0.030\ \dfrac{700}{1000} = 0.021\ \Omega$

Per unit reactance $\quad = (0.021)(0.576) = 0.0121$

RECTIFIER AND FURNACE LOADS

These loads will neither contribute to nor limit the fault current in the system, and so are neglected for the purposes of calculating fault currents.

CURRENT-LIMITING REACTOR

Per unit reactance on the reactor kVA base = 0.06

Reactor through kVA rating $= \sqrt{3}(300)4.16 = 2160$

From Eq. (4):

Per unit reactance on 10 000-kVA base $= 0.06\ \dfrac{10\ 000\ \text{kVA}}{2\ 160\ \text{kVA}} = 0.277$

750-kVA TRANSFORMER

From Table 11:

Per unit reactance on transformer kVA base = 0.055

From Eq. (4):

Per unit reactance on 10 000-kVA base $= 0.055\ \dfrac{10\ 000\ \text{kVA}}{750\ \text{kVA}} = 0.733$

600-HP SYNCHRONOUS MOTORS

From Table 9:

Per unit reactance $X''_d = 0.15$ on motor kVA base
Per unit reactance $X'_d = 0.23$ on motor kVA base
Motor kVA base $= (0.8)(600) = 480$ kVA

Per unit reactance on 10 000-kVA base $X''_d = 0.15\ \dfrac{10\ 000}{480\ \text{kVA}}$

$= 3.13$

$X'_d = 0.23\ \dfrac{10\ 000}{480\ \text{kVA}}$

$= 4.79$

100-HP MOTORS

From Table 10: Per unit reactance $X''_d = 0.20$ on motor kVA base
Motor kVA base = hp rating approximately = 100 kVA

Per unit reactance on 10 000-kVA base $X''_d = 0.20\ \dfrac{10\ 000\ \text{kVA}}{100\ \text{kVA}} = 20.0$

120-FT FEEDER CABLE

From Table 13: $X_a = 0.108 \ \Omega/1000$ ft
From Table 16: $X_d = -0.0729 \ \Omega/1000$ ft for $\frac{1}{2}$-in spacing
                              $X_T = 0.108 -0.0729 = 0.035 \ \Omega/1000$ ft

For 120 ft $X_T = 0.035 \ \dfrac{120}{1000} = 0.0042 \ \Omega$

Per unit reactance $= (0.0042)(43.4) = 0.182$

AIR CIRCUIT BREAKERS

From Table 18:

$$\text{Reactance of 1200-A breaker} = 0.00007 \ \Omega$$

$$\text{Reactance of 150- and 200-A breakers} = 0.001 \ \Omega$$

Per unit reactances:

$$\text{1200-A breaker} = (0.00007)(43.4) = 0.00304$$

$$\text{150- and 200-A breakers} = (0.001)(43.4) = 0.0434$$

CURRENT TRANSFORMERS

From Table 20:

$$\text{Reactance of 1200-5 CT} = 0.00007$$

$$\text{Reactance of 150-5 and 200-5 CTs} = 0.0022$$

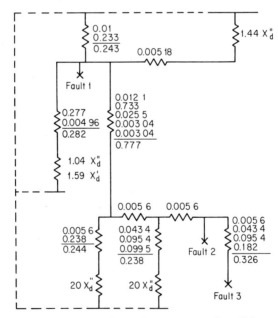

Fig. 28 Simplification of Fig. 20 by combining series- and parallel-reactance values.

Per unit reactances:

$$1200\text{-}5 \text{ CT} = (0.00007)(43.4) = 0.00304$$

$$150\text{-}5 \text{ and } 200\text{-}5 \text{ CTs} = (0.0022)(43.4) = 0.0954$$

The next step is to draw a reactance diagram showing the calculated per unit reactance values in one-line form. Figure 20 shows the values arranged similar to the circuits of Fig. 19.

Many of the reactances can be combined with ease. Series reactances can be added up and represented as a single reactance. Parallel reactances can also be combined into a single value. When motors and generators are represented by their per unit reactances on a common kVA base, all their "neutral" or "center" points are considered to be connected to the same bus as the utility-supply equivalent reactance. Thus the utility-supply equivalent reactance is in parallel with the motor and generator reactances.

The reactance diagram should be simplified as much as possible, retaining the points at which the fault current is to be calculated. Figure 28 illustrates the simplification of Fig. 20 by combining series- and parallel-reactance values. The dotted lines indicate buses of "equal potential" so far as the fault-current calculations are concerned.

Further simplification of the reactance diagram can be made only for a specific fault location. For a fault at location 1, for example, it is no longer necessary to retain fault locations 2 and 3, and further simplifications of the reactance diagram can be made.

FAULT 1

The simplification of the reactance diagram into a single equivalent reactance is shown in Fig. 29a, b, c, and d. Because it is desired to calculate the fault current at both $\frac{1}{2}$ to 1 cycle and at 8 cycles, both X''_d and X'_d values must be included, and separate single equivalent reactances determined. For the momentary-fault-current calculations, the utility-supply reactance and the X''_d values are used. For the interrupting-duty fault-current calculations, the utility-supply reactance, the generator X''_d value, and the synchronous-motor X'_d values are used. The induction-motor reactance values are not used when calculating the interrupting-fault-current values, since their fault-current contribution is negligible after a few cycles. In Fig. 29c and d, the reactances are combined into separate single equivalent values.

From Eq. (7), the rms symmetrical fault current is

$$\frac{10\ 000 \text{ kVA}}{(\sqrt{3})(4.16 \text{ kV})(0.177)} = 7850 \text{ A}$$

From Table 7, a multiplication factor of 1.6 must be applied to account for the effect of the dc component of initial fault current. The rms asymmetrical fault current is then

$$(1.6)(7850) = 12\ 560 \text{ A}$$

From Eq. (6), the rms symmetrical (8-cycle breaker) fault kVA is

$$\frac{10\ 000 \text{ kVA}}{0.187} = 53\ 476 \text{ kVA}$$

From Eq. (7), the rms symmetrical (8-cycle breaker) fault current is

$$\frac{10\ 000 \text{ kVA}}{(\sqrt{3})(4.16 \text{ kV})(0.187)} = 7430 \text{ A}$$

From Table 7, a multiplying factor of 1.0 should be applied to obtain the interrupting requirement for an 8-cycle power circuit breaker.

For proper protection, the 4160-V power circuit breakers would have to be capable of interrupting 53 476 kVA (7430 A at 4160 V) and be capable of withstanding a momentary current of 12 560 A.

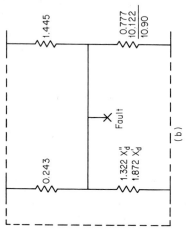

(b)

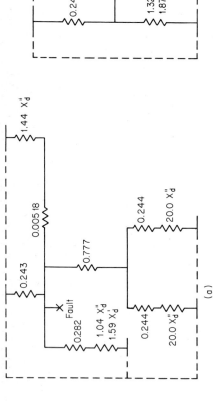

(a)

For momentary ($\frac{1}{2}$ to 1 cycle) fault-current calculation

0.177

$$\frac{1}{X} = \frac{1}{1.322} + \frac{1}{0.243} + \frac{1}{1.445} + \frac{1}{10.90}$$

$$\frac{1}{X} = 5.65$$

$$X = 0.177$$

(c)

For interrupting (8 cycles) fault-current calculation

0.187

$$\frac{1}{X} = \frac{1}{1.872} + \frac{1}{0.243} + \frac{1}{1.445}$$

$$\frac{1}{X} = 5.34$$

$$X = 0.187$$

(d)

Fig. 29 Simplification of reactance diagram by considering only fault 1.

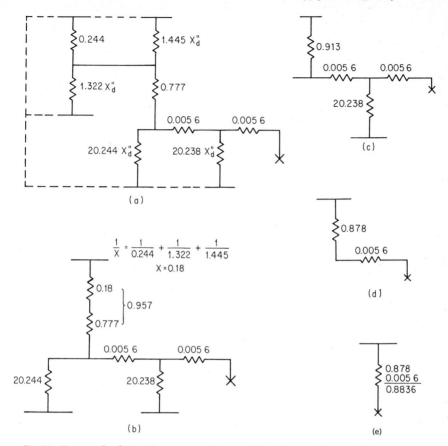

Fig. 30 Process of reducing the reactance diagram of Fig. 19 to single equivalent reactance.

FAULT 2

Figure 28 must be simplified in another manner for the calculation of fault currents at location 2. In systems of 600 V and less, only the momentary ($\frac{1}{2}$ to 1 cycle) fault-current value is of interest. Figure 30a, b, c, d, and e shows the process of reducing the reactance diagram of Fig. 19 to a single equivalent reactance.

From Eq. (7), the symmetrical momentary fault current is

$$\frac{10\ 000\ \text{kVA}}{(\sqrt{3})(0.480\ \text{kV})(0.8836)} = 13\ 629\ \text{A}$$

FAULT 3

Figure 31a shows the simplified reactance diagram for fault location 3. The value 0.8836 is the single equivalent reactance for fault location 2, and the value 0.326 is the reactance between locations 2 and 3 (Fig. 28). The total reactance from source to fault is then

$$0.8836 + 0.326 = 1.2096$$

A check of Table 13 indicates that the resistance of the 120-ft feeder cable is high in comparison with the system equivalent reactance from source to fault. The per unit value of resistance is

$$(0.181) \frac{120}{1000} (43.4) = 0.943$$

This per unit resistance is

$$\frac{0.943}{1.2096} \times 100 = 78\%$$

of the system reactance from source to fault. It now becomes necessary to include the effect of the resistance of the feeder cable to arrive at a realistic value of fault current. Figure 31*a* shows the simplified reactance-resistance circuit.

Account should also be taken of the resistance of the system up to the feeder. Assume a per unit resistance equal to 25% of the corresponding per unit reactance from source to feeder. This value is at best an estimate, but it will hold fairly true for most low-voltage systems not having any generators or a large proportion of motors connected to them.

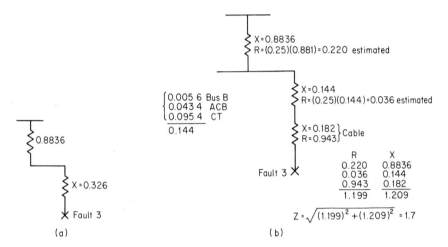

Fig. 31   Simplification of reactance-resistance circuit.

Figure 31*b* shows the calculation of the single equivalent impedance Z for fault 3. From Eq. (7) the rms symmetrical fault current is

$$\frac{10\ 000\ \text{kVA}}{(\sqrt{3})(0.480\ \text{kV})(1.7)} = 7070\ \text{A}$$

The value of rms symmetrical fault current if determined by reactance alone would be

$$\frac{10\ 000\ \text{kVA}}{(\sqrt{3})(0.480\ \text{kV})(1.2096)} = 9956\ \text{A}$$

This value would represent an error of 41% if the resistance were not considered in the calculation of fault 3.

60. Short-Circuit Currents in Low-Voltage Systems (Ref. 7). To determine the interrupting-capacity requirements of low-voltage air circuit breakers, it is necessary to obtain information, at the point of application, of the short-circuit condition of every part of a distribution system. Twelve sets of curves (Figs. 32 to 43) provide graphical means of determining the value of short-circuit currents and the corresponding power factor for various sizes of liquid-filled transformers and various voltages, in combination with different sizes and lengths of cable. In Tables 21 and 22 are given the multiplying factors to use in obtaining the asymmetrical values of current at different power factors.

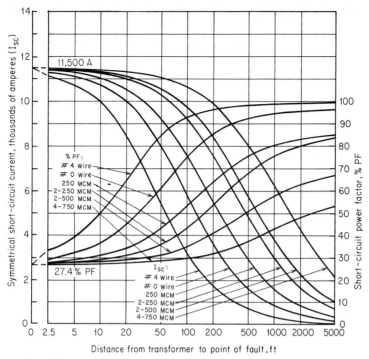

Fig. 32 Values of current and power factor are related to distance from a 150-kVA 208-V liquid-filled transformer.

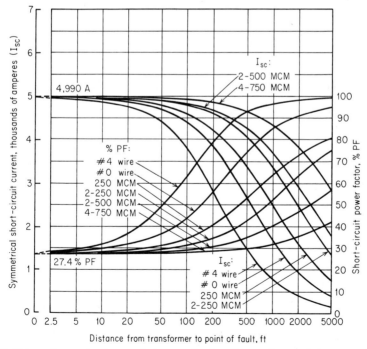

Fig. 33 Values of current and power factor are related to distance from a 150-kVA 480-V liquid-filled transformer.

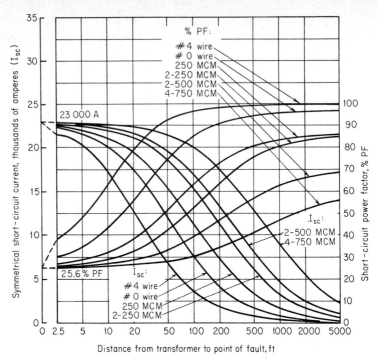

Fig. 34 Values of current and power factor are related to distance from a 300-kVA 208-V liquid-filled transformer.

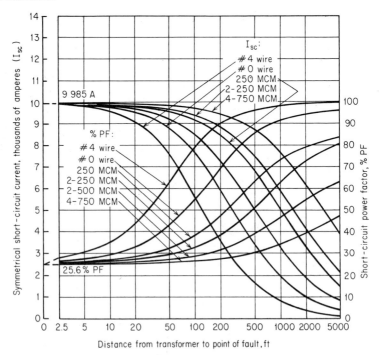

Fig. 35 Values of current and power factor are related to distance from a 300-kVA 480-V liquid-filled transformer.

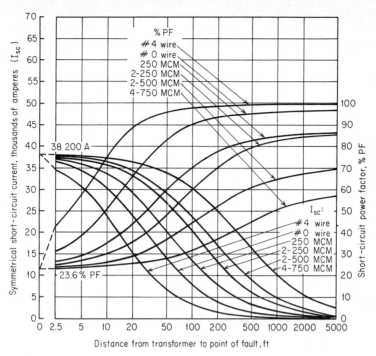

Fig. 36 Values of current and power factor are related to distance from a 500-kVA 208-V liquid-filled transformer.

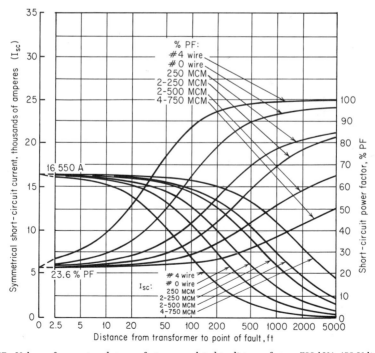

Fig. 37 Values of current and power factor are related to distance from a 500-kVA 480-V liquid-filled transformer.

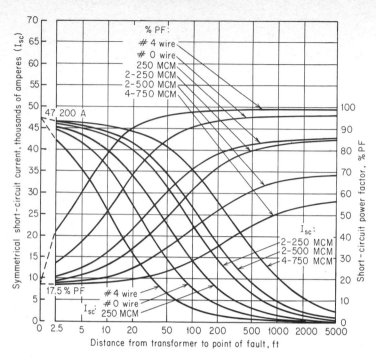

Fig. 38 Values of current and power factor are related to distance from a 750-kVA 208-V liquid-filled transformer.

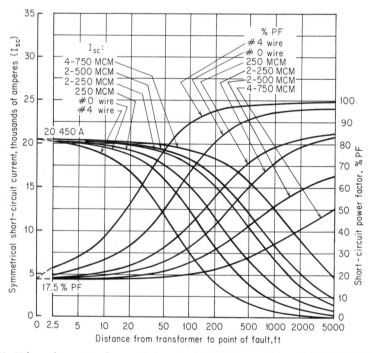

Fig. 39 Values of current and power factor are related to distance from a 750-kVA 480-V liquid-filled transformer.

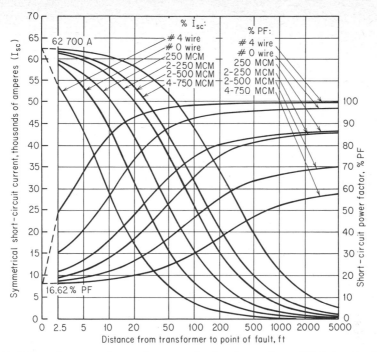

Fig. 40 Values of current and power factor are related to distance from a 1000-kVA 208-V liquid-filled transformer.

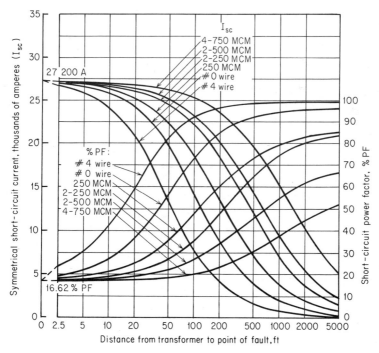

Fig. 41 Values of current and power factor are related to distance from a 1000-kVA 480-V liquid-filled transformer.

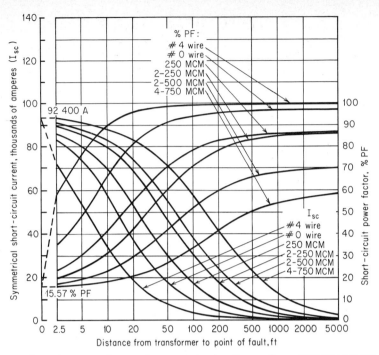

Fig. 42 Values of current and power factor are related to distance from a 1500-kVA 208-V liquid-filled transformer.

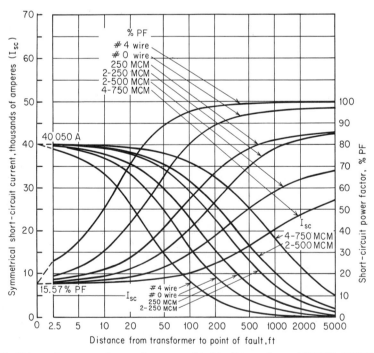

Fig. 43 Values of current and power factor are related to distance from a 1500-kVA 480-V liquid-filled transformer.

In the study diagramed in Fig. 44, these basic characteristics are assumed:

1. Primary source available is 500 MVA at the primary of the transformer, with a source circuit X/R ratio of 25.

2. Transformer kVA covers a range of seven standard sizes: 150, 300, 500, 750, 1000, 1500, 2000.

3. Distribution voltages of 208 and 480 are assumed: 3-phase, 60 Hz.

4. Transformer impedances are 4½% up through 500 kVA and 5½% for 750 kVA and higher, less the standard allowable tolerance of −7½%. X/R ratios are selected from data furnished by numerous transformer manufacturers from which the characteristics in Table 23 are calculated. The data are based on liquid-filled, self-cooled transformers. Dry-type transformers generally have higher reactances, and the results of this study should be conservative for systems in which they are used.

5. Motor impedance is based on an equivalent reactance of 25 and an X/R ratio of 6.

6. Feeder conductors chosen are at least 115% of the size normally required for the standard frame sizes of low-voltage air circuit breakers. Calculations in this study are based on 3-phase conductors, of minimum insulation thickness, in magnetic ducts. Table 24 gives the conductor sizes and alternate arrangements for which this study can be used.

7. Short-circuit duty can be read direct from the set of 12 curves. The short-circuit current available, in symmetrical amperes, and the corresponding short-circuit power

TABLE 21. Multiplying Factors

| Short-circuit power factor, % | Short-circuit X/R ratio | Multiplying factor | | Short-circuit power factor, % | Short-circuit X/R ratio | Multiplying factor | |
		Max 1-phase rms A at ½ cycle (curve M_M)	Av 3-phase rms A at ½ cycle (curve M_A)			Max 1-phase rms A at ½ cycle (curve M_M)	Av 3-phase rms A at ½ cycle (curve M_A)
0	∞	1.732	1.394	29	3.3001	1.139	1.070
1	100.00	1.696	1.374	30	3.1798	1.130	1.066
2	49.993	1.665	1.355	31	3.0669	1.121	1.062
3	33.322	1.630	1.336	32	2.9608	1.113	1.057
4	24.979	1.598	1.318	33	2.8606	1.105	1.053
5	19.974	1.568	1.301	34	2.7660	1.098	1.049
6	16.623	1.540	1.285	35	2.6764	1.091	1.046
7	14.251	1.511	1.270	36	2.5916	1.084	1.043
8	12.460	1.485	1.256	37	2.5109	1.078	1.039
8.5	11.723	1.473	1.248	38	2.4341	1.073	1.036
9	11.066	1.460	1.241	39	2.3611	1.068	1.033
10	9.9501	1.436	1.229	40	2.2913	1.062	1.031
11	9.0354	1.413	1.216	41	2.2246	1.057	1.028
12	8.2733	1.391	1.204	42	2.1608	1.053	1.026
13	7.6271	1.372	1.193	43	2.0996	1.049	1.024
14	7.0721	1.350	1.182	44	2.0409	1.045	1.022
15	6.5912	1.330	1.171	45	1.9845	1.041	1.020
16	6.1695	1.312	1.161	46	1.9303	1.038	1.019
17	5.7967	1.294	1.152	47	1.8780	1.034	1.017
18	5.4649	1.277	1.143	48	1.8277	1.031	1.016
19	5.1672	1.262	1.135	49	1.7791	1.029	1.014
20	4.8990	1.247	1.127	50	1.7321	1.026	1.013
21	4.6557	1.232	1.119	55	1.5185	1.015	1.008
22	4.4341	1.218	1.112	60	1.3333	1.009	1.004
23	4.2313	1.205	1.105	65	1.1691	1.004	1.002
24	4.0450	1.192	1.099	70	1.0202	1.002	1.001
25	3.8730	1.181	1.093	75	0.8819	1.000 8	1.000 4
26	3.7138	1.170	1.087	80	0.7500	1.000 2	1.000 05
27	3.5661	1.159	1.081	85	0.6198	1.000 04	1.000 02
28	3.4286	1.149	1.075	100	0.0000	1.000 00	1.000 00

TABLE 22. Multiplying Factors

Transformer kVA	Short-circuit power factor at terminals, all voltages, %	Multiplying factors to obtain asymmetrical rms A at ½ cycle		Short-circuit current at transformer terminals, A		
		Avg 3-phase	Max offset phase	Symmetrical rms	3-phase avg asymmetrical rms	Max offset phase asymmetrical rms
				208 V		
150	27.4	1.079	1.155	11 550	12 410	13 280
225	26.6	1.083	1.165	17 220	18 650	20 060
300	25.6	1.089	1.174	23 000	25 050	27 000
500	23.6	1.101	1.197	38 200	42 060	45 730
750	17.5	1.148	1.286	47 200	54 180	60 700
1000	16.62	1.155	1.301	62 700	72 420	81 570
1500	15.57	1.165	1.320	92 400	107 650	121 970
2000	14.54	1.176	1.339	121 800	143 240	163 100
				480 V		
150	27.4	1.079	1.155	4 990	5 380	5 760
225	26.6	1.083	1.165	7 470	8 090	8 700
300	25.6	1.089	1.174	9 985	10 870	11 720
500	23.6	1.101	1.197	16 550	18 220	19 810
750	17.5	1.148	1.286	20 450	23 480	26 300
1000	16.62	1.155	1.301	27 200	31 420	35 390
1500	15.57	1.165	1.320	40 050	46 660	52 870
2000	14.54	1.176	1.339	52 800	62 080	70 700
2500	13.53	1.186	1.358	66 700	79 100	90 600

TABLE 23. Transformer Characteristics

Transformer rating, kVA	X/R	R, %	X, %	Z, %
150	3.24	1.23	4.0	4.19
300	3.50	1.14	4.0	4.16
500	3.84	1.04	4.0	4.12
750	5.45	0.94	5.1	5.19
1000	5.70	0.89	5.1	5.19
1500	6.15	0.83	5.1	5.18
2000	6.63	0.77	5.1	5.17
2500	7.18	0.71	5.1	5.15

TABLE 24. Circuit Conductors (Copper)

Circuit rating, A	Conductor size	Conductor insulation	Alternate conductor sizes
50	4 AWG	Type R, T, or TW	
100	0 AWG	Type R, T, or TW	
225	250 MCM	Type RH	
400	2–250 MCM	Type RH	
600	2–500 MCM	Type RH	3–No. 4/0 AWG
		Type RH	4–No. 2/0 AWG
1600	4–750 MCM	Type RH	3–2000 MCM
		Type RH	5–400 MCM
		Type RH	6–300 MCM

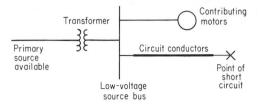

Fig. 44   Basic circuit configuration is used for short-circuit study.

factor are shown for any distance, in circuit feet, from the terminals of the transformer up to 5000 ft. These values include the current contributed by the connected motors.

8. Asymmetrical short-circuit currents can be found from the symmetrical values, shown on the 12 curves, by using a multiplying factor read from the graph in Fig. 45 and corresponding to the short-circuit power factor at the point of fault. The multiplying factor from curve M_A can be used to find the average 3-phase asymmetrical amperes and curve M_M to find the maximum asymmetrical amperes in one phase with the greatest possible offset. For accurate values of M_A and M_M, use the values in Table 21.

9. Short-circuit current at the terminals of transformers may be determined from Table 22. This gives the symmetrical rms amperes, the 3-phase average asymmetrical rms amperes, and the maximum obtainable asymmetrical rms amperes in one phase at the terminals.

VOLTAGE-SURGE PROTECTION

61. Lightning Phenomena. The severity of lightning storms is expressed in the average number of days per year in which thunderstorms occur in various regions of the United States, as shown in Fig. 46. Note that the region in which the severity is the lightest is the west coast, which has only 5 thunderstorm days per year, while the area in the Gulf of Mexico is the most severe and varies between 30 and 90 thunderstorm days per year.

In spite of the great interest in the manner in which electrical charges accumulate in thunderstorms, the question is still controversial. Several theories have been advanced as to how a cloud builds up electrical charges to such a magnitude as to cause lightning strokes. It is not our purpose to discuss these theories in detail. However, it is generally agreed that lightning occurs when a section of cool air meets a section of warm

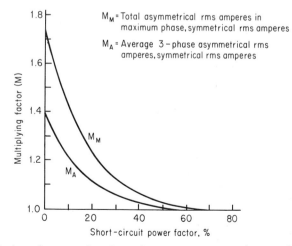

Fig. 45   Multiplying factors used to obtain short-circuit asymmetrical current from symmetrical values at an instant $\frac{1}{2}$ cycle (60-Hz basis) after initiation of a fault.

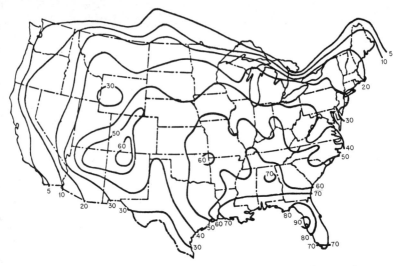

Fig. 46 Annual isoceraunic map of United States shows number of thunderstorm days per year.

air and an updraft results which causes falling raindrops to separate into smaller particles. The separation of the raindrops causes the formation of positive and negative charges. When a sufficient number of positive or negative charges accumulate, this results in lightning discharges between clouds or between the clouds and the earth.

Another theory is that the source of generation of charges is the freezing process in the upper portions of cumulus clouds, which seem to extend above 20 000 ft before electrical discharges occur.

a. Lightning Voltage. The potential between cloud and ground, just prior to a stroke of lightning, has been variously estimated at from 100 million to 1 billion volts. The protection engineer is concerned principally with the potential which appears on the line conductor. If the current in a wave traveling along the line is known, a satisfactory value of voltage is obtained by multiplying the current by the surge impedance Z of the conductor.

The potential which can appear upon the apparatus is limited only by either protective apparatus or flashover of insulating structures, such as bushings or line insulators, plus the impulse strength of wood or other insulating supports, including the effect of ground resistance. Direct strokes must be given consideration on all circuits carried overhead and on underground circuits connected to overhead conductors.

b. Current Amplitudes. The most comprehensive data have been collected from measurements with the magnetic link arranged to read current in tower legs. In some cases current was measured in all four legs. In other cases measurements were made in one leg and extrapolated. The stroke currents were obtained from adding together currents which seemed to be involved in a particular stroke. It is noted that 50% of the stroke currents are in excess of 10 000 A (Table 25). However, only 10% of the tower currents are in excess of 32 000 A.

TABLE 25. Crest Values of Current Peaks (Ref. 8)

% of cases exceeding value given in succeeding columns	Direct strokes, kA		Transmission-line towers, kA	
	Empire State Building	McCann	Lewis and Foust	Waldorf
50	10	5.5	9.5	11
20	18	15	25	23
10	26	25	34	32
Max	58	160	132	114

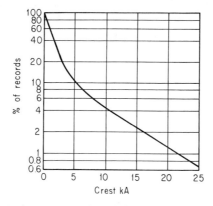

Fig. 47   Lightning current through distribution arresters (Ref. 8).

Experience curves are given showing lightning currents through distribution (Fig. 47) and station arresters (Fig. 48).

62. Protection of Industrial Equipment. *a. Lightning Surges.* A lightning surge is created by nature when a direct lightning stroke strikes an overhead line conductor. It can place thousands of amperes on the line and frequently is followed by long-duration components of several thousand microseconds. If not limited by efficient lightning arresters, such surges can damage severely or totally destroy power apparatus.

b. Induced lightning surges can result from a lightning "near miss." The initial surge is usually of relatively short duration. The principal function of the arrester is to protect the power apparatus during lightning surges without interruption to power on the system.

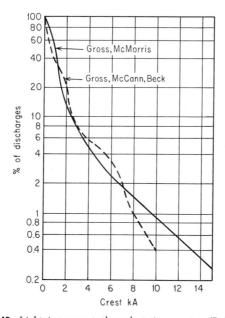

Fig. 48   Lightning currents through station arresters (Ref. 8).

c. Switching surges usually result from opening or closing a breaker, oil switch, or air break or air interrupter switch. They are experienced in nearly all circuit voltages, and can subject apparatus to many times normal line-to-ground voltage. They can originate from a remote point. Switching surges are of long duration—thousands of microseconds.

63. Types of Arrester. There are two general high-voltage arrester designs: valve and expulsion types.

a. Valve-Type Arrester. This type of arrester is the most common and most widely used. Valve-type arresters employ a nonlinear resistance known as a valve element in series with the spark gap. This valve element permits practically unrestricted flow of surge currents at the instant the gap sparks over from a voltage surge, but as soon as the voltage has subsided, it acutely curtails the flow of power-frequency current and assists the series gap in performing its interrupting action. Except as noted here, this arrester is not significantly influenced by the amount of short-circuit power available at the point of installation.

b. Expulsion-type arresters are not used extensively. This type of arrester employs an arc-extinguishing chamber in series with the gaps to interrupt the power-frequency current which flows after the gaps have sparked over. Generally, the power-frequency current is that available for a line-to-ground short circuit at the point of installation on the system, so these arresters are rated in rms amperes to indicate their ability to interrupt within low and high current limits.

Expulsion-type arresters are recommended to be installed on the line for the protection of rotating machines (not dry-type transformers). Their advantage is the low arc drop while passing surge current.

64. Degrees of Protection. Three classes of valve-type arresters are available, each providing a different level of protection. The selection of one of the three arrester classes depends upon the amount of "insurance to be provided" and therefore the amount of money to be spent. Protective levels of station arresters are better than those of intermediates, while intermediates are substantially better than distribution.

a. Station-class valve-type arresters are the most expensive type and are generally used to protect substations or large transformers above 1000 kVA, although they are sometimes used to protect transformers of less capacity in which continuity of service is of paramount importance. Some suppliers have ratings from 3 to 684 kV. Station arresters are also used where the present or future maximum symmetrical system fault current is greater than the rating of intermediate-class arresters. Table 26 shows protective characteristics of a station-class lightning arrester. These characteristics may vary somewhat, depending upon the design and manufacturer.

b. Intermediate-class valve-type arresters are generally used on transformers, 1000 kVA or smaller, indoors, in switchgear where limited space is available or where station-type arresters cannot be economically justified. Some manufacturers provide arresters with a range of ratings from 3 to 120 kV. Since lightning arresters are sealed, and since their internal pressure can vary according to the altitude, the altitude at which they are installed should be checked to be sure it is within the arrester rating.

Intermediate-class arresters can also be used only in locations where the present and future maximum symmetrical system short-circuiting duty does not exceed its rating. If it does, a station-class arrester must be used instead. Table 27 shows the protective characteristics of an intermediate-class lightning arrester.

c. Distribution-class valve-type arresters are smaller and less expensive than the intermediate class. Some manufacturers have lightning arresters with ratings from 3 to 46 kV and are used to protect equipment and transformers supplied by distribution lines. Their location is generally on the distribution lines, and they are often mounted directly on the distribution transformer they protect. Characteristics of distribution-class arresters are ample to provide front-line protection against lightning surges. Table 28 shows electrical characteristics of distribution arresters from one manufacturer.

d. Surge Protection for Rotating Ac Machines (Ref. 8). The basic-impulse-insulation level of rotating machines is limited by the dry type of insulation used, space restrictions, and mechanical and thermal considerations. In comparison with other types of electrical equipment the basic-impulse-insulation level (BIL) of rotating ma-

TABLE 26. Protective Characteristics of Station-Class Lightning Arresters *

Arrester rating, kV rms	Max impulse sparkover ANSI standard front-of-wave † kV crest	Max 100% 1.2 × 50 μs impulse sparkover kV crest	Max switching surge sparkover kV crest	Max discharge voltage for 8 × 20 μs discharge current wave, kV crest				
				1.5 kA	5.0 kA	10.0 kA	20.0 kA	40.0 kA
60	190	178	153	109	130	143	158	185
72	227	216	185	131	155	170	189	222
90	290	255	230	163	194	213	237	277
96	304	270	245	174	207	227	253	296
108	340	300	275	196	233	256	284	332
120	370	325	298	218	259	285	319	370
144	440	342	350	262	311	342	379	444
168	510	400	410	305	362	399	442	518
180	545	428	439	327	388	427	474	555
192	575	456	468	348	414	455	505	592
240	685	560	585	436	518	570	630	740

* McGraw-Edison Company, Power Systems Division.
† On a wave rising 100 kV/μs/12 kV of arrester rating, up to 1200 kV/μs.

TABLE 27. Protective Characteristics of Intermediate-Class Lightning Arresters *

Arrester rating, kV rms	Impulse sparkover ANSI standard front-of-wave kV crest †		Max 100% impulse sparkover kV crest ‡	Max discharge voltage, kV crest §				
	Avg	Max		1.5 kA	3.0 kA	5.0 kA	10 kA	20 kA
9	29	32.5	32	21	22	24	26	28.5
10	34	38	37.5	23	25	27	29	31.5
12	37	42	41	27	29	32	34	37.5
15	44	50	49	34	36.5	39.5	43	47.5
21	59	67	66	47.5	51	56	60	66
24	67	76	74	54	58	64	68	75
30	84	94	91	68	73	79	86	95
36	96	109	107	82	87	95	102	113
39	106	121	120	91	97	106	114	126
48	126	143	139	109	116	127	136	150
60	155	173	168	136	145	159	171	189
72	179	201	196	163	174	191	204	225
90	238	266	282	204	218	239	256	282
96	249	279	299	217	232	254	273	300
108	270	303	328	244	261	286	397	338
120	291	325	361	272	290	319	338	373

* McGraw-Edison Company, Power Systems Division.
† Wave rising 100 kV/μs/12 kV of arrester rating.
‡ Standard 1.2 × 50 μs wave.
§ Standard 8 × 20 μs discharge current wave.

TABLE 28. Protective Characteristics of Distribution-Class Lightning Arresters *

| Voltage rating, kV rms | Sparkover levels | | | Max discharge voltage for 8 × 20 μs discharge current wave, kV crest § | | | |
	Max impulse ANSI standard front-of-wave kV crest †	Rotating machinery front-of-wave kV crest ‡	Max 60-Hz kV crest	1.5 kA	5.0 kA	10.0 kA	20.0 kA
3	12.5	12	12	9	10	11	12
4.5	16.5	16.5	16.5	13.5	15	16.5	17.5
6	22	22	21	18	20	22	23.5
7.5	26	26	25	22.5	25	27	29.5
9	31	30	28.5	27	30	32.5	35
10	35	34	32.5	30	33.5	36	39
12	41	41	36.5	36	40	43	47
15	49	49	44	45	50	54	58.5
18	58	58	52	54	60	65	70
21	66	66	59.5	63	70	75.5	82

* McGraw-Edison Company, Power Systems Division.
† On a wave rising 100 kV/μs/12 kV of arrester rating.
‡ Uniform rate of rise to sparkover in 10 μs.
§ Discharge-current wave (8 × 20 μs) conforms to the latest ANSI Standard C62.1 (newly defined but equivalent to previous 10 × 20 wave).

chinery is very low. For example, the BIL rating of a 5000-V oil-immersed distribution transformer or a 4160-V oil circuit breaker is 60 kV. The BIL rating of a 4000-V motor is considered to be in the order of 13 kV. This level is equal to the peak value of the 1-min high-potential proof test which is at $\sqrt{2}(2 \times \text{rated V} + 1000)$ V.

An ac-motor stator winding can be considered to be a short transmission line. It exhibits the properties of surge impedance, electrical length, and terminal reflection and refraction phenomena normally associated with a transmission line. A surge voltage impressed upon the motor terminals stresses both the ground insulation and the turn-to-turn insulation. In general, the highest stress is across the first few turns of the motor winding, since the surge is attenuated as it passes through the winding. It is possible, however, to have stresses due to reflected waves at ungrounded or high-impedance grounded neutral connections or midway between terminals of delta-connected windings under certain conditions.

Typical surge-impedance values for a transmission line lie between 400 and 500 Ω. Cables have a typical surge-impedance value in the order of 30 to 40 Ω. The surge impedance of ac-motor stator windings varies between approximately 150 and 1500 Ω. The upper end of this range is usually associated with higher-voltage motors having multiturn coils and series-connected windings. Typical velocity-of-propagation values are 1000 ft/μs for transmission lines, 600 ft/μs for cables, and 45 ft/μs for the slot portion of motor stator windings. The velocity in the end-turn region will be much higher than that in the slot portion.

Surge voltages can be generated by lightning strokes, switching operations, fault conditions, forced-current-zero devices such as current-limiting fuses, and simple energization of cable-connected motors. In general, switching surges are less than two or three times rated line-to-neutral peak voltage.

When a surge is impressed on a motor terminal, a portion of the wave is reflected along the incoming conductor back toward the source and a portion is refracted into the motor winding. This effect is most severe when the motor surge impedance is much larger than that of the incoming conductor, such as the case for a cable-connected motor. The refracted voltage can approach twice the incoming voltage level for motors with high surge impedances. Even with a relatively low motor surge-impedance value of 300 Ω and a cable of 30 Ω, the refracted voltage is 1.8 times the incoming surge-voltage value.

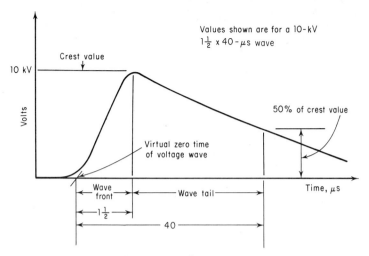

Fig. 49 Typical surge wave.

The magnitude of the turn-to-turn stress caused by this refracted wave is dependent upon the steepness of the wavefront. Figure 49 shows a typical surge waveform. To gain some idea of the voltage impressed across a single turn, the following situation will be considered. Assume that a surge voltage of twice the peak line-to-neutral voltage (6500 V) is refracted into a 4000-V motor with a core length of 25 in. Assume also that the rise time of the wavefront is 1 μs and that the velocity of propagation is 500 in/μs. The electrical length of the iron portion of one turn is therefore 0.1 s. This value corresponds to 650 V on the wavefront and is the turn-to-turn stress.

The wire insulation used in motors is generally a double serving of glass material. The surge voltage between turns should be limited to 600 to 1000 V for this type of insulation, depending upon the thickness used. Repeated exposure to surge voltages above these values may reduce the life of the insulation and result in failure. In practice it has been found that wavefronts with a rise time of 10 μs or longer will not cause excessive stresses between turns of motor stator windings. The use of surge capacitors will reduce the rate of rise of an incoming wave (Table 29). The capacitors must be

TABLE 29. Protective Equipment for Three-Phase Ac Rotating Machines (Ref. 4)

Machine voltage rating (phase-to-phase)	For installation at machine terminals or on machine bus †						For installation 1500 to 2000 ft out on directly connected exposed overhead lines		
	Protective capacitors			Station-type arresters			Distribution-type arresters		
				Voltage rating			Voltage rating		
	Voltage rating	μF/pole	Single-pole units required	Un-grounded or resistance-grounded system	Effectively grounded system	Single-pole units required	Un-grounded or resistance-grounded system	Effectively grounded system	Single-pole units required
0–650	0–650	1.0	3 *	650	650	3	650	650	3
2400	2400	0.5	3 *	3000	3000	3	3000	3000	3
4160	4160	0.5	3 *	4500	3000 ‡	3	6000	3000 ‡	3
4800	4800	0.5	3	6000	4500	3	6000	6000	3
6900	6900	0.5	3	7500	6000	3	9000	6000	3

* A single three-pole unit is commonly used.
† For single-phase machines the same recommendations apply except that only two single-pole units are required if neither line is grounded and only one (on the ungrounded line) if one line is grounded.
‡ The use of 3000-V arresters on a 4160-V system requires an X_0/X_1 ratio less than necessary to make the system "effectively grounded."

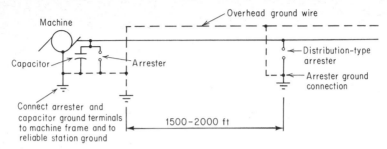

Fig. 50 Lightning-protective equipment for a motor connected to an exposed overhead line (Ref. 10).

connected across each phase to ground at the incoming cable terminals. The leads from this point to the motor should be kept as short as possible, 3 ft or less. Recommended surge capacitor values are 1.0 μF for motors on 600-V systems and below and 0.5 μF for motors on 2400- to 6900-V systems.

 e. Lightning Protection on Exposed Circuits. In addition to surge capacitors to reduce the steepness of the wavefront, machines connected to circuits exposed to lightning require further protection in the form of lightning arresters. Station-type arresters designed specifically for rotating machines are available in ratings of 3 to 25 kV. For protection of motors and other equipment on low-voltage circuits (110 to 125 V), a 175-V arrester is available. The ratings of protective equipment for 3-phase machines are shown in Table 29.

 Motors connected directly to exposed overhead lines can be protected as shown in Fig. 50. This arrangement consists of a distribution-type arrester 1500 to 2000 ft out on the line and a station-type arrester across the surge capacitor for protection of the machine ground insulation. The surge capacitor and the arrester should be located at the machine terminals for maximum effectiveness. If they cannot be located at the machine terminals, they should be connected ahead of the machine on the incoming lines rather than on separate leads from the machine terminals. An overhead ground wire extending for a distance of 2000 ft from the plant will provide additional protection against direct strokes on lines in the immediate vicinity of the plant.

 Figure 51 shows the arrangement of lightning-protective equipment for a motor connected to an exposed overhead line through a transformer. The station-type arrester provides protection for the transformer and eliminates the need for a distribution-type arrester out on the line. The arrester across the motor terminals protects the windings from the electrostatic and electromagnetic transient voltages transmitted through the transformer. The value of transient voltage is highest for wye-wye transformers with both neutrals grounded and autotransformers. Wye-delta-connected transformers are more effective in reducing the transient-voltage magnitude. In smaller installations the less expensive distribution-type arrester may be used on the transformer primary in place of the station-type arrester.

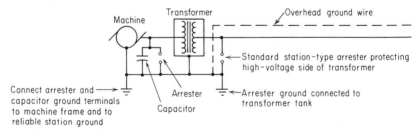

Fig. 51 Lightning-protective equipment for a motor connected to an exposed overhead line through a transformer.

The protective scheme for a motor connected to an exposed overhead line through a reactor or regulator is shown in Fig. 52. This same arrangement can be used where a cable is the connecting link instead of the reactor. In this case the cable sheath should be grounded to the arrester grounds on each end of the cable. Since the surge impedance of a cable is generally so much lower than that of an overhead line, the cable itself will reduce the incoming voltage. The arrester at the line end of the cable may be eliminated if the strength of the cable and the termination equipment is sufficient for the expected voltage. Where the length of cable is considerable, it may be possible to omit the arrester at the machine because of the attenuation of the surge voltage.

Low-voltage motors, 600 V and less, have higher dielectric strengths than higher-voltage motors on a relative basis. Low-voltage motors connected to exposed lines through transformers require no lightning-protective equipment if proper protection has been provided for the transformer. Lightning arresters are required for low-voltage motors connected directly to exposed overhead lines.

f. Surge Protection for Dry-Type Transformers. Dry-type transformers are particularly vulnerable to both lightning and switching surges. Protection is necessary on all dry-type transformers whose primary voltage is 2200 V and higher. The reason for this is that most dry-type transformers have a lower BIL rating than liquid-cooled transformers. However, there is at least one dry-type transformer manufacturer whose transformer has a BIL rating equal to that of a liquid-cooled transformer. The recom-

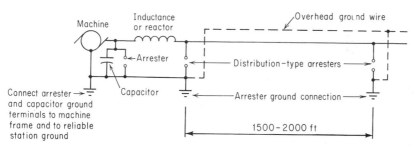

Fig. 52 Lightning-protective equipment for a motor connected to an exposed overhead line through an inductance.

mended lightning arrester for dry-type transformers is the RM type (rotating-machine type). Capacitors for surge protection are not necessary to protect dry-type transformers. The RM-type arrester differs from the conventional valve-type arrester in that it has a lower breakdown voltage to coordinate with the lower BIL rating of the dry-type transformer. This arrester is the same as that used to protect rotating machines. It is recommended that the RM arrester be installed as close to the transformer as possible and possibly in the same transformer enclosure. In all cases, arresters are recommended for protection against switching surges in dry-type transformers even though the transformer location is not vulnerable to lightning surges.

65. Determining Lightning-Arrester Rating. Determination of the proper arrester voltage rating is primarily a matter of being certain the arrester will be able to interrupt power-follow current under any circumstance (Table 30).

A voltage rating at least 25% higher than the maximum continuous phase-to-ground voltage of the system is recommended to minimize the possibility of arrester failure caused by too low a voltage rating.

The probability that power-frequency overvoltages will occur during an arrester operation is very small, but if experience or network analyzer studies indicate maximum system voltages may be exceeded, frequently arresters of a higher voltage rating are recommended.

Arresters are selected and rated so that the maximum permissible power-frequency voltage across the arrester under any condition of system operation including fault conditions is not exceeded.

a. Ungrounded Systems. Three-phase ungrounded and ungrounded neutral systems must have arrester ratings equal to or in excess of the maximum phase-to-phase

**TABLE 30. Recommended Arrester Applications for Distribution Systems
When Using Low-Sparkover Arresters ***

System voltage, kV		Recommended arrester rating, kV		
Nominal circuit voltage	Max tolerable zone voltage	4-wire multi-grounded neutral wye	3-wire solidly grounded neutral wye	Delta and ungrounded wye
2.4	2.6	. . .	. . .	3
4.16 ÷	4.5	3	4.5	4.5
4.8	5.2	4.5	4.5	6
6.9	7.5	6	6	7.5
7.2	7.8	6	7.5	9
8.32 †	9.0	6	9	9
9.0	9.7	7.5	9	9
11.5	12.5	9	10	12
12.0	13.0	9	12	15
12.47 †	13.5	9	12	15
13.2 †	14.3	10	12	15
13.8	15.0	10	12	15
14.4	15.0	12	12	15
16.5	17.8	15	15	18
20.8	22.5	18	21	
21.6 †	23.4	18	21	
23.0 †	24.8	21	21	
24.9 †	27.0	21		

* McGraw-Edison Company, Power Systems Division.
† Usual 4-wire multigrounded distribution systems.
Note. An arrester rating represents the maximum line-to-ground voltage to which the arrester should be subjected. Lightning arresters are inherently sensitive to overvoltages and should never be subjected to 60-Hz voltages above their rated voltages, even during momentary abnormal conditions. If such a condition is likely, a special system study may be necessary, and a higher-rated arrester may be required.

voltage of the system. If system neutrals are not effectively grounded, such as grounded through a resistance, the arrester rating should equal or exceed the highest phase-to-phase system voltage.

b. Grounded Systems. Three-phase, four-wire systems that are effectively grounded to neutral will have a voltage to neutral under fault conditions not exceeding 80% of the highest phase-to-phase voltage. Certain 3-phase, 4-wire distribution systems that are grounded frequently throughout the system under fault conditions will have voltages less than 80% of the highest phase-to-phase voltage. This same criterion is used in selecting the voltage rating of rotating-machine arresters (RM) to protect dry-type transformers, although RM-type arresters have a lower sparkover voltage than the conventional valve-type arrester.

66. Lightning-Arrester Application. Lightning arresters are installed at or near operating equipment to protect the equipment from the destructive effects of direct or induced lightning surges. This protection is effected by providing a short-circuit path to ground for the steep wavefront of the lightning surge and a high-impedance path to ground for the 60-Hz power-follow current.

Figures 53 and 54 are economical applications of lightning arresters for commercial and industrial power systems. Substituting station-class for intermediate-class and intermediate-class for distribution-class arresters will provide better lightning protection, but there may be some question as to whether it is economically justified. More arresters can be installed closer to the equipment that it protects, but again it is questionable whether it is economically justified, and it is up to the engineer to decide on the degree of protection to be provided.

67. Secondary-Class Lightning Arresters. The need for providing lightning-arrester protection on low-voltage circuits is fundamentally the same as for primary-

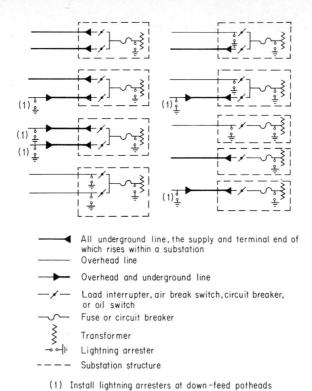

All underground line, the supply and terminal end of which rises within a substation

———— Overhead line

Overhead and underground line

—∕— Load interrupter, air break switch, circuit breaker, or oil switch

—∿— Fuse or circuit breaker

Transformer

Lightning arrester

– – – – Substation structure

(1) Install lightning arresters at down-feed potheads

Fig. 53 Typical lightning-arrester application for outdoor open-type substations.

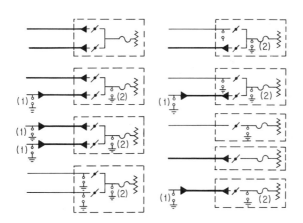

(1) Install lightning arresters at down-feed potheads.

(2) These arresters may be omitted if the transformer is of the dry type, except if the transformer is not located in the substation.

All dry-type transformers are to be protected with valve-type rotating machine (RM) arresters mounted at the transformer bushings or in the transformer enclosure.

Fig. 54 Typical lightning-arrester application for indoor open-type, and all cubicle-type substations 34.5 kV or less.

system protection. While secondary lines usually are shorter than primary lines, they still are subject to lightning surges which can cause stresses beyond the insulation-breakdown strength of connected equipment. The NEC requires arresters in all industrial stations, but there are no requirements for the installation of arresters on industrial, commercial, or residential secondary services. After trouble is experienced from lightning, the tendency is to install lightning arresters. However, insurance companies have experienced considerable losses in certain areas and recommend the installation of secondary lightning arresters. One of the problems is, where to install the arresters? The most effective location is on the source side of the service head, but this does not present a very good appearance.

VOLTAGE FLUCTUATIONS RESULTING FROM SYSTEM CONDITIONS

Power supply which is derived from utility systems provides a stable source of power, generally being devoid of flicker owing mainly to the large generating units as well as interconnections with other systems. Engine-driven generators on isolated systems are probably responsible for most of the rare cases of flicker originating from the power system itself.

Abrupt changes of load on generators produce corresponding changes in the terminal voltages. This voltage fluctuation is the result of two factors: the change in speed and the regulation of the machine. In central-station practice it is very unusual for change in speed to be a significant factor. Sudden load increments are usually too small compared with the total generating capacity to change the speed materially. Even if the speed changes, however, the rate at which the voltage drops is ordinarily so slow that the effect is imperceptible to the eye.

68. Short Circuits and Switching Surges. Short-circuit currents, because of their magnitude, produce large voltage drops and attendant flicker. Reduction in the amount of voltage drop is not feasible without major changes in system layout and large expenditures. The duration of the voltage drop can, however, be markedly reduced in a number of cases by the use of high-speed relays and breakers. Flicker due to short circuits occurs so seldom that no special consideration for this purpose alone is necessary, except of course, that there is a constant vigilance to reduce short circuits. The tendency is toward a gradual reduction in flicker as system improvements are made for other purposes, such as protection of lines against lightning and installation of high-speed relays and breakers. Radial lines and short circuits which produce outages are a distinctly different problem.

Line switching rarely produces flicker unless load is picked up or dropped, or lines with large charging currents are switched. Here again, special provisions to reduce flicker are rarely necessary.

69. Utilization Equipment. Most of the flicker on central-station systems is due to the customer's utilization equipment. However, power companies generally have service rules which cover the starting currents of motors and other utilization equipment, the purpose of which is to eliminate or minimize flicker from this source. The following are some of the more common types of equipment known to cause flicker.

a. Motor Starting. Probably most flicker problems are caused by the starting of motors. For reasons of reduced cost, efficiency, and reliability, and in order to produce sufficient starting torque, commercial general-purpose motors require starting currents several times their full-load running current.

Four general classes of motor installations are of importance in the flicker problem.

1. Single-phase fractional-horsepower motors commonly used in homes and small stores

2. Single-phase integral-horsepower motors operated from secondary distribution circuits and used to serve small office and commercial buildings, and in recent years to serve homes for central air conditioning

3. Integral-horsepower polyphase motors operating from secondary distribution circuits and used to serve small factories, stores, apartments, and small office buildings

4. Large integral-horsepower polyphase motors operated from primary or subtransmission lines by industrial and commercial concerns.

Single-phase fractional-horsepower motors are manufactured in large quantities, and to maintain this extent of usage, they must continue to be low in cost, economical to operate, rugged, and reliable. These requirements have led to several classes of motors, depending upon the service, with one class designed specifically for frequent starting with low starting current. This motor is used in great quantities in domestic refrigerators and oil burners, and the ¼-hp 115-V class usually has a locked-rotor starting current of 20 A or less.

A medium-sized midwestern utility has a starting-current limitation for 115-V single-phase motors operating from a 120/240-V lightning service as follows:

24 A for frequent starting *
48 A for infrequent starting

Single-phase integral-horsepower motors have become increasingly common and popular in recent years to serve central air conditioning in homes. This application is practically the first appliance in the home with an electric motor having a capacity above 1 hp, not including window air-conditioning units. Most central-air-conditioning motors are 2, 2½, or 3 hp.

The same utility also limits the permissible starting current of single-phase motors according to Table 31.

**TABLE 31. Starting Current Limit
for 240-V Motors**

Customer's total demand, kW	Max starting current, A	
	Frequent starting	Infrequent starting
10 or less	30	60
11–20	40	80
21–40	55	110
41–60	70	140
61–80	85	170
81–100	100	200
Over 100	Consult the company	

Under the above limitations, the maximum horsepower ratings of single-phase motors for across-the-line starting for customers with a total demand of 10 kW or less for the same utility are given in Table 32.

If the current drawn by a single-phase motor causes excessive flicker, the recommended method of solving the problem is to wire the motor to a separate outlet so that the utility can isolate a transformer for this service and so that no other services are supplied from this transformer.

Integral-horsepower polyphase motors on secondary circuits are potential sources of flicker. In most cases such motors are used in areas of high load concentration, and the power circuits are correspondingly large. This capacity usually permits ordinary 3-phase squirrel-cage motors to be started directly across the line. In some cases, however, the size of the motor is out of proportion with its supply line. To remedy this problem, a starter can be used to limit the inrush current and thereafter change the current in increments sufficiently small to prevent objectionable lamp flicker.

Supplying large motors from primary power lines is usually not troublesome because such motors are usually located in a "commercial or industrial district" where power-supply lines are inherently heavy. There are nevertheless cases where motor ratings are too high for the power facilities. A suitable motor starter or a special type of motor may be required to remedy these conditions.

Starting currents for both induction and synchronous motors at full voltage vary from 5 to 10 times full load, depending upon the size, number of poles, and other application requirements, such as required starting, pull-in, and pullout torques.

* Frequently started motors are those which normally start more frequently than once every 2 h.

TABLE 32. Code Letter vs. Horsepower Limit

Industry standard code letter *	115-V motors		230-V motors	
	Frequent starting	Infrequent starting	Frequent starting	Infrequent starting
A	$3/4$	$1\frac{1}{2}$	2	3
B	$1/2$	1	$1\frac{1}{2}$	3
C	$1/2$	1	$1\frac{1}{2}$	3
D	$1/2$	1	1	2
E	$1/2$	1	1	2
F	$1/3$	$3/4$	1	2
G	$1/3$	$3/4$	1	2
H	$1/3$	$1/2$	$3/4$	$1\frac{1}{2}$
J	$1/3$	$1/2$	$3/4$	$1\frac{1}{2}$
K	$1/4$	$1/2$	$1/2$	1
L	$1/4$	$1/2$	$1/2$	1
M	$1/6$	$1/3$	$1/2$	1
N	$1/6$	$1/3$	$1/2$	1
P	$1/6$	$1/3$	$1/3$	$3/4$
R	$1/6$	$1/4$	$1/3$	$3/4$
S	$1/8$	$1/4$	$1/3$	$1/2$
T	$1/8$	$1/4$	$1/3$	$1/2$
U	...	$1/6$	$1/4$	$1/2$

* Code letters indicate the locked-rotor kilovoltamperes per horsepower and are shown in NEC Table 430-7 (b). A code letter A motor draws from 0 to 3.14 kVA/hp, and a code letter U from 20.0 to 22.39 kVA/hp. Therefore a code letter U draws much more starting current than a code letter A motor, or about 7.6 times as much.

b. Motor-driven Reciprocating Loads. This type of load usually consists of air compressors, pumps, and refrigerators. The motor load varies cyclically with each power stroke and produces a corresponding variation in the line current. This comparatively small variation of voltage may be objectionable if the pulsation occurs 6 to 12 times per second.

c. Electric Furnaces. There are three general types of electric furnaces: resistance, induction, and arc. The resistance furnace usually causes no more flicker than any other resistance load of comparable size. Because most induction furnaces operate at high frequency and therefore are connected to the power line through a frequency changer or rectifier, they represent a fairly steady load. Induction and arc furnaces are used extensively instead of blast and open-hearth furnaces to minimize air pollution.

Three-phase steel-melting arc furnaces are also used to a considerable extent to make high-grade alloy steel, and they frequently cause voltage flicker. Fluorescent lamps connected to these same lines are generally unsatisfactory, since fluctuations often cause the fluorescent lamps to be extinguished. Most of the time motors operate satisfactorily when connected to these lines.

While the average load factor and power factor of electric-arc furnaces are as good as or better than those of many other industrial devices, the problem of supplying them with power is usually much more difficult. During the melting-down period, pieces of steel scrap will at times more or less completely bridge the electrodes, approximating a short circuit on the secondary side of the furnace transformer. Consequently, the melting-down period is characterized by violent fluctuations of current at low power factor, single-phase. When the refining period is reached, the steel has been melted down to a pool, and arc lengths can be maintained uniform by automatic electrode regulators which move the electrodes up or down so that stable arcs can be held on all three electrodes. The refining period is therefore characterized by a steady 3-phase load of high power factor.

The size of load fluctuations during the melting-down period is influenced by a number of factors, of which the rate of melting is perhaps the most important. The furnace-supply transformers have winding taps for control of the arc voltage and in the smaller sizes (approximately 6000 kVA and below) have separate built-in reactors to limit the current and stabilize the arc. The rate of melting is subject to further control by means of automatic electrode regulators. Sometimes the production of steel is stepped up by raising the arc voltage, reducing the series reactance, raising the regulator setting, or by a combination of several of these procedures. Forcing the furnace in this manner increases both the magnitude and the violence of the load swings. The type of scrap being melted also affects the extent of the load swings, heavy scrap causing wider fluctuations than light scrap.

One authority gives the following figures for maximum instantaneous changes in current: for furnaces 1000 kVA and smaller, 2½ times normal; for 2500-kVA furnaces, 2.0 times normal; for 5000-kVA furnaces, 1.6 times normal; and for 10 000-kVA furnaces, 1.1 times normal.

d. Electric Welders. This class of equipment is of great importance in power-system flicker. Most welders have a smaller on time than off time, and consequently the total energy consumed is small compared with the instantaneous demand. Fortunately, most welders are located in factories, where other processes require a large amount of power and where the supply facilities are sufficiently heavy so that no flicker ordinarily is experienced. In isolated but nonetheless important cases the welder may be the major load in the area, and serious flicker may be imposed on distribution systems adequate for ordinary loads.

The more common type of electric welders are:
1. Flash welders
2. Pressure butt welders
3. Projection welders
4. Resistance welders – spot and seam

In welders the source voltage, usually 230, 460, or 2300 V, is stepped down to a few volts to send high currents through the parts to be welded. Practically all welders in service are single-phase.

With flash welders, one piece is held rigidly, the other is held in quasi-contact with it, and the voltage is applied. An arc is formed, heating the metal to incandescence, and the movable piece is made to follow to maintain the arc. The heating of the metal is partly by the passage of current and partly by burning with the arc. After a sufficient temperature and head penetration have been obtained, the pieces are forced together with great pressure. In some cases, the power is cut off before this "upset"; in others, the power is left on. The current, drawn during the flashing period, is irregular because of the instability of the arc, so that the flicker effect is more obnoxious than if the current is steady at its maximum value. The average power factor during flashing may be as high as 60%. At upset, it is about 40%. The flashing may last up to 20 or 30 s, but 10 s is more common. The duration of power during upset is usually short, about ½ s. This type of welder may draw up to 1000 kVA during flashing, and about twice this at upset.

Pressure butt welders are similar to flash welders, except for the important difference that the parts being welded are kept continuously in contact by pressure. The heating is produced primarily by contact resistance. From a power-supply standpoint the butt welder is more desirable than the flash welder because the welding current, once supplied, is practically steady, and the only flicker produced is at the time power is applied and removed. The range of currents and power factors is about that for flash welders.

Projection welders are similar to pressure butt welders, except that the latter usually join pieces of about equal size, and projection welders usually join small pieces to large ones. The current demand is usually smaller, but the operations are likely to be more frequent.

In resistance welders current is applied through electrodes to the parts to be welded, usually thin sheets of steel or aluminum. The weld is accurately timed to bring the metal just to the welding temperature. The pieces are fused together in a small spot. In the spot welder, one or a few such spots complete the weld. In a seam welder, a long succession of spots produce the equivalent of a single continuous weld or seam. Resistance welders are characterized by large short-time currents.

In spot welders, the current may be applied for only a few cycles (on a 60-Hz basis) with welds following one another in a fraction of a second up to about a minute. Thus, from a flicker standpoint a succession of individual voltage dips occur at objectionable frequent intervals. Seam welders have an on duration of a few cycles, followed by an off duration, also of only a few cycles. The process is a continuous one while a given piece is in the machine, and since the periodicity of the welds is uniform, the flicker can be annoying even for relatively small voltage dips. The essence of good spot and seam welding is accurate control of the heat; consequently accurate magnitude and duration of current are necessary. Electronic controls are being used to a large extent for welder control functions because there are no wearing parts and close and consistent regulation of the heat is possible.

The following are excerpts from Article 630 of the 1975 NEC requirements for welders.

70. Ac transformer and dc rectifier arc welders have special requirements that involve ampacity protection, and disconnect means.

a. Ampacities of Supply Conductors. The ampacities of conductors shall be as follows:

1. Individual welders. The rated ampacities of the supply conductors shall be not less than the current values determined by multiplying the rated primary current in amperes, given on the welder nameplate, and the following factor based upon the duty cycle or time rating of the welders:

Rated % duty cycle of welders	Multiplying factor
20 or less	0.45
30	0.55
40	0.63
50	0.71
60	0.78
70	0.84
80	0.89
90	0.95
100	1.00

For a welder having a time rating of 1 h, the multiplying factor shall be 0.75.

2. Group of welders. The rated ampacities of conductors which supply a group of welders shall be permitted to be less than the sum of the currents, as determined in accordance with the above of the welders supplied. The conductor rating shall be determined in each case according to the welder loading based on the use to be made of each welder and the allowance permissible in the event that all the welders supplied by the conductors will not be in use at the same time. The load value used for each welder shall take into account both the magnitude and the duration of the load while the welder is in use.

Conductor ratings based on 100% of the current, as determined in accordance with the above, of the two largest welders, 85% for the third largest welder, 70% for the fourth largest welder, and 60% for all remaining welders, should provide an ample margin of safety under high-production conditions with respect to the maximum permissible temperature of the conductors. Percentage values lower than those given are permissible in cases where the work is such that a high operating duty cycle for individual welders is impossible

b. Overcurrent Protection. Overcurrent protection shall be as provided in the following paragraph. Where the nearest standard rating of the overcurrent device used is under the value specified in these calculations, or where the rating or setting specified results in unnecessary opening of the overcurrent device, the next higher rating or setting may be used.

1. Each welder shall have overcurrent protection rated or set at not more than 200% of the rated primary current of the welder, except that an overcurrent device is not required for a welder having supply conductors protected by an overcurrent device rated or set at not more than 200% of the rated primary current of the welder.

2. Conductors which supply one or more welders shall be protected by an overcurrent device rated or set at not more than 200% of the conductor rating.

c. Disconnect Means. A disconnect means shall be provided in the supply connection of each welder which is not equipped with a disconnect mounted as an integral part of the welder.

The disconnect means shall be a switch or circuit breaker, and its rating shall be not less than that necessary to accommodate overcurrent protection as specified in these requirements.

71. Motor-Generator Arc Welders. Motor-generator arc-welder installations are covered by the appropriate NEC sections of Chapters 1 to 4, inclusive, applicable to conductors, motors, generators, and associated equipment. Referring specifically to the motor-supply connections, the following sections apply in addition to such other provisions as may be applicable: conductor rating, Sections 430-22 and 430-24; overcurrent protection for motors, Section 430-31; for conductors, Section 430-51; controllers, Sections 430-8 and 430-83; disconnecting means, Section 430-101.

72. Resistance welders have special requirements involving ampacity, protection, and disconnect means.

a. Ampacities of Supply Conductors. The ampacities of the supply conductors necessary to limit the voltage drop to a value permissible for the satisfactory performance of the welder are usually greater than that required to prevent overheating as prescribed in the following. The rated ampacities for conductors for individual welders shall conform to the following:

1. Varying operations. The rated ampacities of the supply conductors for a welder which may be operated at different times at different values of primary current or duty cycle shall be not less than 70% of the rated primary current for seam and automatically fed welders, and 50% of the rated primary current for manually operated (nonautomatic) welders.

2. Specific operation. The rated ampacities of the supply conductors for a welder wired for a specific operation for which the actual primary current and duty cycle are known and remain unchanged shall be not less than the product of the actual primary current and the multiplier given below for the duty cycle at which the welder will be operated.

Duty cycle, %	50	40	30	25	20	15	10	7.5	5.0 or less
Multiplier	0.71	0.63	0.55	0.50	0.45	0.39	0.32	0.27	0.22

The rated ampacities of conductors which supply two or more welders shall be not less than the sum of the value obtained as explained above for the largest welder supplied, and 60% of the values obtained as explained above for all the other welders supplied.

b. Explanation of Terms. (1) The rated primary current is the rated kVA multiplied by 1000 and divided by the rated primary voltage, using values given on the nameplate. (2) The actual primary current is the current drawn from the supply circuit during each welder operation at the particular heat tap and control setting used. (3) The duty cycle is the percentage of the time during which the welder is loaded. For instance, a spot welder supplied by a 60-Hz system (216 000 cycles per hour) making four hundred 15-cycle welds per hour would have a duty cycle of 2.8% (400 multiplied by 15, divided by 216 000, multiplied by 100). A seam welder operating 2 cycles on off would have a duty cycle of 50%.

c. Overcurrent Protection. Overcurrent protection shall be as provided in the following. Where the nearest standard rating of the overcurrent device used is under the value specified or where the rating or setting specified results in unnecessary opening of the overcurrent device, the next higher rating or setting may be used.

Each welder shall have an overcurrent device rated or set at not more than 300% of the rated primary current of the welder, except that an overcurrent device is not required for a welder having a supply circuit protected by an overcurrent device rated or set at not more than 300% of the rated primary current of the welder.

Conductors which supply one or more welders shall be protected by an overcurrent device rated or set at no more than 300% of the conductor rating.

d. Disconnecting Means. A switch or circuit breaker shall be provided by which each welder and its control equipment can be isolated from the supply circuit. The ampacity of this disconnecting means shall be not less than the supply conductor rating. The supply-circuit switch may be used as the welder-disconnecting means where the circuit supplies only one welder.

e. Marking. A nameplate giving the following information shall be provided: name of manufacturer, frequency, primary voltage, rated kVA at 50% duty cycle, maximum and minimum open-circuit secondary voltage, short-circuit secondary current at maximum secondary voltage, and specified throat and gap setting.

REFERENCES

1. Alfred E. Relation, Uninterruptible Power for Critical Loads, *IEEE Trans. Ind. Gen. Appl.*, vol. IGA5, no. 5, September/October 1968.
2. "Electric Power Distribution for Industrial Plants," IEEE Publication 141, 3d ed., October 1964.
3. Lawrence R. Poker, Economical Ground-Fault Protection Available with a Standard Low-Voltage Static Tripping System, *IEEE Trans. Ind. Gen. Appl.*, vol. IGA-6, no. 2, March/April 1970.
4. R. H. Kaufmann and J. C. Page, Arcing Fault Protection for Low-Voltage Power Distribution Systems, *IEEE Trans. Ind. Gen. Appl.*, vol. 79, pp. 160–165, 1960.
5. Eustace C. Soares, Clearing Ground-Fault Currents on Distribution Systems 600 Volts or Less, *IEEE Trans. Ind. Gen. Appl.*, vol. IGA-2, no. 1, p. 53, 1966.
6. A. J. Bisson, "Iron Conduit Impedance Effects in Ground Circuit Systems," AIEE Paper 54–244, 1954.
7. Donald Fink and J. M. Carroll, "Standard Handbook for Electrical Engineers," 10th ed., McGraw-Hill Book Company, New York, 1968.
8. Robert W. Smeaton, "Motor Application and Maintenance Handbook," McGraw-Hill Book Company, New York, 1969.

5

System Considerations
for Ac Control

WILLIAM R. IBACH *

FOREWORD

Before control equipment is specified for any application, the characteristics of the system with relation to the equipment supplied must be considered. A background beyond that given in Sec. 2 with relation to the power supply, voltage conditions, and fault characteristics of the system is given in this section to aid in choosing the types of control that will be suitable on a given system.

POWER SUPPLY

1. General Requirements. Regulatory bodies generally require utilities to regulate the voltage of power services not to exceed 10% above or 10% below normal voltage,

* W. R. Ibach and Associates, Electrical Engineers (formerly Chief Substation Design Engineer, Wisconsin Electric Power Company); Registered Professional Engineer (Wis.); Senior Member, IEEE and Engineers and Scientists of Milwaukee.

or lighting loads not to exceed 3% above or 3% below normal voltage during lighting periods.

If both power and lighting are fed from the same source, the regulation requirements of lighting service generally apply.

Wholesale or primary rates generally include regulation of 10% above and 10% below normal voltage. If this rate is used for shopping centers or other locations where voltage is critical, consideration should be given to use of automatic voltage regulators. These regulators can be purchased as separate units or as a load tap-changer transformer, where the tap changer is an integral part of the transformer.

If computers are being considered, the power source must be checked to determine if the regulation is within the limits set by the computer manufacturer. If this load is fed from an ac network system, the source is generally stable enough for a computer. If the source is not stable, a separate voltage regulator is required (Sec. 4, Art. 8, and Sec. 7).

2. Dc Source. Utility dc systems are being eliminated, reduced in area, or converted to or being fed from ac systems. The original dc systems were fed directly from dc generators, rotary converters, or motor generators. The dc energy from this type of system has been quite expensive, so that most utilities have been seeking means of converting or eliminating this load. The method used is the one that is the most economical to the utility.

One method is to use a large silicon controlled rectifier to feed the present distribution system. The disadvantage of this arrangement is that the dc distribution system remains and must be maintained.

The other alternative is to convert each service from dc to ac. Whether each customer is to be converted or whether individual silicon controlled rectifiers are to be installed at each service is again a question of economics. If the load consists of elevators, fire pumps, or printing presses, or if extensive rewiring is required, individual rectifiers at the service entrance provide the most economical arrangement. If the rewiring is simple and consists of lighting and simple motor replacement, a direct replacement of the utilization equipment will be economical. Silicon controlled rectifiers are considered quite reliable, provided that they are not overloaded or abused. One supplier guarantees his diodes for life if applied within their rating. Such a guarantee is indicative of the reliability of the equipment. By this method of conversion, therefore, the utility's dc distribution system can be removed, the cables, generally copper, salvaged, and duct space freed for ac needs.

3. Ac Source. The ac supply in this country, with few exceptions, has been very reliable, especially if compared with the performance in other countries. The high degree of reliability, according to the splendid past record, makes the public and many public officials less tolerant in cases of rare power failures. The system frequencies that are a part of a coast-to-coast interconnection approach perfection, since the daily number of system cycles are checked at a certain time each day and the system frequency is increased or decreased so that synchronous clocks on the system are practically 100% accurate over a 24-h period.

4. Continuity of Service. Because of the widespread use of electric service for all purposes, including refrigeration for food and drug protection, a great many customers cannot tolerate an outage of the electric service. Some of these occupancies are airports, stadiums, arenas, auditoriums, theaters, shopping centers, large industrial plants, department stores, municipal, state, and federal government buildings, colleges, universities, medical centers, and large apartment buildings. To provide service from a second source, either manual or automatic transfer can be provided. The manual-transfer equipment must be mechanically interlocked so that only one source can be closed at the same time.

Automatic transfer equipment for normal and reserve lines is also mechanically and/or electrically interlocked so that only one source can be closed at a time. The speed of operation generally is so fast that the transfer cannot be detected from the light flicker. Also, in some cases the speed of transfer is so fast that magnetically held motor starters will not drop out.

a. Automatic Transfer, 600 V Maximum. A typical automatic transfer scheme for 600-V maximum service is given in Fig. 1.

b. Automatic Transfer, High-Voltage Circuits.   A typical automatic transfer scheme for higher voltage circuits is given in Sec. 20.

5. Voltage Fluctuations.   The type of motor and motor starter that may be used depends upon the motor starting current and, of course, the maximum voltage drop permitted.

a. Motor Starting.   The polyphase squirrel-cage induction motor is the most rugged, simple, reliable, maintenance-free, and economical motor.   It is rugged and simple because it has a winding in the stator only, the rotor consisting of short-circuiting bars, and since it does not have windings on the rotor or does not have collector rings, commutator, or brushes, it is the most reliable and most maintenance-free of all motors.   Its economy lies in that it can be started by applying full voltage and throwing it across the line.   Its limitation on motor size, therefore, is the maximum voltage drop that will be permitted when thrown across the line.

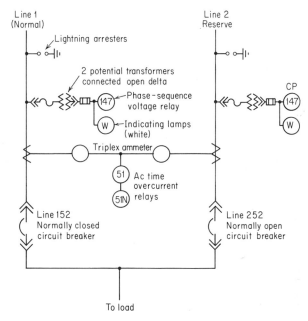

Fig. 1   Automatic transfer scheme and overcurrent protection for normal and reserve lines, using circuit breakers (3-wire, 3-phase service).

b. Estimation of Voltage Drop.   Figure 4 in Sec. 4 shows a series of curves by which the voltage drop can be approximated for various motor sizes supplied through various transformer sizes.   If the starting of the motor will produce a voltage drop that is tolerable, a full voltage, sometimes called "across-the-line starter," is acceptable.   If not, other means must be devised to reduce the starting current of the motor.   To reduce the drop in voltage, a different type of starter, such as reduced voltage, resistance, or autotransformer, reactor, or part winding, may be selected.

c. Full-voltage starters do not have any voltage-reducing elements but essentially consist of a magnetically held contactor for opening and closing of the circuit (Sec. 2).   In addition, the units have overload elements for opening the magnetically held contactor circuit, thus providing running protection for the motor.

If the current drawn by the motor, using a full-voltage starter, cannot be tolerated, other methods of reducing the current must be adopted (Sec. 22).   Some of the characteristics of various starters and motors are shown in Table 1.

d. Light-flicker limitations are given in Sec. 4, Art. 12.

TABLE 1. Comparison of Motor-starting Methods *

Type of starter (settings given are the more common for each type)	Motor terminal voltage, % line voltage	Starting torque, % full-voltage starting torque	Line current, % full-voltage starting current
Full-voltage starter..	100	100	100
Autotransformer:			
80% tap	80	64	68
65% tap	65	42	46
50% tap	50	25	30
Resistor starter, single step (adjusted for motor voltage to be 80% of line voltage)..................	80	64	80
Reactor:			
50% tap..	50	25	50
45% tap..	45	20	45
37.5% tap..	37.5	14	37.5
Part-winding starter (low-speed motors only):			
75% winding...	100	75	75
50% winding...	100	50	50
Wye-delta starter..	58	33	33

* For a line voltage not equal to the motor rated voltage multiply all values in the first column by the ratio

$$\frac{\text{Actual voltage}}{\text{Motor rated voltage}}$$

Multiply all values in the second column by the ratio

$$\left(\frac{\text{Actual voltage}}{\text{Motor rated voltage}}\right)^2$$

and multiply all values in the last column by the ratio

$$\frac{\text{Actual voltage}}{\text{Motor rated voltage}}$$

6. Low-Voltage Protection for Services. This type of protection is not prevalent on services except in connection with automatic transfer or in cases of ac generators located in customer's plants and operating in parallel and synchronism with the utility's system. If the voltage of the utility's system drops, it is necessary for the undervoltage relay to separate the two systems to prevent the customer's generator from exciting or back-feeding into the utility's lines. This relay is in addition to the reverse-power relay which provides a similar function. However, there are times when current flows from the customer's plant to the utility's system, but at a low power factor so that the reverse-power relay would not function on this reverse flow. In this case the customer's generator is not feeding any of the utility's load but instead is charging a utility company's line, and the charging current is sufficient to cause the voltage of the customer's generating system to drop. The undervoltage relay will operate in this case. In addition, experience has shown that the voltage relay is good backup protection to the reverse-power relay when the customer's generators operate in parallel with a utility's system. Low-voltage protection for services, unless absolutely necessary, as described above, should be avoided. Well-intended applications are generally abandoned after nuisance trippings are experienced because of voltage surges on the utility's system.

7. Overvoltage protection is not commonly used because overvoltages are not prevalent. If overvoltage protection is supplied, it is used as a protection against voltage surges. Constant overvoltage can be remedied by bucking transformers, voltage regulators, or tap-changing transformers. Voltage surges due to lightning or switching can be dissipated by means of lightning arresters or surge-protecting capacitors. It is possible, however, to have overvoltages due to a malfunction of a voltage regulator or tap-changing transformer, a wrong connection by the utility, or an accidental falling of a high-voltage wire on a low-voltage wire. However, these cases are not common or probable. Overvoltage-protective equipment for motor circuits is virtually non-existent for standard equipment. Overvoltage protection can be obtained for services, but its application, except in special cases, is questionable and should be avoided and discouraged.

8. Phase Failure or Reversal.   The possibility of a phase failure from a utility system depends on the system or the portion of a system. If the distribution system is entirely underground, the possibilities of failure of one of the phases are rare. If a portion of a distribution system is overhead and/or if the circuit breakers or the reclosures controlling the feeders are single-phase, the possibility of a phase failure is great.

Rural distribution substations generally have single-phase reclosures protecting their feeders because the load, including motors, is largely single-phase and the operation of a single-phase reclosure would result in an outage affecting a smaller area compared with a 3-phase reclosure.

Urban substations generally feed more and larger 3-phase loads, including motors, and the circuit breakers or reclosures are therefore generally 3-phase to minimize and eliminate the single phasing of motors. The possibilities of a phase failure in a feeder supplied from an urban distribution substation, therefore, are not as great as in a rural substation, especially if lines are underground.

a. Reverse-phase voltage relay is a 3-phase contact-making voltmeter which is used to detect reverse-phase connection of lines, transformers, motors, generators, or synchronous condensers. It is commonly used in automatic transfer schemes to assure connection of proper phase rotation. For example, one such relay is used to initiate trip of a source breaker, while another is used to supervise closing of the alternate source breaker.

One set of relay contacts closes on 3-phase normal voltage conditions, whereas the other set of contacts closes on 3-phase undervoltage, loss of voltage, reverse-phase connection, or serious phase unbalance.

When used on motor applications, the relay may operate because of an open phase, such as that caused by an open fuse. The motor will tend to maintain normal voltage across the relay terminals by backfeed unless heavily loaded. However, the relay will operate as soon as the motor is stopped and will prevent the motor starting with one phase open.

b. Motor Single-Phase Protection.   The destruction of polyphase motors due to single phasing caused by the blowing of one fuse protecting a 3-phase transformer or transformer bank, or because an overhead wire falls is quite common. One manufacturer sells a phase-monitoring relay as a protection against economic losses resulting from phase reversal and unbalanced voltages of the electric supply to motors.

Single-phase protection, however, can be obtained by using three running overcurrent devices, one in each of the three legs of the 3-phase circuits in motor starters (Fig. 2).

NEC now requires three running overcurrent units in polyphase motor starters. The three overcurrent units provide better protection against single phasing (Sec. 13).

FAULT CONSIDERATIONS

9. Short-circuit and ground-fault protection of the motor-control circuit, motor starter, motor branch-circuit conductors, and the motor is provided by the motor branch-circuit protection (Sec. 13).

10. Contributions by Other Loads to Faults inside Plant.   Besides the short-circuit capacity provided by the utility's service, there are several other contributing sources inside the customer's plant. These sources include synchronous and induction rotating machines (Sec. 4, Art. 48). These machines may be generators, motors, or synchronous condensers. The current which each of these delivers to a fault at its own terminals is limited by the machine. Each of these rotating machines produces fault current which decreases with time after the initiation of a fault. In other words, these sources exhibit a variable reactance to the flow of fault current.

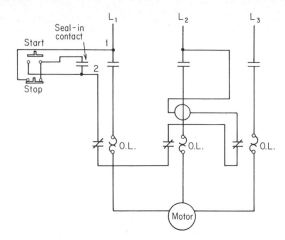

Fig. 2 Typical full-voltage, 3-phase motor starter with three overload relays.

11. Generator faults decrease exponentially from a relatively high initial value to a lower steady-state value some time after the initiation of the fault. Since a generator continues to be driven by its prime mover and to have its field energized from its separate exciter, the steady-state value of fault current will persist unless interrupted by some switching means.

See Sec. 4, Art. 49 for fault-current calculations, and the contributions to the fault current by various pieces of equipment.

12. Backup protection to the motor branch circuit would be by the feeder protection, provided, of course, there is more than one motor (Sec. 13).

The following information indicates the principle of coordination, which is necessary for backup protection.

Today, the circuit designer has available fuses of many characteristics from which a choice can be made to meet his needs. He must consider the abilities of the fuses to interrupt heavy short-circuit currents, but equally important is the ability of the fuses to be selective. Thus under fault conditions, it is highly desirable to have only the fuse

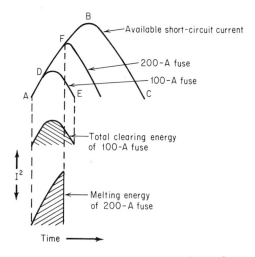

Fig. 3 Comparison of fuse characteristics for coordination.

blow that is in the circuit of origin.   A fault on a branch circuit should be cleared by the branch-circuit fuse without opening the feeder fuse.   Similarly a feeder fault should be cleared by the feeder fuse without opening the service.

To provide data which would allow the selection of fuses to perform selectively, many tests were run wherein fuses were mounted in series, that is, a smaller ampere-rated fuse in series with a larger fuse.   By performing such tests at various available fault currents, data were obtained to size the fuses so that only the smaller fuse blew.

The phenomenon involved is shown in Fig. 3, which displays the blowing of a 100-A fuse in series with a 200-A fuse.   If these two fuses were placed in a circuit set to deliver a short-circuit current as shown by ABC, the same fault current would flow through both fuses.   Since it would take less energy to melt the smaller fuse, the smaller fuse would be expected to melt first and then clear.   With the fault initiated at point A, the current rises to point D, where the 100-A fuse melts and an arc is struck within the fuse and quenched at E.   The crosshatched area shows this melting plus arcing phenomenon as a given amount of energy.   This total clearing energy of the smaller fuse (melting plus arcing) should be less than the energy to just melt the larger fuse.   Under this condition only the smaller one will blow.   It would take longer to melt the 200-A fuse than the 100-A size.   This characteristic is shown by the fact that the current must rise from A to point F in order to melt the 200-A fuse.   At the bottom the melting energy of the 200-A fuse is shown.   In this particular case selectivity existed, as the total clearing energy of the 100-A fuse was less than the melting energy of the 200-A fuse.   A 2/1 ratio in fuse sizes produces selectivity.

6

Dc Supply Considerations

JOHN F. SELLERS
GEORGE F. ECKENSTALER
C. H. LEET

FOREWORD

The type of dc source of power supply affects the selection, application, and care of control equipment. It is also an important factor in the protection required for the system. The three types of sources of dc power presently used in industrial plants — generators, rectifiers, and batteries — are therefore covered in this section to provide the necessary background information.

Part 1

Dc Generators

JOHN F. SELLERS *

GENERATOR CHARACTERISTICS AND REQUIREMENTS

The use of dc generators to supply power to dc motors is now quite limited in applications, but the use of silicon controlled rectifier (SCR) excitation circuits has given extreme flexibility to the generator performance. Secondary protection has also widened in scope for fail-safe operation of larger and higher-rated machines.

* Consulting Engineer, Dc Machines (Retired Chief Engineer, Dc Machines, Allis-Chalmers); Retired Member, Fellow Grade, IEEE.

1. Applications in Greatest Use.   Dc generators are still used in many applications, such as the following:

1. Synchronous motor-generator (MG) sets are necessary for power-factor correction in many large concentrations of induction motors and rectifiers. In some steel-rolling mills the auxiliary 250-V dc power is supplied from synchronous MG sets for this reason.

2. Ability to recover from the effects of high overload or short circuit gives the dc generator an edge over static supplies in the opinion of some users. The fast action of modern circuit breakers has reduced the chance of damage to either type of power supply, when a fault occurs.

3. Relatively ripple-free voltage is available from a dc generator, from residual to 100% rated voltage, which is not the case with a static supply. Some dc motors show extra heating and commutation problems when dc voltage with high ripple content is applied.

4. On reversing metal-rolling mills, a flywheel MG set has often been used to reduce the peak demand from the ac line. However, with the large size of most power systems, the demand factor is not critical.

5. Power for earth-moving vehicles has been from a diesel engine into either a hydraulic torque converter or a dc generator connected to dc motors. In the larger equipment, the dc power system has been almost universally used because of lower maintenance costs.

6. Power for traction on railroads has made use of dc systems when the prime mover is a diesel engine. Flexibility and low cost have been the proved attribute.

7. Dc generators have been chosen when the maintenance personnel are not technically competent. This situation of technical training has been improving with time.

2. Protection of Dc Generators.   The basic consideration of a dc control system is to protect the armature circuit from overload. This protection is usually accomplished with a current relay tripping the circuit breaker (Sec. 18). Fuses may also be used. Protection is also obtained when load increases gradually, by reduction of the shunt-field current, when an overload current relay operates to insert resistance in the dc generator shunt-field circuit. This circuit can produce a continually monitored reduction in excitation dependent on actual current load.

The overload capability of a dc generator varies with the class of design. A general-purpose machine usually has 150% load rating for circuit-breaker or overload-relay setting. Standard industrial-type generators larger than 0.6 kW per r/min have a 125% load rating for 2 h and 200% rating for 1 min. Circuit-breaker setting is usually 10% higher, or 225% of continuous rated A. Special designs for up to 300% of continuous rating are available. They require extra-wide interpoles or pole-face windings and in addition suitable brush area. If the 1-min load capability is higher than the standard 200%, this value is usually shown on the nameplate.

Dc control systems are also required to monitor the operation or conditions of larger generators with respect to:

1. Line voltage to ground
2. Shunt and interpole temperature by embedded detectors
3. Overvoltage across output bus
4. Bearing temperature by detector
5. Oil flow to bearings
6. Space heaters for humidity control when not in operation
7. Air-pressure drop across machine-mounted filters to determine their buildup

3. Factors Affecting Dc Generator Terminal Voltage.   When excitation current is held constant on a shunt-wound dc generator, the terminal voltage will drop from no load to full load, owing to two or more internal factors as shown in Fig. 1.

The IR voltage drop in the windings and brushes subtracts directly from the generated voltage. The windings are the armature and interpole, and series field if used, which will change in resistance by 25 to 35% from ambient to running temperature. Brush-contact voltage will change very little between no load and full load, being in the order of 2 to 2.5 V. The total IR plus brush V is usually from 2 to 5% of maximum rated voltage at continuous rated current.

Magnetic flux is lost from no load to full load because of the armature magnetic field which is at an angle of 90° from the main-pole field. This cross-magnetizing effect adds to the flux density over one-half of the main-pole arc, and subtracts on the other half, producing a net loss of flux, with constant excitation, of 5 to 10% from no load to full load.

A pole face or compensating winding will maintain practically constant main-pole flux regardless of load. This winding is embedded in the pole face, in slots uniformly distributed therein. The turns on this field are as near as possible to that part of the armature winding enclosed by the pole face, about ⅔. The turns are subtractive to the total MMF required on the interpole. The result is that the interpole and compensating windings have nearly the same number of turns on each. The combined total of interpole and compensating turns will be about 115 to 125% of the turns per pole of the armature winding. The same total turns would be used on a noncompensated machine.

The shape of the air gap under the main poles greatly influences the loss of flux from no load to full load in a noncompensated generator. If the pole face is concentric with

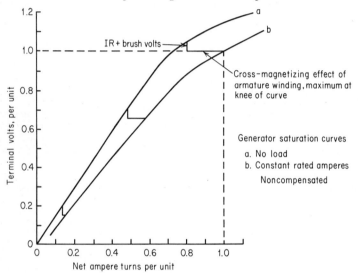

Fig. 1   Saturation curves of shunt-wound noncompensated generator.

the armature, distortion will be large, depending on the ratio of gap and armature teeth ampere-turns to armature ampere-turns. If the pole face has a radius of 1½ times the armature radius, and the gap at the pole tip is 3 times that at the center of the pole, the loss of flux at full load might be as low as 2%. The percent loss of flux increases faster than the percent increase in load with any geometry, owing to saturation of the armature teeth.

Interpole saturation at heavy overloads causes additional voltage droop. When the interpole saturates, commutation of the current in an armature slot is delayed, causing the center of current change to be forward of the center of the brush. In a dc generator this position produces a demagnetizing effect on the main poles, with resulting drooping voltage as load increases.

Correct interpole strength is usually assumed in an explanation of a generator load-voltage characteristic. If the interpole strength is not correct, because of lack of thoroughness in testing, a definite effect is found on the output-voltage characteristic. A weak interpole adjustment will add to voltage droop, as indicated above. On the other hand a strong interpole adjustment will produce a rising component in a generator load-voltage characteristic.

In general, the actual shape or amount of droop in a dc generator load-voltage characteristic is not too important in special applications where the generator voltage is varied continuously during the load cycle.

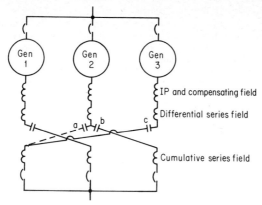

Fig. 2   Paralleling cross-connected series fields.

4. Control of Generator Field for Special Loads.   Flywheel MG sets are often used for power to drive dc motors on reversing metal-rolling mills.   The sets are equipped with one, two, or more dc generators which are arranged either all in parallel, or with one generator supplying power to one armature of a twin-armature dc motor.   The division of current between parallel generators has most often been obtained by the use of series fields which may be either differential to the shunt field or a combination of differential and cumulative series fields.   Figure 2 shows the latter arrangement for three or more generators, which can be adapted to two-generator operation in case one generator must be cut out owing to an armature-winding failure.   The temporary connection a must be made, and series field connections b and c opened, to remove generator 3 from service.

Each series field in Fig. 2 is designed to have 15 to 20% of the net ampere-turns on a main field pole at rated voltage.   Take as an example the no-load and full-load saturation curves of a typical flywheel-type generator, shown in Fig. 3.   The speed of the

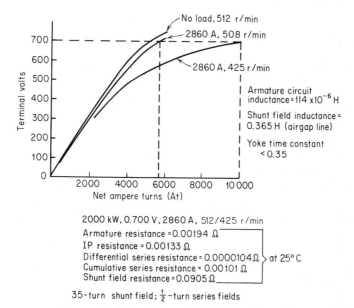

2000 kW, 0.700 V, 2860 A, 512/425 r/min
Armature resistance = 0.00194 Ω
IP resistance = 0.00133 Ω
Differential series resistance = 0.0000104 Ω
Cumulative series resistance = 0.00101 Ω   } at 25° C
Shunt field resistance = 0.0905 Ω

35-turn shunt field; $\frac{1}{2}$-turn series fields

Fig. 3   Saturation curves of compensated cross-connected generators for flywheel MG set.

wound-rotor induction motor driving the dc generator is 512 r/min at no load and 425 r/min at maximum slip. The motor speed is governed by a load-actuated regulator in the wound-rotor circuit. Figure 3 shows that 10 000 At is required at full-load rated voltage at minimum speed, so each series field should have from 1500 to 2000 At. Since the full-load armature current is 2860 A, each differential and cumulative field will be connected in two parallel circuits, giving 1430 At per pole at full-load amperes. Figure 4 shows the arrangement of the three fields on a generator main pole.

In some cases of paralleling dc generators, it is desirable to reduce the cost of the bus-bar connections. Only the differential series field is used to produce a drooping voltage characteristic with increasing load to the extent of 12 to 15% from no load to full load. This droop requires a larger gross shunt-field strength than when a cumulative series field is also used, to give the required ampere-turns shown in Fig. 3. The shunt field then must utilize the coil space that would be taken by a cumulative series field.

5. Solid-State Excitation. Control development using SCR circuitry in the excitation supply to the shunt fields of dc generators provides an alternate method of obtaining the effect of differential or load-balancing series fields. This arrangement greatly simplifies the generator field construction and external bus requirements because all the functions of the series fields are contained in the initial monitoring stage of the

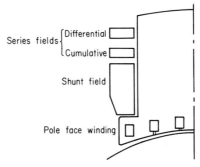

Fig. 4   Typical pole windings.

static excitation supply. Thus simple shunt-wound generators can be used for most applications. However, for backup and fast protection against overloads, the differential series field with 10 to 15% of net field ampere-turns at full load is considered essential by some users, especially in traction and earth-moving applications. The differential series field also improves braking characteristics when reverse current flow occurs. Then the differential series field becomes cumulative to the main shunt field, producing higher volts per armature ampere as compared with a straight shunt-wound generator, and thereby gives a greater braking effect on the connected dc motor.

The use of static excitation supply to a dc generator driven by a diesel engine provides a very flexible means of voltage control to limit engine overload in earth-moving applications.

Figure 5 shows the operating characteristics of a typical dc generator installed on an earth-moving machine. The generator voltage is raised by hand control of the shunt-field amperes, with full field on the dc motor. When maximum rated voltage is reached, the control switches to motor-field weakening for higher vehicle speed. If at any time during the working cycle a limit is reached in engine capability, the governor will act to reduce the generator voltage until the work load has been reduced to within the engine capability. At the same time the operator will usually cut back on the speed control, which increases the motor-field current for more torque per ampere. These comments apply only to shunt-wound motors and not to series-wound motors.

Further consideration shows that it would be desirable to have a combination of motor shunt-field weakening as the generator voltage increases, for then the operator would control only the generator field. Any limit action by the engine governor would automatically increase the motor field for more torque per armature ampere. There is not an exact relation for optimum change of motor field with respect to generator voltage, since two or more factors are present:

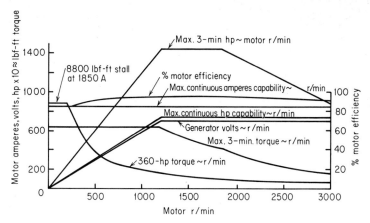

Fig. 5 Characteristic curves of dc generator powering a 360-hp, 350/700-V, 845/425-A, 600/3000-r/min motor for earth moving.

1. If the generator voltage is increased to its maximum value before the connected dc-motor field is weakened, the insulation and brush life, motor commutation, and system efficiency are at an optimum value. Maximum efficiency is gained because the load torque is supplied by the least possible armature current until the operation nears maximum motor speed.

2. With the generator at maximum voltage over more than 50% of the motor-speed range, there is insulation stress and chance of generator flashovers which would not occur if the motor field is weakened while the generator volts increase above the 40 to 50% of maximum voltage level.

The motor field is left at maximum continuous value over the generator voltage range of 0 to approximately 50%, where vehicle operation usually requires the heaviest torque. In fact, the control should be able to force the motor field to at least 150% of its maximum continuous level, when the armature current has increased to 120% of its continuous rating. With usual motor flux densities this field forcing provides a stall torque of 280% of maximum continuous torque, with approximately 220% of continuous ampere rating. By this means, the required stall torque is obtained with minimum stress on the generator.

In view of the above considerations, the actual generator voltage–motor speed curve will be modified to either curve *b* or *c* of Fig. 6, depending on the method of weakening the motor shunt field. Curve *b* is usually considered the most practical in operation, of the three shown.

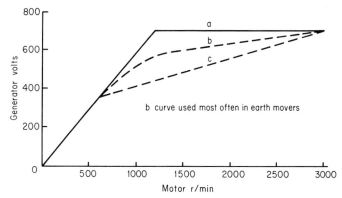

Fig. 6 Optional generator voltage vs. motor r/min curves for 360-hp, 350/700-V, 845/425-A, 600/3000-r/min motor.

In traction (railway) applications the dc generator is connected to one or more dc motors which have series fields instead of the shunt fields covered above. The design characteristics of the generator are no different, however, except that a pole-face winding is nearly always found necessary to reduce generator flashover when switching surges occur or overload is suddenly encountered.

Since a motor with series fields will not automatically provide a reverse flow of armature current for braking, it is customary to disconnect the motor from the dc generator, and connect to braking grids. The series field is then separately excited from the generator, and the braking effect is regulated by the field current of the generator.

6. Armature-Circuit and Shunt-Field Time Constants. The armature-circuit time constant is important in protecting the generator during rapid current change, when flashovers are likely to occur. The magnetic flux in the interpole is not able to keep up with the armature current owing to the damping action of eddy currents in the solid frame. This condition sets up a severe stress on the brushes. Some sparking is likely to occur when the rate of current change exceeds ten times rated value per second, although generators with laminated frames can take twice this value without trouble. The average of many tests of armature circuit inductance shows:

$$L_a = \frac{K \times \text{rated volts} \times 60}{\pi \times \text{rated amperes} \times \text{poles} \times \text{r/min}}$$

where $K = 0.15$ for generators with pole-face windings

$\qquad = 0.60$ for generators without pole-face windings

$\qquad L_a =$ armature inductance, in henrys

The time constant is L_a/R_a s, where R_a is the total armature-circuit resistance at operating temperature, including an allowance for brush resistance based on 2 V at rated current.

The shunt-field time constant is quite variable, owing to saturation in the armature teeth and elsewhere. The higher the saturation, the lower is the inductance. So for a transient shunt-field change from residual to maximum voltage, the transient inductance will change over a ratio of 2/1 or more, with the highest value of inductance at the straight-line part of the no-load saturation curve. The inductance of the shunt-field circuit is

$$L_f = \frac{n \, d\phi}{di_f \times 10^{-8}}$$

where $L_f =$ field inductance, henrys

$\qquad n =$ poles $\times$ turns per pole/number of parallel paths in shunt field

$\qquad d\phi = \dfrac{de \times \text{armature paths} \times 3 \times 10^9}{\text{poles} \times \text{commutator bars} \times \text{r/min}}$

$\qquad di_f =$ incremental change of shunt-field current for incremental change of armature voltage de

$\qquad d\phi/di_f =$ slope of saturation curve at voltage e, the center of de

The time constant L_f/R_f is usually in the range of 2 to 4 s for shunt fields designed for rapid voltage changes.

To the field time constant must be added 0.25 to 0.4 s for the eddy currents induced in a solid field yoke. When the yoke has a laminated construction utilizing ⅛-in or thinner laminations, the extra time constant may be neglected. Figure 3 shows the armature and field time constants for a 2000-kW generator.

It can be seen that for calculation of either time constant, it is necessary to obtain certain design data, along with the no-load saturation curve.

Part 2

Rectifier Power Supplies

GEORGE F. ECKENSTALER *

GENERAL

Many loads in industry require speed and torque characteristics which are inherent features of a dc drive system. In the field of electroplating and electrolytic process work there is no substitute for direct current. The power rectifier, therefore, is a basic piece of equipment in many industrial plants.

Few other types of equipment have experienced such rapid changes in technology as that in the power-conversion field. Originally, conventional rotating machinery such as rotary converters and synchronous MG sets were used. Then, the introduction of the static mercury-arc rectifier tube in the 1940s tended to make rotating machinery obsolete for many applications, and most recently, the development of high-power solid-state silicon semiconductor devices has made the mercury-arc tube and rotating machinery completely obsolete for power supplies. Now only semiconductor-type power rectifiers employing silicon diodes or silicon thyristors (silicon controlled rectifiers) are considered for new installations.

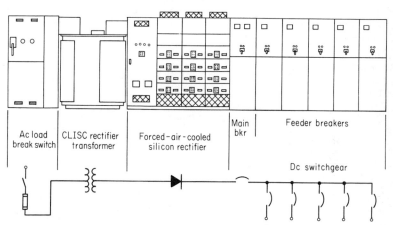

Fig. 7 Typical indoor metal-enclosed heavy-duty industrial silicon-rectifier substation.

A heavy-duty rectifier power supply will ordinarily include an ac incoming line breaker or disconnect switch, a power transformer to reduce the available ac supply voltage to the value needed to obtain the required dc voltage, the power-rectifier assembly, a main dc breaker, and any required dc feeder breakers to distribute the dc power. The proper selection of these components forms a power-rectifier unit capable of handling the desired continuous and short-time load currents (Fig. 7).

APPLICATIONS

The application and ratings of silicon-rectifier power supplies are many and varied. However, in general, the main fields of application can be divided into two basic groups, according to types of load:

* Chief Engineer, Dc Drives, Electronics Operation, Allis-Chalmers; Member, IEEE.

1. Constant-potential dc loads
2. Adjustable potential dc drive loads
Each group has its own specific control methods peculiar to the application.

7. Constant-Potential Dc Loads. Power rectifiers supplying this type of load are designed to maintain a relatively constant voltage over their load range and are not capable of inverting any energy from the load to the ac supply. Typical fields of service are:

1. General industrial
2. Transportation
3. Electrochemical

Typical loads in the industrial-service field are cranes, conveyors, hoists, lifting magnets, and multiple-motor drives. Rectifiers for general industrial service are usually rated for 250 V dc with individual unit capacities up to 2000 kW. Overload ratings are generally 125% for 2 h and 200% for 10 s or 1 min. Transportation rectifiers may be rated up to 1000 V dc with kW ratings sized for the specific job. Overload ratings in the transportation field are generally severe, 150% for 2 h with short-time ratings ranging to 300% for 5 min and 450% for 10 s. Electrochemical-rectifier voltage ratings also may be 1000 V dc but with individual unit capacities in the MW range. Generally there are no overload ratings on these rectifiers.

8. Adjustable-Potential Dc Drive Loads. Typical fields of service in this group are main and auxiliary drives for metal-rolling mills, rotary kilns, and processing lines. These power supplies must provide an adjustable and controllable-on-command dc output voltage from zero to 100% and, in addition, may be required to provide direct current in either direction to the load, and inverting energy from the load to the ac supply. These power supplies may be rated from 1 to many thousands of kilowatts at dc voltages up to 1000 V. Availability of switchgear in the required ratings, the practical limit to the number of semiconductor devices that can be put in parallel, and the necessity of phase multiplication with reference to ac line harmonics are some of the design considerations which dictate a limit to the size of a unit.

BASIC DESIGN CONSIDERATIONS

9. Power Circuit. Except for special applications the two basic circuits used for most industrial applications are:

1. Six-phase double way (3-phase bridge) (Fig. 8)
2. Six-phase double wye with interphase transformer (Fig. 9)

The choice of circuits is based on economic considerations. The basic six-device double-wye circuit requires a more expensive transformer and semiconductor devices with a higher peak voltage rating but has the advantage of being capable of producing twice the current output of the basic six-device double-way circuit. Present economics usually dictate the selection of the double-way circuit for applications requiring dc output voltages of 250 V and higher.

10. Semiconductor Device. The heart of the power supply is the silicon semiconductor cell. When properly applied they exhibit virtually unlimited, trouble-free life. They are, however, both voltage- and temperature-sensitive. Because of their small mass and therefore low thermal capability, they require mounting on a heat sink, which can be either air- or water-cooled, to obtain optimum current output while maintaining temperature within limits (Fig. 10). The "flat-pack" type of package has the advantage that it can be heat-sinked on both sides and therefore provides for better heat transfer to the cooling media.

Although the current ratings of the individual silicon cells have been increasing, heavy-duty industrial applications generally require the paralleling of devices to obtain the desired current ratings. In many applications the short thermal-time-constant characteristic of the semiconductor device makes the required overload or momentary capacity of the equipment the determining factor in the selection of the number of parallel devices (Fig. 11).

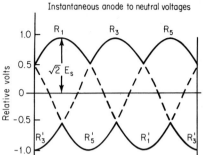

Instantaneous anode to neutral voltages

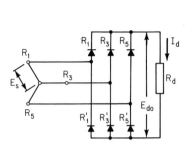

Note: E_{do} = theoretical average direct voltage at no load

Wave shapes shown are theoretical. Load current will cause distortion.

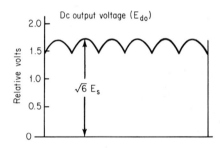

Dc output voltage (E_{do})

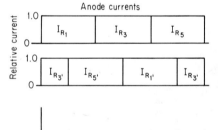

Anode currents

Fig. 8 Six-phase double-way rectifier circuit.

CIRCUIT CHARACTERISTICS:

DC output circuit
$E_{do} = 2.34\ E_s$
% ripple = 4.5
Rectifier element

Average current $= \dfrac{I_d}{3}$

Rms current $= \dfrac{I_d}{\sqrt{3}}$

Peak inverse voltage $= \sqrt{6}\ E_s$

Transformer

DC winding rms current $= \dfrac{\sqrt{2}}{\sqrt{3}}\ I_d$

DC winding kVA rating $= \sqrt{6}\ E_s I_d \times 10^{-3}$
$\qquad\qquad\qquad\quad = 1.05\ E_{do} I_d \times 10^{-3}$

AC winding kVA rating $= \sqrt{6}\ E_s I_d \times 10^{-3}$
$\qquad\qquad\qquad\quad = 1.05\ E_{do} I_d \times 10^{-3}$

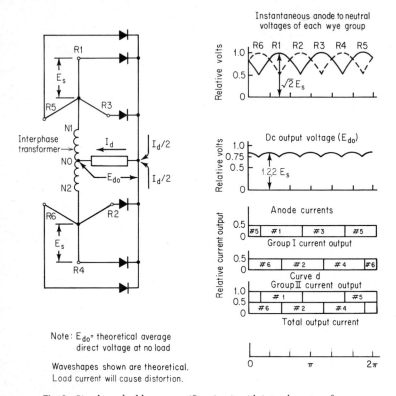

Note: E_{do} = theoretical average
direct voltage at no load

Waveshapes shown are theoretical.
Load current will cause distortion.

Fig. 9 Six-phase double-wye rectifier circuit with interphase transformer.

CIRCUIT CHARACTERISTICS:

DC output circuit
$E_{do} = 1.17 E_s$
% ripple = 4.5
Rectifier element

Average current $= \dfrac{I_d}{6}$

Rms current $= \dfrac{I_d}{2\sqrt{3}}$

Peak inverse voltage $= 2\sqrt{2}\, E_s$ when $I_d = 0$
$\hphantom{Peak inverse voltage} = \sqrt{6}\, E_s$ under normal load

Transformer

DC winding rms current $= \dfrac{I_d}{2\sqrt{3}}$

DC winding kVA rating $= \sqrt{3}\, E_s I_d \times 10^{-3}$
$\hphantom{DC winding kVA rating} = 1.48\, E_{do} I_d \times 10^{-3}$

AC winding kVA rating $= \dfrac{\sqrt{3}}{\sqrt{2}} E_s I_d \times 10^{-3}$
$\hphantom{AC winding kVA rating} = 1.05\, E_{do} I_d \times 10^{-3}$

Fig. 10 Two commonly used types of packages for semiconductor devices, and typical air-cooled and water-cooled heat sinks. (*Power Semiconductors, Inc.*)

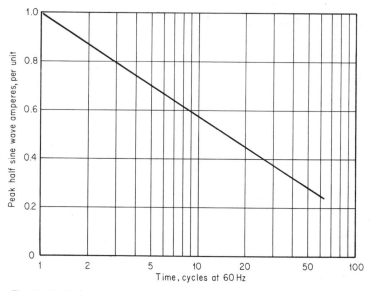

Fig. 11 Typical short-time capacity curve for high-power semiconductor device.

PROTECTION

Basically the control system for a rectifier unit must be designed to (1) protect the power supply and, more specifically, the semiconductor devices from conditions which could cause disastrous results, and (2) control the output flow of power with respect to a given command signal. The effects of the control methods on the ac supply system must also be taken into consideration.

The combination of a small package and a high current density results in a semiconductor-device characteristic which requires special considerations in the protection scheme of a rectifier power supply. Whenever a load current is applied to a semiconductor device, its junction temperature stabilizes to the specific load-condition temperature within a few seconds or less, and under fault-current conditions the junction temperature can rise to self-destruct values within a few milliseconds or less. This low thermal-time-constant characteristic and the inability of a semiconductor device to withstand inverse voltages in excess of their rating, even though this voltage may be impressed across the device for just a small fraction of a cycle, make the device extremely susceptible to damage during any abnormal condition. The degree of sophistication of protection will depend on whether or not the semiconductor devices are considered expendable, which in turn will depend on unit size and the requirements of the application.

11. Classes of Protection. In general, the degree of protection can be divided into three classes which will be arbitrarily called:
1. Minimal protection
2. Basic protection
3. Maximum protection

a. Minimal-Protection Scheme. The semiconductor devices are considered expendable, and no special protective devices or techniques are used to prevent semiconductor-device failure in case of a fault occurrence. Conventional line fuses or circuit breakers are used to clear fault currents and prevent unit destruction. This scheme is generally used only on relatively inexpensive light-duty-type rectifiers employing only one semiconductor device per phase path.

b. Basic Protection Scheme. The semiconductor devices are protected from failures during a fault occurrence by high-speed fuses. Conventional circuit breakers with overload sensing devices are used to provide equipment protection. This class of protection requires that blown fuses be replaced after a fault occurrence before the rectifier can be operated again. Its use is mainly in the medium-duty, medium-power type of rectifier requiring only a limited number of parallel paths per phase leg.

c. Maximum-Protection Scheme. Various protective devices are employed for the specific purpose of preventing, as much as possible, any component failure during a fault occurrence. The scheme is used in heavy-duty, high-power, high-reliability rectifier units employing the paralleling of many semiconductor devices to obtain the necessary power capability. In this class of equipment the idea of shutting down to replace devices or fuses after a fault occurrence is economically unfeasible and cannot be tolerated. This scheme requires that the characteristics of each piece of equipment in the rectifier substation, which include the power transformer, the semiconductor devices, the fuses, the overload and fault-sensing devices, and the protective switchgear devices, all be coordinated to provide the logic of operation for complete protection.

12. Types of Faults. In general, the types of faults that can occur in a rectifier unit are:
1. Internal fault current
2. External overload or fault current
3. Excessive inverse voltage
4. Auxiliary control failure

a. Internal Fault. This type of fault is a short circuit occurring within the rectifier unit. The fault can be caused by the following conditions:
1. Failure of a diode or thyristor to block reverse voltage
2. Failure of a thyristor to block forward voltage during a normal off-state period
3. Insulation failure

A study of the double-wye (Fig. 12a) and double-way (Fig. 12b) rectifier circuits shows that a rectifying element that fails to block reverse voltage in the double-wye circuit, sometimes referred to as a single-way circuit, creates a short circuit on both the ac and dc systems, whereas a failed device in the double-way circuit results in a short circuit on the ac system only. In both circuits the path of the ac system fault current is in a forward direction through the semiconductor devices in each unfaulted phase and into the phase with the failed device. To prevent the good semiconductor devices in the fault feed phases from destroying themselves, the fault current must be interrupted very quickly. The basic protection scheme would employ high-speed current-

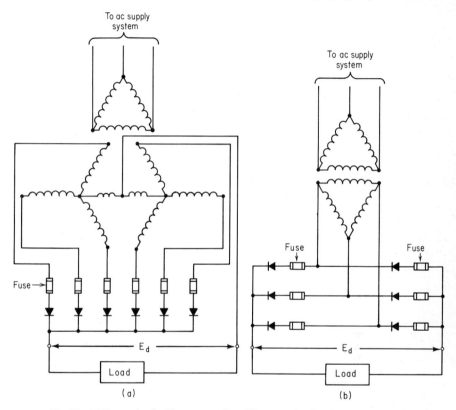

Fig. 12 (a) Fusing for double-wye rectifier. (b) Fusing for double-way rectifier.

limiting fuses placed either in the ac supply lines or directly in series with each semiconductor device (Fig. 12). The fuses, which are specially designed for semiconductor application, operate to limit the first-cycle-peak fault to a value lower than that which would ordinarily flow, and to provide a clearing time in the order of just a few milliseconds. They are coordinated with the semiconductor-device short-time capability so that the fuse will clear before a semiconductor-device failure can occur (Fig. 13).

In heavy-duty high-power applications employing many parallel paths and the maximum-protection concept, the protection scheme should quickly isolate a failed device and prevent failure of any good devices without interrupting the output power flow to the load. To accomplish this result, high-speed semiconductor-type current-limiting fuses are placed in series with each paralleled semiconductor device. Figure 14 shows that when a device failure occurs, a fault current flows in a forward direction through the N parallel paths of the normal devices in the phase arm supplying the fault current,

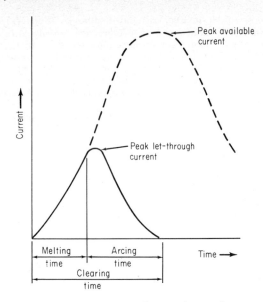

Fig. 13   Limiting action of current-limiting fuse.

and in the reverse to normal direction through the faulted device path. Thus each parallel device in the fault-feed supply phase is subjected to $1/N$ times the current through the faulted device path. The fuses in series with each device must be coordinated with the device short-time capability and the system capacity so that the fuse in the faulted device path isolates this path before any of the semiconductor devices in the fault feed-supply phase fail. In addition, the fuses must have sufficient capacity not to melt during an external overload or fault condition before the external switch-gear-type protective devices can operate.

A thyristor device may fail to block forward voltage because of a premature gate signal or because of excessive rate of forward applied voltage. Ordinarily a device that fails to block forward voltage can recover and is not permanently damaged if resultant fault or overcurrents can be limited to a value within the device's capability.

A properly designed unit will limit dv/dt to within the thyristor capabilities so that only random failures may occur because of device degradation. A premature signal may be the result of a control failure or random unsuppressed noise. When this condition occurs, usually all the devices in a parallel-phase path are turned on. In this failure mode the type of load affects the type of fault. If the load exhibits a relatively constant impedance such as in constant-potential applications, the result may be only a momentary loss of voltage control because the load impedance will tend to limit the breakthrough current to a value just slightly above normal. If, on the other hand, the load impedance varies widely such as in variable-speed-drive applications, a break-through can result in a high fault current which can destroy devices if the current is not interrupted quickly. In units using a basic protection scheme the fuses used for protection against a reverse-blocking failure will also protect the devices in this mode of failure. In high-power rectifiers utilizing the maximum-protection philosophy, the fuse characteristics are coordinated with that of a high-speed dc breaker so that the breaker clears the fault before any semiconductor devices fail or fuses blow. In some applications it may be necessary to use a dc reactor in the load circuit so that proper coordination can be attained. In those cases where a single semiconductor device fails in a parallel-path arm, the entire load current will attempt to flow through that path, and the current-limiting fuse in series with that particular device will blow to isolate the defective path. No other coordination is possible.

An internal insulation failure within the rectifier unit may cause either or both an ac line-to-line fault and a dc bus fault. In this failure mode the ac and dc switchgear devices must protect the rectifier from destruction.

In the parallel-path rectifier, the fault current may at first be impedance-limited by the internal bus and therefore be at a relatively low value, which if not quickly detected will cause considerable damage to the rectifier unit. Conventional switchgear-type overcurrent relays connected to sense alternating current in the input line between the transformer and the rectifier bridge can be set lower than relays in the transformer primary circuit because they do not sense the transformer inrush current. However, they still must be set higher than the overload ratings of the rectifier unit.

In applications requiring a high degree of sensitivity to insulation-type faults, a differential-protection scheme which compares the ac input current with the dc output current can be used. Any slightest deviation in the compared values will cause a tripping signal to be transmitted to the ac and dc breakers.

Where the rectifier is paralleled with other conversion equipment, an internal dc-bus fault will result in a reverse current through the dc breaker due to the contribution of fault current from the parallel equipment. Ordinarily a sensitive reverse-current detector is employed to trip the dc breaker quickly whenever this condition is sensed, thereby limiting the fault contribution from the parallel equipment and providing for the possibility of continuity of dc service from the main dc bus.

b. External Overload or Dc Fault Current. IEEE Standards define an overload current as a current that exceeds the rating of the converter unit in magnitude or time, but the conduction cycles and waveforms remain essentially constant. A fault current is said to exist when the conduction cycles of some phases are abnormal.

Equipment employing the basic protection scheme generally uses an inverse time overcurrent relay to sense overload current and to transmit a signal to the ac and/or dc switching devices and the semiconductor reverse-blocking-failure fuses to provide the fault-current protection. In addition, some thyristor units may use gate suppression to help limit the fault. This system uses a current-sensing device and associated circuitry to remove the thyristor gate signals whenever output current exceeds a predetermined level. Although gate suppression cannot alter the current flowing through the conducting devices for the first cycle of the fault current, it can prevent the nonconducting devices from turning on, and limits the fault-current occurrence time to less than a cycle. Another system closely allied to gate suppression, which is sometimes used instead of gate suppression, is phase-back. In this system the gate pulses are delayed to reduce the dc output voltage whenever a fault current is sensed.

In equipment using the maximum-protection philosophy the dc breaker must clear all overloads and faults without any equipment failure. Therefore the time-current capacity of the semiconductor devices and their isolating fuses must be great enough to carry both the overload and the expected maximum fault current until the dc breaker can clear. Generally conventional switchgear-type inverse time relays with instantaneous attachments are used to match the characteristics of the silicon devices (Fig. 15).

When coordinating the main dc breaker with feeder breakers, the ideal situation would be for the main breaker to be able to distinguish a feeder fault from a main bus

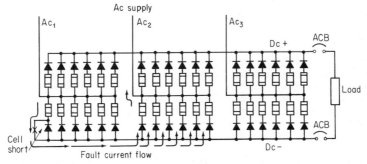

Fig. 14 Fault-current flow in parallel-path rectifier unit under reverse-blocking-failure conditions.

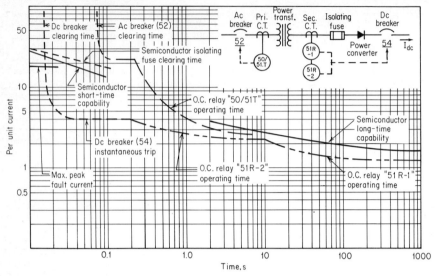

Fig. 15 Typical coordination curve for main dc breaker and feeder breakers.

fault so that selective tripping could be attained. However, selective tripping under all conditions would require a time delay in main-breaker tripping, and since a time delay would require more semiconductor-device capacity in the rectifier unit, the ideal situation is not too practical. In practice, feeder breakers are set for instantaneous trip only, at about 100 to 200% of their rating. Normally this setting is considerably less than the main-breaker instantaneous current setting, and therefore a degree of selectivity by magnitude only is attained.

   c. Excessive Inverse Voltage.   One of the important concerns with semiconductor devices is that of externally generated overvoltages. These voltages may be extremely short in duration with a steep wavefront. They may be caused by lightning, by the opening or closing of a breaker or switch in the ac supply line, by the tripping of a dc breaker, or by the commutation of current between the semiconductor devices in each phase leg. The overvoltage may, however, be of a relatively long duration with a slowly rising wavefront caused by the type of load on the dc bus. These loads include an overhauling load on motor equipment, a motor-field-changing scheme, or magnetic load switching by polarity reversal. In any case, an overvoltage in excess of the device rating can puncture the junction and thereby destroy the ability of the device to block voltage.

   For these reasons, voltage-surge-suppression circuits and devices must be employed to suppress any overvoltage. The short-time transient voltages can be suppressed by the use of *RC* circuits and surge-suppression devices such as thyrite, selenium, or metal varistor devices. However, since these devices have only limited energy-absorbing capability, they are incapable of suppressing any long-duration overvoltage condition without destroying themselves. When long-time overvoltages are known to occur, the use of a regenerative load-resistor bank with its associated voltage-sensing device and dc contactor arranged to connect the resistor across the dc bus whenever an overvoltage occurs is generally the accepted practice. If more sophistication is desired, the converter can be made to invert the dc energy back into the ac supply.

   To provide an additional safety margin for coordination with the surge devices, the semiconductor devices will generally have a voltage rating between 2 and 2.5 times the normal crest working voltage of the circuit. However, as better surge-suppression devices are developed, it is reasonable to expect that the device safety-voltage factor will be reduced.

   d. Auxiliary Control Failure.   The basic auxiliary control requiring monitoring consists of the cooling system and the circuitry necessary to gate the thyristor devices.

In the case of forced air-cooled equipment, loss or impairment of airflow is monitored by an airflow or pressure-sensing device and by temperature detectors. Similarly with water-cooled equipment, loss of water flow is monitored by a flow or pressure relay and temperature detectors. In equipment using the maximum-protection philosophy, temperature detection in two stages is employed, the first stage sending an alarm signal and the second stage a breaker-tripping signal. Complete loss of air or water flow should result in an immediate tripping signal because of the previously mentioned low-thermal-capacity characteristic of the semiconductor device.

A low supply voltage or the loss of a phase voltage to the gating circuit may result in erratic firing and semiconductor-device failure and therefore should be monitored. Operation of the monitoring relay should transmit a trip signal to the protective switchgear device.

VOLTAGE CONTROL

13. Inherent Regulation. In some constant-potential industrial-service applications, no control of the output voltage other than the proper choice of impedance of the rectifier transformer to provide the desired volt-ampere droop characteristic is necessary. The output voltage of the nonregulated unit will vary with load in accordance with the inherent regulation of the unit (Fig. 16). It should be noted that any supply-line system-voltage variations will cause the dc output voltage to vary in direct proportion to the ac voltage change. No-load taps, usually in the range of ±5%, are generally provided on the primary of the rectifier transformer so that the primary-secondary winding ratio can be matched to the ac supply voltage actually existing at the jobsite. The basic limitations to the application of nonregulated rectifiers for industrial service are that when paralleling with other equipment, the regulation curves may not match and therefore proportionate load sharing cannot be attained, and that ac line-voltage fluctuations and load changes may adversely affect the operation of the equipment on the dc bus.

14. Automatic Continuous Control. For applications requiring automatic continuous control of voltage, a number of basic methods are available. Some control schemes may incorporate the use of several basic methods to attain the desired result in the most economic manner. The methods include the following:

1. Step-type voltage regulators
2. Induction regulators
3. Saturable reactors
4. Self-saturating reactors
5. Thyristors

a. Step-Type Voltage Regulators. The step regulator employs load-tap-changing equipment to vary the voltage applied to the ac winding of the rectifier transformer without interrupting the load current. It acts to control the level of the rectifier regulation curve in discrete steps as shown in Fig. 17. Two tap-changing mechanisms may be coordinated in a single regulator to provide a wide range of control with a narrow bandwidth between steps. Its use is limited to those applications where the coarseness of discrete steps is not objectionable and fast response is not a requirement. It is a

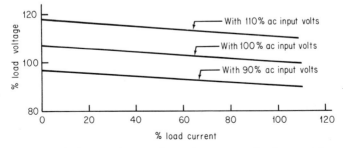

Fig. 16 Typical inherent voltage-regulation characteristic of a silicon-rectifier unit.

mechanical device and therefore requires routine maintenance and periodic inspection for trouble-free operation. Its main use is in the electrochemical field because it can be obtained in large sizes and can therefore be used to control a large group of rectifiers feeding a single load.

b. Induction Regulators. The 3-phase induction regulator has the general construction of a wound-rotor induction motor and consists of a rotor which is excited from the source and a stator which is in series with the source. The output voltage is dependent on the relative position between rotor and stator, and is the vector sum of the rotor and stator voltages. Like the step regulator, the induction regulator acts to control the level of the rectifier regulation curve by changing the voltage applied to the ac winding of the rectifier transformer. The control is stepless, but the speed of response is relatively slow owing to the inertia of the rotor. Because the rotor is a moving piece of equipment, the regulator is subject to mechanical wear and therefore requires routine maintenance. Because of its motorlike construction it is both current- and voltage-limited, seldom being used at voltages higher than 15 kV. Its use is limited to special applications where voltages and currents are moderate, the load is relatively steady, and speed of response is not a factor.

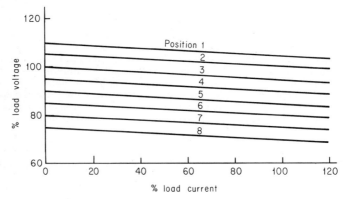

Fig. 17 Typical voltage characteristic of a rectifier unit with step-type voltage regulator or induction regulator control.

c. Saturable Reactors. The saturable reactor is a variable-impedance device which has a dc control winding and a set of ac windings on a common magnetic core. On a single-phase basis the saturable reactor consists of two identical ferromagnetic cores wound with identical ac gate windings and a common dc control winding. It is well known that the impedance of an iron-cored coil can be varied by changing the degree of saturation of the core. In the saturable-reactor device the saturation level of the core, and hence the reactance, is altered by changing the direct current in the control winding. Maximum reactance occurs when the control current is zero, and minimum reactance occurs when the control current is maximum. Stepless control is obtained by modulating the control current.

The reactor is located in the ac supply line to the rectifier bridge on either the primary or secondary side of the transformer and acts to control the dc output voltage by varying the ac circuit reactance (Fig. 18). Note that the control acts to shape the regulation curves, giving them an inherent constant-current characteristic for each specific value of control current. Although the saturable-reactor control can provide a full range of control, it is generally not acceptable because of its associated low power factor, sluggish response time, and high control-current requirements. Because of these adverse characteristics its use in industrial service is normally limited to those applications not demanding high speed of response and requiring only a limited range of control in the order of 10% or less (Fig. 19). Note that an inherent characteristic of this control is that at light load currents, the dc output voltage of the rectifier unit tends to rise sharply to the open-circuit voltage value. In applications where this characteristic is objection-

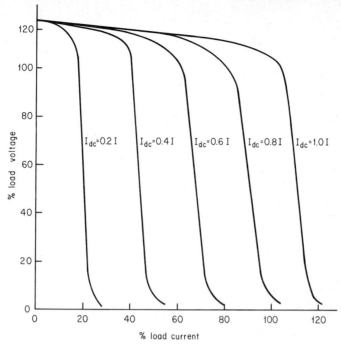

Fig. 18 Typical voltage characteristic of a rectifier unit with full-range saturable-reactor control.

able, a loading resistor may be automatically connected across the dc bus whenever the load current falls below a predetermined value.

The device is completely static and therefore requires no more maintenance than a transformer.

d. Self-saturating Reactors. These devices are basically single-phase saturable reactors applied in the rectifier bridge circuit so that an ac gate winding is in series with the main rectifier diodes in each phase leg (Fig. 20). The circuit arrangement permits load current to flow only unidirectionally through each ac gate winding, hence the term "self-saturating." Actually the entire rectifier unit becomes a magnetic amplifier acting to control the dc output voltage by the gating action of the cores (Fig. 21). This

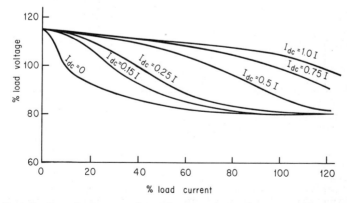

Fig. 19 Typical voltage characteristic of a rectifier unit with partial-range saturable-reactor control.

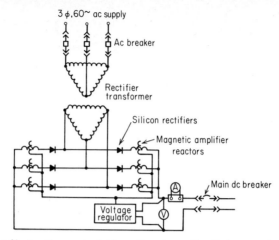

Fig. 20 Self-saturating reactor bridge with main rectifier diodes in each phase leg.

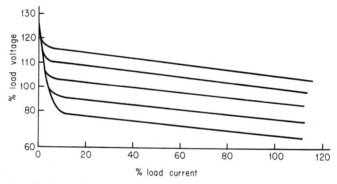

% load current

Fig. 21 Typical voltage characteristic of a rectifier unit with partial-range magnetic amplifier.

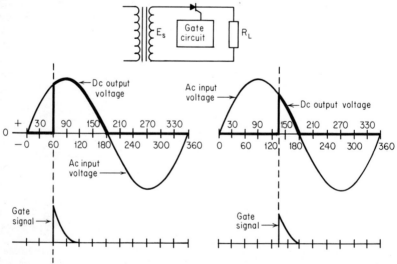

Fig. 22 Phase shift controls dc output.

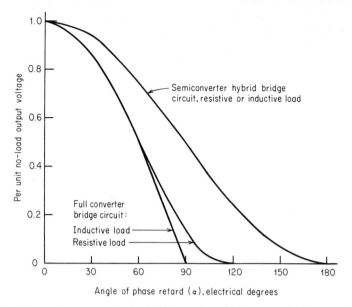

Fig. 23 Output voltage vs. angle of retard for six-phase double-way or double-wye circuits.

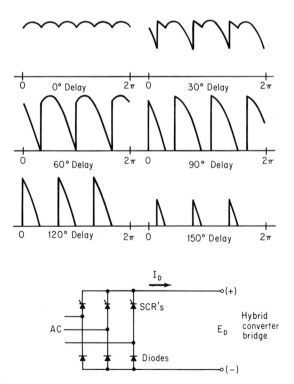

Fig. 24 Three-phase double-way circuit with phase control on the positive side of the bridge. Voltage E_D and current I_D waveforms are shown with resistive load.

voltage-control method has the advantage over the simple saturable reactor of higher speed of response and much lower control power requirements, but it still exhibits the poor power-factor characteristic of the saturable reactor.

  e. Thyristor Control.   The use of the reverse-blocking, forward-conducting triode-type semiconductor device, known as a thyristor or silicon controlled rectifier (SCR), as the rectifying element in a power-rectifier unit permits output-voltage control without the need for external power-type control equipment.   The thyristor device not only blocks reverse voltage (cathode to anode) similar to a conventional diode, but it also has the ability to block forward voltage (anode to cathode) until an appropriate signal

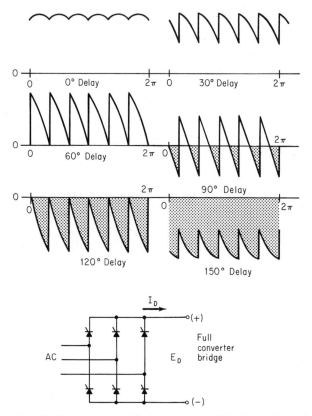

Fig. 25   Three-phase double-way circuit with phase control on both the positive and negative side of the bridge.   Unshaded area above zero line is E_D for a resistive load.   Shaded area and unshaded area on both sides of zero line is E_D for an active load.

is applied to its third element called the "gate."   The thyristor then conducts until the anode voltage becomes negative or its anode current is caused to decrease below its holding current value.   By utilizing a time-controlled signal to the gate with respect to the applied anode voltage to delay the start of conduction in each cycle, the load voltage can be controlled smoothly from maximum available to zero.   This mode of voltage control is known as "phase control" (Fig. 22).   The amount of phase control may be expressed either as the percent reduction in dc voltage obtained from maximum, or by the angle of retard, which is the angle in electrical degrees by which forward conduction is delayed by the control means.   The effect of phase control is to control the level of the rectifier regulation curve.   The dc voltage appearing across the load for a given phase-retard angle depends on the rectifier circuit and the type of load.   Figure

23 shows how the output voltage varies with angle of retard for resistive and inductive loads on six-phase circuits. Figures 24 and 25 illustrate the theoretical voltage-output waveforms produced with the hybrid semiconverter circuit and the full converter circuit. Note that in the full converter circuit with inductive loading, angles of retard beyond 90° permit the power supply to act as a dc to ac inverter.

The advantage of the hybrid circuit employing diodes on one side of the double-way bridge and thyristors on the other side is one of economics, since in general diodes are less expensive than thyristors and only three gate-firing circuits are needed. However, the circuit has the disadvantages that it cannot invert dc power, it has a high 180-Hz ripple voltage, and on high-power equipment it exhibits an inherent commutation-failure problem at high angles of retard. Because of these characteristics its use is basically limited to constant-potential applications requiring partial range control and to adjustable potential applications in the low-power range. Thyristor-controlled power supplies have the advantage over other methods of voltage control of a higher speed of response, high efficiency, and the prospect of extremely low maintenance. They have the disadvantage of the saturable-reactor and magnetic-amplifier power supplies in that the ac supply-system power factor decreases with phase control approximately equal to the per unit voltage reduction caused by phase control. However, because of their advantages, thyristors are used almost exclusively in all adjustable-potential power supplies. Where operation at reduced voltage for extended periods of time is expected, tap-changing equipment on the power-rectifier transformer can be used so that the amount of phase control is always limited by the range between taps.

REFERENCES

1. D. W. Borst, E. J. Diebold, and F. W. Parrish, Voltage Control by Means of Power Thyristors, *IEEE Trans. Ind. Gen. Appl.*, vol. IGA-2, March/April 1966.
2. L. E. Nickels, "Power Control and Conversion," IEEE Education and Lecture Series, Chicago, Ill., May 1964.
3. H. Winograd and John B. Rice, Conversion of Electrical Power, Rectifiers and Other Static Converters, in Donald Fink and J. M. Carroll, "Standard Handbook for Electrical Engineers," 10th ed., chap. 12, McGraw-Hill Book Company, New York, 1968.

Part 3

Batteries

C. H. LEET *

GENERAL

Before an engineer specifies a storage-battery system, it is best for him to review these basic concepts. Better knowledge of battery design and operation can prevent embarrassing and costly consequences.

15. Primary and Secondary Batteries. A battery is an electrochemical device that is a source of dc electricity. Some batteries allow recharging; some do not. Primary battery cells are designed for only one discharge and are not intended to be recharged. A typical example is the carbon-zinc "dry-cell" design that is familiar to users of flashlights and sports lanterns. Primary batteries vary in size from the tiny mercury cells used in electric wristwatches to the "wet-cell" air-depolarized batteries in large sizes. Primary batteries are not discussed in detail in this section.

Secondary batteries are designed for repeated discharging and recharging, or "cycling," without appreciable decrease in capacity per cycle. Secondary batteries are

* Marketing Manager, Exide Lightguard, ESB Incorporated; Registered Professional Engineer (Pa); Senior Member, IEEE.

usually capable of hundreds, even thousands of cycles when properly maintained. This type of battery is normally a "wet-cell" design and is commonly used in automobiles, boats, submarines, fork-lift trucks, railroads, and utility and industrial facilities.   Secondary batteries are often known as electric storage batteries, and they are the subject of this part of this section.

When serviced by the proper charging system and correctly maintained, an electric storage battery is the fastest, most reliable source of dc power available today.   Electronic inverters have increased battery flexibility, allowing the storage battery to serve loads requiring ac power as well.

16. Stationary and Motive Power.   Applications of electric storage batteries are defined as being either stationary or motive-power.   The two types of service require drastically different types of battery and charger designs.

A stationary battery is designed to serve as an auxiliary and/or emergency source of power to the load.   This battery is normally mounted on racks and is continuously charged except for intermittent discharges of varying times and power.

A motive-power battery is designed to serve as the prime power source for the load. This battery is mounted in some form of electric vehicle and commonly has been rated to supply about 6 h of operating power before the discharge phase of the cycle is ended. This discharge is followed by about 8 h of recharging to complete the cycle.   Most motive-power batteries are cycled daily.

The cells of a motive-power battery are the prime source of energy needed to move the electric vehicle.   Battery discharge is more or less continuous whenever the driver operates the vehicle.   There are probably millions of battery-powered electric vehicles in use today throughout the world.   Most are industrial vehicles such as fork-lift trucks, mining and warehouse tractors, and floor sweepers.

17. Lead-Acid and Nickel-Iron-Alkaline Cells.   Regardless of the type of application, all industrial storage batteries have one thing in common.   Each battery is made up of a group of cells electrically connected in series to supply an electrical load with a given voltage.   The voltage rating of the fully charged battery depends on the number of cells and the chemical structure.

Commercially available industrial storage-battery cells are either lead-acid or nickel-alkaline in chemical structure.   Each of these two electrochemical couples offers a variety of plate designs and materials.   But the basic voltage rating of the fully charged lead-acid design is normally a nominal 2 V per cell, while the nominal rating of the fully charged nickel-alkaline design is always 1.2 V per cell.

Obviously a lead-acid battery requires fewer cells than a nickel-alkaline battery to meet the voltage requirements of a particular load.   Voltage specified for the load is obtained by connecting a sufficient number of cells in series.   For example, a 36-V load requires 18 lead-acid cells or 30 nickel-alkaline cells.   Electrical-load specifications normally identify a minimum and maximum voltage range, a factor which must be considered when the battery is specified.

18. Voltage Decline during Discharge.   Battery voltage gradually declines during discharge and should not be permitted to drop below the minimum tolerated by the load plus the line drop.   The use of preset low-voltage detection relays is recommended to protect the battery against harmful overdischarge.

The rate of voltage decline depends on these factors:

1. Amperage or current demand of the load
2. Duration of the discharge
3. Chemical type and design of the cell
4. Number and size of plates in each cell
5. State of charge of the battery at start of discharge
6. Age of the battery cells
7. Temperature of the cells

The point at which the low-voltage relay disconnects the battery is known as the final voltage of the system, which differs from that of a cell.   Final voltage of the system is the lowest voltage accepted by the electrical equipment connected to the battery. Final voltage of the cell is the lowest recommended discharge voltage.   Only a small percentage of the cell's rated capacity can be obtained beyond that value.

19. Battery-Capacity Ratings.   Capacity of the battery is basically its ability to supply a given amperage for a given period of time at a given initial cell temperature while maintaining voltage above a given minimum level. Both stationary and motive-power batteries are rated in ampere-hours (Ah) at a given discharge rate. Stationary batteries are commonly rated for 8-h, 3-h, 1-h, or 1-min discharges but may be rated for any desired time interval, including minutes or seconds. Motive-power batteries are usually rated for 6-h discharges, considered a realistic average of the total actual discharge during an 8-h working shift.

The ampere-hour rating is simply the product of the discharge in amperes multiplied by a given time period. For example, a cell may be rated to deliver a total of 300 Ah at a 3-h discharge rate. Thus, when a load of 100 A is imposed on the fully charged battery, the cells will carry it for 3 h if necessary, provided that the initial temperature of the cells is 77°F (25°C), an industry standard.

But that same cell will not be able to carry a 300-A load for 1 h. A given cell delivers fewer total ampere-hours when the discharge rate is increased, i.e., when more amperes per hour are taken from the cell. On the other hand, the same cell will probably be able to carry a 50-A load longer than 6 h. A given cell size usually delivers more total ampere-hours when the discharge rate is decreased, under similar conditions.

Manufacturers of batteries commonly list several discharge ratings for a given cell size. This rating is given to help engineers make comparisons and select the proper battery sizes. The reference temperature almost universally used is 77°F (25°C). Battery capacity is increased at higher temperatures, but life expectancy is shortened. Conversely, battery capacity is reduced at lower temperatures, while life expectancy is prolonged (unless the electrolyte freezes, which can damage the lead-acid cell components). Care should be taken to keep all cells of a battery at about the same temperature at any given time. Battery location should avoid exposing cells to direct sunlight or mounting cells next to radiators or outlets of heating systems.

20. Battery-Cell Components. There is some similarity between the components of the lead-acid and nickel-alkaline cells. Each uses an element consisting of positive and negative plates sandwiched together alternately and insulated from each other by separators or spacers between each pair of plates (Fig. 26). The element is immersed in electrolyte, a liquid solution held in a container or jar. A cover is sealed to the container or jar. The positive plates of the element are connected to a positive post which usually protrudes through the cover, while the negative plates are connected to a negative post that also protrudes through the cover. When high discharge rates are desired from the cell, four posts may be used—two positives and two negatives.

21. Importance of Battery Watering. Almost all storage-battery cells have a vent well in the cover. The cover allows water to be added to the electrolyte as needed or during scheduled maintenance checks. This procedure is the most important single maintenance factor in a storage-battery installation.

Water is lost from the electrolyte during the charging process. Charging is actually the reversal of electric-current flow through the battery cell, causing a reversal of the electrochemical processes that created power for the load. High charging currents that are acceptable at the start of the charge will gradually cause the water molecules in the electrolyte to break down into hydrogen and oxygen gases, which bubble up through the electrolyte and are vented from the cells as the charge progresses. This phenomenon, called gassing, can be controlled, but not eliminated, by careful downward adjustment of the charging current or voltage. Gassing occurs in both lead-acid and nickel-alkaline cells, but the rate of gassing increases as the ratio of charging amperes to battery ampere-hour capacity increases. For example, 1 A of charge creates less gas than 10 A of charge on a fully charged 100-Ah battery.

Water loss from the electrolyte, if uncorrected, will eventually cause plate damage and lower the cell's rated capacity by exposing the plates of the element to air. The electrolyte should completely cover the element at all times. This requires the periodic addition of water (see Battery Maintenance). No other chemical or additive should be used unless authorized by the battery manufacturer. Municipal water supplies are normally within the impurity limits specified in Table 1, but a periodic water analysis should be requested from the water supplier. When there is doubt about the suitability of the water supply, distilled water should be used.

Since it is impractical to monitor the electrolyte level of a stationary battery continuously, these cells are designed with a substantial reservoir volume of electrolyte above the element. The electrolyte level should be checked on a monthly or semi-

Fig. 26 A typical multitubular lead-acid motive-power battery cell is shown disassembled, illustrating the various component parts. Microporous rubber separators are used between each pair of positive and negative plates. Liquid electrolyte, not shown, is poured into the fully assembled cell before shipment. (*a*) Transparent plastic cell jar. (*b*) Positive plates (multitubular shown in photograph). (*c*) Negative plates. (*d*) Microporous separators. (*e*) Cell cover with identifying nameplate. (*f*) Two seal nuts with rubber ring gaskets. (*g*) Standard vent plug. (*h*) Optional explosion-resistant vent plug. (*i*) Two connector bolts.

annual basis, if convenient. In addition, the stationary battery should be checked after an extended discharge. These precautionary checks are the most effective means of attaining maximum battery life and reliability.

Motive-power batteries lose water much more rapidly than do stationary batteries, since the electric vehicle battery is usually cycled every day. Electrolyte level should be checked when approximately 2 h of the charging process remain, and water should be added to all cells as necessary. Gassing action during the remaining charging process will thoroughly mix the water with the electrolyte, eliminating uneven dilution of the solution.

LEAD-ACID BATTERY DESIGNS

The most popular type of storage battery in use today is the lead-acid family. Well over 95% of modern stationary and motive-power batteries use some kind of lead-acid design.

22. Three Positive-Plate Designs. Three basic designs of positive plates are offered in lead-acid battery cells. These include the pasted or Fauré plate, the multitubular plate, and the Planté plate. Variations of the pasted plate and multitubular plate are used in both stationary and motive-power batteries, but the Planté design is considered suitable only for stationary applications.

a. Pasted plates (Fig. 27), also called flat plates, use a basic construction consisting of a latticework metallic grid with openings filled with lead-oxide paste, which is referred to as the active material. The grid may be cast of either lead-antimony or lead-calcium

Fig. 27   Typical pasted positive plate and pasted negative plate for lead-acid battery cell.

alloy. Lead-antimony is preferred for installations where frequent cycling occurs. Lead-calcium is superior for installations requiring longer intervals between watering maintenance. The calcium alloy is unsuited for motive-power applications where the battery is cycled daily. Frequent cycling of the lead-calcium battery drastically shortens the life expectancy because the positive plates will "grow" in size until they either break up or short-circuit the element.

b. Multitubular plates (Fig. 28) use porous tubes to contain the active material. The lead-antimony grid is basically a row of *spines* extending down from a top bar. A porous tube is slipped over each spine and filled with the powdered lead oxide. A cap bar across the bottom of the plate seals the active material into the tubes. This design offers the highest power density among lead-acid batteries, providing more ampere-hours of capacity per cubic foot of battery volume at moderate rates of discharge.

Fig. 28   Typical multitubular positive plate and pasted negative plate for lead-acid battery cell.

Fig. 29   Manchester type of Planté positive plate use in lead-acid battery cell.

c. Planté plates (Fig. 29) are considered to have the longest life expectancy of all stationary lead-acid storage-battery designs.   Installations that have served for more than 30 years are common.   Basic construction of the Planté positive consists of a formed lead-antimony plate of large area, the active material of which is formed in thin layers at the expense of the lead itself.   Such plates have been (1) cast with complex ridges or grooves, (2) mechanically furrowed to obtain greater surface area, and (3) cast with circular openings in which separate corrugated lead ribbons, rolled into spiral buttons or rosettes, have been inserted as shown in Fig. 29.

23. Negative plates (Fig. 29), regardless of the type of positive-plate design in the cell, are basically built with the pasted-plate design.   Metallic sponge lead is used as the active material.   In multitubular and Planté cells, the negative grid must be a lead-antimony alloy.   In pasted-plate cells, the negative grid normally uses lead-antimony but can be made of a lead-calcium alloy, particularly if the positive-plate grid uses lead-calcium.

24. Grid Alloys.   Antimony or calcium is needed to give structural strength to the soft lead in the grid, which serves as both a physical support and an electrical conductor. Active materials alone have no rigid mechanical form or strength and, particularly the positive materials, are very poor conductors of electricity.   The grid achieves and retains a physical shape and conducts the current to all parts of the material.   The advantages and disadvantages of antimony and calcium are covered in Art. 22.

Some battery manufacturers now offer a selection of sealed maintenance-free lead-acid cells of limited capacity.   Pasted plates with lead-calcium-alloy grids are used in these batteries, which require no watering during the normal projected life expectancy, which may be shorter than that of conventional stationary batteries.   At present, these maintenance-free cells are restricted to small emergency lighting systems.

25. Electrochemical Reactions – Lead-Acid.   The basic electrochemical reactions are similar in all lead-acid cells, despite the variations in plate design (Fig. 30).   Lead peroxide and metallic sponge lead are the active materials on which the electrolyte, a solution of dilute sulfuric acid (H_2SO_4), acts.   When the battery is fully charged, the active material of the positive plate is lead peroxide (PbO_2) and the material of the negative plate is sponge lead (Pb).

As the cell is discharged, the electrolyte (H_2SO_4) divides into H_2 and SO_4.   The H_2 combines with some of the oxygen formed at the positive plate to produce water (H_2O), which reduces the percentage of acid in the electrolyte.   The SO_4 combines with the lead (Pb) of both plates, forming lead sulfate ($PbSO_4$).

When a cell is being charged, this chemical action is reversed.   The lead sulfate ($PbSO_4$) on the positive and negative plates is converted to lead peroxide (PbO_2) and sponge lead (Pb), respectively.   The concentration of acid in the electrolyte increases as the SO_4 from the plate combines with the hydrogen from the water to form H_2SO_4.

The chemical formula is as follows:

$$\underset{\text{Charged}}{PbO_2 + Pb + 2\ H_2SO_4} \rightleftharpoons \underset{\text{Discharged}}{2\ PbSO_4 + 2\ H_2O}$$

In less technical terms, the active material of the positive plates in a fully charged battery is lead peroxide, while that of the negative plates is pure sponge lead. All the acid is in the electrolyte, and the specific gravity is at maximum. As the battery discharges, some of the acid separates from the electrolyte which is in the pores of the plates, forming a chemical combination with the active material. This changes the active material to lead sulfate and produces water. As the discharge continues, more acid is drawn from the electrolyte by the active material, forming more sulfate and water. As this process continues, the specific gravity of the electrolyte will gradually decrease, because the proportion of acid is decreasing while the water content is increasing.

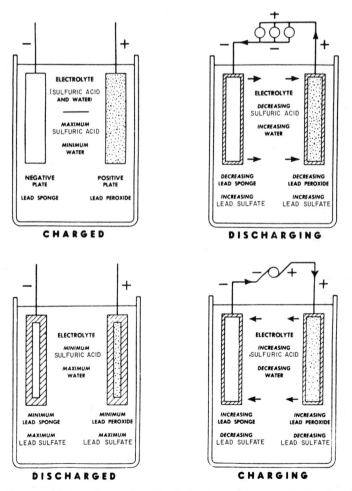

Fig. 30 Electrochemical reactions produced by discharging and charging of lead-acid battery cell.

When a battery is placed on charge, the reverse action takes place. The sulfate (SO_4) in the active material of the plates is driven out and back into the electrolyte by the charging current, changing into acid. Sulfate in the plates is gradually reduced by its return to the electrolyte; the acid produced by the process raises the specific gravity of the electrolyte. The specific gravity will continue to rise until all the sulfate is driven out of the plates and back into the electrolyte as acid. There will then be no sulfate in the plates.

After all the acid is returned to the electrolyte, more charging will not raise the gravity any higher unless the electrolyte level is lowered by water loss caused by overcharging. All the acid in the cells is now in the electrolyte, and the battery is said to be fully charged. The material of the positives is again lead peroxide, the negatives are sponge lead, and the specific gravity is at its maximum reading if the solution level has been properly adjusted. Briefly speaking, on discharge the plates absorb acid from the electrolyte; on charge, the plates return the acid to the electrolyte.

26. Specific Gravity. The volume of acid in the electrolyte of a lead-acid battery· is measured by specific gravity. Since the acid content of the solution decreases lineally during discharge, the specific-gravity reading is a good indication of the lead-acid battery-cell state of charge.

The specific-gravity value recommended for a full charged battery is a matter of design. Selection of the gravity value depends on many factors. On the one hand, the gravity must be at least high enough to supply enough sulfuric acid to meet the minimum chemical requirements needed to achieve lead-acid reaction. On the other hand, too much acid will adversely affect certain parts of the cell. Between these two extremes, other factors such as capacity rating, operating temperature, and battery life expectancy must be considered.

Heavily worked batteries, such as those used in electric lift trucks, are normally supplied with the lowest volume of acid in the electrolyte, but in the highest concentration. Lightly worked batteries, such as those in standby or emergency service, are usually assigned the highest volume of acid, but in lower concentrations.

The most commonly used full-charge specific-gravity values (usually acceptable within a range of ±10 points) are listed below for lead-acid batteries:

1.275 – Motive power batteries for lift trucks, etc.
1.260 – Automotive batteries
1.250 – Partially cycled batteries used for railway-car lighting, engine starting, etc.
1.210 – Stationary batteries for switchgear, controls, emergency lighting, etc.

Higher-specific-gravity electrolyte has a lower electrical resistance which better maintains the terminal voltage of the cell during discharge. Minimum resistance is encountered at 1.216 specific gravity. The degree to which specific gravity affects the cell-capacity rating will vary considerably with different types of designs. A common rule of thumb is that a difference of 25 points in gravity will change the capacity by 8 to 10%. Battery manufacturers have established certain standards of specific gravities for particular applications.

Where battery room temperatures average above 90°F more than 30 days in the year and drop below 40°F only a few days per year, a full-charge specific gravity of 1.170 ±10 points should be considered. Longer battery life will be attained with this low gravity at such "tropical" temperatures.

The choice of full-charge gravity value depends on a compromise between many factors. Some effects of difference in gravity (in varying degree), with other features constant, are:

Higher gravity	Lower gravity
More capacity	Less capacity
Shorter life	Longer life
Less space required	More space required
Higher momentary discharge rates	Lower momentary discharge rates
Less adaptable to "float" charging	More adaptable to "float" charging
More standing loss	Less standing loss

The specific gravity is measured with a hydrometer syringe. However, the reading is meaningless unless combined with a temperature reading of the electrolyte. Both the hydrometer syringe and the thermometer should be specially designed for use with lead-acid batteries. *Tools used with alkaline electrolyte must never be inserted into acid electrolyte, since they will "poison" the acid solution.* The specific-gravity reading is taken first and noted; then the temperature reading is taken. A scale on the battery thermometer will indicate the number of points to be added or subtracted from the specific-gravity reading if the cell temperature is above or below the standard of 77°F (25°C).

To determine the state of charge, the gravity reading is compared with the full-charge value and the published "specific-gravity drop" of a particular cell size at a particular discharge rate. For example, a partially discharged lead-acid cell is checked with a hydrometer syringe that shows a specific-gravity reading of 1.193. The thermometer reading shows 83°F and indicates that two points should be added to the specific-gravity reading to correct it to 77°F. The corrected gravity reading is now 1.195. Assume that this type of battery has an average full-charge gravity of 1.245 and a published gravity drop of 125 points during the 8-h discharge rate. The 1.195 reading is about 50 points below full charge, indicating that the battery is 40% (50/125) discharged.

In addition to temperature, another factor may influence the specific-gravity reading. When the electrolyte level drops below the recommended level in the cell, the volume of acid in the solution decreases (if the fall in level is due to the loss of water during charging). In a certain type of cell, for example, the gravity may rise 15 points with each ½-in drop of electrolyte level. After water is added to the electrolyte and is mixed with it by gassing, the specific gravity will return to its normal value.

Specific-gravity readings should not be attempted immediately after water is added to electrolyte. Acid is heavier than water and will tend to concentrate at the bottom of the cell, where the hydrometer syringe will not be able to reach it. Charging action will create gassing that will mix the water into the electrolyte. Water should therefore be added to electrolyte before the end of charging. Water should not be added to motive-power batteries before the charge is started, since the solution could overflow before the charge is completed.

27. Cell Final Voltage. Although specific-gravity readings are an accepted method of measuring the state of charge of the lead-acid battery, the factor that determines the maximum length of discharge is the cell's final voltage. This term refers to the minimum useful and accepted voltages at various rates of discharge. These are the values at which the maximum number of ampere-hours have been delivered before the cell voltage begins its rapid decline as the point of exhaustion is approached. Final voltages decrease with increasing or higher discharge rates.

Voltage supplied by a battery gradually declines from the start of a discharge and continues to decline as the discharge proceeds. Eventually, the voltage reaches a critical point known as the "knee" of the discharge curve. After this point, the voltage drops at an ever faster rate—unless the load is disconnected from the battery. This point is known as the cell's final voltage and in the lead-acid battery may occur anywhere between 1.75 and 1.50 V per cell, depending on the application and other factors.

Since a repeated uncontrolled discharge could damage the lead-acid battery, it is advisable to disconnect the load at the cell's final voltage. Another factor that may affect the selection of a final voltage is the lowest voltage acceptable to the load.

Lead-acid batteries used in motive-power applications can be protected against overdischarge by a voltage-discharge indicator. This indicator is simply a warning lamp mounted on the electrical vehicle's dashboard. When the battery's voltage drops to a critical level, the warning lamp is activated, signaling the vehicle operator to return to the battery-charger area.

In stationary applications, the lead-acid battery can be protected by low-voltage relays (LVR). While on charge, the battery's cell voltage can be between 2.15 and 2.33, but when a load is imposed on the battery, the cell voltage almost immediately drops to about 2 V. If a low-voltage relay is preset for 2 V per cell and is installed on the battery's circuit, it can be used to activate a visual or audible warning system that will alert the facility staff to an emergency that is draining power from the battery. A second low-voltage relay, preset to 1.75 or 1.70 V per cell, would automatically disconnect the load from the battery.

28. Charging Lead-Acid Batteries. A motive-power lead-acid battery requires a different charging technique from that used with a stationary lead-acid battery. Motive-power batteries are normally charged with either a two-rate or a tapered charging current for about 8 h. A stationary battery is float-charged while on a standby status and is charged with a constant-potential current-limited unit after a discharge.

a. Motive-Power-Battery Charging. The battery charger will recharge the motive-power lead-acid battery within 8 h if the charger's ampere-hour rating is matched to that of the battery. In normal operation of electric-vehicle fleets, the battery-charging area will have several chargers of different ratings. The operator must make certain that the battery is attached to the proper charger before activating the charger.

The operator must check three basic factors, consisting of (1) charger dc voltage vs. battery voltage, (2) charger dc ampere-hour capacity vs. battery ampere-hour rating, and (3) whether the battery and charger are matched for acid or alkaline types. If a charger rated for 18 lead-acid cells is connected to a battery with only 12 cells, internal damage will occur to the battery. If a charger rated for 12 lead-acid cells is connected to an 18-cell battery, internal damage may occur to the charger, and the battery will not charge. And if a high-ampere-hour battery is connected to a low-ampere-hour charger, a substantially longer charging time will be required. When a low-ampere-hour battery is connected to a high-ampere-hour charger, the charging process will be cut short, and the battery will be undercharged.

A lead-acid battery that has been partially or completely discharged within its ampere-hour capacity rating can safely absorb high-ampere charging currents at the start of the charge, which may last as long as 4 or 5 h, depending on the current and the depth of discharge. But the *finishing* rate during the last 3 or 4 h of the charge must be much lower, perhaps only one-fifth as much current as the start of the charge.

A total of about 8 h is economically required to recharge completely a lead-acid battery that has been discharged to its full rated capacity. If charger cost was not a factor, the battery could be recharged in as little as 3 h and 20 min. A two-rate charge (considered obsolete) feeds high current into the battery for about 4 h, then automatically drops the current to a low level for the remaining 4 h. A tapered charge feeds a moderately high current into the battery for about 5 h and then curves down to the low final rate.

Most modern motive-power chargers, properly matched to the battery by voltage, number of cells, and ampere-hour rating, can recharge about 80% of the discharge ampere-hours in only 4 h. But the battery is then only partially recharged and should not regularly be returned to duty until it has been completely recharged. The final charging rate is more important to the charging process than the high starting rate. The charger usually has to feed the lead-acid battery about 10% more ampere-hours than the battery discharged. The 10% additional charge is known as "overcharge."

When properly cycled and maintained, a motive-power lead-acid battery will achieve a normal life of about 6 years, if cycled daily, 5 days per week. Well-maintained batteries frequently exceed this life expectancy, but improper care and abuse can substantially shorten service life.

b. Stationary Battery Charging. Charging of a stationary lead-acid battery is normally done by a continuous *float* charge until the battery experiences a discharge. Then the charging voltage is raised to a higher value, which depends on the recharge time specified for the installation, increasing and sustaining a higher charge rate in amperes. The cost of the charger is directly related to its high-current capacity. As the capacity is increased in the specifications, the recharge time is shortened—and the charger's cost is increased.

In a typical stationary battery installation, the charger and battery are located in a building, near the load. The charger may be either a motor-generator set or a static rectifier, which is preferred. Its function is to change alternating current into direct current suitable for charging the battery and maintaining a constant voltage throughout its load range.

If a stationary battery had no charger connected to it, low-level electrochemical reactions taking place inside the cells would slowly drain the discharge capacity of the battery, causing "standing loss" or "self-discharge." To prevent this discharge, the charger maintains a float charge that continuously monitors and corrects these internal losses of the battery.

In the majority of float-charge installations, each cell of the battery has a "low" gravity electrolyte, specified as 1.210 ± 10 points at a full-charge specific gravity. Under these conditions, the charger and battery are usually operated at a voltage of 2.15 V per cell, or 120 V for a typical 60-cell lead-acid battery with lead-antimony grids.

Lead-acid battery cells equipped with lead-calcium alloy grids require a slightly higher float voltage, usually 2.17 to 2.25 V per cell, depending on the application. The reason for the higher floating voltage of the calcium alloy cells is that it requires more float voltage to restore current which was taken from the battery by its occasional use.

Current drawn by the fully charged floating battery with lead-antimony grids and 1.210 nominal specific gravity will vary somewhat with the battery's age and condition. It will be in the general range of 50 to 100 mA (0.05 to 0.10 A) per 100 Ah of the battery rated capacity at the 8-h discharge rate. For fully charged calcium-alloy batteries, the float current will be 10 to 20% of that needed by a lead-antimony battery of the same capacity.

Temperature variations from the standard 77°F (25°C) will also affect the battery current demand. Current drawn by a fully charged lead-acid battery will approximately double for each 15°F rise in cell temperature, and be reduced to about one-half for a 15°F temperature decrease from the standard. Most chargers automatically compensate for variations in float-current demands if the charger is preset for the correct float voltage.

Still another factor in the floating battery's current demand is the floating voltage from the charger. Under otherwise similar conditions, current input to a battery increases or decreases with a rise or fall in charger or bus voltage. This current change will be considerable, even with relatively minor voltage changes, depending on rate of voltage change. Close control of floating voltage is therefore necessary. For lead-acid batteries, the current drawn will approximately double for each increase of 0.06 V per cell (2.15 to 2.21, etc.) and be reduced to one-half by a similar decrease. When the speed of voltage change is measured in fractions of a second, the current demand will be five or even ten times as high as that imposed by a slow voltage change. The charger should be able to provide the floating voltage with a variation of no more than ±1% throughout its ampere rating.

A voltmeter is essential for on-the-spot indication of the floating voltage across the battery. It should be easy to read, and its accuracy should be verified once each year by comparison with another meter. Voltmeter indication is the quickest way to detect too high float voltage; another method is to notice excessive watering requirement, but water usage would take many weeks to be disclosed. Too high float voltage shortens the life expectancy of the battery.

Too little float voltage is indicated by the specific-gravity reading, which would be below the full gravity range. This condition is undesirable because it decreases the capacity available from the battery in the event of an emergency and may permanently damage the negative plates.

c. Equalizing Charge. All lead-acid batteries used for motive power should be given periodic equalizing charges. These are actually only a prolonged charge at a higher voltage per cell, designed to correct small variations among the cells of the battery. Motive-power batteries are given an equalizing charge once a week, usually on weekends, for about 3 to 4 h longer than a normal charge. The extra charge is a continuation of the finishing rate. If it is known that the lead-acid battery was not fully charged at the start of the equalizing charge, the charge should be extended until the lowest-gravity cell is restored to within the full-charge gravity range.

29. Motive-Power Lead-Acid Batteries. These batteries consist of a group of cells in a steel tray. The cell "jars" are made of high-impact rubber compound or plastic. Many different voltages are available in motive-power lead-acid batteries, ranging from only 6 V (three cells) for small materials-handling equipment to 240 V (120 cells) for large mine tractors. The most popular sizes are 12 V (six cells), 24 V (12 cells), 36 V (18 cells), and 48 V (24 cells).

Smaller (12-V, 6-cell) batteries are usually housed in a steel tray with a hinged cover. Larger (rider-type) batteries are normally assembled in an open-top steel tray. All trays have holes or lifting tabs that should be used as attachment points when the battery is lifted from the electrical vehicle. Some vehicles also allow smaller batteries to be slid from the battery compartment using rollers.

Connection of the battery to the electrical vehicle and the charger is by means of a plug or receptacle on the ends of electrical cables leading from the battery's main negative and positive terminals. A matching plug or receptacle must be provided on the vehicle and on the end of the charger's dc cables. The length of the cables attached to the battery terminals can be specified on the purchase order but should not be excessively long.

An almost limitless variety of capacity ratings are available in motive-power lead-acid batteries. The physical size of a battery does not necessarily indicate its capacity. The size and number of plates and the type of positive plate inside each battery cell are the governing factors—plus, of course, the amperage of the electrical load.

Selection of the proper ampere-hour rating should be based on a load analysis plus a 20% safety factor. Naturally, the specified battery must also match the space limitations of the electrical vehicle battery compartment.

Analyzing the load to be imposed on the lead-acid battery requires that the following questions be answered:

1. How many round trips per shift will the vehicle be required to make?
2. How many feet does the vehicle travel in each round trip?
3. What is the average weight of the vehicle load?
4. What distances will the load be carried?
5. How many ramps will the vehicle have to climb in each round trip?
6. How long and how high are the ramps?
7. If the vehicle is a lifting type, how high and how often will lifts be made?
8. What are the electrical specifications of the vehicle and its attachments?

The load-analysis data should be provided to the battery manufacturer, whose representatives will be able to match them to the proper battery. If temperatures are a problem, such as outdoor operation in tropical or arctic climates or indoor operation in freezers or foundries, the thermal loads should be identified.

At the same time, it is advisable to coordinate the battery purchased with the purchase of new chargers matched to the battery type and capacity rating. A new battery can easily be ruined by obsolete, inaccurate charging equipment. A motive-power battery charger should be capable of automatically changing its rate of charge to match the battery state of charge, and should automatically terminate its charge before excessive gassing occurs.

30. Stationary Lead-Acid Batteries. These batteries are usually supplied as individual cells in transparent plastic jars. A rack assembly is normally required in order to properly mount the cells. The cells are connected in series on the jobsite by a contractor, using intercell, intertier, and interrack connectors normally supplied by the battery manufacturer. The most popular lead-acid battery sizes are 24-V (12-cell), 48-V (24-cell), 120-V (60-cell), and 240-V (120-cell) systems. Occasionally, some other number of cells may be specified, usually for voltage control on charge for one of the above-listed systems.

a. Battery Racks. Most battery manufacturers are prepared to supply a wide variety of steel-rack configurations for stationary lead-acid battery systems. Typically, these include a single-row one-tier, a single-row two-tier, a two-step, and a three-step rack. In areas where earthquake tremors or other anticipated disturbances could cause cell jars to slip from the racks, optional seismic-shock-protected racks should be specified. Plastic channels are available to protect the steel rails against corrosion and to provide electrical insulation.

Lead-acid storage batteries are inherently heavy in weight. This weight factor increases with larger cell sizes. Individual cell weights may range as high as 400 lb or more, which means that a 120-cell lead-acid battery could weigh more than 24 tons plus rack and charger. Floor construction must be capable of supporting the battery and rack weight.

A stationary battery may range from a voltage of only 6 V (a three-cell emergency-lighting unit battery) to 240 V (120 cells) or more. Cell containers may vary in size from smaller than a cigar box to as large as a shower stall. Selection of the proper cell is usually too complex for an engineer to undertake without assistance from manufacturers familiar with their many cell dimensions and capacities.

b. Plotting the Load. Before the battery manufacturer can recommend a particular cell, the project engineer *must* prepare a "discharge profile" indicating the duty cycle of the battery in the event of a power failure. Most stationary batteries supply power to several loads during an emergency. The *discharge profile* should indicate (1) the amperage, (2) the duration, and (3) the sequence of each load. Accompanying data must include the minimum sustained operating temperature and the minimum allowable dc system voltage.

Total load imposed on the battery should be divided into three classes:

1. Momentary currents required for the operation of connected equipment, particularly the closing and tripping of switchgear.

2. Steady or continuous load of indicating lamps, holding coils for relays, supervisory control equipment, and other control devices.

3. Emergency light and power load.

As an example, a battery may be selected to furnish a continuous load for several hours, after which it must have sufficient capacity for circuit-breaker operation.

If the battery is expected to carry some other emergency load of shorter duration, the capacity rating required will depend on whether the additional load occurs earlier or later in the emergency period. Peak capacities can be minimized by sequencing additional loads whenever possible. Two 50-A loads that follow each other count as one 50-A load, but two 50-A loads that occur simultaneously must be counted as a 100-A load.

When determining the maximum momentary demand for circuit-breaker operation, it is usually planned to close no more than one circuit breaker at a time, although there may be an occasional exception. The tripping load per breaker is normally much less than the closing load. However, automatic protective devices may trip an entire bank of circuit breakers, which may result in a momentary demand larger than that required to close one circuit breaker.

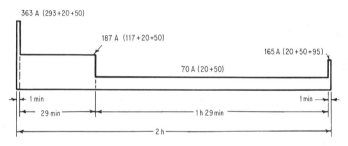

Fig. 31 Sample discharge profile for the duty cycle described in the text.

A simple discharge profile would show a 3-h emergency period, with a 20-A emergency-lighting load imposed throughout. In addition, the specifications call for the closing of a circuit breaker rated at 100 A at any time during the 3-h period. The discharge profile should always show the random load imposed at or near the end of the discharge period, since this load requires the greatest battery capacity.

A more complex discharge profile would show a 2-h emergency period with five distinct loads:

1. Eleven circuit breakers, each drawing 6 A, tripped simultaneously for a total of 66 A

2. A continuous load of 20 A throughout the emergency

3. An emergency-lighting load of 50 A throughout the emergency

4. Startup and operation of an emergency-power electric motor, requiring 293 A inrush and 117 A running current for a 30-min period

5. Closing 11 circuit breakers one at a time, each requiring 95 A

Loads 2 and 3 impose a base load of 70 A throughout the emergency. The first additional load on the profile would normally be the 66 A required for tripping current. But this load is of such short duration (1 s or less) and is so closely followed by the high inrush motor current of 293 A, the profile would have to show a momentary demand of 359 A imposed on top of the 70-A base load. Since the 66-A and 293-A loads follow each other, only the inrush motor current and the base demand of loads 2 and 3 are shown in the first minute of the load cycle (Fig. 31).

When the stationary battery is used with an inverter, the battery must be sized to handle the load profile imposed on the inverter plus the electrical load of the inverter itself. Static-inverter systems as small as 1 kVA are in use with battery systems. Other static-inverter systems, made up of individual inverter units usually connected in parallel, may total 1000 kVA or more.   A battery serving an inverter should not serve any other loads.

This section is not intended to review the fine points of matching an inverter to a load. But when sizing the stationary battery within a static uninterruptible ac power supply (UPS) system, the engineer should remember that the inverter is part of the load and must be considered a steady load on which the fluctuations of an ac load, such as a computer or a hospital intensive-care unit, are added (Secs. 4 and 7).

While manufacturers of industrial motors and other dc equipment are accustomed to defining inrush-current demands, manufacturers of other equipment such as electronic systems have only limited data available on the inrush characteristics of such components.   Engineers should carefully evaluate any load intended for a stationary battery, making certain that current demands are accurately defined for all stages of operation.

Certain types of loads are not compatible with other loads.   For example, electrical "noise" generated by some loads could interfere with other loads.   The engineer must exercise caution when assigning battery loads.   Sometimes two or more battery systems may be advisable.

c. Unscheduled Current Demands.   Discharge capacity of a stationary lead-acid battery is usually a carefully calculated factor.   In a properly designed system, every load connected to the battery is completely defined in terms of its inrush and operating amperage, its duration, and its sequence position in the discharge profile.   Unscheduled loads should be avoided if possible.   When this procedure is impracticable (for example, if dc wall outlets are needed for special equipment that may or may not be operating when an emergency occurs), the engineer must calculate the load on the high side.

Surplus capacity may add to the costs of installation, but inadequate capacity will cause problems during emergencies.   The battery that has an unexpected demand imposed on it may not be able to meet the scheduled loads later in the discharge cycle.

d. Matching Load to Cell Type.   Engineers unfamiliar with the fine points of lead-acid cell designs should not attempt to specify the cell type and size for a particular load until they have consulted with technical representatives of the battery manufacturer.   Once the engineer has defined the load in every detail, including individual amperages and load sequences, the manufacturer's staff can match the load to the proper cell.

Factors involved in matching load to cell include:

1. Type of positive plate design
2. Dimensions of positive and negative plates
3. Number of positive plates per cell
4. Specific gravity of the electrolyte
5. Number of posts on each cell
6. Design of the charger serving the battery

Manufacturers can approach the specification of a battery with one of two methods, both involving calculating formulas.   One method gives an answer in terms of the number of positive plates required; the second (and most modern) method provides an answer directly in ampere-hours.

The positive-plate method uses the principle which states that the ampere capacity of a battery for a given time is always directly proportional to the number of positive plates contained in the cell (Ref. 1).   Another method (Ref. 2) analyzes the ampere-hours being used up to the end of each step in the duty cycle.   The discharge-rate capacity required to provide these ampere-hours at the end of each step is calculated using the relative capacity factor for the relative time.   In this formula, the relative time for any given step of the discharge profile is the product of the accumulated ampere-hours to the end of that step divided by the current for that step.

Replacing the customary 8-h ampere-hour rating of stationary batteries with a 3-h reference rating has been proposed (Ref. 1).   The 8-h rating of batteries as a basis for switchgear and control-bus operation was established many years ago, when 8 h of protection was regarded as necessary to allow the facility's staff to diagnose the problem, gather forces, and get the station back in operation.   Battery sizes steadily increased as time went by, owing to the growth of power demands that must be provided during an ac outage.   In most applications, it was found that 3 h of protection time was all that was usually required, and it is now general practice to size a switchgear battery for a 3-h duty cycle.   A battery sized for 3 h will be approximately half the capacity rating of a battery sized for 8 h.

31. Installing Stationary Batteries.   Specifying engineers should be aware of problems that sometimes develop during the installation of stationary-battery systems. These problems occasionally arise because of vague drawings, oral instructions, or last-minute changes not shown in the drawings provided to the installing contractor. They can also be caused by contractors who allow sloppy craftsmanship during installation, or who make on-site changes without checking with the specifying engineer.

Such problems can be resolved at the start if the engineer arranges to be on hand during the installation and knows what should be avoided.

For example, the battery's main positive and negative terminals require special care. Lead-acid batteries use lead posts which sometimes contain copper inserts for better conductivity.   If the cable conductors to the charger and the load are improperly mounted, they will impose a physical load on the soft posts which can either damage the posts or eventually cause a fatigue rupture of the cell cover.   Conductors connected to the terminals should be supported by some other physical system so that minimum weights and torques are applied to the posts.

The engineer's plans should indicate exactly how many conductors will be connected to the battery and what the size of the cables will be.   When these data are properly prepared, appropriate connectors can be ordered from the battery manufacturer at the time of the battery order.   Intercell connectors should be long enough to permit about 1/2-in spacing between cell jars.

The battery should be located as close to the load as possible to avoid excessive voltage drop along the conductors.   Positive and negative conductors between the battery and any individual system component should be run through the same conduit.   Engineers should consult equipment manufacturers to establish sensitivity of hardware to the variables of cable runs.

As mentioned earlier in this text, individual cell weights can be as heavy as 400 lb or more.   Adequate lifting equipment must be used at the site to prevent accidents. Any cell that is dropped must be considered to be damaged, even if no visible damage is indicated.   A battery manufacturer's representative should be called in to inspect the cell.   Such accidents can cause extensive delays in installation schedules and should therefore be avoided at all costs.

All cells in the battery should be numbered, with labels either on the jars or on the racks.   A record book should be supplied for the installation, and personnel should be assigned to take periodic cell-voltage readings and check electrolyte levels in selected cells.   Ideally, the procedure should alternate the cell inspections so that each cell is inspected once or twice a year, while the battery is checked once a month.   Some battery manufacturers will perform this maintenance service for a fee, relieving the facility personnel from an unfamiliar task.

A plastic water jug should be provided for adding water to the electrolyte.   Other necessary tools are a hydrometer syringe marked for lead-acid electrolyte, plus a voltmeter for periodic cell readings and a battery thermometer.

A ventilation system should be provided to prevent buildup of hydrogen inside the battery room.   When the battery is mounted in an area in which the air-exchange rate to the outdoors is normally once an hour or more frequent, no auxiliary air exhaust is necessary.   Explosion-resistant vent plugs are advisable as a means of preventing external flame or sparks from igniting the gases concentrated inside the cell, between the cover and the electrolyte.   *Smoking and open flames should always be prohibited in any storage-battery room.*

The configuration of the battery rack must be clearly specified or shown on the drawings, including the number of cells per row, the number of rows, the location of the main positive and negative terminals, and the location of the charger.   The engineer designing the installation must keep in mind the need for free access to all cells, necessary to allow maintenance personnel to add water periodically to the electrolyte of each cell.

Temperature controls for the battery room should be set at 77°F (25°C) if a thermostat is used.   All cells of the battery must be kept at the same temperature.   Battery cells should not be installed next to hot water or steam radiators, heating ducts, or in the path of hot air or radiant-heat energy from a heating system.   *Battery cells should never be installed where sunlight can fall on them.* Clear glass windows in the battery room should be painted to prevent sunlight from entering the room.

The battery charger should never be mounted against a wall unless all its components can be serviced through the front panel. If the charger is located near a wall, the designer should allow enough space to permit the front and rear panels of the charger to be fully opened for maintenance work. The same factor applies to the static-inverter cabinets of such systems.

NICKEL-ALKALINE BATTERY DESIGNS

Nickel-alkaline batteries were invented about 1900, some years after lead-acid batteries were commercially introduced. But the relatively high cost per ampere-hour of capacity, plus the greater space requirements, have limited the appeal of the nickel-alkaline designs. However, their superior reliability, long life expectancy, and extreme sturdiness have contributed to an increasing popularity in recent years.

32. Two Basic Nickel-Alkaline Types. In stationary power systems, design engineers can specify one of the nickel-cadmium alkaline battery designs. In motive-power systems, the nickel-iron alkaline battery is sometimes preferred over the lead-acid battery.

a. Nickel-cadmium alkaline batteries are supplied in two basic designs, pocket-type plates and sintered plates. Commercial nickel-cadmium batteries in the United States are commonly the pocket-plate type.

Pocket-plate nickel-cadmium alkaline batteries are available in three basic designs built to meet the varying discharge rates of stationary battery systems, including high rate, medium or general-purpose rate, and low rate. High rate is defined as service for discharges shorter than 1 h. Medium rate is designed to feed loads shorter than 4 h. Low rate will carry loads for up to 20 h. Naturally there is overlapping between the three ratings, so selection must be based on a number of pertinent factors as well as the discharge rate.

Discharge rates for nickel-cadmium alkaline batteries are different for the various kinds of duty cycles, which are usually defined as switchgear rates, emergency-lighting rates, and engine-starting rates. The 5-s rate can safely be used when the duty cycle consists primarily of switchgear tripping and closing circuits. The 1-s rate to 0.65 V per cell for engine breakaway and the 30-s rate to 0.85 V per cell for engine rolling are generally used for engine-starting duty. Emergency-lighting duty is normally defined as a constant-current discharge, with the discharge rate based on the maximum length of discharge desired. If the length of recharge time specified is less than 12 h following a maximum discharge, the emergency-lighting battery may have to be oversized.

Small-capacity nickel-cadmium alkaline battery cells are usually encased in plastic cell containers, while large-capacity cells are housed in steel cell containers. The dividing point is about 150 Ah, depending on the discharge rate and cell design. Rates vary from as low as 5 Ah to as high as 550 Ah, depending on the plate size and the number of plates. Only a few manufacturers supply nickel-cadmium alkaline batteries, but there are differences in plate designs, which are explained in literature available from the manufacturers.

Basic construction of the positive and negative plates of the nickel-cadmium alkaline battery is similar, both being pocket types. The pockets are formed from thin strips of nickel-plated steel, finely perforated. The active material is sandwiched between two perforated strips, which are then crimped together to form a long, narrow pocket. The crimps of the pockets are locked to one another to form a chain of individual pockets, as many as are needed to produce the required plate height and width. This assembly is called a plaque.

A U frame is placed around the edges of the plaque and pressed into place. At the same time, insulator-pin grooves are pressed into the face of the now completed plate.

Negative plates are attached to their terminal pole by a steel bolt secured by a locknut; steel washers assure proper spacing of the plates. The positive plates are attached to their terminal pole in the same manner. The groups of positive and negative plates are then intermeshed and pin insulators put in place, forming an element. After the element is placed in either a plastic or steel cell container, the cell cover—with vent cap and appropriate holes for terminal poles—is installed, completing the cell assembly.

Cell containers are available in a variety of sizes to allow specification of 2- or 4-inch electrolyte height above the plate tops. When the higher electrolyte level above the cell tops is specified, the watering frequency for a given application can be prolonged. Watering intervals can be as long as 5 years in some nickel-cadmium alkaline applications.

b. Nickel-iron alkaline batteries are designed with tubular positive plates and pocket-type negative plates. Each positive plate consists of several perforated steel tubes, filled with alternated layers of nickel hydrate and nickel flake. The tubes are welded together, then welded to a tab or headpiece across the tops of the tubes.

The negative plate of the nickel-iron alkaline battery resembles the construction of nickel-cadmium alkaline plates. Each nickel-iron negative plate consists of rows of interlocked pockets of perforated steel that encase the powdered active material. However, the active material of the nickel-iron battery plates is different from that of the nickel-cadmium battery.

Nickel-iron alkaline-battery cell elements are always installed in steel cell containers. These cells are then assembled into a battery tray, since nickel-iron alkaline batteries are intended primarily for motive-power applications. The tray consists of hardwood

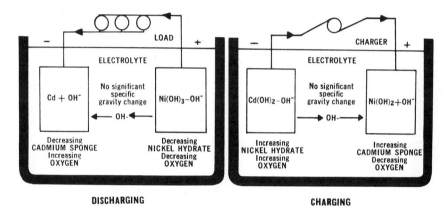

Fig. 32   Chemical action in a nickel-cadmium alkaline-battery cell.

slats into which hard-rubber buttons have been recessed. Steel suspension bosses spot-welded to the sides of each cell container fit the rubber buttons, holding the cells to the sides of the tray.

Unlike the lead-acid motive-power battery, the nickel-iron battery can be used for awhile, then placed in indefinite storage after a few easy steps have been taken. The battery can later be reactivated after months or years of storage.

33. Electrochemical Reactions – Nickel-Cadmium. The active material of the positive plate is nickel hydrate, with graphite added to help current conductivity. The negative plate's active material is cadmium sponge with additives to help conductivity. The electrolyte is a solution of potassium hydroxide (KOH) diluted in water with a normal specific gravity of 1.160 to 1.190 (depending on the type of cell) at 77°F. A small amount of lithium hydroxide is usually included to improve capacity.

When the nickel-cadmium alkaline battery is fully charged, the nickel hydrate of the positive plate is highly oxidized, while the active material of the negative plate is metallic cadmium sponge. During discharge, oxygen ions are transferred from one set of plates to the other, the electrolyte acting as transfer agent for the oxygen. When the battery is discharged, the positive-plate active material is reduced to a lower oxide, while the metallic cadmium sponge in the negative plate is oxidized.

The chemical formula of the nickel-cadmium alkaline battery is as follows (Fig. 32):

$$\underset{\text{Charged}}{2\ \text{Ni(OH)}_3 + \text{Cd}} \rightleftharpoons \underset{\text{Discharged}}{2\ \text{Ni(OH)}_2 + \text{Cd(OH)}_2}$$

34. Electrochemical Reactions – Nickel-Iron. The active material of the positive plate is nickel hydrate, with layers of nickel flake alternated in the perforated steel tubes to help electrical conductivity. The nickel hydrate changes to an oxide of nickel after the formation treatment, which consists of several cycles of charge and discharge after the cell is assembled. Formation is needed to stabilize the nickel-iron cell's electrochemical characteristics, and is conducted at the factory. Active material of the nickel-iron cell negative plates is finely divided iron oxide, contained in pockets of perforated steel strip. The electrolyte is a solution of potassium hydroxide (KOH) in water with a normal specific gravity of 1.210 to 1.215 at 77°F. A small amount of lithium hydroxide is usually included to increase the cell's capacity and life expectancy.

When the nickel-iron alkaline cell is fully charged, the positive active material is essentially nickel oxyhydrate (NiOOH), and the negative material is metallic iron sponge (Fe), while the electrolyte holds potassium hydroxide (KOH) and water with the addition of lithium hydroxide. Although certain intermediate transitory compounds

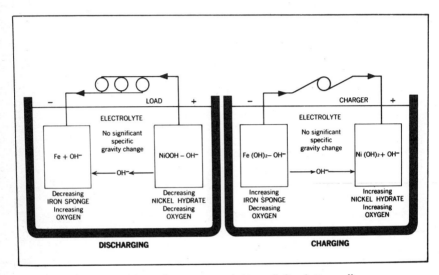

Fig. 33 Chemical action in a nickel-iron alkaline-battery cell.

are produced during discharge, the end result is a transfer of oxygen from the positive to the negative. In the discharged nickel-iron battery, the active material of the positive is nickel hydroxide $Ni(OH)_2$ and the negative is iron hydroxide, $Fe(OH)_2$. Ignoring the intermediate reactions, the end formula for the nickel-iron alkaline cell is (Fig. 33)

$$\overset{Charged}{2\ NiOOH + H_2O + Fe} \rightleftharpoons \overset{Discharged}{2\ Ni(OH)_2 + Fe(OH)_2}$$

35. Specific Gravity. The volume of potassium hydroxide (KOH) in the electrolyte of a nickel-alkaline battery is measured by specific gravity. None of the constituents of alkaline electrolyte combine with the active material of the plates during charge or discharge. Ions pass from positive to negative plates and back again without affecting the chemical structure of the alkaline electrolyte. Therefore, the specific gravity of the alkaline electrolyte cannot be used as an indication of the state of charge of a nickel-alkaline battery.

The only exception to the above statement is when the nickel-iron alkaline battery is discharged beyond its normal limits, down to or near zero voltage, as when the cells are being prepared for storage. This procedure drives much of the lithium from the plates into the electrolyte, causing the specific gravity to rise 25 to 35 points (0.025

to 0.035). This action is reversed when the cells are charged again at the time the battery is reactivated.

However, specific-gravity readings will vary from the normal rating when the electrolyte temperature is higher or lower than 77°F, when the solution level drops below the correct level within the cell, or when the battery grows older.

The specific-gravity reading is taken with a hydrometer syringe designed for use with alkaline electrolytes. *An alkaline hydrometer syringe should never be used with acid electrolyte, and vice versa.* The specific-gravity reading is meaningless unless combined with a temperature reading of the electrolyte and a note as to the state of the electrolyte level within the cell. An example of a properly noted reading of alkaline specific gravity from a nickel-cadmium cell is

Specific gravity 1.180 at 107°F, solution level normal.

This reading indicates a temperature-corrected gravity of 1.190 at 77°F, with no correction for solution level necessary.

The temperature correction for nickel-cadmium and nickel-iron alkaline specific-gravity readings is about 1 point (0.001) of gravity for every 3°F above or below 77°F. Add points for higher temperatures; subtract points for lower temperatures.

The specific-gravity correction for variations in nickel-cadmium solution level is different among various manufacturers. Pertinent literature of the manufacturer should be consulted. The need for correction is due to water losses that result in concentration of the potassium hydroxide with the electrolyte, raising the specific gravity above normal. Adding water will dilute the electrolyte, lowering its specific gravity while raising the solution level. However, hydrometer syringe readings should not be taken immediately after water is added, since the water will not immediately mix with the rest of the electrolyte.

Variations in the electrolyte level of nickel-iron cells will also affect specific-gravity readings.

However, these readings can be corrected to indicate the proper state of the specific gravity by subtracting 5 points of gravity for every ¼ in that the level is below its normal position, or adding 5 points for every ¼ in above normal.

For example, a nickel-iron alkaline-battery pilot cell produces a reading of 1.220 specific gravity at 89°F and ½-in low level; 89°F minus 77°F equals 12°F, which divided by 3 equals 4 points to be added for temperature correction. The ½-in low level equals 10 points to be subtracted. The net result is that 6 points should be subtracted from the 1.220 reading, producing a corrected reading of 1.214 specific gravity at 77°F and proper electrolyte level within the cell.

Temperature of the electrolyte in alkaline batteries should be measured with a thermometer designed for use with alkaline electrolyte. *Thermometers for lead-acid batteries should never be used with alkaline electrolytes, and vice versa.*

Naturally, any alkaline electrolyte solution drawn into a hydrometer syringe for the measurement of specific gravity should be returned to the cell from which it was drawn. *The alkaline electrolytes are extremely caustic, and care should be taken to avoid contact with the skin or membrane areas.*

Specific gravity of alkaline electrolyte will gradually decrease over the years, owing to loss from spray, evaporation, and the tendency to carbonate in contact with air. If the nickel-cadmium battery's specific gravity declines to 1.130, renewal of the electrolyte is necessary. When the nickel-iron battery's gravity falls to about 1.160, the capacity and operation of the battery are adversely affected. The battery can usually be restored to its proper performance rating by renewing the entire electrolyte solution in all the battery's cells. This procedure may be necessary several times during the life of the nickel-iron alkaline battery.

Solution level within the steel containers of a nickel-cadmium or nickel-iron alkaline battery can be checked by using a plastic test tube supplied by the manufacturer. Both ends of the tube are open; the operator closes the top opening with a fingertip and lowers the tube into the battery until the tube contacts the plate tops. Keeping the finger tightly pressed over the top opening, the operator removes the tube from the cell, carrying the electrolyte within it. After noting the level of the solution within the tube and its relation to the markings on the tube, the operator returns the electrolyte to the cell by inserting the tube into the vent well and taking his fingertip off the top end.

Solution level within plastic cell containers of nickel-cadmium batteries can usually be checked visually, since the electrolyte is seen as a shadow on the plastic. The container has the proper level of electrolyte indicated by markings imprinted on the plastic cell container.

36. Cell Final Voltage. As the cells of any storage battery continue to supply power to a load, the voltage produced by each cell declines. Eventually the voltage reaches a point at which it is no longer useful to the load, or the battery cells have delivered as much power as they can before a rapid decline of cell voltage brings the cell to exhaustion. Final voltage is the minimum useful and accepted voltage at the specified rate of discharge. The cell final-voltage rating decreases with increasing or higher discharge rates.

At normal discharge rates, a nickel-cadmium alkaline battery's cell final voltage is normally 1.14 V per cell. But at high current levels, a final voltage of 0.85 or even 0.65 V per cell may be specified. Ampere-hour capacity ratings decrease for a given cell size as the current rates are increased. This decrease does not represent an accompanying loss of capacity; it means only that the final voltage is reached in less time.

Many stationary battery applications involve a series of loads that are imposed on the battery in steps. If the first load requires a high rate of discharge, and each of the succeeding loads is rated for a lower and still lower rate of discharge, the nickel-cadmium alkaline battery offers an interesting advantage. The cell voltage will recover if a high rate of discharge is followed by a lower rate of discharge. The process of lowering the discharge rate and getting an accompanying rise in voltage can be repeated several times. Where this practice is feasible, the design engineer can plan on getting almost full ampere-hour capacity at each consecutive lower rate, regardless of higher-than-average discharge rates at the start of the discharge (Fig. 34). However, this characteristic will not hold true if high rates are used at the end of the discharge.

A nickel-iron alkaline battery is normally used in an electrical vehicle; if properly sized to the vehicle's load demands, the battery will power the vehicle through an 8-h work shift before reaching a final voltage of about 1.0 V per cell. The voltage-discharge indicator mentioned for use with lead-acid batteries in electrical vehicles is not suited for use with nickel-iron alkaline batteries.

The most practical method of noting that a nickel-iron battery has reached final-voltage level is to watch for signs of sluggishness in the operation of the electric vehicle. This condition is not necessarily an indication that the battery voltage has dropped to final-voltage level, but it certainly should cause the driver to return the vehicle to the battery-charging room.

The only practical means of reading the voltage of a nickel-iron alkaline battery is to attach a voltage test fork to contact points on the battery pilot cell in the battery-charging room. The test fork consists of a resistor drawing a high current and a voltmeter to measure the cell voltage at the time. This device can provide a rough indication of the nickel-iron battery state of charge by comparing the reading to a charge indicator scale matched to the type of nickel-iron battery being tested.

Although both the nickel-cadmium and nickel-iron alkaline batteries are capable of withstanding abuse better than the lead-acid battery, repeated overdischarge of either type should be avoided whenever possible. The nickel-cadmium battery should be protected by a low-voltage relay (LVR) as described in the lead-acid battery section on stationary applications. The LVR can be preset for 1.2 V per cell to indicate that the nickel-cadmium battery has begun discharging. A second LVR, preset for the specified final voltage, would automatically disconnect the nickel-cadmium battery from the load to avoid overdischarge.

37. Charging Nickel-Alkaline Batteries. A motive-power nickel-iron alkaline battery requires a different charging technique from that used with a stationary nickel-cadmium alkaline battery. Nickel-iron batteries are normally charged with a nominally constant charge rate throughout the entire charge, which lasts about 8 h if the discharge was about 6 h in length. Nickel-cadmium batteries are float-charged while on a standby status; after a discharge, a higher charge value is used for a prolonged period, after which float-charge rates are resumed until the next discharge.

a. Nickel-Iron Battery Charging. The battery charger will recharge the motive-power nickel-iron alkaline battery within 8 h if the charger voltage and ampere-hour rating is matched to that of the battery. In normal operation of electrical-vehicle fleets, the battery-charging area will have several chargers of different voltage and ampere-hour ratings. The operator must make certain that the battery is attached to the proper charger before activating the charger.

Three basic factors must be checked, including charger dc voltage vs. battery voltage, charger dc ampere-hour capacity vs. battery ampere-hour rating, and whether the battery and charger are matched for acid or alkaline types. All factors must be properly matched before the battery can be safely recharged.

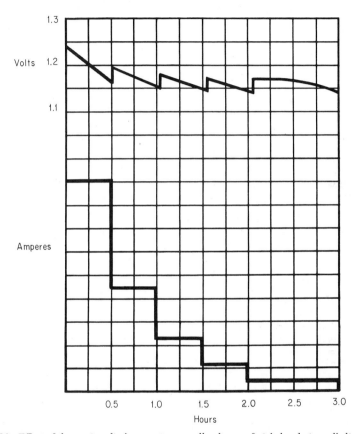

Fig. 34   Effect of decreasing discharge rates on cell voltages of nickel-cadmium alkaline battery.

Unlike the lead-acid motive-power battery, the nickel-iron battery does not require a finishing rate of charge that is substantially lower than the starting rate. The starting and finishing rates for the nickel-iron battery charge are substantially the same throughout the period of the charge. But the recharging of a nickel-iron battery requires about 25% more ampere-hours of current than were drawn from the battery during discharge, while a lead-acid battery needs only an extra 10% more ampere-hours than were taken from it.

When properly cycled and maintained, a motive-power nickel-iron battery will achieve a normal life of about 10 years if cycled daily, 5 days per week. Well-maintained batteries frequently exceed this life expectancy, but improper care and abuse can substantially shorten service life.

  b. Nickel-Cadmium Battery Charging.   Charging of a stationary nickel-cadmium battery is normally done by a continuous-float charge until the battery experiences a discharge. Then the charging current is raised to a higher value equal to at least 5% of the 8-h capacity of the battery.

Float rate for nickel-cadmium batteries is about 1.40 to 1.45 V per cell at 77°F. After a discharge, the higher charge value is supplied until voltage rises to 1.55 V per cell, which is maintained for 15 to 30 h. Float voltage is then resumed until the next discharge occurs.

The size of the charger ampere rating is normally matched to the load demand, plus the maximum high charging rate specified for the battery. The charger may be either a motor-generator set or a static rectifier, of the constant-potential type. The only time the battery discharges is when the load peaks exceed the capacity of the charger—or during an emergency when the normal ac power to the charger is lost.

c. *Equalizing Charge.* Periodic equalizing charges are necessary for lead-acid batteries used in motive-power service. But nickel-iron alkaline batteries should never be given an equalizing charge. Stationary batteries of all types, whether acid or alkaline, require periodic equalizing charges, except for lead-acid batteries with calcium-reinforced grids when maintained at specified float voltage. Equalizing of nickel-cadmium battery cells feeds 1.55 V per cell to the battery for 5 to 15 h past the normal float-charge condition, depending on the relationship of the charger current capacity to the battery ampere-hour rating.

38. Motive-Power Nickel-Iron Batteries. These batteries usually consist of a group of cells in a hardwood tray. The cell containers are made of steel coated with an electrically insulating paint. Many different voltages are available in motive-power nickel-iron alkaline batteries, ranging from only 12 V (10 cells) for small materials-handling equipment to 64 V (54 cells) for large tractors. The most popular sizes are 12 V, 24 V (20 cells), 36 V (30 cells), and 48 V (40 cells).

All the factors pertaining to sizing and use of lead-acid batteries in electric vehicles, described in Art. 29, are pertinent to the use of nickel-iron batteries.

39. Stationary Nickel-Cadmium Batteries. These batteries are usually supplied in translucent plastic jars or steel containers. A rack assembly is normally required to mount the cells properly. The cells are connected in series on the jobsite by a contractor. The most popular nickel-cadmium battery sizes are 12-V systems (using 9 to 10 cells), 24-V systems (with 18 to 20 cells), 48-V systems (using 36 to 40 cells), and 120-V systems (having 86 to 95 cells). The number of cells is determined by the charging-voltage limitations and discharge final voltages.

a. *Battery Racks.* Specifications for rack assemblies for nickel-cadmium batteries are essentially the same as for lead-acid batteries, except that the individual cell weights of the nickel-cadmium battery are substantially lower than those of the lead-acid battery. The largest nickel-cadmium cells weigh scarcely 75 lb, while the smallest weigh less than 2½ lb apiece.

b. *Plotting the Load.* The discharge profile described under lead-acid batteries for stationary duty is basically the same for nickel-cadmium battery installations. The nickel-cadmium battery is suitable for the same loads as the lead-acid battery, except those loads exceeding the maximum ampere-hour ratings of nickel-cadmium alkaline battery types, which rarely exceed 550-Ah ratings.

Installation of the nickel-cadmium battery is also similar to that of the lead-acid stationary battery, except that the weight of the alkaline-battery cells is substantially less (as mentioned in an earlier paragraph).

BATTERY SAFETY

A storage battery is constantly "live" electrically and is therefore a source of electrical shock. Care must be taken to avoid simultaneous contact with the battery positive and negative posts. Tools should never be laid down on the top of the battery, where they can cause a dangerous short circuit. Any employee experiencing electrical shock should be given medical attention as soon as possible.

Smoking should never be permitted in a battery-charging room, since all storage batteries (lead-acid, nickel-cadmium, nickel-iron) generate hydrogen gas during the charging process. This gas should never be permitted to reach a level of 4% by volume in the room atmosphere. Adequate ventilation is required in any battery-charging room.

Acid electrolyte used in lead-acid batteries is highly corrosive. If spilled on the floor or on other objects, it should be neutralized with a solution of baking soda (bicarbonate of soda) mixed with water. The normal ratio is 1 lb of soda to 1 gal of water. When the spilled electrolyte and soda stop foaming, the spill can be flushed safely down a floor drain with clear water.

If the acid solution is accidentally splashed on the skin, soda solution can be used to neutralize the acid before washing the skin with soap and water. A doctor should be called immediately if pain continues.

If acid or alkaline electrolyte solution is accidentally splashed in the eyes, only clear water should be used to flush the affected areas. Employees filling water into battery cells should always wear protective goggles or face shields—the minor inconvenience is nothing compared with possible loss of sight if a drop of electrolyte is sprayed into one or both eyes. Medical attention should be called for immediately if there is any suspicion that electrolyte has affected the eyes.

If alkaline electrolyte used in nickel-cadmium or nickel-iron batteries is spilled on the floor or other objects, it should be neutralized by flooding the spill with a solution of vinegar diluted with water in 50/50 proportions. The spill can be flushed down the floor drain with clear water after the electrolyte and vinegar solutions stop reacting to each other.

Clothing is easily stained and damaged by electrolyte. A rubber apron is recommended when an employee is servicing batteries, and rubber gloves are desirable to protect the hands against electrolyte burns or electrical shock.

BATTERY MAINTENANCE

There are no moving parts in a battery, so wear and tear are minimal. The most important factor in battery maintenance is proper addition of water to correct electrolyte-solution levels.

Motive-power batteries, both lead-acid and nickel-iron alkaline types, should be checked daily for possible low electrolyte levels. Typical volume of water required for a lead-acid battery in motive-power service is about 1/10 p per cell per week for each 100 Ah of battery capacity. Typical volume of water required for a nickel-iron battery is about 1/4 to 1/2 p per cell per week for each 100 Ah of battery capacity. Water level in each stationary battery cell should be checked once or twice a year, or following recharge after any prolonged outage.

Water from the local water supply is normally good enough for battery use, provided it meets the maximum impurity limits shown in Table 1. Where there is any doubt, distilled water is the safest medium. Impurities in the water supply can cause short circuiting of the plates in the battery cell or can adversely affect the acidity or alkalinity of the electrolyte.

Tools used with electrolyte must be restricted to either acid or alkaline electrolyte. No tool used in acid solution should be dipped into an alkaline solution, and vice versa.

TABLE 1. Impurity Limits for Water Used to Replenish Electrolyte of Acid and Alkaline Battery Cells

	Parts per 100,000	Parts per million	%
Total solids	35	350	0.035
Fixed solids	20	200	0.020
Organic and volatile	15	150	0.015
Calcium	4	40	0.004
Iron	0.4	4	0.0004
Nitrates	4	40	0.004
Chlorine, bromine (combined fluorine, iodine), total	4	40	0.004

Impurity limits should be compared with routine tests supplied by the water-supply company. If impurity levels are higher than noted above, distilled water may be required instead of tap water.

STORAGE BATTERY REPORT — Battery in Floating Service

Company...   Batt. Type ....   New ....

Location...   Pilot Cell No. ....   (rotate as needed)

Battery No..   Full Charge Gravity (Range)..

DATE & INITIALS OF READER	BUS VOLTS	PILOT CELL HYDROMETER READING	TEMP. ADJACENT CELL	DATE →					ADDING WATER
				CELL	VOLTS	VOLTS	VOLTS	HYD. RDG.	
				1					Add water after completing hydrometer readings.
				2					
				3					
				4					
				5					
				6					
				7					Date Quantity
				8					
				9					...................................Qts.
				10					
				11					...................................Qts.
				12					
				13					
				14					Equalizing Charges
				15					
				16					Amps. or
				17					Date Volts Length
				18					
				19					
				20					
				21					
				22					
				23					
				24					
				25					
				26					
				27					
				28					
				29					
				30					REMARKS:
				31					
				32					
				33					
				34					
				35					
				36					
				37					
				38					
				39					
				40					
				41					
				42					
				43					
				44					
				45					
				46					
				47					
				48					
				49					
				50					
				51					
				52					Readings Reviewed
				53					
				54					
				55					By:................................
				56					
				57					Date:..............................
				58					
				59					
				60					

Cell voltage readings should be recorded each month, on the right side of the report form. This may be either on float or 10 to 20 minutes after starting the equalizing charge. Repeat readings if more than 0.10 volt spread is observed.

Cell hydrometer readings should be recorded every three months. Record as read, do not correct for temperature.

NOTE: Previous users of this form will see a rearrangement. This was done to make it appropriate also for unattended locations visited every 3 to 10 days. These readings, as well as daily readings, are recorded on the left side.

IF COMMENTS ARE DESIRED SEND TO THE NEAREST EXIDE REPRESENTATIVE OR TO EXIDE INDUSTRIAL DIVISION, THE ELECTRIC STORAGE BATTERY CO., PHILA. 2, PA.

Fig. 35(a) Sample maintenance form for motive-power and stationary batteries.

BATTERY No._____

BATTERY RECORD

Start of CHARGE							End of CHARGE				
DATE	HOUR	FROM TRUCK No.	GRAVITY PILOT CELL	CIRCUIT No.	REMARKS	OPER-ATOR	DATE	HOUR	GRAVITY PILOT CELL	TEMP.	OPER-ATOR

INSERT W IN REMARKS WHEN WATER IS ADDED.
" E " " " EQUALIZING CHARGE IS GIVEN.
" C " " " BATTERY IS CLEANED OR WASHED.

BATTERY No._____

Fig. 35(b) Sample maintenance form for motive-power and stationary batteries.

INDIVIDUAL CELL HYDROMETER READINGS

BATTERY NO. . . . .   DATE . . . .

(Every three months, record gravity after equalizing charge.)

Cell No.	Hydrometer Reading	Cell No.	Hydrometer Reading	Remarks
+ 1		19		
2		20		
3		21		
4		22		
5		23		
6		24		
7		25		
8		26		
9		27		
10		28		
11		29		
12		30		
13		31		
14		32		
15		33		
16		34		
17		35		
18		36		

Temp. of inside cell . . . .   Electrolyte level . . . .

Readings by . . . .

Fig. 35(c) Sample maintenance form for motive-power and stationary batteries.

Motive-power batteries should be washed down regularly, to clean off the dust and grime that they usually acquire in their working day. Washdowns with warm water flowing from a hose at moderate pressure can be scheduled twice a year or more frequently if necessary. Hand brushes with plastic-filament fill can be used to scrub stubborn deposits from the tops of the batteries when necessary.

All batteries, both motive-power and stationary, whether lead-acid, nickel-iron, or nickel-cadmium alkaline, should be provided with a permanent file. Each motive-power battery should have a number painted on its tray. Each cell of a stationary battery should be identified by a number. Scheduled checks of electrolyte specific gravity, voltage levels, temperature, etc., should be noted on forms. Water-consumption rates should also be kept on file. Such records are the best help possible when it becomes necessary to analyze battery performance. The file should be maintained in its entirety as long as the facility has the battery (Fig. 35).

REFERENCES

1. E. A. Hoxie, "Some Discharge Characteristics of Lead-Acid Storage Batteries," AIEE Paper 54–177, January 1954.
2. E. A. Wagner, IEEE Working Group on Station Auxiliary Apparatus, IEEE Station Design Subcommittee of IEEE Power Generation Committee, April 1971.

7

Emergency Power—
Standby or Uninterrupted

H. K. EVERSON *

* Senior Application Engineer, Motor Generator Division, Allis-Chalmers Corporation (Retired).

FOREWORD

Many of today's activities, both commercial and governmental, require power supplies in addition to or in place of the usual utility source. Typical users who are confronted with the need of standby power, or quality of power that is better than normal, include users of the following equipment or applications:

Communication systems
Nuclear-power generators
Computer installations
Telemetering
Vital instrumentation and recorders
Critical lighting
Boiler controls
Microwave systems
Fuel processing and control systems
Data-processing computers
Pipeline instrumentation
Process-control computers
Flame-safeguard systems
Continuous-process machinery
Air-traffic control
Hospital operating rooms and intensive-care facilities
Military installations
Ship and aircraft navigational aids
Fire-alarm systems
Military fire control
Microwave repeaters
Electronic switching centers
Utilities

Anyone planning installation or operation of equipment included in the above list should carefully review his power requirements and determine in advance the quality of electric power he will require to operate his installation successfully, efficiently, and economically.

GENERAL

1. **Standby vs. Uninterrupted Power.** In applying emergency power equipment, an understanding of the difference between the two types of equipment is essential.

a. Standby power is that power which is standing by but is not necessarily immediately available. In some instances, the standby power may actually be running idle but must be switched into the system. This operation requires a time interval resulting in an interruption of power. Generally speaking, most standby systems are idle and must receive a signal to start and accelerate to operating speed, voltage, and frequency before they can be connected to the load.

b. Uninterrupted power implies continuous voltage and frequency within the limits of the equipment that is using it. One type of equipment might supply what could be termed uninterrupted power, even though there might be a perceptible change in voltage and frequency, but for a very short period of time. In another system, there may be little or no discernible change in voltage or frequency.

It is important to realize that there is probably no perfect continuous power from any source, including the best-regulated utility. There will always be instances in which, although there may be no discontinuity of power, there will be changes in voltage or frequency unacceptable to the equipment.

Even though power is continuous, intermittent changes in voltage of 10% may occur. Such a change can be acceptable to most equipment but not acceptable to some. Spikes occurring on the voltage would cause no problem for most equipment but could be disastrous in other equipment. Usually on a normal utility system no change in frequency will occur that is unacceptable to any type of equipment. A complete loss of power for a very short period of time can occur, however. This period of time is important, because it can determine the type of uninterrupted-power system that should be selected.

STANDBY POWER

If the equipment and its power requirements have been analyzed and can tolerate a power interruption of 10 to 15 s, either an idle-standby system or a continuously running system can be selected.

2. Idle-Standby System. The first system is most commonly applied and consists of an engine-generator set standing by in an idle condition (Fig. 1). Power is supplied to the load through the transfer switch for all normal conditions. Upon a power failure, the engine is started and brings the generator up to operating speed and voltage, at which time the transfer switch disconnects the load from the power line and connects it to the generator. The functions of sensing a power-line failure, starting the generator, and switching the transfer switch can all be accomplished by automatic controls and can therefore be unattended. This system is the least expensive arrangement in initial costs and operating expense. Also, because the unit operates only intermittently, the maintenance problems are accordingly reduced.

3. Continuously Running System. The second system is similar to the previous one, with the exception that engine and generator run continuously. Figure 1 is also applicable in this case. With this system the engine-starting period is eliminated and

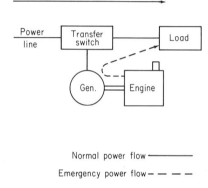

Normal power flow ———

Emergency power flow — — — —

Fig. 1 Idle-standby engine generator must be brought up to speed before it can supply emergency power. This time delay is an important consideration and may not be permissible with a critical load.

after the switching transient has died down, power is supplied to the load from the engine-generator set.

This system, however, has the disadvantage that the engine runs constantly and is inefficient because of fuel consumption and higher maintenance. It also has the disadvantage of the first system that when the transfer switch is closed, voltage and frequency transients appear on the line for a short period of time.

One characteristic of standby power as defined here is that in all cases a transfer switch is used and there is a period of time, from a fraction of 1 s to 10 or 15 s, depending upon the system used, when power is interrupted.

It is also important to note that the prime mover need not be a reciprocating engine but can be any other type of supply, such as a turbine. Other than the speed, size, type of prime mover, and complexity of controls, the above systems all fall in the category of standby power, and the user must determine which of the approaches best suits his situation.

UNINTERRUPTED POWER

As previously stated, standby power does not satisfy the requirements of the equipment in numerous applications.

4. Computer Power. Uninterrupted power supplies are used in increasing numbers for computer installations. A computer often needs its own power supply for reasons other than uninterrupted power, such as eliminating spikes or transients on the voltage from the utility line. Thus an uninterrupted power supply also acts as a buffer in these installations.

The computer, to operate correctly, has a limited frequency and voltage tolerance, which it can accept without introducing error into the output. In addition to the normal variations, such as light flicker or a momentary reduction in voltage, the computer is also sensitive to pulses or spikes of millisecond or even microsecond duration. With equipment generally available, it is not possible to detect interruptions or spikes, and the first indication that such conditions may be occurring is that errors appear in the computer output.

5. Power for Data Relaying. In high-speed communications or data-relaying centers, interruption of a transmission would result in the loss of costly information, or possibly the value of the complete transmission. This loss could not only be costly, but in national defense systems, could even be disastrous. Because of the type of equipment used, especially in wireless transmissions, a short failure of power could disrupt services not only for that period of time, but often for a half hour or longer owing to the nature of the transmitting equipment and the restarting cycle.

6. Power for Special Functions. In a modern utility, it is essential to have continuous, uninterrupted power to such devices as flame guards, for if there is an interruption of power the result is the same as if there were a signal indicating a loss of flame, and they would shut down the fuel supply. Also, in any nuclear installation it is important to have complete control of the reactors at all times, as well as to be able to monitor all the control equipment during a period of a power outage.

7. Power Requirements. If a study of an installation indicates some type of uninterrupted power is required, the first thing to determine is the quality of power needed. This information will determine the cost. A number of questions should be answered. Is a voltage accuracy within $\pm 1\%$ necessary? Is there a maximum allowable transient-voltage value that the equipment will tolerate, and how soon does the voltage have to recover and be within its steady-state band? Are there any requirements regarding frequency? How long are the power shortages that must be protected against? It may not be necessary to protect against power outages that occur infrequently, such as once or twice a year. For the majority of installations, interruptions of this type might be acceptable. On the other hand, for continuous-process computers, communication equipment, and the like, it may be mandatory to protect against any type of outage.

8. Interruption Duration. The duration of interruption greatly influences the type and complexity of power supply to be selected.

If a study of requirements shows that the maximum power interruption may be in order of 1 s or less, it will be possible in many instances to apply a flywheel "ride-through" type of set to supply these requirements. On the other hand, if the power interruption will be more than a few seconds, depending upon the frequency drop permitted, it will be necessary to go to a more elaborate type of uninterrupted set in which a second source of energy is available. It is not necessary to consider only one form of energy for a second source. The energy could be in the form of kinetic energy stored in a flywheel, potential energy such as hydraulic energy, chemical energy from a battery, or other forms of energy.

9. Buffer MG Set. The simplest and least expensive power supply, although not normally considered an uninterrupted power supply, might be a straight motor-generator set, referred to as a "buffer" MG set. The primary purpose of this unit is to isolate the load from the normal power supply, so that transients occurring in the normal power supply are not fed through to the equipment. In many instances, such a unit is the simplest form of motor-generator set with no complicated equipment required. It must be remembered, however, that the output power will decay in frequency upon an interruption of primary power, and although voltage may be maintained within the limits of the equipment, there may be deviation on a frequency basis.

In any type of uninterrupted power supply, although the output power is electrically isolated from the input power, complete isolation may not exist for other than the utility frequency. It is possible that high-frequency transients may exist which are created by switching or other equipment on the utility line. Even such innocent equipment as office machinery could produce transients which would feed through the ground path and into the equipment. To protect against these transients, it is usual practice to insulate the generator completely, both on its base and at its shaft, so that these high-frequency transients will also be blocked.

10. Flywheels. To improve the characteristics of the buffer type of motor-generator set, a flywheel is added to release kinetic energy to the generator to help maintain frequency for a short period of power interruption.

In a typical application this type of unit would be used to maintain output frequency within 1 to 5% for a fraction of a second to a period of 1 or more seconds. This unit is basically a straightforward MG set with a flywheel, except that the control becomes more complicated and in some cases quite involved. Time delay under voltage protection must be provided on the motor starter to prevent drop-out for the period of power interruption anticipated. Depending upon the size of flywheel, some means of bypassing the normal motor-overload relays may be provided while the flywheel is accelerated from rest up to full speed. Frequently, depending upon the size of the set, the size of the flywheel, and the capabilities of the normal power supply, it will be necessary to provide some form of reduced-voltage starting. In many cases the motor generator is a complete independent system, including input and output circuit breakers, and input and output instrumentation. Often a frequency-sensing relay is connected to the generator output to signal a low-frequency condition or to reduce the generator output to zero should low frequency occur. The latter feature is incorporated to prevent the possibility of a ride-through on an abnormally low frequency. Such an occurrence might result in equipment damage.

In addition to the above, an input-voltage-sensing device may be added to signal an abnormally low voltage condition and thus avoid a potential problem. In addition, many of these units are connected with high- and low-output-voltage indication or protection, again to signal a potentially dangerous condition or to shut down the unit should it be necessary.

Another basic, simple type of uninterrupted power set consists of a single generator driven by two electric motors, either induction or synchronous. This arrangement is suitable for applications with two primary sources of power, which are separated and would normally not be subject to simultaneous failure. In this application, both motors are sized to carry the full generator load, and under normal operation each carries only half load. Upon failure of one or the other primary power sources, the remaining motor takes over full load and continues to produce an uninterrupted power output.

An improved version of the flywheel MG set has a prime mover with a flywheel that operates at a higher speed than the generator and is connected to the generator through a torque-controlled coupling. The motor and flywheel might be operating at 1800 r/min normally, with the generator operating at 1200 r/min. If with this type of system a prime-power failure occurs and the coupling is automatically controlled by the generator output frequency, considerably more flywheel and motor kinetic energy is utilized than if the unit were direct-connected. In this system, the unit can continue to operate and produce uninterrupted power until the motor and flywheel have reached a speed of 1200 r/min and the carry-through or ride-through feature can be extended to 15 or more seconds. When power is reapplied, the motor reaccelerates the flywheel, and all conditions return to normal. As with the flywheel MG set, there is a practical limit to the length of time that this unit can operate without input power, and if longer periods of time are required another type of system must be selected. A disadvantage of this system is low efficiency resulting from the coupling loss during normal operation.

11. Engine-Type Units. In an internal-combustion-engine type of unit the motor-generator combination runs as a synchronous motor on the line at all times (Fig. 2). The engine is equipped with special valve-lifting means and a fuel cutoff so that no fuel is used during normal operation. At the time of the power failure, the switch is opened, the engine is started, and the motor becomes a generator, supplying power to the load. The advantages of this unit are the elimination of the starting period and the reduction of fuel consumption and maintenance on the engine. However, voltage and frequency transients occur at the time the engine-generator set starts supplying power to the load.

12. Flywheel-Clutch System. A set similar to the one just described has a flywheel and clutch added to the system to allow the engine to be idle under normal conditions (Fig. 3). Under normal operation, the motor runs as a synchronous motor, driving

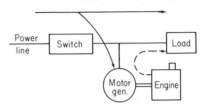

Fig. 2 The motor generator runs continuously as a synchronous motor under normal conditions in this arrangement. Valve lifters and fuel cutoff eliminate engine compression and fuel consumption when the motor is driving it. In emergencies the engine operates normally and drives the generator to supply power.

the flywheel. Upon power failure, the transfer switch disconnects the power line and energizes the clutch, with the result that the engine is cranked and started. The motor turns into a generator and the engine-generator set now provides power to the load for the duration of the emergency. Here again, as with all the other systems, all the functions of opening and closing the switch and starting and stopping the engine can be accomplished automatically so that the unit can be unattended. The main purpose of the flywheel in this system is to help accelerate the engine quickly and to minimize the frequency transient. This system is therefore an improvement over the system in Art. 11. The arrangement represents an approach to uninterrupted power with a minimum investment and reduced maintenance because the engine is normally idle. On the other hand, the engine is subjected to rather high stresses during the starting sequence and there is some distortion of the voltage and frequency supplied to the load at the time the switch opens and the motor converts to a generator. This system and the previous system require reverse current and/or reverse power relays. They also require voltage-sensing relays to sense loss of power and cause the switch to open. The relay arrangement also prevents the motor-generator-engine combination from pumping power back into the source.

It is important to remember that uninterrupted-power systems are designed to carry only the critical loads. The critical load is usually on one bus and can be isolated from the normal supply by a transfer switch.

13. Generator-to-Load System. This arrangement supplies power from the generator to the load at all times (Fig. 4). The source power drives the motor, generator, and flywheel, supplying power to the load. Upon failure of input power the clutch is energized, starting the engine, which supplies the load for the emergency period. At the same time that the clutch is energized, the power line to the motor is interrupted to prevent any feedback or unwanted contributions from the power system. This system, like the one described in Art. 12, can run for a considerable period of time, depending upon the fuel supply to the engine. It has less transient than the previous system when the motor-generator combination changes from a motor to the generator, because the generator in this system is connected at all times. A possible disadvantage of this system is that even though a flywheel is used, there is a small voltage and frequency transient at the time primary power is lost and during the period when the engine is cranked and comes up to speed. For a prolonged period of operation, the last two sys-

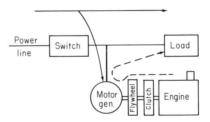

Fig. 3 Engine is cranked through the clutch with flywheel energy when emergency power is required. Engine wear during normal operation is eliminated.

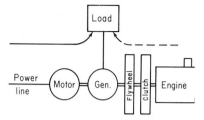

Fig. 4 The generator supplies power to the load under both normal and emergency conditions. Flywheel energy is used to crank the engine in emergencies.

tems described perhaps have a minimum initial cost and supply the longest emergency running condition of any of the systems previously described. They require less space and less complex control than systems to be described later.

14. Refined System. The following arrangement is basically the same as the one described in Art. 13, except for the addition of a torque-controlled coupling and a flywheel running at a higher speed than the motor or generator (Fig. 5). The flywheel operating at a higher speed can release a considerably greater amount of kinetic energy to the motor and generator over a longer period of time. Thus the engine can be started by normal means brought up to speed before the electric clutch is energized. Once the engine is up to speed and the clutch is energized, the complete system is driven at its nominal speed by the engine for the complete period of the emergency. When primary power is reapplied, the flywheel support motor reaccelerates the flywheel to its higher nominal speed. After a predetermined time delay the synchronous motor is reenergized and the engine is shut down.

An important point in all uninterrupted units in which a second prime mover is brought into play is that a time delay is inserted in the system to prevent the transfer of the unit back to the normal prime power immediately upon its restoration.

During the time delay the prime power is monitored for a given period of time, usually 5 to 15 min, to ascertain that the power is within limits and is stable. The monitor prevents recycling of the unit back and forth, if the prime power is restored only momentarily.

15. Dc Drive System. One arrangement consists of a motor-generator set utilizing a dc motor and an ac generator which is directly connected to the critical load (Fig. 6). The power supply for the dc motor consists of a static-type rectifier which obtains its

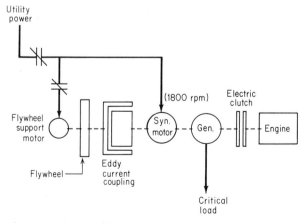

Fig. 5 Flywheel runs at higher speed than the motor and generator. Eddy-current clutch governs speed at which the motor, generator, and engine are driven while cranking the engine through the electric clutch.

power from the prime source. A battery is floated between the rectifier and the motor. The battery acts as the emergency source of energy when prime power fails. The rectifier in this instance must be designed to produce sufficient dc power for the dc motor and for charging the batteries so that they are kept at full charge while floated on the line.

The power flow is through the rectifier into the dc motor. It is mechanically connected to the generator, which supplies power to the load. In the event of primary-power failure, the motor keeps running with the battery as its source of power and does not react to the power failure. To apply this system, it is necessary to determine the length of time that an emergency might exist. The size of the battery bank is based on this information. The next consideration is to determine how often power failures may occur. The size of the rectifier battery-charger combination must be sufficient to recharge the battery completely before a second prime-power failure. Normal time to operate on battery alone is in the order of 5 to 30 min, although some installations have a battery backup sufficient to carry the unit for as much as 8 h. A running time of 8 h is generally applicable only to small units because the size of the battery bank and the cost would be prohibitive for larger units. Battery-recharging time can be in the order of 12 h, and the rectifier size is based on this time element.

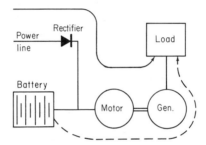

Fig. 6 Battery supplies energy for dc motor to drive ac generator in emergencies. Normally, motor is supplied power through the rectifier. With this system power is uninterrupted during transitions.

Depending upon the requirements of the load, the control for a system of this type can become quite involved. The rectifier must be protected from incoming transients and the maximum current withdrawal from the rectifier must also be controlled to prevent excessive current output when the rectifier is supplying the motor after the battery has been completely discharged. As on other systems, alarms can be included to signal power failure, rectifier failure, or low battery voltage, which will indicate that a shutdown is imminent. In addition to the above controls, a motor-speed regulator must be included to maintain the motor speed within desired tolerances for all conditions of load, machine temperature, and battery potential. Voltage-regulator controls on the generator and output instrumentation are also included, so that the voltage and load can be monitored.

When the application of units involves batteries, the unit and battery bank are selected to run under emergency conditions and only until standby power can be supplied in lieu of the primary source. Generally, as soon as prime-power failure has occurred and the unit is operating on batteries, a signal is sent to the standby-power supply to start and accelerate up to rated speed and voltage. When the standby unit is ready to accept power, the rectifier is transferred from primary power to standby power and continues to operate in the normal mode, supplying the motor with dc power and recharging the battery. With this system, an uninterrupted power supply is therefore available for several hours or days, depending upon the fuel supply to the standby unit.

If high reliability is required, several units are paralleled with excess capacity in the generating units, so that failure of any one unit will not adversely affect the load. Automatic-control equipment can be included which will automatically isolate and shut down the faulted unit and, if required, automatically accelerate and bring a standby idle

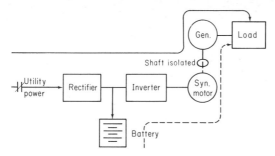

Fig. 7 Rectifier-inverter system provides uninterrupted power during transition from normal to emergency power. System eliminates dc machine.

unit into operation and synchronize and parallel it with the other running machines. If an absolute requirement for continuous power is necessary, almost any combination is possible to produce the desired results. It must be remembered, however, that the equipment becomes considerably more expensive and more complicated.

16. Ac Drive System.   Another variation of the system in Art. 15 uses a static inverter to provide ac power (Fig. 7). This system is similar in that a rectifier is used to convert ac power to dc power, which in turn is fed into an inverter that converts the dc into a voltage wave of the desired frequency. In one such system, the square-wave output of the inverter is fed into a synchronous motor, which in turn drives a synchronous generator that is connected to the load. This method eliminates the costly dc motor with its speed regulator, as well as the attendant problems with brushes and commutation. The system is a modification of the basic system and serves to illustrate the many varieties and approaches that are available to the power user.

17. Static System.   A system in which all components are static, utilizing no rotating equipment, is another approach to uninterrupted power. The system uses a rectifier to convert alternating current to direct current which supplies a static inverter and charges a battery bank (Fig. 8).

The static inverter changes the dc into a square wave of the desired frequency. Many of these inverters are used, each producing a square wave of different lengths, and these square waves are superimposed to form a staircase approximating a sine wave. This wave is in turn filtered to produce a high-quality sine wave. This system is similar to those previously described in that it will run on the battery bank for a limited amount of time, after which that unit must be supplied from a standby power until normal primary power has returned. Pulse-width control can also be used in these systems.

Like those systems using a dc motor and ac generator, power is normally supplied from a primary source with the battery floating fully charged between the rectifier and the inverter. Upon loss of input power, the inverter draws power from the battery and the output is unaffected. These units, while providing uninterrupted power, also serve as isolators from any power-line transients in voltage or frequency, as well as from high-frequency interference. They also act as a line-voltage regulator to supply constant voltage to the load in either a normal or the emergency state. One advantage of this type of installation is that the output frequency is independent of load or line variations and can be fixed at a precise value within approximately 0.1% or better. The units are also designed to have adjustable output frequency when required. They can also be designed to vary output frequency to synchronize with the input power.

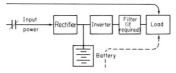

Fig. 8 Completely static system supplies ac power directly to the load for both normal and emergency operation. Inverter is supplied from either rectifier or battery.

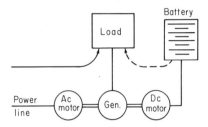

Fig. 9 Generator of three-machine set is driven by ac motor under normal operation and by the dc motor for emergency power. System provides uninterrupted power through transfer from ac to dc drive.

This arrangement makes it possible to remove the unit from service for normal maintenance.

18. Three-Machine Set. The main power in this arrangement is fed to an ac induction or synchronous motor (Fig. 9). It is mechanically connected to the synchronous generator and dc motor which is supplied from a battery bank. With this type of set, the ac driving motor normally drives the generator. Upon failure of prime power, the dc motor is energized and supplies the mechanical torque to the generator. Here again, the control can be more or less involved, depending upon the requirements of the user. There are some basic requirements such as an input-voltage monitor to determine at what point the switch to the ac motor should be opened and the dc motor energized. In small units of this type, a brush-lifting mechanism is used on the dc motor to eliminate brush wear when the unit is in the normal mode of operation. These brushes are held away from the commutator by ac solenoids. When the input power fails, the solenoids release and permit the brushes to contact the commutator. In this type of system, the fields of the dc motor are energized at all times so that the motor is up to temperature and stabilized when power is allowed to flow through the armature.

As in other types of installations, the motor can be an induction or synchronous type, depending upon the exactness of output frequency required by the load. Also, as in other installations, the input monitor has a time delay of about 0.25 s to avoid unnecessary transfers from ac to dc and back. There is also a time delay on transfer back to ac, permitting time to determine definitely that the ac power has returned and is stable. When prime power has returned and is stable, the control allows the unit to transfer back to ac, at which time the brushes are lifted from the dc motor and the battery charger recharges the battery.

Instead of using a static battery charger, the dc motor can be used as a generator to charge the batteries during normal operation. If this system is used, however, the brush-lifting feature, generally considered advantageous with a static battery charger, cannot be utilized. Generally these units incorporate a flywheel to smooth out frequency variations during the transfer from one prime mover to the other.

The descriptions of the various arrangements provide a brief résumé of some of the types of uninterrupted power supplies that are available and have been used in the past. Details of the many features of each have been omitted. Some systems are quite complex and require extensive knowledge of both the equipment and the application.

19. Communications Power. One of the more complex systems was constructed for a government continuous-communication facility (Fig. 10). Only one unit is shown.

A second unit was also in continuous operation, supplying power to the load, and a third unit was available as an idle standby, to be accelerated and to replace any faulted unit. Functions including start-up and shutdown were automatic.

a. Principles of Operation. These units are rated 125 kW. In addition to being an uninterrupted power supply, they also serve to convert the frequency from 50 Hz to 60 Hz. The result is that their rotational speed is 600 r/min. This system includes most of the devices normally encountered in no-break (uninterrupted) sets and many that are usually not included. It is described to show what can be accomplished, not what should be done in all systems.

The basic system consists of three rotating units of the in-line construction. The in-line construction is definitely desirable to eliminate any speed-changing devices such as belted or geared arrangements which would tend to reduce the reliability of the equipment. In supplying uninterrupted power, no vital functions or vital pieces of equipment should be eliminated to reduce cost. The specification for equipment should be carefully reviewed with an eye to providing the maximum reliability within a reasonable cost.

Whether the normal drive motor and the generator are furnished as two separate units or as a combined unit must be determined by the user of the equipment. If a synchronous motor-generator combination is used, it is necessary to add only a dc drive motor, making a two-unit mechanical set. There are, however, many drawbacks to the use of a combination motor generator, primarily from the load standpoint.

In one type of system the motor-generator combination runs as a motor idling on the normal power supply, and upon failure of primary power the motor is converted into a generator and then is driven by the dc motor. There are several problems with converting the motor to a generator without affecting some of the generator operating characteristics. The first problem is to determine the type of detection equipment. There is a choice of whether to detect changes in voltage, frequency, or power direction. In this system the critical load is fed directly from the utility and upon failure it is sometimes difficult to separate the critical and noncritical loads. As a result the noncritical load sometimes causes transients on the critical load. These transients can cause both voltage and frequency deviations which may be objectionable to the critical load. In a properly designed system these transients are minimized, but they still may not be acceptable to some of today's critical power-consuming devices. The ultimate in no-break power sets should therefore consist of three units wherein the critical load is driven by the generator of the no-break set under all conditions of operation. It is also important not to size the generator for critical and noncritical loads, because the additional noncritical loads will only tend to increase the cost of installation.

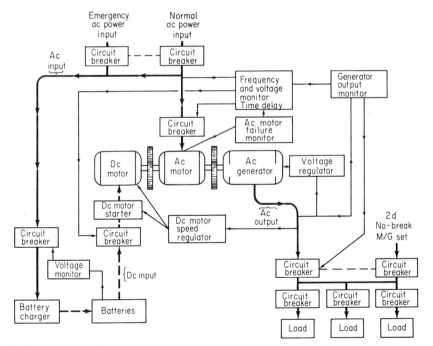

Fig. 10 Complex no-break system is used to assure continuity of power for continuous-communication system. In application this system is backed up by one running and one idle no-break set to carry power in case of equipment failure.

b. Input-Power Protection.   The incoming power to the MG set should be supplied through a disconnecting means which includes overload and short-circuit protection based on the capacity of the incoming power system.   Two incoming lines are indicated in Fig. 10, one being the normal supply and the other the emergency supply.

The normal power supply can be obtained from a utility or some other local power system.   The emergency supply is usually obtained from diesel generators.   In these arrangements the dc-motor emergency drive is normally only an interim source of drive power until such time as emergency ac power can be supplied from diesel engines or other means.   Only in the case of very small no-break sets is it practical to run the entire emergency period on dc batteries because of the cost of supplying an adequate battery bank and the space that it would require.

The emergency supply and the normal supply must be interlocked in such a manner that they cannot be paralleled.   The means for determining whether the breaker on the normal power supply or the breaker on the emergency power supply should be closed can be accomplished manually or automatically.   If automatic control is used, a monitoring device is required to sense a failure of the normal power supply.   If a failure occurs, the monitoring device opens the breaker of the normal supply, and when the emergency power reaches proper voltage and frequency, it closes the emergency breaker.   The emergency breaker remains closed until the normal power supply is properly restored.   The emergency breaker is then opened and the normal power-supply breaker is again closed.   When one breaker is open prior to closing the second breaker, the no-break set will automatically cycle from ac drive to dc drive, and back to ac when the proper input power has been restored.

c. A No-break transfer monitor is a device which continually monitors the power input to the ac motor ahead of the motor controller.   Its main purpose is to determine and cause either the ac motor or the dc motor to be selected as the driving means. This monitor contains, as a minimum, two voltage monitors for over- and undervoltage, and two frequency monitors for over- and underfrequency.   These monitors may be relays commercially available for this purpose or may be specifically constructed static circuits.   It is important for maximum flexibility of the system that the frequency and voltage monitors be adjustable in both their upper and lower limits.

These monitors must perform two complete functions:

1. Retain the ac motor on the line as long as the proper power is available, and then transfer to the dc motor when ac power is outside prescribed limits.   For this purpose, the limits are usually relatively wide, with the voltage range being as much as plus or minus 20% and the frequency range plus or minus 5%.   It is desirable to keep the unit functioning on the ac supply as long as the output power from the generator is of acceptable quality.

2. Following a transfer from the ac drive to the dc drive, the monitor must determine when normal ac power is again within proper limits and transfer the unit from the dc drive back to the ac drive.   The frequency and voltage limits must be closer when transferring back to ac to ensure that power has been correctly restored.   As a result, the re-transfer relay will usually be plus or minus 5% on voltage, and plus or minus 1% or better on frequency, and again these relays should be adjustable.

If the possibility exists that all voltage phases may not be equally disturbed, it is necessary to use three voltage monitors on a 3-phase circuit.   However, one frequency monitor is sufficient.

Another function of the monitor is to introduce an artificial time delay into the signals that it originates for transfer and retransfer.   The purpose of this delay is to eliminate unwanted transfers caused by momentary changes in the input power supply.   In other words, if a large motor is started on the same input power line, causing a voltage dip of 100 ms or thereabouts, and power is immediately restored to its proper level, the no-break set should not transfer to dc and again back to ac.   If the monitoring is of the instantaneous type, there can be many unnecessary transfers per day or even possibly per hour.   This condition should be avoided whenever possible.

d. Ac Drive Motor.   The normal drive motor for a no-break set can be of the induction or synchronous type.   The choice of which to use should be determined by the frequency requirements of the generator output.   This is a point which should be weighed and studied carefully for the following reasons: First, the standard ac induction motor, being the workhorse of industry, is by far the most dependable, reliable, and economical. This motor will operate over the greatest length of time with minimum maintenance and maximum reliability owing to the elimination of the slip rings and secondary excitation required by synchronous motors. In addition, the ac induction motor usually has the capability of accelerating from rest with much higher moments of inertia than a synchronous motor.

The ac induction motor, in addition, can be designed for very low slip and thus under normal operating conditions can provide a generator output frequency that is as little as a fraction of 1% below that which would be supplied by a generator driven by a synchronous motor.

The synchronous motor, on the other hand, will provide a constant output frequency with a constant input frequency. The biggest drawback to the synchronous motor, in addition to its initial cost, is its need for secondary excitation and the resulting lower reliability and higher maintenance associated with the exciter and other devices such as slip rings. If a brushless synchronous motor is used, it is still necessary to have the exciter, diodes, and a voltage control for the brushless exciter, all of which are not required by an induction motor. The biggest problem when using a synchronous motor is its reduced ability to accelerate the required flywheel into step, as might be required by the performance specification. It is quite possible that, if an unusually large flywheel is used, a synchronous motor having several times the horsepower required to drive the load might be needed to accelerate the flywheels and pull its Wk^2 into synchronous speed. In addition to a higher initial cost of the larger motor is the fact that under normal running conditions the motor would be running at very light loads with attendant inefficiency. It is also important to note that if a synchronous motor is used, the starting device will be more complex and again will reduce the reliability of the no-break set.

e. Motor Control. The device for controlling the ac motor can be a conventional starter or a circuit breaker with proper overcurrent relays to match the characteristics of the ac motor. A major point of consideration in the control of a no-break set is time. Therefore a minimum number of control devices should be installed. This practice will also contribute to higher reliability. It is normally not recommended to have a circuit breaker in addition to a motor control. The control may dictate that the breaker and control must operate in sequence. This sequencing will introduce additional time delays and cause greater frequency dips during transfer from one drive motor to the other. This function of the motor control can usually best be accomplished by a single device providing all the protection that the motor needs and giving maximum speed of operation.

f. Ac Motor-Failure Monitor. In a completely instrumented no-break set requiring maximum reliability, it is desirable to introduce a device between the motor control and the ac motor to detect failure which could occur either in the motor control or in the ac motor itself. This device is necessary because the no-break transfer monitor is measuring only the ac power and not the actual performance of the motor-generator set. Should a failure occur in the motor control or the ac motor, it is necessary to have a signal transmitted to start the dc motor. The control line (Fig. 10) which runs from the ac generator-output monitor to the input-frequency and voltage monitor serves to supply a signal to the dc motor in the event that the unit is on ac drive when the output frequency or voltage goes outside predetermined limits. Whether or not the ac motor-failure monitor is needed as well as a monitor on the output of the ac generator depends on personal preference and the degree of reliability required. Often the ac motor-failure monitor can be omitted and complete control will still be provided by the no-break transfer monitor and the generator-output monitor. In a highly complicated system of the larger sizes of uninterrupted sets, in addition to having complete automatic control, it is desirable to know why a unit failed when it failed. The ac motor-failure monitor can provide this signal to tell why the unit transferred over to dc, so that proper corrective maintenance can be made without recycling the MG set a second time.

g. Dc Motor. The emergency drive motor can be a standard dc type with the addition of auxiliary fields for speed control, usually required by the specifications. This motor can usually be of the type having an intermittent rating, in that it will usually run for something less than 15 min. The only time this motor might be used for continuous duty would be on small no-break sets where the dc motor might run for 1 or 2 h or possibly longer.

The choice of whether the dc motor should act as a battery charger when the set is under normal ac drive is one that should be carefully considered. First, is the motor of a size in which a brush-lifting device can be employed? This feature is desirable to reduce maintenance, brush wear, and commutator wear. Since a dc motor running with very light loads does not provide sufficient brush lubrication, brush wear is greater under this mode of operation than under full-load operation. If brush lifting is not used, a regular program of dc-motor brush replacement may be necessary.

If the dc motor is to be utilized as a battery charger, it will, to a small extent, reduce the brush wear as the result of the charging cycle. However, if the motor operates in the ac mode for 99% of the time, it will perhaps be charging 1% of the time, and the remaining 99% will be running with very light brush current, again with excessive brush wear.

If the batteries must be recharged in a very short duty cycle, the dc motor may not have the necessary capability. It would be better, in this instance, to supply a separate battery charger, which could accomplish this charging recycle in the time required. If the brush-lifting technique is used, it will be necessary to provide a separate device for charging the batteries.

h. Dc-Motor Speed Regulator. Whether or not a speed regulator is required will be determined by the basic specification and the quality of power required from the ac generator output. A dc motor can normally be designed to run within fairly close limits for a constant load and a constant battery supply voltage. However, the ac generator loads will vary and will affect the motor demand. Wide variation in dc-motor supply voltages may also cause change in dc-motor speed. The relatively wide variation in dc-motor supply voltage cannot be avoided, because it is normal to run the batteries from their full charge down to a minimum level.

If batteries are selected to minimize the voltage drop, considerably more money will be expended for additional batteries than would be required for the dc-motor speed regulator. Therefore, most no-break sets of this type require a dc-motor speed regulator.

i. Batteries. The battery selected must, of course, be based on the ampere demand that will be required by the motor, taking into consideration the length of time it will have to supply this power. Another point to consider is the voltage difference between full charge and no charge, and the effect it will have on the size and design of the dc drive motor, the speed regulator, and other components in the dc system. Other items such as shelf life, charging efficiency, hazardous atmosphere, initial investment, and desired life should be carefully discussed with battery manufacturers. Battery characteristics must be carefully coordinated between the no-break set manufacturer and the battery supplier (see Sec. 6, Part 3, Art. 30).

j. A battery voltage monitor is a voltage-detecting device which can be arranged to sound an alarm or shut down the dc motor in the event of low voltage, so as to protect the batteries (see Sec. 6, Part 3, Arts. 27 and 36).

k. A battery charger can be either the rotating or the static type and may be composed of one charger or one main charger and one trickle charger. When the battery-charging load is at its minimum level, the batteries will normally be maintained by a trickle charger. If the no-break set has been running on the battery bank and ac power is reapplied, the charging demand will be heavy and the main charger can be switched on and the trickle charger off by proper automatic controls. When the battery charge has been reestablished and only the trickle charge is required, the main charger can be switched off and the trickle charger on. All these operations can be accomplished automatically. The battery-charger supply is through a circuit breaker which is used as an automatic disconnecting means.

l. Ac Generator. The selection of the generator is another item which must be given careful consideration so that it performs as required once it is put into service.

The generator must of course provide voltage within established limits, proper waveform, acceptable deviation factor, acceptable harmonic content, and acceptable telephone interference factor (TIF) and must meet other criteria established by the load. The frequency is a function not of the generator but rather of the drive motors. An important item, often overlooked, is that the transient performance can be specified for full-load application or rejection, or for some partial load. If possible, the proper performance should be coordinated with the actual generator-load conditions.

It is not sufficient to specify standard generator performance for an average power supply. If as in Fig. 10 the total load is comprised of three separate loads and they are equal in size, the generator performance should be dictated by the effect that one of these loads will have on the system, assuming only one load will be applied at a time. In determining the maximum voltage drop and recovery time that is permissible, the effect of the transient on an unloaded generator or a partially loaded generator when a second load is applied should be considered. If it is necessary that all loads be connected simultaneously rather than sequentially, generator performance must be based on full-load conditions.

Generators should not be deliberately oversized as in any other application but should be selected and specified for anticipated conditions. It is generally not advisable to buy an oversized no-break set on the general assumption that additional power may be needed at some later date. This additional power can be costly. Unless a definite expansion is planned at the time the original uninterrupted power supply is purchased, the equipment may not be adequate when the new load is added.

Transient performance of the generator is a function of the generator reactance, the generator exciter, the voltage-regulator characteristics, and the size and power factor of the load to be applied. Consequently, transient performance affects generator size. However, if some of the load equipment will be affected by voltage transients in excess of a predetermined value, there is no choice but to size the generator or reduce its reactance to minimize the transient effects. The proper relationship of the generator, exciter, and voltage regulator is the responsibility of the generator manufacturer. It is therefore necessary that he be made acquainted with all the critical operating conditions and parameters as part of the original specifications.

m. Generator Exciter. The choice of exciter may be based on the performance required. If minimum disturbances in voltage and minimum time of response are required, one might select a static exciter and voltage regulator combination. If the voltage deviation can be controlled by generator reactance and response of time is not critical, a brushless exciter will probably be the better choice from the standpoint of reliability, since slip rings and brushes are eliminated. If a brushless exciter is chosen, it is important that all diodes be considerably derated and that adequate diode protection be provided against transients that might be reflected from the load back into the generator. Another item which might affect the choice of exciter is the rotational speed at which the units are to run. Low rotational speeds require a physically large exciter with its attendant long time delays and relatively slow response characteristics. Therefore, for low speeds a static exciter might be selected.

n. Generator Voltage Regulator. In choosing between a static exciter and a rotating exciter for long life and high reliability, a static voltage regulator should be selected. However, static components in the regulator must be selected on a derated basis.

o. Generator and Load Protection. As is true of any distribution system, the individual loads are normally protected by individual circuit breakers which may be of the manually operated molded-case type. Because the main circuit breaker is under the control of an automatic device, it must be electrically operated. It is installed primarily to protect the generator and to complete the necessary automatic functions.

p. Ac-Generator Output Monitor. Usually the loads connected to a noninterrupted power supply require more than normal protection. The generator output is therefore continually monitored for overvoltage, undervoltage, overfrequency, underfrequency, and any other requirements the specifications indicate. If a voltage or frequency condition exists which is outside predetermined tolerances, the monitor will open the output circuit breaker or shut down the MG set, thereby protecting the load. While this arrangement seems contrary to the uninterrupted-power concept, it may be only a matter of time before the load will be damaged if the output power is not of proper quality, and consequently the power should be disconnected.

A noninterrupted power supply cannot inherently protect against equipment failure. The ac-generator output monitor can send a signal back to the incoming-line monitor and cause the unit to be transferred from ac drive to dc drive in the event that the output power is outside frequency requirements. If the voltage is outside predetermined limits, there is no advantage in changing the prime mover. In this case, the generator-output circuit breaker should be opened and the MG set stopped.

q. Flywheel or Flywheels. After all steady-state operating conditions which will dictate the size of the no-break unit, the type of ac drive motor, the type of dc drive motor, and the type of ac generator, exciter, and voltage regulator have been determined, it is necessary to establish transfer conditions. These conditions are important during transfer from ac drive to dc drive, or from dc drive to ac drive.

The load must tolerate, at least for a short period of time, some frequency less than the steady-state value. It is impossible to make transfer from one drive motor to a second drive motor without having a perceptible change in frequency, regardless of how small.

The less the limitations regarding transfer frequency, the less the cost of the no-break set. If the frequency requirements during transfer are very rigid, the price will increase accordingly. If the frequency dip allowed is small, the flywheels are large and larger drive motors will be required. Consequently, larger motor-starter devices, larger circuit breakers, and larger power supplies are required to back up the system.

In determining final requirements, it is necessary to establish two parameters. The first is the minimum tolerable frequency disturbance. The second is the length of time this frequency disturbance can persist. It may even be necessary to define the frequency during transfer by a third parameter of some intermediate frequency with a

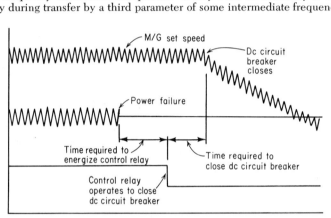

Fig. 11 Time delays are involved in closing dc breaker after power failure. Flywheel energy maintains speed of set during these delays.

stated period of time. Specifications for one set permitted a frequency as low as 57 Hz for 5 s, or an absolute instantaneous minimum frequency of 54 Hz. This precise definition of the frequency permissible during transfer is of the utmost importance to the system and motor-generator-set designer, if the greatest value is to be realized for a minimum cost. This monitoring of output frequency, both instantaneous and that referred to 57 Hz for a period of 5 s, is the function of the ac-generator output monitor. In some cases, this function can be accomplished with a few devices, and in other cases, many are required to provide proper automatic control.

The above specifications represent a somewhat normal to lower value commonly required of noninterrupted power supplies. The most critical frequency requirement might be in the range of 60 Hz plus or minus 0.5%.

20. Uninterrupted Power-System Operation. In an uninterrupted power system, a dc motor can be used as the emergency prime mover. First, in the normal mode of operation, the ac power system supplies power to an ac motor, which in turn drives the ac generator that provides power at the proper voltage and frequency for the load. In the event of normal power failure, the no-break transfer monitor disconnects the ac motor from the power supply and connects the dc motor to the batteries. During the interval required for switching, the flywheels provide the necessary energy to drive the ac generator.

A time-delay relay in the no-break transfer monitor prevents its operation on short transients. This delay is in the order of 250 ms. A second delay of about 100 ms occurs while the dc-motor circuit breaker closes (Fig. 11). A third delay sometimes occurs as a result of residual voltage produced by the ac drive motor following a power failure (Fig. 12). The voltage decay delays the monitor signal. A fourth delay occurs from the

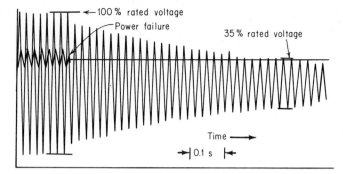

Fig. 12 Monitor signal is delayed by residual voltage decay after motor power source fails.

time the dc motor is energized until it has returned the complete rotating unit from some subnormal speed back to the normal required speed.

The dc motor field in this arrangement is continually energized, either by the battery or by the dc motor running as a generator. Energizing the fields continually has two advantages. First, it eliminates the time required for the field to build up, and second, it keeps the field structure at a stabilized temperature, thus minimizing the speed change resulting from temperature change.

It is not always sufficient to maintain a fixed potential to the dc motor fields. Sometimes the potential must be compared with the battery potential. This comparison is necessary because the difference between the voltage that the motor is putting out as a generator being driven by the flywheels and the battery voltage will determine the inrush current to the dc motor. In turn, the inrush current to the dc motor will determine its ability to accelerate the heavy inertial load at the moment the circuit breaker closes, rather than waiting for the dc speed regulator to operate. If the battery voltage and the motor voltage as a generator are equal, the speed of the motor-generator set may decrease as much as 2% before the speed regulator is able to provide the proper signal to the dc motor and make necessary correction.

If, however, a predetermined voltage difference is set into the speed regulator and comparative circuits, so that the dc motor is allowed to take larger currents at the time the dc breaker is closed, the motor will immediately produce an accelerating torque to the motor-generator set and will continue to do so until the speed regulator has an overriding effect. In other words, the speed regulator connected to the dc motor is a speed regulator only when the dc motor is driving, and is a voltage regulator when the dc motor is being driven as a generator.

It is evident that the time following deenergization of the ac motor and energization of the dc motor can be from 200 to 400 ms (Fig. 13). The flywheels must be calculated

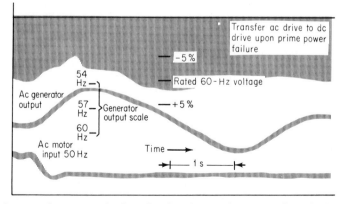

Fig. 13 Generator frequency and voltage drop less than 5% during typical transfer from ac to dc.

to drive the whole rotating unit until the dc motor can become effective. In addition to the above time delays of monitoring and switching, improper coordination of the dc speed regulator–voltage regulator can introduce as much as 300 ms additional time before the dc motor is actually putting energy back into the system.

Precise engineering is required, and very close attention to detail is needed if the system is to meet transfer-frequency requirements (Fig. 14). If the dc-motor circuit breaker is replaced with a dc-motor starter, an additional delay of 200 to 400 ms will be introduced in the case of a large dc-motor starter having a long time constant. A dc circuit breaker is therefore recommended in preference to a dc-motor starter.

The dc-motor starter is an accessory which is only occasionally required. Figure 10 includes a dc-motor starter. It is not necessarily fundamental to the operation of the no-break set. In many instances, a dc-motor starter is provided to bring the rotating unit from rest to synchronous speed, and then switch it to ac. The necessity of adding the dc-motor starter is determined by two requirements. First, if the ac supply, normal or emergency, has insufficient capacity to start the unit from rest, it is necessary to start from the battery, and a dc-motor starter with time-delay acceleration is required. Second, a dc-motor starter might be required if the ac motor was unable to start the set, to accelerate the rotating mass from rest to synchronous speed.

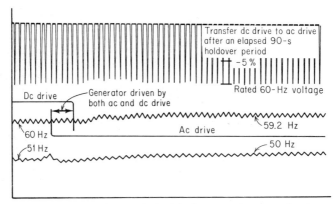

Fig. 14 Smooth transfer is made from dc drive to ac drive in no-break set.

Once the dc-motor starter has served its purpose of accelerating the unit to synchronous speed, it is not used during normal operation. The dc-motor starter must be connected in the circuit in such a manner that it serves as a starting means only and after starting the main contactor remains closed. Connection and disconnection of the motor with the battery is accomplished through the circuit breaker only.

If the unit can normally be started from the ac motor, the cost of the dc-motor starter can be eliminated. Transfer to and from dc is made by means of the circuit breaker only. If the dc starter is removed, positive-interlock protection must be provided to prevent the dc breaker from closing until the MG set is running at rated speed.

It is not possible to run a no-break set without the incoming power monitor, while relying only on the generator-output monitor. It is, however, possible to cause the MG set to transfer from ac motor drive to dc motor drive by means of a signal furnished from the ac generator output. The MG set could be transferred to dc, without the ac input monitor, but there would be no signal to cause the MG set to transfer back to ac, and consequently the system would be incomplete.

21. Multiple No-Break-Set Installation. The only method by which complete uninterrupted power can be supplied is by installing "backup" uninterrupted power sets for the units normally supplying power. This backup can be accomplished by various methods, depending upon the degree of insurance required and whether or not an interruption can be tolerated in the event of equipment failure.

The first system, while not truly uninterrupted power, is one in which it is assumed that power interruption can be tolerated in the event of equipment failure. The method normally used to provide this protection is to have a second auxiliary noninterrupted power-supply unit, a duplicate of the first unit, running at all times. In the event of an equipment failure of the set carrying the load, the automatic control, consisting of the generator output monitor, will open the circuit breaker to the faulted unit, and all load will be transferred to unit 2.

In some installations it is possible to back up as many as three running and loaded noninterrupted MG sets with one running-unloaded MG set, all under automatic control. In some installations a second auxiliary idle MG set is connected into the automatic control in such a manner that should the running-unloaded set be switched in to take the place of a faulty machine, the idle set is immediately accelerated from rest and then becomes the running standby.

One installation of this type of system consists of three 125-kW MG sets, all under the automatic control of one switchboard. Of the three units, two run continuously supplying separate loads and the third set remains idle as a standby in case of equipment failure of either of the other two sets. In the event of equipment failure of either of the running sets, the idle set will immediately be started and brought up to speed. When proper voltage and frequency are obtained, the load will be switched from the faulty set to the incoming set. In the event of a complete generator failure, an outage of power would result for approximately 10 s, the time required to accelerate the idle set up to speed and obtain proper voltage and frequency.

22. Automatic-Control Considerations. The control which monitors one or several rotating units can, in addition to the functions previously described, determine and select the proper drive motors for each unit. The system becomes more complex with the addition of automatic battery-charging control and an unattended standby set that is automatically accelerated and switched onto the load in case of equipment failure. The whole system must be designed for a fail-safe operation. Failure protection for the speed regulator for the dc motor, for example, must cause the motor to slow down in case of regulator failure rather than permit it to run away.

When a complete automatic control exists which is properly adjusted and running as designed, it at first gives an operator some concern when circuit breakers start closing and opening on all areas of the control board, with lights signaling and alarms ringing. It is only after the system stabilization that it becomes apparent the automatic control performed as designed.

23. Redundancy for Reliability. For improved reliability, many units may be operated in parallel. This is true whether they are the motor-generator type, the engine-generator type, the static type, or a combination of any of the above. Admittedly the efficiency of such a system is reduced, but reliability is considerably improved as a result of redundancy.

This approach to increased reliability is to divide the load among several units in such a manner that the failure of any unit will have no adverse effect on the load. Four 25-kW units might carry a 75-kW load or five 100-kW units carry a 400-kW load.

In all installations where a multiple number of units are operated, admittedly the control becomes considerably more involved, including automatic synchronizing equipment and automatic starting and shutdown equipment, but even though the controls become more involved, the net result is improved reliability.

The possibility that flywheel sets can also be operated in parallel to provide increased reliability in the event of a malfunction in one of the units should be remembered. Another system that is used is supplying the motor of a flywheel MG set with two primary sources of power, connecting the motor to either of these by means of a transfer switch and voltage monitors. Under normal operation the MG set would be connected to the preferred source, and should this supply fail, the transfer switch will connect the motor to the second source.

Another addition to an installation can be a transfer switch on the output of the generator. In the event of a malfunction of equipment, the load can be connected to the primary source, provided that the voltage and frequency are the same. This arrangement does not provide uninterrupted power, but in some instances such a transfer is desired to provide increased reliability and operating flexibility. When this transfer switch is not used as described, it can be used as a maintenance tool wherein the output of the MG set is synchronized with the input power and the load is transferred to the source so that the MG set can be shut down for routine maintenance.

24. Motor Selection.   Either a synchronous or an induction motor may be used for these units.   Quite often, if an induction motor is used, a specially designed low-slip motor is applied to maintain nominal output frequency as close to 60 Hz as possible. On the other hand, where exact 60-Hz output frequency is required for normal operation, a synchronous or a synchronous-reluctance type of motor must be used, so that under normal operating conditions the output frequency will be constant.

While induction-type MG sets can be paralleled, and generators driven by dc motors can also be paralleled with no problem, the parallel operation of synchronous-driven machines does present a problem and is generally not recommended.

In applying any type of motor-generator set, using an induction or synchronous motor, another point which must be kept in mind is the reconnecting of the motor to primary power following a failure, taking into consideration the residual voltage of the driving motor.   At the moment of reapplication of power, if the residual voltage of the motor is of such a magnitude and out of phase with the primary power, one can expect inrush current in excess of locked-rotor.   Consequently the motor must be designed to stand this type of operation.   In the case of synchronous motors, it might even be desirable to have a lockout feature to prevent reenergization until the residual voltage has decayed to a value at which it will cause no problem.

8

Installation and
Service Considerations

KENNETH L. PAAPE[*]

[*] Director of Industrial Control Development, Allen-Bradley Company, Milwaukee, Wis.;
Registered Professional Engineer (Wis.); Senior Member, IEEE; Chairman of NEMA Subcommittee on Enclosures and Service Conditions.

FOREWORD

In planning for or specifying switchgear and control equipment, there are many considerations and precautions regarding the installation and ambient conditions in which the equipment must operate. These conditions determine how the equipment shall be installed and protected by proper enclosures, and may restrict its rating or operations. This section provides an understanding of the standards and codes relating to switchgear and control-installation requirements.

INSTALLATION CONSIDERATIONS

1. General. Industrial control and switchgear apparatus may be obtained in a variety of configurations and enclosures to match the requirements of the location. These requirements must be given to the suppliers of the apparatus so that the manufacturer can provide equipment suitable for the application and in such construction that it can be properly installed, inspected, operated, and serviced.

2. Defining Service Conditions. The International Electrotechnical Commission (IEC), an international body organized to facilitate the coordination and unification of national electrotechnical standards, has developed a classification of service conditions which affect the performance of switchgear and control equipment (Technical Committee TC65). These classes can be used to describe the service conditions applying to control or switchgear installations.

Three major groups of service conditions are outlined. Each group is separated into specific types. Typically, vibration is a type of service condition in the group of service conditions related to mechanical effects. Classes of severity are described for each type. Each type of service condition should be considered independently, since predictable relationships do not exist between most types. For instance, one area may be subject to severe atmospheric temperature variation, while another area may have corrosive agents as well as extreme temperatures.

CLASSIFICATION OF SERVICE CONDITIONS

3. Group I, Type A, Service Conditions with Respect to Temperature and Humidity. Temperature and humidity conditions are divided into four major classes according to severity.

a. Class 1, Air-conditioned Areas. Air-conditioned areas are those with temperature maintained within ±2°C of a selected base temperature which is not less than 20°C (68°F) and not more than 25°C (77°F). Temperature is measured at one selected representative location, near the control equipment, but not where it is affected by heat developed therein. Humidity does not exceed 75% relative humidity. Air-conditioned areas may assume temperature and humidity equal to sheltered outdoor conditions (Class 3a) in case of failure of the air-conditioning system.

Areas with air conditioning which do not maintain temperature and humidity within the limits specified above are classified as heated areas (Class 2).

b. Class 2, Heated Areas. Heated areas are intended for continuous human occupation without outdoor clothing. A minimum temperature is maintained. Ventilation is sufficient so that maximum temperature is not unduly greater than outdoor shaded-area air temperature. This class includes air-conditioned areas where temperature and humidity control are not maintained within the limits specified in Class 1.

The temperature of heated areas is not less than 10°C (50°F) or more than 5°C (9°F) above outdoor shaded-area temperature when outdoor temperatures exceed 30°C (86°F). The upper limit does not apply to space within control or switchgear enclosures whose temperature is significantly above area temperature. Within such enclosures, temperatures may rise to extreme levels. No levels are specified for humidity. Heated areas may assume temperatures equal to outdoor sheltered areas (Class 3a) in case of failure of the heating or ventilating systems.

Areas with heating and ventilation which do not maintain temperatures within the limits specified for heated areas are classified as outdoor sheltered areas.

c. Class 3, Outdoor Areas. Equipment exposed to outdoor temperature and humidity is considered as being in an outdoor area, whether or not sheltered from direct sunlight, wind, rain, etc. Outdoor service conditions may be specified by a temperature range and a range of relative humidity or according to geographic areas, such as temperate, tropical, desert, subarctic, and polar.

Environmental temperature and humidity may be further subdivided into Class 3a, Outdoor Sheltered Areas, and Class 3b, Outdoor Exposed Areas.

d. Class 3a, Outdoor Sheltered Areas. Outdoor sheltered areas are those sufficiently enclosed to protect equipment from wind, rain, snow, and direct sunlight. This class includes shelters such as corrugated-steel "shacks" protecting equipment. Temperatures within these shelters when exposed to direct sunlight may exceed outdoor shaded-area air temperature by as much as 30°C (54°F). Also included are storage areas in warehouse-type buildings which, with their contents, respond slowly to changes in outdoor air temperature. Rapid rise in outdoor air humidity may result in exposure of equipment to air whose dew point exceeds the equipment temperature, resulting in water condensing on and in the equipment. Sheltered areas are generally less subject to extremely rapid changes in temperature.

This class includes areas with insufficient heating and/or ventilation to meet Class 2 (Heated Areas) specifications.

e. Class 3b, Outdoor Exposed Areas. Equipment in outdoor exposed areas may be subject to direct sunlight, wind and wind-driven particulate material, rain, snow, hail, and sleet. Temperature may change rapidly, for instance, when a rain shower immediately follows direct sunshine. Maximum temperature which equipment may reach when exposed to direct sunshine is not considered as a service condition, since it depends on the design, surface finish, and other factors, as well as the air temperature and solar-radiation intensity.

f. Class 4, Extreme Service Conditions. Conditions more severe than outdoor areas are considered extreme temperature and humidity service conditions.

4. Group I, Type B, Service Conditions with Respect to Solids, Liquids, Vapors, and Gases in the Atmosphere Other than Water Vapor and Clean Air. Combinations of materials which may be present in air surrounding control equipment and affect its performance are almost unlimited. Although no precise separation between classes of service conditions is possible, three classes of service conditions are listed. In addition, temperature and humidity play a major part in the effects of corrosion, electrical leakage, etc.

a. Class 1, Clean Areas. Clean areas are areas with controlled ventilation including suitable air filtering to remove particulate material above 1 μm in size. The atmosphere should not contain any significant quantities of gas or vapor not present in pure air.

b. Class 2, Average Areas. Average areas are areas subject to normal but not extreme atmospheric dust and air pollution. One criterion is that the worst short-term conditions are not beyond human tolerance levels for continuous exposure without detectable adverse effects. In addition, the combination of gas and vapors which may be present, including water vapor, shall not be substantially corrosive. For example, a usual electroplating area, while apparently not detrimental to human health, is sufficiently corrosive so that it does not qualify as a Class 2 area.

c. Class 3, Severe Dust, Corrosion, or Pollution Conditions. This class covers a variety of areas in many industries, including chemical plants, cement mills, and similar areas subject to corrosive vapors, dust, sandstorms, etc. The degree of severity varies greatly. Severe conditions often increase maintenance requirements, and may decrease the life expectancy of equipment.

5. Group I, Type C, Service Conditions with Respect to Air Circulation and Ventilation. Air circulation has important effects as a service condition, since it is the primary means of removing heat developed in switchgear and controls. It can convey (and in some cases remove) dust and other materials. Wind can have significant effects on operation of equipment in outdoor areas.

a. Class 1, Enclosed Areas with Continuous Controlled Ventilation. Class 1 areas are enclosed with continuous ventilation that is controlled and directed to provide at least one complete air change per minute. There is no limit on the size of Class 1 ventilated areas. Included are small areas with individual circulating fans, or a complete building. The criteria are continuity and distribution of air circulation, so that no significant dead-air pockets exist at any time in proximity to equipment. This result is obviously dependent upon the design of the ventilating system with respect to the size, shape, and location of equipment to assure elimination of dead-air pockets.

b. Class 2, Enclosed and Semienclosed Ventilated Areas. Class 2 ventilated areas comprise enclosed and semienclosed areas provided with positive air circulation. Circulation must provide a complete air change an average of once in 5 min or less. Ventilation may be by forced-air input, a positive exhaust system, or both. The system must provide reasonable circulation through the entire area, but pockets with very limited circulation may exist, especially where equipment located in the area impedes circulation.

c. Class 3, Enclosed and Semienclosed Areas with Natural Air Circulation. Class 3 areas are enclosed and partially enclosed areas without continuously operating positive ventilating systems. The significant characteristic of this class of area is that during extended periods, air circulation may be limited to circulation induced by thermal effects.

Class 3 areas differ from Class 4 (totally enclosed areas) in that natural circulation is encouraged and often supplemented by fans, vents, etc. However, in a Class 3 area, with all doors and windows closed, and with no wind, circulation may be minimal.

d. Class 4, Totally Enclosed Areas with No Normal Circulation. Class 4 areas are those within enclosures with no provision for ventilation or circulation. Class 4 areas are not intended for normal human occupancy. Heat generated by equipment in the enclosed area is transferred out by thermal conduction through the enclosure walls. Since these enclosures are usually not completely dust- and vaporproof, they breathe with changes of temperature and barometric pressure.

Many switchgear rooms with doors normally closed, and without vents or fans, and underground manholes are Class 4 areas. When work is required in these areas, doors or covers are left open and forced ventilation is often used.

e. Class 5, Outdoor Areas. Class 5 areas are subject to natural air circulation from both wind and thermal effects. In congested areas, with no wind, circulation may be negligible. In open areas, outdoor-mounted equipment may be subjected to considerable mechanical forces from high wind, and to erosive effects of wind-driven dust and sand. Direct sunlight with no wind can cause substantial temperature rise, particularly in low latitudes if thermal circulation is impeded by nearby surfaces.

Shelters for outdoor equipment are often used. The space inside the shelters may be Class 1, 2, 3, or 4, depending upon the enclosure.

6. Group II, Service Conditions with Respect to Mechanical Effects. The mechanical effects considered as service conditions are shock, vibration, and mechanical stress.

7. Group II, Type A, Mechanical Shock as a Service Condition. Significant mechanical shock is an abnormal condition. In operation or storage, significant shock occurs only because of accidental or unpredictable (e.g., earthquakes) occurrences. In operation, shock may occur because a moving object falls on or strikes the equipment or support structure. In storage conditions, there is also the possibility that the equipment may be dropped. Shock hazards, being unpredictable, are difficult to define or classify. No attempt is made to classify shock in terms of measurable parameters.

a. Class 1, Minimal Mechanical-Shock Hazard. Minimal shock hazard exists for equipment operated in control rooms where it is protected from vehicular traffic, crane-carried equipment, etc. Freedom from shock in this environment results in part from location and in part from minimal contact by persons unfamiliar with the apparatus.

b. Class 2, Normal Mechanical-Shock Hazard. This class includes the usual field or shop operating conditions with the normal hazard of being struck by moving objects, and general-purpose storage areas, where untrained persons may be handling the equipment.

c. Class 3, Abnormal Mechanical-Shock Hazard. Severe shock hazards exist in areas where equipment is located in the field or shop in heavy-industry operations, such as mining, steelmaking, and the like, where heavy equipment is in constant use for moving large objects and there is a significant probability of violent physical contact with the equipment or its supports.

8. Group II, Type B, Vibration as a Service Condition. Switchgear and control apparatus may be subjected to wide ranges of amplitude and frequency of vibration which are difficult to predict. In addition, there is the possibility of resonance in structures, particularly pipe mountings and the like, used for field-mounted equipment. Vibration may vary significantly between similar locations quite close together. It is recommended that vibration as a service condition be based on measurements at specific locations rather than in a general area. Vibration frequency can vary from a few cycles per second to as high as a kilohertz. Available data indicate no significant correlation between amplitude and frequency. The classification is in terms of acceleration measured in g. No attempt is made to classify by frequency.

a. Class 1, Negligible Vibration. Vibration with less than 0.1 g acceleration is considered to be negligible. In general, vibration which cannot be sensed by hand contact with the equipment is considered negligible.

b. Class 2, Moderate Vibration. Vibration with acceleration between 0.1 and 1 g is considered moderate.

c. Class 3, Severe Vibration. Vibration with acceleration between 1 and 5 g is considered severe.

d. Class 4, Extreme Vibration. Vibration with acceleration of more than 5 g is considered extreme.

9. Group II, Type C, Mechanical Stress as a Service Condition. Mechanical stress is distinguished from shock in that mechanical stress is applied for an interval at least 1 s in duration. Continuous mechanical stress can result from mounting, conduit, and other connections to the equipment.

a. Class 1, Installations of Minimal Mechanical Stress. Installations where connections are made through tubing and flexible electrical cable or lightweight conduit are considered Class 1, with little likelihood of any appreciable mechanical stress.

b. Class 2, Installations with Equipment Subject to Normal Mechanical Stress. Class 2 installations include all other situations where devices are mounted in accordance with manufacturer's instructions.

10. Group III, Power Supply as a Service Condition. For the purposes of classification, a power source is defined as the primary source, usually ac mains, from which the system power supply is derived. A power supply is defined as the power applied to the control system.

11. Group III, Type A, Ac Power-Supply Systems. *a. Class 1, Precisely Regulated Ac Power Supplies with Constant Frequency.* This class provides a voltage varying not more than ±2% from the designated value. Frequency varies not more than 0.2% from the stated value. Waveform distortion is less than 2% from sinusoidal. Short-term variations (more than 0.1 s) beyond the prescribed limits are not acceptable. Variations due to transients (source disturbances with duration of less than 0.1 s) do not exceed double the stated values.

b. Class 2, Voltage-regulated Ac Power Supplies. Class 2 ac power-supply systems provide a regulated ac voltage. Mechanisms for providing this regulation include electromechanical regulators and many solid-state devices. All are subject to power-source voltage variation in varying degrees. This class includes systems in which voltage is maintained to $\pm3\%$, with source voltage variations from ±10 to -15% of specified source voltage, output (load) variations from 0 to 100% of rated load, and $\pm1\%$ frequency variation. This class is subject to the same frequency variations as the source voltage. Waveform is not specified; it is generally not better than the source and with some types of induction and solid-state regulators may be severely distorted.

Transient performance of the regulators is not specified. The frequency response of various types varies from a period of 0.5 s or more on some electromechanical regulators to one-half cycle on some electronic devices.

c. Class 3, High-Grade Commercial Power Sources. Most control systems are operated directly from a commercial power source which has a steady-state voltage which does not deviate more than ±10 or -15% from stated value. Short-term voltage variations (more than 0.1 and less than 10 s) do not exceed $+15$ or -25% of the stated voltage. Absolute peak value of voltage, regardless of duration, is not more than 30% above the peak value corresponding to stated rms value. Steady-state frequency does not vary more than $\pm2\%$ of the stated value. Short-term frequency variations (less than 10 s) do not exceed 5% of the reference value. Waveform does not deviate from sinusoidal by more than 5%.

d. Class 4, Large-Variation Power Supplies. Class 4 power supplies are all those which exceed permitted variations as stated in Class 3.

12. Group III, Type B, Dc Power-Supply Systems. *a. Class 1, Precision-regulated Dc Power Supplies.* Class 1 supplies have a voltage regulation of $\pm1\%$ throughout the normal range of source voltage and load power. Regulation is sufficiently responsive so that, for short-term variations in the source or for sudden changes in load, voltage variation is maintained to within $\pm2\%$. The system is equally effective against pulses and other extreme short-duration transients in the supply. Ripple in Class 1 dc systems is less than 1% (measured as the ratio of the rms value of all ac components to the dc voltage).

b. Class 2, Voltage-regulated Dc Power Supplies. Lead-acid storage-battery power supplies are typical of Class 2 systems. Steady-state voltage is maintained to $+5$ to -10% of rated value. Short-term voltage variations (10 s or less) are not greater than $+10$ to -15% rated voltage. Class 2 systems provide a major filtering effect on transients in the source voltage. Transient limits are not included in specifications for this class. The ac ripple in Class 2 system does not exceed 3%.

c. Class 3, Unregulated Dc Power Supplies. A Class 3 dc power supply is typically a rectifier and filter operating directly from an ac power source. Class 3 voltage will normally vary in direct proportion to the power source and is affected by changes in load. A common situation which is often adequate and economical where the power source is subject to severe variations and reasonable stability of dc voltage is important is the combination of a Class 2 regulated ac voltage feeding an unregulated dc system.

AMBIENT CONSIDERATIONS

13. Ambient Temperature. Ambient temperature is the temperature of the medium such as air, water, or earth into which the heat of the equipment is dissipated. For self-ventilated equipment, the ambient temperature is the average temperature of the air in the immediate neighborhood of the equipment. For air- or gas-cooled equipment with forced ventilation or secondary water cooling, the ambient temperature is taken as that of the ingoing air or cooling gas. Table 1 gives ambient-temperature limitations (Ref. 1).

Ambient-temperature limitations are imposed on the storage and application of industrial control and switchgear apparatus to prevent damage, failure, or rapid deterioration. Devices such as those containing semiconductors which are very temperature-

TABLE 1. Ambient-Temperature Limitations

Situation	Minimum	Maximum
Operation of air-cooled low-voltage apparatus (Ref. 1)	0°C, 32°F	40°C, 104°F
Operation of high-voltage controllers (Ref. 1)	0°C, 32°F	40°C, 104°F
Operation of high-voltage switchgear with copper contacts (Ref. 3)	0°C, 32°F	40°C, 104°F
Operation of high-voltage switchgear with silver contacts (Ref. 3)	0°C, 32°F	55°C, 131°F
Operation of water-cooled low-voltage apparatus (Ref. 1)	5°C, 41°F	40°C, 104°F
Temperature of cooling water (Ref. 1)	5°C, 41°F	30°C, 86°F
Apparatus in storage (Ref. 1)	−30°C, −22°F	65°C, 149°F

sensitive will fail if their maximum allowable temperature is exceeded. Insulated parts will deteriorate with exposure to excessive temperatures. In the case of enclosed apparatus, the designer must select materials which have adequate life when operated at a temperature which is the sum of the temperature rise of the material under operating conditions plus the temperature rise of the air within the enclosure (usually 15 to 25°C above the air outside the enclosure) plus the maximum allowable ambient temperature surrounding the enclosure, usually 40°C.

14. Supply Voltage. The variations in supply voltage considered usual or normal will vary with the switchgear and control manufacturer and the country of origin. In addition, certain kinds of control systems have a narrower range of voltage limitations. In the absence of more specific information the limits shown in Table 2 may be considered to apply. If the application is outside any of the limits shown in Table 2, the manufacturer must be consulted.

15. Altitude. Equipment that depends on air for its insulating and cooling medium will have a higher temperature rise and a lower dielectric strength when operated at higher altitudes. High-voltage air switches must be derated in accordance with ANSI C37.30 when operated at altitudes above 3300 ft (1000 m). Regulated systems and adjustable-speed drives are limited to altitudes of 3300 ft for full rating unless for every 330-ft increase in altitude over 3300 ft there is a compensating 1°C decrease in maximum allowable ambient temperature. Motors having a service factor of 1.15 (115% of rated load) will operate satisfactorily at rated load at an ambient temperature of 40°C at altitudes between 3300 and 9000 ft. Industrial controls need not be derated for altitudes up to 6600 ft if they consist of manual or electromagnetic devices. Industrial controls which include power semiconductors usually require derating when operated above 3300 ft.

16. Contaminants. Typical contamination problems are shown in Table 3 along with recommended solutions. For a list of hazardous atmospheres see Table 4.

17. Unusual Service Conditions (Ref. 1). Where unusual service conditions exist, they should be called to the manufacturer's attention, since unusual construction or protection may be required. Enclosures are available to provide protection against

TABLE 2. Supply-Voltage Limitations (Ref. 1)

Supply	Minimum *	Maximum†
Alternating current ‡	90% of rated voltage §	110% of rated voltage§
Direct current	80% of rated voltage	110% of rated voltage

* Measured under worst-case load conditions, e.g., motor starting.

† Measured open circuit.

‡ Must have a sinusoidal waveform and if system is polyphase, it must be a balanced system. The supply frequency must not deviate more than 1% from rated.

§ Regulated systems, adjustable-speed drives, and other industrial systems using semiconductor power converters require that the supply voltage not deviate more than 10% above or 5% below rated voltage and any changes in excess of 5% occur at a rate less than 0.5% per cycle.

TABLE 3. Solutions to Contamination Problems

Contaminant	Primary action	Alternate action
Contaminants originating outside the control enclosure, such as dust; dirt; rain; snow; sleet; ice; water by dripping, hosedown, or flooding; oil; coolant; corrosive agents, salt air, hazardous materials	Select appropriate enclosure from application tables. Provide conduit system equivalent to enclosure	Relocate control in sheltered, possibly pressurized, area away from contaminants
Contaminants originating within the control enclosure, such as condensation	Add drain and/or breather	Add space heater
Moisture and fungus	Have manufacturer supply moisture- and fungus-proofing	Add space heater
Leakage via the conduit system	Add seals at conduit openings	Relocate conduit to provide drain

TABLE 4. Hazardous Atmospheres (Ref. 2)

Class	Group	Explosive atmosphere containing	
I	A	Acetylene	
	B	Butadiene Ethylene oxide Hydrogen	Manufactured gases containing more than 30% hydrogen (by volume) Propylene oxide
	C	Acetaldehyde Cyclopropane Diethyl ether	Ethylene Unsymmetrical dimethyl hydrazine
	D	Acetone Acrylonitrile Ammonia Benzene Butane 1-Butanol (butyl alcohol) 2-Butanol (secondary butyl alcohol) n-Butyl acetate Isobutyl acetate Ethane Ethanol (ethyl alcohol) Ethyl acetate Ethylene dichloride Gasoline Heptanes Hexanes Isoprene Methane (natural gas) Methanol (methyl alcohol) 3-Methyl-1-butanol (isoamyl alcohol)	Methyl ethyl ketone Methyl isobutyl ketone 2-Methyl-1-propanol (isobutyl alcohol) 2-Methyl-2-propanol (tertiary butyl alcohol) Octanes Pentanes 1-Pentanol (amyl alcohol) Petroleum naphtha Propane 1-Propanol (propyl alcohol) 2-Propanol (isopropyl alcohol) Propylene Styrene Toluene Vinyl acetate Vinyl chloride Xylenes
II	E	Dust of metals such as aluminum or magnesium	
	F	Carbon black, coal, or coke dust	
	G	Flour, starch, or grain dust	
III		Easily ignitable fibers or flyings (lint)	

For further information, see NFPA 70, National Electrical Code; NFPA 30, Flammable and Combustible Liquids Code; NFPA 32, Dry Cleaning Plants; NFPA 35, Manufacture of Organic Coatings; NFPA 36, Solvent Extraction Plants; NFPA 58 (ANSI Z106.1), Storage and Handling of Liquefied Petroleum Gases; and NFPA 59, Storage and Handling of Liquefied Petroleum Gases at Utility Gas Plants. NFPA is the National Fire Protection Association, 60 Batterymarch Street, Boston, Mass. 02110.

moderate pollution by dust, moisture, corrosion, oil, etc. (Sec. 16). Contaminants beyond the capability of the standard enclosures are considered unusual.

1. Unusual environmental conditions:
   Excessive moisture
   Excessively corrosive conditions
   Steam
   Sudden changes in temperature
   Fungus, insects, vermin, etc.
   Abnormal nuclear radiation
2. Unusual supply-voltage characteristics
   Unbalanced polyphase system
   Nonsinusoidal waveform or severe transient voltage surges
3. Other unusual service conditions
   Excessive vibration and shock
   Unusual shipping or storage conditions
   Installation or operation in accordance with electrical or safety codes other than
       the National Electrical Code (NEC)
   Unusual space limitations
   Limitations on radio interference generated by the equipment
   Unusual operating duty
   Restriction of ventilation
   Radiated or conducted heat from other sources

HAZARDOUS LOCATIONS (Ref. 2)

18. Identification.   An area or location is deemed "hazardous" under the NEC if it has an atmosphere containing an explosive vapor or an explosive dust (Table 4). Hazardous locations are identified by class, group, and division according to the nature of the hazardous material and the extent to which the material is present.

a. Classes.   Class I hazardous locations are those where an explosive vapor is present, as in a petroleum refinery. Class II hazardous locations are those where explosive dust is present, as in a grain mill. Class III hazardous locations are those where loose fibers and flyings (lint) are present, as in a textile mill.

b. Divisions.   Division 1 locations are those where the probability of an explosive atmosphere's existing is high. Division 2 locations are those where the probability of an explosive atmosphere's existing is low, yet the potential hazard under certain conditions is so great that appropriate steps must be taken to ensure a safe installation.

19. Class I, Division 1.   Specifically, Class I, Division 1 locations are those (*a*) in which hazardous concentrations of flammable gases or vapors exist continuously, intermittently, or periodically under normal operating conditions; (*b*) in which hazardous concentrations of such gases or vapors may exist frequently because of repair or maintenance operations or because of leakage; or (*c*) in which breakdown or faulty operation of equipment or processes which might release hazardous concentrations of flammable gases or vapors might also cause simultaneous failure of electrical equipment.

This classification includes locations where volatile flammable liquids or liquefied flammable gases are transferred from one container to another; interiors of spray booths and areas in the vicinity of spraying and painting operations where volatile flammable solvents are used; locations containing open tanks or vats of volatile flammable liquids; drying rooms or compartments for the evaporation of flammable solvents; locations containing fat- and oil-extraction apparatus using volatile flammable solvents; portions of cleaning and dyeing plants where hazardous liquids are used; gas-generator rooms and other portions of gas-manufacturing plants where flammable gas may escape; inadequately ventilated pump rooms for flammable gas or for volatile flammable liquids; the interiors of refrigerators and freezers in which volatile, flammable materials are stored in open, lightly stoppered, or easily ruptured containers; and all other locations where hazardous concentrations of flammable vapors or gases are likely to occur in the course of normal operations.

20. Class I, Division 2.   Specifically, Class I, Division 2 locations are those (*a*) in which volatile flammable liquids or flammable gases are handled, processed, or used, but in which the hazardous liquids, vapors, or gases will normally be confined within closed containers or closed systems from which they can escape only in case of accidental rupture or breakdown of such containers or systems, or in case of abnormal operation of equipment; (b) in which hazardous concentrations of gases or vapors are normally prevented by positive mechanical ventilation, but which might become hazardous through failure or abnormal operation of the ventilating equipment; or (c) which are adjacent to Class I, Division 1 locations, and to which hazardous concentrations of gases or vapors might occasionally be communicated unless such communication is prevented by adequate positive-pressure ventilation from a source of clean air, and effective safeguards against ventilation failure are provided.

This classification includes locations where volatile flammable liquids or flammable gases or vapors are used, but which, in the judgment of the authority having jurisdiction, would become hazardous only in case of an accident or some unusual operating condition. The quantity of hazardous material that might escape in case of accident, the adequacy of ventilating equipment, the total area involved, and the record of the industry or business with respect to explosions or fires are all factors that should receive consideration in determining the classification and extent of each hazardous area.

Piping without valves, checks, meters, and similar devices is not normally deemed to introduce a hazardous condition even though used for hazardous liquids or gases. Locations used for the storage of hazardous liquids or of liquefied or compressed gases in sealed containers are not normally considered hazardous unless subject to other hazardous conditions also.

21. Class II, Division 1. Specifically, Class II, Division 1 locations are those (a) in which combustible dust is or may be in suspension in the air continuously, intermittently, or periodically under normal operating conditions, in quantities sufficient to produce explosive or ignitable mixtures; (b) where mechanical failure or abnormal operation of machinery or equipment might cause such mixtures to be produced, and might also provide a source of ignition through simultaneous failure of electrical equipment, operation of protection devices, or from other causes; or (c) in which dusts of an electrically conducting nature may be present.

This classification includes the working areas of grain-handling and -storage plants; rooms containing grinders or pulverizers, cleaners, graders, scalpers, open conveyors or spouts, open bins or hoppers, mixers or blenders, automatic or hopper scales, packing machinery, elevator heads and boots, stock distributors, dust and stock collectors (except all metal collectors vented to the outside), and all similar dust-producing machinery and equipment in grain-processing plants, starch plants, sugar-pulverizing plants, malting plants, hay-grinding plants, and other occupancies of similar nature; coal-pulverizing plants (except where the pulverizing equipment is essentially dusttight); all working areas where metal dusts and powders are produced, processed, handled, packed, or stored (except in tight containers); and all other similar locations where combustible dust may, under normal operating conditions, be present in the air in quantities sufficient to produce explosive or ignitable mixtures.

Combustible dusts which are electrically nonconducting include dusts produced in the handling and processing of grain and grain products, pulverized sugar and cocoa, dried egg and milk powders, pulverized spices, starch and pastes, potato and woodflour, oil meal from beans and seed, dried hay, and other organic materials which may produce combustible dusts when processed or handled. Electrically conducting nonmetallic dusts include dusts from pulverized coal, coke, and charcoal. Dusts containing magnesium or aluminum are particularly hazardous, and every precaution must be taken to avoid ignition and explosion.

22. Class II, Division 2. Specifically, Class II, Division 2 locations are those in which combustible dust will not normally be in suspension in the air, or will not be likely to be thrown into suspension by the normal operation of equipment or apparatus, in quantities sufficient to produce explosive or ignitable mixtures, but (a) where deposits or accumulations of such dust may be sufficient to interfere with the safe dissipation of heat from electrical equipment or apparatus; or (b) where such deposits or accumulations of dust on, in, or in the vicinity of electrical equipment might be ignited by arcs, sparks, or burning material from such equipment.

Locations where dangerous concentrations of suspended dust would not be likely, but where dust accumulations might form on, or in the vicinity of electrical equipment, include rooms and areas containing only closed spouting and conveyors, closed bins or hoppers, or machines and equipment from which appreciable quantities of dust would escape only under abnormal operating conditions; rooms or areas where the formation of explosive or ignitable concentrations of suspended dust is prevented by the operation of effective dust-control equipment; warehouses and shipping rooms where dust-producing materials are stored or handled only in bags or containers; and other similar locations.

23. Class III, Division 1.   Specifically, Class III, Division 1 locations are those in which easily ignitable fibers or materials producing combustible flyings are handled, manufactured, or used.

Such locations include some parts of rayon, cotton, and other textile mills; combustible fiber-manufacturing and -processing plants; cotton gins and cottonseed mills; flax-processing plants; clothing-manufacturing plants; woodworking plants; and establishments and industries involving similar hazardous processes or conditions.

Easily ignitable fibers and flyings include rayon, cotton (including cotton linters and cotton waste), sisal or henequen, istle, jute, hemp, tow, cocoa fiber, oakum, baled waste kapok, Spanish moss, excelsior, and other materials of similar nature.

24. Class III, Division 2.   Specifically, Class III, Division 2 locations are those in which easily ignitable fibers are stored or handled (except in process of manufacture).

25. Code Check.   For additional information relative to equipment-installation requirements for hazardous locations, see national and local electrical codes.

INSPECTION CONSIDERATIONS

26. Electrical Codes.   The installation of industrial switchgear and control apparatus done as part of construction covered by a building permit is usually inspected and judged under local building codes.   Local electrical codes are based largely or entirely on the National Electrical Code (NEC) sponsored by the National Fire Protection Association (NFPA).   The NEC is revised and republished approximately every 3 years. Appropriate action by federal authorities under the Occupational Safety and Health Act (OSHA) gives the NEC and revisions to the NEC the status of federal law.   In the absence of local inspection, installation should be in accordance with the NEC.

27. Laboratory Approval.   Local inspectors rely heavily on labels, markings, and listings to identify equipment which has been evaluated, found to comply with applicable standards, and thus recognized by a testing laboratory, most often Underwriters' Laboratories, Inc. (UL).   This "recognition" is often miscalled "approval" in spite of the fact that only the authority enforcing an electrical code may "approve" an installation. Switchgear and certain items of control apparatus have an attached label indicating UL recognition, under label service.   The majority of control apparatus is governed by UL reexamination service, and the equipment recognized by UL carries a UL listing mark or is merely listed in a UL publication as a recognized component.

Many items of control apparatus are manufactured to UL standards but are not listed, marked, or labeled because the manufacturers have chosen not to spend the time and money required to obtain UL recognition of these items.   Manufacturers who design and build to UL and National Electrical Manufacturers Association (NEMA) standards are usually willing to certify compliance with their interpretation of these standards if such certification will assist in obtaining local approval.

28. Inspector's Scope.   Local codes, the NEC, and all building inspectors are concerned only with the safety of the electrical installation.   They are not concerned with the efficiency or the adequacy of the installation in terms of the work the equipment is expected to do.   For example, a local electrical inspector will check to see if the components in the motor branch circuit for a 10-hp motor are proper for that 10-hp motor.   He does not judge if the 10-hp motor is adequate for the load it is attached to.

WIRING AND TERMINATIONS

29. Ampacity.   National and local codes specify the ampacity of conductors housed in raceways, conduit, or cables and used to interconnect the various parts of an industrial-control installation.   These ampacities, however, do not apply to the internal wiring of a controller installed by the control or switchgear manufacturer.   This internal wiring is selected by the manufacturer on the basis of not exceeding the maximum allowable temperature for the wire insulation involved.

30. Aluminum Wire. Industrial switchgear and control is usually provided with terminals for the installer to use in wiring. These terminals may not be used with aluminum wire unless the terminals are marked AL. Aluminum wire requires more preparatory steps for proper installation than does copper to remove and penetrate the aluminum oxide (a high-resistance material) which forms as soon as aluminum is exposed to air. Special compounds in the form of greases are available to prevent or penetrate the aluminum oxide. Crimp-type lugs provide more reliable termination for aluminum wire than do pressure wire connectors or lugs with setscrews. With screw-type terminals the aluminum creeps or flows under mechanical pressure, vibration, and the expansion and contraction caused by heating and cooling. In the matter of a few days, this creep creates a joint which is not as tight as it was originally unless some form of continuing pressure (such as from a conical washer) is provided. Once a high-resistance joint is created, it is rapidly self-expanding, creating more aluminum oxide, which creates more heat, which creates more aluminum oxide, etc., until the device fails.

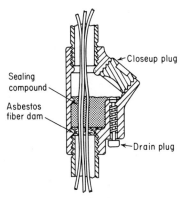

Sealing compound

Asbestos fiber dam

Closeup plug

Drain plug

Fig. 1 Typical seal-off fitting.

Aluminum wire properly installed has proved to be satisfactory for distribution-equipment applications, whereas to date, aluminum wire has produced a disproportionate number of termination problems in control equipment for the amount of aluminum wire used, probably for one or more of the following reasons:

1. The initial installation was improper and/or joint maintenance was nonexistent.

2. Control equipment cycles on and off many more times per day than does distribution equipment.

3. Control equipment carries current closer to its maximum rating more often than does distribution equipment.

4. The temperature excursion of the terminals from no-load to full-load conditions is greater for control equipment than for distribution equipment. (Industrial-control equipment is designed to standards which permit a 50°C temperature rise at field wiring terminals, and distribution equipment is designed to standards which permit only a 30°C temperature rise.)

5. Control equipment is more apt to be mounted where it is exposed to vibration than is distribution equipment.

The potential material cost savings of aluminum wire applied to industrial-control equipment should be weighed against the cost of obtaining proper termination and installation, additional maintenance, and the performance to date.

Aluminum alloys containing a small amount of iron have been developed which improve ductility and greatly reduce the tendency to creep and yet match the resistivity of other aluminum alloys used for wire. As a greater percentage of aluminum wire is manufactured from the alloys containing iron, termination problems related to creep can be expected to be correspondingly reduced.

TABLE 5. Equipment Arrangements for Motor Branch Circuits

Arrangement	Equipment	Advantages	Disadvantages
I	Disconnect device (enclosed switch) and motor starter in separate enclosures	Starters may be built into a machine cavity or mounted with motor	(a) Disconnect device cannot be mechanically interlocked* with enclosure cover of starter (b) Ampere rating of fuse clips may not exceed ampere rating of enclosed switch, thereby often requiring either time-delay fuses or an oversize switch. (c) Conduit and wiring between disconnect device and starter must be provided by installer. (d) Physical size of two enclosures exceeds that of combination starter
II	Disconnect device and motor starter combined in a common enclosure, i.e., a combination starter	(a) Wiring between starter and disconnect device is provided by control manufacturer. (b) Disconnect device is mechanically interlocked* with the enclosure door of the motor starter. (c) Single enclosure requires less mounting area than two separate enclosures. (d) Ampere rating of fuse clips may exceed ampere rating of disconnect switch	Physical size of combination starter exceeds that of either a motor starter or an enclosed disconnect device (but not both)
III	Combination starter unit (insert) in a motor-control center (see Sec. 23)	(a) All the advantages of arrangement II. (b) No separate feeder wiring to line terminals of disconnect device is required. (c) Unit can be easily disconnected from power and load and replaced. (d) Installation appearance is improved	All motor controllers are concentrated in a single location, some distance from the motors they control. Enclosure required may not be available in control-center construction

* Mechanically interlocked means that the disconnect device must be in the off position before the door permitting access to the motor starter can be opened, except where the disconnect-operating mechanism is provided with a defeater, an inconspicuous bypass requiring at least one additional deliberate step to operate.

31. Conduit Seals. Seals must be provided in conduit and cable systems in Class I, Division 1 and 2 hazardous locations to prevent the passage of gases, vapors, or flames from one portion of an electrical installation to another by means of the conduit (Ref. 2). These seals are usually constructed with wadding and sealing compound in seal-off fittings as shown in Fig. 1. The sealing-compound thickness must be not less than the trade size of the conduit and not less than $5/8$ in. Splices and taps should not be made in fittings intended only for sealing with compound, nor should other fittings in which splices or taps are made be filled with sealing compound.

Seals should also be provided in conduit systems exposed to cutting oils or coolants used in manufacturing processes in order to maintain the oiltight integrity of control devices such as limit switches or push buttons.

32. Grounding. Most electrical codes and industrial practices require a continuous grounding circuit from each metallic enclosure to earth via the conduit or cable system. Where the conduit or cable covering is nonmetallic, an additional conductor (ground wire) must be provided to provide this path to ground. When nonmetallic enclosures are used, ground wires or plates must be provided to maintain the continuity of the ground circuit between the conduits or cables entering and leaving the enclosure. Metallic parts mounted through the nonmetallic enclosure must be shielded internally from accidental contact with live parts and/or tied internally to the grounding system.

EQUIPMENT GROUPING AND IDENTIFICATION

33. Motor Branch Circuits. Safety considerations require that motor branch circuits (see Sec. 16) have a disconnect means such as a switch or circuit breaker within 50 ft of and visible from the equipment it disconnects. For small motors, 1 hp and less, the disconnect means may be part of a manual motor controller known as a fractional-horsepower starter. For all magnetic controllers and larger manual controllers, the disconnect means is a separate device and may be provided for in one of three ways (Table 5).

34. Nameplates. Industrial switchgear and control equipment can be identified from the manufacturer's nameplate, which usually gives a catalog number and rating information. The catalog number may be expressed as a bulletin, class, type, folio,

TABLE 6. Controller Checklist

1. Determine motor data:
   a. Full-load current (FLA)
   b. Voltage rating
   c. Horsepower and/or locked-rotor current (LRA)
2. Determine if load or local power company requires that motor be started at reduced voltage, and specify accordingly
3. Determine if the load is such that the motor must run in both a forward and a reverse direction, and specify accordingly
4. Determine if the motor is wound for more than one speed, and specify accordingly
5. Determine if the starter will be so inaccessible that automatic-reset overload relays are required. If so, specify automatic-reset overload relays and design the circuit and equipment safeguards so that automatic restarting either will not occur (use 3-wire control) or if 2-wire control is used, automatic restarting will not create a dangerous situation
6. Determine the type of enclosure required to best match the environmental conditions where the controller is to be located (see Sec. 16)
7. Determine the control-system considerations (see Sec. 5) including the maximum fault current which might occur
8. Select the method of providing fault protection for the motor branch circuit, fuses, circuit breaker, or motor short-circuit protectors, ensuring that the interrupting capacity of the fuse or circuit breaker is not less than the maximum available fault current. If fuses are specified, they must be designated as time-delay or non-time-delay, and the appropriate class
9. Specify the equipment required (manual or magnetic) (nonreversing or reversing) (single or multispeed) (full or reduced voltage), the enclosure, type and size of wire to be used if other than copper, and the equipment arrangement desired (Table 5)
10. Provide the controller manufacturer with complete motor and branch-circuit data and any ambient limitations

form, series, mark, or similar number plus a date code or stamp. In correspondence with the manufacturer, all the information shown on the nameplate should always be provided.

35. Legend Plates. In addition to nameplates, control apparatus should have legend plates which identify the function of the controller. Examples of the latter are Roof Exhaust Fan, Conveyor No. 1, and Motor-Generator Set No. 2. Legend plates for push buttons and selector switches are usually furnished with these devices by the manufacturer. Controller-function legend plates for controllers mounted as units in a control center (see Sec. 23) or switchboard are usually supplied by the manufacturer. All other function legend plates must be supplied by the installer or user of the control equipment.

CONTROLLER SPECIFICATIONS

36. Checklist. Each motor requires a controller which may be manual or magnetic, rated in either (1) horsepower at a given voltage or (2) full-load current (FLA) and locked-rotor current (LRA) at a given voltage. Use Table 6 as a checklist in writing specifications for motor control.

REFERENCES

1. "Industrial Controls and Systems," NEMA Standards Publication, ICS-1970, New York, 1970, Reaffirmed 1975.
2. "Hazardous Locations," National Electrical Code, 1975, National Fire Protection Association, Arts. 500, 501, NFPA Publication 70-1975, Boston, 1974.
3. "High Voltage Air Switches, Insulators and Bus Supports," ANSI Standards Publication C37.30-1962, New York, 1962.

9

Seismic Requirements

KI S. JOUNG, Ph.D.*

FOREWORD

The present seismic withstand criteria of electrical equipment for nuclear-power plants have evolved through the development and growth of nuclear power-generating stations for more than a decade. The human loss and extensive property damage caused by the 1964 Alaska earthquake aroused widespread attention to proper seismic design of buildings and structures. More recently, heavy damage sustained by electrical equipment during the 1971 San Fernando earthquake strongly influenced West Coast utilities to initiate seismic requirements even for non-nuclear-related electrical equipment.

* Manager, Engineering Analysis and Computer Applications, Advanced Technology Center, Allis-Chalmers Corporation; Registered Professional Engineer (Wis.).

Therefore seismic design and qualification of switchgear and motor controls is relatively new for suppliers of this equipment; furthermore, in the present state of the art, it is often a difficult and time-consuming process because of the very complex nature of both earthquakes and the electrical equipment. Accordingly, the present seismic technology for this equipment is in the process of rapid development and improvement.

The principal purpose of this section is therefore to summarize some fundamentals associated with seismic analysis and testing of electrical equipment to assist engineers who specify or apply switchgear or control equipment that might be subjected to seismic disturbances and also to aid design, system, and construction engineers to initiate or continue detailed investigation of other electrical equipment.

A brief background on the nature and measurement of earthquakes is followed by a historical development of the present seismic criteria and the equipment classification for nuclear-plant applications.

1. Earthquakes.   Of interest in designing or applying switchgear and control equipment are the nature of earthquakes, their measurement, and their characterization.

a. Nature of Earthquakes.   An earthquake may be described as the vibratory and often violent movement of the earth's surface which is preceded by a release of energy in the earth's crust. A majority of the destructive quakes are caused by a sudden dislocation of segments of the crust. When the crust is subjected to forces whose origins and nature are mostly unknown, it bends and deforms, accumulating its energy in the process. However, if the stresses exceed the strength of the crust, it breaks and "snaps" to a new position. During this breaking and snapping process, the stored energy is released, generating the accompanying vibrations called "seismic waves." These waves travel from the source of the earthquake to distant locations along the earth's surface and through the earth at different speeds depending upon the medium of their paths.

A fault is a fracture in the earth's crust along which a relative movement of the crust has occurred. A fault with a horizontal relative movement is called a strike-slip fault, an example being the crustal movement along California's San Andreas fault, which is predominantly horizontal. Geologists have learned that earthquakes tend to reoccur along faults, which represent zones of weakness in the earth's crust.

The focal depth of an earthquake is the distance (in depth) between the earth's surface and the regions (focus) of the origin of the earthquake energy. The epicenter of an earthquake is the point on the earth's surface directly above its focus. The location of an earthquake is usually described by the geographic position of its epicenter and its focal depth.

b. Measurement of Earthquakes.   Earthquakes produce two general types of vibrations known as surface waves and body waves; the former travel along the earth's surface and the latter through the earth.

Surface waves generally have the strongest vibrations, causing most of the damage by earthquakes. Body waves consist of compressional and shear waves, both of which travel through the earth's interior from the focus, but only compressional waves travel through the earth's molten core. Compressional waves, which push tiny particles of earth material directly ahead of them, travel at high speeds and usually reach the surface first. Hence they are often referred to as "primary" waves, or P waves. Shear waves travel less rapidly and usually reach the surface later than compressional waves; therefore, they are known as "secondary" waves, or S waves. Shear waves displace material in the directions perpendicular to their path and are often called transverse waves.

The arrival times of compressional and shear waves at selected seismograph stations throughout the world can indicate the location and time of an earthquake and often its focal depth. The amount of released energy can be estimated from the recorded amplitudes of seismic waves.

The Richter scale and the modified Mercalli scale are the two common ways to express the severity of an earthquake. The Richter scale indicates the magnitude of an earthquake, which is measured by the amplitude of the seismic waves and related to the amount of energy released. The modified Mercalli scale, expressing the intensity of an earthquake, is a subjective means to describe severity of shock felt at a given location.

The Richter scale is logarithmic, so that each higher number on the scale represents an excitation with ground motion ten times as large as that of the preceding number (or about thirty times the energy). Hence, the 1906 San Francisco earthquake (magnitude 8.3) released more than a billion times the energy of the smallest earthquakes (magnitude 2) normally felt by human beings (Ref. 1).

The intensity of an earthquake at a particular location is measured by the modified Mercalli scale ranging in values from I ("not felt except by very few, favorably situated") to XII ("damage total, lines of sight disturbed, objects thrown into air"). The maximum intensities experienced in the 1964 Alaska earthquake and the 1906 San Francisco earthquake were X and XI, respectively.

c. Characterization of Earthquakes. Earthquakes may be characterized in various ways, depending upon the interests and objectives of the person concerned. To a seismologist the motion of solid bedrock due to an earthquake may be very important, since he is interested in a study of the earth's internal geophysical structure. However, for an electrical-equipment designer whose primary objective is to design equipment that must survive earthquakes, the ground motion representative of one or more earthquakes at a given site is sufficient to characterize earthquakes for that site.

2. Earthquakes Experienced by Equipment. To perform a good seismic design, an equipment designer must know the earthquake intensity the equipment will experience. In general, seismic waves in equipment housed in a building are different from those experienced by the ground or the foundation.

Beginning their journey from the focus, the seismic waves of an earthquake reach a particular location on the earth's surface after traveling through different layers of the earth's crust. They then travel through the building foundation, the building itself with its damping and structural members, and finally the floor supporting the equipment.

Because of its mass, stiffness, and damping characteristics, each medium responds differently to the seismic waves. Therefore, the characteristics of waves alter as they pass through each medium, and the intensity and frequency experienced by the equipment can be quite different from what is experienced by the building foundation.

The process described above may be simulated by the following series of analytical computations involving certain assumptions and approximations. The earthquake ground motion in the form of an acceleration-time history is applied to a mathematical model of the foundation-building system. In this model the soil stiffness is approximated by equivalent springs and dampers, while a simplified building model may consist of a series of masses, each representing a floor level (Fig. 1). A dynamic analysis is performed on this model, and the computed acceleration response of the floor supporting the equipment is the level which the equipment will experience.

The following observations are made based on Arts. 1 and 2:

1. The soil-foundation-building system tends to filter out certain frequencies present in earthquakes.

2. The ground acceleration may be magnified depending upon the dynamic characteristics of the building and the foundation.

3. A degree of uncertainty in the accuracy of computed acceleration values is inevitable because of approximations made in modeling the soil-building-foundation system and estimating its damping.

3. Brief Historical Background. Although earthquakes have been known to man for thousands of years, their recorded history is fairly short; detailed records of earthquakes are available only for those which occurred in the last three or four decades. Accordingly, the art of seismic design and analysis is relatively new and still being developed.

There were no specific seismic-design requirements for buildings before the beginning of this century. In the United States, the 1925 Santa Barbara earthquake resulted in the inclusion of an earthquake-resistant design code in the first edition of the Uniform Building Code in 1927. The United States code for building seismic shear loads based on static g-force level depends on seismic probability zones dividing the United States into four regions (Ref. 2).

Generally speaking, seismic-design requirements for nuclear-power-plant facilities existed only on the West Coast before 1964; the Uniform Building Code or equivalent local codes were used for design in other regions of the United States.

Until 1963, seismic design was the applying of a static horizontal load at an equipment mass center, regardless of its location in the plant or plant building (Ref. 2).

However, licensing requirements of the Atomic Energy Commission prompted the adoption of dynamic analysis and testing methods by utilities and architect/engineers for seismic evaluation of equipment and structures. These requirements have, in turn, been passed on to equipment suppliers.

From 1964 to 1967, seismic loads on equipment were determined by assuming that the ground seismic motion was transmitted unchanged by the rigid building to the equipment or that the equipment was rigid and experienced the inertia load of the building at its point of attachment (Ref. 2).

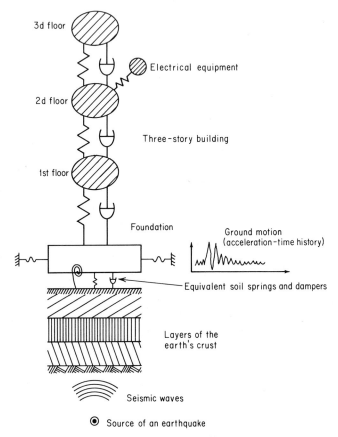

Fig. 1   Schematic of soil-foundation-building-equipment system.

Since 1967, the dynamic effects resulting from resonance between the building and the equipment have been considered. Accessibility to digital programs based on the finite-element technique has brought to reality an efficient dynamic analysis of complex structures considering their damping and resonance effects.

The response of equipment and structures to vertical ground motion can be as significant as their response to horizontal motions. However, it has been ignored in the past, with the assumption that structures were rigid in the vertical direction and that the vertical-acceleration component was less than the horizontal components. The vertical-acceleration components of the 1971 San Fernando earthquake were greater than the horizontal components; therefore seismic requirements in the vertical direction are likely to receive careful consideration in the future.

For the last 5 to 8 years, there have been increasing requirements for equipment suppliers to perform dynamic analysis and/or testing to verify the seismic adequacy of their equipment for nuclear as well as nonnuclear applications. The development of these requirements has been accelerated by the two major United States earthquakes in that period (1964 Alaska and 1971 San Fernando) and the public awareness of the human loss and property damage that can be inflicted by an earthquake.

In view of these rapidly developing requirements, a seismic analysis and testing guide for electrical-equipment suppliers has been prepared by the Joint Committee on Nuclear Power Standards of the IEEE to assist them in meeting seismic requirements (Ref. 3). Other organizations such as NEMA are also in the process of drafting a seismic-requirement guide.

4. Equipment Classification. Equipment classification depends upon various industrial applications. The following three seismic design classes are for the equipment associated with nuclear-power plants and are based on the design functions of that equipment (Ref. 2):

Class I. Those structures and components, including instruments and controls, failure of which might cause or increase the severity of a design-basis accident or result in an uncontrolled release of a substantial amount of radioactivity

Class II. Those structures or components which are important to reactor operation, but not essential to safe shutdown and isolation of the reactor, and whose failure could not result in the release of a substantial amount of radioactivity

Class III. Those structures not directly related to reactor operation or safety in the event of a design-basis accident and whose failure would not result in the release of a significant amount of radioactivity

Class I equipment is usually designed for earthquakes of two different magnitudes called the operational-basis earthquake (OBE) and the design-basis earthquake (DBE). The OBE is the greatest earthquake which can be expected to occur during the life of the plant and for which the equipment must remain operational or be able to shut down and start up again. The DBE is the greatest earthquake that is likely to occur during the life of the plant and during which the equipment must be able to perform its safety function. In general, the equipment is not required to function normally during or after a DBE.

Class II equipment is investigated for the OBE only and is required to meet the same criteria as Class I equipment. Class III equipment is designed for normal building or equipment code requirements in force at the site, where there may even be no earthquake-design provisions.

For nonnuclear applications in general, no distinction is made between the OBE and the DBE.

FUNDAMENTAL CONCEPTS

Seismic design, analysis, and testing require a knowledge of the theory of vibrations of elastic structures. Therefore, some fundamental concepts and terminology in mechanical vibrations are now discussed for a simple spring-mass system and then for a more complex system and the response spectrum. A somewhat more detailed presentation is made in Arts. 11 and 12. A more exhaustive treatment on structural dynamics may be found in the literature (Refs. 4, 6, 7, and 8).

5. Fundamental Concepts of Simple System. Consider a simple spring-mass system (Fig. 2) consisting of a mass m (weight W) suspended by a spring of stiffness k. It is assumed that the values of the weight W, the spring constant k, and the magnitude of motion are such that the force developed in the spring is proportional to its deformation; in other words, a linearly elastic spring is assumed.

a. Free Vibration. The motion of the mass, considered only in the vertical direction, is completely described by specifying the location of the mass at any instant of time.

Assume the mass is released after being displaced by a certain initial distance (small enough so as not to cause a permanent deformation in the spring) in the vertical direction. It will then oscillate up and down continuously with a certain frequency, if friction or resisting forces are assumed to be absent in the system. This oscillatory motion is called the free vibration, since it is sustained by the action of the forces inherent in the system and not by the influence of external forces. The frequency of the motion, known as the natural frequency of the system, is given by

$$f_n = \frac{1}{2\pi}\left(\frac{k}{m}\right)^{1/2} \qquad \text{Hz} \tag{1}$$

if k and m are in lb/in and lb-s²/in, respectively. The circular natural frequency of the system is

$$\omega_n = 2\pi f_n = \left(\frac{k}{m}\right)^{1/2} \qquad \text{rad/s} \tag{2}$$

while the natural period or period of the motion given by

$$\tau_n = \frac{1}{f_n} = 2\pi\left(\frac{m}{k}\right)^{1/2} \tag{3}$$

is the time required for the mass to undergo a complete cycle of motion. It is seen in Eq. (3) that for a given spring it will take longer for a heavier mass to complete a cycle of motion.

If W and k of the system shown in Fig. 2 are 100 g lb and 400 lb/in, respectively, the system natural frequency is

$$f_n = \frac{1}{2\pi}\left[\frac{k}{(W/g)}\right]^{1/2} = \frac{1}{2\pi}\left(\frac{400}{100}\right)^{1/2}$$

$$= 0.318 \text{ Hz}$$

The system circular natural frequency [Eq. (2)] is

$$\omega_n = \left(\frac{400}{100}\right)^{1/2} = 2.0 \text{ rad/s}$$

and the system natural period is

$$\tau_n = \frac{1}{f_n} = 3.14 \text{ s}$$

The free vibration of this system in the vertical direction can be initiated by displacing the mass any arbitrary distance, provided the spring was not stretched beyond its elastic limit. This condition implies that oscillation amplitudes of its free vibration can be different depending upon the amplitudes of initial displacement. Therefore, the amplitudes of a free vibration are arbitrary and relative values. The absolute magnitudes of an actual motion are determined only when known external forces are considered in the analysis (Art. 11).

The natural frequency of a system depends on its mass and stiffness [Eq. (1)] as well as boundary conditions; this conclusion is true even for more complex systems.

b. *Transient and Steady-State Vibration.* Suppose now the mass in Fig. 2 is excited in the vertical direction by a periodically fluctuating force acting on it. If monitored and plotted as a function of time, the motion of the mass will appear irregular for a certain time after the force is applied. The behavior of this motion, called the transient vibration or transient response, depends on the initial conditions as well as the mass, stiffness, and damping properties of the system.

Fig. 2 Simple spring-mass system.

The irregularities in the motion disappear or "die out" soon because of damping, and the remaining motion appears regular or "steady"; this motion, known as the steady-state vibration or response, is sustained by the disturbing force. The steady-state response of a system depends on its dynamic characteristics including damping as well as the amplitude and frequency of the disturbing force.

c. Damping. When the disturbing frequency of a harmonically oscillating force approaches the natural frequency of a system, its vibration amplitude builds up, approaching infinity in the absence of damping. This condition is known as resonance. In reality, however, every system has some damping, which prevents vibration amplitude from becoming infinitely large.

Damping is an extremely complex phenomenon. Of the many types of damping, viscous damping is not only the simplest but the most useful and mathematically manageable. Viscous damping is found where the resisting force is due to viscous resistance in a fluid medium such as the fluid motion in an ideal dashpot. Viscous-damping force is given by the product of the velocity and the proportionality factor c, called the coefficient of viscous damping.

Shown in Fig. 3 is a simple spring-mass system with a dashpot representing viscous damping. Suppose this system is subjected to the same initial displacement and released each time while its damping is increased, starting from the no-damping condi-

Fig. 3   Spring-mass system with viscous damping.

tion. The system will execute its free vibration indefinitely if it has no damping at all. As the damping is increased, however, the oscillating motion tends to die out. It can be deduced intuitively that at a certain value of damping coefficient c the motion will become aperiodic or dead-beat motion. The value of c at which this phenomenon occurs is referred to as the critical damping c_{cr} and is given, for this simple system, by the expression

$$c_{cr} = 2(km)^{1/2} = 2m\omega_n = \frac{2k}{\omega_n} \tag{5}$$

The dimensionless ratio c/c_{cr}, called the viscous-damping factor ζ, is given by

$$\zeta = \frac{c}{c_{cr}} = \frac{c}{2}(km)^{-1/2} = \frac{c}{2m\omega_n} = \frac{c\omega_n}{k} \tag{5a}$$

The damping as a percent of the critical damping, denoted by β, is determined from the expression

$$\beta = 100\,\zeta$$

As an example, 5% damping means that $\beta = 5$ or $\zeta = 0.05$. The viscous-damping factor of a system provides a convenient measure of its damping magnitude.

For a single-degree-of-freedom linear system with low damping, damping is related to the response curve—i.e., acceleration-magnification factor (the ratio of output acceleration to input acceleration) vs. excitation frequency (see Fig. 4)—by the relationship

$$Q_o = \frac{1}{2\zeta} = \frac{f_n}{f_2 - f_1} \tag{6}$$

in which Q_o = peak magnification factor at resonant frequency f_n
      ζ = viscous-damping factor
  f_1 and f_2 = frequencies at which $Q = Q_o/\sqrt{2}$
Equation (6) may be written as

$$\zeta_p = \frac{1}{2Q_o} \qquad \text{(resonant peak)} \tag{6a}$$

$$\zeta_b = \frac{f_2 - f_1}{2f_n} \qquad \text{(bandwidth)} \tag{6b}$$

which are two possible methods of computing damping. The first method, given by Eq. (6a), is referred to as the resonant-peak method and is concerned with only the value of the peak magnification. The second method [Eq. (6b)], known as the bandwidth

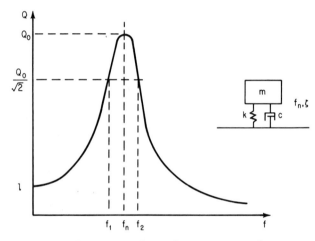

Fig. 4 Typical acceleration-magnification factor vs. excitation-frequency curve.

method, takes into account the peak amplification and the shape of the response curve near resonance. The method is particularly useful for a system with between 1 and 10% of the critical damping (Ref. 5). Below this range difficulties are encountered owing to the narrow resonant bandwidth, while above this range resonances become less clearly defined.

Equation (6) is, strictly speaking, valid only for the displacement-response curve of a linear, single-degree system with low viscous damping. However, the bandwidth method based on Eq. (6) has been widely used and accepted as a reasonable method to estimate damping for multidegree systems (Ref. 5). Although a simpler method to use than the bandwidth method, the resonant-peak method has limited validity in estimating damping for multi-degree-of-freedom systems.

d. Degrees of Freedom. The motion or configuration of the vibrating systems shown in Figs. 1 and 2 is completely described by only one coordinate—i.e., the vertical coordinate specified as a function of time. Such systems are said to be single-degree-of-freedom systems. A vibrating system for which more than one coordinate is necessary to describe its motion is called a multi-degree-of-freedom system. The number of degrees of freedom of a vibrating system is equal to the minimum number of coordinates necessary to specify its configuration at any time.

A system whose configuration is described by an infinitely large number of coordinates is known as a continuous system. An example is the transverse vibration of a cantilever beam requiring an infinitely large number of coordinates to describe its motion completely. The beam may, however, be idealized, for example, as a five-mass lumped (or discrete) system. If only a transverse displacement is of interest at each of the five masses, the system is said to have five degrees of freedom. The system has, however, ten degrees of freedom if a transverse displacement and a rotation at each mass are used to describe its configuration.

6. **Fundamental Concepts of Multidegree System.** Free and forced vibrations for multidegree and continuous systems are discussed briefly using an example, and some useful conclusions are summarized without proof. A somewhat detailed discussion on the vibrations of multidegree linear systems is presented in Art. 12.

a. Free Vibration. Consider the free vibration of a two-mass system (Fig. 5), assuming that the masses are on a frictionless horizontal surface (hence no resisting forces

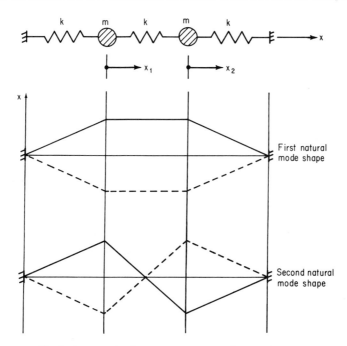

Fig. 5   An example of a two-mass system and its mode shapes.

present) and considering their motion only in the x direction. The displacements x_1 and x_2 of the masses are measured from their equilibrium positions.

If the equations of motion were derived using, for example, Newton's second law of motion, two interrelated equations, one for each mass, would be obtained. Disturbing external forces are not included for the free-vibration analysis.

The solution of the equations of motion would indicate that the resulting motion of the system is harmonic, and there are two harmonic motions, each with a frequency. For example, if $m = 1$ and $k = 2$ in the example, the two natural circular frequencies are computed to be $\omega_1 = \sqrt{2}$ rad/s and $\omega_2 = \sqrt{6}$ rad/s. Corresponding to each of these frequencies there is a "shape" of the motion called the natural mode or mode shape of the system. The mode shape corresponding to ω_1 is $x_1 = 1$ and $x_2 = 1$, while that associated with ω_2 is $x_1 = 1$ and $x_2 = -1$. In the first natural mode both masses oscillate in phase with a natural frequency ω_1, while they oscillate out of phase with another natural frequency ω_2 in the second mode. These mode shapes are shown in Fig. 5. The magnitudes x_1 and x_2 of the free vibration are only relative and determined only when external disturbing forces are considered.

It is known that the number of natural frequencies of a linear lumped-mass system equals its degrees of freedom, and there is, associated with each natural frequency, a natural mode of vibration or mode shape.

An example of a continuous system now considered is the free transverse vibration of a flexible finite-length string supported at both ends (Fig. 6). Theoretically, an infinitely large number of coordinates is needed to describe the motion of the string completely. Therefore the system has an infinite number of natural frequencies and the associated mode shapes. The solution of the equation of motion indeed results in the natural frequencies of the form $\omega_n = np$, where $n = 1, 2, 3, \ldots$ and p is a constant, while the mode shapes are of the form $\phi_n(x) = \sin (n\pi x/l)$, with l the length of

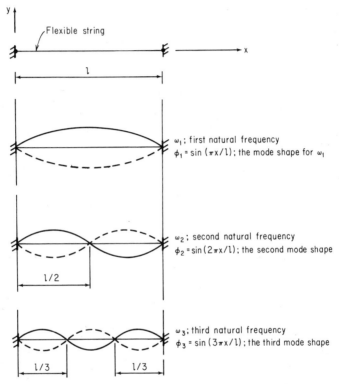

Fig. 6   An example of a continuous system—transverse vibration of a finite-length, supported, and flexible string and some mode shapes.

the string. Some of these mode shapes are shown in Fig. 6. Note that the first (or fundamental) natural mode ϕ_1 oscillates with the lowest frequency $\omega_1 = p$, and as the wave number increases, so do the natural frequency values.

b. Transient and Steady-State Vibration. A descriptive treatment of the forced vibration is presented here, using as an example the vibrating string discussed above.

Assume that a harmonically oscillating concentrated force begins to act on the midpoint of the string; its disturbing frequency Ω is assumed to be different from any of the string natural frequencies. Assume further that the transverse-response displacements at certain points on the string are monitored and plotted as functions of time. Then, the displacements would appear irregular during a time duration following the initial application of the force. The response displacements depend on the string mass and stiffness characteristics as well as its initial conditions (i.e., initial displacement and velocity conditions) at the time of the force application. If damping is present in the string, it damps out this irregular motion in a relatively short time, and only the steady-state portion of the response motion remains.

It can be shown analytically that the dynamic response of a system to disturbing forces consists of the modal-response contributions of all its natural modes. In this example each natural mode of the string responds to the concentrated force and the final dynamic response can be determined by summing these modal contributions in an appropriate manner.

The relative magnitude of each modal contribution of a system depends upon the disturbing frequencies, the magnitudes and distribution of disturbing forces, and the system characteristics and damping. If the disturbing frequency Ω equals the fundamental (lowest) frequency ω_1 in the example, the first-mode response goes to infinity (this is the resonance in the absence of damping). It can be shown that the second natural mode makes no contribution to the overall dynamic response, since the disturbing force, acting at the node (point of zero displacement) of the second mode, cannot

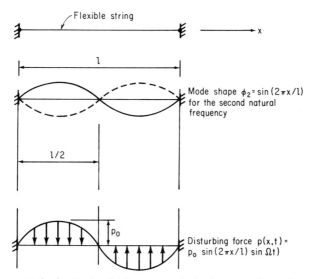

Fig. 7 The magnitude distribution (with two waves) of a harmonically oscillating force on a finite-length flexible string.

impart any energy into the mode. On the other hand, if the magnitude distribution of a harmonically oscillating force is of sine-wave form with two waves (Fig. 7)—i.e., $p(x,t) = p_o \sin(2\pi x/l) \sin \Omega t$—it can be shown that regardless of the disturbing frequency Ω, only the second natural mode contributes to the system dynamic response.

The above discussion for an undamped system indicates that a knowledge of natural frequencies and mode shapes of a system is essential in computing its dynamic response to disturbing forces.

c. Damping. If the disturbing frequency approaches one of the natural frequencies of a system, resonance condition exists. However, the damping present in the system prevents its response from becoming large. Hence even an approximate knowledge of the system damping is very important and useful.

Although the bandwidth method discussed earlier is valid only for the displacement-response curve of a linear, single-degree-of-freedom system with low damping, it has been widely used and accepted as a reasonable method to estimate damping for multi-degree systems (Ref. 5).

d. Modeling Continuous Systems. In practice many structures are continuous systems with arbitrary geometry, boundary, and loading conditions. Structural portions of switchgear and motor-control units are continuous systems, while some components (or masses) may be considered as lumped-mass systems.

Closed-form or exact dynamic analysis for a complex structure is often impossible. However, an approximate analysis of a continuous system may be performed using its equivalent lumped-mass model and the finite-element technique (Refs. 6 to 8). This method is, in essence, a matrix structural-analysis technique, wherein the dynamic response of a continuous system is computed only at a finite number of masses of its idealized lumped-mass model. Digital-computer programs based on the finite-element technique are powerful tools for the natural frequency, mode shape, and forced-vibration analyses of multidegree systems.

7. Response Spectrum. Although the dynamic response of a single-degree system to an earthquake acceleration-time history can be determined, computing its entire history of forces and displacements during the earthquake is a time-consuming process. Furthermore, for most practical and design purposes, the maximum-response (dis-

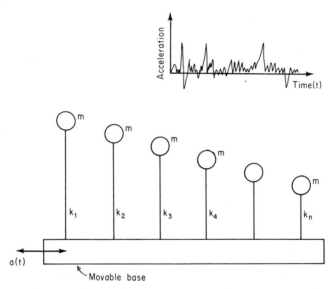

Fig. 8 A series of conceptual, single-degree-of-freedom cantilever pendulums fixed on a movable base.

placement, velocity, and acceleration) values are quite useful. It is therefore convenient to plot these maximum values as functions of the natural frequencies of a series of single-degree-of-freedom systems subjected to an acceleration-time history record. The resulting curves are called response spectra.

a. Ground Response Spectrum. Shown in Fig. 8 is a series of conceptual cantilever pendulums supported on a movable base. If the pendulums are assumed to have equal masses but different stiffness values, the natural-frequency value is lowest for the leftmost pendulum and increases from left to right—i.e., $\omega_1 < \omega_2 < \cdots < \omega_n$, where $\omega_1 = \sqrt{k_1/m}$, $\omega_2 = \sqrt{k_2/m}$, etc. Suppose the base is now subjected to the ground motion of an earthquake as recorded by a strong motion accelerometer. Then the computed maximum response (displacement, velocity, or acceleration) of each pendulum represents the maximum response of a single-degree-of-freedom system during the earthquake. Generally, a maximum-velocity value called the spectral velocity (or more correctly the spectral pseudo-velocity, since it is not exactly equal to the maximum velocity in a damped system) is computed. The maximum displacement, known as the spectral displacement, is obtained by dividing the spectral velocity by the circular frequency ω. Similarly, the spectral acceleration (or more accurately the spectral pseudo-acceleration, since it is, in general, not equal to the peak acceleration value) is computed by multiplying the spectral velocity by the circular frequency.

The computed spectral velocity for a given pendulum depends on its natural frequency, damping ratio, and ground motion (usually, the acceleration-time history). Therefore, for a given earthquake time history and a specified damping ratio, the spectral velocity, displacement, or acceleration can be computed as a function of the pendulum natural frequency or period. The curves generated by plotting these maximum-response quantities as functions of the natural frequencies or periods of a series of pendulums (or single-degree systems) are referred to as the response spectra or response-spectrum curves. The spectral displacement, velocity, or acceleration may be plotted, resulting in the displacement, velocity, or acceleration response spectrum, respectively. The response may be in either the horizontal or the vertical direction. The curves developed by computing the maximum response of single-degree systems supported on the ground are called the ground response spectra.

Shown in Fig. 9 is the velocity response spectrum for El Centro (California) earthquake of May 18, 1940. The curves show clearly the influence of damping on the maximum-response values; the response is quite sensitive for low damping values (i.e.,

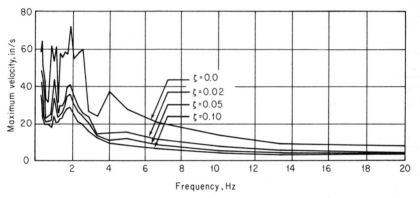

Fig. 9 Velocity response spectrum for El Centro earthquake, May 18, 1940 (north-south component) (Ref. 5).

local peaks and valleys are distinct and pronounced), while it is "smoother" for higher damping values. The presence of the sharp peaks and valleys is also indicative of local resonances in the ground-motion record.

b. Floor Response Spectrum. To develop a floor response spectrum for a floor of a building, a dynamic analysis must be performed for the building mathematical model subjected to a ground-motion record, and the maximum response of a series of pendulums supported on the floor computed and plotted against their natural frequencies or periods.

Floor-response-spectrum curves are developed for three to five different pendulum-damping values for the same building and its damping. It is seen in Fig. 10, representing typical acceleration response spectrum for a hypothetical building, that damping can significantly affect design requirements, since a slight increase in damping results in a considerable reduction in load requirements in the resonance regions of equipment. Although 0.5 to 1.5% damping is commonly used for analysis purposes, higher values may be used if justified by test or experience.

The floor-response-spectrum curves for a building show, in general, pronounced peaks at its dominant natural frequencies. For example, the high peaks near 12 Hz in Fig. 10 suggest that one of the building natural frequencies is near 12 Hz. It should, however, be emphasized that the entire spectrum is greatly affected by the building model and its damping used to develop the spectrum.

Floor response spectra are affected significantly by variations in soil properties

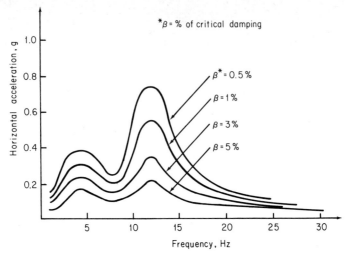

Fig. 10 Floor-acceleration response spectrum of a hypothetical building with floor elevation of 905 ft.

(estimated linear and angular stiffness, and damping values), since these variations strongly influence the natural frequencies of a model of a soil-foundation-structure system. Because of the relatively unknown nature of the soil properties, a range of these values is usually used to generate floor response spectra. This method tends to produce broader peaks in the curves, subjecting equipment items to high acceleration values over a wider frequency range.

To verify the seismic adequacy of equipment through testing, the floor motion must be known. In most cases the floor motion may be regarded as the maximum combined-mode acceleration of the floor in the frequency range above 20 to 30 Hz as determined from the floor-response-spectrum curve. The floor motion is not the same as the floor response spectrum.

SEISMIC QUALIFICATION METHODS

Analysis and testing, two common seismic qualification methods, are discussed here. Although an integral part of a seismic qualification program, seismic design is not, in the present state of the art, fully developed in the form of design formulas or procedures. Hence seismic design is given a rather brief treatment.

8. Seismic-Analysis Methods. Seismic analysis of electrical equipment such as switchgear and motor controls requires a knowledge of proper mathematical modeling of the equipment and its damping, and perhaps the use of digital programs based on the finite-element technique.

Two common dynamic-analysis methods are the time-history analysis and the response-spectrum technique.

a. Time-History Analysis. In this method a multi-degree-of-freedom model of equipment on a floor is subjected to the time-history response of the floor; the floor time-history response may be in acceleration, velocity, or displacement and is determined from the dynamic analysis of the building-foundation-soil system subjected to a given earthquake record. Hence, each time-history analysis is valid only for a given earthquake record and cannot guarantee the equipment withstand capability against other quakes; this is particularly true, since there is a tendency for each earthquake to be different from others. Because of this, the time-history method may need to be performed with three or four different earthquake histories.

Results of this analysis are the equipment responses as functions of time. Computer time required can be considerable. The advantage of the method is that some marginal equipment may be qualified through this method. See Arts. 11 and 12 for a further technical discussion.

b. Response-Spectrum Method. It was noted earlier that each natural mode of a multidegree system responds to any given excitation, and that the system dynamic response consists of its modal contributions. This analysis is known as the normal-mode method.

In the response-spectrum method a multidegree model of equipment or a structure is used in conjunction with the modal analysis and response spectra to compute loads or stresses in each of its natural modes. Since response spectra are used, each modal response represents a pseudo-maximum response and the time-phase relationship among the natural modes is eliminated. Therefore total or combined mode response is usually computed on a square-root sum-of-squares basis.

Consider the two-mass example (Fig. 5) subjected to the velocity response shown in Fig. 9. It can be seen that the spectral-velocity value associated with the second natural frequency of $\sqrt{6}$ rad/s (or 0.389 Hz) is approximately 42 in/s for 5% damping. This value multiplied by certain parameters gives a modal contribution toward displacement, velocity, or acceleration. Since the phase relationship between the two natural modes has been lost, for the final results each modal contribution is added either absolutely (which usually results in a conservative analysis) or on a square-root sum-of-squares basis.

Digital programs based on the finite-element technique may be used to perform the response-spectrum analysis. This method is perhaps the best compromise between rigorous analysis and design conservatism.

9. Seismic Tests. In the present state of the art, verification by testing may often be desirable in evaluating complex structures such as switchgear and motor controls with moving or active parts. This assumption is true, in part, because of the difficulty involved in developing an adequate mathematical model and accounting for possible structural nonlinearity.

Especially for the initial certification of equipment, it appears prudent to test it and gather valuable information on its damping, natural frequencies, mode shapes, and nonlinear behavior or response. The accumulated test data may then be utilized to meet as many seismic specifications as possible or to verify analytical results of the same or similar equipment.

Seismic testing is performed on a shaker table, test bed, or suitably constructed test fixture. A knowledge of the floor motion is necessary for equipment seismic testing. However, floor-response-spectrum curves usually furnished to the equipment supplier do not directly provide information on the actual floor motion. Hence care must be exercised in extracting required test input levels from the response-spectrum curves.

Since stiff single-degree systems tend to oscillate with the floor, the peak acceleration level of a floor can be estimated as the asymptote at high frequencies of the floor-acceleration response spectrum. But the frequency content of the peak is not conveyed by the asymptote. A floor response spectrum exhibits maximum single-degree oscillator response near building resonant frequencies.

Some seismic specifications call for a simultaneous input of the vertical and horizontal accelerations. Many tests are, however, performed separately in each of the three mutually orthogonal directions — i.e., front to back, side to side, and up and down.

Measuring equipment response during an actual earthquake would constitute a true seismic test for the equipment. However, this type of test is neither easy nor practical. Various types of seismic tests are based on different input motions.

a. Time-History Input. An actual floor motion in terms of time-history record is applied to the equipment base. This motion reproduces an actual earthquake most realistically. But, because of the tendency for an earthquake motion to be unique, a test based on one particular earthquake does not assure seismic adequacy against future ones.

b. Broad-Band Random Input. A randomly generated simulated history of appropriate frequency content is applied.

c. Narrow-Band Random Input. A time-varying motion of essentially one frequency is applied. A special case is sine-beat testing, in which the amplitude envelope is sinusoidal (Refs. 9, 10).

In sine-beat testing, sinusoidal beats capable of reproducing floor response spectrum similar to an actual earthquake motion can be used to simulate reasonably well a typical earthquake motion. The buildup and decay of the beat tend to reduce amplification and number of stress cycles. Hence this method is particularly suitable for equipment with low or little damping. On the other hand, care should be used in exciting equipment only in the vicinity of its apparent cabinet resonances, because a simple analysis can show that resonant frequencies of smaller components, which are often away from those of the cabinet, may not be excited.

The motion of the sinusoidal beat test consists of sinusoidal resonant frequency with an amplitude producing the maximum base acceleration modulated by a sinusoid of a lower frequency. The modulated beat contains a certain number of resonant-frequency cycles. A time duration is allowed between beats to eliminate any sizable superposition of motion.

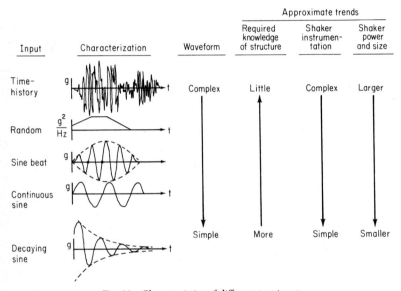

Fig. 11. Characteristics of different test inputs.

d. Continuous Sine Test. A constant-amplitude motion of a single frequency is applied to the equipment base. Its advantage is the ease with which the input motion can be generated, while disadvantages are a high output buildup near system resonances and possible fatigue failure due to a large number of fluctuations. However, only three to five sine waves with an appropriate amplitude are used, since the duration of peak earthquake acceleration is relatively short.

e. Decaying Sine Test. The input motion is of a single frequency with amplitude decreasing with time, which may be generated as a transient response of a suspended table from an initial displacement (Ref. 11).

These input waveforms are shown in Fig. 11 (Ref. 12). It is apparent that some waveforms are different from actual seismic excitations. However, the use of these waveforms may be justified if they can simulate important characteristics of seismic excitations.

The waveforms shown may be divided into the following two groups: the first group (time-history and random) contains multiple frequencies in its input, and the second consists essentially of a single frequency.

It has been shown (Refs. 5, 9) by analysis and actual recording during earthquakes that dynamic characteristics of a building tend to filter much of the frequency contents of the ground motion, and amplify acceleration components with frequencies close to its resonant frequencies. Therefore, the floor motion corresponds more closely to that of a single frequency with time-varying random amplitude than does the ground motion.

For switchgear and motor controls supported on a floor, therefore, a single-frequency input excitation can reasonably simulate the floor motion and is valid, provided there is a sufficient separation among the structure resonant frequencies in the frequency range of interest (usually up to 25 or 30 Hz).

10. Seismic Design.   No detailed and specific seismic design guide for switchgear and motor controls is available in the present state of the art.   The objective is therefore to outline briefly some general steps or considerations to be taken in these equipment designs.

The most important consideration is to design the equipment so that resonance is avoided.   If the equipment is on a floor of a building, the equipment resonances should not coincide with the building resonant frequencies.   Similarly, components must not be in resonance with the cabinets or other support structure.

If resonance or near resonance is unavoidable, damping must be increased either in the equipment itself or in its support (in the form of a damper system) to reduce vibration amplitudes.

Establishing dynamic amplification levels for cabinet structures, and acceleration and frequency withstand levels for components and instruments, can aid immeasurably in design processes.   This consideration is particularly true for a line of products with similar designs but different sizes and weights.

Since the 20- to 35-Hz frequency range is considered a dividing line or cutoff frequency for a rigid body, every effort should be made, if possible, to design the equipment so that its fundamental resonant frequency is above 20- to 35-Hz range.   It should be noted in this regard that tightening or stiffening the equipment support greatly influences the equipment resonant frequency and its dynamic-response characteristics.

SUMMARY OF ANALYSIS OF ELASTIC SYSTEMS

An earthquake response analysis is in essence a classical problem in the theory of structural dynamics, wherein a multi-degree-of-freedom structure or system is subjected to a time-varying excitation at its foundation.   Hence, the evaluation of the structural physical properties (mass, stiffness, and damping characteristics) and the use of valid earthquake motions are extremely important factors for any seismic analysis.   However, these critical factors are assumed to be accounted for.   The techniques of dynamic-response analysis are summarized — first for simple systems, followed by multidegree systems.

11. Analysis of Single-Degree-of-Freedom System.   The equation of motion for a single-degree system and its solution are summarized, followed by a summary of results for undamped and damped cases.   Some results of the base or support motion are also presented.

a. Equation of Motion.   The equation of motion for a single-degree system with viscous damping (Fig. 12) is

$$m\ddot{y} + c\dot{y} + ky = F_0[f(t)] \tag{7}$$

where y, $\dot{y}$, and $\ddot{y}$ are, respectively, the displacement, velocity, and acceleration of the mass m; c and k are the viscous-damping coefficient and spring constant, respectively; and F_0 and $f(t)$ are, respectively, a constant force amplitude and a nondimensional time function.

b. Undamped System.   For the free vibration with no damping, $c = 0$ and $F_0[f(t)] = 0$ so that the equation of motion becomes

$$m\ddot{y} + ky = 0 \tag{8}$$

and its solution is

$$y(t) = \frac{\dot{y}_0}{\omega} \sin(\omega t) + y_0 \cos(\omega t) \tag{9}$$

where y_0 and $\dot{y}_0$ are, respectively, the initial displacement and velocity, and ω is the circular natural frequency given as

$$\omega = \sqrt{\frac{k}{m}} \qquad \text{rad/s} \tag{10}$$

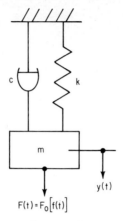

Fig. 12 A single-degree system with viscous damping subjected to a disturbing force.

The dynamic response of an undamped, linearly elastic single-degree system subjected to an arbitrary forcing function $F(t) = F_0[f(t)]$ and initial conditions y_0 and $\dot{y}_0$ is given by the expression

$$y = y_0 \cos(\omega t) + \frac{\dot{y}_0}{\omega} \sin(\omega t) + y_s \omega \int_0^t f(\tau) \sin[\omega(t - \tau)]\, d\tau \qquad (11)$$

in which $y_s = F_0/k = F_0/(m\omega)$ is the static displacement due to F_0. In Eq. (11) the first two terms account for initial conditions, while the last term, representing the response due to the forcing function, may be determined by computing the system response to a unit impulse and adding the contribution of all infinitesimal impulses between zero and the time of interest t (superposition principle for linear systems). Hence the evaluation of the integral in Eq. (11) involving the function $f(t)$ yields the dynamic response.

c. Damped System.   The equation of motion for free vibration is [see Eq. (7)]

$$m\ddot{y} + c\dot{y} + ky = 0 \qquad (12)$$

and its solution [for the case of $\beta < \omega$ where $\beta = c/(2m)$ is a measure of damping given as a percent of the critical damping] is

$$y = e^{-\beta t}\left[\frac{\dot{y}_0 + \beta y_0}{\omega_d} \sin(\omega_d t) + y_0 \cos(\omega_d t)\right] \qquad (13)$$

in which

$$\omega_d = \sqrt{\omega^2 - \beta^2} \qquad (14)$$

is the system damped natural frequency. The response in Eq. (13) may be considered as a harmonic function with decaying amplitudes; the rate of the decay is related to β or damping of the system.

Consider now a special case when β approaches ω. Then ω_d approaches zero (which implies there is no frequency of oscillation), $\cos(\omega_d t) \to 1$, and $\sin(\omega_d t) \to (\omega_d t)$. Hence Eq. (13) reduces to

$$y = e^{-\omega t}[\dot{y}_0 t + (1 + \omega t)y_0] \qquad (15)$$

The motion described by Eq. (15) is no longer periodic; the system, instead of oscillating, merely returns to its neutral position. The amount of damping at which this phenomenon occurs is known as the critical damping. For critical damping,

$$\omega = \beta = \frac{c_{cr}}{2m} \qquad (16)$$

so that

$$c_{cr} = 2m\omega = 2\sqrt{km} \tag{17}$$

where c_{cr} is the critical-viscous-damping coefficient.

It is seen from Eq. (14) that the difference between ω and ω_d is quite small even for a system with $\beta = 10$ or 15% of the critical damping. Therefore, the decrease in natural frequency due to damping may be neglected for most practical applications.

The dynamic response of the damped system (Fig. 11) at any time t including initial conditions and subjected to a forcing function $F(t) = F_0[f(t)]$ is

$$y(t) = e^{-\beta t}\left[\frac{\dot{y}_0 + \beta y_0}{\omega_d}\sin(\omega_d t) + y_0\cos(\omega_d t)\right] + \frac{y_s\omega^2}{\omega_d}\int_0^t f(\tau)e^{-\beta(t-\tau)}\sin[\omega_d(t-\tau)]\,d\tau \tag{18}$$

The first two terms in Eq. (18) representing the contribution of the free vibration become negligible (because of damping) after some cycles, while the last term is the forced-vibration response. Equation (18) reduces to Eq. (11) if $\beta = 0$.

If the disturbing force is of the sinusoidal form

$$F(t) = F_0[f(t)] = F_0\sin(\Omega t) \tag{19}$$

then y_{ss}, the steady-state part of the response, is given by

$$y_{ss} = \frac{F_0}{k}\frac{(1 - r^2)\sin(\Omega t) - 2(\beta\Omega/\omega^2)\cos(\Omega t)}{(1 - r^2)^2 + (2\beta\Omega/\omega^2)^2} \tag{20}$$

where $r = \Omega/\omega$. The maximum value of Eq. (20), known as the dynamic magnification factor, is

$$\tau = \frac{(y_{ss})_{\max}}{y_s} = \left[(1 - r^2)^2 + \left(\frac{2\beta\Omega}{\omega^2}\right)^2\right]^{-1/2} \tag{21}$$

At resonance (i.e., $\Omega = \omega$), Eq. (21) becomes

$$\tau_r = \frac{1}{2(c/c_{cr})} = \frac{1}{2\zeta} \tag{22}$$

Equation (22) states that the ratio of dynamic response to static response, at resonance, of a single-degree damped system subjected to a sinusoidally fluctuating force [Eq. (19)] is inversely proportional to the damping expressed as a percent of its critical damping.

d. Base Motions. For earthquake analysis and design of structures and equipment, the determination of their response to movement of the base or support becomes necessary.

For the undamped single-degree system (Fig. 13) subjected to a base motion $y_b = y_B f_B(t)$, the equation of motion is

$$m\ddot{y} + k(y - y_b) = 0 \tag{23}$$

and the forced-vibration part of the general solution given by Eq. (11) becomes

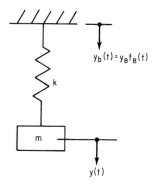

Fig. 13 An undamped single-degree system subjected to a base motion.

$$y = y_B\omega \int_0^t f(\tau) \sin\left[\omega(t - \tau)\right] d\tau \tag{24}$$

When the base acceleration is used in the equation of motion, it is convenient to introduce the relative displacement defined as

$$u = y - y_b \tag{25}$$

Then Eq. (23) becomes

$$m\ddot{u} + ku = -m\ddot{y}_b = -m\ddot{y}_B f_B(t) \tag{26}$$

in which $f_B(t)$ is the time function of the base acceleration. Using Eq. (11), the general solution for the relative motion is

$$u = -\frac{\ddot{y}_B}{\omega} \int_0^t f_B(\tau) \sin\left[\omega(t - \tau)\right] d\tau \tag{27}$$

for no initial conditions.

For damped single-degree systems, it is more convenient to write the equation of motion in terms of the base acceleration. Hence Eq. (26) becomes (adding the viscous-damping term)

$$m\ddot{u} + c\dot{u} + ku = -m\ddot{y}_b = -m\ddot{y}_B f_B(t) \tag{28}$$

The general solution of Eq. (28) is

$$u = -\frac{\ddot{y}_B}{\omega_d} \int_0^t f_B(\tau) e^{-\beta(t-\tau)} \sin\left[\omega_d(t - \tau)\right] d\tau \tag{29}$$

if $u_0 = \dot{u}_0 = 0$ (i.e., no initial conditions). Equation (29) is useful in the following discussion of earthquake analysis of multidegree systems.

12. Analysis of Multi-Degree-of-Freedom System (Lumped-Mass System). The finite-element technique of structural analysis mentioned earlier makes possible a good engineering analysis of complex continuous systems subjected to arbitrary loadings and boundary conditions. The most important step in this technique is to idealize a structure by assuming it to be an assemblage of a finite number of "meshes" called elements interconnected at a finite number of points called nodes or joints. A detailed treatment of this topic is beyond the purpose of this section. It can be simply assumed that this structural idealization has been achieved and the masses are lumped at nodes so that the resulting mass matrix is diagonal. Furthermore, these masses are assumed to be connected by massless, equivalent springs representing stiffness of the structure; these springs are expressed by the stiffness matrix discussed below.

a. Determination of Natural Frequencies and Modes. The equations of motion for a system with N masses, N degrees of freedom, and no external forces are

$$\begin{aligned}
m_1\ddot{y}_1 + k_{11}y_1 + k_{12}y_2 + \cdots + k_{1N}y_N &= 0 \\
m_2\ddot{y}_2 + k_{21}y_1 + k_{22}y_2 + \cdots + k_{2N}y_N &= 0 \\
\cdots \cdots \cdots \cdots \cdots \cdots \cdots \cdots \cdots & \\
\cdots \cdots \cdots \cdots \cdots \cdots \cdots \cdots \cdots & \\
\cdots \cdots \cdots \cdots \cdots \cdots \cdots \cdots \cdots & \\
m_N\ddot{y}_N + k_{N1}y_1 + k_{N2}y_2 + \cdots + k_{NN}y_N &= 0
\end{aligned} \tag{30}$$

in which the k's and the y's are, respectively, the stiffness coefficients (spring constants) and the displacements of the lumped masses m's. In matrix form Eq. (30) may be written as

$$\begin{bmatrix}
m_1 & 0 & 0 & \cdots & 0 \\
0 & m_2 & 0 & \cdots & 0 \\
\cdot & \cdot & \cdot & \cdots & \cdot \\
\cdot & \cdot & \cdot & \cdots & \cdot \\
\cdot & \cdot & \cdot & \cdots & \cdot \\
0 & 0 & 0 & \cdots & m_N
\end{bmatrix}
\begin{Bmatrix}
\ddot{y}_1 \\ \ddot{y}_2 \\ \cdot \\ \cdot \\ \cdot \\ \ddot{y}_N
\end{Bmatrix}
+
\begin{bmatrix}
k_{11} & k_{12} & \cdots & k_{1N} \\
k_{21} & k_{22} & \cdots & k_{2N} \\
\cdot & \cdot & \cdots & \cdot \\
\cdot & \cdot & \cdots & \cdot \\
\cdot & \cdot & \cdots & \cdot \\
k_{N1} & k_{N2} & \cdots & k_{NN}
\end{bmatrix}
\begin{Bmatrix}
y_1 \\ y_2 \\ \cdot \\ \cdot \\ \cdot \\ y_N
\end{Bmatrix}
=
\begin{Bmatrix}
0 \\ 0 \\ \cdot \\ \cdot \\ \cdot \\ 0
\end{Bmatrix} \tag{31}$$

or, more compactly,

$$[m]_d\{\ddot{y}\} + [k]\{y\} = \{0\} \tag{32}$$

where $[m]_d$ is the diagonal mass matrix and $[k]$ the stiffness matrix. The displacement vector $\{y\}$ describes vibratory motion of this system.

In Eq. (31) the mass matrix is diagonal, so that the equations are said to be dynamically decoupled. However, the stiffness matrix is not diagonal and the equations are said to be coupled statically or elastically. If both the mass and the stiffness matrices are diagonal, the equations will be completely decoupled and each equation may be solved independently of the others.

The free-vibration equation represented by Eq. (31) or (32) may be solved by assuming the solution displacement vector $\{y\}$ to be harmonic with the frequency parameter ω—i.e., $\{y\} = \{\phi\}e^{-i\omega t}$. Substitution of this assumed solution into Eq. (31) results in

$$\begin{bmatrix} (k_{11} - m_1\omega^2) & k_{12} & \cdots & k_{1N} \\ k_{21} & (k_{22} - m_2\omega^2) & \cdots & k_{2N} \\ \cdot & \cdot & \cdots & \cdot \\ \cdot & \cdot & \cdots & \cdot \\ \cdot & \cdot & \cdots & \cdot \\ k_{N1} & k_{N2} & \cdots & (k_{NN} - m_N\omega^2) \end{bmatrix} \begin{Bmatrix} \phi_1 \\ \phi_2 \\ \cdot \\ \cdot \\ \cdot \\ \phi_N \end{Bmatrix} = \begin{Bmatrix} 0 \\ 0 \\ \cdot \\ \cdot \\ \cdot \\ 0 \end{Bmatrix} \tag{33}$$

The nontrivial solution of Eq. (33) exists only if the determinant of its matrix is zero:

$$\begin{vmatrix} (k_{11} - m_1\omega^2) & k_{12} & & k_{1N} \\ k_{21} & (k_{22} - m_2\omega^2) & \cdots & k_{2N} \\ \cdot & \cdot & \cdots & \cdot \\ \cdot & \cdot & \cdots & \cdot \\ \cdot & \cdot & \cdots & \cdot \\ k_{N1} & k_{N2} & & (k_{NN} - m_N\omega^2) \end{vmatrix} = 0 \tag{33a}$$

The polynomial equation resulting from expanding the above determinant is called the secular or frequency equation. It has a set of N roots, which are related to the system natural frequencies. Substituting each of these roots into Eq. (33), a displacement vector, whose amplitudes are relative, may be determined. In other words, associated with each eigenvalue ω_i, there is an eigenvector $\{\phi\}_i$. A set of N eigenvectors represents all N-mode shapes of the system. The problem as posed by Eq. (31) is known as an eigenvalue problem and the parameters ω its eigenvalues.

Consider now the two-mass system shown in Fig. 14. The equations of motion for the free vibration are

$$\begin{aligned} m_1\ddot{y}_1 + k_1 y_1 - k_2(y_2 - y_1) &= 0 \\ m_2\ddot{y}_2 + k_2(y_2 - y_1) + k_3 y_2 &= 0 \end{aligned} \tag{34}$$

or, in matrix form,

$$\begin{bmatrix} m_1 & 0 \\ 0 & m_2 \end{bmatrix} \begin{Bmatrix} \ddot{y}_1 \\ \ddot{y}_2 \end{Bmatrix} + \begin{bmatrix} (k_1 + k_2) & -k_2 \\ -k_2 & (k_2 + k_3) \end{bmatrix} \begin{Bmatrix} y_1 \\ y_2 \end{Bmatrix} = \begin{Bmatrix} 0 \\ 0 \end{Bmatrix} \tag{35}$$

If $k_1 = k_2 = k_3 = k$ and $m_1 = m_2 = m$, then Eq. (35) becomes

$$\begin{bmatrix} m & 0 \\ 0 & m \end{bmatrix} \begin{Bmatrix} \ddot{y}_1 \\ \ddot{y}_2 \end{Bmatrix} + \begin{bmatrix} 2k & -k \\ -k & 2k \end{bmatrix} \begin{Bmatrix} y_1 \\ y_2 \end{Bmatrix} = \begin{Bmatrix} 0 \\ 0 \end{Bmatrix} \tag{36}$$

Assuming the solution in the form $\{y\} = \{\phi\}e^{-i\omega t}$, or

$$\begin{Bmatrix} y_1 \\ y_2 \end{Bmatrix} = \begin{Bmatrix} \phi_1 \\ \phi_2 \end{Bmatrix} e^{-i\omega t} \tag{37}$$

and substituting into Eq. (36),

$$\begin{bmatrix} (2k - m\omega^2) & -k \\ -k & (2k - m\omega^2) \end{bmatrix} \begin{Bmatrix} \phi_1 \\ \phi_2 \end{Bmatrix} = \begin{Bmatrix} 0 \\ 0 \end{Bmatrix} \tag{38}$$

The nontrivial solution of Eq. (38) exists only if [see Eq. (33a)]

$$\begin{vmatrix} (2k - m\omega^2) & -k \\ -k & (2k - m\omega^2) \end{vmatrix} = 0 \tag{39}$$

or, expanding Eq. (39) and solving for ω^2,

$$\omega_1 = \sqrt{\frac{k}{m}} \quad \text{and} \quad \omega_2 = \sqrt{\frac{3k}{m}} \tag{40}$$

which are the two natural frequencies of the system. Substituting $\omega_1 = \sqrt{k/m}$ into Eq. (38),

$$\begin{bmatrix} k & -k \\ -k & k \end{bmatrix} \begin{Bmatrix} \phi_1 \\ \phi_2 \end{Bmatrix} = \begin{Bmatrix} 0 \\ 0 \end{Bmatrix}$$

or

$$\phi_1 - \phi_2 = 0$$
$$-\phi_1 + \phi_2 = 0$$

both of which yield the result that $\phi_1 = \phi_2$, but do not give the absolute magnitudes of ϕ_1 or ϕ_2. The eigenvector (or mode shape) corresponding to ω_1 is written as

$$\{\phi\}_1 = \begin{Bmatrix} \phi_{11} \\ \phi_{21} \end{Bmatrix} = \begin{Bmatrix} 1 \\ 1 \end{Bmatrix} \tag{41}$$

where ϕ_{11} and ϕ_{21} are, respectively, the relative amplitudes at mass 1 and 2 for the first natural frequency. Similarly, the eigenvector corresponding to the second natural frequency ω_2 is

$$\{\phi\}_2 = \begin{Bmatrix} \phi_{12} \\ \phi_{22} \end{Bmatrix} = \begin{Bmatrix} 1 \\ -1 \end{Bmatrix} \tag{42}$$

The orthogonal modal transformation matrix denoted by $[\phi]$ is formed by combining the eigenvectors given by Eqs. (41) and (42) as follows:

$$[\phi] = [\{\phi\}_1 \{\phi\}_2] = \begin{bmatrix} \phi_{11} & \phi_{12} \\ \phi_{21} & \phi_{22} \end{bmatrix} = \begin{bmatrix} 1 & 1 \\ 1 & -1 \end{bmatrix} \tag{43}$$

The modal transformation matrix $[\phi]$ is useful in transforming the equations of motion expressed in the present coordinate system $\{y\}$ into those written in a new coordinate

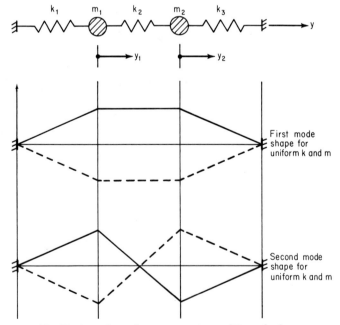

Fig. 14 An undamped two-mass system and its mode shapes.

system [see Eq. (50)]. When expressed in this new coordinate system, the equations of motion are completely decoupled. In other words, the mass and the stiffness matrices are diagonal in the new coordinate system. The diagonalization is possibly due, among other factors, to the following two important orthogonality relationships (proof not given).

The first orthogonality relationship for normal modes is expressed as

$$[\phi]^T[m]_d[\phi] = [M]_d \qquad (44)$$

where $[\phi]^T$ is the transpose of $[\phi]$ and $[M]_d$ is called the generalized mass matrix. The second orthogonality condition is

$$[\phi]^T[k][\phi] = [K]_d = \lceil \omega^2 \rceil_d [M]_d \qquad (45)$$

in which $[K]_d$, known as the generalized stiffness matrix, is a diagonal matrix. For the above example, Eq. (44) yields

$$\begin{bmatrix} 1 & 1 \\ 1 & -1 \end{bmatrix}\begin{bmatrix} m & 0 \\ 0 & m \end{bmatrix}\begin{bmatrix} 1 & 1 \\ 1 & -1 \end{bmatrix} = m\begin{bmatrix} 2 & 0 \\ 0 & 2 \end{bmatrix} \qquad (46)$$

and Eq. (45) gives

$$\begin{bmatrix} 1 & 1 \\ 1 & -1 \end{bmatrix}\begin{bmatrix} 2k & -k \\ -k & 2k \end{bmatrix}\begin{bmatrix} 1 & 1 \\ 1 & -1 \end{bmatrix} = k\begin{bmatrix} 2 & 0 \\ 0 & 6 \end{bmatrix} = [K]_d \qquad (47)$$

b. Normal-Mode Method. An important advantage of the normal-mode method in seeking the solution for a forced-vibration problem is that the coupled equations of motion can be decoupled; this enables each equation to be solved separately one at a time.

Consider the matrix equations of motion for a multidegree system with disturbing external forces (Refs. 13, 6, 7):

$$[m]_d\{\ddot{y}\} + [k]\{y\} = \{F(t)\} \qquad (48)$$

$$\text{where } \{F(t)\} = \begin{Bmatrix} F_1 \\ F_2 \\ \cdot \\ \cdot \\ \cdot \\ F_N \end{Bmatrix} f(t) \qquad (49)$$

and N is the number of degrees of freedom, $F_1, F_2, \ldots, F_N$ the force magnitudes, and $f(t)$ a nondimensional force function. The modal transformation relationship is introduced through the orthogonal modal transformation matrix $[\phi]$:

$$\{y\} = [\phi]\{\eta\} \qquad (50)$$

where $\{\eta\}$, the generalized displacement vector, is a function of time and may also be viewed as the new coordinate system. Note in Eq. (50) that the orthogonal modal transformation matrix $[\phi]$ provides a linkage between the original ($\{y\}$) and the new $\{\eta\}$ coordinate systems. Substituting Eq. (50) into Eq. (48),

$$[m]_d[\phi]\{\ddot{\eta}\} + [k][\phi]\{\eta\} = \{F(t)\}$$

Premultiplying in the above equation by $[\phi]^T$,

$$[\phi]^T[m]_d[\phi]\{\ddot{\eta}\} + [\phi]^T[k][\phi]\{\eta\} = [\phi]^T\{F(t)\} \qquad (51)$$

or, utilizing Eqs. (44) and (45),

$$[M]_d\{\ddot{\eta}\} + [\omega^2]_d[M]_d\{\eta\} = [\phi]^T\{F(t)\} \qquad (52)$$

or, dividing by $[M]_d$,

$$\{\ddot{\eta}\} + [\omega^2]_d\{\eta\} = \frac{[\phi]^T\{F(t)\}}{[M]_d} \qquad (53)$$

A comparison of Eq. (52) with Eq. (48) shows that the equations in Eq. (52) are decoupled when expressed in a new coordinate system $\{\eta\}$ related to the original coordinate system $\{y\}$ through the transformation matrix $[\phi]$ according to Eq. (50). The nth modal equation of Eq. (52) is

$$M_n\ddot{\eta}_n + K_n\eta_n = P_n(t) \tag{54}$$

where M_n, the generalized mass for the nth mode, is determined from

$$M_n = \{\phi\}_n^T[m]_d\{\phi\}_n \tag{55}$$

while K_n, the generalized stiffness for the nth mode, is

$$K_n = \{\phi\}_n^T[k]\{\phi\}_n \tag{56}$$

and the generalized load $P_n(t)$ is

$$P_n(t) = \{\phi\}_n^T\{F(t)\} \tag{57}$$

It is important to note that Eq. (54) is identical in form to the equation of motion for a single-degree system and hence most of the solution technique for the simple system can be utilized to solve Eq. (54).

Following the solution of η for each mode, the final solution $\{y\}$ is evaluated from Eq. (50).

   c. *Response to Base Motion.*   Consider the equations of motion for a damped multi-degree system written in terms of the relative displacement u:

$$[m]_d\{\ddot{u}\} + [c]\{\dot{u}\} + [k]\{u\} = \{F_B(t)\} \tag{58}$$

where $\{u\} = \{y\} - \{y_b\}$ is the relative displacement with respect to the base displacement $\{y_b\}$, and $\{F_B(t)\}$ is the effective-force vector at the base.

If the viscous-damping matrix is assumed to be proportional to the mass matrix as

$$[c] = 2[\zeta]_d[\omega]_d[m]_d \tag{59}$$

where ζ represents the damping ratio, then

$$[\phi]^T[c][\phi] = 2[\phi]^T[\zeta]_d[\omega]_d[m]_d[\phi] = 2[\zeta]_d[\omega]_d[M]_d \tag{60}$$

in which Eq. (44) was used.   Hence, introducing the transformation

$$\{u\} = [\phi]\{\eta\} \tag{61}$$

into Eq. (58), premultiplying by $[\phi]^T$, and using Eqs. (44), (45), and (60),

$$[M]_d\{\ddot{\eta}\} + 2[\zeta]_d[\omega]_d[M]_d\{\dot{\eta}\} + [\omega^2]_d[M]_d[\eta] = [\phi]^T\{F_B(t)\} = \{P(t)\} \tag{62}$$

where $\{P(t)\} = [\phi]^T\{F_B(t)\}$ is the generalized effective-force vector. The nth modal equation of Eq. (62) is

$$M_n\ddot{\eta}_n + 2\zeta_n\omega_n M_n\dot{\eta}_n + \omega^2 M_n\eta_n = \{\phi\}_n^T\{F_B(t)\} = P_n(t) \tag{63}$$

The effective-force vector $\{F_B(t)\}$ is given as

$$\{F_B(t)\} = [m]_d[I]_d a_g(t) \tag{64}$$

where $a_g(t)$ is the acceleration-time history of the base or ground and $[I]_d$ is an $(N \times 1)$ unit vector with all components equal to unity.   Hence the generalized effective-force vector is

$$\{P(t)\} = [\phi]_n^T\{F_B(t)\} = [\phi]_n^T[m]_d[I]_d a_g(t) \tag{65}$$

and its nth component is

$$P_n(t) = \mathscr{L}_n a_g(t) \tag{66}$$

where $\mathscr{L}_n$, known as the earthquake-participation factor for the nth mode, is given as

$$\mathscr{L}_n = \{\phi\}_n^T[m]_d[I]_d \tag{67}$$

Substitution of Eq. (66) into (63) results in

$$\ddot{\eta}_n + 2\zeta_n\omega_n\dot{\eta}_n + \omega_n^2\eta_n = \frac{\mathscr{L}_n a_g(t)}{M_n} \tag{68}$$

The solution of Eq. (68) is

$$\eta_n(t) = \frac{\mathscr{L}_n}{M_n\omega_n} \int_0^t a_g(\tau)e^{-\zeta_n\omega_n(t-\tau)} \sin\left[\omega_n(t-\tau)\right] d\tau \tag{69}$$

After the solutions $\eta_n(t)$ are evaluated from Eq. (69), the relative displacements are determined from Eq. (61).

d. Response-Spectrum Analysis. Note that the relative-displacement response history $\{u\}$ of a damped multidegree system is computed from Eq. (61) following the evaluation of its modal responses $\{\eta\}$ from Eq. (69). Since the response expression for any mode given by Eq. (69) is identical in form to that presented previously for a single-degree system, the maximum response of any mode can be determined with the use of the earthquake response spectra.

Introducing the spectral velocity for the nth mode S_{vn}, defined as the maximum value of the integral in Eq. (69), i.e.,

$$S_{vn} = \left[\int_0^t a_g(\tau)e^{-\zeta_n\omega_n(t-\tau)} \sin\left[\omega_n(t-\tau)\right] d\tau\right]_{max} \tag{70}$$

into Eq. (69) (Refs. 5, 13),

$$\eta_n|_{max} = \frac{\mathscr{L}_n}{M_n}\frac{S_{vn}}{\omega_n} \tag{71}$$

in which $\mathscr{L}_n/M_n$, known as the modal participation factor of the nth mode, may be viewed as a measure of the nth modal contribution to the total system response. Defining the spectral displacement for the nth mode S_{dn} as

$$S_{dn} = \frac{S_{vn}}{\omega_n} \tag{72}$$

Eq. (71) may be written as

$$\eta_n|_{max} = \frac{\mathscr{L}_n S_{dn}}{M_n} \tag{73}$$

Finally, the relative maximum-displacement vector is

$$\{u\}_{max} = [\phi]\{\eta\}_{max} \tag{74}$$

where the vector $\{\eta\}_{max}$ consists of the elements $\eta_n|_{max}$ defined in Eq. (73).

It should be noted that the determination of the maximum response of each normal mode as given by Eq. (73) does not lead to a complete solution of the system dynamic response, because the time-phase relationship has been eliminated owing to the use of the response spectrum. However, an approximation to the total maximum response, based on probability considerations, may be achieved by the root-mean-square procedure involving the modal maximum responses.

This response-spectrum technique makes an approximate determination of the earthquake response of multidegree systems possible through the use of the earthquake response spectra and without performing a complete time-history analysis.

REFERENCES

1. "Earthquakes," U.S. Department of the Interior, Geological Survey, 1971.
2. J. D. Stevenson, "Engineering and Marketing Guide to Seismic Design Requirements for Nuclear Plant Equipment," Lecture Notes, 1970.
3. "Guide for Qualification of Class I Electrical Equipment for Nuclear Power Generating Stations," IEEE Document 344–1971, 1971.
4. W. C. Hurty and M. F. Rubinstein, "Dynamics of Structures," Prentice-Hall, Inc., Englewood Cliffs, N.J., 1964.
5. L. R. Wiegel (ed.), "Earthquake Engineering," Prentice-Hall, Inc., Englewood Cliffs, N.J., 1970.

6. J. S. Przemieniecki, "Theory of Matrix Structural Analysis," McGraw-Hill Book Company, New York, 1968.
7. O. C. Zienkiewicz and I. K. Cheung, "The Finite Element Method in Engineering Science," 2d ed., McGraw-Hill Book Company, New York, 1971.
8. M. F. Rubinstein, "Structural Systems — Statics, Dynamics and Stability," Prentice-Hall, Inc., Englewood Cliffs, N.J., 1970.
9. E. G. Fischer, "Sine Beat Vibration Testing Related to Earthquake Response Spectra," 42d Shock and Vibration Symposium, Department of Defense, Naval Research Laboratories, Key West, Fla., Nov. 2, 1971.
10. A. P. Colaiaco and W. S. Albert, "Seismic Testing of Metal Clad and Metal Enclosed Switchgear Using Sine Beat Vibrations," IEEE Power Engineering Society Winter Meeting, Jan. 30–Feb. 4, 1972.
11. C. R. Gallant et al., A Testing Program for Qualification of Switchgear to Be Subjected to an Earthquake Experience, IEEE *Trans. Power Apparatus Syst.*, vol. PAS-90, no. 1, January/February 1971.
12. K. S. Joung, K. E. Rouch, S. H. Telander, and D. S. Totten, "Seismic Testing of Switchgear and Control Equipment," IEEE Paper T72 537-9 presented in San Francisco, Calif., July 1972.
13. J. M. Biggs, "Introduction to Structural Dynamics," McGraw-Hill Book Company, New York, 1964.

GENERAL REFERENCES

P. Barkan, "Some Observations on Seismic Failure and the Specification of High Voltage Electrical Apparatus," IEEE Summer Power Meeting, July 1971.

P. Beutler, Seismic Considerations for High Voltage Sub-Station Facilities, *IEEE Trans.*, 70TP529-PWR, July 1970.

J. M. Biggs and J. Roesset, "Seismic Analysis of Equipment Mounted on a Massive Structure," November 1968, private communication, June 1971.

J. J. Herrera, "Earthquake Considerations in Substation Design — Sylmar HVDC Converter Station," Panel Presentation, IEEE Summer Power Meeting, July 1971.

H. L. Holland, "City of Los Angeles AC Substations," Panel Presentation, IEEE Summer Power Meeting, July 1971.

T. E. Johnson and R. J. McCaffrey, "Current Techniques for Analyzing Structures and Equipment for Seismic Effects," undated internal report, Bechtel Corporation, San Francisco, Calif., private communication, October 1970.

C. H. Norris et al., "Structural Design for Dynamic Loads," McGraw-Hill Book Company, New York, 1959.

F. Novoa, "Earthquakes and the Substation Equipment — Arrangement and Specification," CIGRE Paper 23-02, 1970.

"Nuclear Reactors and Earthquakes," TID 7024, U.S. Atomic Energy Commission, Division of Reactor Development, Washington, D.C., 1963.

E. G. Pereboom, "Earthquake Considerations in Substation Design," Panel Presentation, IEEE Summer Power Meeting, July 1971.

W. A. Sells, "Introductory Remarks — A Brief Summary of the Causes of Earthquakes," Panel Presentation, IEEE Summer Power Meeting, July 1971.

T. Ushio et al., "Security of Aseismic Construction of High Voltage Switchgear," IEEE Paper T72 110-0, IEEE Winter Power Meeting, New York, Jan. 30–Feb. 4, 1972.

10

Load Considerations

W. G. LEFFERTS *

FOREWORD

For every load there is a choice of drive motor and control. Often, the motor and control are considered simultaneously as an electrical drive system. For example, in some cases the motor selected to drive a specific load may be either an induction- or synchronous-type motor. The type of control selected must correspond to the type of motor used. The combination chosen may be based on economics or on many other factors (Ref. 1).

The load characteristics not only affect the motor selection; they are also important in the choice of control equipment. Such items as enclosures, protection, starting equipment, speed control, and braking are largely based on the load characteristics and environment.

* Senior Engineer, Allis-Chalmers Corporation; Member, IEEE (retired).

Table 1 gives the load characteristics of a large variety of common types of machines found in industrial and processing plants. A careful reference to this table prior to the selection of control equipment for a particular type of machine may indicate problems to be considered, such as special ambient or torque conditions. While the conditions do not hold true in all cases, if attention is given to the possible severity of the load conditions, application problems may be avoided.

If the exact load is not given in the table, a similar load may be used as a rough guide in considering control equipment. However, further investigation of the load conditions in this case may be indicated.

CONSTANT-SPEED LOADS

1. Motor Load. The term "motor load" usually means horsepower required by the driven machine. "Shaft horsepower" is a mechanically oriented synonym. Since 1 hp = 33 000 ft-lb/min motor load in hp = ft-lb/33 000 and ft-lb for rotating machinery is the force required to turn the shaft multiplied by the number of feet through which the force moves, the ft-lb term becomes torque in lb-ft = radius × 2π × r/min × lb. Therefore

$$\text{Motor load in hp} = \frac{\text{radius} \times 2\pi \times \text{lb} \times \text{r/min}}{33\ 000} = \frac{\text{radius} \times \text{lb} \times \text{r/min}}{5250}$$

The relationship is established by the following formula:

$$\text{Horsepower} = \frac{\text{lb-ft} \times \text{r/min}}{5250}$$

where lb-ft is torque or effort required to turn the load. The term "motor load" is therefore best described as the torque required by the load. Additionally, the torque requirement is usually dependent upon the speed at which the torque is required. To describe a motor load accurately, consideration must be given to the torque requirements under the following conditions.

2. Breakaway torque is normally defined by motor manufacturers as "locked-rotor" torque. It is the torque required to start a shaft turning. Methods of bearing lubrication and types of lubricants have a pronounced effect on this torque requirement. Some loads are harder to start turning than others. Consider, for example, the case of grinding mills of the types called ball mills, rod mills, or pebble mills. For many years these grinding mills were built with grease-lubricated bearings. After standing idle overnight, they were difficult to start on cold mornings. The motor had to supply at least 150% of "full-load" running torque to start the mill. Nowadays mill bearings are pressure-lubricated to float the trunnions on a film of oil before the motor is energized. As a result the torque required to start a modern mill may be 70 to 90% of full-load torque. Another case where breakaway torque must be carefully considered is in crushers whose crushing chambers are full of material. To allow a crusher shaft to start turning when its bearings are dry and its chambers are full of material, more than 200% locked-rotor torque may be required. High-torque-jog reversing capabilities have been provided to permit starting some crushers under such conditions.

3. Accelerating or pull-up torque, usually expressed in percent of running torque, is required to accelerate the load from standstill to full speed. It is the torque required not only to overcome friction, windage, and product loading but also to overcome the inertia of the machine. The torque required by a machine may not be constant after the machine has started to turn. This type of load is particularly characteristic of fans and centrifugal pumps and also of certain machine tools.

Another type of load different from pumps and fans but which also demands high accelerating torque is that imposed by vibrating screens whose vibration results from rotating eccentric weights. The ratio of accelerating to full-speed running torque may reach 3/1 in some screen designs.

In considering the torque required by a machine during acceleration, its maximum torque required is the most significant. The minimum accelerating-torque capability of the driving motor must exceed the maximum accelerating torque required by the machine. Special consideration must be given to the selection of the motor to assure that it will have the necessary thermal capacity to bring the machine to full speed. During the period of acceleration one-half of the energy input to the motor is absorbed by the motor rotor circuit while the other half is stored in the driven machine. In larger cage motors of conventional design the temperature rise of the rotor may limit its ability to accelerate a high-inertia load to its full speed.

4. Peak Torque. The peak torque is the maximum momentary torque that a machine may require from its driving motor. The peak torque required by a load is directly related to the "breakdown" or "pullout" torque for its driving motor. High peak-torque requirements for brief periods of time are available from the breakdown torque of an induction motor assisted by the inertia of the rotating system. If the peak-torque requirement is of any appreciable duration, however, it is necessary that the breakdown torque of the motor exceed the peak-load requirement. The term "pullout" torque usually refers to the maximum running torque of a synchronous motor. Inertial energy cannot assist the pullout torque in carrying a peak-torque requirement of the load unless the motor is of the non-salient-pole or reluctance type. It is therefore necessary in practically every instance to consider specifying a synchronous motor whose breakdown or pullout torque exceeds the peak running torque of the load.

For some basic-materials-processing machinery, such as rock or ore crushers, the peak-torque requirements may be 300% or more of the full-load running-torque capabilities of their drive motors. In the case of gyratory crushers the breakdown torques of the driving motors should exceed those peaks to prevent stalls since the inertias of their rotating systems are relatively small. Jaw crushers, like punch presses, store most of their energy in flywheels to supply their peak-torque requirements. Their driving motors are of quite small horsepower, but they should be of the high-slip induction type to permit their flywheels to deliver their energy upon demand.

It is apparent from these considerations that once a load has been brought up to speed, the motor must be sized both on a thermal basis and on a torque basis to match the load. The control selected must then match the motor thermal capability.

5. Inertia Ratio. This column in Table 1 provides an indication of the accelerating time required to bring the load to full speed. With a cage motor this value gives the control designer an approximate time during which abnormally large starting currents will flow.

$$t = \frac{Wk^2(N_1 - N_2)}{308\ T_a}$$

where t = accelerating time,

N_2 = final speed of load, r/min

N_1 = initial speed of load, r/min

T_a = accelerating torque available at load shaft, lb-ft, which is the motor torque times motor speed divided by load speed

Reference 2 lists normal-load-inertia (Wk^2) capabilities of cage motors of 250 to 10 000 hp. Normal-load inertias (Wk^2) for smaller motors (1 through 200 hp) have been estimated in arriving at the tabulated inertia ratios.

a. Load Wk^2 for large polyphase squirrel-cage motors is shown in Table 2. Table 2, also generally applied to 60°C rise motors, provides a norm for standardization. Motors for applications having load Wk^2 in excess of values shown in the table or deviating therefrom in other respects can be designed. High-load Wk^2 or multiple successive starts may require special designs or larger motor frames and thus depart from standard-motor design. Such motors are also more expensive to build than those for which standard Wk^2 applies.

Load inertia Wk^2 is calculated knowing the weight of the rotating parts in pounds. k is the radius of rotation in feet of the center of gravity of the parts. To compare load inertia with motor inertia capability, the load inertia must be divided by the motor speed squared divided by the load speed squared.

b. Number of successive starts of squirrel-cage induction motors initially from ambient or rated-load temperature is indicated in Table 2. The same number of starts may also be specified for load Wk^2 greater than Table 2. However, a price penalty may apply over standard-motor designs.

Frequent motor starting, when required for an application, should definitely be made

TABLE 1. Load Characteristics of Various Machines

Load description	Load torques, % full-load drive torques			Inertia ratio*	Ambient†	Environment‡	Cage motor stall or locked-rotor inrush kVA/hp or motor rating	Types of motors usually used§	Adjustable-speed range, max¶	Remarks
	Breakaway	Accelerating	Peak running							
Actuators:										
Screw-down (rolling mills)	200	150	125	1	A	D	....	DC	2/1	Could be intermittently operated. Reversing required. Stall possible
Positioning	150	110	100	1	3C	5	3.5–4.0	C, DC	6/1, S	Reversing required. Stall possible
Agitators:										
Liquid	100	100	100	2	AW	FJ	5.0–6.0	C	....	Steady load
Slurry	150	100	100	2	AW	CDJ	5.0–6.0	C	....	Settling of solids when idle may cause difficult restarting. Steady load
Barrels, tumbling (foundry)	50	150	100	3	A	DJ	5.0–6.0	C	....	Steady load
Bars, boring, rotary kiln	75	125	100	4	A	DJ	5.0–6.0	C4, DC	4/1	Reversing required, steady load
Beaters:										
Standard	110	120	100	4	AW	DJ	5.0–6.0	C	....	Steady load
Breakers	100	120	120	4	AW	5	5.0–6.0	C	....	
Blowers, centrifugal:										
Valve closed	30	50	40	60	3	5	5.0–6.0	C, C2	3/1	Some applications would require constant speed. Steady load
Valve open	40	110	100	60	3	5	5.0–6.0	C, C2	3/1	
Aircraft	20	110	100	2	AHC	....	3.0–5.0	C	3/1	High-frequency drive usually used
Blowers, positive-displacement, rotary, bypassed	40	40	100	2	AW	DJ	5.0–6.0	C	....	Steady load
Breakers, flake, starting loaded	150	110	100	2	A	DEJ	5.0–6.0	C	....	
Calenders, textile or paper	75	110	100	3	AW	J	5.0–6.0	C	....	
Cars, mining, traction drive	200	200	200	50	3	E	....	DC	....	Reversing required. Overhauling load
Centrifuges (extractors)	40	60	125	50	AW	DJ	5.0–6.0	C	....	Starting unloaded
Chippers, wood, starting empty	50	40	200	100 max	AW	D	6.0–8.0	S	....	Fluctuating load

Equipment										Remarks
Compactors, solids	100	110	110	1	A	DEJ	5.0-6.0	DC, C	2/1	Constant speed may be used, fluctuating load
Compressors:										
Air, shop-type	40	100	100	1	AHC	D	4.0-6.0	C	....	Size determines type of motor, fluctuating load
Axial-vane, loaded	40	100	100	8	AW	DJ	4.0-6.0	C, S, WR	....	Steady load
Reciprocating, start unloaded	40	50	100	10	A	DJ	4.0-6.0	C, S	....	Fluctuating load
Converters:										
Copper, loaded	150	150	125	8	AH	D	....	WR, DC	4/1, S	Reversing required, overhauling load
Conveyors:										
Belt (loaded)	110	130	100	4	A	CDJ	5.0-6.0	C, DC	10/1, S	Inertia depends on load. Constant-torque adjustable-speed multidrives may be required. Antirollback and/or forward or regenerative braking may be required
Drag (or apron)	100	150	100	2	AH	D	5.0-6.0	C	....	Starting loaded. Inertia depends on load. Constant-torque adjustable-speed drive may be required. Steady load
Miscellaneous small	75	150	125	2	3	DF	4.0-6.0	C	....	Fluctuating load
Mining	100	150	150	2	3	CDE	5.0-6.0	C	....	Starting loaded. Inertia depends on load.
Coolers:										
Screw (loaded)	150	100	100	1	AH	CDEJ	5.0-6.0	C	....	Steady load
Shaker-type (vibrating)	50	150	75	6	AH	DJ	5.0-6.0	C	....	Rapidly fluctuating load
Hot solids, rotary (loaded)	175	140	100	2	AH	DJ	5.0-6.0	C	....	Steady load
Grate, reciprocating (loaded)	50	125	75	1	A	DJ	5.0-6.0	C, DC	3/1	Springs cause overhauling load each half cycle
Grate, oscillating (loaded)	50	100	40	1	A	DJ	5.0-6.0	C, DC	1/5/1	Steady load
Grate, traveling (stoker-type)	100	110	100	1	AH	D	5.0-6.0	C, DC	6/1	Starting loaded—torque-limiting drive is desirable. Dripproof motors have been used
Cotton gin	90	100	100	4	A	E	5.0-6.0	C	....	Steady load
Cranes, traveling:										
Bridge motion	100	300	100	4	AH	CDJ	....	WR, DC	10/1 ⎤	Drives must be suited to duty cycle and service. Hoisting inertia depends on load. Reversing required. Shock and vibration encountered, jogging required
Trolley motion	100	200	100	4	AH	CDJ	....	WR, DC	10/1 ⎬	
Hoist motion	50	200	100	....	AH	CDJ	....	WR, DC	10/1 ⎦	
Crushers:										
Gyratory:										
Starting unloaded	50	60	300	2	AC	DJ	5.0-6.0	C	....	Widely fluctuating load. Stall possible
Choke-fed	100	200	300	2	AC	DJ	4.0-6.0	WR, C	....	Reverse jogging with maximum torque may be necessary to start. Stall possible
With feeder	100	150	150	2	AC	DJ	5.0-6.0	C	....	Stall unlikely, load steadier because of feeder

TABLE 1. Load Characteristics of Various Machines (continued)

Load description	Load torques, % full-load drive torques			Inertia ratio*	Ambient†	Environment‡	Cage motor stall or locked-rotor inrush kVA/hp or motor rating‡	Types of motors usually used§	Adjustable-speed range, max¶	Remarks
	Breakaway	Accelerating	Peak running							
Crushers (continued)										
Jaw:										
Starting unloaded	50	100	200	10	AC	DJ	4.0–6.0	C	....	Usually started unloaded. Oscillatory loading can cause line-voltage unsteadiness. Stall possible
Choke-fed	....	....	200	10	AC	DJ	4.0–6.0	C, WR	....	
Pulverizing (hammer-mill)	50	100	150	25	A	CDJ	4.0–6.0	C	....	Usually started unloaded. Stall possible. May be a fluctuating load
Roll:										
Starting unloaded	50	50	....	10	A	DJ	4.0–6.0	C	....	Stall possible. Load can fluctuate widely
Choke-fed, loaded	200	200	150	10	A	DJ	4.0–6.0	C	....	
Cutter bars, balling drum	50	150	150	25	A	DJ	....	DC	2/1	Cyclically overhauling load
Cutter heads, dredge	50	125	150	2	W	....	....	WR	S	Stall possible. Fluctuating load
Dampers:										
Fan, centrifugal, cold	200	200	100	1	AC	CDEJ	3.0–5.0	C, DC	S, 10/1	Reversing required. Stall possible
Fan, centrifugal, hot	400	300	100	1	AH	CDJ	4.0–5.0	C, DC	S, 10/1	Reversing required. Stall possible
Drawbridges	100	125	100	10	AW	....	5.0–6.0	C	....	Reversing required
Draw presses (flywheel)	50	50	200	10	A	D	4.0–6.0	C	....	High inertia
Drill presses	25	50	150	2	A	D	5.0–6.0	C, DC	4/1	Constant-horsepower drive may be required. Fluctuating load
Drums, balling (ore)	50	125	100	4	A	D	....	DC	2/1	Steady load. 10% braking for rollback desirable
Dryers:										
Rotary (rock or ore)	50	150	100	4	AH	D	5.0–6.0	C, WR	....	Steady load. Constant-torque drive
Grain	50	100	90	2	AH	DJ	5.0–6.0	C	....	Steady load
Edgers (starting unloaded)	40	30	200	10	A	DE	5.0–6.0	....	....	High inertia. Fluctuating load

Load										Remarks
Elevators:										
Bucket (starting)	150	175	150	2	AHC	CDJ	5.0–6.0	C		Magnetic brake may be used. Stall possible
Freight (loaded)	100	125	100	4	ACW	DFJ	5.0–6.0	C	S	Reversing required. Overhauling load.
Man lift	50	125	100	1	AH	5	5.0–6.0	C		Brake necessary. Steady load
Personnel (loaded)	110	150	100	4	AHC	5		DC	10/1	Speed range required during acceleration and deceleration. Brake necessary. Reversing required. Controlled acceleration may be required
Escalators, stairways (starting unloaded)	50	75	100	2	A		5.0–6.0	C		Steady load
Extractors (press-type)	50	150	150	1	3	5	4.0–6.0	C		Cyclically variable load
Extruders (rubber or plastic)	100	150	100	1	3	5	5.0–6.0	C		Fluctuating load
Fans:										
Centrifugal, ambient:										
Valve closed	25	60	50	25	ACW	5	4.0–6.0	C	2/1	Variable-torque adjustable-speed may be specified. Steady load
Valve open	25	110	100	25	ACW	5	4.0–6.0	C	2/1	
Centrifugal, hot gases:										
Valve closed	25	60	100	60	AH	DJ	4.0–6.0	C	3/1	Peak running overload torque occurs when handling colder gases. Steady load
Valve open	25	200	175	60	AH	DJ	4.0–6.0	C	3/1	
Propeller, axial-flow	40	110	100	25	AH	DJ	4.0–6.0	C		High inertia
Feeders:										
Belt (loaded)	100	120	100	2	3	5	4.0–6.0	C, DC	10/1, S	Steady load. Constant-torque characteristic
Distributing, oscillating drive	150	150	100	4	AW	CDJ	5.0–6.0	C, DC	6/1, S	Cyclically variable load
Screw, compacting rolls	100	100	100	1	AW	CDEJ	5.0–6.0	C	S	Stall possible. Torque-limited coupling desirable
Screw, filter-cake	100	100	100	1	3	DJ	5.0–6.0	C	3/1, S	Eddy-current coupling desirable
Screw, dry	150	100	100	1	AH	CDEJ	5.0–6.0	C	3/1, S	Eddy-current coupling usual. Constant torque
Slurry, ferris-wheel	110	100	75	2	AW	DJ		DC	3/1	Steady load. Constant torque
Table	125	110	100	2	A	DJ	5.0–6.0	C, WR, DC	6/1, S	Steady load
Vane-type	150	80	75	1	AW	CDEJ	5.0–6.0	C	6/1, S	Steady load. Magnetic coupling desirable. Constant torque
Vibrating, magnetic	100	100	100		AH	CDEJ			3/1, S	Feeder manufacturer selects controller
Vibrating, motor-driven	50	150	100	4	AH	CDEJ	5.0–6.0	C	3/1, S	Feeder manufacturer selects controller
Forge presses	25	50	150	10	AH	D	4.0–6.0	C		Fluctuating load
Frames, spinning, textile	50	125	100	2	A	E	3.0–5.0	C		
Furnaces, holding, copper	150	125	100	4	AH	D		WR	4/1, S	Overhauling load. Reversing required
Gates:										
Diverting, solids	200	125	100	1	3	CDJ	5.0–6.0	C	S	Reversing required
Locks, hydraulic	25	200	200	2	W		5.0–6.0	C	S	

10-7

TABLE 1. Load Characteristics of Various Machines (continued)

Load description	Load torques, % full-load drive torques — Break-away	Load torques, % full-load drive torques — Accelerating	Load torques, % full-load drive torques — Peak running	Inertia ratio*	Ambient†	Environment‡	Cage motor stall or locked-rotor inrush kVA/hp or motor rating	Types of motors usually used §	Adjustable-speed range, max ¶	Remarks
Generators:										
Electric, flywheel-type	50	100	400	100	A		....	WR	....	High inertia. High peak running overloads
Electric, general use	25	30	150	3	AH	D	5.0–6.0	C	....	Steady load
Electroplating	25	30	100	3	AW	F	5.0–6.0	C	....	Steady load
Welding	30	50	200	3	AHC	D	5.0–6.0	C	....	Peak torque required when arc is struck. Fluctuating load
Grates:										
Indurating (preheater)	100	110	200	3	AHC	DJ	5.0–6.0	DC, C	4/1, S	Torque-limiting drive desirable. Steady load. Eddy-current coupling required for cage motor
Stoker (furnace)	75	110	100	1	AH	D	5.0–6.0	DC, C	2/1, S	Steady load
Grinders:										
Metal	25	50	100	2	A	D	5.0–6.0	C	....	Starting unloaded
Pulp or meat	40	50	150	2	AW	F	5.0–6.0	C	....	Starting unloaded
Pulp, magazine-type	50	50	150	5	AW	CJ	5.0–6.0	C	....	Starting unloaded
Pulp, pocket-type	40	30	150	5	AW	CJ	5.0–6.0	C	....	Starting unloaded
Hammers, power, flywheel	50	50	150	10	A	D	5.0–6.0	C	....	High inertia, fluctuating load
Hoists:										
Skip	100	150	100	10	A	D	....	WR, DC	6/1	Regenerative braking. Reversing required
Small shop	100	150	125	3	AHW	DJ	5.0–6.0	C	S	Reversing required
Hydropulpers	125	125	150	1	W	....	5.0–6.0	C	....	Reversing required
Indexers	150	200	150	2	A	D	5.0–7.0	C	S	
Ironers, laundry (mangles)	50	50	125	1	3	....	5.0–6.0	C	....	Fluctuating load
Jointers, woodworking	50	125	125	4	A	E	5.0–6.0	C	....	Fluctuating load
Jordans, plug out	50	50	150	9	AW	....	5.0–6.0	C	....	Fluctuating load

Kilns, rotary (loaded)	200	125	125	4	AH	DJ	5.0–6.0	DC, WR, C	6/1, S	Approx. 10% reverse dynamic braking may be required
Knife, cane, sugar	75	100	100	5	H	J	5.0–6.0	C		Load may vary
Log washers, rock or ore (loaded)	75	125	150	1	AW	J	5.0–6.0	C		Steady load
Looms, textile, without clutch	125	125	150	2	A	E	5.0–6.0	C		Cyclically overhauling load
Machines:										
Boring (loaded)	150	150	100	2	A	D	5.0–6.0	DC, C	6/1	Constant-horsepower drives may be required on large machines. Fluctuating load
Bottling	50	50	100	2	AW		5.0–6.0	C		Steady load
Briquetting	100	125	150	2	A	CDEJ	5.0–6.0	C		
Buffing, automatic	50	75	100	2	A	DJ	5.0–6.0	C		Frequent starting and plugging required
Cinder-block vibrating	50	150	70	4	AW	DJ	5.0–6.0	C		Cyclically varying load
Keyseating	25	50	100	2	A	D	5.0–6.0	C		Fluctuating load
Kneading	50	150	175	1	AW	J	5.0–6.0	C		Fluctuating load
Mining, loading, and cutting	50	150	200	4	AH	E		WR		Fluctuating load
Polishing	50	75	100	2	A	DJ	5.0–6.0	C		High frequencies may be used
Shipboard, navy	Depends on driven machine				HW	5	5.0–6.0	DC, C	6/1	Cage motors preferred
Spinning, metal	100	150	150	4	3	D	5.0–6.0	C		Fluctuating load
Textile, card, with clutch	50	75	200	10	A	E	5.0–6.0	C		
Textile, card, without clutch	75	300	200	10	A	E	5.0–6.0	C		Slip coupling may be required
Textile, draw-frame	90	100	125	2	A	E	5.0–6.0	C		
Textile, roving frame	60–100	100	100	2	A	E	3.0–5.0	C		
Textile, twister	100	160	150	2	A	E	4.0–6.0	C		Maximum torques determined by trial
Textile, warper or slasher	100	110	100	2	A	E	5.0–6.0	C	10/1	Adjustable-speed coupling with inching and torque control
Wire-drawing	50	125	100	2	A	D	5.0–8.0	C		Multispeed cage
Mills:										
Attrition (starting unloaded)	100	60	120	10	AW	CDEJ	5.0–7.0	C, S		High inertia. Small mills will use cage motors
Autogenous, grinding (prelubricated)	90	140	100	10	A	DJ	3.0–7.0	S		High inertia. Inching required. Low-starting-inrush motors require clutches, steady load
Ball, grinding (prelubed bearings)	90	130	100	6	AW	CDJ	3.0–7.0	S		High inertia. Inching required. Low starting-inrush motors require clutches, steady load
Ball, grinding (dry bearings)	140	130	100	6	AW	CDJ	1.5–7.0	C, WR, S		High inertia. Horsepower requirements will dictate type of motor
B&W, coal (loaded)	150	110	100	3	A	CDE	5.0–6.0	C		
Bowl, Raymond, coal (loaded)	130	120	100	5	A	CDE	5.0–6.0	C		Grinding mill and exhauster coupled together, steady load

TABLE 1. Load Characteristics of Various Machines (continued)

Load description	Load torques, % full-load drive torques			Inertia ratio*	Ambient†	Environment‡	Cage motor stall or locked-rotor kVA/hp or motor rating	Types of motors usually used §	Adjustable-speed range, max ¶	Remarks
	Breakaway	Accelerating	Peak running							
Mills (continued)										
Bradley-Hercules (loaded)............	150	110	100	3	A	CDEJ	C	5.0–6.0	....	....	High inertia, steady load
Flour, grinding..................	50	75	100	6	A	E	C	5.0–6.0	....	....	Steady load
Pan............................	125	125	150	4	A	D	C	5.0–6.0	....	....	
Rod or tube, grinding (prelubed bearings)	90	120	100	6	AW	CDJ	C, WR, S	1.5–7.0	....	....	Horsepower requirements dictate type of motor. Low-starting-inrush motors require clutches, steady load
Rod or tube, grinding (dry bearings) ...	140	130	100	6	A	CDJ	C, WR, S	3.0–8.0	....	....	Horsepower requirements dictate type of motor. Low-starting-inrush motors will require clutches, steady load
Rolling metal:										
Billet, skelp and sheet, bar.........	50	30	200	1	A	D		....	DC, WR	1.5/1	Accurate speed control required, fluctuating load
Brass and copper finishing..........	120	100	200	1	A	D		....	DC, WR	1.5/1	Fluctuating load
Brass and copper roughing..........	40	30	200	1	A	D		....	WR, DC	1.5/1	Fluctuating load
Drawbench.........................	75	100	200	2	A	D		5.0–7.0	C, WR, DC	....	Multispeed cage used, fluctuating load
Merchant mill trains...............	50	30	200	1	A	D		....	WR, DC	1.5/1	Fluctuating load, automatic speed control
Plate..............................	40	30	250	1	A	D		....	WR, DC	1.5/1	
Reels, wire or strip...............	100	100	100	2	A	D		5.0–6.0	DC, C	20/1	Torque-controlled cage motor with speed-controlled coupling can be used
Rod...............................	90	50	200	1	A	D		....	WR, DC	1.5/1	Fluctuating load, automatic speed control
Rolls, roughing...................	90	100	250	4	A	D		....	WR	....	Fluctuating load
Scalper...........................	60	90	200	4	A	D		5.0–6.0	C	....	Fluctuating load

Application									Remarks
Sheet and tin (cold rolling)	150	110	200	1	A	D	……	1.5/1	WR, DC	Fluctuating load, automatic speed control
Strip, hot	40	30	200	1	A	D	……	1.5/1	WR, DC	Fluctuating load
Structural and rail finishing	40	30	200	1	A	D	……	1.5/1	WR, DC	Fluctuating load, automatic speed control
Structural and rail roughing	40	30	250	1	A	D	……	1.5/1	WR	Fluctuating load
Tube	50	30	200	1	A	D	……	1.5/1	DC	Fluctuating load, automatic speed control
Tube piercing and expanding	50	30	250	1	A	D	……	10/1	WR, DC	Torque and speed control, cyclically varying load
Tube reeling	50	30	200	1	A	D	……		DC	Fluctuating load
Rubber	100	100	200	1	A	CE	5.0–7.0	……	C	Fluctuating load
Saw, band	50	75	200	100	A	D	5.0–7.0	……	C	High inertia, fluctuating load
Stamp	50	75	150	10	A	D	5.0–7.0	……	C	High inertia, cyclically varying load
Wash	25	30	100	1	A	……	5.0–6.0	……	C	Steady load
Mixers:									
Banbury	125	100	250	1	A	CE	5.0–7.0	……	C, WR	Multispeed cage may be used, fluctuating load
Concrete	40	50	100	2	AW	D	4.0–7.0	……	C	Steady load
Dough	100	125	100	1	AW	E	5.0–7.0	……	C	Fluctuating load
Liquid	100	100	100	2	AW	EFJ	5.0–7.0	……	C	
Sand, centrifugal	50	100	100	4	A	D	5.0–7.0	……	C	
Sand, screw	100	100	100	1	A	D	5.0–7.0	……	C	
Slurry	100	100	100	1	A	DJ	5.0–6.0	……	C	
Solids (mullers)	100	125	175	1	A	CDJ	5.0–7.0	……	C	Steady load
Pans, pelletizing, ore	50	100	100	6	A	DJ	5.0–7.0	2/1	DC	
Planers:									
Metalworking	50	150	150	4	A	D	……	4/1	DC	Plugging and reversing service, cyclically varying load
Woodworking	50	125	150	10	A	D	5.0–7.0	……	C	High inertia, varying load
Plasticators	125	100	250	1	A	5	5.0–7.0	……	C	Multispeed may be required, fluctuating load
Plows, conveyor, belt (ore)	150	150	200	1	A	DJ	5.0–7.0	……	C	Reversing and jogging required
Positioners, indexing (machine tool)	50	200	100	4	A	D	5.0–7.0	……	C	Reversing and jogging required
Presses:									
Brick	100	175	150	4	A	DJ	5.0–7.0	……	C	Cyclically varying load
Drill, production, automatic	50	60	125	1	A	D	5.0–7.0	……	C	Frequent starts with plugging, fluctuating load
Pellet (flywheel)	50	75	150	10	A	D	5.0–7.0	……	C	High inertia, fluctuating load
Printing, production-type	100	150	150	4	A	……	5.0–7.0	20/1	DC, C	Reversal required for setup, cage motor is used with speed-adjusting coupling
Punch (flywheel)	50	75	100	10	A	D	5.0–7.0	……	C	High inertia, fluctuating load
Punch (no-flywheel)	10	40	150	10	A	D	4.0–7.0	……	C	Fluctuating load
Pug mill (solids mixing)	150	125	100	1	A	CDJ	5.0–7.0	……	C	Solids may "set up" on emergency shutdown
Puller, car	150	110	100	25	A	D	5.0–7.0	……	C	Inertia depends on number of cars. Unloaded reversals required, fluctuating load

TABLE 1. Load Characteristics of Various Machines (continued)

Load description	Load torques, % full-load drive torques			Inertia ratio *	Ambient †	Environment ‡	Cage motor stall or locked-rotor inrush kVA/hp or motor rating	Types of motors usually used §	Adjustable-speed range, max ¶	Remarks
	Break-away	Accelerating	Peak running							
Pumps:										
Adjustable-blade, vertical...	50	40	125	1	AW	.....	5.0-7.0	C	.....	Unloaded start, steady load
Centrifugal, discharge open...	40	100	100	1	AW	FJ	4.0-7.0	C	.....	Loaded start, steady load
Hydraulic, aircraft...	100	100	100	1	AW	E	5.0-7.0	C	.....	Loaded start. Reversing may be required, fluctuating load
Oil-field, flywheel...	50	200	200	10	AC	D	5.0-7.0	C	.....	Cyclically varying load
Oil, lubricating...	40	150	150	1	AC	D	5.0-7.0	C	.....	Cold oil can cause drive overloads, steady load
Oil, fuel...	40	150	150	1	AC	D	5.0-6.0	C	.....	Peak torque caused by more viscous oils, steady load
Propeller...	40	100	100	1	AW	F	4.0-6.0	C	.....	Handling nonviscous fluids, steady load
Reciprocating, positive-displacement...	40	30	150	1	AW	.....	5.0-6.0	C	.....	Starting dry, handling nonviscous fluids, cyclically varying load
Reciprocating, positive-displacement...	40	30	20	1	AW	.....	5.0-6.0	C	.....	Bypassed, handling nonviscous fluids, cyclically varying load
Reciprocating, positive-displacement...	100	100	150	4	AW	.....	5.0-6.0	C	.....	3-cylinder, not bypassed, handling a nonviscous fluid, cyclically varying load
Screw-type, started dry...	40	30	100	1	AW	F	5.0-6.0	C	.....	Handling nonviscous fluids, steady load
Screw-type, primed, discharge open...	40	100	100	1	AW	F	5.0-6.0	C	.....	Handling nonviscous fluids, steady load
Slurry-handling, discharge open...	100	100	100	1	AW	D	5.0-6.0	C	.....	Steady load
Submersible, deep-well...	40	100	100	1	.....	.....	5.0-7.0	C	.....	
Turbine, centrifugal, deep-well...	50	100	100	2	AW	.....	5.0-6.0	C	.....	
Vacuum (paper, min. service)...	60	100	150	4	AW	.....	5.0-6.0	C	.....	Cyclically varying load
Vacuum (other applications)...	40	60	100	4	A	.....	5.0-6.0	C	.....	Cyclically varying load
Vacuum, reciprocating...	40	60	150	10	A	.....	4.0-6.0	C	.....	Starting unloaded, cyclically varying load
Vane-type, positive-displacement...	100	150	150	1	A	DJ	5.0-6.0	C	.....	Viscous fluids may overload drive, steady load

Application										Remarks
Refiners, disk-type, starting unloaded.....	50	50	150	20	A	D	4.0–6.0	C	.....	Steady load
Rolls:										
Bending.....	150	150	100	2	A	D	5.0–7.0	C	.....	Reversing required, fluctuating load
Compacting (loaded).....	100	110	125	1	A	DJ	5.0–6.0	C	.....	Usually steady load
Crushing (sugarcane).....	50	110	125	2	AW	J	4.0–6.0	C	.....	Fluctuating load
Flaking.....	30	50	100	2	A	E	4.0–5.0	C	.....	Steady load
Rubber.....	75	75	100	1	.....	.....	4.0–7.0	C	.....	Multispeed cage, approx. 6.3 kVA/hp
Sanders, woodworking.....	30	50	100	1	A	D	4.0–6.0	C	.....	
Saws:										
Band, metalworking.....	30	50	100	4	A	D	5.0–6.0	C	.....	Fluctuating load
Circular, metal cutoff.....	25	50	150	6	A	D	5.0–6.0	C	.....	Fluctuating load
Circular, wood, production.....	50	30	150	10	A	E	5.0–7.0	C	.....	High inertia, fluctuating load
Edger (see Edgers)										
Gang.....	60	30	150	10	A	D	5.0–7.0	C	.....	High inertia, fluctuating load
Trimmer.....	40	30	150	10	A	D	5.0–7.0	C	.....	High inertia, fluctuating load
Screens:										
Centrifugal, paper-mill.....	50	100	100	50	AW	FJ	5.0–7.0	C	.....	High inertia
Centrifugal (centrifuges).....	40	60	125	50	AW	DJ	5.0–7.0	C	.....	High inertia
Rotary, stone (trommel).....	70	100	100	1	A	DJ	4.0–6.0	C	.....	
Vibrating.....	50	150	70	6	3	5	5.0–6.0	C	.....	
Separators, air (fan-type).....	40	150	100	15	A	5	5.0–7.0	C	.....	Plug stops may be required
Shakers, foundry or car.....	50	150	70	6	AHC	CJ	5.0–7.0	C	.....	High inertia
Shears, flywheel-type.....	50	50	120	10	A	D	5.0–7.0	C	.....	
Shovels:										
Dragline, hoisting motion.....	50	150	100	4	A	CDJ	.....	WR	6/1	Similar to cranes. Reversing service required
Dragline, platform motion.....	50	100	100	4	A	CDJ	.....	WR	4/1	Similar to cranes. Reversing service required
Large, digging motion.....	50	200	200	3	ACW	CDJ	.....	WR, DC	10/1	Similar to cranes. Reversing service required
Large, platform motion.....	50	100	100	4	ACW	CDJ	.....	WR, DC	4/1	Similar to cranes. Reversing service required
Shredders (see crushers, pulverizing).....										
Sifters, shaker-type.....	50	100	70	3	A	EJ	5.0–6.0	C	.....	
Stokers:										
Small, screen-type.....	50	100	100	1	AH	CD	5.0–6.0	C	.....	
Traveling-grate type.....	50	110	100	1	A	CD	5.0–6.0	C	.....	Torque-limiting drive is desirable
Swagers.....	100	110	150	1	A	C	4.0–7.0	C	.....	Fluctuating load
Tension-maintaining drives.....	100	100	100	1	AW	DE	5.0–6.0	C	10/1	Cage motors to drive through torque- and speed-controlled couplings
Tools:										
Machine.....	100	150	100	2	A	D	4.0–7.0	C, DC	4/1	Cage motors use speed-controlled couplings, load usually fluctuates
Machine, broaching, automatic.....	50	150	150	1	A	D	4.0–7.0	C	.....	Frequent starts and plugged stops, fluctuating load
Machine, grinding.....	50	150	100	2	A	D	5.0–7.0	C	.....	

TABLE 1. Load Characteristics of Various Machines (*continued*)

Load description	Load torques, % full-load drive torques			Inertia ratio*	Ambient†	Environment‡	Cage motor stall or locked-rotor inrush kVA/hp or motor rating	Types of motors usually used§	Adjustable-speed range, max¶	Remarks
	Break-away	Accel-er-at-ing	Peak run-ning							
Tools (*continued*):										
Machine, lathe, metal, production......	50	200	200	2	A	D	5.0–7.0	C	....	Frequent starts, plug stops, and changes in depth of cut, fluctuating load
Machine, mill, boring, production, metal....	100	125	100	4	A	D	5.0–7.0	C, DC	20/1	Inertia depends on work on table. Cage motors use speed-controlled coupling, fluctuating load
Machine, milling, production...........	100	100	100	1	A	D	5.0–7.0	C, DC	4/1	Cage motors require speed-controlled coupling, fluctuating load
Machine, shaper, metal automatic......	50	75	150	2	A	D	....	DC	10/1	Reversing may be required with plug stops, fluctuating load
Portable, hand...........	300	200	250	10	AHC	D	2–3	SW	4/1	Usually universal type, intermittent-duty motor, widely fluctuating load
Vehicles:										
Freight...........	200	200	200	50	AW	DJ	1–3	SW	3/1	Special engineering required. Reversing required. Inertia depends on load, fluctuating load
Passenger...........	100	400	200	25	AW	DJ	1–3	SW	10/1	Special engineering required. Inertia depends on load, fluctuating load
Walkways, mechanized...........	50	50	100	2	A	....	5.0–6.0	C	....	Steady load usual
Washers, laundry...........	25	75	100	4	AW	....	5.0–6.0	C	....	Steady load usual
Winches...........	125	150	100	4	A	D	4.0–6.0	C, WR, DC	S, 10/1	Drive must be coordinated with service, frequent starts required
Wood hogs...........	60	100	200	30	AW	....	5.0–6.0	C, WR	....	High inertia. Drive choice depends on size and location, fluctuating load

* Load inertia compared with normal inertial capability of its drive motor.

† A = high altitude, which for motors may be in excess of 3300 ft above sea level

H = high temperatures, which for motors may be in excess of 104°F

C = extreme cold. 40° below 0°F or lower. Such temperatures can affect lubricants to prevent starts

W = high humidity, continuous exposure to atmospheres of 100% relative humidity and/or frequent "hose-downs"

3 = where A, H, and W may exist either simultaneously or individually at different times

‡ C = atmosphere heavy with carbon dust such as occasionally exists in rubber plants or coal- and coke-handling facilities

D = "Dirty" or atmospheres containing abrasive dusts

E = explosive atmospheres (dust or gas)

F = atmospheres containing relatively high concentrations of acid fumes

J = atmospheres containing quantities of chemical dusts which may be corrosive or gummy after exposure to high humidity

5 = all of above

§ C = squirrel-cage induction motor (single-speed)

C2 = squirrel-cage induction motor (two-speed)

C4 = squirrel-cage induction motor (four-speed)

WR = wound-rotor induction motor

DC = dc motor

SW = series-wound (dc or ac) motor

S = synchronous

¶ Ratios indicate maximum ranges employed on equipment indicated. S indicates that starting and stopping may replace adjustable-speed range.

Note. Any one application may not meet all the conditions listed.

---

a part of motor specification. During initial plant start-up, etc., additional motor starts beyond Table 2 standard may be desired. They should not be made without consulting the motor manufacturer. The life of the motor is affected by the number of successive motor starts.

Normal motor-overload protective relays will not protect a motor against damage resulting from too frequent successive motor starts.

ADJUSTABLE-SPEED LOADS

Additional mechanical requirements of motor loads must be considered when machines require adjustable-speed drives. Load characteristics require further definitions to describe them adequately.

6. Constant torque refers to loads whose horsepower requirements vary linearly with changing speeds during normal operation. Most machinery for processing basic materials falls in this category. Conveyors of nearly all types, some crushers, some coolers, some rock-grinding machines, rotary kilns, and a host of other machines require practically the same torque whether running fast or slowly when operating normally. Constant-displacement pumps delivering liquid against a constant head at varying speeds are another prime example of a constant-torque load.

7. Constant Horsepower. Such a load absorbs the same amount of horsepower regardless of its speed during normal operation. Machine tools normally used for production purposes are prime examples of constant-horsepower loads. Reel motors also fall into this class under certain operating conditions, and some types of earth-moving equipment may also be constant-horsepower loads. For this type of load, torque requirements decrease as the speed increases. Traction loads frequently approach the characteristics of constant horsepower.

8. Variable Torque. A load whose torque requirements vary with speed in any fashion other than those mentioned in Arts. 6 and 7 is in this class. Centrifugal pumps and blowers, whose torque requirements increase approximately as the square of the speed, are examples of this type of load. Some mixers are also variable-torque loads.

The type of solid-materials feeder having a rotating vibration generator that operates near its critical speed presents a torque ratio at low speed compared with that at high speed that is approximately the fifth root of the speeds.

Most loads, when abnormally operated, present variable-torque characteristics to their drives.

TABLE 2. Load Wk^2 (Exclusive of Motor Wk^2), lb-ft^2

hp	Speed, rpm						
	3 600	1 800	1 200	900	720	600	514
250	210	1 017	2 744	5 540	9 530	14 830	21 560
300	246	1 197	3 239	6 540	11 270	17 550	25 530
350	281	1 373	3 723	7 530	12 980	20 230	29 430
400	315	1 546	4 199	8 500	14 670	22 870	33 280
450	349	1 714	4 666	9 460	16 320	25 470	37 090
500	381	1 880	5 130	10 400	17 970	28 050	40 850
600	443	2 202	6 030	12 250	21 190	33 100	48 260
700	503	2 514	6 900	14 060	24 340	38 080	55 500
800	560	2 815	7 760	15 830	27 440	42 950	62 700
900	615	3 108	8 590	17 560	30 480	47 740	69 700
1 000	668	3 393	9 410	19 260	33 470	52 500	76 600
1 250	790	4 073	11 380	23 390	40 740	64 000	93 600
1 500	902	4 712	13 260	27 350	47 750	75 100	110 000
1 750	1 004	5 310	15 060	31 170	54 500	85 900	126 000
2 000	1 096	5 880	16 780	34 860	61 100	96 500	141 600
2 250	1 180	6 420	18 440	38 430	67 600	106 800	156 900
2 500	1 256	6 930	20 030	41 900	73 800	116 800	171 800
3 000	1 387	7 860	23 040	48 520	85 800	136 200	200 700
3 500	1 491	8 700	25 850	54 800	97 300	154 800	228 600
4 000	1 570	9 460	28 460	60 700	108 200	172 600	255 400
4 500	1 627	10 120	30 890	66 300	118 700	189 800	281 400
5 000	1 662	10 720	33 160	71 700	128 700	206 400	306 500
5 500	1 677	11 240	35 280	76 700	138 300	222 300	330 800
6 000	1 673	11 690	37 250	81 500	147 500	237 800	354 400
7 000	1 612	12 400	40 770	90 500	164 900	267 100	399 500
8 000	1 484	12 870	43 790	98 500	181 000	294 500	442 100
9 000	1 294	13 120	46 330	105 700	195 800	320 200	482 300
10 000	1 046	13 170	48 430	112 200	209 400	344 200	520 000

The table applies to standard polyphase squirrel-cage motors having locked-rotor torques equal to 60% of full-load torque and a rated temperature rise of 40°C. Motors can accelerate without injurious temperature rise under the following conditions:

1. Rated voltage and frequency applied.

2. During the accelerating period, the connected load torque should be equal to, or less than, a torque which varies as the square of the speed and is equal to 100% of full-load torque at rated speed.

3. Two starts in succession (coasting to rest between starts) with the motor initially at ambient temperature or one start with the motor initially at ambient temperature not exceeding its rated-load operating temperature.

Note. Data from Ref. 2. Also, see Ref. 2 for values of Wk^2 at lower speeds.

OTHER LOAD CONDITIONS

9. Duty Cycle. Duty cycle includes frequent starts, plugging stops, reversals, or stalls. These characteristics are usually involved in batch-type processes and may include tumbling barrels, certain cranes, shovels and draglines, dampers, gate- or plow-positioning drives, drawbridges, freight and personnel elevators, press-type extractors, some feeders, presses of certain types, hoists, indexers, boring machines, cinder-block machines, keyseating, kneading, car-pulling, shakers (foundry or car); swaging and washing machines, and certain freight or passenger vehicles. The list is not all-inclusive. The drives for these loads must be capable of absorbing the heat generated during the duty cycles. Adequate thermal capacity would be required in slip couplings, clutches, or motors to accelerate or plug-stop these drives or to withstand stalls. It is the product of the slip speed and the torque absorbed by the load per unit of time which generates heat in these drive components. All the events which occur during the duty cycle generate heat which the drive components must dissipate.

Since motors are rated by horsepower, designs have been standardized to include locked-rotor, accelerating, and breakdown torques (Arts. 2 to 4). It is seen that each of the three torques listed must be sufficiently large or a motor cannot handle its load. To assure that a motor is not overpowered, the duty cycle of the load must be considered. A load whose torque requirement varies in accordance with a repetitive cycle would have its motor horsepower determined on the basis of the internal heat generated in the motor after it has reached full speed. Core loss, windage, and friction losses are considered constant no matter what load the motor is carrying. The copper losses, however, are a function of the torque output of the motor, and so determine its horsepower rating. They are proportional to the square of the line current to the motor. The rms horsepower concept is therefore generally used to calculate the motor rating where the speed remains essentially constant.

Duty cycle not only affects the selection of the motor but also is an important factor in selecting such items as acceleration control, overload protective devices, and braking control.

10. Repetitive loads include banbury, dough mixers, plasticators, large shovels (digging), tension-maintaining drives, some textile machinery, and machine tools. Machines operating on a batch basis usually have definitely predictable duty cycles. During parts of these duty cycles, torque requirements may increase considerably. The drives, of course, must be able to supply the maximum torque but do not have to bear a continuous rating at this maximum torque. Because of this characteristic, an economic advantage can be taken in selecting suitable drives for these machines.

11. Overhauling Loads. Copper converters, cranes, cutter bars, elevators, escalators, hoists, conveyor-belt plows, vehicles, and winches are examples of overhauling loads.

The drives for these machines must be able to brake their loads during parts of their duty cycles. Braking may be accomplished with friction, magnetic, or fluid brakes as separate units, if the prime mover does not have the means of braking. Some of these machines are also required to hold their loads stationary. Cutter bars act as pendulums. The overhauling load on the downswing must not cause an appreciable increase in speed. A full copper converter beginning to pour its charge will require more torque to hold the load than that required when the converter is almost empty.

REFERENCES

1. Robert W. Smeaton, "Motor Application and Maintenance Handbook," McGraw-Hill Book Company, New York, 1969.
2. NEMA Standards Publication MG1 for Motors and Generators, New York.

11

Motor Types and Characteristics

R. C. MOORE *

* Adjunct Professor, Milwaukee School of Engineering; Registered Professional Engineer (Wis.); Senior Life Member, IEEE.

FOREWORD

Since many motor-application considerations have a direct bearing on the selection and application of both switchgear and control equipment, this section was written to provide information and background data that relate to motor control and system protection.

Motors treated in this section are ac single-phase and polyphase types. Two-phase motors, because of their application rarity, will be discussed only in relation to their terminal markings. Polyphase commutator-type motors are not treated, nor are single-phase railway-type commutator motors, which are, of course, nonindustrial and of special construction and control.

STANDARDS

Standards provide guidance for motor manufacturer, user, service shops, etc., for proper and safe motor usage. For example, standards identify torque values, frequency of motor starts, load inertias, and efficiency determination. Thus standards are beneficial for mutual understanding of all concerned with motor design, fabrication, use, servicing, etc. Accordingly standards are frequently referenced.

1. **Motor Classifications.** Classifications listed in this section are taken from NEMA Standards (Ref. 1), which represent the thinking and experience of industries and engineers, and provide a common ground for mutual understanding between manufacturer and user. In the following, only points relating to the nature and purpose of this section, as it fits in with other sections, will be emphasized. Complete details relating to the classifications may be found in the reference cited.

CLASSIFICATION BY SIZE

2. **Fractional-horsepower motors** are built in small sizes in a frame smaller than the frame of an integral-horsepower motor which has a continuous rating of 1 hp, open construction, at 1700 to 1800 r/min (Table 1).

A two-digit frame number assigned to fractional-horsepower motors is the dimension of the foot height (bottom of foot mounting to centerline of shaft) in inches times the number 16.

Fractional-horsepower motors may be built as either single-phase or polyphase types. Most applications are of the single-phase cage type, although polyphase squirrel-cage motors are also offered.

3. **Integral-horsepower motors**, ac type, are built in a 680 frame, or smaller, with an open continuous rating of 1 hp at 1700 to 1800 r/min or in a larger frame—but no larger than the frame required for the ratings shown in Table 2.

TABLE 1. Horsepower and Synchronous-Speed Ratings, r/min, of Single-Phase Integral-Horsepower Motors Rated 115 and 230 V

hp	60 Hz				50 Hz			
1/2	......	......	......	900	......	......	1000	750
3/4	......	......	1200	900	......	1500	1000	750
1	......	1800	1200	900	3000	1500	1000	750
1½	3600	1800	1200	900	3000	1500	1000	750
2	3600	1800	1200	900	3000	1500	1000	750
3	3600	1800	1200	900	3000	1500	1000	750
5	3600	1800	1200	900	3000	1500	1000	750
7½	3600	1800	1200	900	3000	1500	1000	750
10	3600	1800	1200	900	3000	1500	1000	750
15	3600	1800	1200	900	3000	1500	1000	750
20	3600	1800	1200	900	3000	1500	1000	750
25	3600	1800	1200	900				

In the three-digit frame-number assignment to integral-horsepower motors the first two digits equal the foot height in inches times the number 4. The third digit is related to the (side view) distance from the motor centerline to a foot-mounting hole.

Integral-horsepower motors may be built as either single-phase or polyphase types. For example, a 25-hp ac squirrel-cage induction motor may be offered at the higher speeds for either single-phase or polyphase application. In the higher range of horsepower, integral-horsepower motors are generally of the polyphase type.

TABLE 2. Horsepower and Synchronous-Speed Ratings, r/min, of Polyphase Integral-Horsepower Induction Motors

hp	60 Hz							50 Hz			
1/2	....	......	......	900	720	600	514	....	......	......	750
3/4	....	......	1200	900	720	600	514	....	......	1000	750
1	....	1800	1200	900	720	600	514	....	......	1000	750
1½	3600 *	1800	1200	900	720	600	514	3000 *	1500	1000	750
2	3600 *	1800	1200	900	720	600		3000 *	1500	1000	750
3	3600 *	1800	1200	900	720	600	514	3000*	1500	1000	750
5	3600 *	1800	1200	900	720	600	514	3000*	1500	1000	750
7½	3600 *	1800	1200	900	720	600	514	3000 *	1500	1000	750
10	3600 *	1800	1200	900	720	600	514	3000 *	1500	1000	750
15	3600 *	1800	1200	900	720	600	514	3000 *	1500	1000	750
20	3600 *	1800	1200	900	720	600	514	3000 *	1500	1000	750
25	3600*	1800	1200	900	720	600	514	3000 *	1500	1000	750
30	3600 *	1800	1200	900	720	600	514	3000 *	1500	1000	750
40	3600 *	1800	1200	900	720	600	514	3000 *	1500	1000	750
50	3600 *	1800	1200	900	720	600	514	3000 *	1500	1000	750
60	3600 *	1800	1200	900	720	600	514	3000 *	1500	1000	750
75	3600 *	1800	1200	900	720	600	514	3000 *	1500	1000	750
100	3600 *	1800	1200	900	720	600	514	3000 *	1500	1000	750
125	3600 *	1800	1200	900	720	600	514	3000 *	1500	1000	750
150	3600 *	1800	1200	900	720	600	.....	3000 *	1500	1000	750
200	3600 *	1800	1200	900	720	.....	.....	3000 *	1500	1000	750
250	3600 *	1800	1200	900	.....	.....	.....	3000 *	1500	1000	750
300	3600 *	1800	1200	.....	.....	.....	.....	3000 *	1500	1000	
350	3600 *	1800	1200	.....	.....	.....	.....	3000 *	1500	1000	
400	3600*	1800	......	.....	.....	.....	.....	3000 *	1500		
450	3600*	1800	......	.....	.....	.....	.....	3000 *	1500		
500	3600 *	1800	......	.....	.....	.....	.....	3000 *	1500		

* Applies to squirrel-cage motors only.

4. Voltages for ac fractional- and integral-horsepower motors are (Ref. 1):
Universal motors, 115 and 230 V
Single-phase motors, Table 2
1. 60 Hz, 115 and 230 V
2. 50 Hz, 110 and 220 V
Ac polyphase motors
1. 60 Hz, 115,* 200, 230, 460, 575, 2300, 4000, 4600, and 6600 V
2. Three-phase, 50 Hz, 220 and 380 V
5. Frequencies
1. Ac motors. The frequency shall be 50 and 60 Hz (Ref. 1).
2. Universal motors. The frequency shall be 60 Hz/dc (Ref. 1).

CLASSIFICATION BY APPLICATION

6. General-purpose ac motors are induction motors of the open type continuously rated at 200 hp or less, with a service factor (Arts. 24 and 40) and Class A or B insulation.

The general-purpose-type motor is designed in standard ratings with standard operating characteristics and mechanical construction for use under usual service (Art. 23) conditions without restriction to a particular application or type of application.

7. Definite-purpose motors are any motors designed in standard ratings with standard operating characteristics or mechanical construction for use under service conditions other than usual or for use in a particular type of application (Ref. 1).

8. Special-purpose motors are motors with special operating characteristics or special mechanical construction, or both, designed for a particular application and not falling within the definition of general-purpose or definite-purpose motors.

CLASSIFICATION BY ELECTRICAL TYPE

Ac motors are of three general types—induction, synchronous, and series.

9. Induction motors have their primary winding (usually the stator) connected to the power-supply source. The secondary winding (usually the rotor) carrying induced currents may be of the squirrel-cage or polyphase winding type. These motors during normal operation run at speeds less than synchronous.

10. Synchronous motors run at synchronous speed during normal operation and transform electrical into mechanical power.

11. Series-wound motors have their field and armature circuits connected in series. The motors are used in single-phase ac applications.

CLASSIFICATION BY ENVIRONMENTAL PROTECTION AND METHODS OF COOLING

Under this category various types of motor construction satisfy definite requirements, such as the open machine and the totally enclosed machine.

12. Open machines have ventilating openings which permit passage of external cooling air over and around the windings of the machine. The term "open machine," when applied to large apparatus without qualification, designates a machine having no restriction to ventilation other than that necessitated by mechanical construction (Ref. 1).

Open-machine construction includes dripproof, splashproof, semiguarded, guarded, dripproof fully guarded, open externally ventilated, open pipe-ventilated, and weatherproof types.

13. Totally enclosed machines are so enclosed as to prevent the free exchange of air between the inside and the outside of the case but not sufficiently enclosed to be termed airtight.

Totally enclosed machine construction includes totally enclosed fan-cooled, explosionproof, dust-ignition-proof, waterproof, totally enclosed pipe-ventilated, totally

---

*60 Hz, 115 V applies to motors rated 15 hp or smaller. Also, it is not practical to build motors of all horsepower ratings for all the standard voltages.

enclosed water-cooled, totally enclosed water-air-cooled, totally enclosed air-to-air-cooled, and totally enclosed fan-cooled guarded types (Ref. 1).

CLASSIFICATION BY VARIABILITY OF SPEED

14. Constant-speed motors operate at either a constant or practically constant speed during normal operation, such as a synchronous motor or low-slip induction motor.

15. Varying-speed motors operate with speed varying with load, such as a series-wound or repulsion motor where the speed decreases with load increase.

16. Adjustable-speed motors may have their speed varied gradually over a considerable range. At a particular adjusted condition the speed remains practically unaffected by the load.

17. Adjustable varying-speed motors may have their speed adjusted gradually. When an adjustment is made and set, the speed may vary considerably with load, such as a wound-rotor motor under secondary resistance control.

18. Multispeed motors may operate at any one of two or more definite speeds, each being practically independent of load, for example, an induction motor with windings capable of various pole groupings. In multispeed permanent-split capacitor and shaded-pole motors the speeds are dependent upon the load.

SINGLE-PHASE MOTOR TYPES

19. Single-phase induction motors are not self-starting. A single stator winding excited from a single-phase power source produces a pulsating magnetic field in the motor air gap. With the rotor at standstill no "breakaway" torque is developed. Various methods have been devised to produce useful torque so that a single-phase motor will start and bring its connected load to full speed. Some of these methods have resulted in important single-phase motor types such as split-phase, capacitor, repulsion, and universal motors.

a. Split-phase motors are single-phase induction motors equipped with an auxiliary winding displaced in magnetic position from, and connected in parallel with, the main winding (Ref. 1).

A schematic diagram (Fig. 1) of a split-phase motor shows (Ref.1) two windings, main and auxiliary. The axis of the auxiliary winding is displaced in space with respect to the axis of the main winding, and the current in the auxiliary winding is out of phase in time with the main winding current. To accomplish the phase displacement, the main winding has lower resistance and greater reactance than the auxiliary winding, which is of fine wire placed in the top of the slots. With the same line voltage placed across each winding, the current in the auxiliary winding (high resistance and low reactance) is more in phase with the voltage than the current in the main winding, as shown in Fig. 2. The time-phase displacement between the currents is small; hence the locked-rotor torque is not large. The locked-rotor current is the vector sum of the winding currents and is high, as the vector currents add up almost directly.

During starting, at about 75 to 80% synchronous speed, a centrifugal switch disconnects the auxiliary winding from the power line, and the motor may then accelerate on the main winding only to attain full speed (Fig. 3).

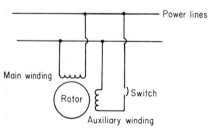

Fig. 1 Elementary schematic diagram of a split-phase induction motor.

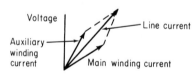

Fig. 2 Winding and line currents of a split-phase induction motor.

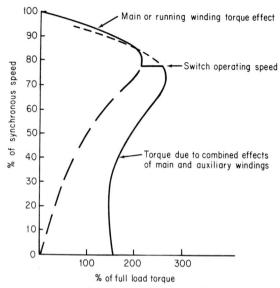

Fig. 3 Speed-torque curve characteristic of one type of split-phase motor design.

It is important that the auxiliary winding not be left on the power-supply source for any extended time. The thermal capacity of the winding is limited. This limitation also implies that high-inertia loads must be avoided.

If the auxiliary-winding leads are brought out, external resistance can be inserted in the auxiliary winding circuit during starting. A relay can be used instead of a shaft-mounted centrifugal switch for auxiliary-winding switching.

Motor rotation can be reversed by reversing the leads of either main or auxiliary winding but not both simultaneously.

b. Capacitor motors are single-phase induction motors with a main winding arranged for direct connection to a source of power and an auxiliary winding connected in series with a capacitor. There are three types of capacitor motors:

1. A capacitor-start motor is a capacitor motor in which the capacitor is in the circuit only during the starting period. This type of motor is widely used in general-purpose applications. As in the split-phase motor, the capacitor-start motor has two windings, a main and auxiliary (Fig. 4), displaced in space and their currents in time. A centrifugal switch (or relay) is used to cut out the auxiliary winding at an appropriate speed during the starting period. As shown in Fig. 5, there is a large time-phase displacement of currents with resulting high torque and low locked-rotor current. Note that the currents add up vectorially for a low locked-rotor current.

2. A permanent-split capacitor motor is a capacitor motor having the same value of capacitance for both starting and running condition.

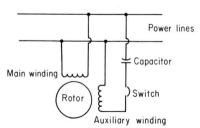

Fig. 4 Elementary schematic diagram of a capacitor-start induction motor.

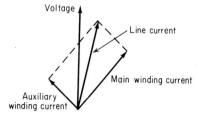

Fig. 5 Winding and line currents of a capacitor-start induction motor.

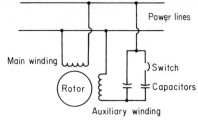

Fig. 6 Elementary schematic diagram of a two-value capacitor induction motor.

3. A two-value capacitor motor uses two starting capacitors as in Fig. 6; one capacitor in the auxiliary winding is permanently installed. Thus, in this type of capacitor motor there are different values of effective capacitance for the starting and running conditions.

20. Single-phase wound-rotor motors, such as the repulsion motor, repulsion-start induction motor, and repulsion-induction motor, are in less demand, mostly for economic reasons.

a. Repulsion motors are single-phase motors which have a stator winding arranged for connection to a source of power and a rotor winding connected to a commutator. Brushes on the commutator are short-circuited and are so placed that the magnetic axis of the rotor is inclined to the magnetic axis of the stator winding. This type of motor has a varying speed characteristic like a dc series motor.

b. Repulsion-start induction motors are single-phase motors having the same windings as a repulsion motor, but at a predetermined speed the rotor winding is short-circuited or otherwise connected to give the equivalent of a squirrel-cage winding. This type of motor starts as a repulsion motor but operates as an induction motor with constant-speed characteristics. The motor develops high torque per ampere at starting compared with other single-phase motor types.

c. Repulsion-induction motors are a form of repulsion motors which have a squirrel-cage winding in the rotor in addition to the repulsion-motor winding. A motor of this type may have either a constant-speed or varying-speed characteristic.

21. Universal motors are suitable for operation on single-phase 60 Hz or dc. They have dc series-motor-type characteristics and construction with commutator, brushes, etc. Many universal motors are of this fractional-horsepower type and may operate at several thousand r/min (Ref. 3).

POLYPHASE-MOTOR TYPES

22. Polyphase motors cover the range of horsepower ratings from fractional- to integral-horsepower to large-apparatus types.

a. Fractional- and integral-horsepower types are generally squirrel-cage induction motors. However, secondary (rotor) data in volts and amperes for wound-rotor induction motors are given in standards (Ref. 1) as low as 1 hp. Synchronous reluctance-

TABLE 3. Horsepower Ratings of Large-Apparatus Induction Motors (Ref. 1)

100	600	2 500	9 000	19 000	45 000
125	700	3 000	10 000	20 000	50 000
150	800	3 500	11 000	22 500	55 000
200	900	4 000	12 000	25 000	60 000
250	1 000	4 500	13 000	27 500	65 000
300	1 250	5 000	14 000	30 000	70 000
350	1 500	5 500	15 000	32 500	75 000
400	1 750	6 000	16 000	35 000	80 000
450	2 000	7 000	17 000	37 500	90 000
500	2 250	8 000	18 000	40 000	100 000

**TABLE 4. Synchronous-Speed Ratings
of Large-Apparatus Induction
Motors (Ref. 1)**

r/min at 60 Hz *

3600	720	400	277
1800	600	360	257
1200	514	327	240
900	450	300	225

* At 50 Hz, the speeds are ⅚ of the 60-Hz speeds.

type motors with good starting characteristics are popular with improved design developments.   Horsepower ranges are given in Table 2.

b. Large-apparatus-type motors are of the induction squirrel-cage and wound-rotor design, and synchronous motors are of the salient-pole and cylindrical-rotor types. The cylindrical-rotor design may be of synchronous-induction motor construction (see Art. 88) or turbo type for high-speed 3600-r/min operation at 60 Hz.   The turbo type is not widely applied and may require a separate starting motor.

c. Standard horsepower and speed ratings are shown in Tables 3 and 4 for induction and Tables 5 and 6 for synchronous motors, respectively.

d. Standard voltages for 60 Hz only are 200, 230, 460, 575, 2300, 4000, 4600, 6600, and 13 200 V.   It is not practicable to build motors of all horsepower ratings for all these voltages.

e. Standard frequencies are 50 (for export) and 60 Hz for induction motors.

23. Usual service conditions for fractional-horsepower, integral-horsepower, and large-apparatus motors are:

a. Fractional- and Integral-Horsepower Motors

1. Exposure in an ambient temperature in the range of 0 to 40°C.   When water cooling is used, the range is 10 to 40°C.

2. Exposure to an altitude not exceeding 3300 ft (1000 m).

3. Installation on a rigid mounting surface.

4. Located or installed in supplementary enclosures which do not seriously interfere with motor ventilation.

5. Voltage variation up to ±10% under rated load at running conditions.   Synchronous motors have rated excitation current unchanged.   With the voltage change, performance may not be in accordance with rated voltage values.

6. Frequency variation up to ±5% under rated load at running conditions.   Synchronous motors have rated excitation unchanged.   With the frequency change, performance may not be in accordance with rated frequency values.

**TABLE 5. Horsepower Ratings of Large-Apparatus
Synchronous Motors (Ref. 1)**

20	450	4 000	17 000	60 000
25	500	4 500	18 000	65 000
30	600	5 000	19 000	70 000
40	700	5 500	20 000	75 000
50	800	6 000	22 500	80 000
60	900	7 000	25 000	90 000
75	1 000	8 000	27 500	100 000
100	1 250	9 000	30 000	
125	1 500	10 000	32 500	
150	1 750	11 000	35 000	
200	2 000	12 000	37 500	
250	2 250	13 000	40 000	
300	2 500	14 000	45 000	
350	3 000	15 000	50 000	
400	3 500	16 000	55 000	

TABLE 6. Synchronous-Speed Ratings of Large-Apparatus Synchronous Motors (Ref. 1)

Speed Ratings, r/min at 60 Hz *

3600	514	277	164	100
1800	450	257	150	95
1200	400	240	138	90
900	360	225	129	86
720	327	200	120	80
600	300	180	109	

* At 50 Hz, the speeds are $\frac{5}{6}$ of the 60-Hz speeds.

7. Combined voltage and frequency variation up to ±10% provided the frequency range is within ±5%. Synchronous motors have rated excitation current unchanged. Performance may not be in accordance with that established for rated voltage and frequency.

b. Large-Apparatus Induction and Synchronous Motors
1. Ambient temperature not less than 10°C or greater than 40°C
2. Altitude not exceeding 3300 ft (1000 m)
3. A location and supplementary enclosure such that there is no serious interference with motor ventilation

24. Important Terms in Motor Usage
a. Efficiency is power output/power input. This ratio times 100 gives the efficiency in percent.

b. Power Factor
1. Single-phase-motor power factor is watts input divided by voltamperes input. This ratio times 100 gives power factor in percent.
2. Polyphase-motor power factor is watts input divided by voltamperes input. Voltamperes for three-phase motors is the product $\sqrt{3}\, E_L I_L$, where E_L is line-to-line voltage and I_L is line current to the motor. The ratio of watts input/voltamperes times 100 gives power factor in percent. If kilowatts are used instead of watts, kilovoltamperes must be used instead of voltamperes.

c. Service factor for ac motors is a multiplier which, when applied to rated horsepower, indicates a permissible horsepower loading which may be carried (MG1-1.43 of Ref. 1).

d. The secondary voltage of wound-rotor motors is the voltage across the open-circuited slip rings when rated voltage is applied to the primary winding at rated frequency.

e. The rating of a motor consists of the output and any other characteristic such as speed, current, voltage, or frequency assigned by the manufacturer.

f. Torque is a turning or twisting effort. In very small motors it may be expressed or measured in ounce-force-in (ozf-in) or ounce-force-ft (ozf-ft). In large-motor ratings torque is usually expressed in pounds-force-ft (lbf-ft) or as a percent of full-load torque. Occasionally torque may be expressed in ft-lb. This terminology may be found in past engineering literature but has been dropped because of conflict with work units also expressed in foot-pounds.

Many kinds of torque are described by qualifying modifiers applied to the term. Some torque terms are vague, some broad or inclusive, and some very definite. The following torque terms are commonly used (Fig. 7):
1. The full-load torque of a motor is the torque necessary to produce its rated horsepower at full-load speed. In pounds-force foot (pounds force at 1-ft radius) it is equal to the horsepower times 5252 divided by the full-load speed.
2. Locked-rotor (static torque) is the minimum torque which it will develop at rest for all angular positions of the rotor, with rated voltage applied at rated frequency.
3. The pull-up torque of an ac motor is the minimum torque developed by the motor during the period of acceleration from rest to the speed at which breakdown torque occurs. For motors which do not have a definite breakdown torque, the pull-up torque is the minimum torque developed up to rated speed.

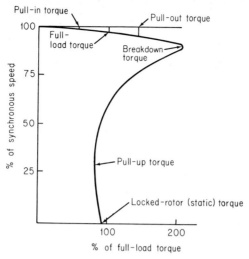

Fig. 7 Types of motor torques.

4. The breakdown torque of a motor is the maximum torque which it will develop with rated voltage applied at rated frequency, without an abrupt drop in speed.

5. The pullout torque of a synchronous motor is the maximum sustained torque which the motor will develop at synchronous speed with rated voltage applied at rated frequency and with normal excitation (Ref. 1).

6. The pull-in torque of a synchronous motor is the maximum constant torque under which the motor will pull its connected inertia load into synchronism, at rated voltage and frequency, when its field excitation is applied.

The speed to which a motor will bring its load depends on the power required to drive it, and whether the motor can pull the load into step from this speed depends on the inertia of the revolving parts, so that the pull-in torque cannot be determined without having the Wk^2 as well as the torque of the load (Ref. 1; also Sec. 10).

g. The *locked-rotor current* of a motor is the steady-state current taken from the line with the rotor locked and with rated voltage and rated frequency applied to the motor.

h. *Ambient temperature* is the temperature of the surrounding cooling medium, such as gas or liquid, which comes into contact with the heated parts of the apparatus. Ambient temperature is commonly known as "room temperature" in connection with air-cooled apparatus not provided with artificial ventilation.

TERMINAL MARKINGS

25. Terminal Markings of Single-Phase Motors. Industry conforms to standards (Ref. 1) in the marking of terminals of many types of single-phase motors. The simple terminal marking to be described illustrates elementary features of the marking system for single-phase motors (Art. 28).

Dual-voltage windings of the reconnectable series-parallel type have terminal markings as shown in Fig. 8. Terminals numbered 1 through 4 refer to the main winding and those numbered 5 through 8 to the auxiliary winding.

For the high-voltage series connection, even to odd connections are made, T_2 to T_3 and T_6 to T_7. For standard direction of rotation, terminals T_4 and T_5 are connected together. Standard direction of rotation is counterclockwise at the end opposite the drive. For the lower-voltage or parallel connection, odd- to odd-subscript terminals are connected together, T_1 to T_3 and T_5 to T_7. With T_4 connected to T_5 the motor operates with standard direction of rotation.

For a single-voltage single-phase motor or where either winding is intended for one

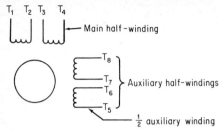

Fig. 8 Dual-voltage terminal-marking arrangement for single-phase motors.

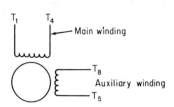

Fig. 9 Single-voltage terminal-marking arrangement for single-phase motors.

voltage, Fig. 9 applies.   Standard direction of rotation is obtained when T_4 and T_5 are joined to one line and T_1 and T_8 to the other.

26. Motor auxiliary devices may be internally or externally connected to the motor windings.

a. Motor internal self-contained devices, such as thermal protectors, capacitors, or starting switches, may be wired within the motor structure and no terminal be brought out from the internal-connected junctions.   Any such device interposed between the marked terminal and the part of the winding to which it ultimately connects does not affect the marking.

If a terminal is brought out from the internally connected junction mentioned above, the junction terminal marking is determined by the part of the winding to which it is connected.   Other terminals are identified by letters indicating the auxiliary device within the motor to which the terminal is connected.   The letter P, for example, refers to a thermal protector (Fig. 10).

b. Auxiliary devices external to the motor may be located separately from the motor. The terminal-marking letters are determined from the device to which they connect directly.   For example, the letter P refers to a thermal protector, J to a capacitor, etc. The subscript number for the terminal markings is the same as the subscript numbers of the letter T of the motor terminals to which they connect for the standard direction of rotation.   For dual-voltage motors subscripts for the auxiliary-device terminals are determined with the motor connected for the higher-voltage value.   Other special conditions may arise as, for example, when the auxiliary device joins to a junction of more than one motor terminal.   Then the lowest motor terminal subscript is used for the auxiliary-device terminal subscript.   It is suggested that the latest publication of Ref. 1 be consulted for a complete discussion of terminal markings for single-phase motors in complex or general cases.

27. Terminal Color Markings.   When single-phase motors use lead colors instead of letter markings to identify the leads, the color assignment is determined from the following:

T_1............	Blue	T_5............	Black
T_2	White	T_6	Red
T_3............	Green	T_7............	No assigned color
T_4............	Yellow	T_8............	Brown

28. Terminal Markings—General Principles.   Three simple rules provide a general guide for the terminal markings of a single-phase motor:

1. Main-winding terminal designations are T_1, T_2, T_3, and T_4, and auxiliary-winding designations are T_5, T_6, T_7, and T_8 (Fig. 8).   This arrangement distinguishes single-phase- from quarter-phase-motor terminal markings, which use odd numbers for one phase and even numbers for the other phase.

2. Odd- to odd-numbered terminals of each winding are joined for lower-voltage (parallel) connection and odd- to even-numbered terminals for the higher-voltage (series) connection.

3. Rotors of single-phase motors are represented by a circle.   This distinguishes a single-phase rotor from the quarter-phase motor rotor which is never represented.

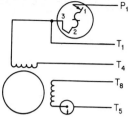

Fig. 10 Schematic diagram for split-phase motors, single-voltage-reversible, with thermal protector, motor starting switch shown in running position.

Line leads	L_1	L_2
Counterclockwise rotation............	T_1, T_8	T_4, T_5
Clockwise rotation	T_1, T_5	T_4, T_8

29. Terminal markings and connections for the following single-phase-type motors are shown in the schematic diagrams of the figures indicated.

Note that on terminal boards, where used, the studs are identified by numbers only. Terminal-board numbers are not necessarily related to winding-lead terminal numbers. For terminal-board stud marking and internal devices permanently connected to terminal-board studs see Ref. 1.

  a. Split-Phase Motors
  1. Without thermal protector (Fig. 11)
  2. With thermal protector (Fig. 10)
  b. Capacitor-Start Motors, Reversible Single-Voltage
  1. Without thermal protector (Fig. 12)
  2. With thermal protector (Fig. 13)
  c. Capacitor-Start Motors, Reversible Double-Voltage
  1. Without thermal protector (Fig. 14)
  2. With thermal protector (Group I, Fig. 15)
  3. With thermal protector (Group II, Fig. 16)
  4. With thermal protector (Group III, Fig. 17)
  d. Two-Value Capacitor Motor, Single-Voltage Reversible (Figs. 18 and 19).
  e. Permanent-Split Capacitor Motor, Single-Voltage Reversible (Fig. 20)
  f. Universal Motors
  1. Single-voltage, nonreversible (Fig. 21)
  2. Single-voltage, reversible (Fig. 22)
  g. Repulsion, Repulsion-Start Induction, Repulsion-Induction Motors
  1. Reversible by brush shifting, single-voltage (Fig. 23)
  2. Reversible by brush shifting, double-voltage (Fig. 23)
  3. Single-voltage, externally reversible (Fig. 24)

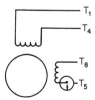

Fig. 11 Schematic diagram for split-phase motors, single-voltage-reversible, without thermal protector, motor starting switch shown in running position.

Line leads	L_1	L_2	Join
Counterclockwise rotation............	P_1	T_4, T_5	T_1, T_8
Clockwise rotation	P_1	T_4, T_8	T_1, T_5

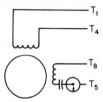

Fig. 12 Schematic diagram for capacitor-start motors, reversible, single-voltage, without thermal protector, motor starting switch shown in running position.

Line leads	L_1	L_2
Counterclockwise rotation............	T_1, T_8	T_4, T_5
Clockwise rotation	T_1, T_5	T_4, T_8

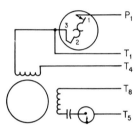

Fig. 13 Schematic diagram for capacitor-start motors, reversible, single-voltage, with thermal protector, motor starting switch shown in running position.

Line leads	L_1	L_2	Join
Counterclockwise rotation............	P_1	T_4, T_5	T_1, T_8
Clockwise rotation	P_2	T_4, T_8	T_1, T_5

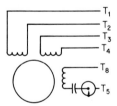

Fig. 14 Schematic diagram for capacitor-start motors, reversible, double-voltage, without thermal protector, motor starting switch in running position.

Line leads	L_1	L_2	Join
Higher nameplate voltage:			
Counterclockwise rotation	T_1	T_4, T_5	T_2, T_3, and T_8
Clockwise rotation.....................	T_1	T_4, T_8	T_2, T_3, and T_5
Lower nameplate voltage:			
Counterclockwise rotation	T_1, T_3, T_8	T_2, T_4, T_5	
Clockwise rotation.....................	T_1, T_3, T_5	T_2, T_4, T_8	

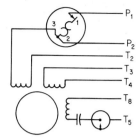

Fig. 15 Schematic diagram of capacitor-start motors, reversible, Group I, double-voltage, with thermal protector, motor starting switch in running position.

Line leads	L_1	L_2	Join	Join
Higher nameplate voltage:				
Counterclockwise rotation .... P_1		T_4	P_2, T_8	T_2, T_3, T_5
Clockwise rotation .... P_1		T_4	P_2, T_5	T_2, T_3, T_8
Lower nameplate voltage:				
Counterclockwise rotation .... P_1		T_2, T_4, T_5	P_2, T_3, T_8	
Clockwise rotation .... P_1		T_2, T_4, T_8	P_2, T_3, T_5	

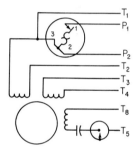

Fig. 16 Schematic diagram of capacitor-start motors, reversible, Group II, double-voltage, with thermal protector, motor starting switch in running position.

Line leads	L_1	L_2	Join	Join	Insulate separately
Higher nameplate voltage:					
Counterclockwise rotation.......... P_1		T_4, T_5	T_2, T_3, T_8	....	P_2, T_1
Clockwise rotation.................... P_1		T_4, T_8	T_2, T_3, T_5	....	P_2, T_1
Lower nameplate voltage:					
Counterclockwise rotation......... P_1		T_2, T_4, T_5	P_2, T_3	T_1, T_8	
Clockwise rotation.................... P_1		T_2, T_4, T_8	P_2, T_3	T_1, T_5	

30. Three-phase schematic marking method for motor-winding terminals provides important information about a motor winding, especially when the motor is enclosed and only the phase-winding leads are visible. For example, in the circuit diagram of two circuits per phase, shown in Fig. 25, a "clockwise rotating spiral" describes the terminal numbers starting from terminal T_1. In the inverted Y of Fig. 25, the terminal markings are arranged to give the correct polarity of the circuits.

Assume the circuits of Fig. 25 are arranged in multiple (parallel) as shown in Fig. 26 and the two circuits of each phase are then permanently connected together as shown in Fig. 27. Note that the highest numbers of Fig. 26 are dropped and only the lowest numbers retained in Fig. 27.

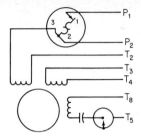

Fig. 17 Schematic diagram of capacitor-start motors, reversible, Group III, double-voltage, with thermal protector, motor starting switch in running position.

Line leads	L_1	L_2	Join	Insulate separately
Higher nameplate voltage:				
Counterclockwise rotation............... P_1	T_4, T_5	T_2, T_3, T_8	P_2	
Clockwise rotation .... P_1	T_4, T_8	T_2, T_3, T_5	P_2	
Lower nameplate voltage:				
Counterclockwise rotation............... P_1	T_2, T_4, T_5	P_2, T_3, T_8		
Clockwise rotation .... P_1	T_2, T_4, T_8	P_2, T_3, T_5		

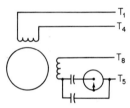

Fig. 18 Schematic diagram for two-value capacitor motors, reversible, single-voltage, without thermal protector, motor starting switch shown in running position.

Line leads	L_1	L_2
Counterclockwise rotation............. T_1, T_8	T_4, T_5	
Clockwise rotation .... T_1, T_5	T_4, T_8	

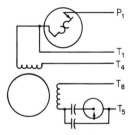

Fig. 19 Schematic diagram for two-value capacitor motors, reversible, single-voltage, with thermal protector, motor starting switch shown in running position.

Line leads	L_1	L_2	Join
Counterclockwise rotation............. P_1	T_4, T_5	T_1, T_8	
Clockwise rotation .... P_1	T_4, T_8	T_1, T_5	

Fig. 20   Schematic diagram for permanent-split capacitor motors, reversible, single-voltage.

Line leads	L_1	L_2
Counterclockwise rotation............	T_1, T_8	T_4, T_5
Clockwise rotation	T_1, T_5	T_4, T_8

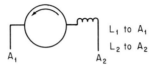

Fig. 21   Schematic diagram for universal motors, nonreversible, single-voltage.

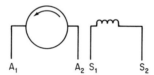

Fig. 22   Schematic diagram for universal motors, reversible, single-voltage.

Lined leads	L_1	L_2	Join
Counterclockwise rotation............	A_1	S_2	A_2, S_1
Clockwise rotation	A_1	S_1	A_2, S_2

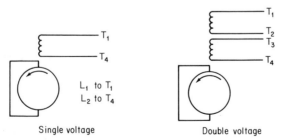

Fig. 23   Schematic diagrams for repulsion, repulsion-start induction and repulsion-induction motors, reversible by shifting brushes.

Line leads	L_1	L_2	Join
Higher nameplate voltage............	T_1	T_4	T_2, T_3
Lower nameplate voltage............	T_1, T_3	T_2, T_4	

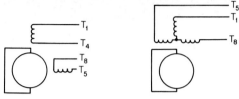

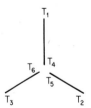

Fig. 24 Schematic diagrams for repulsion, repulsion-start induction, and repulsion-induction motors, single-voltage, externally reversible.

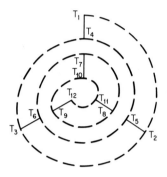

Fig. 25 Terminal markings for two circuits per phase determined by clockwise-rotating-spiral method.

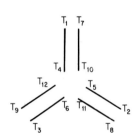

Fig. 26 Terminal markings for two circuits in multiple per phase.

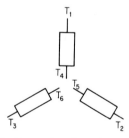

Fig. 27 Terminal markings for two circuits in multiple per phase, permanently connected.

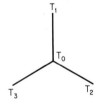

Fig. 28 Terminal markings for 3-phase single-circuit, wye connection, all phase ends brought out.

Fig. 29 Terminal markings for 3-phase single-circuit, wye connection, neutral brought out.

Fig. 30 Terminal markings for 3-phase single-circuit, wye connection, neutral formed but not brought out.

Further, if the neutral leads of the three phases of Fig. 27 are permanently connected together, the phase lead ends thus connected drop their numbers. A neutral lead brought out is marked T_0.

The simple wye-connected winding is shown in Figs. 28, 29, and 30 for all phase ends brought out, neutrals joined with neutral lead out, and for three terminal leads with neutral phase ends joined permanently.

The method of clockwise rotating spiral can be applied to more than two circuits and to 2-phase motors. However, the method is unsuited for 3-phase motors having two synchronous speeds obtained from a reconnectable winding. Rotation will not be in the same direction for both speeds if the power lines are connected to each set of terminals in the same sequence.

A delta-connected two-circuit-per-phase winding uses the two-circuit-per-phase wye-connected diagram of Fig. 25 rotated 30°, and the delta terminal-marking diagram is then constructed as shown in Fig. 31.

Dual-voltage 3-phase motor terminal markings and connection arrangements are shown in Figs. 32 through 34. Figure 34 also shows wye connections for starting with final delta connections for running.

Multispeed 3-phase motor terminal markings and connection arrangements are shown in Figs. 35 through 42. As stated previously, the "clockwise spiral rotating" system is unsuited for reconnectable two-speed 3-phase induction motors.

31. Terminal-marking diagrams and connections for many of the diagrams are shown in the following listing for 3-phase motors.

1. Two circuits in multiple per phase, all phase circuit ends brought out (Fig. 26)
2. Two circuits in multiple (parallel) per phase, permanently connected, all phase ends brought out (Fig. 27)
3. Single circuit, wye connection, all phase ends brought out (Fig. 28)
4. Single-circuit, wye connection, neutral brought out (Fig. 29)
5. Single-circuit, wye connection, neutral formed but not brought out (Fig. 30)
6. Two circuits per phase, delta connection (Fig. 31)
7. Wye connection, dual-voltage (Fig. 32)
8. Delta connection, dual-voltage (Fig. 33)
9. Wye-connection start, delta-connection run, single-voltage or wye-delta-connected, dual-voltage, voltage ratio 3/1 (Fig. 34)
10. Variable-torque motors (Fig. 35)
11. Constant-torque motors (Fig. 36)
12. Constant-torque motors (Fig. 37)
13. Constant-horsepower motors (Fig. 38)
14. Constant-horsepower motors (Fig. 39)
15. Three-speed motor using three windings (Fig. 40)
16. Four-speed motor using two windings (Fig. 41)
17. Three-speed motor using two windings (Fig. 42)

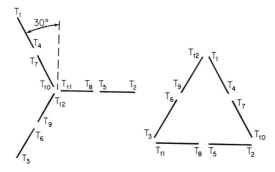

Fig. 31   Terminal markings for two circuits per phase, delta-connected.

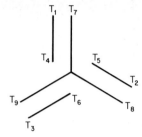

Fig. 32 Wye-connected, dual-voltage.

Voltage	L_1	L_2	L_3	Tie together
Low............	T_1, T_7	T_2, T_8	T_3, T_9	(T_4, T_5, T_6)
High............	T_1	T_2	T_3	$(T_4, T_7) (T_5, T_8) (T_6, T_9)$

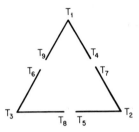

Fig. 33 Delta-connected, dual-voltage.

Voltage	L_1	L_2	L_3	Tie together
Low.................	(T_1, T_6, T_7)	(T_2, T_4, T_8)	(T_3, T_5, T_9)	
High.................	T_1	T_2	T_3	$(T_4, T_7) (T_5, T_8) (T_6, T_9)$

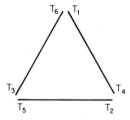

Fig. 34 Wye-connected start, delta-connected run, single-voltage.

	L_1	L_2	L_3	Tie together
Start......................	T_1	T_2	T_3	(T_4, T_5, T_6)
Run	(T_1, T_6)	(T_2, T_4)	(T_3, T_5)	

Wye-delta-connected, dual-voltage (voltage ratio $\sqrt{3}/1$)

	L_1	L_2	L_3	Tie together
High voltage............	T_1	T_2	T_3	(T_4, T_5, T_6)
Low voltage.............	(T_1, T_6)	(T_2, T_4)	(T_3, T_5)	

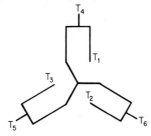

Fig. 35 Variable-torque motors.

Speed	L_1	L_2	L_3	Insulate separately	Tie together
Low............	T_1	T_2	T_3	T_4, T_5, T_6	
High...........	T_6	T_4	T_5	....	(T_1, T_2, T_3)

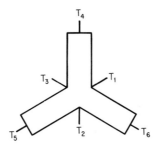

Fig. 36 Constant-torque motors.

Speed	L_1	L_2	L_3	Insulate separately	Tie together
Low............	T_1	T_2	T_3	T_4, T_5, T_6	
High...........	T_6	T_4	T_5	....	(T_1, T_2, T_3)

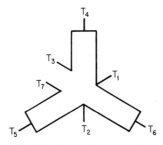

Fig. 37 Constant-torque motors.

Speed	L_1	L_2	L_3	Insulate separately	Tie together
Low............	T_1	T_2	(T_3, T_7)	T_4, T_5, T_6	
High...........	T_6	T_4	T_5	....	(T_1, T_2, T_3, T_7)

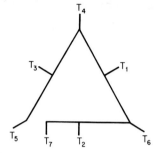

Fig. 38 Constant-horsepower motors.

Speed	L_1	L_2	L_3	Insulate separately	Tie together
Low............	T_1	T_2	T_3	....	(T_4, T_5, T_6, T_7)
High............	T_6	T_4	(T_5, T_7)	T_1, T_2, T_3	

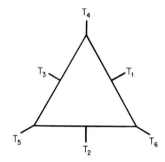

Fig. 39 Constant-horsepower motors.

Speed	L_1	L_2	L_3	Insulate separately	Tie together
Low............	T_1	T_2	T_3	....	(T_4, T_5, T_6)
High............	T_6	T_4	T_5	T_1, T_2, T_3	

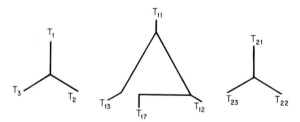

Fig. 40 Three-speed motor using three windings.

Speed	L_1	L_2	L_3	Insulate separately
Low...............	T_1	T_2	T_3	$T_{11}, T_{12}, T_{13}, T_{17}, T_{21}, T_{22}, T_{23}$
Second............	T_{11}	T_{12}	(T_{13}, T_{17})	$T_1, T_2, T_3, T_{21}, T_{22}, T_{23}$
High...............	T_{21}	T_{22}	T_{23}	$T_1, T_2, T_3, T_{11}, T_{12}, T_{13}, T_{17}$

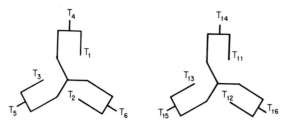

Fig. 41   Four-speed motor using two windings.

Speed	L_1	L_2	L_3	Insulate separately	Tie together
Low	T_1	T_2	T_3	T_4, T_5, T_6, T_{11}, T_{12}, T_{13}, T_{14}, T_{15}, T_{16}	
Second	T_{11}	T_{12}	T_{13}	T_1, T_2, T_3, T_4, T_5, T_6, T_{14}, T_{15}, T_{16}	
Third	T_6	T_4	T_5	T_{11}, T_{12}, T_{13}, T_{14}, T_{15}, T_{16}	(T_1, T_2, T_3)
High	T_{16}	T_{14}	T_{15}	T_1, T_2, T_3, T_4, T_5, T_6	(T_{11}, T_{12}, T_{13})

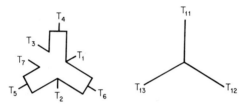

Fig. 42   Three-speed motor using two windings.

Speed	L_1	L_2	L_3	Insulate separately	Tie together
Low	T_1	T_2	(T_3, T_7)	T_4, T_5, T_6, T_{11}, T_{12}, T_{13}	
Second	T_6	T_4	T_5	T_{11}, T_{12}, T_{13}	(T_1, T_2, T_3, T_7)
High	T_{11}	T_{12}	T_{13}	T_1, T_2, T_3, T_4, T_5, T_6, T_7	

32. Two-Phase Terminal Markings.   Two-phase terminal marking uses the clockwise-rotating-spiral method shown in Fig. 43. Note that one phase has odd-numbered markings and the other phase even-numbered markings. The method follows procedures previously covered for 3-phase motors in Art. 30.

33. Two-Phase Diagrams and Connections.   Terminal-marking diagrams and connections for one of the diagrams are shown in the following list for 2-phase motors.

1. Three circuits per phase, all leads brought out (Fig. 44)
2. Three circuits per phase, all circuits paralleled permanently, phase end leads only brought out (Fig. 45)
3. Two-speed, single-winding, variable-torque induction motor (Fig. 46)
4. Two-speed, two-winding induction motor (Fig. 47)
5. Four-speed, two-winding induction motor (Fig. 48)

34. Three-Phase Wound-Rotor Terminal Markings.   Terminal markings for 3-phase and 2-phase three-slip-ring rotor windings of induction motors are shown in Fig. 49.

35. Pole-Amplitude-Modulation (PAM) Markings.   Multispeed winding terminal markings and connection arrangement are not given in standards. Generally they are provided to the control manufacturer by the motor builder for the particular type of induction motor.

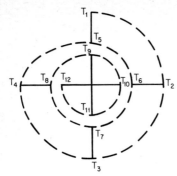

Fig. 43 Two-phase terminal markings using clockwise rotating-spiral method.

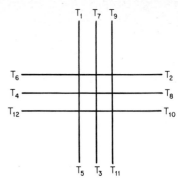

Fig. 44 Terminal markings for three circuits per phase, all circuit leads brought out.

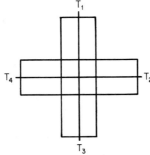

Fig. 45 Terminal markings for three circuits per phase connected in parallel permanently inside the motor.

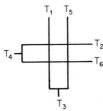

Fig. 46 Terminal markings for 2-speed, 2-phase, single-winding, variable-torque induction motor.

Speed	L_1	L_3	L_2	L_4	Insulate separately
Low............	T_1	T_5	T_2	T_6	
High............	(T_1, T_5)	T_3	(T_2, T_6)	T_4	T_3, T_4

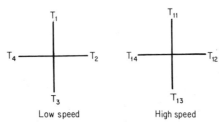

Low speed   High speed

Fig. 47 Terminal markings for a 2-phase, 2-winding, 2-speed induction motor.

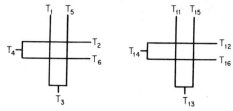

Fig. 48 Terminal markings for a 2-phase, 4-speed induction motor using two windings each of which is reconnectable to give two synchronous speeds.

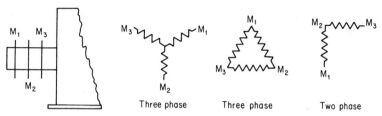

Three phase   Three phase   Two phase

Fig. 49 Terminal markings of the rotors of wound-rotor induction motors.

PERFORMANCE

36. Design Letters—Polyphase. Design letters applying to polyphase squirrel-cage integral-horsepower induction motors indicate a degree of performance relative to locked-rotor torque, breakdown torque, and locked-rotor current. Hence, these letters may be used in specifications to indicate certain design characteristics desired. All design letters A through D listed below apply to squirrel-cage motors which must withstand full-voltage starting.

a. Design A motor design develops locked-rotor and breakdown torques adequate for general application as specified in Tables 7 and 8, draws a locked-rotor current higher than the values shown in Tables 9 and 10 for 60 and 50 Hz, respectively, and has a slip at rated load of less than 5% except that for motors of 10 or more poles the slip may exceed 5% slightly. For pull-up torque see Table 11.

b. Design B motor design develops locked-rotor and breakdown torques adequate for general application as specified in Tables 7 and 8, draws a locked-rotor current not to exceed values in Tables 9 and 10 for 60 and 50 Hz, respectively, and has a slip at rated load of less than 5%, except that for motors of 10 or more poles the slip may exceed 5% slightly.

c. Design C motor design develops a locked-rotor torque for special high-torque applications up to values shown in Table 12, breakdown torque up to values shown in Table 13, locked-rotor current not to exceed the values shown in Tables 9 and 10 for 60 and 50 Hz, respectively, and has a slip at rated load of less than 5% slightly.

d. Design D motor design develops high locked-rotor torque. The locked-rotor torque for 60 and 50 Hz, 4-, 6-, and 8-pole, single-speed, polyphase squirrel-cage motors rated 150 hp and smaller, with rated voltage and frequency applied, should be 275%, expressed in percent of full-load torque, which represents the upper limit of application for these motors. Locked-rotor current for Design D motors is not greater than shown in Tables 9 and 10 for 60 and 50 Hz, respectively. Rated-load slip is less than 5%.

37. Design Letters—Single-Phase Integral. Design letters applying to single-phase integral-horsepower motors have the same purpose stated in Art. 36 relative to certain performance quantities. The design letters listed below apply to single-phase, integral-horsepower motors which must be able to withstand full-voltage starting.

a. Design L has a locked-rotor current not to exceed the values shown in Table 14 and develops a breakdown torque listed in Section 10.33, Part 10 of Ref. 1. This list is extensive and can be consulted for actual values when desired.

TABLE 7. Locked-Rotor Torque of Single-Speed Polyphase Squirrel-Cage Integral-Horsepower Motors with Continuous Ratings (Ref. 1)

		Synchronous speed, r/min						
	60 Hz	3600	1800	1200	900	720	600	514
hp	50 Hz	3000	1500	1000	750			
½		.....	.....	.....	140	140	115	110
¾		.....	.....	175	135	135	115	110
1		.....	275	170	135	135	115	110
1½		175	250	165	130	130	115	110
2		170	235	160	130	125	115	110
3		160	215	155	130	125	115	110
5		150	185	150	130	125	115	110
7½		140	175	150	125	120	115	110
10		135	165	150	125	120	115	110
15		130	160	140	125	120	115	110
20		130	150	135	125	120	115	110
25		130	150	135	125	120	115	110
30		130	150	135	125	120	115	110
40		125	140	135	125	120	115	110
50		120	140	135	125	120	115	110
60		120	140	135	125	120	115	110
75		105	140	135	125	120	115	110
100		105	125	125	125	120	115	110
125		100	110	125	120	115	115	110
150		100	110	120	120	115	115	
200		100	110	120	120	115	115	
250		70	80	100	100			
300		70	80	100				
350		70	80	100				
400		70	80					
450		70	80					
500		70	80					

Values are in percent of full-load torque.   Rated voltage and frequency implied.   Table applies to Design A and B motors.

b. Design M has a locked-rotor current not to exceed the values shown in Table 14 and develops a breakdown torque listed in the same reference for Design L.

38. Design Letters—Single-Phase Fractional.   Design letters applying to single-phase fractional-horsepower motors have the same purpose stated in Art. 36.   The design letters listed below apply to single-phase, fractional-horsepower motors which must be able to withstand full-voltage starting.

a. Design N has a locked-rotor current not exceeding the values shown in Table 15.

b. Design O has a locked-rotor current not exceeding the values shown in Table 15.

39. Code Letters—Fractional and Integral.   Code letters for fractional- and integral-horsepower ac motors (except wound-rotor motors) rated 1/20 hp and larger are specified in standards (Ref. 1) to be marked on the motor nameplates.

Letter designations for locked kVA/hp as measured at full voltage and rated frequency are shown in Table 16.   To clarify different types of applications, the following data are specified:

a. Multispeed motors are marked with the code letter designating the locked-rotor kVA/hp for the high speed at which the motor can be started, except constant-horse-power motors, which are marked with the code letter for the speed giving the highest locked-rotor kVA/hp.

b. Single-speed motors starting on wye connection and running on delta connection

TABLE 8. Breakdown Torque of Single-Speed Polyphase Squirrel-Cage Integral-Horsepower Motors with Continuous Ratings

| | Synchronous speed, r/min | | | | | | |
| | 60 Hz | 3600 | 1800 | 1200 | 900 | 720 | 600 | 514 |
hp	50 Hz	3000	1500	1000	750			
½		.....	.....	.....	225	200	200	200
¾		.....	.....	275	220	200	200	200
1		.....	300	265	215	200	200	200
1½		250	280	250	210	200	200	200
2		240	270	240	210	200	200	200
3		230	250	230	205	200	200	200
5		215	225	215	205	200	200	200
7½		200	215	205	200	200	200	200
10–125		200	200	200	200	200	200	200
150		200	200	200	200	200	200	
200		200	200	200	200	200		
250		175	175	175	175			
300–350		175	175	175				
400–500		175	175					

Design A values are in excess of values shown above. Values are in percent of full-load torque. Rated voltage and frequency implied.

are marked with a code letter corresponding to the locked-rotor kVA/hp for the wye connection.

c. Dual-voltage motors which have a different locked-rotor kVA/hp on the two voltages are marked with the code letter for the voltage giving the highest locked-rotor kVA/hp.

d. Motors with 60- and 50-Hz ratings are marked with a code letter designating the locked-rotor kVA/hp on 60 Hz.

TABLE 9. Locked-Rotor Current at Rated Voltage and Frequency of Single-Speed (Constant-Speed) Induction Motors Rated at 230 V

hp	Locked-rotor current, A	Design letters	hp	Locked-rotor current, A	Design letters
½	20	B, D	60	870	B, C, D
¾	25	B, D	75	1085	B, C, D
1	30	B, D	100	1450	B, C, D
1½	40	B, D	125	1815	B, C, D
2	50	B, D	150	2170	B, C, D
3	64	B, C, D	200	2900	B, C
5	92	B, C, D	250	3650	B
7½	127	B, C, D	300	4400	B
10	162	B, C, D	350	5100	B
15	232	B, C, D	400	5800	B
20	290	B, C, D	450	6500	B
25	365	B, C, D	500	7250	B
30	435	B, C, D			
40	580	B, C, D			
50	725	B, C, D			

For other voltages, the current shall be inversely proportional to the voltage. Values apply to 3-phase, 60-Hz, integral-horsepower squirrel-cage motors.

TABLE 10. Locked-Rotor Current at Rated Voltage and Frequency of Single-Speed 3-Phase 50-Hz Integral-Horsepower Squirrel-Cage Constant-Speed Induction Motors Rated at 380 V

hp	Locked-rotor current, A	Design letters	hp	Locked-rotor current, A	Design letters
1 or less	20	B, D	30	289	B, C, D
1½	27	B, D	40	387	B, C, D
2	34	B, D	50	482	B, C, D
3	43	B, C, D	60	578	B, C, D
5	61	B, C, D	75	722	B, C, D
7½	84	B, C, D	100	965	B, C, D
10	107	B, C, D	125	1207	B, C, D
15	154	B, C, D	150	1441	B, C, D
20	194	B, C, D	200	1927	B, C
25	243	B, C, D			

For other voltages the current shall be inversely proportional to the voltages.

TABLE 11. Pull-up Torque of Single-Speed Polyphase Squirrel-Cage Integral-Horsepower Motors Having Continuous Ratings, Designs A and B

Column 1	Column 2
Locked-rotor torque from Table 8	Pull-up torque, %
110% or less	90% of Column 1
More than 110% but less than 145%	100% of full-load torque
145% or more	70% of Column 1

TABLE 12. Locked-Rotor Torque at Rated Voltage and Frequency of Design C, 60- and 50-Hz, Single-Speed, Polyphase Squirrel-Cage Integral-Horsepower Motors with Continuous Ratings

		Synchronous speed, r/min		
	60 Hz	1800	1200	900
hp	50 Hz	1500	1000	750
3		.....	250	225
5		250	250	225
7½		250	225	200
10		250	225	200
15		225	200	200
20–200		200	200	200

Values are in percent of full-load torque and represent the upper limit of the range of application for these motors (Ref. 1).

e. Part-winding-start motors are marked with a code letter designating the locked-rotor kVA/hp that is based upon the locked-rotor current for the full winding of the motor.

40. General-Purpose Motor Definitions. The general-purpose motor defined in Art. 6 is an ac induction motor of 200 hp or less of open construction and continuously rated.

a. A Class A insulated motor operating in a 40°C ambient under usual service conditions (Art. 23) has a winding temperature rise at rated load which is specified by stand-

TABLE 13. Breakdown Torque at Rated Voltage and Frequency of Design C, 60- and 50-Hz, Single-Speed, Polyphase Squirrel-Cage Integral-Horsepower Motor with Continuous Ratings

hp		Synchronous speed, r/min		
	60 Hz	1800	1200	900
	50 Hz	1500	1000	750
3		.....	225	200
5		200	200	200
7½–200		190	190	190

Values are given in percent of full-load torque and represent the upper limit of the range of application for these motors (Ref. 1).

TABLE 14. Locked-Rotor Current of Single-Phase Integral-Horsepower, 60 Hz, Design L and M Motors of All Types

hp	Locked-rotor current, A		
	Design L motors		Design M motors
	115 V	230 V	230 V
½	45	25	
¾	61	35	
1	80	45	
1½	...	50	40
2	...	65	50
3	...	90	70
5	...	135	100
7½	...	200	150
10	...	260	200
15	...	390	300
20	...	520	400

Data for rated voltage and frequency.   Values should not be exceeded for these designs.

TABLE 15. Locked-Rotor Current of Single-Phase, Fractional-Horsepower, 60 Hz, Design N and O Motors

2-, 4-, 6-, and 8-Pole, 60 Hz Motors, Single-Phase

hp	Locked-rotor current, A			
	115 V		230 V	
	Design O	Design N	Design O	Design N
⅙ and smaller	50	20	25	12
¼	50	26	25	15
⅓	50	31	25	18
½	50	45	25	25
¾	...	61	...	35
1	...	80	...	45

Values should not be exceeded for these designs.

TABLE 16. Code Letters for Locked-Rotor kVA

Letter designation	kVA/hp	Letter designation	kVA/hp *
A	0–3.15	L	9.0–10.0
B	3.15–3.55	M	10.0–11.2
C	3.55–4.0	N	11.2–12.5
D	4.0–4.5	P	12.5–14.0
E	4.5–5.0	R	14.0–16.0
F	5.0–5.6	S	16.0–18.0
G	5.6–6.3	T	18.0–20.0
H	6.3–7.1	U	20.0–22.4
J	7.1–8.0	V	22.4 and up
K	8.0–9.0		

* Locked kVA/hp range includes the lower figure up to but not including the higher figure. For example, 3.14 is designated by letter A and 3.15 by letter B.

ards not to exceed 40°C by thermometer or 50°C by resistance. Either method of temperature-rise measurement is acceptable for demonstrating the meeting of standards. Temperatures are measured in accordance with the latest revisions of Refs. 6 and 7.

Service factors for the Class A insulated general-purpose motor are shown in Table 17. When the voltage and frequency are maintained at nameplate value, the motor may be overloaded up to the horsepower obtained by multiplying the rated horsepower by the service factor shown on the nameplate.

At any service factor greater than 1, motor temperature will be higher than the rated-load value and the efficiency, power factor, and speed may differ from it, but the locked-rotor torque and current and the breakdown torque will remain unchanged.

b. A *Class B insulated motor* operating in a 40° C ambient has a winding-temperature rise when tested at the service-factor load, which is specified by standards not to exceed 90°C rise by resistance. Service factors are shown in Table 18. Temperatures are measured in accordance with the latest revision of Refs. 6 and 7.

In its application the motor may be operated at its rated load under usual service conditions (Art. 23). At nameplate voltage and frequency the motor may be overloaded up to the horsepower obtained by multiplying the rated horsepower by the service shown on the nameplate.

c. *Performance characteristics* of interest for general-purpose motors are:

1. Locked-rotor torques of single-phase fractional-horsepower motors are shown in Table 19.

2. Locked-rotor torques of single-phase integral-horsepower motors are shown in Table 20.

3. Locked-rotor amperes of fractional-horsepower motors must not exceed the values shown in Table 15 for Design N motors.

TABLE 17. Service Factor of General-Purpose Ac Motors when Operated at Rated Voltage and Frequency

hp	Service factor	hp	Service factor
1/6	1.35	1	1.25*
1/4	1.35	1 1/2	1.20 *
1/3	1.35	2	1.20 *
1/2	1.25 *	3–200	1.15 *
3/4	1.25 *		

* In the case of polyphase squirrel-cage integral-horsepower motors, these service factors apply only to Design A, B, and C motors.

TABLE 18. Service Factor for General-Purpose Ac Single-Phase and Polyphase Induction Motors Having a Temperature Rise of 90°C by Resistance

hp	Synchronous speed, r/min						
	3600	1800	1200	900	720	600	514
$1/20$	1.4	1.4	1.4	1.4			
$1/12$	1.4	1.4	1.4	1.4			
$1/8$	1.4	1.4	1.4	1.4			
$1/6$	1.35	1.35	1.35	1.35			
$1/4$	1.35	1.35	1.35	1.35			
$1/3$	1.35	1.35	1.35	1.35			
$1/2$	1.25	1.25	1.25	1.15 *			
$3/4$	1.25	1.25	1.15 *	1.15 *			
1	1.25	1.15 *	1.15 *	1.15 *			
$1^{1}/_2$–125	1.15 *	1.15*	1.15 *	1.15 *	1.15*	1.15*	1.15 *
150	1.15 *	1.15*	1.15 *	1.15 *	1.15 *	1.15 *	
200	1.15 *	1.15 *	1.15 *	1.15 *	1.15 *		

* In the case of polyphase squirrel-cage integral-horsepower motors, these service factors apply only to Design A, B, and C motors.

4. For breakdown torque of single-phase fractional- and integral-horsepower motors, see Section 10.33, Part 10 of Ref. 1.

5. Breakdown torque of a general-purpose, polyphase, squirrel-cage, fractional-horsepower motor at rated voltage and frequency is not less than 140% of the breakdown-torque values of a single-phase, general-purpose, fractional-horsepower motor of the same horsepower and speed rating given in the reference in the preceding paragraph.

6. Pull-up torque of single-phase integral-horsepower motors is not less than the rated-load torque.

d. *General-purpose, 60-Hz, polyphase, integral-horsepower induction motors* of 2-, 4-, 6-, and 8-pole type are capable of satisfactory operation on 50 Hz with adjustments as follows:

1. Reduce 60-Hz voltage to $5/6$ value for 50-Hz operation.
2. Reduce 60-Hz horsepower to $5/6$ value for 50-Hz operation.
3. Speed on 50 Hz will be $5/6$ of 60-Hz value. Slip will be $6/5$ of the 60-Hz slip.
4. Rated-load torque in lbf-ft will be approximately the same for either frequency.
5. Locked-rotor torque in lbf-ft will be approximately the same at either frequency.

TABLE 19. Locked-Rotor Torque of Single-Phase Fractional-Horsepower General-Purpose Induction Motors

hp	60-Hz			50-Hz		
	Speed, r/min					
	3600 3450	1800 1725	1200 1140	3000 2850	1500 1425	1000 950
	Min locked-rotor torque, ozf-ft					
$1/8$	...	24	32	...	29	39
$1/6$	15	33	43	18	39	51
$1/4$	21	46	59	25	55	70
$1/3$	26	57	73	31	69	88
$1/2$	37	85	100	44	102	120
$3/4$	50	119	.....	60	143	
1	61	.....	.....	73		

TABLE 20. Locked-Rotor Torque of Single-Phase Integral-Horsepower General-Purpose Induction Motors

	r/min		
	3600	1800	1200
hp	Min locked-rotor torque, lbf-ft		
¾	......	......	8.0
1	......	9.0	9.5
1½	4.5	12.5	13.0
2	5.5	16.0	16.0
3	7.5	22.0	23.0
5	11.0	33.0	
7½	16.0	45.0	

6. Breakdown torque in lbf-ft will be approximately the same for either frequency.

7. Locked-rotor amperes will be approximately 5% less than the 60-Hz value. The code letter on the nameplate to indicate the locked-rotor kVA/hp applies only to the 60-Hz rating.

8. Service factor will be 1.0.

9. Temperature rise will not exceed 90°C.

41. Temperature Rise. All rises are based on an ambient of 40°C and an altitude not exceeding 1000 m (3300 ft). Temperature measurements are to be taken in accordance with Ref. 6 or 7. See Tables 1 and 2 for applicable ratings.

1. Temperature rises by thermometer for motors having continuous ratings are shown in Table 21.

2. Temperature rises by thermometer for short-time rated motors are shown in Table 22. Short-time ratings are based upon a corresponding short-time load test which shall commence only when the windings and other parts of the machine are within 5°C of the ambient temperature at the time of the starting of the test.

42. Continuous or Short-Time Ratings. Temperature rises of ac fractional-horsepower motors having a continuous or short-time rating are shown in Table 23 (Art. 41). Where two methods of temperature measurement are shown, either method fulfills the meeting of the standards.

Temperature rises of single-phase and polyphase induction integral-horsepower motors having continuous or short-time ratings are shown in Table 24. Where two methods of temperature measurement are shown, either method fulfills the meeting of standards.

TABLE 21. Temperature Ratings, Rise of Windings by Thermometer for Continuous Rated Ac Fractional- and Integral-Horsepower Motors

Enclosure	Insulation, temp rise, °C		
	Class A	Class B	Class H *
Open general-purpose	40		
Dripproof	40 †		
Splashproof, guarded, semiguarded, and dripproof fully guarded	50	70	110
Totally enclosed	55	75	115
All others	50	70	110

Ambient 40°C; if higher, decrease rise by amount in excess.

* Applies only to polyphase integral-horsepower induction motors.

† May also be rated 50°C.

TABLE 22. Temperature Rise of Windings by Thermometer for Short-Time Rated Fractional- and Integral-Horsepower Motors

			Insulation, temp rise, °C		
Enclosure	hp rating up to and including	Time rating, min	Class A	Class B	Class H *
Totally enclosed	30	5	55	75	115
	50	15	55	75	115
	60	30	55	75	115
	All ratings	60	55	75	115
Open	All ratings	5, 15	50	70	110
	All ratings	30, 60	50	70	110

Ambient temperature 40°C; if higher, decrease rise by amount in excess.
* Applies only to polyphase integral-horsepower induction motors.

TABLE 23. Temperature Rise of Coil Windings above 40°C Ambient and under Rated Conditions of Ac Fractional-Horsepower Motors

	Insulation class	
Motor	A	B
Totally enclosed; see Art. 13:		
Thermometer............................	55	75
Resistance................................	65	85
Motors in frame smaller than		
No. 42 frame, resistance	65	85
Motors other than above:		
Thermometer............................	50	70
Resistance	60	80

Time rating may be continuous or short-time. Short-time ratings are 5, 15, 30, and 60 min.

TABLE 24. Temperature Rise, °C, above 40°C Ambient of Single-Phase and Polyphase Induction Integral-Horsepower Motors

	Insulation		
Windings	A	B	H *
Totally enclosed motors, including variations:			
Thermometer..	55	75	115
Resistance..	65	85	135
Totally enclosed fan-cooled, including variations:			
Thermometer..	55	75	115
Resistance..	60	80	125
Motors other than above:			
Thermometer..	50	70	110
Resistance..	60	80	125
Commutator and collector rings, by thermometer, insulation adjacent thereto:			
Class A, rise 65°C (55°C for general-purpose only)			
Class B, rise 85°C (75°C for general-purpose only)			
Class H, rise 125°C			

Time ratings continuous and short-time duty. Short-time ratings are 5, 15, 30, and 60 min.
* For polyphase induction motors only.

TABLE 25. Temperature Rise, °C, above 40°C Ambient of Single-Phase and Polyphase Induction Motors

	Insulation class		
	B	F *	H †
1. Windings, fractional-horsepower motors:			
a. Open motors, service factor 1.15 or higher, resistance	90	115	
b. Open motors except item 1*a* or 1*d* resistance......................	80	105	125
c. Totally enclosed nonventilated or fan-cooled			
including variations thereof, resistance..............................	85	110	135
d. Any motor in a smaller than No. 42 frame, resistance	85	110	135
2. Windings, integral-horsepower motors:			
a. All motors, service factor 1.15 or higher, resistance.............	90	115	
b. Totally enclosed fan-cooled motors including			
variations thereof, resistance..	80	105	125
c. Totally enclosed nonventilated motors including			
variations thereof, resistance..	85	110	135
d. Motors with encapsulated windings, service factor 1.0,			
all enclosures, resistance...	85	110	

Time ratings are continuous or short-time. Short-time ratings are 5, 15, 30, and 60 min. All rises are at full-load rating except for motors with service factor greater than 1.0 for which the rise is at the service-factor load.

  * Applies to integral-horsepower and polyphase fractional-horsepower motors.

  † Applies to polyphase induction motors only.

43. Temperature Rise vs. Service Factor. Temperature rises of single-phase and polyphase induction motors having continuous or short-time ratings are shown in Table 25. See comments on short-time ratings in Art. 41. Note that for motors having a service factor greater than 1.0, the rise is at the service-factor load.

44. Performance Data listed below apply to appropriate ac motors listed in Table 26.

45. Number of starts which a squirrel-cage integral-horsepower motor rated 250 to 500 hp, as shown in Table 1, shall be capable of, are two in succession, provided:

1. Motor coasts to rest between starts.

2. Motor is initially at ambient temperature, or one start with motor at a temperature not exceeding rated-load operating temperature.

3. Motor is designed for (1) Wk^2 of the load, (2) load torque during acceleration, (3) applied starting voltage, and (4) method of starting.

46. Overspeed values which squirrel-cage and wound-rotor motors (except crane

TABLE 26. Ac Motor Ratings

Synchronous speed, r/min	Squirrel-cage and wound-rotor hp	Synchronous hp	
		Power factor	
		Unity	0.8
3600	500	200	150
1800	500	200	150
1200	350	200	150
900	250	150	125
720	200	125	100
600	150	100	75
514	125	75	60

Ac motors covered in this table are motors up to and including ratings built in frames corresponding to the continuous open-type ratings.

TABLE 27. Overspeed for Squirrel-Cage and Wound-Rotor Induction Motors

hp	Synchronous speed, r/min	Overspeed, % of synchronous speed
200 and smaller	1801 and over	25
	1201–1800	25
	1200 and below	50
250–500, inclusive	1801 and over	20
	1800 and below	25

motors) will withstand without mechanical injury in an emergency of short duration are shown in Table 27.

47. Variations from rated speed, nameplate, or otherwise indicated value which ac single-phase and polyphase integral-horsepower motors will withstand are specified in standards not to exceed 20% of the difference between synchronous and rated speed when measured at rated voltage, frequency, and load in a 25°C ambient (Ref. 1).

48. Service factor of 250- to 500-hp motors, open-type ac, operated at rated voltage and frequency is 1.0.

49. Variations of voltage and frequency or combined voltage and frequency under which motors will successfully operate are indicated in Art. 23a, paragraphs 5, 6, and 7.

50. Wound-rotor-motor breakdown torques for continuous-rated polyphase integral-horsepower types are shown in Table 28.

APPLICATION

51. Direction of Rotation. All ac single-phase motors, all synchronous motors, and all universal motors are stated in standards (Ref. 1) to have counterclockwise rotation when facing the end of the machine opposite the drive. Where two or more motors are mechanically coupled together, the standard may not apply to all units.

The above does not apply to polyphase induction motors where either or both directions of rotation may be required. Also the phase sequence of the power source may not be known.

52. Altitude Operation. Temperature rises are generally based on altitude of 1000 m (3300 ft) or less. Applications may be made at higher altitude, in which case the following comments apply:

a. Motors having Class A or Class B insulation and temperature rises according to

TABLE 28. Wound-Rotor Motor Breakdown Torques for Continuous-rated Polyphase Integral-Horsepower Types

hp	Breakdown torque, % of full-load torque		
	Speed, r/min		
	1800	1200	900
1	.....	.....	250
1½	.....	.....	250
2	275	275	250
3	275	275	250
5	275	275	250
7½	275	250	225
10	275	250	225
15	250	225	225
20–200	225	225	225

Tables 23, 24, and 25 will operate satisfactorily at altitudes above 3300 ft in locations where the decrease in ambient temperature compensates for the increase in temperature rise as follows:

Ambient temp, °C	Max altitude, ft
40	3300
30	6600
20	9900

b. Motors with a service factor of 1.15 or higher will operate satisfactorily at unity service factor at an ambient temperature of 40°C at altitudes above 3300 ft up to 9000 ft.

c. Motors applied at altitudes exceeding 3300 ft in an ambient of 40°C should have temperature rises at sea level not exceeding values calculated as follows:

$$T_2 = T_1 \left(1 - \frac{\text{alt} - 3300}{33\,000}\right)$$

where T_2 = test temperature rise, °C at sea level

T_1 = temperature rise in appropriate Table 23, 24, or 25

alt = altitude, ft above sea level, at which motor operates

53. Voltage operation above or below nameplate value may shorten motor life, for example:

a. Undervoltage operation, such as a 230-V motor operated on a 208-V system, is not recommended. As indicated in Art. 49, the 208-V system could drop below 208 V and thus the 230-V motor could operate considerably below the standard 10% value. Such operation reduces torques appreciably and generally causes overheating when line current increases with voltage decrease.

b. Overvoltage operation, such as a 575-V rated motor operated at 10% overvoltage or 630 to 635 V, was indicated to be satisfactory (Art. 49). However, where the supply voltage exceeds, say, 635 V, motor operation may be unwise because of the injury to insulation. Thus such operation should be avoided in the interest of good engineering application.

54. Unbalanced-voltage operation of polyphase induction motors may seriously affect their performance and characteristics. Motor line-current increases may be much larger than the voltage-unbalance ratio.

a. Unbalance Defined. The voltage unbalance in percent may be defined as follows:

$$\% \text{ voltage unbalance} = 100 \times \frac{\text{max voltage deviation}}{\text{avg voltage}}$$

Thus, for voltages of 220, 215, and 210 the average is 215 V, the maximum deviation is $220 - 215 = 5$ V, and the percent voltage unbalance is $100(5/215) = 2.3\%$.

b. Unbalance Effects. With balanced voltages, the air-gap magnetic field rotates in a positive direction. The rotor rotates in the same direction. With voltage unbalance, the positive rotating field is slightly reduced but the voltage unbalance introduces a negatively rotating magnetic air-gap field which rotates in a direction opposite to rotor rotation and tends to produce high currents.

c. The motor full-load speed is reduced somewhat owing to the unbalanced voltage system. The negatively rotating magnetic air-gap field induces rotor currents and hence reverse or backward torque and thus reduces the motor speed somewhat.

d. Locked-rotor and breakdown torques are reduced with voltage unbalance owing to the reverse torque of the negatively rotating air-gap magnetic field.

e. Temperature rise may be considerably increased even with a relatively small voltage unbalance. In the phase with the highest current the percentage increase in temperature rise will be approximately two times the square of the volt unbalance. Thus a 3.5% voltage unbalance will produce an increase of approximately 25% in temperature rise. The increase in losses and the increase in average heating of the complete winding will be somewhat lower than the winding with the highest current.

f. Locked-rotor current will be unbalanced to about the same degree as the voltages. Locked kVA will increase only slightly.

The currents at normal operating speed with unbalanced voltages will be greatly unbalanced, about six to ten times the voltage unbalance. Selecting and setting overload devices may present a problem, as settings for one amount of voltage unbalance may not be suitable for a different voltage unbalance. Increasing overload-protection-device size is not a solution, as heating from overload and from single-phase operation is lost (Ref. 1).

55. Effects of voltage variations on performance within standards stated variations (Art. 23a, paragraphs 5, 6, and 7) can alter the motor performance appreciably from rated full-load conditions.

As may be expected, for example, greater departures from rated voltage can cause greater departures from rated voltage performance. The following are some results applying to integral-horsepower-type polyphase motors.

1. A 10% increase in voltage *above* base value may cause an increase in motor heating for a given horsepower load. Low-speed motors especially may have a considerable reduction in power factor.

2. A 10% decrease in voltage below base value may cause an increase in motor heating and in motor power factor.

3. Locked-rotor and breakdown torque are proportional to the square of the voltage.

4. Motor speed at rated load is altered when voltage departs from base value. Thus, if the voltage increases 10% the slip in either percent of synchronous speed or revolutions per minute, decreases by $1 - (1/1.10)^2 = 1 - 0.83 = 0.17$ or 17%. If the voltage decreases by 10%, the slip in either percent of synchronous speed or revolutions per minute increases by $(1.1)^2 - 1 = 0.21\%$. Thus a 5% slip motor at full load would have 5×1.21 or 6.05% slip at full-load with a voltage depressed 10% below base (rated) value.

5. A frequency change operates opposite to a voltage change. For example, if frequency is greater than base (rated) value, the effect is similar to a voltage reduction. Thus for a 5% increase in frequency, the locked-rotor torque is reduced to approximately $(0.95)^2$, or about 90% of base value. A decrease in frequency below base value operates to increase starting or breakdown torque.

An increase in frequency above base value increases motor base speed by the same ratio and also increases windage and friction loss. A decrease in frequency below base value produces opposite effects.

6. Voltage and frequency variations may occur simultaneously. Effects are then superimposed. A high voltage and low frequency produce additive torque effects. If both voltage and frequency increase by the same amount, the results compensate and torque remains unchanged.

56. Part-winding Starting of Polyphase Induction Motors. Polyphase induction motors may be designed so that only a portion of the motor winding is initially connected to the supply lines. The purpose is to reduce the initial starting current. One type of connection, for example, results in about 60% of normal (full-winding) locked-rotor current. Torque is also reduced, in this example, to about 50% of full-winding locked-rotor torque.

The nature of the part winding used, even in the best connection, results in the speed torque's being adversely affected by harmonic torques. Thus, care must be taken in the application on the part-winding connection that the load torque is such that the motor will start and accelerate. Acceleration may not be up to full speed. After a few seconds on the part-winding connection, however, the remainder of the winding is inserted and the motor, if properly applied, will accelerate to full speed.

During the operation on the part-winding connection the motor may be noisier than on its full winding.

57. Load Wk^2. Polyphase squirrel-cage induction motors listed in *a* below can accelerate without injurious temperature rise. Under certain conditions listed in *b* below, the load Wk^2 is given in Table 29.

a. Motor Definition

1. 200 to 500 hp as shown in Table 1.

2. Locked-rotor torque as shown in Table 7.

3. Breakdown torque as shown in Table 8.

4. Class A or B insulation system with temperature rise in accordance with Table 24 or 25.

TABLE 29. Load Wk^2 for Polyphase Squirrel-Cage Induction Motors (Ref. 1)

The following table lists load Wk^2 which polyphase squirrel-cage motors having performance characteristics in accordance with Part 20 of Ref. 1 can accelerate without injurious temperature rise under the following conditions:
1. Rated voltage and frequency applied.
2. During the accelerating period, the connected load torque shall be equal to, or less than, a torque which varies as the square of the speed and is equal to 100% of full-load torque at rated speed.
3. Two starts in succession (coasting to rest between starts) with the motor initially at ambient temperature or one start with the motor initially at a temperature not exceeding its rated load operating temperature.

						Speed, r/min						
	3600	1800	1200	900	720	600	514	450	400	360	327	300
Hp						Load Wk^2 (Exclusive of Motor Wk^2), lbf-ft^2						
100	......	....	....	....	....	....	....	12670	16830	21700	27310	33690
125	......	....	....	....	....	....	....	15610	20750	26760	33680	41550
150	......	....	....	....	....	....	13410	18520	24610	31750	39960	49300
200	......	....	....	....	....	12060	17530	24220	32200	41540	52300	64500
250	210	1017	2744	5540	9530	14830	21560	29800	39640	51200	64400	79500
300	246	1197	3239	6540	11270	17550	25530	35300	46960	60600	76400	94300
350	281	1373	3723	7530	12980	20230	29430	40710	54200	69900	88100	108800
400	315	1546	4199	8500	14670	22870	33280	46050	61300	79200	99800	123200
450	349	1714	4666	9460	16320	25470	37090	51300	68300	88300	111300	137400
500	381	1880	5130	10400	17970	28050	40850	56600	75300	97300	122600	151500
600	443	2202	6030	12250	21190	33110	48260	66800	89100	115100	145100	179300
700	503	2514	6900	14060	24340	38080	55500	76900	102600	132600	167200	206700
800	560	2815	7760	15830	27440	42950	62700	86900	115900	149800	189000	233700
900	615	3108	8590	17560	30480	47740	69700	96700	129000	166900	210600	260300
1000	668	3393	9410	19260	33470	52500	76600	106400	141900	183700	231800	286700
1250	790	4073	11380	23390	40740	64000	93600	130000	173600	224800	283900	351300
1500	902	4712	13260	27350	47750	75100	110000	153000	204500	265000	334800	414400
1750	1004	5310	15060	31170	54500	85900	126000	175400	234600	304200	384600	476200
2000	1096	5880	16780	34860	61100	96500	141600	197300	264100	342600	433300	537000
2250	1180	6420	18440	38430	67600	106800	156900	218700	293000	380300	481200	596000
2500	1256	6930	20030	41900	73800	116800	171800	239700	321300	417300	528000	655000
3000	1387	7860	23040	48520	85800	136200	200700	280500	376500	489400	620000	769000
3500	1491	8700	25850	54800	97300	154800	228600	319900	429800	559000	709000	881000
4000	1570	9460	28460	60700	108200	172600	255400	358000	481600	627000	796000	989000
4500	1627	10120	30890	66300	118700	189800	281400	395000	532000	693000	881000	1095000
5000	1662	10720	33160	71700	128700	206400	306500	430800	581000	758000	963000	1198000
5500	1677	11240	35280	76700	138300	222300	330800	465600	628000	821000	1044000	1299000
6000	1673	11690	37250	81500	147500	237800	354400	499500	675000	882000	1123000	1398000
7000	1612	12400	40770	90500	164900	267100	399500	565000	764000	1001000	1275000	1590000
8000	1484	12870	43790	98500	181000	294500	442100	626000	850000	1114000	1422000	1775000
9000	1294	13120	46330	105700	195800	320200	482300	685000	931000	1223000	1563000	1953000
10000	1046	13170	48430	112200	209400	344200	520000	741000	1009000	1327000	1699000	2125000
11000	......	13010	50100	117900	220000	366700	556200	794000	1084000	1428000	1830000	2291000
12000	......	12670	51400	123000	233500	387700	590200	844800	1155000	1524000	1956000	2452000
13000	......	12160	52300	127500	244000	407400	622400	893100	1224000	1617000	2078000	2608000
14000	......	11470	52900	131300	253600	425800	652800	934200	1289000	1707000	2195000	2758000
15000	......	10620	53100	134500	262400	442900	681500	983100	1352000	1793000	2309000	2904000

The values of Wk^2 of connected load given in the foregoing table were calculated from the following formula:

$$\text{Load } Wk^2 = A\left[\frac{Hp^{.95}}{\left(\frac{r/min}{1000}\right)^{2.4}}\right] - 0.0685\left[\frac{Hp^{1.5}}{\left(\frac{r/min}{1000}\right)^{1.8}}\right]$$

Where A = 24 for 300 to 1800 r/min, inclusive, motors
       A = 27 for 3600 r/min motors

5. Service factor of 1.0 shall apply for item ratings operating with temperature rises of item 4.

b. *Operating Conditions*

1. Rated voltage and frequency.
2. Connected load torque during the accelerating period should not exceed 100% of full-load torque at rated speed and should vary as the square of the speed.
3. Two successive starts (coasting to rest between starts) with motor initially at ambient temperature or one start with the motor initially at a temperature not exceeding its rated full-load operating temperature.

58. Encapsulated windings of integral-horsepower ac machines. This type of wind-

ing is referred to in Table 25, part 2d. The windings are suitable for exposure to the following environmental conditions (Ref. 1):
1. High humidity
2. Water spray and condensation
3. Detergents and mildly corrosive chemicals
4. Mildly abrasive nonmagnetic airborne dust in quantity insufficient to impede proper ventilation or mechanical operation.

For other environmental conditions motor manufacturers should be consulted.

LARGE-APPARATUS INDUCTION MOTORS

The following articles pertain to large-apparatus induction motors having horsepower and speed ratings built in frames larger than those listed in Table 26. The service factor at rated voltage and frequency and in accordance with the temperature rise shown in Table 30 is 1.0 for these motors.

Horsepower and speed ratings are in standard step values for large-apparatus induction motors and are shown in Tables 3 and 4. Note that it is not practical to build motors of all horsepowers at all speeds.

59. Induction-motor types commonly used in larger sizes are (1) squirrel-cage and (2) wound-rotor machines. Most widely used is the squirrel-cage-type motor. To a small degree and for special applications the solid-rotor type is also applied.

a. Squirrel-cage induction motors are simple and rugged. The rotor is a completely self-contained type of mechanical construction with no insulation, slip rings, externally brought out connections, etc. Large power systems can tolerate switching very large cage-type induction motors directly across the line. However, reactors, autotransformers, or other voltage-reducing devices can be used in many applications in series with the motor to reduce its starting kVA demand.

b. Wound-rotor induction motors were widely used when systems were unable to tolerate the kVA demand in starting large squirrel-cage motors across the line and where squirrel-cage-motor design and construction had features unsuitable to starting large Wk^2 loads. Improvements in system capability and squirrel-cage-motor design and construction have to a large degree relegated wound-rotor induction-motor usage to special cases. A further contribution toward their reduced present-day application is economic. The wound-rotor motor is more expensive to build than the cage-type motor.

TABLE 30. Observable Temperature Rise of Large-Apparatus Induction Motors over 40°C Air Entering the Ventilating Openings of the Motor

		Temp rise, °C			
		Insulation class system			
Machine part	Method of temp determination	A	B	F	H
1. Insulated windings					
a. All hp ratings........	Resistance	60	80	105	125
b. 1500 hp and less................	Embedded detector *	70	90	115	140
c. Over 1500 hp					
(1) 700 V and less...........	Embedded detector *	65	85	110	135
(2) Over 7000 V.....	Embedded detector *	60	80	105	125
2. Cores, squirrel-cage windings, and mechanical parts, such as collector rings and brushes, may attain such temperatures as will not injure the machine in any respect					

Altitude 1000 m (3300 ft) or less applies. Temperatures are to be measured in accordance with Ref. 6. See Art. 60 for comments on ventilating-air temperature differing from 40°C.

* For motors equipped with embedded detectors, this method will be used to demonstrate conformity with the standard. Detectors may be either resistance elements or thermocouples.

60. Observable temperature rises for several insulation systems are shown in Table 30.   Comments applicable to Table 30 provide for certain thermal variations, for example:

a. Totally Enclosed Water-Air-Cooled Motors.   Temperature of the cooling air is the temperature of the air leaving the coolers.

1. For 30°C cooling water, air leaving the coolers should not exceed 40°C.

2. For higher cooling-water temperature, the air leaving the cooler may exceed 40°C but the rises in Table 30 must be reduced by the number of degrees that the air leaving the coolers exceeds 40°C.

b. Altitude.   For motors operating under prevailing barometric pressure and designed not to exceed the specified temperature rise at altitudes from 1000 m (3300 ft) to 4000 m (13 000 ft) the temperature rises as checked by tests at low altitudes must be less than those listed in Table 30 by 1% of the specified temperature rise for 100 m (330 ft) of altitude in excess of 1000 m (3300 ft).

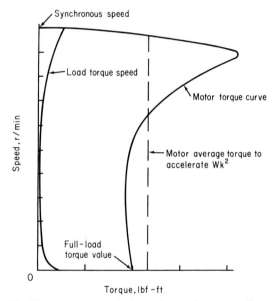

Fig. 50   Motor torque is much greater than load torque at all speeds.

c. Ambient Exceeds 40°C.   For ambients 40°C to and including 50°C, rises given in the table should be reduced by 10°C.   For ambients 50°C to and including 60°C, rises in the table should be reduced by 20°C.

61. Torques   (see Art. 24f for definitions).   All torque values are given for rated voltage and frequency.

1. Locked-rotor torque shall not be less than 60% of the rated full-load torque.

2. Breakdown torque shall not be less than 175% of the rated full-load torque.

3. Pull-up torque shall be at least 1.3 times the torque obtained from a curve which varies as the square of the speed and is equal to 100% of full-load torque at rated speed. In no case will the pull-up torque be less than 60% of full-load torque.

62. Load Acceleration.   For load Wk^2 which polyphase squirrel-cage motors can accelerate under specified conditions.   (See Table 29.)

63. Acceleration Time.   Calculation of time to accelerate Wk^2 to speed.   The time to accelerate an inertia from standstill to full speed may be calculated from the following formula:

$$t = \frac{Wk^2 \times \text{r/min}}{308 \times T} \qquad \text{s} \qquad\qquad (1)$$

where Wk^2 is in lbf-ft^2 (pounds-force-feet squared) and includes all rotating parts

r/min = revolutions per minute

T = torque in lbf-ft (pounds force at 1 ft) applied to the Wk^2

= motor torque minus load torque

1. If the load torque is small compared with the motor-developed torque, a rough estimate may be made from Fig. 50 of the average motor torque applied to the Wk^2 over the entire speed range from standstill to full speed.   Thus, with this single value of average torque entered in Eq. (1) and assumed acting over the entire speed range (r/min = synchronous speed, a fair approximation), the time to accelerate the Wk^2 is quickly computed.

2. When the load torque is appreciable, the torque available to accelerate the Wk^2 must be obtained by subtracting the load torque from the motor torque.

A quite accurate determination of acceleration time to full speed may be made using a step-by-step method.   For example, for step 1, in Fig. 51 the average torque T_1 is

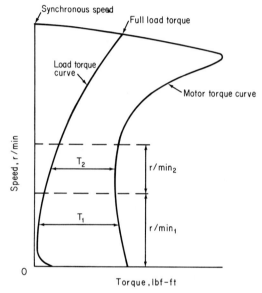

Fig. 51 Load torque is appreciable and must be subtracted from motor torque to obtain torque available to accelerate Wk^2.

simply determined by inspection as an average value for the speed interval r/min$_1$. Using Eq. (1), t_1, the time in seconds, to accelerate the Wk^2 over r/min$_1$, can be made as follows:

$$t_1 = \frac{Wk^2 \times \text{r/min}_1}{308 \times T_1}$$

A second interval of speed r/min$_2$ is taken as shown in the figure, T_2 is estimated, and t_2 is estimated, and t_2 calculated:

$$t_2 = \frac{Wk^2 \times \text{r/min}_2}{308 \times T_2}$$

The process is repeated several times until full speed is reached, and all the time values are added together to get the total acceleration time.

Note that where the torque-curve curvature changes rapidly, the r/min intervals must be small so that the average torque T can be closely determined.

64. Overspeed. Standards (Ref. 1) specify that squirrel-cage and wound-rotor induction motors shall be so constructed that, in an emergency, they will withstand without mechanical injury overspeeds above synchronous speed in accordance with the following:

Synchronous speed, r/min	Overspeed, % of synchronous speed
1801 and over	20
1800 and below	25

65. Operation with Variations from Rated Voltage and Frequency.

a. Running. Motors shall operate successfully under running conditions at rated load with a variation in the voltage or frequency up to the following:

1. Plus or minus 10% of rated voltage with rated frequency
2. Plus or minus 5% of rated frequency, with rated voltage
3. A combined variation in voltage and frequency of plus or minus 10% (sum of absolute values) of the rated values, provided the frequency variation does not exceed plus or minus 5% of rated frequency

Performance within these voltage and frequency variations will not necessarily be in accordance with standards established for operation at rated voltage and frequency.

b. Starting. Motors shall start and accelerate to running speed a load which has a torque characteristic and an inertia value not exceeding that listed in Table 29 with the voltage and frequency variations specified in *a* above. For loads with other characteristics, the starting voltage and frequency limits may be different.

66. Noise (Sound Level). Polyphase induction motors, either squirrel-cage or wound-rotor types, may have air- or magnetic-generated noise. In a well-designed motor ventilation (air) noise may be reduced in part at the source of generation and in part by using acoustic silencers. Complete reduction of noise at its source is very difficult in high-speed motors. In low-speed machines windage noise is generated in a minor degree unless the diameter and peripheral velocity of the rotor are large. Magnetic noise can be present in any polyphase induction motor unless due care in design is taken. This type of noise, if present, is very difficult to reduce because the frame of the motor vibrates and transmits the vibration to motor parts and supports, attached ductwork, piping, etc.

Motor methods of noise measurement are described in IEEE 85 (Ref. 5). Methods used are:

1. The hemisphere octave-band sound-power-level method
2. The five-position octave-band pressure-level frequency analysis with the microphone 5 ft from the machine surface

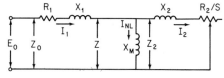

Fig. 52 Induction-motor equivalent circuit. Quantities are per phase for 2-phase motors and per phase wye for 3-phase motors. Nomenclature:

E_0 = voltage (volts), for 3-phase use line to neutral
I_1 = line or stator current, A
I_2 = rotor current referred to stator, A
I_{NL} = no-load current at rated voltage and frequency
R_1 = stator resistance, Ω, at temperature $t_s = 25°C$ + temperature rise of winding-coil end at rated load
R_2 = rotor resistance, Ω, referred to stator at temperature t_s
X_M = magnetizing reactance, Ω
X_1 = stator leakage reactance, Ω
X_2 = rotor leakage reactance, Ω, referred to stator
Z_0 = motor circuit impedance, Ω, at input terminals
Z_2 = rotor impedance, Ω, referred to stator
Z = paralleled rotor and magnetizing impedance, Ω, referred to stator
S = slip in per unit value (at standstill $S = 1$)

The Walsh-Healey Act of 1969 has set limits of noise tolerance or exposure. The requirements apply to performance of federal-contract work. Failure to meet the requirements may cause severe penalties to be applied such as contract cancellation or blacklisting for 3 years.

Walsh-Healey regulations permit sound-pressure-level exposures listed below. Should the daily exposure be made up of two or more periods of different sound levels, their combined effect is obtained by the exposure formula or index, which should not exceed unity. Thus,

$$\frac{C_1}{T_1} + \frac{C_2}{T_2} \cdots \frac{C_n}{T_n} = 1$$

where C is the exposure time at a given sound level and T is the allowable exposure time at that sound level.

Permissible Noise Exposure

Duration per day, h	Sound level, dBA [*]
8	90
6	92
4	95
3	97
2	100
1–1½	102
1	105
½	110
Less than ½	115

[*] Sound-pressure level measured on the "A-weighted" scale.

67. Equivalent Circuit of Induction Motor. Calculations to determine power factor, efficiency, starting and breakdown torques, starting current, and many other characteristics of the polyphase squirrel-cage or wound-rotor induction motor are conveniently made using the equivalent-circuit method of computation. The equivalent circuit is shown in Fig. 52. This circuit is used in the United States for evaluating much induction-motor performance data. Some countries, in their standards, use the circle diagram in determining many induction-machine characteristics.

The so-called "constants" in the equivalent circuit must be known to solve the circuit for motor characteristics. These constants are determined by the motor designer, of course, since he must solve the circuit from the design he makes to determine if specifications have been met. Hence the constants can be obtained only from the motor manufacturer. In many designs the rotor resistance R_2 and the rotor reactance X_2 will vary with the slip S at which the motor operates. Thus certain so-called "constants" really may be somewhat variable.

68. Starting methods for polyphase squirrel-cage induction motors. Complicated methods of starting induction motors are usually avoided wherever possible. Generally such methods increase overall costs and require additional space and maintenance care. The following methods of stating squirrel-cage induction motors may be considered: full-voltage, reactor, autotransformer, wye-delta, and part-winding.

a. Full-voltage starting is desirable wherever possible. It provides the minimum of cost. Absence of starting accessory equipment means no additional space requirements or additional maintenance care. The size of motor which may be thrown across the line depends on many factors. Careful consideration must be given to line drop as it affects other connected equipment and as it affects the motor-developed torque to provide adequate ability to accelerate the load to full speed under the worst starting conditions of the application. Squirrel-cage motors as large as 15 000 hp have been thrown across the line.

b. Reactor starting is one of the simplest methods of reduced-voltage starting of polyphase squirrel-cage induction motors. Standard reactor types are given in Table 31 showing the percent voltage which may be expected across the motor at standstill.

TABLE 31. Reactors and Autotransformers for Squirrel-Cage Motors

Duty	Motor hp rating	Approx % of line voltage at start
Heavy.............	All	As required
Medium..........	50 hp or less	65–80
Medium..........	Above 50 hp	50–65–80
Light..............	All	30–37.5–45

Reactors may be connected in series with each motor line or in the neutral phase ends of the motor winding as shown in Fig. 53.

Computations to determine line current and motor torque using a series reactor are not difficult. For example:

1. Refer to Fig. 52 and for an assumed slip S calculate the motor impedance Z_0.
2. $Z_0 = R_0 + jX_0$ = motor resistance + j (motor reactance).
3. $\qquad jX_r = +j$ (reactor ohms).
4. Add items 2 and 3; Z_t = total impedance = $R_0 + j(X_0 + X_r)$.
5. Line current = motor current = $E_0/Z_t = I_L$.
6. Motor current without reactor at full voltage = $E_0/Z_0 = I$.
7. Motor torque at full voltage without reactor = T.
8. Motor torque with reactor = $(I_L/I)^2 \times T$.

Assuming a reactor in series with the induction motor (Fig. 53) supplies 50% voltage to the motor at standstill when the series combination is connected to full line voltage, the line kVA will be 50% of the line kVA of the motor alone thrown across the line without benefit of a reactor. The same reasoning applies to other percentages of reactor values which may be used.

As the motor accelerates from zero speed to full speed, the voltage across the motor increases automatically as long as the reactor is in the circuit. The reason is that the reactor value in ohms stays constant during motor acceleration but the motor ohms Z_0 increase, as may be expected, since the value R_2/S increases as the slip becomes smaller (Fig. 52). Hence motor impedance Z_0 increases as speed increases and becomes a larger part of Z_t, the sum of motor and reactor impedance. Therefore, only at standstill or in the vicinity of standstill does a reactor tap of 50%, for example, supply 50% of line voltage to the motor. Within a few percent of full speed with the reactor connected in the motor circuit the voltage across the motor may be 80% or more. Generally the reactor is switched out of the circuit and full voltage is applied to the motor before full speed is attained.

c. Autotransformer starting is depicted in a single-phase equivalent-circuit diagram

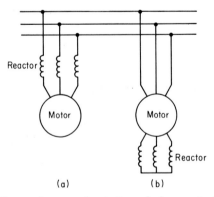

(a)   (b)

Fig. 53 (*a*) Line and (*b*) neutral reactors for starting polyphase squirrel-cage induction motors.

in Fig. 54. Standard voltage taps are the same as those percentages of line voltage at start in Table 31 for reactors. Special taps are available, of course, but with a price penalty.

Tap starting from an autotransformer reduces the starting kVA a motor takes from the supply lines. The kVA reduction from the full-voltage starting method varies as the square of the tap times motor full-voltage current. Unlike the reactor-type starting method the voltage across the motor does not increase as the motor speed increases.

d. Wye-delta-type starting is not widely applied to starting large 3-phase induction motors. This type of starting requires that the motor be designed for normal full-load operation on the delta connection and be started on the wye connection. Thus 73% more conductors are required over designs for a normal wye-operating connection at full load. More conductors per slot for the wye-delta-type design means more slot insulation, since conductors are insulated from one another. For a standard slot generally used for standard wye-connected motors, less slot space is therefore available for copper. In addition to this, coil-loop winding requires more turns to be wound and hence a more expensive coil.

Started on the wye connection, the motor develops current and torque which are one-third the values which would apply on the delta connection.

e. Part-winding starting for large squirrel-cage induction motors is the same as for integral-horsepower polyphase squirrel-cage induction motors described in Art. 56. The same comments and values apply.

f. Other methods of starting squirrel-cage induction motors to obtain large torque with low current are not commonly used for large motors. Low-frequency starting, for example, can produce large torque per ampere of current. However, special equipment is required for producing a low-frequency source of power. Developments in electronic switching show promise for future high power production of variable frequency generally now produced by rotating equipment.

Step starting is used in some instances. The method involves switching direct current between winding terminals in such a fashion that in the motor air gap the magnetic field moves from one position forward to another as a different set of winding terminals are energized. Thus the rotor moves forward. Switching may be accomplished by mechanical (contactors) or solid-state (electronic) means.

69. Single-phase operation of a three-phase motor alters normal motor performance considerably. One mode of single-phase operation is shown in Fig. 55. Under single-phase operation the motor current produces a pulsating magnetic field in the motor air gap instead of the rotating field produced by 3-phase currents. However, the pulsating-air-gap magnetic field may, by a mathematical concept, be replaced for most practical purposes by two equal magnetic fields each rotating at synchronous speed but in opposite directions. Each field acts like a normal 3-phase rotating field.

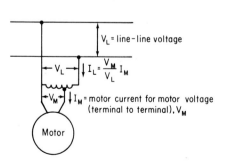

Fig. 54 Line diagram showing autotransformer starting arrangement.

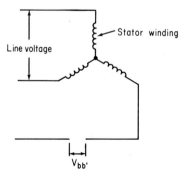

Fig. 55 Three-phase motor operates single-phase. Voltage V'_{bb} is shown between disconnected line and motor terminal.

Using the mathematical opposite-rotating-field concept, the voltage V_{bb}' may be shown to be equal to three times the negative-sequence voltage,

$$V_{bb}' \times 3\, I_2\, Z_2$$

where $I_2 =$ negative-sequence current
$Z_2 =$ impedance to negative-sequence current

a. At locked rotor each of the oppositely rotating fields generates identical current in the rotor to produce identical torque but in opposite directions.   Hence the net torque is zero and there is no tendency for the motor to start.   The effect is like two identical mechanically coupled motors supplied with the same voltage but of opposite sequence, causing opposite rotating tendencies.   In fact an analysis of much of the performance of a 3-phase motor on single-phase can be made using the two-identical-motor idea (Ref. 7).

At standstill single-phase motor current is

$$I = \frac{3 \times E_2}{Z_1 + Z_2} = \frac{3 \times E_0}{2Z_1} = 0.866\,\frac{E_0}{Z_1} = 0.866 \times 3\text{-phase locked-rotor current}$$

where $Z_1 = Z_2$, the impedances to positive- and negative-sequence currents, respectively, of the two-motor idea, $E_0 =$ line voltage $\sqrt{3}$.

b. An idle-running (unloaded) motor will continue to run on a single-phase supply if one power line of a 3-phase system is opened up.   The voltage V_{bb}' (Fig. 55) across the disconnection may be very low and the idle-running current on single-phase will be about $\sqrt{3}$ times the 3-phase balanced-voltage idle-running current.

During single-phase operation voltages measured at the terminals of the motor at no load will be fairly well balanced 3-phase voltages.   Thus the motor, unloaded, acts as a single-phase to 3-phase converter.   Small 3-phase induction motors (that is, smaller than the motor operating single-phase), connected to the terminals of a 3-phase motor operating on single-phase, will start and run up to speed and carry load.

c. Under loaded conditions a 3-phase motor on single-phase will have, as speed decreases, an increase in the voltage V_{bb}'.   The general shape of the voltage V_{bb}' is shown in Fig. 56 for a slow-speed motor.

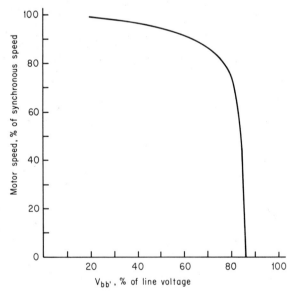

Fig. 56   Three-phase motor operates single-phase.   Voltage V_{bb}', between disconnected line and motor terminal, is shown.   Motor is a low-speed induction motor.

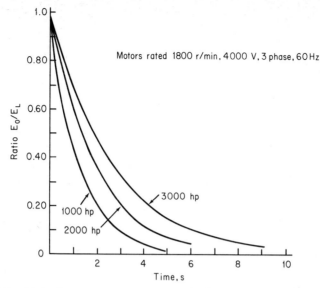

Fig. 57 Ratio of induction-motor open-circuit terminal voltage E_0 to line voltage E_L vs. seconds elapsed after disconnection from supply. T_o, time constant, occurs at 0.368 E_0/E_L.

70. Residual voltage exists at the terminal of an induction motor removed from its power supply. The fact may be deduced from a study of Fig. 52, the equivalent circuit of the induction motor. After the motor is disconnected from its power supply, the rotor circuit is still closed. From the theorem of constant flux linkages, the linkage associated with the inductances in the closed circuit is the same the instant after as it was before the disconnection. Thus the voltage across the magnetizing branch X_M is the same after as it was before line disconnection. The voltage may therefore be assumed equal to line voltage because the voltage drop across X_1, the stator leakage reactance, is generally small compared with the drop across X_M in large motors. As time passes, the motor open-circuit voltage will not remain constant. In the closed rotor circuit resistance losses cause the motor terminal voltage to decrease in accordance with the circuit time constant, which from an inspection of Fig. 52 is

$$T_o = \frac{X_M + X_2}{\omega R_2} \quad \text{s} \tag{2}$$

The stator terminal voltage decreases in accordance with T_o the motor open-circuit time constant, and is expressed by

$$E_0 = E_L e^{-t/T_o} \tag{3}$$

where E_0 = motor terminal voltage after removal from power supply
$\quad E_L$ = power-supply line voltage
$\quad t$ = time
$\quad e$ = base of logarithm = 2.718
The following is pertinent to induction-motor open-circuit voltage phenomena and induction-motor application:
1. For a given motor speed, large horsepower motors generally have longer time constants than smaller-horsepower motors (Fig. 57).
2. For a given horsepower, higher-speed motors generally have longer time constants than low-speed motors (Fig. 58).
3. For the decrement curves of Figs. 57 and 58 in about $0.7T_o$ (actually $0.693T_o$)

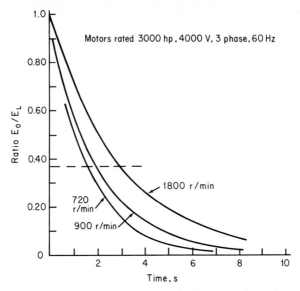

Fig. 58 Ratio of induction-motor open-circuit terminal voltage E_0 to line voltage E_L vs. seconds elapsed after disconnection from supply. T_o, time constant, occurs at 0.368 E_0/E_L.

the voltage decays to $\frac{1}{2}$ value; in another $0.7T_o$ the voltage decays to $\frac{1}{4}$ value; in another $0.7T_o$ down to $\frac{1}{8}$ value, etc.

4. High transient torques and currents will occur if the motor is switched from one power supply to another when the motor still has substantial residual terminal voltage 180° out of phase with the new power supply. The phenomenon is precisely the same as a generator being synchronized 180° out of phase. In such switching cases practice is often to permit a deferred time reconnection to the new power supply of about $2 \times 0.7T_o$ or $1.4T_o$ which permits the residual terminal voltage to drop to 25% of line value.

5. By means of phase-angle switching, thus avoiding out-of-phase reclosure, the motor may be rapidly switched and synchronized to the new power bus. In such cases the motor may have substantial residual voltage when switched.

6. Residual voltage at the motor terminals will decay faster than shown in Eq. (3) if the motor speed is not sustained at the value corresponding to the time of disconnection.

71. Number of Starts with Large-Apparatus Motors. Number of starts which a squirrel-cage induction motor (large apparatus) shall be capable of making are the same as described for integral-horsepower motors and under the same conditions specified (Art. 45) (Fig. 29).

72. Fault Contributions. Fault (short-circuit) contribution of 3-phase induction machines. Immediately following a terminal short circuit, an induction motor may, for a short time, contribute a considerable current. Large high-speed machines have longer time constants associated with these short-circuit currents than small low-speed machines.

At the instant of short circuit at the terminals of a 3-phase machine, the magnetic flux linking the closed stator winding becomes trapped or fixed to the winding. This condition implies that the stator magnetic-flux poles no longer rotate. Similarly the rotor flux is trapped in and fixed to the closed rotor winding with which it then rotates. The flux for either machine part, stator or rotor, is maximum at the instant of short circuit and starts decaying immediately in accordance with machine time constants. According to Faraday's law, voltage induces short-circuit currents in the respective machine parts, stator and rotor, as the flux decreases.

Stator induced current due to decaying stator trapped flux is a unidirectional or direct current (Fig. 59). The equation showing the current decay may be written

$$i_{dc} = \sqrt{2}\,\frac{E}{x'}\,e^{-kt} \qquad \text{A} \qquad\qquad (4)$$

where E = terminal to neutral rms supply voltage before short circuit
$\quad x'$ = transient reactance of the induction machine, Ω per phase
$\quad e$ = base of natural (Napierian) system of logarithms
$\quad k = 1/T_{dc}$
$\quad T_{dc}$ = dc time constant, s
$\qquad = x'/\omega\,r_1$, s
$\quad t$ = time, s, during continuous-current decay
$\quad \omega = 2\pi$ times supply frequency
$\quad r_1$ = stator winding resistance, Ω per phase
$\quad$ Note that x' may be expressed as

$$x' = x_1 + \frac{x_2 x_M}{x_2 + x_M} = x_1 + x_2 \qquad \text{approx}$$

where x_1, x_2, x_M are, respectively, the stator, rotor, and magnetizing reactances in ohms per phase as used in the induction-machine equivalent circuit (Art. 67, Fig. 52).

Stator current is also produced by voltage induced by the flux trapped in the revolving rotor. If the rotor speed is maintained, for example, by a large inertia-driven apparatus and the apparatus loading is light, the stator induced-current frequency is likely to be just slightly lower than the frequency of the power source which supplies the motor during normal operation. Thus the stator current induced by the rotor flux only will be alternating current (Fig. 60). The decrease of this current due to the decrease of rotor flux will introduce an exponential decline of the alternating component of current in the stator, or

$$i_{ac} = -\sqrt{2}\,\frac{E}{x'}\,e^{-mt}\cos \omega t \qquad \text{A} \qquad\qquad (5)$$

where $m = 1/T_{ac}$
$\quad T_{ac}$ = ac time constant

$$T_{ac} = \frac{x'}{x_1 + x_M}\,\frac{x_2 + x_M}{\omega r_2} \qquad \text{s} \qquad\qquad (6)$$

$$= \frac{x'}{x_1 + x_M}\,T_o \qquad\qquad (6a)$$

where T_o is the machine open-circuit time constant from Art. 70.

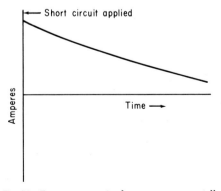

Fig. 59 Dc component i_{dc} decreases exponentially.

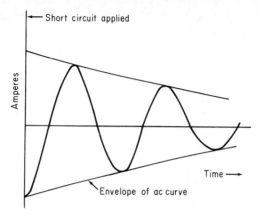

Fig. 60 Ac component i_{ac} of induction-machine short-circuit current.

When the ac and dc components are considered together, the decrement curve for stator short-circuit current has the form shown in Fig. 61 and is expressed by Eq. (7),

$$i_{sc} = i_{dc} + i_{ac} = \sqrt{2}\,\frac{E}{x'}\,(e^{-kt} - e^{-mt}\cos \omega t) \qquad \text{A} \tag{7}$$

Note that the information in this article pertains generally to cage-type induction machines, although the method is valid for wound-rotor motors where r_2 (Fig. 52) would include any external rotor resistance.

73. Wound-rotor induction motors in some applications have significant performance advantages over cage-type motors. Some advantages are a high torque per ampere, an ability to start high-inertia loads (much higher than cage motors do), an ability to operate at reduced speeds, etc. A disadvantage of the wound-rotor motor compared with cage-type induction motors is economic. The motor costs more than cage-type motors.

a. Starting torque and current performance characteristics of a wound-rotor motor constitute a chief advantage in their application. For example, assume rated line voltage is applied to the stator terminals and sufficient resistance is inserted in the rotor circuit, via the slip rings, so that rated full-load stator current flows. Assume it

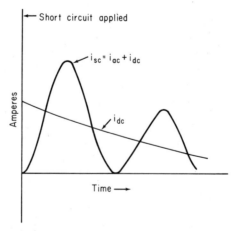

Fig. 61 Induction-machine short-circuit current. Asymmetrical (completely offset) wave.

is desired to calculate the value of this resistance. The motor nameplate provides the necessary data for the computation.

Nameplate data give the open-circuit voltage E_{22}, terminal-to-terminal value, across any two of the three slip rings (3-phase rotor winding is assumed) at standstill, when full line voltage at rated frequency is applied to the stator terminals. Nameplate data also supply the actual rotor slip-ring current I_{22} at full load; hence,

$$r_{22} = \frac{E_{22}}{3I_{22}} \qquad \Omega, \text{ terminal to neutral, in rotor terms} \qquad (8)$$

This value is the total actual resistance terminal to neutral in the rotor circuit at standstill so that the motor will develop full-load torque at standstill with full-load current taken from the power-supply line. Neglecting the rotor winding resistance, the external resistance, terminal-to-neutral value, in the rotor circuit is also expressed by Eq. (8).

Referring to the equivalent circuit (Fig. 52) with $r_2/S = $ a constant, the line current is unchanged. Thus at 50% speed ($S = 0.50$) the value of r_2 is 50% of the standstill value to obtain motor full-load current at half speed. It may be shown that the motor also develops full-load torque with full-load line current at half speed. The value of r_{22} calculated from Eq. (8) is, at half speed, one-half the value calculated at standstill.

To obtain full-load torque with full-load current at, say, 25% speed ($S = 0.75$), the rotor external resistance must be 0.75 times the value calculated at standstill. Other speed and r_{22} values for the speed chosen may be readily calculated for the full-load torque and current condition (Ref. 2).

If at any speed the motor draws full-load current and develops full-load torque (r_{22}/S is constant), the motor power factor has the same value as at full load, full speed with the slip rings short-circuited.

b. Frequency Converter. A wound-rotor motor may be used as a frequency converter if the rotor speed is properly controlled. If the stator is connected to a constant-frequency supply source, the rotor output frequency at standstill (rotor speed zero) is the same as the stator supply frequency. At half speed the rotor output frequency is one-half the stator supply frequency. If the rotor is driven at synchronous speed backward against the stator rotating magnetic field, the frequency from the rotor slip rings will be double the stator supply frequency (Fig. 62). A dc machine coupled to the wound-rotor motor shaft can be used to hold the rotor speed at the value desired for the required slip-ring-frequency output. The dc machine will act as a generator to recover power for slip-ring output frequency less than the stator supply frequency or

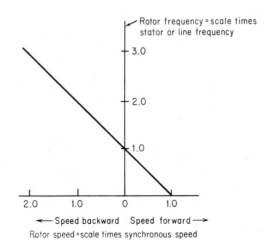

Fig. 62   Rotor frequency vs. rotor speed.

as a motor to drive the wound-rotor motor for frequency output at the slip rings greater than the stator supply frequency.

c. A wound-rotor motor with one slip ring open-circuited will operate with resistance across two slip rings only. The motor will start from rest if the load torque is not excessive but may and generally will lock at half synchronous speed approximately. This phenomenon is called the "Görges effect." If one slip ring is open-circuited when the motor is running at full load and full speed, the motor will continue to operate, but the unbalance in the rotor circuit will be reflected in the motor performance and line-current unbalance.

d. Phase Converter. If the rotor of a wound-rotor motor is wound for a different number of phases than the stator, a wound-rotor motor may be operated as a phase converter. Thus a wound-rotor motor with a 3-phase stator may have a 2-phase, 6-phase, etc., rotor winding, and loads may be operated from the slip rings.

e. Selsyns. Wound-rotor motors may be operated as synchronous ties, or selsyns. Two identical motors, for example, would have their rotors suitable connected together electrically so that if one of the rotors is shifted mechanically and thus electrically with respect to the other, circulating currents are produced between the two rotors to cause a synchronizing effect and both rotors tend to keep the same speed. Applications have been made on lift bridges. A motor is located at each end of the bridge with only an electrical tie or connection between the rotors. The motors lift the bridge or leaves of the bridge at the same speed. Mechanical gearing or other such connections between ends of the bridge is thus avoided.

f. Short-circuited Slip Rings. Starting a wound-rotor motor with the slip rings short-circuited like a cage motor is rarely successful. The motor will take currents from the line in excess of designed value and will develop little, if any, starting torque. If the connected load inertia is large, the motor, if it starts at all, may overheat excessively. There is great danger of seriously damaging a wound-rotor motor trying to start it as a cage-type motor. In rare cases where the load inertia is small and the load torque is light during the accelerating period, successful starting may be achieved.

74. Solid-rotor induction motors have a rotor of solid steel or steel rim of suitable magnetic qualities. The rotor steel carries both magnetic flux and rotor current. In carrying rotor current, the steel acts as a high-resistance conductor. There is also a "skin effect" in the rotor—that is, current tends to flow more near the outer rotor surface as the rotor induced frequency increases with higher rotor slips (lower speeds).

Motors of this type have high slip and low power factor and efficiencies for a given horsepower and speed rating compared with squirrel-cage motors. Since the rotor is a homogeneous body without bars, cage end rings, brazing, etc., an important application of this type of machine is for starting very high-inertia (Wk^2) loads and for short-time operating duty (unless the motor is made somewhat oversized). A good application for this type of motor is as a starting motor for a large synchronous condenser. The motor is disconnected from the power supply when synchronism has been reached by the condenser. The number of motor poles may be two less than the condenser so that condenser synchronous speed may be reached.

Because of the extreme ruggedness of the motor and its ability to withstand high temperature in the rotor, consideration should be given to its possible application for continuous operating service despite the motor low power factor and efficiency. For a given horsepower rating the motor may be slightly larger than a standard cage motor, but its starting torque will be greater and starting current less than those of the standard cage motor.

LARGE-APPARATUS SYNCHRONOUS MOTORS

Large-apparatus synchronous-motor horsepower and speed ratings are in standard step values (Ref. 1) as shown in Tables 5 and 6. Note that it is not practical to build motors of all horsepowers at all speeds.

75. Standard voltages and frequencies are given in Art. 22.

76. Standard power factor is 1.0 or 0.8 (overexcited) (Ref. 1).

77. Service Factor. The service factor of synchronous motors is 1.0 when operated at rated voltage and frequency and power factor and in accordance with the temperature rise given in Table 32.

TABLE 32. Observable Temperature Rise of Large-Apparatus Synchronous Motors over 40°C Maximum Air Entering the Ventilating Openings of the Motor

Machine part	Method of temp determination	Temp rise, °C Insulation class system			
		A	B	F	H
1. Insulated windings					
a. All horsepower ratings........................	Resistance	60	80	105	125
b. 1500 hp and less......................	Embedded detector *	70	90	115	140
c. Over 1500 hp					
(1) 7000 V and less.................	Embedded detector *	65	85	110	135
(2) Over 7000 V	Embedded detector *	60	80	105	125
2. Field winding					
a. Salient pole by resistance.......................................		60	80	105	125
b. Cylindrical rotor by resistance.................................. ...			85	105	125
3. Cores, amortisseur windings, and mechanical parts such as collector rings, brushholders, and brushes may attain such temperatures as will not injure the machine in any respect					

Altitude 1000 m (3300 ft) or less applies. Temperature to be measured in accordance with Ref. 8. See Art. 60 for comments on ventilating-air temperature differing from 40°C.

* For motors having slot embedded detectors, resistance elements, or thermocouple types, this method shall be used to demonstrate conformity with the standard.

78. Excitation voltages for field windings are (Ref. 1), 62½, 125, 250, 375, and 500 V dc. It is impractical to design all horsepower ratings for all excitation voltages shown. The excitation voltages do not apply to brushless-type excitation methods.

79. Observable temperature rise for rated-load conditions is shown in Table 32. Note that all comments in Art. 60 apply to this table.

80. Standard torques (Ref. 1), such as locked-rotor, pull-in, and pullout at rated voltage and frequency, are shown in Table 33.

81. Normal load Wk^2 is important because it has a considerable effect on the motor pulling into synchronism (Art. 24f, paragraph 6). Wk^2 values for many horsepower and speed ratings of synchronous motors are tabulated in Ref. 1. The tabulated values are calculated from the following formula:

$$\text{Normal-load } Wk^2 = \frac{0.375 \, (\text{hp})^{1.15}}{(\text{r/min}/1000)^2}$$

where r/min is motor speed in revolutions per minute and Wk^2 is on the motor-speed basis.

When applications of synchronous motors are made to loads having greater than normal Wk^2, the motor designer must have all information pertaining to the apparatus to be driven. This information includes not only the load Wk^2 but also any speed changes which might reflect a greater inertia at the motor shaft, load torque-speed curve for the starting period under the most severe conditions, method of starting, including final voltage at the motor terminals, possibility of voltage variations during the starting period, etc.

82. The number of starts which the motor shall be capable of making, provided that the Wk^2 of the load, the load torque during acceleration, the applied voltage, and the method of starting are those for which the motor was designed, are:

1. Two starts in succession, coasting to rest between starts, with the motor at ambient temperature

2. One start with the motor initially at a temperature not exceeding its rated-load operating temperature

Additional starts other than those listed should not be made without consulting the

TABLE 33. Torques of Minimum Value for Large-Apparatus Synchronous Motors

| Speed, r/min | hp | Power factor | Torque, % of rated full-load torque | | |
			Locked rotor	Pull-in* torque	Pullout † torque
500–1800	200 and below	1.0	100	100	150
	150 and below	0.8	100	100	175
	250–1000	1.0	60	60	150
	200–1000	0.8	60	60	175
	1250 and larger	1.0	40	60	150
		0.8	40	60	175
450 and below	All ratings	1.0	40	30	150
		0.8	40	30	200

* Based on normal Wk^2 of load (Art. 81). Rated excitation current applied.
† Rated excitation current applied.
Note. Motors are capable of delivering the pullout torque for at least 1 min.

motor manufacturer and without considering motor-control devices protecting the motor parts, especially the amortisseur starting winding. Since the life of the motor is affected by the number of starts, they should be kept to a minimum.

83. Overspeed. Salient-pole synchronous motors are so constructed that in an emergency they will withstand an overspeed of 25% without mechanical injury.

84. Operation at Other than Rated Power Factor

a. A motor rated, say, 1000 h, 80%, operates at greater armature current and excitation current than at a loading of 1000 hp, 100% PF. Full-load current I_{fl} is calculated as

$$I_{fl} = \frac{\text{hp output}}{\text{PF}} \frac{0.746}{\text{eff} \times \sqrt{3}\, E} \tag{9}$$

where PF and eff are power factor and efficiency, respectively, expressed in decimal form, and E is line-to-line voltage at the motor terminals.

Based on no change in armature current, I_{fl} in Eq. (9) indicates that the motor output at 100% instead of 80% PF could be raised approximately 25% or to $1.25 \times 1000 = 1250$ hp (efficiency assumed unchanged).

Locked-rotor torque and pull-in torque in terms of pounds-feet or turning torque is, of course, unaffected by a horsepower consideration. If the torque values are 100% each on a 1000-hp basis, torque values will be $(1000/1250) \times 100\%$ or 80% each on a 1250-hp basis.

Pullout torque of the 1000 hp, 80% PF motor is 200% or 2000 hp. On a 1250-hp basis this calculates to be $2000/1250 \times 100$ or 160% pullout torque. However, this value must be reduced to approximately 120% because field current must be reduced so that at the new 1250 hp, 100% PF loading the current I_{fl} does not exceed the value of the original 1000 hp, 80% PF rating.

The following tabulation applies.

80% PF Motor

Horsepower   = 1000
Locked-rotor torque = 100%
Pull-in torque   = 100%
Pullout torque   = 200%

This motor operated at 100% PF has the following approximate characteristics:

Horsepower   = $1.25 \times 1000 = 1250$
Locked-rotor torque = $0.8 \times 100\% = 80\%$
Pull-in torque   = $0.8 \times 100\% = 80\%$
Pullout torque   = $0.6 \times 200\% = 120\%$

b. A 1000-hp, 100% PF synchronous motor when operated at 80% PF will carry less than 1000 hp on the basis of no I_{fl} increase; see Eq. (9). Assuming no change in efficiency, the equation indicates 800 hp permissible loading at 80% PF. This is not the case, however. A reduction below 800 hp at 80% PF is necessary because field current is not adequate for 80% PF at 800 hp. Field current cannot be increased or field heating will become excessive. The horsepower must therefore be dropped, in this case, to 350 hp at 80% PF with the original field-current value (at 1000 hp, 100% PF).

The following table applies for the 80% PF loading of a 1000-hp, 100% PF rated motor:

Horsepower	$= 0.35 \times 1000$	$= 350$ hp
Locked-rotor torque	$= 2.85 \times 100\%$	$= 285\%$
Pull-in torque	$= 2.85 \times 100\%$	$= 285\%$
Pullout torque	$= 2.85 \times 100\%$	$= 285\%$

85. Variation from Rated Voltage and Frequency

a. With a voltage variation of $\pm 10\%$ from rated value, a synchronous motor, at rated load and frequency and with rated exciting current maintained, will operate successfully in synchronism. Operation may not be in accordance with standards established for operation at rated voltage.

b. With a frequency variation of $\pm 5\%$ from rated value, a synchronous motor, at rated load and voltage and with rated exciting current maintained, will operate successfully in synchronism. Operation may not be in accordance with standards for operation at rated frequency.

c. With a combined voltage and frequency variation of $\pm 10\%$ (sum of absolute values) from rated value, provided the frequency variation does not exceed 5%, and with rated exciting current maintained, a synchronous motor will operate successfully in synchronism. Operation may not be in accordance with standards established for operation at rated voltage and frequency.

86. Rapid braking of synchronous motors is necessary in emergencies and in fact may be required by law for certain applications such as rapid stopping of rubber mills for safety to personnel. In other applications, rapid braking may be desirable for economic production purposes.

Dynamic braking of synchronous machines is generally accomplished by quickly disconnecting the stator winding from the power system and then immediately connecting the winding to a resistor bank as shown in Fig. 63. Field excitation current of the synchronous-machine field is generally unaltered. In some instances, however, the field may be forced or boosted by shorting out a resistor in series with the field winding, thus applying higher excitation voltage and current to the field winding. Fewer stopping revolutions of the rotating masses are thus achieved. As connected in Fig. 63, the synchronous machine acts as a generator to supply resistor losses. The torque required to drive this generator comes from the rotating inertia masses such as the synchronous-machine rotor and connected apparatus. The resistor energy loss is obtained from a reduction in the kinetic energy given up by the rotating masses slowing down. The slowing down is the braking process.

The optimum dynamic-braking single-resistor value in ohms, terminal to neutral, wye-connected, for a 3-phase machine is

$$R = \frac{x'_d}{\sqrt{3}} \quad \Omega \tag{10}$$

where x'_d is the machine direct-axis transient reactance. Resistor values larger or smaller than the optimum single-valued R of Eq. (10) will allow more revolutions to stop than the formula shows. The shape of the curve shown in Fig. 64 indicates the effect of varying the resistance value on the number of revolutions the rotating masses take to stop.

Resistance thermal dissipation is mostly due to storage, as convection cooling has very little time to be effective during stopping of the rotating elements. The thermal-storage capacity of the resistors may be based on the calculation of the kinetic energy stored in the rotating masses at full speed. This calculation provides a good approximation because the kinetic energy removed from stopping the rotating system is

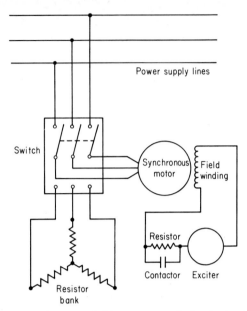

Fig. 63 Elementary circuit showing switching arrangement of motor to other power supply or dynamic-braking resistors.

assumed converted into resistor heat losses (windage, friction, etc., losses in the generator are neglected).

Kinetic energy of the rotating system is given by the expression

$$\text{Kinetic energy} = 0.231 \frac{\text{r/min}}{(1000)^2} \times Wk^2 \qquad \text{kW-s} \tag{11}$$

where r/min is full speed of the rotating system in revolutions per minute; r/min is usually the machine speed and all Wk^2 connected apparatus are converted to the synchronous-machine speed base. Wk^2 is in pounds-feet squared.

Calculations based on synchronous generator operation (Ref. 9) show that the minimum revolutions to stop are approximately

$$\text{Min revolutions to stop} = \frac{N_o H x'_{du}}{26(e'_d)^2} \tag{12}$$

where N_o = synchronous speed of motor, r/min
H = kinetic energy [Eq. (11)]/synchronous-machine kVA rating
x'_{du} = unsaturated value of synchronous-machine direct-axis transient reactance per unit
e'_d = per unit voltage behind the transient reactance
= $1.0 + jix'_{du}$, where i is per unit armature current prior to dynamic braking

Resistor current during the braking period may be calculated from the following equation (Ref. 4):

$$i = e'_d \frac{\sqrt{(r/n)^2 + x_q^2}}{(r/n)^2 + x'_d x_q} \tag{13}$$

where n = per unit speed (except synchronous speed = 1.0, 50% speed = 0.50 per unit, etc.)
x_q = synchronous-machine quadrature-axis synchronous reactance, per unit value
r = per unit resistance in stator circuit

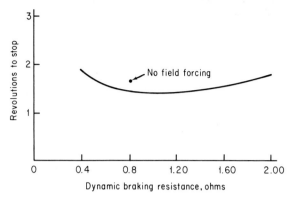

Fig. 64   Revolutions to stop vs. ohms dynamic-braking resistance with field forcing, 800 hp, 120 r/min, 3-phase, 60 Hz. 80% PF, 2300-V synchronous motor.

87. Noise specifications for synchronous motors come under the same control as induction machines, as stated in Art. 66.

88. Nonsalient-Pole Synchronous and Induction Motors. Synchronous-induction motors of the cylindrical or round-rotor (nonsalient-pole) type are particularly applicable for low-speed drives where high starting torque with the low line-current capability of the wound-rotor motor is desired and where, in the same machine, the full-load running good power factor of the synchronous-type motor is obtainable.

The stator of the synchronous-induction motor differs little from the standard induction or synchronous motor. The rotor of the synchronous-induction motor resembles the wound-rotor induction motor closely in both construction and winding type. The polyphase-type rotor-winding leads may be brought out (generally) to three or five slip rings, as shown in Fig. 65. Also shown are rotor resistor connections for starting purposes and exciter insertion into the rotor circuit for synchronous-motor operation.

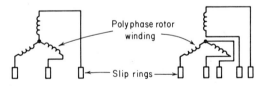

(a) Two types of synchronous-induction-motor rotor windings

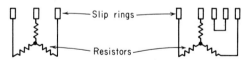

(b) Rotor external resistors for wound-rotor induction-motor-type starting

(c) Exciter supplies direct current to rotor winding for synchronous-motor operation

Fig. 65   Synchronous-induction motor connection arrangements of rotor for induction and for synchronous-type operation.

During starting, the motor operates as a wound-rotor induction motor with resistance (Fig. 65*b*) inserted into the rotor circuit. As the motor approaches synchronous speed and the slip becomes small, a source of dc excitation is connected into the rotor winding (Fig. 65*c*). Then as in the salient-pole-type synchronous motor pull-in to synchronism occurs provided the load torque and inertia and power-line voltage are correct values for which the motor is designed.

During synchronous-speed operation the motor operates as a synchronous motor with unity or other prescribed and designed power factor.

The synchronous-induction motor may require a special low-voltage, high-current exciter. Certain other characteristics of the motor, such as efficiency and pullout torque, may not be quite as good as in the salient-pole-type motor. However, the wound-rotor induction-motor-type starting characteristics are distinct advantages as well as the synchronous-motor full-load operating PF characteristics.

89. Brushless Synchronous Motors. Modern solid-state circuitry has enabled designers to build synchronous motors (and generators) which require no collector slip rings or brushes.

At least two types of excitation methods for brushless motors are in general use.

1. Voltage for excitation is generated by an exciter by motor rotation. The exciter has a stator with field winding supplied by direct current to produce a nonrotating flux. The direct current may be from a battery, rectifier, etc., supply. A wound exciter rotor, mounted on the synchronous-motor shaft, turns in the exciter magnetic field, thereby having voltage generated within the rotor winding. An air gap separates the stator from the rotor. The exciter rotor voltage is alternating current, which is rectified to direct current for the main motor field winding. Rectifier elements are mounted on the motor shaft and rotate with the motor rotor. Also mounted on and rotating with the motor shaft is suitable circuitry for field application and inserting the motor field resistors for starting purposes.

During motor full-speed normal operation an increase in exciter stator direct current produces an increase in generated voltage in the exciter rotor and thereby increases the motor field current. In this manner motor field current may be widely varied.

2. Another method of brushless excitation uses a rotating transformer in which the primary is stationary and the secondary, mounted on the rotor shaft, rotates. Primary and secondary are separated by an air gap. The primary may be supplied with alternating current and the rotor secondary by voltage transformer action also produces alternating current which is changed to direct current by rectifier elements mounted on and rotating with the synchronous-motor shaft. The direct current provides the motor normal excitation. A variation of the stationary primary ac voltage of this type of exciter will alter the secondary winding voltage, which when rectified to direct current in turn increases the motor field-winding excitation. Circuitry and switching elements are mounted on and rotate with the motor rotor to insert starting resistance in the rotor field winding during the starting period and to switch from the starting to pull-in sequence.

An advantage of this method is that excitation voltage is available at all speeds from standstill to full speed. Thus, for example, for rotor positioning at near standstill the exciter output is fully available when switching is done on the motor stator.

An obvious difference in the two methods is that in method (1) power for the motor field is produced by motor rotation which drives the exciter rotor. In method (2) the power for the synchronous-motor field excitation is obtained by the exciter transformer action derived from a source, single-phase, for example, outside the motor. Controls, etc., are somewhat different for the two methods.

90. Polyphase Synchronous-Reluctance Motors. A synchronous-reluctance-type motor resembles a cage-type induction motor but runs at synchronous speed at its full-load rating.

A salient-pole motor starting on its squirrel-cage-type winding accelerates its connected load to just under synchronous speed and slips poles before excitation is applied. The synchronously rotating air-gap mmf wave produced by the stator winding finds a low-reluctance path as it slips over a pole piece. Hence, for the condition of the center or axis of the mmf wave directly over the center axis of the pole piece, the reluctance is least for the flux path. Maximum flux is produced with this alignment. It is called the "direct axis."

If the center axis of the mmf wave is midway between poles (called the "quadrature axis"), the flux is greatly reduced from that of the direct axis. The reason is that the quadrature-axis path is mostly through air, while the direct-axis path is almost entirely through magnetic material. The result is that when the rotor is close to synchronous speed, there is a strong tendency for the rotor and load to be pulled into the maximum-flux condition. The rotor will "lock in" with the synchronously rotating air-gap field and then run at synchronous speed and resist a tendency to pull out of step. Many synchronous machines, lightly loaded, will pull into synchronism in this manner without excitation being applied.

The synchronous-reluctance motor operates on the principle described. However, the direct- and quadrature-axis paths are specially designed and differ in mechanical configuration from the salient-pole structure. The design also incorporates a "squirrel-cage" type of starting winding. In fact, the rotor may have the round appearance of the common cage-type motor, yet it runs at synchronous speed by "reluctance torque."

Synchronous-reluctance motors of 100 hp or more are available. They have lower power factor and efficiency and larger starting currents than their dc-excited salient-pole cousins. Their advantages are synchronous-speed operation with no rotor excitation required and simple cage-type motor control (Ref. 3).

REFERENCES

1. "Motors and Generators," NEMA Standards Publication MG1-1967, National Electrical Manufacturers Association, New York.
2. "Autotransformers and Reactors," NEMA Industrial Control Publication IC-1959, Part 14, Table 14-1.
3. Robert W. Smeaton, "Motor Application and Maintenance Handbook," McGraw-Hill Book Company, New York, 1969.
4. C. E. Kilboutne and I. A. Terry, Dynamic Braking of Synchronous Machines, *Elec. Eng.*, December 1932.
5. "IEEE Test Procedures for Air Borne Sound Measurements on Rotating Electrical Machinery," IEEE 85, The Institute of Electrical and Electronics Engineers, New York, 1973.
6. "Test Procedure for Polyphase Induction Motors and Generators," IEEE 112A, The Institute of Electrical and Electronics Engineers, New York.
7. "Test Procedure for Single Phase Induction Motors," IEEE 114, The Institute of Electrical and Electronics Engineers, New York.
8. "Test Procedure for Synchronous Machines," IEEE 115, The Institute of Electrical and Electronics Engineers, New York.

12

System Protection and Coordination

ROBERT L. STOUPPE, JR.[*]

*Senior Engineer, System Relaying, Detroit Edison Company; Registered Professional Engineer (Mich.); Member, IEEE.

FOREWORD

This section provides information on the equipment and devices that can be used to provide protection against short circuits (faults) on electrical systems. It covers the coordination of these devices for sustaining system integrity and provides some guides to effect a coordinated protective system.

The complex and automated nature of industrial and commercial businesses makes a reliable supply of electrical energy mandatory. A properly designed and coordinated power system can go a long way to help achieve this required reliability.

GENERAL

1. Reliable Electric Service. There are three basic methods of obtaining electrical energy. The first and probably the most common is to purchase electric power from the local electric utility. The utility can generally supply electric power at a cheaper overall rate than privately produced power can be supplied. This type of system generally has the least complicated protective scheme, with the accompanying simplicity of coordination between the various equipment utilized in the system. A second source of power is on-site private generation. This type of energy source leads to more complex and more difficult coordination problems because of the relatively low magnitude of short-circuit currents that are available from small isolated generators. The available short-circuit current varies over wide extremes depending upon how many generators are running. Since the electric utility generally has a more constant source impedance at a given customer, the short-circuit current magnitudes do not vary, making the coordination of the protective equipment much more reliable. The third common method is a combination of utility power and private generation. This form of electric service is the most complicated and requires the greatest amount of protective equipment with a corresponding complexity in the system coordination.

There are synchronizing problems, reverse-power problems, varying fault-current magnitudes, and many other areas where difficulties may arise with the third method.

2. Rapid Removal of Faulted Equipment. Since everything man makes is imperfect in one way or another, it ultimately fails. In large higher-capacity systems the amount of available short-circuit current is so great that major damage can occur to expensive equipment in split seconds.

The job of protective equipment is to operate and remove a faulted segment of the system or a piece of major equipment from the rest of the system as quickly and safely as possible. The speed with which these devices operate has a direct bearing on the continuity of service to the remainder of the system. Rapid fault clearing helps to keep large motors and generators stable and thus to ride through a momentary disturbance. Rapid clearing of short circuits can markedly reduce the amount of damage done to a piece of equipment, such as a transformer, and will in turn reduce the time necessary to repair the equipment so that downtime or production loss can be kept to a minimum.

The protective equipment must protect the system against uncontrolled short circuits, but it must not hinder the load-carrying capability of the system.

In higher-voltage systems (over 600 V) the protective equipment is to sense and remove short circuits, but it is not necessarily to be used for overload protection of equipment such as cables and transformers. If a protective device is limited by being required to sense overloads as well as faults, the pickup setting will be so low that adequate coordination cannot be achieved and system reliability will be greatly reduced.

The thermal capacity of equipment in higher-voltage systems must be engineered into the system. It is not the function of protective equipment to monitor system loads to prevent overloading of cables, transformers, and bus work.

The use of automatic transfer equipment can reduce the downtime for some manufacturing processes because the electric service can be reestablished via an alternate source after a momentary service outage.

Some systems and loads that cannot tolerate the slightest dip in voltage or interruption of service may require some form of "uninterruptible power" supply to ensure that the power is not interrupted even for an instant under most normal circumstances (see Secs. 4 and 7). General-purpose utility power is usually not suitable for highly sophisticated machines such as computers without special devices. The cost for such service would be prohibitive and may not be feasible because of other system characteristics. Even a small isolated private power system would have problems that could arise from insulation failures in the connected equipment.

A reliable electric system must be planned, and the equipment used must be compatible. Equipment characteristics must be chosen after the system requirements are established to ensure that the proper devices are put into the system and coordination can be achieved.

3. Coordination is the selective operation of the various protective devices so that the device closest to the fault operates first and the smallest possible portion of the system is interrupted for a given short circuit. The selective operation of overcurrent devices is achieved by setting or choosing the protective devices with current pickups and operating times so that the device closest to the fault operates first. The successive devices back toward the power source have a progressively higher ampere pickup and a progressively longer time of operation. Coordination is most easily achieved when the time-overcurrent characteristics of the various devices are plotted on a single sheet of log-log graph paper. This graphical representation of the characteristics of overcurrent relays, fuses, and direct-acting trip devices helps in the choice of the proper device to achieve the desired selectivity.

Curves for the various devices are obtainable from the manufacturer. Most of these characteristics are available on "industry-standard" paper so that curves may be traced or compared with the aid of a light box or a tracing table. The plots should be made at one voltage level. If several voltage levels are involved and coordination is between devices on the primary and secondary of a transformer, one voltage level is chosen and the other currents are referred to the chosen voltage level using the appropriate plotting ratio. The plotting ratio consists of three items: (1) The transformer voltage ratio of primary to secondary line-to-line voltages. (2) The transformer winding connection primary to secondary (wye-wye, wye-delta, delta-delta, or delta-wye). The manner in which the fault currents are transformed through a transformer is very important to proper coordination. (3) The regulation available on the transformer and any external regulator.

For proper selectivity the various protective devices are coordinated even under the most severe operating conditions (Art. 22).

Many systems employ automatic reclosing of the protective devices to help restore service after a transient fault has occurred. If this reclosing is done a few seconds after the initial fault, the overcurrent relay may not have completely reset, and if the fault is permanent rather than transient, the overcurrent relay may be "inched" out and the contacts closed to trip a device upstream of the device nearest the fault. This same type of operation may occur to fuses and other thermal devices. The heat buildup within the fuse must be dissipated before the fault current is reapplied to prevent a false operation of an upstream device.

Preloading of fuses, relays, and other overcurrent devices must be taken into consideration when protective devices are coordinated in series. Load currents and fault currents both cause protective devices to operate, and the usual overcurrent device cannot distinguish between the two. Heavy load currents should be considered as aiding the upstream device to operate on faults beyond a downstream device. This preloading is sometimes represented on the plot of the characteristics by using what is called a load triangle. This load triangle in effect moves the downstream-device curve to the right, and the safety margin is measured from this new location to ensure selective operation (Fig. 1).

Once a selective scheme is achieved on paper, the various settings must be applied to the actual devices. It is sometimes the practice to adjust the tap and time dial of the relays and hope the settings will allow the scheme to operate as planned. A better way is to have the relays and trip devices tested and the devices calibrated to the precise curves that were drawn on the coordination study. The relays can be proved if three points are prescribed and these points are checked by passing current through the relay coils and measuring the time of operation with a cycle counter.

Fuse characteristics must be taken on faith because no adjustments are possible.

Direct-acting trips can be field-checked if an adequate high-current test set is available. The new static-type direct-trip devices can be checked with secondary current because they use current transformers.

Table 1 gives suggested minimum safety margins to act as guides in achieving a coordinated system.

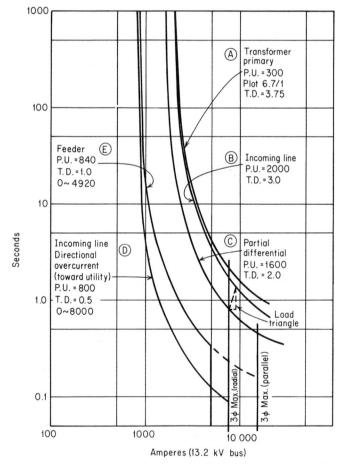

Fig. 1 Plot of phase-protective devices. P.U. is pick up and T.D. is time dial.

TABLE 1. Margins for System Coordination

Devices	Safety margin, s*
Relays over relays	0.4–0.5
Relays over trip devices	0.1–0.2
Relays over fuses	0.1–0.2
Fuses over relays	0.3–0.4
Fuses over fuses	0.1
Trip devices over fuses	Clear space on plot

*The coordination margins above assume that the devices are operating in a time-delay area and are not in an area where the device is practically instantaneous for the current magnitude involved.

a. Relays over Relays. A time margin of between 0.4 and 0.5 s at the maximum available fault current should be maintained between overcurrent relays in series. This time includes the breaker operating time (about 0.13 s), relay overtravel or coasting time, variations in the manufacture of the devices used including current-transformer performance, and a small safety factor. An examination of the time-current characteristic operating curves for the various devices will show that overcurrent relays have a "line" characteristic. The direct-trip device and the fuse have a "band" within which the device will operate. This band is necessary to accommodate the tolerances required for the manufacture of the devices. When a relay is coordinated *over* a band, the upper limit of the band is used. When a device is coordinated *under* the band, the lower limit is used. This practice will ensure selectivity even if the upper device is operated near the bottom of its band and the lower device is operated near the top of its band. If calibration equipment is not available, relay characteristics should be shown as a "band" to allow for manufacturing tolerances.

Several curve shapes are available for overcurrent relays. Each of these shapes has its area of being the "best" for the application. Relays within an electric system should be compatible, e.g., have the same time-current characteristic, so that adequate selectivity can be achieved when relays are in series. Some combinations of overcurrent-relay curve shapes cannot be made to coordinate. The most common and most versatile characteristic is called the "very-inverse" relay. It will provide time delay over most fuses and other very-inverse relays. This characteristic is found in the General Electric Type IAC53 relay and the Westinghouse Type CO-8 relay or any equivalent device.

Overcurrent relays should also include an instantaneous attachment if the relays are to be used in applications other than incoming lines. This instantaneous element, in conjunction with the time-delay element, can help achieve overall selectivity with several steps of protection in series and still not have excessively high operating times on the devices nearest the power source.

b. Relays over Trip Devices or Fuses. A time margin of between 0.1 and 0.2 s at the maximum available fault current should be maintained between the relay curve and the trip device or fuse under the relay. The entire operating time including manufacturing tolerances is included in the operating characteristic. The safety margin is 0.1 to 0.2 s.

It is assumed that the various devices being coordinated are pickup-selective, with the device closer to the power source having the higher pickup value.

When the devices are on different sides of a power transformer, the total plotting ratio should be utilized (the winding ratio, winding connections, and regulation).

c. Fuses over Relays. A time margin of between 0.3 and 0.4 s should be maintained between the fuse and relay curves. This time includes the breaker operating time, relay overtravel time, and a safety factor.

d. Fuses over Fuses. The time margin between two fuses in series should be about 0.1 s at the maximum current at which the two must coordinate. The available fault current will then be low enough so that the larger fuse, at least, is operating in the time-delay portion of the curve and not in the instantaneous or current-limiting area of the fuse characteristic. The downstream device may operate in the instantaneous or current-limiting area if desirable. Sometimes the use of a current-limiting fuse makes an otherwise impossible coordination problem workable and acceptable. Preloading should be considered in all these applications to ensure that large load currents or starting-inrush currents do not cause unnecessary operation of the protective devices.

e. Trip Devices over Fuses. A clear space between the device characteristics on the plot is usually sufficient to ensure selectivity. The tripping characteristic of both these types of devices includes the arcing time, manufacturing tolerances, and safety factors.

f. Instantaneous Device over Instantaneous Device. The term "instantaneous" means that the action of the device is not intentionally delayed in any way. These devices are actuated only by the magnitude of current that passes through the operating coil. The time of operation for these devices is essentially the same for all and is determined by the physical properties of the individual devices. In order that any selectivity between instantaneous devices in series may be achieved, there must be a sufficient difference in the current magnitude at the two locations. This current-magnitude difference is achieved by having impedance in the circuit. The upstream device must be set at least 150% of the current available at the downstream protection location. The downstream device should be set no higher than about 60% of the current for which it is expected to operate. The apparently high margin of safety is necessary because instantaneous devices tend to "overreach" and operate on currents that are apparently lower than the relay setting. Since no time delay is involved, the instantaneous device tends to operate on the instantaneous asymmetrical current values. The time-delay relays are restrained until the asymmetrical current has dissipated, but the instantaneous relay will operate on this offset current and hence cause the "overreach" problem.

g. Time-Delay Relay over Time-Delay Relay with an Instantaneous Attachment. The time margin at the maximum available fault current can be shortened and the total system operating time reduced by the judicious use of instantaneous elements which cut short the time-delay curve so that any current higher than the pickup of the instantaneous unit will cause the unit to operate before the time-delay element. This allows the coordination point to be moved back to the current value represented by the instantaneous element pickup. The safety margin between devices can be applied at this point instead of at the maximum current value with great advantage because of the inverse nature of the time-delay overcurrent relay (Fig. 1).

4. The Required Information to do a coordination study includes the following:

1. System layout that includes all transformer sizes with their self-impedances (impedance volts) and winding connections; motor horsepower including nameplate data for full-load amperes, locked-rotor amperes, accelerating time to attain 99% synchronous speed, and service factor; cable sizes and lengths; voltages; type of automatic switching desired; and the expected short-term and long-term loads.

2. Short-circuit study of the electrical system to determine the maximum and minimum fault-current magnitudes for 3-phase, phase-to-phase, and phase-to-ground faults on both the primary and secondary of all connected transformers. The secondary faults should be reflected back to the primary of the transformer and the winding connections, such as delta-wye, wye-delta, and delta-delta, taken into consideration.

3. Knowledge of the operating characteristics and principles of operation of the various devices to be used in the protective schemes.

4. Knowledge of how the system is to be operated, such as a radial system, an automatic transfer to an alternate feed, and parallel operation of two or more sources.

5. Knowledge of the interrupting ratings and ampacities of the devices in the system. The protection engineer can be the overseer of the entire system, because quite often he is the only one who must have the complete picture in order to do a thorough job. It is his responsibility to make the system function as a coordinated entity.

SENSING EQUIPMENT

5. General Description. Three basic types of equipment are used to sense short circuits. Two of these, the protective relay and the current sensor for a direct-tripping breaker, are sensors only and must cause another device to function to actually interrupt the short-circuit current. The third, a fuse, is both a sensing and an interrupting device contained within one housing. The three devices will be covered separately to provide background on how they work and what they will do in coordinating individual protective devices within an electrical power system. Functions and terminology for protective devices are given in Ref. 1.

These devices have some common attributes:

1. They all have an inverse time-current characteristic. Their operation is therefore predictable within certain limits, and the time to operate decreases as the current magnitude increases until the mechanical and arc-quenching limits of the device are reached.

2. They all have a definite pickup value—again within the tolerance limitations for the device. The protective relay is the most accurate of the various devices. The solid-state sensors used in low-voltage "direct-acting" equipment are generally the next most accurate. The fuse is next, and magnetic direct-acting trip devices are the least accurate.

3. Each of these protective devices has a time-current characteristic so that a definite time to function or operate for each unit of current can be determined. These devices will repeat this characteristic provided that proper maintenance is performed on the devices. These last statements apply to fuses also, provided that a new fuse is used each time. The manufacturing tolerances are such that a given fuse of the same manufacture and ampacity will have similar characteristics fuse after fuse.

These operating characteristics are the building blocks for establishing a co-ordinated system.

These devices and their characteristics can be assembled in such a manner that only the device closest to the faulted equipment will operate and the remainder of the system can continue to supply electric power to the other equipment and loads.

6. Limitations. Each of these devices (fuses, direct-acting trips, and relays) has some restriction or limitation that prevents its indiscriminate use. The devices are covered separately below.

7. Fuses. Present-day fuses can accomplish most of the protection required for good system operation and design. The fuse is described in Ref. 1 as "An overcurrent protective device with a circuit-opening fusible part that is heated and severed by the passage of current through it." A fuse is a thermal device, and heat, no matter how it is generated, helps to cause a fuse element to melt.

Fuses have several other functional features such as:

1. A fuse combines the sensing and interrupting elements in one unit.

2. A fuse is a single-phase device. Only the element in the affected phase will melt to deenergize the faulty phase. Sometimes this is an advantage. Single-phase loads can be maintained. Three-phase motors will continue to run but may overheat and be damaged if run on single-phase power for extended periods.

3. A fuse responds to a combination of current magnitude and the duration of time the current exists (a function of I^2T). It has an inverse-time current characteristic—the higher the current the faster the fuse blows.

4. One of the chief disadvantages of fuses is that the device requires considerably more current than its ampere rating to cause the fusible element to melt.

NEMA standards require that E-rated fuses of 100E and below melt in 300 s at 200 to 240% of their rating and fuses above 100E must melt in 600 s at 220 to 264% of their rating. These times are extremely long in terms of short-circuit protection. Fuses should be applied so that there is coordination with downstream devices and also so that faults are cleared within reasonable times like 1 to 5 s or faster. This range of operating times generally requires a fault magnitude of five or more times the ampere rating of the fuse. This current-magnitude requirement makes the application of fuses a little difficult at times.

When the application is critical, some other protective scheme must be employed, usually at a higher expense.

Fuses come in many sizes, shapes, characteristics, voltage ratings, and current-interrupting capabilities. In broad terms there are two basic types of fuses. These are the silver-sand fuses called the current-limiting fuse and the expulsion-type fuse. The expulsion-fuse category also includes the solid-material-type fuse.

a. Current-limiting fuses are designed to achieve circuit opening in less than ¼ cycle

(60-Hz basis) provided the threshold-current magnitude is exceeded. This means that a current-limiting fuse will react to low and medium magnitudes of fault current like any other fuse. Only when high magnitudes of current occur will there be any current-limiting action. A current-limiting fuse produces no external arcing. The fault energy is expended inside the fuse container to transform the special quartz sand therein into glass, which creates an insulating material to effect a circuit opening. The current-limiting fuse has the highest short-circuit-interrupting rating of any fuse available (Fig. 2).

Because the current-limiting fuse is very critical as to the voltage applied to it, it must be applied at the voltage for which it is designed.

A voltage surge is associated with the operation of a current-limiting fuse which develops to drive the fault current to zero. Proper application of these fuses will keep this voltage surge to a minimum to prevent insulation damage to the other equipment connected to the circuit.

The time-current characteristic of current-limiting fuses is usually rather "fast" or extremely inverse. Some manufacturers produce a medium-voltage (4.0 to 14.4 kV) current-limiting fuse that has a time-current characteristic closely following that of expulsion-type fuses. These "slower" fuses can provide more equipment protection than faster fuses (Fig. 3).

A comparison of the TCC (time-current characteristic) of a fuse of a given ampacity will show this difference in characteristic. The two fuses shown are both E-rated fuses at the 5.0-kV level. The fuse on the right is a "slower" fuse and is more adaptable to transformer primary protection than the "fast" fuse on the left.

Consider a 500-kVA transformer connected delta-wye on a 4800-V system. The full-load current is 60 A. The magnetizing inrush must be considered to be between ten and twelve times full load and lasting for approximately 0.1 s. At a current of twelve times full load, or 720 A, the "fast" fuse on the left in Fig. 3 may have blown on magnetizing inrush current. The fuse on the right had ample time to withstand the inrush current. A second consideration would be to ensure that a secondary phase-to-ground fault could be sensed by the primary fuse.

The magnitude of fault current is dependent upon the source impedance and the transformer impedance, and these quantities vary widely throughout an electrical system. For the sake of an example, a primary fault of 2400 A and a transformer impedance of 5.75% on its own base were chosen. The reflected 3-phase fault produced 730 A, and the primary phase current for a secondary phase-to-ground fault is about 470 A. The 470 A is barely adequate to blow these 100E fuses, and any larger-sized fuse in this situation would not provide complete transformer protection.

The choice would have to go to the fuse on the right in Fig. 3, the "slower" fuse, if complete protection were to be provided for this transformer. Each application of fuses should be studied so that the proper fuse is applied to do the whole job and not just a part of it.

Current-limiting fuses are fully rated devices that can usually sustain moderate overloads without damage or a change in characteristic.

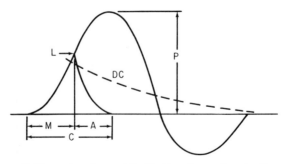

Fig. 2 The current-limiting action of current-limiting-type fuses. *A*, arcing time; *M*, melting time; *C*, total clearing time; *P*, available peak current; *L*, let-through current; *DC*, dc component.

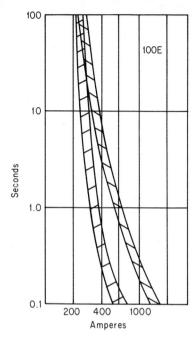

Fig. 3 Plot of two different types of current-limiting fuses showing differences in time-current characteristics.

Current-limiting fuses are generally used in motor starters and low-voltage circuit breakers to protect the motor contactor or circuit breaker from destruction when subjected to fault-current magnitudes in excess of their interrupting rating. This type of fuse, however, is not restricted to this usage, and current-limiting fuses may be used very effectively as transformer and circuit protection within the limits determined by the magnitude of available fault current.

b. Expulsion Fuses—Fiber-lined Tube. This fuse type is limited to outdoor use because of the explosive nature of the emitted ionized gases and flame. These fuses get their interrupting capability from the tube that contains the fuse link.

Low-magnitude faults cause the rapid destruction (burning) of the tube that surrounds the fusible element and the resulting pressure buildup is released down the tube, which causes the arc to be elongated and thus extinguished (Fig. 4).

Operation at the higher fault magnitudes involves not only the consumption of the tube but also the burning of some of the fiber tube, and the resulting blast blows out the arc caused by the parting fuse element. There are single-vented and double-vented fuse holders. The double-vented type has the higher interrupting rating.

This type of fuse uses the inexpensive replaceable links but also has the lowest fault-interrupting rating of the fuse types considered.

These expulsion-type fuses cannot be safely used indoors. They must be used outdoors in the open and in a location away from personnel.

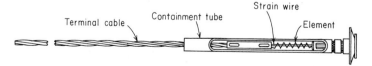

Fig. 4 Cross section of a typical universal-link-type fuse.

c. Expulsion Fuses—Solid-Material. Solid-material fuses overcome some of the shortcomings of the fiber-tube type. The solid material is generally a boric acid compound which decomposes when subjected to heat to form steam. This steam (water vapor) is expelled from the end of the fuse to elongate and thus to extinguish the arc caused by parting of the fuse element.

Since the expelled gases are steam, these fuses can be utilized indoors within an enclosure, provided that a discharge filter is connected to the fuse to reduce the velocity of the expelling gases and hence contain the "explosion." These filters are called "mufflers," "condensers," "snufflers," or "discharge filters," depending upon the manufacturer.

Solid-material-type fuses have the widest range of application because of the many time-current characteristics available. They can be made into larger ampacities without employing the two- and three-barrel units required on current-limiting fuses with a resulting lower cost. Fuse ampacities up to 400E can be obtained in the 14.4-kV rating.

Solid-material fuses have respectable interrupting capabilities, but the rating is less than that available in some current-limiting fuses. The X/R ratio (phase angle) of the fault impedance must be considered when the maximum interrupting-current value is being approached because the higher the X/R ratio is, the lower the available interrupting capability is. The manufacturer's literature should be consulted before any of these devices are applied.

d. Application. Each of the fuse types discussed has its own area of usage where it is the "best" for the application. Fuses within the broader category can be further singled out by manufacturer on the basis of time-current characteristics. The differences between the various fuse time-current characteristics are greater than the differences between other protective devices such as relays and direct-trip sensors.

The protection engineer should study these various characteristics and choose the TCC that best serves his purposes. Substitution of a fuse with a more desirable TCC is not always possible because of economic factors.

The final decision on the electrical-system layout and equipment to be utilized should not be made until the protection engineer has had an opportunity to study the system and make his recommendations to ensure a system that is both adequate and coordinated.

Since these fuses have a stable operating characteristic, they can be precisely coordinated with other fuses as well as other overcurrent devices, provided that the available fault-current magnitude does not approach a value that causes all the devices in series to operate simultaneously. There may be some advantage to mixing the current-limiting and other types of fuses in an electrical system to achieve coordination. The current-limiting fuse would usually be closest to the fault to allow the upstream devices time to provide the desired selectivity.

8. Direct-tripping Devices. Direct-tripping devices are those which sense an overcurrent and function to cause a circuit-opening action. These are usually attachments to some form of circuit breaker except in the molded-case breaker where the sensor is an integral part of the device.

There are two general classifications: (1) under 600 V and (2) over 600 V.

a. Direct-acting circuit breakers under 600 V are further divided into power circuit breakers and molded-case breakers.

1. Low-voltage power circuit breakers are small, compact, rugged circuit-protective devices that contain heavy-duty contacts, arc-quenching chambers, a stored-energy operating mechanism, and an overcurrent-sensing device that initiates the trip of the breaker. The operating energy is obtained from the fault current itself either by passing the current through the sensing-unit coil to effect a trip or by the use of current transformers so that the sensing units will have smaller magnitudes of current in the sensor units and to supply tripping power for the trip coil.

Two types of sensing devices are presently on the market: magnetic and solid-state. Both these overcurrent trip devices have three basic characteristics: long time delay, short time delay, and instantaneous which can be used singly or in combination. They also have three operating characteristics—maximum, intermediate, and minimum time delay. The general shape of the operating characteristic is the same for each band, but the band is shifted vertically to help achieve coordination among similar devices.

The devices achieve pickup adjustment by sliding the tripping characteristic horizontally to provide a more accurate setting and to help achieve coordination.

The magnetic devices are readily made selective on the same voltage level. Some difficulty is encountered when attempting to coordinate a transformer primary fuse and a secondary magnetic-trip device (Fig. 5).

The solid-state overcurrent tripping devices could have a more favorable operating characteristic than the magnetic variety depending upon the manufacturer. The solid-state sensor is a more accurate device, and its operating characteristic has a much narrower tolerance band than does the magnetic variety. The slope of the time-current characteristic is smooth without the troublesome "hump" that is typical of the magnetic device (Fig. 6).

If the solid-state device is applied to the transformer (circuit of Fig. 5) the improvement is obvious (Fig. 7).

2. Molded-case breakers are small, compact circuit breakers that contain a thermal-magnetic sensing device that operates the spring-loaded contacts.

These molded-case devices have time-current characteristics and can be coordinated one with another.

b. Direct-acting circuit breakers over 600 V encompass all oil and air circuit breakers. It is a type of trip initiation that uses the energy from the fault current to actuate the tripping mechanism of the breaker. Several forms are in use:

1. Instantaneous trip coil
2. Time-delay dashpot trip-coil device
3. Time delay using circuit-opening overcurrent relays
4. Time delay using circuit-closing overcurrent relays in conjunction with either a tripping reactor or an autotransformer

Each of these methods utilizes current transformers as the operating energy source. These current transformers along with the associated breaker trip coil, relay, and reactor or autotransformer must be selected with deliberate care to ensure that the components are compatible and will function when called upon to do so. The current transformer is the most vital component in the scheme and should have an accuracy classification of 10C100 or better to be considered for this type of application (see Sec. 15).

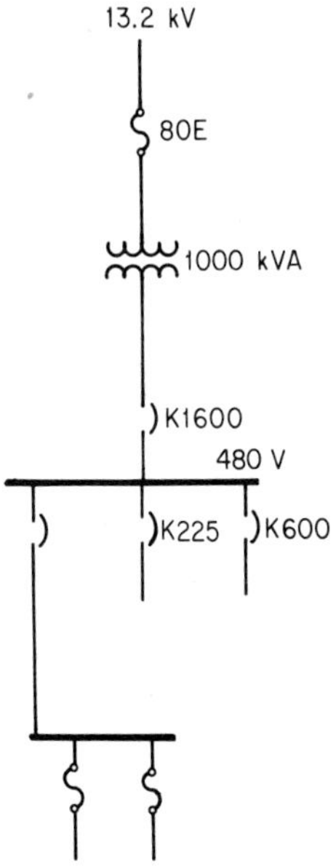

13.2 kV

80E

1000 kVA

K1600

480 V

K225   K600

Fig. 5 One-line diagram of a typical single service using direct-acting breakers.

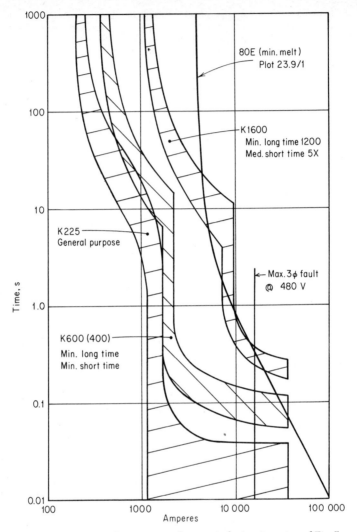

Fig. 6 Time-current characteristics of magnetic devices in system of Fig. 5.

The current-transformer iron must not become saturated when the high-impedance burden is imposed upon the secondary leads.

c. Instantaneous trip coil is the simplest form. The current-transformer secondary is connected directly to a breaker trip coil. The trip-coil plungers are usually pickup-adjustable to provide a range of setting. When the current through the breaker exceeds the setting of the trip coil, the breaker trips instantaneously (Fig. 8).

d. Time-delay via dashpot tripping devices are essentially the same as the previously mentioned device, but the trip-coil plunger must push against an oil-dashpot diaphragm that slows down the action and allows some downstream device an opportunity to function should the fault be beyond that device.

e. Time-Delay Tripping Using Circuit-opening Overcurrent Relays. The current from the current transformer passes through the coil of an overcurrent relay and then through the normally closed contact (*b* contact) of the relay, which shorts out the breaker trip coil.   When the pickup value of the overcurrent relay is exceeded, the *b* contact will open and the current is then forced through the breaker trip coil to unlatch the mechanism and open the circuit breaker (Fig. 9).

f. Time-delay tripping using a circuit-closing overcurrent relay is the most widely used scheme where a battery is not available.   The scheme utilizes an overcurrent relay, a tripping reactor or autotransformer, and the breaker trip coil (Fig. 10).

This scheme should have exclusive use of the current transformers that are in the circuit.   The current-transformer secondary circuit has a rather high burden introduced when it operates.   Any additional burden may cause the current-transformer iron to saturate, which would prevent the current transformer from producing sufficient secondary current to operate the scheme.

The trip-coil plunger is usually adjustable and should be calibrated to operate at a current that is less than 80% of the pickup value of the overcurrent relay.   This setting

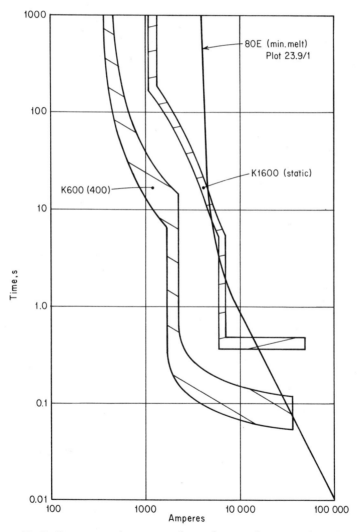

Fig. 7   Time-current characteristic of static device in the system of Fig. 5.

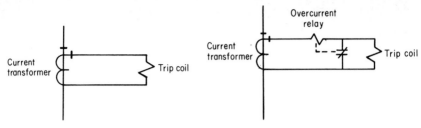

Fig. 8 Schematic of a simple ac trip scheme.   **Fig. 9** Schematic of time-delay ac trip scheme.

is necessary to ensure that when the relay contacts close, sufficient energy is available to operate the trip latch on the breaker.

g. *Important rules* to follow when designing or specifying trip circuits include:

1. Choose adequate current transformers as to accuracy classification and ratio in light of the available fault-current magnitudes.

2. Choose a relay in the 4-A range (this range has the lowest internal impedance).

3. Choose a trip-coil ampere pickup and impedance that are compatible with the tripping reactor or autotransformer to be used.

4. Above all, perform an operational test on the completed assembly by passing current through the primary of the current transformer and allowing the scheme to operate as it would under actual service.

h. Short-circuit magnitudes are calculated for the example installation shown in Fig. 5.

Transformer 1000 kVA, 13.2 kV, 480/277 V

Impedance volts = 6.0% at 1000 kVA

The K1600 is set to operate at a pickup of 1200 A. The feeders have K225 and K600 devices with the tripping characteristics as shown in Fig. 6.

Maximum available short circuit = 2500 A at 13.2 kV or 57.2 MVA

Calculation of short-circuit-current magnitudes for 3-phase and phase-to-ground faults on the secondary of the 1000-kVA transformer (10-MVA base):

3-phase faults
480-V side:

$$I_{3\phi} = \frac{12\ 000}{0.755} = 15\ 500\ \text{A} \qquad \begin{array}{l} \text{Source} = 0.175\ \text{per unit} \\ \text{Transformer} = \underline{0.600}\ \text{per unit} \\ \phantom{\text{Transformer} = }0.775 \end{array}$$

13.2-kV side:

$$I_{3\phi} = \frac{437.5}{0.775} = 565\ \text{A}$$

Phase-to-ground faults
480-V side:

$$\sqrt{3}I_0 = \frac{(3)(12\ 000)}{2.150}\ 16\ 750\ \text{A} \qquad \begin{array}{l} Z_1 = 0.775 \\ Z_2 = 0.775 \\ Z_3 = \underline{0.600} \\ 2.150\ \text{per unit} \end{array}$$

13.2-kV side:

$$3I_1 = \frac{\sqrt{3}(437.5)}{2.150} = 353\ \text{A}$$

Primary fuse plotting ratio:

Phase-to-phase fault = 23.9/1
Phase-to-ground fault = 47.5/1

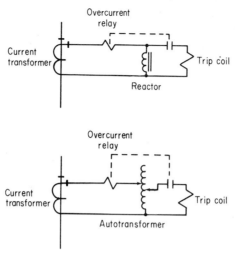

Fig. 10 Schematics of two time-delay ac trip schemes using reactors or autotransformers in the current-transformer circuit.

The 80E primary fuse must blow for a current magnitude of 353 A in order to sense secondary phase-to-ground faults. A fuse must have current that is at least 166% of the maximum fuse-blowing current at 100 s to be considered adequate to ensure the melting of the fuse.

The 80E fuse chosen is the largest fuse that can be applied and still have some protection against secondary phase-to-ground faults. This fuse choice has caused the incoming-line K1600 breaker to be set down to 1200-A pickup, which restricts the maximum load that the 1000-kVA transformer can carry without tripping on load. The downstream devices were chosen to be selective under the K1600 (1200) breaker.

This example is a typical situation—the installation has a restricted load because of the amount of available fault current at the particular location. The use of a delta-delta-connected transformer or a transformer of lower impedance would improve the selectivity problems because more current would be available with which to operate the protective devices.

The inselectivity shown in Fig. 6 between the primary fuse and the secondary device is somewhat typical. The inselectivity exists for phase faults only. The primary fuse is selective with the secondary devices for phase-to-ground faults because of the higher plotting ratio available.

9. Protective Relays are either electromechanical or static (solid-state) devices that are connected to the secondary of instrument transformers that operate on current, potential, or a combination of both to sense abnormal system conditions and to operate contacts or a solid-state switching device to cause a circuit breaker or other device to function to remove the abnormal condition from the electrical system.

There are relays that are sensitive to high current or high voltage, low current or low voltage, reverse power, ratio of volts to amperes (impedance), frequency, voltage or current balance, and nearly every other measurable electrical phenomenon that can occur on an electric-power system (Ref. 1; also see Sec. 1, Table 3).

Protective relays are used in conjunction with potential transformers, current transformers, circuit breakers, and buses to provide the building blocks that comprise the schemes that achieve overall system coordination and equipment protection.

10. Current Transformers. Proper selection of adequate current transformers is vital to the protective-relay scheme (see Sec. 15). The relays must have fairly accurate reproductions of the primary current over the entire range of operation to ensure coordination and system protection.

The current transformers used for relaying purposes are usually the bushing type. This type of current transformer should have at least 40 turns that are distributed around the iron core. Bushing current transformers with fewer turns and/or with the turns not distributed uniformly around the core do not perform adequately over the range of fault currents usually found in industrial electric systems.

Here are some guides to help choose current transformers for relaying purposes.

1. Choose a turns ratio so that a 5-A secondary current is 110 to 125% of the expected maximum load through the circuit expressed in secondary amperes.

2. Choose a ratio that will produce no more than 100A in the secondary windings with maximum primary current through the current transformer.

Some compromise may be necessary between 1 and 2 above to get a ratio that will satisfy system requirements.

3. Choose a current transformer that has a relay accuracy classification high enough to supply adequately the burden imposed by the relays, meters, and other devices in the circuit.

Do not skimp on adequate current transformers, because these devices are the heart of the protective-relay scheme. Do not specify an accuracy class that greatly exceeds the present or known future needs. The higher the accuracy class of the current transformer the more it will cost.

a. Ground-sensor current transformers are a special application of current transformers. These devices usually have a large internal diameter, a small cross section of iron, and 10 to 20 turns around the core. These "sensors" are designed to be used with a specific relay or static device, and the overall performance of the combined components must be considered. The current transformer by itself is a very poor current transformer but when it is used as a unit with its relay, it makes a satisfactory fault-sensing device for this specific area of system protection.

The ground sensor is applied to incoming lines and feeders as very sensitive ground-fault protection. The 3-phase conductors all pass through the ground-sensor current transformer. The flux from the current in these phase leads has a zero resultant for loads and phase-to-phase faults because all the currents involved pass through the current transformer and there is a magnetic-flux balance. If a phase-to-ground fault occurs, the current leaves the phase conductors and returns to the source through some other path. The ground sensor will detect this leaked-off current and operate because the flux is no longer at a balance and the unbalanced flux produces current to operate the sensing device.

11. **Fused load-break Switches** are becoming popular as primary voltage-switching devices. A fused load-break switch is usually a 3-phase set of disconnects that are group-operated by an external operating handle and are equipped with interrupters capable of opening load currents. There are either current-limiting fuses or boric acid fuses with expulsion restrictors included in the unit to provide short-circuit protection. These load-break switches can be packaged to provide remote operation and to provide a load-transfer scheme either automatically or manually.

An adequate two-line service with manual transfer can be achieved when three load-break switch assemblies are connected together. This type of installation is becoming popular for use in high-rise buildings.

The configuration in Fig. 11 is adaptable to many two-line-service situations. The load-break disconnects are on the source side to provide the means for load transfer

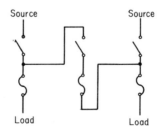

Fig. 11   One-line diagram of load-break switches in two-line service with load transfer.

and the fuses are located on the load side. When the fuse is located as shown, adequate transformer protection can usually be achieved even where one incoming line is serving the entire load. The fuse in the tie-switch position should be as large as the largest fuse or larger if the utility circuits can tolerate a larger fuse. When the tie-switch fuse and the feeder fuse are the same size, there is no selectivity, but there is no real penalty other than the fact that two fuses will blow instead of only one for a fault condition.

The configuration shown in Fig. 12 would be adequate and selectivity would be achieved if the available fault current were low enough so that the fuses at A, B, and C have time clearance between fuses and the utility circuit can tolerate the required large-sized fuses at A and B. Sometimes the fuse and B can be replaced in function by the cable-pole fuse and thereby eliminate one step of fusing. It becomes quite difficult and at times nearly impossible to achieve adequate coordination under high available fault-

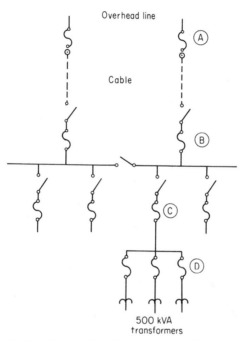

Fig. 12   One-line diagram of load-break switches in large office complex.

current situations. Sometimes a mixture of current-limiting and the slower boric-acid-type fuses can achieve what one type of fuse alone is unable to do.

Figure 13 is a plot of the fuses installed in Fig. 12. A careful study of the curves reveals that at a current of 5000 A maximum available fault current, the incoming lines have about 3 cycles of selectivity with the feeder fuses for feeder faults. The feeder fuses are not selective with the 500-kVA transformer primary fuses for primary faults. Any fault current higher than 5000 A will cause the fuses to melt practically simultaneously. The selectivity is further reduced when the upstream devices are preloaded. Preloading of fuses tends to move the curves to the right, thus reducing the time interval between the different fuse characteristics at a given current magnitude.

Figure 14 shows the time-current plot with the current-limiting fuse installed at D. The 40E fuse is operating in the current-limiting region at the 5000-A point so there is adequate selectivity between the feeder fuse and the transformer primary fuse.

When high maximum fault currents cause serious inselectivities to exist, breakers with relays may be required (Fig. 15) at location B in order to obtain the desired coordination. A preliminary fault study must be made and the coordination problem worked before any equipment is purchased if the scheme is to be a fully coordinated system.

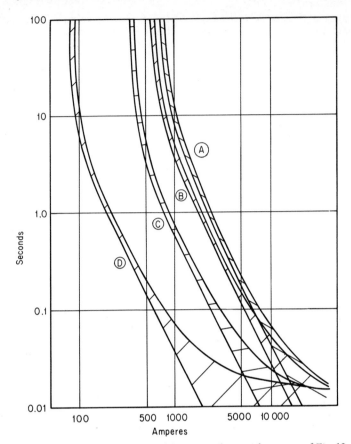

Fig. 13 Plot showing coordination of the various fuses in the system of Fig. 12.

Figure 16 is a plot of the relay and fuse characteristics of the scheme in Fig. 15 to show the ample coordinating time available when overcurrent relays are used on the main breakers and a high magnitude of short-circuit current is present.

BASIC PROTECTIVE-RELAY SCHEMES

There are many combinations of the various types of relays, but each complicated relaying scheme consists of the basic types of relays as simple building blocks interconnected to achieve the desired result.

12. Overcurrent Protection is self-explanatory. The overcurrent relay may have either an inverse-time or an instantaneous operating characteristic and is connected to the secondary of a current transformer of suitable ratio. The contacts of the overcurrent relay are connected to actuate the trip mechanism of the circuit-opening equipment. This type of scheme is the most commonly used throughout industry. It is nondirectional; that is, the direction of power flow has no influence on the operation of the device.

13. Directional Overcurrent Protection is a modification of the nondirectional type. There is a device that senses watts or a voltampere relationship within the relay that allows only the inverse-time overcurrent element of the relay to operate if the power flow is in the desired direction. In the normal inductive system there is no problem with the direction-determining element getting confused because of the phase angle

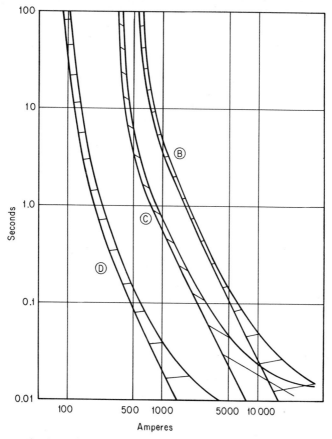

Fig. 14 Plot showing coordination of the system of Fig. 12 with current-limiting fuse.

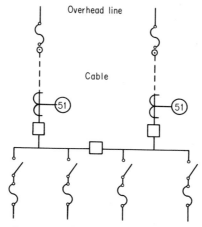

Fig. 15 One-line diagram showing relayed breakers at incoming line positions to achieve coordination.

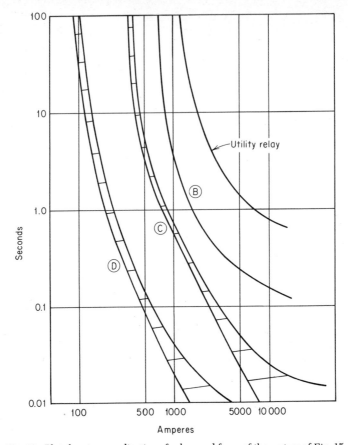

Fig. 16 Plot showing coordination of relays and fuses of the system of Fig. 15.

of the current involved. If a large amount of capacitive vars is flowing through the system, the directional element could be confused and sense a fault or an overcurrent in the reverse direction to that which was intended. The directional element is not a true watt element. Its function here requires maximum torque at the fault-current power-factor angle.

14. Total Bus Differential in its simplest form is shown in Fig. 17. This scheme algebraically sums up all the currents entering or leaving the bus section. Normal loads and short circuits outside the bus section cause currents to circulate among the various current transformers so that each is satisfied. The system is balanced, and no

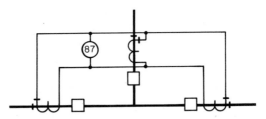

Fig. 17 One-line diagram showing simple bus-differential relaying scheme.

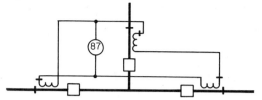

Fig. 18 One-line diagram showing simple bus-protective scheme using linear-coupler devices.

current should flow through the relay operating coil. However, some current may flow because the iron cores of the current transformers tend to saturate when over-excited by high-magnitude fault currents and the current transformers do not reproduce the primary currents at the secondary terminals.

Short circuits that occur inside the zone of protection cause the currents on the secondary of the current transformers to add and are forced through the 87-relay operating coil. The 87-relay coil actuates an auxiliary relay that causes all the breakers on the bus section to open, thus clearing the short circuit from the system.

The use of relays designed especially for bus-differential application helps to mini-mize the effects of saturated current transformers and should be used when this type of protection is required.

Another method of bus protection is the "linear-coupler" system, which is basically a voltage scheme (Fig. 18). The linear coupler is an ironless "current transformer" that produces a voltage proportional to the primary current flowing through it. This scheme removes the problems of saturated iron cores but introduces minor problems in application.

15. Partial Bus Differential is a name given to a scheme that sums the source cur-rents to a bus. It is essentially a bus-overcurrent relay scheme that functions on the total load on a bus. This scheme is used mostly as feeder-breaker backup and as bus protection when no total differential exists and there are two or more sources for fault current.

The partial-differential scheme is essential when the power system consists of two incoming lines and a tie breaker that are all operated normally closed (Fig. 19).

16. Over-under Voltage Relays can be either single-phase or 3-phase, instantaneous or time delay. The relay is used to cause an automatic transfer scheme to function or trip a motor breaker if the supply voltage is too low to maintain adequate motor speed that would cause the motor to overheat.

The 3-phase voltage relays also provide a phase-sequence check. This arrangement is useful if the industrial system is supplied by an overhead line in which the phase wires might be rolled during restoration after the wires were felled by a motorist or a storm.

17. Over-under Frequency Relays are usually used on large motors or on systems that have local generation. The normal 3-phase voltage relay is not sensitive to minor fre-

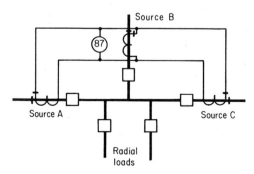

Fig. 19 One-line diagram showing a partial-differential or bus-overload relaying scheme.

quency variations and cannot distinguish between the voltage supplied by a large synchronous motor coasting to a stop after the electric service was interrupted and the normal utility voltage. A frequency relay would detect a change in system frequency and operate to open the motor circuit breaker. The removal of the motor voltage would allow the automatic transfer scheme to operate and would allow the alternate power source to close to reestablish normal voltage and to pick up the other loads. The large synchronous motor would have to be restarted.

18. Power-directional relays are either single-phase or 3-phase devices that operate on very small magnitudes of watts and are inherently directional. They can be either a disk type to produce some time delay or a high-speed induction-cup design. These relays are generally used as power-flow sensors to control backfeed or reverse power flow. They can be used to operate other devices or to cause a circuit breaker to operate.

The *power-flow-type relay* is generally one that will operate on normal voltage and low current.

The *fault-sensing-type relay* will operate on the reduced voltage and high currents that would be prevalent during fault conditions.

19. Single-phasing Protection is necessary when 3-phase motors are served from a system that is protected by single-phase devices, either fuses or line reclosers.

The normal 3-phase voltage relay is not usually adequate for this service, because the running motor will tend to supply the missing phase voltage, thus preventing the 3-phase relay from operating.

Some voltage relays are designed to operate when the voltage triangle is distorted. These devices combined with the motor contactor or circuit breaker could prevent motor burnouts caused by single phasing of a 3-phase motor.

20. Transfer Tripping is a method of achieving system protection by operating a circuit breaker at a location remote from the protected equipment. This type of scheme saves the expense of local breakers or other circuit-opening devices and still gives protection for the low-grade faults that are not sensed by devices at the remote breaker location.

21. High-speed Grounding Switch is a device that performs the function of a transfer-trip scheme but does not use a wire line to transmit a signal. A phase-to-ground fault is purposely applied to the system. On resistance-grounded systems, this intentional fault would not be severe, and it could actuate some sensitive ground-fault relaying to clear the affected equipment from the system in the shortest possible time.

MAJOR EQUIPMENT PROTECTION

22. Transformer Protection.   There are several types of protective schemes that can be utilized for transformer protection. These are:

1. Primary fuses
2. Primary overcurrent relays
3. Transformer-differential scheme
4. Fault-pressure device

a. Primary fuses are the least expensive type of transformer protection. They provide local sensing and fault clearing as well as creating a minimal amount of system disturbance (Fig. 20).

Fuses that are intended to protect transformers have several criteria to meet:

1. The fuse must be able to withstand the transformer inrush current of ten to twelve times full load for 0.1 s (6 cycles) without melting or damaging the fusible element.

2. The fuse must be capable of interrupting the maximum available fault current at the application location.

3. The fuse should be able to carry up to 200% of nameplate full load in an emergency. This capability may, however, have to be sacrificed to fulfill some other requirement.

4. The fuse must be able to sense secondary phase-to-ground faults on a solidly grounded delta-wye-connected transformer bank. If it is unable to comply with this requirement, some other equipment should be installed to effect a transfer trip of the feeding breaker.

5. The fuse must be selective with the secondary protective device.

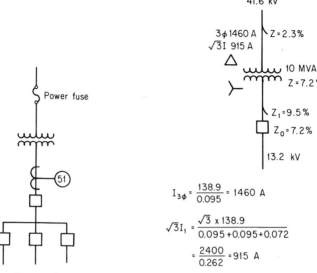

$$I_{3\phi} = \frac{138.9}{0.095} = 1460 \text{ A}$$

$$\sqrt{3}I_1 = \frac{\sqrt{3} \times 138.9}{0.095 + 0.095 + 0.072}$$

$$= \frac{2400}{0.262} = 915 \text{ A}$$

Fig. 20 One-line diagram showing a primary fuse for transformer protection.

Fig. 21 Problem sheet showing calculation of primary current available for a secondary phase-to-ground fault.

Item 4 is usually the most difficult requirement to fulfill. The reflected current for the secondary phase-to-ground fault is dependent upon the source impedance and the transformer self-impedance. The sample calculation of Fig. 21 computes to 62% of the reflected 3-phase secondary fault.

If the wye connection of the transformer is grounded through a grounding resistor, the ground-fault current is essentially governed by the phase-to-neutral voltage divided by the resistance, in ohms, of the resistor. This current when reflected to the primary of the transformer is so small that a phase fuse cannot possibly sense it and still fulfill the other requirements above.

b. Primary overcurrent relays afford about the same degree of protection as do fuses, but the relays allow higher loading capabilities than fuses because of the fixed inherent time-current characteristics of fuses vs. the variable time-current relationship of an overcurrent relay. There are at least two disadvantages to primary overcurrent relays (Fig. 22).

1. The device must cause some other equipment to function to clear the fault. This device might be a local primary breaker or another switching device. If there is no local device, some remote equipment must operate to clear the fault. Some form of transfer tripping may be required such as a pair of wires from the relay contacts to a breaker if the distance is not too great, a tone frequency-shift scheme, a high-speed grounding switch to fault one of the primary phase leads to ground intentionally, or some other scheme that might be devised.

2. The use of primary overcurrent relays as well as primary fuses inserts another step of protection into the coordination scheme that forces the upstream devices to be set with longer operating times so that time selectivity can be maintained.

c. Transformer differential is a relaying scheme that provides fast, sensitive protection for faults within its zone of protection. Its zone of protection extends to the area or the equipment between the sets of current transformers that make up the differential scheme. The secondary cable and circuit breaker could be included. The differential requires some sort of primary switching device or transfer-tripping scheme to effect the circuit opening to remove the short circuit (Figs. 23 and 24).

The transformer-differential scheme generally senses most faults within the transformer (3-phase, phase-to-phase, and phase-to-ground faults) provided the neutral of

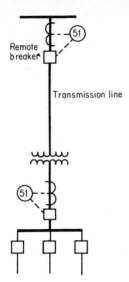

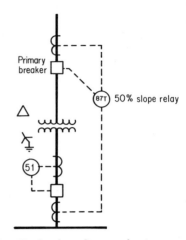

Fig. 22 One-line diagram showing transformer protection using a remote breaker and relaying.

Fig. 23 One-line diagram showing transformer differential with a local primary breaker.

the transformer is solidly or effectively grounded. If the transformer is grounded through some current-restricting impedance, other equipment is required in the neutral connection to sense these low-grade phase-to-ground faults.

The phase angle as well as current magnitude must be considered in designing a differential. The usual delta-wye transformer connection introduces a 30° phase shift between the primary and secondary phase to neutral voltages, and this phase shift must also be provided in the secondary circuit of the current transformers. Care

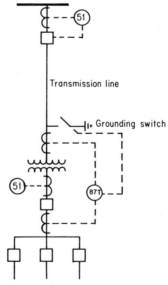

Fig. 24 One-line diagram showing transformer differential and high-speed ground switch to transfer trip to a remote breaker.

TABLE 2. Current-Transformer Connections

Transformer connection		Current-transformer connection	
Primary	Secondary	Primary	Secondary
Delta	Delta	Wye	Wye
Wye	Delta	Delta	Wye
Wye	Delta	Wye + wye-delta aux.	Wye
Delta	Wye	Wye	Delta
Delta	Wye	Wye	Wye + wye-delta aux.
Wye*	Wye	Delta	Delta

* Assume delta tertiary winding.

The main current transformers are assumed to be in the line conductors and not inside any delta-connected windings.

should be taken to connect the current-transformer secondary leads in a delta that duplicates the power-transformer winding connections. The necessary phase-angle correction can be achieved by either connecting the secondary leads of one set of the main current transformers in delta or inserting a set of auxiliary current transformers to provide phase angle as well as ratio correction.

Table 2 shows the recommended current-transformer connections for the various power-transformer winding connections.

If these connections are followed, the transformer wye-connected windings can be grounded at the neutral and the differential scheme will not be affected. The delta connection in the secondary of the current transformers filters out the zero-sequence currents supplied to the system by the grounded wye-connected winding.

Care must be taken to ensure that proper phase-angle relationship is maintained in making up the delta connections.

d. A fault-pressure device is the most sensitive sensing device employed for transformer protection. It is either a gas-operated device mounted on top of the tank that senses the sudden increase in gas pressure that exists over the oil in a transformer or a liquid-operated device mounted low in the oil on the side of the tank that operates from the pressure that results from an arc within the transformer tank. These devices sense faults and operate before the fault develops into a damaging short circuit that would destroy transformer windings and core iron. The zone of protection is limited to the transformer tank, and any faults external to the transformer tank and up to the secondary breaker must be sensed by some other equipment.

The type of protection that is employed for transformers depends upon the size of the unit and the importance of the equipment to overall operation. Several questions should be answered. How much downtime can be tolerated? Is there an alternate supply of electric power for the process in question?

These things and others must be considered and a compromise reached—good, fast, reliable, equipment protection costs money, and the amount of protection that is installed depends upon the situation.

23. Motor Protection can range from the simplest situation of a low-voltage plug fuse to a very sophisticated scheme utilizing several kinds of relays and protective devices.

Figure 25a shows the simplest arrangement for low-voltage motors. The fuse can be either a cartridge or plug-type dual-element slow-blowing fuse sized for motor load in amperes. The low ampere rating provides some motor-overload protection, and the fast portion of the dual fuse provides circuit protection. The motor may or may not have an internal overcurrent or thermal-cutoff device.

Fig. 25 One-line diagram of simple fuse protection for low-voltage motors.

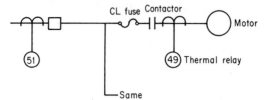

Fig. 26 One-line diagram of current-limiting fuse and motor controller with a thermal device.

Figure 25*b* shows a common arrangement for low-voltage single- and 3-phase motors. The circuit is protected by a molded-case breaker or fuse of sufficient capacity for the circuit load and wire size. The motor contactor provides the switching and contains thermal elements sized for the particular motor to provide thermal protection from over-currents.

The use of current-limiting fuses in motor-starting equipment allows the application of a contactor that is rated to carry and open currents equal to motor loads (Figs. 26 and 27). The short-circuit currents are handled by the fast-acting current-limiting fuses, thereby removing the interrupting duty from the motor contactor. On medium-voltage levels, there is usually some form of thermal relay with its characteristic time-current curve to protect the motor from overloads. The current-limiting fuse protects the motor

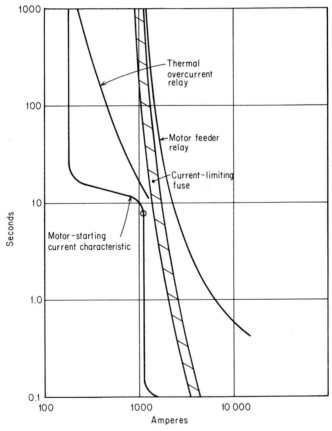

Fig. 27 Plot of protective devices of equipment of Fig. 26.

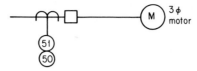

Fig. 28 One-line diagram of overcurrent-relay protection for large motors.

contactor and provides fast-clearing circuit protection which allows the upstream devices to be coordinated without unduly long clearing times.

A somewhat similar effect can be accomplished by using overcurrent relays that have
instantaneous attachments (Fig. 28).

a. Typical Settings. A few guidelines may help to achieve adequate motor protection and system coordination. Figure 29 is a plot of the various devices on a medium-
voltage motor with the following characteristics: 1500 hp at 4800 V, full-load amperes
= 180 A, locked-rotor amperes = 6.5 × full-load amperes = 1170 A, service factor
= 1.0, and accelerating time 8 s. The guides are:

1. The motor characteristics must be known: full-load amperes, locked-rotor amperes, accelerating time, and service factor.

2. The thermal-overcurrent device is set to pick up at 125% of the full-load amperes
and the time dial is set to provide several seconds over the normal accelerating time of
the motor.

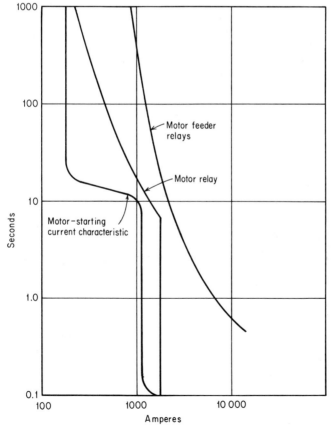

Fig. 29 Plot of relay characteristics for scheme of Fig. 28.

3. The circuit-protecting current-limiting fuse is chosen to carry the normal full-load amperes of the motor, but it must also provide time clearance over the accelerating time at 110% locked-rotor current. This guide is useful in obtaining the smallest adequate current-limiting fuse. Upstream protective devices should not be forced into excessively long clearing times.

4. The motor-feeder overcurrent relays must carry the entire load of all connected motors as well as start the largest motor without having the overcurrent relay close its contacts.

5. When a long-time-delay overcurrent relay with an instantaneous attachment is used, the basic guides still apply. The instantaneous attachment, however, takes the place of the current-limiting fuse in function. The instantaneous overcurrent attachment is usually set about 150% of the locked-rotor current to ensure motor starting.

b. Range of Protection. Large motors are usually provided with nearly all the available types of protective devices. Figures 30 and 31 are essentially the same, but each shows a slightly different approach to provide a motor-differential scheme.

Figure 30 shows the typical motor-differential scheme utilizing a set of matched current transformers on each side of a motor. The percentage differential relay (87) is the disk-type relay having a 10% slope.

The motor is equipped with an under-over frequency relay (81) and a phase-sequence voltage relay (47). There are overcurrent relays (51/50) of the long-time-delay variety

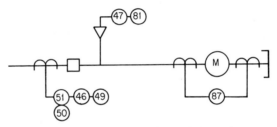

Fig. 30   One-line diagram showing typical protection for large motors.

with an instantaneous attachment, phase-balance relays (46) to protect against single phasing, and thermal-replica relays (49) for overloading.

Figure 31 shows another method of providing a very sensitive differential scheme using only one window-type current transformer per phase and an instantaneous overcurrent relay. This scheme utilizes the principle of flux balancing. The normal current passes through the current transformer twice in opposite directions, thus producing zero flux in the current-transformer iron. The rest of the protective devices are the same. It is not always necessary to provide all the different forms of protection shown in the figures. The amount of protection to be applied is determined by the protection engineer or the motor manufacturer and is based on the size, cost, and importance of the unit under consideration (see Sec. 13).

24. Grounding.   System neutral grounding is an important consideration in system protection and coordination. There are several accepted methods of achieving a grounded-neutral system, and each method has its own advantages.

The common connection of wye-connected transformer windings supplies a ready-made system neutral which can be connected to ground or the earth either solidly or through a resistor or some other impedance to supply ground current under fault conditions, provided that there is a delta winding on the transformer. The closed delta windings provide a path for zero-sequence currents to flow so that the transformer ampere-turns for each winding are satisfied. It is recommended that a delta-wye transformation be used to isolate the ground systems in a multivoltage installation to minimize coordination problems.

A solidly grounded transformer neutral will provide the most fault current for a phase-to-ground fault of all the systems considered. This current magnitude is generally somewhat greater than the 3-phase fault current at the point of grounding. The $3I_0$

or phase-to-ground fault current diminishes faster than the 3-phase fault current as the fault is moved away from a source bus because the cable or line impedances are introduced into the circuit. The zero-sequence impedance of cables and lines is considerably larger than the positive-sequence impedance.

The solidly grounded system is usually applied to voltage levels of 600 V and under but is not restricted to these voltage levels. Utilities will usually solidly ground their medium- and high-voltage systems and reserve the impedance grounding for the main-station-generator neutrals.

a. Resistance grounding can be applied as either high-resistance or low-resistance. High-resistance grounding uses a relatively high ohmic value of grounding resistor, and the current is limited to less than 0.1% of the 3-phase fault current. There usually is not sufficient current available to utilize overcurrent relaying, and consequently the system should either initiate an alarm to indicate there is a ground on the system or use some sort of voltage sensor to operate a tripping device. The small amount of available fault current prohibits any time or current coordination between series-protective devices.

Low-resistance grounding is the more popular method of system neutral grounding. A current magnitude between 5 and 20% of the 3-phase fault is usually chosen as a design value. At least 400 A should be available for relaying. The grounding resistor for all practical purposes determines the magnitude of fault current that will be avail-

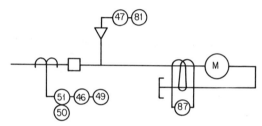

Fig. 31   Protection similar to Fig. 30 but using a different method for motor differential.

able. The phase-to-neutral system voltage divided by the ohms of the resistor will be the maximum value of amperes.

Actually, there will be something less than this maximum value because of the addition of the source and transformer impedances to the resistance of the grounding resistor.

The grounding resistor is usually the controlling factor, and the ground-fault current is essentially constant throughout the system.

b. Motor and generator neutrals can also be grounded to supply additional ground-fault-current sources. Motor neutrals are not recommended as ground sources because of the backfeed supplied to other system faults, especially if reduced-voltage starting equipment is utilized. Large circulating currents can flow between a running motor and a motor being started if open-delta starting transformers are used on a grounded wye-connected motor.

c. Grounding Transformers. Several transformers on the market are designed to supply ground current to a normally ungrounded system. Two of the more common are the zigzag-connected transformer and the wye-delta grounding transformer. Both these devices will supply zero-sequence currents for ground faults that are essentially limited by the impedances of the transformers used.

d. Why Ground? There are several reasons why an electrical system should be grounded:

1. To detect insulation failure. Phase-to-ground insulation failure is the most common type of fault.

2. Fault location is made easier because the faulted equipment is removed from the system automatically by the protective device.

3. Transient overvoltages are kept at safe limits to prevent a sparkover of the con-

nected lightning arresters and to prevent insulation punctures in the transformers and motors.

4. Less expensive equipment may be applied to a solidly grounded system than can be used on an ungrounded system because of the "graded" insulation used on some equipment.

5. To improve service reliability by preventing other equipment from being damaged by overvoltages and by excessively high currents flowing between two different pieces of equipment that have failed to ground on different phases — a phase to phase through ground. There could be insufficient current in this type of fault to operate the protective devices, but there could be more than enough current to burn up the windings of expensive equipment.

6. Personnel safety. Rapid clearing of equipment that has failed to ground reduces the possibility of electrocution from voltage rises in the ground paths that the current follows to return to the source.

7. Better protection for the system and equipment by isolating the faulted device from the system in as short a time as possible to reduce burning and other damage.

   e. Guides for Grounding

1. Ground only at the power sources, such as main transformers connected delta-wye or wye-delta-wye (three winding units) or generators. The ground systems (zero sequence) of the various voltage levels should be kept isolated from each other.

2. If the system is to be low-resistance-grounded, choose a resistor ohmic value to limit the current to between 5 and 20% of a 3-phase fault.

3. If the system is to be high-resistance-grounded, choose a resistor ohmic value to limit the current to less than 0.1% of a 3-phase fault.

4. Only in the solidly grounded system can other than fully insulated equipment be used. In any system that has an impedance grounded neutral, fully insulated equipment must be used because of the overvoltages that could be present on the unfaulted phases during a phase-to-ground fault.

5. Additional ground sources on a solidly grounded electrical system other than the main sources will increase the available fault current above the 3-phase fault magnitude and may require the circuit-interrupting devices to have a higher interrupting capability than would be required otherwise.

6. Grounding resistors have a "time" rating. This rating should be kept in mind when determining the tripping time and system coordination for ground faults.

INTERRUPTING RATINGS

25. Equipment-interrupting Capability may seem to be outside the area of system protection, but actually it is a very important factor that must be considered when designing or applying protective equipment and schemes to an electrical system.

The proper application of circuit breakers, fuses, motor starters, and other circuit-opening devices must be based on the available short-circuit current and the projected future fault-current magnitudes as well as those factors particular to the specific equipment.

Utility electrical systems are growing rapidly, and the concentration of generators and use of underground cable lines with their inherent low impedances have raised fault currents to such a level that some old equipment is no longer adequate to handle the fault currents.

New equipment may even be very near the top of their ratings when installed in some present installations.

The designer for an industrial complex should be aware of the fault magnitude and if necessary should install series reactors or high-impedance transformers or utilize some other scheme to keep the available fault current within the limits of the equipment to be used.

If the protective equipment is old but still serviceable except for the interrupting rating, it is necessary to provide some means of protecting it from being destroyed if it should attempt to open a current in excess of its rating. Cascade tripping might be employed if there is at least one adequate piece of equipment that can interrupt the fault current on the system. The incoming-line breaker usually serves this purpose.

This device would be made to operate first either to reduce the fault current, as in a multiple-feed system, or to clear the fault before the inadequate gear can operate. Sometimes the inadequate equipment can be made to open after the current is removed to isolate the fault and then the main breaker automatically recloses to reestablish the remainder of the load.   It is impossible to have a fully coordinated system if the equipment is unable to handle the fault-current magnitudes available within a particular system.

If a system contains inadequate equipment, the method of operating the system must be tailored to circumvent its deficiencies.   Some types of industrial processes may not be able to tolerate this restricted mode of operation and therefore require that the equipment has interrupting capabilities commensurate with the duty to which it will be subjected.

The economies of the particular situation will dictate how the system will be built, but the protection engineer's duty is to inform those involved what the situation is and how it may affect the overall operation.

26. Short-circuit Data required to establish whether or not protective equipment is adequate come from a short-circuit study.   The short-circuit study is a basic tool for the application and coordination of protective schemes.   It supplies the means to:

1. Determine the adequacy of the equipment.

2. Determine which equipment to use if a choice exists.   Sometimes a fault study will determine whether a choice is available.

3. Determine whether fuses or relays and circuit breakers may be used.

4. Determine coordination and selectivity for the various devices installed on the system.

5. Determine the fault contributions within a complex multisource system.

6. Determine the maximum and minimum values of fault current under the various operating conditions.

The short-circuit study can be conducted in several ways.   Once a one line has been established and major equipment is chosen, at least as to size, a fault study can be conducted.   If actual system-component parameters are not available, assumptions should be made to get at least approximate answers.

If it appears that high fault-current magnitude might become a problem, a transformer impedance that will give a worst-case situation (or lower impedance) should be chosen. If a minimum situation is required, a worst-case impedance for that situation (or higher impedance) should be chosen.

Maximum-fault situations must include motor contributions from synchronous and induction motors.   Since the operating times for circuit breakers and current-limiting fuses must also be taken into consideration, the proper impedance, either transient or subtransient, must be used to provide a more realistic current magnitude.

A maximum fault should truly be a maximum and include all the contributions applicable to the operating time of the equipment involved.

There are several methods of producing a fault-current study.

1. The longhand method with calculations made using a calculator.   This type is generally limited to small radial systems.   It is not impossible to work a more complex system, just time-consuming.

2. Analog computers such as ac and dc calculating boards are useful for the more complex systems, but few of these machines are still available.

3. Digital computers produce many of the fault studies being conducted today. Some digital computers are coupled to an analog computer to provide a ready means of altering a program or system to simulate breaker openings or different operating conditions.

The usual fault study includes maximum and minimum source conditions for 3-phase and phase-to-ground faults on the primary and secondary of characteristic transformers connected to the system while taking the transformer-winding connections into consideration.

A ready tabulation is received from the digital-computer study, but a problem sheet or map is still required to identify the nodes.   If the system is large enough and complicated enough to cause confusion, it may be advantageous to make a schematic layout in the form of a problem sheet for ready reference.   Sometimes the problem sheet can present an overall picture of the system that helps to achieve a coordinated electrical system.

EXAMPLES

27. Single Radial Feed. *a. Single-Line Feed—The High-Rise Building.* This example of a high-rise building is essentially a vertical 13.2-kV 4-wire distribution circuit. The high-rise building is served from two different utility circuits to a 3-unit, 3-phase fused load-break switch arrangement (Fig. 32). One set of three switches serves a 3-phase transformer that is used for general building power.

The middle unit is used as a manually operated tie switch between the two sources. The third set of switches is used to serve three single-phase circuits up through the building. The three risers have provisions at the top of the building so that a riser could be backfed from an adjacent cable under emergency conditions. Any one riser may be required to carry the load of two risers, and the fuse in the fused load break must be sized accordingly. Each 167-kVA dry-type transformer is connected phase to neutral through a 40E current-limiting fuse. The secondary fuse is a 600-A low-voltage current-limiting fuse chosen to ensure that the secondary molded-case circuit breakers will not be subjected to more than 10 000-A interrupting duty. The current-limiting fuse limits the let-through current to less than 10 000 A. The current-limiting fuses in the transformer primary are used in an attempt to achieve coordination with the feeder fuse in the basement. Under high current magnitudes, two similar fuses will melt in approximately the same time. The use of a current-limiting fuse and an expulsion fuse in series allows the slower expulsion fuse to remain intact while the current-limiting unit melts to clear the fault. The current-limiting fuse is downstream from the expulsion fuse.

Figure 33 shows the plot of the various fuse characteristics to show the fuse coordination between the steps of protection.

The 200E expulsion-type fuse in the load-break compartment must be equipped with expulsion restrictors to prevent equipment damage when the fuse blows. The 40E current-limiting fuse is chosen so that a secondary fault can be sensed and adequate coordination clearance can be attained over the 600-A secondary fuse.

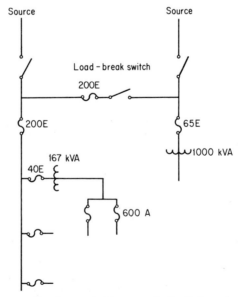

Fig. 32   Fused load-break switches as applied to high-rise buildings.

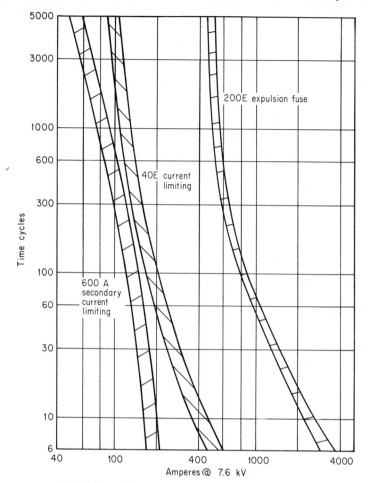

Fig. 33 Plot of fuses and coordination for system of Fig. 32.

b. Single-Line Feed. The example in Figs. 34 and 35 is an installation of a 2000-kVA transformer served from a 13.2-kV line. The primary of the main transformer is fused with a 125E expulsion fuse. The fuse must be able to sense both 3-phase and the reflected phase-to-ground faults on the secondary side of the transformer. The primary fuse must also be able to carry at least transformer full-load current and must withstand the magnetizing inrush current of ten to twelve times full load for 0.1 s.

The secondary breaker in the example is a relayed breaker. The use of relays on the secondary side allows better coordination with both the upstream and downstream devices.

The motor is protected by the thermal-overcurrent devices in the motor controller. The class J current-limiting fuse provides short-circuit protection for the contactor and motor insulation.

The lighting transformer is protected by a fuse sized to carry the full load of the transformer. The secondary and branch circuits are protected with magnetic circuit breakers.

Equipment-interrupting capabilities must be considered to ensure that a safe system is installed to prevent equipment damage and hazard to personnel.

The installation should also conform to the local code and the National Electric Code.

28. Multiple Radial Feed with Automatic Transfer.   *a. Multiple Radial Feed with Emergency Generator.*   This example, shown in Fig. 36, is a primary-voltage service with secondary automatic transfer and a limited amount of on-site emergency generation.

This installation consists of two 750-kVA, 13.2-kV, 480Y/277-V transformers with an automatic transfer between their secondary breakers and a normally open tie breaker. The low-voltage breakers are K-type switchgear. The transformer secondary, the tie breaker, and the emergency-generator breaker are all electrically operated to open and close. The feeder breakers have direct-acting trip devices and are manually operated. All the critical loads, which are mainly computers, are served from one bus that has an emergency generator that will start if both incoming lines are open. If one of the incoming lines returns, the generator is shut down and the load is automatically transferred to the outside source. The computers have an uninterruptible power supply so that they are not disturbed because of the power loss (see Sec. 7).

The 750-kVA transformers are served through 50K universal fuse links mounted on the cable poles. These fuses provide adequate protection for phase and phase-to-ground faults on the secondary of the transformers. The fuse size (50K) was chosen to be able to sense the reflected secondary phase-to-ground fault, but the 50K restricts the permissible loading to less than twice transformer nameplate. The fuse curve also restricts the secondary overcurrent-relay setting. Another fuse with a slower time-current characteristic could have been used to provide more coordination clearance between the fuse and the secondary overcurrent relay but at a much greater expense.

If a larger-ampacity universal-link fuse is used, some form of transfer trip is required. The size of this load did not warrant extra expense.

Since there are no large synchronous motors at this location, the low-voltage or single-phase sensing device for the automatic-transfer scheme can be a 3-phase undervoltage

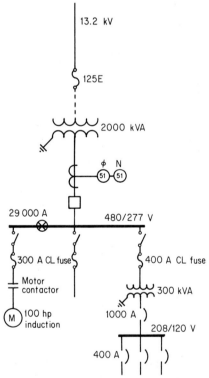

Fig. 34 One-line diagram of single-feed system using primary fuses and a relayed secondary breaker.

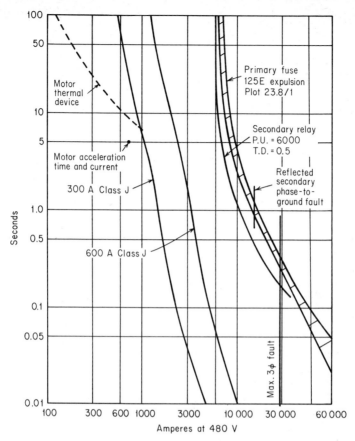

Fig. 35 Plot of the protective devices for system of Fig. 34.

relay such as a type CP or ICR relay. A negative-sequence voltage relay with sufficient time delay could be used to initiate the automatic transfer if motor backfeed presents any problems.

Calculations:

Primary fuse: 50K, plot ratio 24/1
Secondary relay: inverse time pick up 1600, time dial = 1.0, no instantaneous trip
Feeder breakers: K225 100% pick up universal device

Primary 3-phase fault: 1300 A $Z = 0.337$ 10-MVA base
Transformer $Z = 5.75\%$ at 750 kVA $Z = 0.765$
                                       Total $Z_1 = 1.102$ per unit

Secondary $I_{3\phi} = \dfrac{12\ 000}{1.102} = 10\ 900$ A   Summation   1.102
                                       of   1.102
                                                   0.765
                                       $Z_1 + Z_2 + Z_0 = 2.969$ per unit

Secondary $3I_0 = \dfrac{3(12\ 000)}{2.969} = 12\ 100$ A

Reflected secondary $3I_0 = \dfrac{\sqrt{3}\ (437.5)}{2.969} = 255$ A

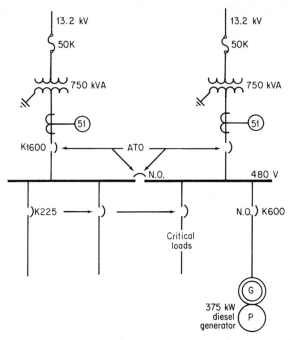

Fig. 36   One-line diagram of primary-voltage service with secondary automatic load transfer.

The total clearing-time curve of a 50K fuse at 100 s = 120 A with 255 A available for a bolted fault. The fuse will protect for secondary phase-to-ground faults. (There must be at least 200 A of available fault current for adequate protection.)

The maximum loading is limited by the overcurrent-relay pickup setting adjusted by a suitable safety factor to ensure that load current will not cause any unwanted tripping. One such safety factor is 80%. The loadability will be 80% of the relay pickup current. This factor limits the load to 1280 A (0.8 × 1600) or 1065 kVA.

The setting cannot be made larger because the primary fuse size is limited to 50K to ensure clearing for secondary phase-to-ground faults. If the available primary short-circuit current were higher, larger primary fuses and hence higher overcurrent-relay settings would allow higher total loads to be carried on one incoming line under emergency conditions (Fig. 37).

b. Multiple Radial Feeds—One Load-carrying, One Standby for a Pumping Station (Fig. 38). This installation is a storm-water pumping station consisting of three 2750-hp synchronous motors and a 500-kVA house-service transformer. Each motor has a reduced-voltage starting device consisting of an open-delta-connected autotransformer. The wye-connected motors are not grounded for two reasons:

1. The 4800-V utility system is an ungrounded delta system, and no ground sources can be tolerated.

2. The starting equipment is an open-delta-connected autotransformer which tends to shift the system neutral when starting a motor. If one motor were running and a second unit were started, the resulting circulating current that would flow from having two system "neutrals" would cause the protective relaying to operate and trip off one or both of the motors.

This installation has many varied forms of protective equipment.

1. The 4800-V bus has a total differential using 1500/5 current transformers and a variable percent differential relay.

2. The 2750-hp motors have: (*a*) thermal-overload relays (49), (*b*) phase-balance relays (46), (*c*) motor-differential relays (87), and (*d*) undervoltage relays (47).

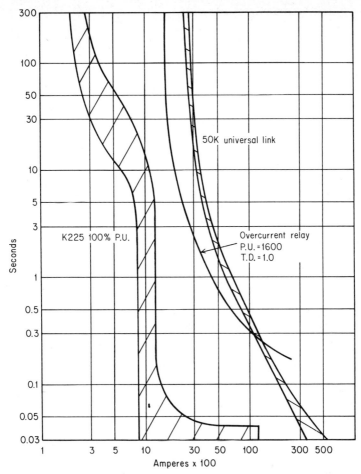

Fig. 37 Plot of protective devices for system of Fig. 36.

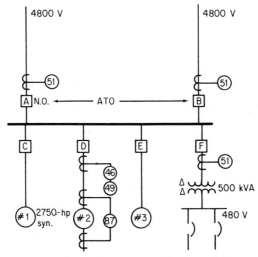

Fig. 38 One-line diagram of primary-voltage service with primary-voltage automatic load transfer.

The thermal-overcurrent relays are set at full-load amperes, and the instantaneous attachment is set at approximately ten times full-load amperes. The ten times value is usually large enough to prevent unnecessary trip-outs on starting inrush or motor backfeed for short circuits on the other feeders or the utility system.

The phase-balance relay is used to detect unbalanced loadings caused by single-phase voltage applied to the motor.

The motor differential utilizes a 10% slope motor-differential relay to provide sensitive fault detection for insulation failures in the motor windings.

The undervoltage relay is used to prevent the motor from being started if there is insufficient voltage to have the motor come up to synchronous speed.

3. The incoming lines have phase overcurrent relays (51) and the automatic-transfer-control undervoltage and phase-sequence relay (47).

4. The house-service transformer has overcurrent relays (51) only.

Most of these protective devices and schemes are differentially connected or balanced quantities that do not introduce a step of time or current to the overall coordination scheme. This arrangement makes a much simpler selectivity problem (Fig. 39).

The overcurrent relay for the house-service transformer can be set low enough to be no problem to the incoming-line overcurrent relays.

The system has two services, but one is strictly a standby that is utilized only for the loss of the normal feed.

29. Two Incoming Lines in Parallel. The example shown in Fig. 40 is an installation that utilizes two 120–13.2-kV step-down transformers that are fed from separate 120-kV lines, and the secondaries are operated in parallel.

The 120-kV lines are tie lines with these transformers tapped to them. The 120-kV lines are protected by pilot-wire differential relaying using metallic wires between

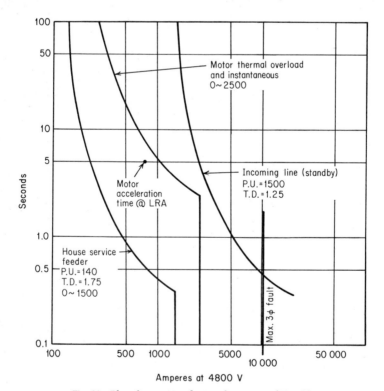

Fig. 39 Plot of protective devices for system of Fig. 38.

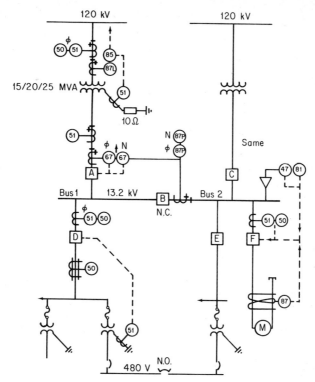

Fig. 40 One-line diagram of system where two services are operated in parallel at the 13.2-kV level.

the various terminals. There is also a transfer trip scheme on this pilot wire for use with the transformer protection.

The main power transformers, which are resistance-grounded, are protected with primary phase and neutral overcurrent relays, a fault-pressure device, and an over-current relay in the transformer neutral-to-ground connection. All these devices initiate a transfer tripping function to the remote line terminals. The transformer-neutral overcurrent relay also trips the transformer secondary breaker. There are directional phase and neutral overcurrent relays on the transformer secondary that sense faults toward the transformer and trip the secondary breaker. The phase relays (67) are backup for the 120-kV line relaying, and the neutral relay is necessary for relay coordination with the neutral overcurrent relay (51) on the other transformer for ground faults between the transformer and its secondary breaker. The phase (51) over-current relays on the transformer secondary are somewhat redundant but do provide backup protection should the bus partial-differential (87P) relays fail to function.

The 13.2-kV buses are protected by a partial bus-differential or bus-overcurrent scheme that sums the currents from the two principal feeds. This scheme acts as the feeder breaker backup as well as the bus protection. These relays operate to open the transformer secondary breaker involved and the normally closed tie breaker. If the tie breaker is open, the partial-differential relays act as transformer over-current relays.

The feeder breakers *D* and *E* serve several fused transformers that are connected delta-wye grounded. The reflected secondary phase-to-ground fault does not produce sufficient current for the phase fuses to operate reliably. A neutral overcurrent relay is used to sense the secondary phase-to-ground fault, and the tripping function is trans-

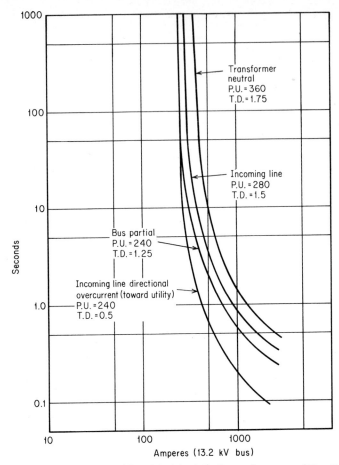

Fig. 41   Plot of the ground or neutral protective devices for system of Fig. 40.

ferred to the feeder breaker. The zero-sequence device and the instantaneous relay operate to provide very sensitive ground-fault protection on the feeder.

Plots of the relay characteristics are given in Fig. 1. Curve *A* is for the primary overcurrent relay, which must be selective in both pickup and time-dial setting with curve *B*, which is for the secondary-transformer overcurrent relay. Curve *A* is plotted at the 13.2-kV level using a ratio of 6.7/1. The two curves *A* and *B* have the same pickup at this extreme plotting ratio. The time dial was chosen to provide a minimum of time over the secondary relay *B*. However, under normal operating situations since the two transformers are operated in parallel, the primary relay must coordinate with the partial-differential relays, curve *C*, which have ample time clearance.

The load triangle is a coordinating help for the situation in which one main transformer unit is out of service and a fault occurs on bus 2, which is normally served by the missing transformer. The load is carried through the tie switch. The partial differential on bus 2 now acts as an overcurrent for bus 2. To save the load on bus 1, the relays of curve *C* must be under the relay curve *B* at the magnitude of current available from one main transformer.

For faults on a feeder just off the bus, adequate selectivity must be maintained. The feeder breaker-relay curve *E* must operate first before the partial-differential relay. The instantaneous element helps to achieve the necessary coordination by cutting short the inverse-time curve as shown dotted. The extended curve provides only 0.3 s time clearance between the relay curves where 0.5 s should be maintained.

The directional overcurrent relay shown as curve D represents the relay curve that must coordinate with a relay with a curve B on the opposite transformer. There is ample time between the curves. The short time-dial setting is used to get as fast a trip as possible for faults in the transformer or the secondary cable so that the system disturbance is kept to a minimum.

A similar presentation could be made for the neutral relays that are shown in Fig. 41.

30. Multiple Feed with In-Plant Generation. *a. Limited vs. Unlimited Backfeed Generation.* There are at least two types of service contracts that could be used when on-site generation is involved. The on-site generation referred to is normally running and operated in parallel with the utility system. The generation may be a by-product from process-steam requirements, or it may be diesel equipment or others.

One method of parallel operation would allow the in-plant generation to backfeed the utility and thus be an energy source for the utility system.

The other method would not allow any backfeed into the utility system except for momentary swings for disturbances on the utility system.

The unlimited-backfeed scheme requires considerably more automatic protective equipment to be installed to protect both the utility system and the plant equipment than the no-backfeed system. The master breaker must be relayed so that the maximum allowable power transfer can occur but still enable the incoming-line breaker to be opened for all faults on the utility line serving the plant. The plant load should be maintained via the local generation if sufficient generating capacity is available. The generator should not become unstable because of the utility-system fault. The generation protection should include a voltage-controlled overcurrent relay that will continue to time out even though the generator is slowing down and the current from the machine is dropping off owing to the decaying action of the short-circuit response of the generator.

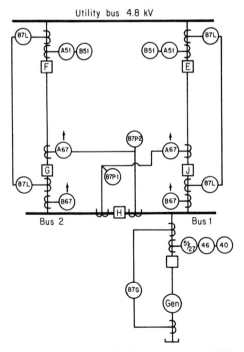

Fig. 42 One-line diagram of system with two feeds operating in parallel and on-site generation and with backfeed allowed.

The limited-backfeed system would require only a sensitive reverse-power relay with an external timer to help ride over any transient faults on the utility system. The generator protection has only to provide protection for any local faults and not reach out into the supply line to provide backup or other protection.

b. Parallel Operation with On-Site Generation. The system in Fig. 42 has a two-line service from the utility with the lines operated in parallel and a generator located on one of the two plant buses and with backfeed allowed. The 4800-V lines are fairly short cable lines that use a rather simple line-differential relaying scheme with an instantaneous overcurrent relay at each terminal to trip its respective breaker. The A51 and B51 overcurrent relays at breakers *E* and *F* are backup for the line-differential relays. The plots for the coordination are shown in Fig. 43. The A51 relays are set to pick up at 1600 A and trip their respective breakers. The B51 relays are set to pick up at 1000 A and have the contacts of the relays on position *E* in series with the contacts of the relays on position *F*. The A51 relays are functional at all times but especially when one line is out of service. The B51 relays, in effect, sum the current toward the plant bus to sense faults downstream when the system is normal.

The B67 directional overcurrent relays are set to pick up at 640 A and the time-dial is set to coordinate with the opposite-line A51 relays for faults on the protected line. The A51 and B67 relays have compatible time-current characteristics to allow coordination.

The A67 directional overcurrent relays were chosen so that the characteristic more or less follows the generator decrement curve. The relays were calibrated to guard against faults just off the bus at the utility substation. These relays must function whether both lines are in service or not.

The 87P-1 and 87P-2 relays are partial-differential or bus overcurrent relays that are set to provide clearance under the A51 and B51 relays at the utility as well as provide

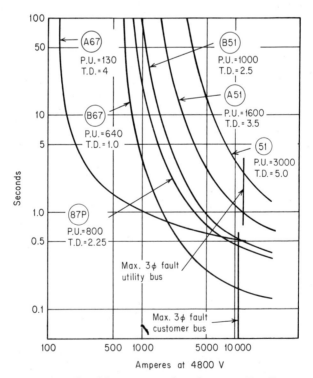

Fig. 43 Plot of the protective relays of system in Fig. 42.

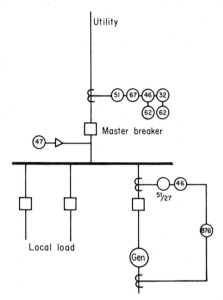

Fig. 44 One-line diagram for system with a single feed and on-site generation and no backfeed allowed.

time clearance over plant feeder breakers. The use of the partial-differential scheme provides backup protection for the feeder breakers and also prevents a total shutdown of the facility in event of a bus fault.

The generator has the usual complement of relays that include a generator differential (87G), voltage-controlled overcurrent relays (51/27), negative-sequence protection (46), and loss-of-excitation protection (40). This protection will allow power flow either into the plant load or out to the utility system, provided that the load is balanced. Any uncontrolled fault on the incoming lines or on the utility system operates the voltage-controlled overcurrent relays (51/27) to trip the generator breaker to isolate the fault. This arrangement provides a backup function because some other protective device should have removed the fault from the system long before the generator relays are required to function.

The system of Fig. 44 is a single-line feed with on-site generation that has limited backfeed to the utility system.

The master breaker is equipped with the following types of protective equipment:

1. Overcurrent relays (51) for faults on the plant bus.

2. Directional overcurrent relays (67) toward the utility.

3. Current-balance relay (46) plus a 10-s timer to ride through transient conditions.

4. Power-directional relays (32) plus a 2-s timer directional toward the utility to trip the master breaker in event of power flow toward the utility.

5. Undervoltage and phase-sequence relay (47) to trip the master breaker in event of low bus voltage and to ensure proper phase rotation.

The generator in this installation is equipped with the following protective equipment:

1. Voltage-controlled overcurrent relays (51/27). This device has a variable-ampere pickup as a function of the voltage applied to it. This arrangement helps to maintain the timing out of the device even though the machine may be slowing down because of a short-circuit condition.

2. Negative-sequence relay (46) to protect the generator from overheating and damage from single-phase currents.

3. Generator differential relays (87G) that have a 10% slope to provide sensitive protection for faults in the generator.

REFERENCES

1. ANSI Standard C37.100-1966.
2. Frank L. Cameron, The Application of High Voltage Power Fuses, *Westinghouse Eng.,* vol. 23, pp. 90–93, May 1963.
3. T. L. Bourbonnais II, The Coordination and Testing of Relays in Industrial Plants, *AIEE Trans.,* vol. 78, pt. II, A, pp. 1–10, April 1959.
4. J. F. Hower, A. C. Lordi, and A. A. Regatti, Keep Protective Devices Coordinated, *Power,* vol. 107, no. 2, pp. 65–67, February 1963.
5. E. R. Perry and A. M. Frey, Transformer Protection with High-Speed Fault Initiating Switches, *Allis-Chalmers Eng. Rev.,* vol. 30, no. 4, pp. 4–8, 1965.
6. V. C. Cook, Coordinating Protective Device Settings Gives Maximum Freedom from Electrical Outages, *Westinghouse Eng.,* vol. 26, no. 6, pp. 183–187, November 1966.
7. C. A. Hatstat, Industrial Power System Design and Protection, *Proc. Am. Power Conf.,* vol. 28, pp. 988–1003, 1966.
8. When Do You Use Fused Interrupters Instead of Metal-Clad Switchgear, *Power,* vol. 110, no. 5, pp. 86–88, May 1966.
9. A. J. Petzinger, Today's Power Systems Require Modern Relay Protection, *Westinghouse Eng.,* vol. 26, no. 1, pp. 12–13, January 1966.
10. P. Katzaroff, Designing Power Systems for Data Processing, *Elec. Construction Maintenance,* January 1966.
11. A. Kusko and F. E. Gilmore, "Application of Static Uninterruptible Power Systems to Computer Loads," IEEE-IGA Paper SPC-Tue-4, September 1969.
12. Westinghouse Electric Corp., "Electrical Transmission and Distribution Reference Book," 1950.
13. "Industrial Plants," IEEE 141 (Red Book).
14. "Commercial Buildings," IEEE 241 (Gray Book).
15. "Industrial Grounding," IEEE (Green Book).
16. "Protection Book," IEEE (Buff Book).
17. C. F. Wagner and R. D. Evans, "Symmetrical Components," McGraw-Hill Book Company, New York, 1961.
18. Edith Clarke, "Circuit Analysis of A.C. Power Systems," vol. 1, John Wiley & Sons, Inc., New York, 1943.

13

Motor and Motor-branch-circuit Protection

FRANK W. KUSSY, D.Eng.*

* Director of Engineering, Control and Instrumentation Division, I-T-E Imperial Corporation;
Registered Professional Engineer (Pa.); Member, IEEE.

FOREWORD

Many choices are available for motor and motor-branch-circuit protective devices. In selecting or specifying motor or circuit protection, many precautions should be observed. Since safety of personnel and equipment is involved, proper selection and application of protective devices are extremely important.

Various codes and standards have been developed through the years to aid in the use of protective devices. This section interprets these codes and standards and explains why they were established.

Under certain special conditions it may be good engineering to exceed the requirements of present standards. Some of these conditions are also covered.

APPROPRIATE STANDARDS

1. Basic Standards. Standards for protection of motors and motor branch circuits are treated in Articles 430 and 440 of the National Electrical Code (NEC), in Underwriters' Laboratories (UL) Standards 508, in ANSI C19, in NEMA Standards, "Industrial Controls and Systems," and in NEMA "Motor and Generator Standards." It is mandatory to meet the NEC provisions; and since the OSHA legislation, it is required to meet the UL standards and highly desirable to meet the NEMA standards. "American Standards for Industrial Control Apparatus," published by the Institute of Electrical and Electronic Engineers (IEEE), does not deal with standards for motor protection.

In addition to those general standards, a great number of special standards are published which have to be met when the motor-protection system is intended for special purposes. An example of such standards is the JIC Standards, which have widespread application in the automotive industry. There are also special standards for protecting hermetically sealed motors for the air-conditioning and refrigeration industry, which are often developed by the manufacturers of this equipment. There are military standards, standards for the Navy, specifications for maritime purposes (AIEE 45), and many others.

The basic circuit for motor protection is shown in Article 430-1A of the NEC (Fig. 1).

2. Motor Information. Motor-nameplate data are necessary for selection of proper protection equipment. The NEC and NEMA standards state the minimum requirements for the information which must be given on the nameplate of the motor. These requirements include the rated voltage, the rated full-load motor current (FLMC), the secondary voltage and the full-load rotor current for wound-rotor motors, the time rating, the code letter, the rated temperature of the insulation class system, the rated ambient temperature, and the rated horsepower for motors 0.125 hp and larger.

On nameplates for dc motors, the FLMC must be marked for the base speed, the time rating, and the horsepower rating. Furthermore, the nature of the winding arrangement must be marked as shunt, stabilized-shunt, compound, or series winding. In many cases the motor is marked with a service factor. Some additional information is required on the nameplate of the motor, but these items are not important for the selection of the proper protective device.

The FLMC is most important and must be known to select the proper thermal-protective devices for the motor. These protective devices detect overcurrents that may become dangerous for the motor and disconnect the motor before a dangerous temperature is reached. Often the FLMC is not known at the time the overload device is selected. The NEC in Article 430-150 has tables for estimated average values for

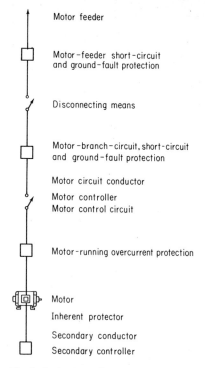

Motor feeder

Motor-feeder short-circuit
and ground-fault protection

Disconnecting means

Motor-branch-circuit, short-circuit
and ground-fault protection

Motor circuit conductor

Motor controller
Motor control circuit

Motor-running overcurrent protection

Motor

Inherent protector

Secondary conductor

Secondary controller

Fig. 1 Basic circuit for motor protection.

3-phase induction motors, which are indicated in Table 1.   Other tables in the NEC have been developed for estimating the FLMC for dc single-phase and 2-phase motors. Table 2 shows the average running kVA/hp for induction motors based on Table 1. If the overload relay is selected, based on these tables, it is recommended that the FLMC in the table be compared with the FLMC found on the nameplate as soon as the motor is available.   A very important item of information is the code letter (Sec. 11, Table 16).   By comparing Tables 1 and 2, the average ratio of the inrush current to the FLMC can be estimated.

Standard motors are built with Class A or Class B insulation.   The permitted temperature rise for 3-phase induction motors is shown in Table 3.

The definition of a general-purpose motor is a motor having open construction, continuously rated, having a service factor of 1.15 for motors of 3 to 200 hp; 1.2 for motors of 1.5 to 2 hp, and 1.25 for motors of 0.5 to 1 hp, and being Class A or Class B insulated. It is important to know the proper insulation class and cooling system of the motor because, based on this knowledge, the overcurrent vs. time characteristic of the selected motor protector is determined.

If the insulation class permits a higher temperature rise for the winding, the motor itself is smaller for the identical horsepower rating and has less mass.   The mass becomes still smaller if the motor is artificially cooled and hermetically sealed.   If the mass of a motor is small, in case of a motor overload or locked rotor the temperature limit above which the motor winding will be permanently damaged is reached faster. A running overcurrent-protective device must therefore trip faster for a motor with a higher insulation class or a better cooling system than for a motor with Class A insulation.

3. Ultimate Trip Current. The service factor is also important for the selection of the proper protective device.   The service factor is defined in the NEMA "Motor and

TABLE 1. Full-Load Current,* 3-Phase AC Motors

	Induction type				Synchronous type			
	Squirrel-cage and wound-rotor, A				Unity power factor, A †			
hp	115 V	230 V	460 V	575 V	2300 V	220 V	440 V	550 V	2300 V
1/2	4	2	1	0.8					
3/4	5.6	2.8	1.4	1.1					
1	7.2	3.6	1.8	1.4					
1 1/2	10.4	5.2	2.6	2.1					
2	13.6	6.8	3.4	2.7					
3	......	9.6	4.8	3.9					
5	......	15.2	7.6	6.1					
7 1/2	......	22	11	9					
10	......	28	14	11					
15	......	42	21	17					
20	......	54	27	22					
25	......	68	34	27	...	54	27	22	
30	......	80	40	32	...	65	33	26	
40	......	104	52	41	...	86	43	35	
50	......	130	65	52	...	108	54	44	
60	......	154	77	62	16	128	64	51	12
75	......	192	96	77	20	161	81	65	15
100	......	248	124	99	26	211	106	85	20
125	......	312	156	125	31	264	132	106	25
150	......	360	180	144	37	.....	158	127	30
200	......	480	240	192	49	.....	210	168	40

For full-load currents of 208- and 200-V motors, increase the corresponding 230-V motor full-load current by 10 and 15%, respectively.

The voltages listed are rated motor voltages. Corresponding nominal system voltages are 110 to 120, 220 to 240, 440 to 480, and 550 to 600 V.

* These values of full-load current are for motors running at speeds usual for belted motors and motors with normal torque characteristics. Motors built for especially low speeds or high torques may require more running current, and multispeed motors will have full-load current varying with speed, in which case the nameplate current rating shall be used.

† For 90 and 80% PF the above figures shall be multiplied by 1.1 and 1.25, respectively.

TABLE 2. Average Running kVA/hp for 3-Phase Induction Motors

hp	Running kVA	Running kVA/hp
1/2	0.8	1.6
3/4	1.12	1.48
1	1.44	1.44
1 1/2	2.08	1.39
2	2.72	1.36
3	3.84	1.28
5	6.08	1.22
7 1/2	8.80	1.17
10	11.2	1.12
15	16.8	1.12
30	32.0	1.07
60	61.6	1.03
100	99.2	0.99
200	192.0	0.96

TABLE 3. Temperature Rise, °C, of Winding of Integral-Horsepower Induction Motors (Measured by Increase of the Resistance of the Winding)

Motor	Insulation			
	Class A	Class B	Class F	Class H
General-purpose motors	50*	90		
Motors having service factor 1.15 or higher	50	90	115	
Totally enclosed motors:				
Nonventilated	65	85	105	135
Fan-cooled	80	85	105	125
Encapsulated motors with 1.0 service factor	...	85	110	
All other motors	60	80	105	125

* Measured with a thermometer, the value would be 40°C.

Generator Standards" as a multiplier which, when multiplied by the rated horsepower of the motor, gives a permanent rating for which the temperature will be within those limits shown in Table 3. If that factor is more than 1, it is usually stamped on the nameplate. The NEC states that an overcurrent-protective device must eventually disconnect the motor at 125% of the FLMC if the motor is marked with a service factor of not less than 1.15 and with a temperature rise not over 40°C. It is becoming more common to use motors with Class B insulation with a service factor of 1. Motors with a temperature rise of 40°C are usually general-purpose motors with Class A insulation.

At present, many exceptions exist to the general NEC rule; e.g., under certain conditions the Code allows an increase of the maximum tripping current from 125 to 140% FLMC. These exceptions are treated in Article 430-34 of the NEC. A motor circuit being protected by a device selected according to Article 430-34 would not necessarily have the proper branch-circuit protection (Art. 4). For hermetically sealed compressor motors, there are special rules dealing with the selection of their protective devices.

The NEC, Part 430-32 (A) (2) and (c) (2), treats the selection of inherent motor protectors. Inherent motor protectors are protective devices which disconnect the motor when its temperature rise reaches a dangerous level. The inherent protector does not protect the branch circuit because the protector is placed after the supplying branch circuit.

The ultimate trip current of a thermally protected motor may not exceed the following percentage of FLMC given in the NEC (Table 1):

FLMC not exceeding 9 A: 170%
FLMC not exceeding 20 A: 156%
FLMC exceeding 20 A: 140%

Since the NEC now allows higher values for the ultimate trip current for inherent motor protectors than for standard relays, in some cases the controller which could carry 125% of the largest heater rating for many hours at 40°C (ultimate trip current) must now carry the ultimate trip current of the thermal protector even if this value is more than 125% of the FLMC of the largest heater for a particular controller. The user of the motor with inherent protector must check with the controller manufacturer to determine if a chosen controller can carry the increased ultimate trip current for the time necessary to trip the inherent protector.

Inherent protectors may be selected in four ways (NEMA "Motor and Generator Standards," MG 1-52/53).

1. If the motors are marked "THERMALLY PROTECTED," (a) they must not exceed under continuous load the temperature rise in Table 4; (b) the protector shall cycle to limit the winding temperature to the values given in Table 5; (c) the ultimate trip currents must not exceed the values required by NEC 430-32.

2. If the motors are marked "OVER TEMP PROT 1," conditions 1A and 1B must be fulfilled.

**TABLE 4. Maximum Permitted Winding
Temperature of Thermally Protected Motors**

Insulation system class	Temperature °C*
A	140
B	165
F	190
H	215

*Temperature based on 25°C ambient temperature.

3. If the motors are marked "OVER TEMP PROT 2," condition 1A must be fulfilled.

4. If the motors are marked "OVER TEMP PROT 3," the motor manufacturer must be consulted for details of protected conditions.

Based on the NEC requirements for running overcurrent protection, UL has standardized in UL 508 the ultimate trip current for an acceptable thermal-overload relay. It must trip at 125% or less of the FLMC. UL will not approve a higher tripping value. If a manufacturer rates an overload relay or a heater for a relay 1 to 1.1A FLMC, UL requires that the relay with the selected heater must trip at 1.25 or less at 40°C ambient.

Based on the service factor for most Class A insulated motors, UL did require in its standards that the proper heater must cause tripping of the relay at six times its lowest tripping current, within 30 s and at twice its lowest tripping current within 8 min. The tripping time of 30 s is too high for Class B insulated motors; thus, most manufacturers recommend that an overload relay for general-purpose Class B insulated motors should trip at six times its lowest FLMC within 20 s or less. This problem is recognized by new UL and NEMA Standards (Art. 5). If the service factor of the motor is 1, the value of 1.25 for the ratio of the lowest tripping current to FLMC is too high because a motor could be indefinitely subjected to 56% (1.25^2 is 1.56) higher temperature rise than permitted. The maximum tripping current in this case may not be higher than 115% FLMC. Most manufacturers indicate this value in their heater tables.

There are several other heater characteristics required for special-purpose motors or for other than 40°C ambient conditions; however, most standards for motor and motor overcurrent-protective devices are based on 40°C ambient conditions. They must be corrected if the relay and the motor are in different locations which have approximately consistent temperature differences if the protective device is not ambient-compensated. The lowest trip current must be higher than 100% of the FLMC to avoid nuisance tripping. Each heater must be used for a certain range of FLMC. Most heater tables show methods used to determine the proper relay or heater for other than 40°C ambient temperature.

TABLE 5. Winding Temperature Under Locked-Rotor Conditions

Type of protection	Maximum temperature °C* insulation class				Average temperature °C* insulation class			
	A	B	F	H	A	B	F	H
Automatic reset								
During first hour	200	225	250	275	. . .	. . .	. . .	. . .
After first hour	175	200	225	250	150	175	200	225
Manual reset								
During first hour	200	225	250	275	. . .	. . .	. . .	. . .
After first hour	175	200	225	250	. . .	. . .	. . .	. . .

*Temperature values are based on 25°C ambient temperature.

4. Branch-Circuit Protection.   It is required that the motor-running protection disconnect the motor in case a hazardous condition occurs during normal operation. Such conditions might be single phasing or a locked rotor.   In case the branch-circuit conductor is overloaded, the branch-circuit protector must disconnect the branch circuit; e.g., if a fault occurs in the motor winding or phase to phase, or if a ground fault in the motor or motor branch circuit occurs, the branch circuit must be disconnected before a fire hazard or extensive damage occurs.

The motor-overcurrent-running protection might disconnect the branch circuit and it might even be possible that the motor-overcurrent-running protection may take over the branch-circuit protection in certain cases.   The branch short-circuit protection may protect the branch circuit and the motor only at currents higher than locked-rotor current.   Therefore, the locked-rotor current should be known (Sec. 11, Table 16). NEMA Standards have been developed for the maximum permitted locked-rotor current for general-purpose motors.   Table 6 shows the maximum permitted motor inrush current for 220- and 230-V rated motors.   Standard utilization voltage for motors is 230 V.   The ratio between the maximum permitted locked-rotor current and the FLMC is also found in Table 6.

The FLMC is based on the estimated values in Table 1.   The locked-rotor current values are taken from NEMA "Motor and Generator Standards."   The ratio shown for 220- and 230-V motors is identical to the same horsepower rating for other standard voltages.   In the NEMA "Motor and Generator Standards," the maximum locked-rotor current is standardized.   Tables can also be found in these standards relating locked-rotor current rate to the FLMC for other than general-purpose 3-phase induction motors. Table 6 indicates the motor design letter for which it is valid.   These design letters concern torque vs. speed characteristics and have no relationship to the insulation class of the motors.

TABLE 6. Maximum Permitted Locked-Rotor Current for Motors at 230 V *

hp	Max locked-rotor current, A	Locked-rotor current/FLMC, A	Design letter
1/2	20	10.0	B, D
3/4	25	8.9	B, D
1	30	8.3	B, C
1 1/2	40	7.6	B, D
2	50	7.3	B, C, D
3	64	6.6	B, C, D
5	92	6.0	B, C, D
7 1/2	127	5.8	B, C, D
10	162	5.8	B, C, D
15	232	5.5	B, C, D
20	290	5.4	B, C, D
25	365	5.4	B, C, D
30	435	5.4	B, C, D
40	580	5.4	B, C, D
50	725	5.6	B, C, D
60	870	5.6	B, C, D
75	1085	5.7	B, C, D
100	1450	5.8	B, C, D
125	1815	5.8	B, C, D
150	2170	6.0	B, C, D
200	2900	6.0	B, C
300	3650	6.0	B
400	5800	6.0	B

*The ratio locked-rotor current/FLMC is identical for other standard voltages for the identical horsepower.

Unfortunately the same letters are standardized for quite different purposes; for example, in insulation Class B, the permitted temperature rise for the motor winding is based on usage of Class B insulated wire. Except for switching or instantaneous currents, which last for only a few cycles, a motor current higher than locked-rotor current can flow only if the motor is already damaged; e.g., by a short circuit in the motor winding, by flashing from phase to phase on the terminal, or by flashing to ground on the terminals. At least under these conditions the branch overcurrent protection not only protects the branch but also prevents more extensive damage to the motor.

Motor-branch-circuit protection is provided by a fuse, a thermal-magnetic circuit breaker, adjustable magnetic circuit breaker, overload relays, or motor short-circuit protectors. The branch-circuit conductor for motor loads shall have an ampacity of not less than 125% FLMC. When UL tests motor starters, they select a cable size up to 100 A, chosen according to the table for the allowable ampacity for insulated conductors for 60°C temperature rating. For 100-A FLMC and above, UL selects a wire of 75°C temperature rating for their test. Today, cables with higher temperature ratings are often used.

If the motor is protected by an inherent protector, the branch-circuit protection must be provided by another branch-circuit overcurrent device. If an oversized fuse or thermal-magnetic circuit breaker is used for that purpose, only the overload relays provide running overcurrent protection, because only they would trip if the wire selected for 125% full-load motor current-carrying capacity is overloaded. If the motor is protected by an inherent protector, it may carry, depending on the FLMC rating according to the NEC 1975, 170, 156, or 140% full-load motor current for an indefinite time without being disconnected by the inherent protector. Therefore the wire and all other components of the branch circuit should be able to carry these loads for an indefinite time. This ampacity may, in some cases, be higher than the sizes selected for 125% full-load motor current. Proper branch-circuit protection for running overcurrents is provided only by overload relays or properly sized fuzes or circuit breakers. The NEC 1975 permits the selection of the same size wire and same branch-circuit components for motors being protected by inherent protectors or thermal-overload relays. This method of selection seems to be illogical. The installer must therefore check to determine whether the selected wire and other branch-circuit components can really withstand the increased current in case of a running overcurrent for an indefinite time.

A wire should not be overheated; i.e., its temperature rise may never be higher than the permitted value for an indefinite time if protected by a properly selected branch-circuit device. No tables have been standardized to indicate the time that a wire can be subjected to a higher temperature than its temperature rating for a short time. The standardized ambient temperature for wire is given in NEC 310 as 30°C. Article 430-24 requires the ampacity of a conductor supplying a single motor to be at least 1.25 FLMC. An overload relay trips at 40°C ambient temperature at 1.25 FLMC or less. The Code does not require larger wire for motors with inherent protector than for motors protected by separate overload relays.

Assume the lowest tripping temperature of an overload relay is 90°C, the ambient temperature is 40°C, and the temperature rise of the eutectic solder or the bimetal at the tripping temperature is 50°C. A 60°C wire allows a temperature rise of 30°C at 30°C ambient temperature. A 75°C wire allows a 45°C temperature rise at 30°C ambient. Actually, at 30°C ambient temperature, this overload relay allows a temperature rise of the tripping element of 60°C before the overload relay trips. The temperature rise of the tripping element is approximately proportional to the heat produced in the heater. This heat is proportional to I^2. It is apparent that I_{30}^2 is 60/50 or 1.2 I_{40}^2. (I_{30} is the maximum tripping current of the overload relay at 30°C ambient temperature.) Since I_{40} is 1.25 FLMC, I_{30} is $\sqrt{1.2} \times 1.25$ FLMC, or 1.36 FLMC.

If the selected wire is 75°C, it may actually operate at a temperature of

$$\left(\frac{1.36}{1.25}\right)^2 \times 45 + 30 = 88°C$$

This slight overheating is accepted practice. If the ambient temperature for the wire is higher than 30°C, a correction factor for the permitted maximum current (compare

Article 310 of the NEC) is included in the Code; for example, for 40°C ambient, this factor is equal to 0.82 for 60°C wire and equal to 0.88 for 75°C wire.

The branch-circuit protective device must be selected differently if the motor is exposed to specific short-time, intermittent, or periodic duty. The duty cycle is shown on the motor's nameplate. The size of the branch-circuit device is then determined by NEC 430-22.

Branch-circuit running-overcurrent protection is, under certain code-specified conditions, provided by the overload relays. Under these circumstances, the overload relays must provide protection in each phase as required since 1971. Earlier installations may therefore be incorrect. Conductors may be protected by fuses or nonadjustable thermal-magnetic circuit breakers according to the ampacity of the Code. If no standardized devices exist for the exact value, the next higher standard fuse or standard nonadjustable thermal-magnetic circuit breaker may be selected (NEC 430-34).

The standard values are given in the Code. For motor circuits, a dual-element fuse should have a rating not higher than 175% FLMC, and a one-time fuse not higher than 300% FLMC. If the fuse or circuit breaker selected causes nuisance interruption under normal inrush current, its rating may be increased to values specified in the Code.

The maximum permitted value for dual-element fuses may be increased in case of nuisance blowing to 225% FLMC and for one-time fuses to 400% FLMC. For motors having low inrush currents, these values should be decreased. The NEC assumes that fuses or circuit breakers chosen to these values in combination with overload relays protect the branch circuit. In reality, the branch-circuit protection is, in some respects, now provided by the overload relay. If a fuse or thermal-magnetic circuit breaker has a rating of 225, 250, or even 400% FLMC, the branch circuit, with the exception of very small motors, is not properly protected.

An instantaneous-trip-type adjustable circuit breaker or a motor short-circuit protector (MSCP) disconnects the motor branch circuit at a value of less than 700% FLMC. This value can be increased to 1300% FLMC according to the Code, Article 430-52. UL requires in their newest standards (especially for combination starters with instantaneous-trip-type circuit breakers) that starters must be self-protected up to ten times FLMC. Since 1968 the NEC has required that an adjustable magnetic circuit breaker be combined with three overload relays to protect the wire against overheating due to small overloads in each phase. The same condition is established for MSCP since the 1971 NEC. The motor short-circuit protector or instantaneous-trip-type circuit breaker provides short-circuit and ground-fault protection only.

At the present time, UL demands that a starter protected by the largest permissible fuse may not be damaged when subjected to a 5000-A short-circuit current at 600 V. The only permitted damage at this short-circuit current would be welding (but not disintegration) of the starter contacts. Heater burnout would not be permitted. The fault must be cleared. UL may allow heater burnout if starters protected by an instantaneous adjustable trip circuit breaker are subjected to a fault of 5000 A available fault current. The combination starter, however, must be marked to show that in the event of heater burnout the overload relay must be replaced. The reason for this precaution is that an overload relay may lose its calibration after a heater burnout. Larger starters, 50 hp and up, are additionally exposed to a 10 000-A short-circuit current. Motor controllers having a rating in excess of 200 hp must withstand a short-circuit current of 18 000 A up to 400 hp; 30 000 A up to 600 hp; 42 000 A up to 900 hp; and 85 000 A up to 1600 hp for 3 cycles. If tested with more than 5000 A available short-circuit current, some damage is permitted (UL Standards 508). Under 3.75-A FLMC, a heater in an overload relay may burn out if the protective fuse is the smallest code fuse (15 A), although the burnout of the heater must not change the calibration of the overload relay. This requirement may be dropped, but the overload relay must be then replaced after heater burnout. Small controllers (1 hp and below at maximum 240 V) are tested at a 1000-A available short-circuit current at maximum rated voltage.

The increased short-circuit capacity in many networks requires that new standards be developed for the protection of controllers and motor branch circuits at higher available short-circuit currents. These developments are not yet reflected in the NEC. UL has developed performance and test standards. NEMA has standardized some performance criteria at higher short-circuit currents in the "Standards for Industrial Control and Systems," Part ICS 2-322, "AC General Purpose Motor Control Centers." After a short-circuit current is interrupted by the branch-circuit protective device, dielectric tests must be passed on the line side of the disconnect device. But a complete replacement of the motor control-center unit would not constitute a failure as long as the door of the unit which was subjected to the fault in the motor control center remains closed and no fire hazard outside the unit occurs. According to UL Standard, a high fault (over 5000 or 10 000 A) must not blow open the controller door. It is not necessary to place cotton outside the enclosure during the withstandability test. These standards do not reflect all the presently available protection possibilities. The user may request more than the minimum standardized conditions. Nearly complete branch-circuit protection is available.

Since the overload relay is part of the controller, UL has established safety standards for this device in UL Standards 508. UL tests the controller at six times FLMC. NEMA, and in some UL Standards, requires some additional tests at ten times FLMC. The controller must function properly after these tests in addition to the short-circuit tests mentioned above.

The NEC 1968 and before had different requirements about the number of necessary overload relays per circuit. Generally two thermal-overload relays were permitted for 3-phase motors. The only exceptions were where the overload relay is the only branch-circuit protective device for running overloads (the adjustable-trip-type instantaneous circuit breaker is a protective device for short-circuit currents and ground faults and must be combined with three thermal-overload relays) or where the motor is in an inaccessible location. The Codes, 1971 and following, change the requirements and request the usage of three overload relays for the protection of 3-phase motors without exception. Two overload relays are required for 2-phase application and one overload relay for single-phase and dc application.

At the present time, under certain conditions, the Code permits group motor protection, i.e., that two, three, or four sets of overload relays are protected by one fusible switch or circuit breaker. A number of conditions are specified for group motor protection. The most important one is that branch-circuit protective devices should have a rating not larger than the one determined for the largest motor plus the FLMC for all other motors. The approval for all group installation can be obtained even when the controller is completely destroyed if destruction does not create a fire hazard. For test purposes, cotton is placed around all openings of the enclosure. The cotton may not burn.

OVERLOAD RELAYS AND THEIR CHARACTERISTICS

5. Thermal-Overload-Relay Features. Overload relays are standardized by NEMA since 1974. The time-current characteristic (tripping time vs. current) at 40°C ambient shall be given in loglog scale, the tripping time in seconds, the current in multiples of ultimate trip current. These curves shall not be extended beyond the limit of self-protection. The limit of self-protection is the maximum current value that the relay can respond to without sustaining damage that will impair its function (tripping). These relays must be self-protected at least up to ten times FLMC. Two classes are standardized: Class 20 must trip within 20 s or less at 600% of its current rating (ultimate trip current); Class 30 must trip within 30 s or less at 600% of the current rating: A Class 10 (tripping within 10 s or less at 600% of its current rating) is suggested as a standard for future design. The tripping time at 200% of its current rating is 8 min or less at 40°C ambient temperature. The most important protective devices are the thermal-overload relays, which protect motors against overheating due to running overloads. These devices are current-sensitive. A thermal element is heated either directly or by heat transfer by the same current which flows through the motor. The relay trips and disconnects the motor from the line when this thermal element reaches a predetermined tripping temperature, which happens if the overload prevails for a dangerous time. The relay should not trip while the FLMC flows through the heating element. The relay has an inverse time characteristic, which means that the higher the overload is, the faster the relay trips.

Current-responsive elements are installed either in two (still in use in the United States) or in three (required in new installations since 1971) phases of a 3-phase circuit. In case of tripping, the relay causes interruption of the coil circuit of a magnetic starter. Smaller motors, up to 10 hp, may also be protected by manual starters in which the thermal-responsive elements, in case of tripping, unlatch a tripper bar. The manual starter mechanism then opens the circuit. A handle or push button may be used to reset the latching mechanism. After resetting, the manual starter can be reclosed. The manual starter, like the thermal-overload relay for a magnetic starter, is trip-free.

Manual starters also are combined with fuses or MSCPs in one housing. This combination has been available since 1974. This combination is used in applications similar to those of combination controllers when remote control or a large number of operations (more than 40 000 to 80 000) are not required. The combination has all the necessary door interlocks (the door may not be opened when the switch is in the on position and the switch may not be closed when the door is open but both safety requirements may be defeated by authorized personnel); only one set of main contacts is used. The short-circuit protective device is placed between the starter contacts and the overload relay. The characteristics of various fuses and the MSCPs are covered in articles 6 to 11 in this section.

A trip-free manual starter or circuit breaker cannot be held closed with the push button or operating lever when the overload relay trips because the overload device is free to trip the starter or breaker. Its tripping is independent of the starter push button or circuit-breaker operating handle.

The most common thermal-responsive relays are of the bimetallic or eutectic-solder types. A bimetal consists of two metals which are usually bonded together. The two metals have different expansion coefficients. Under the influence of heat, the bimetal moves in the direction of the metal with the lower expansion coefficient. Usually one end of the bimetal is fixed and the other can move freely. It is also possible to restrain the movement of the bimetal and to create internal forces due to the heat. When the temperature of the bimetal reaches a predetermined value, it causes a spring action which opens or closes a relay contact or releases the trip latch.

The most common relays in the United States have heater elements of resistive material in series with the main load current which warms the heat-responsive elements by heat conduction, radiation, and convection. In some cases, predominantly in Europe, load current flows directly through the bimetal or through the bimetal and heating element which is wound around the insulated bimetal. This heat warms the bimetal mainly by heat conduction. A typical American relay is shown in Fig. 2. As soon as the bimetallic element (1) moves a certain distance, because of heat, an over-the-center spring (2) opens, by snap action, a normally closed contact (3). In a magnetic starter this contact is connected in series with the coil circuit. The coil circuit is interrupted, and therefore, the main contacts of the starter open. The heater (4) warms the bimetallic element (1). At low currents heat transfer occurs mainly by heat convection and conduction. At higher currents, more than three times FLMC, heat is transferred predominantly by radiation.

The heater is mounted in mounting straps (5 and 6). The over-the-center mechanism is actuated by a tripper bar (7) which depends on the movement of the bimetal. The contact can be reset by the reset slide (8). The reset slide can move the contact back into the normally closed position when the bimetal is no longer overheated. With an adjustment knob (9), the position of the bimetal in relationship to the heater can be changed slightly. By this positioning, the tripping current can be changed to a certain degree (between 85 and 115% of its rated current). The heaters are formed differently for various motor currents and usually change in current steps of about 10%; for example, one particular heater can be used from 10 to 11 A FLMC. The normally closed contact can lead either to terminals or to male parts which plug into female parts of a plug-in base (Fig. 2). The female parts for all relays are then connected in series to the starter coil (Fig. 3).

If the reset slide is held down, the normally closed contact should open when the bimetal is overheated. This action makes a relay trip-free. If the reset slide is pushed when the normally closed contact is closed and the contact remains closed, the relay is called tamper-proof. If the reset slide can be held in an in-between position, between fully on and off, and the normally closed contact opens only if the bimetal is

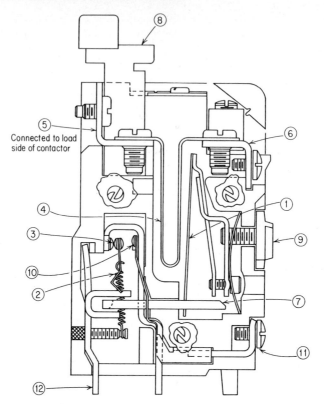

Fig. 2 Thermal-overload relay studs (12) plug into plug-in base of Fig. 3. (*I-T-E Imperial.*)

overheated, and returns automatically to the on position when the bimetal cools, then the relay is said to have an automatic-reset position.

The automatic-reset position is not permitted for a two-wire control circuit. If in such a case the contact is open, the starter deenergizes automatically owing to the overload, but as soon as the relay is cool again, the starter closes automatically. This is called "pumping." If the starter has 3-wire control, the automatic-reset position has some advantages, especially if the relay is not easily accessible because the remote on button must be pushed before the starter can be closed. Three-wire control and two-wire control are shown in Fig. 4.

Some maintenance engineers do not wish to have a relay with automatic-reset position because they want to force the electrician to reset the relay manually in the hope that he will check the cause of the tripping. Some plant engineers prefer to use a reset slide with an emergency stop for the starter, in which case they use a reset slide where the normally closed contact opens every time the reset slide is pushed completely down. After releasing the slide, the normally closed contact closes again in the mid-position of the slide. In the relay shown in Fig. 2, part 12 can be discarded to eliminate the automatic-reset position. The relay in Fig. 2 further shows a normally open indicating contact 10, which can be provided. Terminal 11 is connected to a lamp or other indicating load to indicate that the relay has tripped when contact 10 closes. The indicating contact is optional.

It is also possible to build three overload relays in one block. A bimetal in each phase pushes a crossbar under overload condition. If the crossbar moves far enough, it opens the normally closed contact by pushing the contact spring over the center. This arrangement saves the electrical connection between the relay contacts; only one control contact is used for all phases.

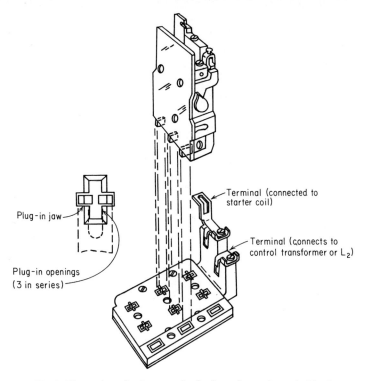

Fig. 3 Plug-in base for three overload relays of type shown in Fig. 2.

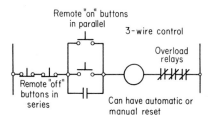

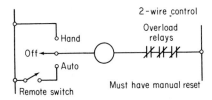

Fig. 4 Two- and three-wire controls.

A eutectic-solder relay (Fig. 5) has an alloy which melts between 80 and 150°C. Usually in this device a small ratchet is under torsion force of a spring, but kept from turning by the solidified eutectic metal. This eutectic metal melts under the influence of a heater if the overcurrent flows too long. The heat transfer occurs by conduction, convection, and radiation to the eutectic metal. When the eutectic metal melts, the ratchet with the shaft turns under the influence of the spring. At the same time, the contact lever, under the influence of the contact spring, opens the normally closed contact of the relay. The normally closed contact can be closed in a similar way as the bimetallic relay with the exception that the automatic-reset position for the reset slide is not possible. The relay may be made tamper-proof or with emergency-stop action depending on how the reset slide is developed. A slight adjustment for various ambient temperatures as indicated for the bimetallic relay cannot be obtained easily. But one manufacturer (Cutler-Hammer in Milwaukee) developed a relay in which a small adjustment of the tripping current is possible by allowing a different position of the heater in relationship to the eutectic solder. Also, 3-phase solder-pot relays are built with one control contact.

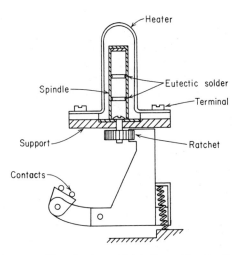

Fig. 5 Eutectic-solder relay (principle similar to Allen-Bradley relays).

6. Overload-Relay Application and Characteristics. The most common motor, the squirrel-cage motor, was developed for continuous duty. If such a motor is overloaded, its temperature increases approximately proportionally to KI^2 if the speed of the motor does not change. The time necessary to produce a dangerous winding temperature decreases proportionally to I^2. Actually the time decreases somewhat more because the time necessary to transfer heat is shorter when the current increases. If the motor ventilation fails or if the rotor is locked, a dangerous temperature rise is reached much faster than when the motor is running. With a failing ventilation system, a motor can be protected only by a device which senses the motor temperature or its rise. Such a device is an inherent motor protector. If the motor is overloaded, it may run slowly. In such a case a dangerous temperature is reached in a shorter time. This problem may occur during the starting period. The time-current characteristic of an induction motor is dependent on its design. Normal motors with Class A insulation can withstand a locked-rotor current for a longer time than motors with Class B insulation; and those with Class B insulation for a longer time than hermetically sealed motors. A small motor has a lower time constant; therefore, the temperature increases in a smaller motor at a higher rate for the same current. Only three maximum points of the tripping characteristics are standardized for thermal-overload relays and only for Class A and Class B insulated motors. This range is insufficient for many applications.

Figure 6 indicates various tripping-characteristic ranges. The maximum-trip-time curve of relays having a characteristic standard trip 1 may be sufficient for 40°C motors having Class A insulation. The maximum-trip-time curve standard trip 2 may be sufficient for T-frame motors. Quick-trip and ultra-quick-trip heaters have still faster time-overcurrent characteristics. The choice shown in Fig. 6 might be controversial, but the reader has a fair idea about the required time-current characteristics of thermal-overload relays. The characteristic for running-overcurrent protection of induction motors is necessary only up to the locked-rotor current. A higher current can flow through the motor only under exceptional cases—either for a very brief period on plug reversing or jogging, or when a two-speed motor is switched from the higher speed to the lower speed without any interval; but those periods are extremely short. A higher

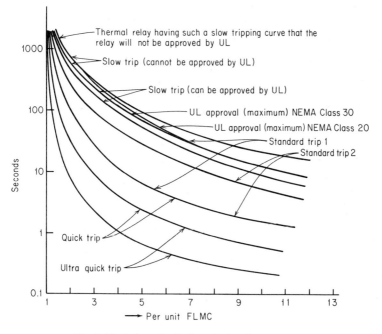

Fig. 6 Typical overload-relay tripping characteristic.

current can flow through the motor for a longer time only if the motor is already damaged.

The overload relay should respond faster to higher than the locked-rotor current, since overload relays often protect the branch circuit. If the motor has been damaged, for example, if some stator windings are shorted, a higher current than locked-rotor current flows. It still would be advisable to remove the motor from the source as quickly as possible to restrict the damage.

It is generally assumed that a heater withstands short-circuit current better if its mass is greater; however, its tripping time will increase with its mass. Heaters which protect both Class A and B motors must have a smaller mass than heaters which protect Class A motors only. Their withstandability (I^2t) may therefore be lower under short-circuit condition. The formula for the temperature rise of a heater is complex and depends on the time the current flows through it and on the transfer of the heat to the bimetal or eutectic allow. To understand the principle, it is sufficient to assume that the temperature of the heater increases according to the formula

$$\theta = \theta_{max}(1 - e^{-t/T})$$
$$T = \frac{Cw}{\alpha A} \qquad \theta_{max} = \frac{I^2R}{\alpha A}$$

where A = dissipating surface, in^2
$\quad \alpha$ = heat-dissipating constant, W/in^2/°C
$\quad R$ = resistance of the heater, Ω
$\quad T$ = time constant, s
$\quad w$ = weight, g
$\quad C$ = specific heat of the heater material, Ws/°C/g
$\quad I$ = rms current, A
$\quad \theta$ = temperature increase, °C
$\quad \theta_{max}$ = maximum temperature increase, °C
$\quad t$ = time, s

This formula is an oversimplification. The thermal end effects at the heater terminals are neglected. The formula should not be used for actual calculations but is included to show some of the main factors which determine the temperature change and the

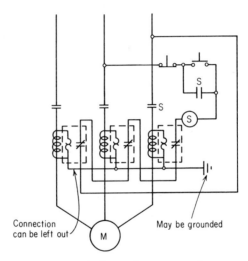

Fig. 7 Overload relays with current transformers.

dependence of the time constant on the mass of the heater. For more extensive analysis see Ref. 6.

Larger relays, for size 5 starters and larger, use a current transformer feeding a small standard overload relay (Fig. 7). Without a current transformer, the resistance of a heater for these currents would be so low that the resistance would be influenced by the torque with which the heater is attached to its base. Such current transformers must have enough iron so that they transfer the current proportionally at least up to locked-rotor current. If the overload relay is also used for branch-circuit protection, it should produce a current proportional to the line current up to the current at which another protective device takes over the protection of the branch circuit.

Hermetically sealed motors must be protected by quick-trip heaters. Quick-trip heaters may be developed in one of the following ways:

1. By reducing the mass. The reduction of the mass has its limitations, because if the heater mass becomes too small, the heater cannot withstand the minimum standard short-circuit currents (Art. 3).

2. By placing the heater very close to the heat-sensitive element. This method also has limitations. Essentially, as a heater is placed very close to the heat-sensitive ele-

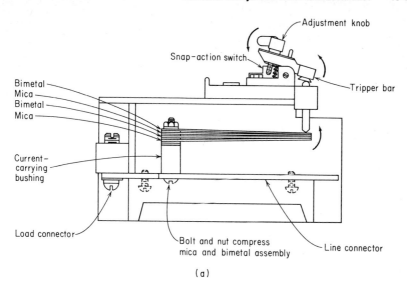

-Adjustment knob

Snap-action switch

-Tripper bar

Bimetal
Mica
Bimetal
Mica

Current-
carrying
bushing

Load connector

Bolt and nut compress
mica and bimetal assembly

Line connector

(a)

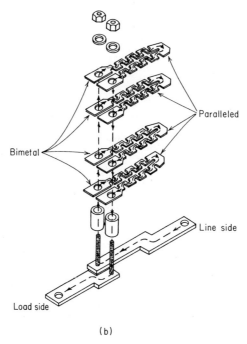

Paralleled

Bimetal

Line side

Load side

(b)

Fig. 8 (*a*) Directly heated bimetal relay—mica spacers used for series connectors, metal used for parallel connectors. (*b*) Assembly of bimetal actuator for relay in *a*. (*I-T-E Imperial.*)

ment, the heat is transferred through the air by heat conduction. The air forms a small, thin film (about 0.07 to 0.18 in thick) and heat is conducted through this film in the same manner as through a solid body. The heat transfer now becomes very dependent on tolerances. If the heater is combined with a heat-sensitive element (eutectic solder) in such a way that they form an integral unit and are interchanged together for various full-load motor currents, their tolerances may be controlled more precisely because the heat can be conducted to the eutectic solder through a better heat conductor than air.

3. By directly heated or partly directly heated bimetals. While some such relays are used in the United States, they are predominantly used in Europe. They use directly heated bimetals or a combination of heat transfer by heat conduction through an insulator to the directly heated bimetal through which the current also flows (Fig. 8). If the heat-sensitive element is a bimetal, a different arrangement for different FLMC or for small ranges of FLMC must be selected. A different relay must therefore be available for each range. The disadvantage is that a large number of complete overload relays must be stocked by a distributor. The accuracy of an adjustable directly heated bimetal is not easily obtainable, since a small adjustable knob cannot easily be read

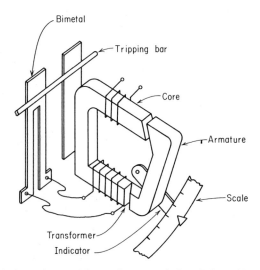

Fig. 9 Principle of adjustable saturable transformers with directly heated bimetal as burden. In other designs heater may be the burden.

exactly. Another disadvantage is that the tripping temperature of the heat-sensitive element varies with its position because it has to move a different distance depending on the position of the adjustment knob. A larger movement of the bimetal can be obtained only with a higher tripping temperature.

4. Another solution for obtaining quick-trip action is to use a heater made of a highly resistive bimetal or a bimetal insulated from a Nichrome wire. This bimetallic heater heats another bimetallic tripping element in the same way as a standard heater on FLMC. On small overcurrents, the bimetallic heater approaches the bimetallic tripping element; thus the heat transfer improves and the tripping time decreases. Under locked-rotor conditions, the heater itself acts as a tripping element and trips the relay directly. Since at ultimate trip current the heater does not trip the overload relay, the relay with such a quick-trip heater can be calibrated like any standard device.

A number of methods are also available to obtain slow tripping:

1. By increasing the mass of the heater. This method has its limitations if the device must have UL approval.

2. By using a saturated transformer. If the current increases (to locked-rotor current, for example) over a certain value, the transformer magnet becomes saturated

(Fig. 9); thus the current on the secondary side of the transformer will not proportionally increase with the primary current of the transformer. This characteristic may even occur on currents higher than two or three times FLMC. On the secondary side of the transformer, the current would be proportionally lower; therefore, the tripping time would be longer than for a standard relay. If a current transformer is used for standard applications as mentioned above; for example, for size 5 starters and larger, the transformer must not saturate at locked-rotor current. For slow-tripping characteristics, the transformer should partly saturate at this current. Such transformers may also be built with changing air gaps, thus obtaining a range of time-current characteristics of the secondary overload relay.

3. By using an Inducto-Therm relay (Art. 25). Wound-rotor motors are usually protected by placing the overload relays in the stator circuit. The characteristic of the relay is different from standard squirrel-cage-motor protectors since the starting current is lower, usually not more than $2\frac{1}{2}$ times FLMC. A standard wound-rotor motor cannot withstand locked-rotor current for any length of time without a starting resistor. Therefore, special magnetic relays must be used to protect the motor if the motor is stalled without the starting resistor.

If motors have a lower code letter, the standard inrush current is lower than six times FLMC, and special relays which trip faster are required. If a motor is designed for reduced-voltage starting and therefore can withstand the locked-rotor current for only an extremely short period of time, precautions must be taken. In such a case, it might be advisable to use an electromagnetic instantaneous-trip-type relay.

MOTOR-BRANCH-CIRCUIT PROTECTION

7. **Protection with One-Time Fuses and Thermal-Overload Relays.** The thermal-overload relay can be combined with various devices for protection of the motor branch circuit. Each has its advantages and disadvantages; the simplest device is the one-time fuse.

A one-time fuse which may have an ampere rating up to 400% of the FLMC may be selected. Since a standard fuse must ultimately blow at a current of 135% (UL Standards 198, "Fuses") of its ampere rating, the branch circuit is not protected with a one-time fuse against low current faults and running overloads. A branch-circuit conductor is usually selected for an amperage of 125% of the FLMC. Only a fuse sized for the amperage of the conductor gives full branch-service protection. If the fuse is selected for a higher amperage, as is here the case, the overload relay functions practically as the branch-circuit protective device at low overcurrents. The coordination of standard heaters with one-time fuses may be based on 300 to 400% FLMC (Fig. 10).

Time-current characteristics are usually expressed on a loglog scale (Tripping-response blowing time of the overcurrent devices vs. FLMC). The coordination curves of Fig. 10 and following are, however, drawn on a semilog scale because the coordination at low fault current between various protective elements can be more easily evaluated.

When a one-time fuse in combination with overload relays is used to protect the motor circuit, however, it is advisable to have overload protection in each conductor, since low-current ground faults could not otherwise be detected if the ground fault occurs between the ground and the phase not protected by the overload relay (Art. 16). This statement is valid despite the fact that before 1971 the usage of two overload relays was allowed for 3-phase systems in combination with one-time fuses. The previous NEC allowed this protection when two overload relays were sufficient for the protection of the motor, but the protection of the motor branch circuit must also be considered.

Another important requisite not part of any test requirements is that the overload relay should be self-protecting up to the crossover point between the time-current characteristics of the overload relay and the one-time fuse, and that a starter may be able to interrupt any current below that point. At least the permitted damage level should be standardized up to this crossover point.

As stated previously, UL presently tests overload relays at six times tripping current with each heater and at a 5000-A available short-circuit current at maximum rated

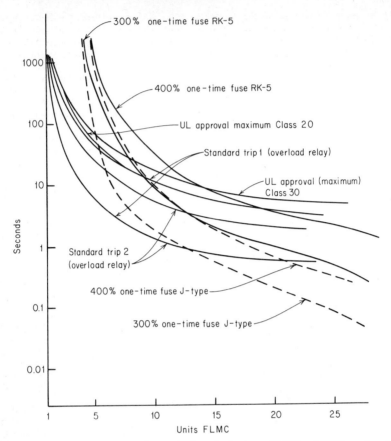

Fig. 10 Time-current characteristics for standard overload relays with 300 and 400% one-time RK-5 and J-type fuses, FLMC = 10 A (fuse mean curves plus or minus 10% accuracy).

voltage and 70 to 80% PF. Standards do not cover the permitted maximum damage level between six times FLMC and a 5000-A available short-circuit current. UL requests an additional test for control equipment over 50 hp, and another test for motor controllers for motors over 200 hp. A test at a 10 000-A available short-circuit current at maximum rated voltage with 70 to 80% PF is to be conducted. The test unit, for example, the combination starter, may be severely damaged and still be acceptable if the fuse between the enclosure and the live pole does not rupture and no obvious fire hazard occurs during this test.

Since 1971–1972, UL also lists combination starters withstandable to higher faults than 10 000 A. The permitted damage level is high. Heater burnout is allowed. Since, after heater burnout, the overload relay may change the calibration, it is required that any approved device must be marked with the requirement that the replacement of the overload relay is necessary after heater burnout.

The IEC Standards recognize three damage levels:

Type A. Any kind of damage to the starter is allowed if the enclosure remains externally undamaged.

Type B. The overload-relay characteristic may have changed permanently. Welding of contacts is permitted.

Type C. Only welding of contacts is permitted.

Requirements beyond Type C are not standardized; the manufacturer must have specifically agreed upon such standards.

At present the installer is not obliged to use a smaller fuse than 400% FLMC or at least the next smaller standard-sized fuse unless the controller is specially marked 300% or the next smaller-sized fuse. No standardized fuses are available smaller than 15 A. The control manufacturer either recommends that a user install 15-A rated one-time fuses as the short-circuit protective device for FLMC of 3.75 A and less or requests the use of smaller-sized nonstandardized fuses rated at lower currents. These fuses should not have a rating less than 400% FLMC.

If the manufacturer recommends the protection of his equipment with such a small fuse, he must indicate on the heater table the size of the fuse selected and he must certify that the overload relay was not damaged at a 5000-A available short-circuit current tested at maximum rated voltage. If a manufacturer recommends the use of standard 15-A one-time fuses for motors rated smaller than a 3.75-A FLMC, the heater may burn out when tested at 5000 A at maximum rated voltage if the relay does not change its calibration after the short-circuit test has been performed and no fire hazard occurs. In the future, UL may require the replacement of the overload relay after heater burnout.

One-time fuses include H, RK-5, RK-1, RK-9, J, or T-type fuses. T fuses have the same characteristics as J fuses but are considerably smaller. ·In the future, UL requires the usage of RK fuses instead of K fuses. For available short-circuit currents over 10 000 A, RK-5 fuses have the same characteristics as K-5 fuses, but H fuses cannot be placed in the same holders as RK fuses. RK fuses can, however, be placed in H fuse holders. RK-9 and RK-1 fuses are no longer common. They also could not be interchanged with H fuses. H-type fuses are not tested with short-circuit currents higher than 10 000 A and therefore must not be used for more than 10 000 A available short-circuit current. The question of what may happen if a controller protected by one-time fuses is tested at higher faults than 5000 A is dependent on the selection of the specific protective device. Generally it is recognized that some damage to the control equipment is permitted at higher fault currents. However, means to prevent this damage are now available. It is necessary to indicate on the heater table the maximum permitted fuse size and type (J, RK). Different values of withstandabilities are standardized. Generally heaters with known resistance R have a withstandability under short-circuit conditions, which can be expressed either in I^2t or in I^2Rt. The I^2t withstandability of the heater increases with its mass, and the permitted I^2t value decreases proportionally to the increase of the resistance. A heater must develop a minimum watt loss at ultimate trip current. The maximum permitted watt loss is determined by the overload-relay design and the various standards. At FLMC, the permitted maximum temperature rise at the terminals of a controller may not be higher than 50°C. The ultimate trip current may not be higher than 125% FLMC for standard heaters for motors with a service factor of 1.15 or larger. As a general rule, it is assumed that at ultimate trip current each heater may produce the following heat energy in watts depending on the construction of the relay:

Starter size	Energy, W, produced in the heater
1	2–4
2	2½–5
3	3½–7
4	4–8

Larger starters usually use current transformers with overload relays on the secondary side of the current transformer.

The watt loss at FLMC may be 64% or more of the watt loss at ultimate trip current if the ultimate trip current is equal to 1.25 FLMC and $0.64 I^2 = (1/1.25)^2 I^2$. The resistance of the heater may be approximately 1.94 Ω at a 1-A FLMC for a relay using a heater with 1.25-A ultimate trip current if the relay must produce 3 W at ultimate trip current. If such a heater is placed in each phase, the resistance of this heater alone in case of short-circuit current of infinite available energy restricts the let-through current at 600 V to

$$\frac{600}{1.94 \times \sqrt{3}} = 180 \text{ A}$$

We must consider that this restriction of the let-through energy is not as remarkable for heaters for a higher FLMC. For a motor with a 10-A FLMC, a restriction of the let-through current by the heater alone would be 18 000 A at infinite available energy and for a 100-A FLMC, even considering that the permitted watt loss may be twice as high (6 W at tripping current), the let-through current at infinite available energy would be 90 000 A. The withstandability of a properly designed heater at various short-circuit currents can be calculated by the following formula:

$$\theta^1 = \frac{I_L^2 t R}{wC}$$

$$RI_L^2 t = Q^1 C w$$

where w = weight, g, of resistance material of the heater
$I_L^2 t$ = let-through energy of the heater in A^2s
C = specific heat, Ws/g/°C of resistive material
R = resistance of the heater, Ω
t = time, s
θ^1 = maximum permitted temperature rise of the heater material

A properly designed heater has its weakest spot in the resistance material of the heater. The formula assumes that all heat during a short circuit is used to increase the temperature of the heater.

R and C are assumed constant over a wide range for most heater materials. Nichrome is such a material.

The permitted let-through energy in the heater is independent of the available short-circuit current. The calculation gives an indication of the type of protective devices required if a heater is not to be damaged owing to short-circuit current. Generally, for small FLMC the maximum let-through current through a starter is mainly dependent on the resistance of the heater. For larger FLMC, approximately 3 to 12 A at 480 or 600 V, it becomes more difficult to design a protective device, since the heater does not appreciably restrict the let-through current at higher faults, especially not at 600 V.

On still larger FLMC, the contactor part of the control equipment may not withstand the let-through energy which the heater can withstand. The contactor may become the weakest member in the circuit. If a smaller than permitted maximum J-type fuse is selected to protect the control equipment, the protection for short-circuit conditions is improved. Normally the fuse should not blow at locked-rotor current. This limitation becomes especially important when J-type fuses are selected for protection. They are silver-sand fuses and are comparatively expensive. It is generally assumed that the replacement of fuses at running overcurrents is uneconomical, not only because of the cost of the fuses but because of the time involved in looking for and installing replacement fuses.

Precaution should be taken when a combination of a fuse of an amperage rating of three to four times FLMC and overload relays are used to protect the cable. Since the overload relay trips comparatively slowly, care should be taken to assure that the wire withstands currents thermally up to the locked-rotor current or up to the crossover point between fuse and overload relay until the overload relay has tripped. Sometimes a plasticizer of smaller wire, No. 12 or 14, may soften under these conditions if exposed repeatedly or even only once to locked-rotor current during the time the overload relay takes to trip. This repeated softening causes a breakdown in the insulation of the wire.

Unfortunately, at the present time, no curves are standardized showing how long a wire can be overloaded thermally without being damaged. Wire must thermally withstand approximately six times its rating for 20 to 30 s if protected by standard heaters. Figure 11 shows the time a thermoplastic cable can be overloaded without softening. Proper cables for motor circuits should be chosen not alone by the NEC, but the installer should be sure that they do not become soft before an overload relay trips at locked-rotor current.

Assuming it might be possible that an overload relay trips at 7.5 times FLMC in 30 s (permitted by UL) or 20 s (highest UL standard permissible value for T-frame motors) it is possible to choose the proper wire size from Fig. 11, but the wire size must be chosen for 125% FLMC. For example, No. 14 wire is too small for a 15-A FLMC current and No. 12 wire for a 20-A FLMC.

Experience shows that small wires are more easily damaged if they are used for motors rated 80% of the wire ampacity. Therefore it is recommended that larger-diameter wires than those permitted by the NEC be used for such small loads. It may be necessary to make tests with individual relays and heaters if repeated locking of the rotor occurs, since the reset time of the relay may be important in determining what happens to the wire. For motors rated not larger than 10-A FLMC, it is advisable to use AWG 14 copper and AWG 12 aluminum wire. AWG 12 copper wire should be used for motors with not larger than a 13-A FLMC and No. 10 aluminum wire for motors rated not over 16.5 A.

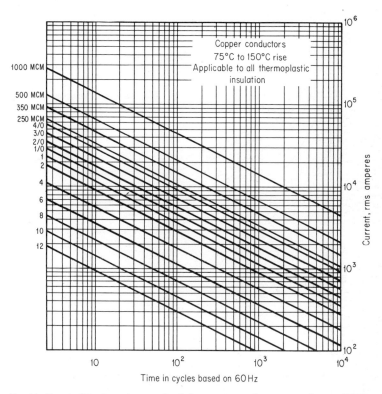

Fig. 11 Permissible short-time overload characteristics of copper conductors (75°C).

8. Branch-Circuit Protection with Dual-Element Fuses and Overload Relays. Another common way to protect branch motor circuits is through the use of a combination of time-delay fuses and overload relays. The NEC, 1968, in Tables 430-152 and 430-153 did not differentiate between the permitted maximum fuse size for one-time fuses and time-delay fuses. The NEC, 1971 and later, has changed this position and put a limit of 225% FLMC for the selection of time-delay fuses. However, higher-rated long-time fuses will still be in use. A problem that often arises is that the time vs. current characteristic of a time-delay fuse is similar to the characteristic of an overload relay in a wide current range. Both devices have manufacturing tolerances of at least 10 to 15% of the time vs. current characteristics. It is therefore very difficult to obtain a clear crossover point between the time vs. current characteristic of the overload relay and the fuse. When selecting a fuse at 150% FLMC or less, predicting if the overload relay or the fuse would cause a disconnection in case of running overcurrents is difficult, especially at locked-rotor currents (Fig. 12) (Ref. 7).

As pointed out earlier, it is undesirable to have the fuse blow in case of running over-currents because the exchange of fuses is time-consuming and expensive. If the fuse selected is larger, i.e., at 175, 200, or 225% FLMC, the two characteristics intersect at higher currents. If the crossover point is so high that it is certain that at locked-rotor currents the overload relay trips before the fuse blows, the question arises whether the overload relay protects the circuit at lower fault currents.

The combination of a starter and a fuse requires that the overload relay must be self-protecting up to the value at which the fuse characteristics intersect the overload-relay

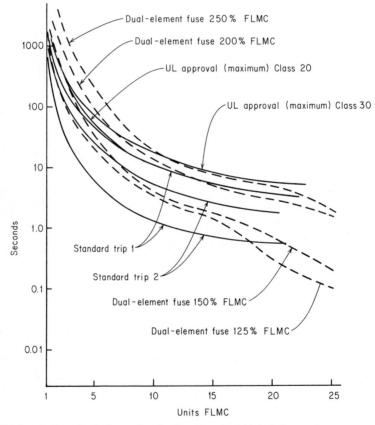

Fig. 12 Standard overload relay with 125, 150, 200, and 250% dual-element fuse RK-5, FLMC = 24 A (fuse mean curves plus or minus 10% accuracy).

characteristics. At least the damage level in this case should be standardized; however, this is not the case.

A dual-element fuse is usually an RK-5 fuse, which is not as limiting as a J-type fuse. If we assume that a 50-A motor may be protected with a 100-A dual-element RK-5 fuse, then the let-through energy, according to UL Standards for RK-5 fuses, can be 500 000 A^2s (Ref. 2). For a 200-A J-type fuse, which is the maximum permitted one-time fuse size for the same motor circuit, the let-through energy is 300 000 A^2s and a 150-A J-type fuse has a let-through energy of only approximately 150 000 A^2s (Ref. 2). Since the J-type fuse provided much better protection than dual-element-type fuses at high current faults, the one-time J-type fuse is superior.

The actual let-through energy values of current-limiting dual-element and one-time fuses may be slightly lower than the permitted maximum values by UL. However, the manufacturer has no control over what fuse the user actually selects. At the above proportion the J-type one-time fuse is, on high faults, superior to the dual-element fuse. Of course, based on 300 or 400% FLMC, the protection by a one-time RK-5 fuse would be even more inferior than the dual-element fuse at available short-circuit currents over 5000 A.

Proper wire protection can be obtained only with a dual-element fuse of 125% FLMC rating if the wire rating is 125% of the FLMC. But it may be expected that if such a low-rated dual-element fuse is selected, the fuse will blow on running overcurrents before the overload relay trips.

9. Branch-Circuit Protection with Circuit Breakers and Overload Relays. Thermal-magnetic molded-case (time-limit) circuit breakers for standard motors are selected so that they must have a rating of 250% or less of FLMC, or 400% if nuisance tripping occurs.

The crossover point between their tripping characteristic and the overload-relay tripping time-current characteristic is higher than six times FLMC. If the breaker size is selected at 125% FLMC, the circuit breaker will protect the wire, but the time-delay circuit breaker will trip too fast at locked-rotor current. Generally it is not desirable for a time-delay circuit breaker to interrupt running overcurrents when a starter is provided. Circuit breakers have contact material which may overheat after a circuit breaker has frequently interrupted the locked-rotor current. The overload relay should guard against running overcurrents.

To obtain the proper coordination of circuit breaker and overload relay, the circuit breaker must be rated between 175 and 250% FLMC. But if the rating is 250% FLMC, the crossover point between the characteristics of the overload relay and the circuit breaker is at about 18 to 27 times FLMC. A time-delay circuit breaker may protect the branch circuit at high fault currents, because at this higher current the circuit breaker interrupts magnetically. The magnetic tripping current is adjustable only on larger, more expensive thermal-magnetic molded-case breakers. It is difficult to avoid damage to the starter below the value at which the magnetic trip unit becomes effective. Even at fault currents which are sensed by the magnetic tripping device of the thermal-magnetic breaker, damage often occurs. Figure 13 shows the coordination between a starter and time-limit circuit breakers.

The time in which a thermal-magnetic molded-case circuit breaker interrupts the arc is usually between $\frac{1}{2}$ and $1\frac{1}{2}$ cycles, rarely more than 12 ms, but a shorter interrupting time than 1 to $1\frac{1}{2}$ cycles cannot be guaranteed. Larger molded case circuit breakers may interrupt in 2 or 3 cycles. These breakers, rated 600 A and higher, are rarely used in motor branch circuits. Therefore, the let-through energy to the breaker is considerably higher at short-circuit currents than if the controller is protected by a J or an RK-5 type of fuse.

Often damage, even on a 5000-A available short-circuit current, cannot be avoided with thermal-magnetic circuit breakers. At higher fault currents, the time-limit circuit breaker must be protected with current-limiting fuses. Usually fuses which protect the breaker have too high a let-through energy to give protection to the starter. Only if a certain damage level is agreed upon, combination starters with molded-case thermal-magnetic circuit breakers with the current-limiting fuses with their high let-through energy may be used for protection of starters.

The problem with most thermal-magnetic breakers is that the impedance in the breaker is too low and the limitation of the let-through energy due to the breaker is therefore not sufficient to avoid damage to the starter. An exception is a combination for small FLMC in which the heater in the overload relay restricts the let-through energy so much that the overload-relay impedance need not be supported by the impedance in the breaker to limit the let-through current. Devices for larger FLMC must, however, be checked thoroughly.

Another problem with the time-limit thermal-magnetic circuit breaker is the protection of small starters, since there are no standardized breakers rated smaller than 15 A. If the FLMC is small (1 to 2 A), the let-through current may not be high enough to trip the breaker magnetically. For starters having higher FLMC, the let-through energy at

5000-A available short-circuit current is often far too high to protect the starter. If with heaters for very small motors, the breaker does not trip magnetically at short-circuit currents, the tripping time of the overload relay becomes too long and heater burnout occurs.

If the circuit breaker is not combined in a single unit with the controller, it must be a time-limit circuit breaker because the NEC does not allow the usage of a separate adjustable magnetic-trip-type circuit breaker and a controller.

An adjustable magnetic-trip-type circuit breaker is a better protector than a time-limit circuit breaker. Such a breaker works with the overload relays of the starter. The overload relays protect the branch circuit up to the crossover point between the ad-

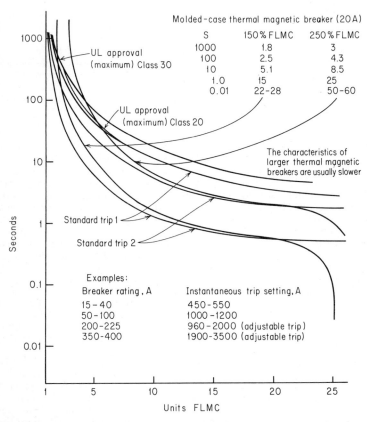

Fig. 13   Standard overload relay with 150 and 250% thermal-magnetic molded-case breaker, FLMC = 20 A (molded-case breaker plus or minus 15% mean curve shown).

justable magnetic-trip-time characteristic and the overload-relay characteristic. This point can be clearly defined and should be between 7 and 13 times FLMC. This value is a maximum range permitted by the NEC; therefore, the nominal values should be between 7 ½ and 11 times FLMC, since the overload relays (Fig. 14) and the magnetic-trip-type circuit breakers require these values because of tolerances.

If the maximum permitted crossover point is less than 13 times FLMC, UL requires that a standard controller protected by an adjustable trip-type circuit breaker must be self-protecting up to 10 times FLMC. This value may be higher for quick-trip heaters. The code requires that the combination of the instantaneous trip-type circuit breaker and controller must be approved for the purpose.

Combination starters with adjustable magnetic-trip-type circuit breakers and overload relays having current transformers on their primary side should have a crossover point of the two tripping characteristics at 6 to 9 times FLMC. Current transformers usually saturate by a current slightly higher than six times FLMC, and at 10 to 13 times FLMC, the current transformer would therefore not transform the overcurrent proportional to the overload relay, and the tripping time would be too long. UL finds it acceptable to allow damage to the controller between 10 times FLMC and crossover point between the breaker and the starter characteristic, if no hazardous condition occurs at rated 600 V or lower. The duration of the short-circuit current, as in a ther-

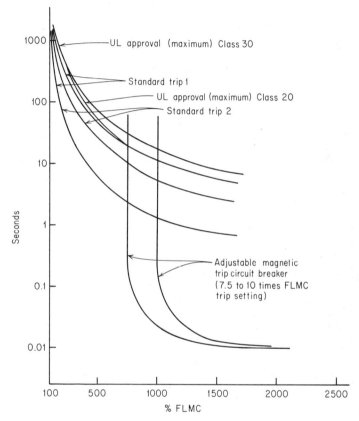

Fig. 14 Standard overload relays with adjustable magnetic-trip-type circuit breaker (trip setting between 7½ and 11 times FLMC). Breaker accuracy minus 30 plus 10%.

mal-magnetic breaker, is usually ½ to 1½ cycles. A shorter time than 1 to 1½ cycles cannot be guaranteed, since the interruption always occurs at current zero. However, in an adjustable-trip-type circuit breaker, the magnet can trip at a comparatively low current and may be built with enough impedance that the let-through breaker current through the starter at a 5000-A available short-circuit current can be restricted so that the damage is small. UL Standards allows considerable damage, especially to the starter. Above a 5000-A available short-circuit current, the damage to the controller permitted by UL may require complete replacement of the units if the door of the enclosure does not blow open and the line side of the terminals of the circuit breaker pass a dielectric test with twice rated voltage or 900 V, whichever is higher.

NEMA has set some standards for the usage of standard combination starters consisting of a circuit breaker and a starter in motor-control-center units. At high faults, severe damage of a motor-control-center unit is permitted if a dielectric test can successfully be performed with twice the rated system voltage at the line side of the breaker after a short circuit. Also damage in one compartment due to fire must not influence the performance of an adjacent unit. It might not be necessary to permit a high damage level in any case; but as already explained, it is probably safest to replace the unit if the short-circuit current has caused damage to the starter and the protective device. It would be difficult for an electrician in the field to determine what repair work must be done if damage occurs.

By combining an adjustable-type circuit breaker with a current-limiting fuse which may have characteristics similar to a J-type fuse, the interrupting range of the adjustable-type circuit breaker can be extended to higher fault levels. A current-limiting fuse which is selected to protect the breaker, however, usually does not have enough current limitation to protect the starter in any case. The combination of an adjustable-trip-type circuit breaker with a current-limiting fuse and starter at high faults will act at those high fault values somewhat like a thermal-magnetic breaker in combination with current-limiting fuses.

Large air breakers, with or without fuses, are not used for the protection of industrial-control equipment as frequently as molded-case breakers. They have the same advantages and disadvantages as a molded-case-type breaker, but they usually have long-time magnetic- and instantaneous-trip devices instead of a thermal-tripping device. An air breaker has a characteristic similar to a thermal-magnetic molded-case breaker. A disadvantage of large air breakers in comparison with a molded-case breaker may be that the duration of the short circuit is often not limited to $1/2$ to $1\frac{1}{2}$ cycles; therefore, under fault conditions, the let-through energy is higher. Large air breakers are more expensive than molded-case breakers. They may be used in special applications when a tripping characteristic is required that cannot be obtained with molded-case breakers or with draw-out switchgear applications. They are more frequently used at voltages above 600 V, but in recent years molded-case circuit breakers and starters of standard-type design with some adaptations have been developed for voltages up to 820 and 1000 V for the petroleum and mining industry. The required short-circuit capacity in these applications is usually comparatively low.

10. Branch-Circuit Protection with Motor Short-Circuit Protectors and Overload Relays. Based on the fact that a properly designed heater can withstand a fixed number of watt-seconds times the weight of its resistance material without being damaged, the maximum let-through energy which an overload relay can withstand without damage can be determined for a motor short-circuit protector (MSCP).

The contactor part of a starter can also withstand without damage certain maximum values of I^2t and I_p (peak let-through current) which can be determined by tests. It is therefore possible to find parameters for protectors which can completely protect starters at the highest available short-circuit currents. Such a protector is a fuselike device which has an extremely steep time vs. current characteristic steeper than those found in most J-type fuses (Fig. 15a). If a protector is placed in each phase and overload relays are provided for the motor-running-overload protection and for wire protection at low overloads, full protection can be obtained.

The MSCP must cut the overload-relay characteristics beyond the locked-rotor current, but below the maximum current which the overload relay and the contactor can withstand without being damaged. The NEC, since 1971, allows the use of such a combination if the whole system consisting of switch, MSCP, and starter with three overload relays is approved for the purpose. One condition is that the crossover point between the overload-relay characteristic and the MSCP characteristic is not more than 13 times FLMC. This maximum permitted value should be used only if the starter is completely self-protected up to this point; otherwise, the crossover point must be selected at a lower current because the purpose of the MSCP is to protect the controller and the wire under any condition. A lower crossover point probably must be selected in each case where a hermetically sealed motor with quick-trip relays is protected by an MSCP.

For normal motors, this crossover point will usually be between 7 and 11 times FLMC. It must not be forgotten that NEMA requires that starters interrupt 10 times

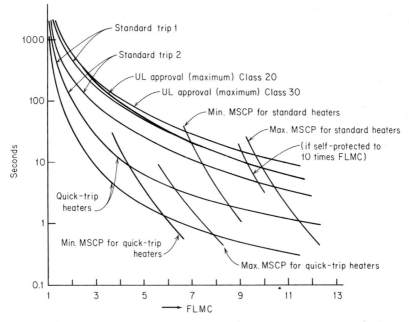

Fig. 15 Overload relay with motor short-circuit protector (MSCP) with plus or minus 10% accuracy.

FLMC. The value of the crossover point can be as low as 4 times FLMC if the protector is combined with a starter having three quick-trip overload relays. The protector must always carry overcurrents for the time it is necessary for the overload relay to trip. The average I^2t and I_p values for MSCP ratings under short-circuit condition depending on FLMC for a standard and quick-trip overload relay are shown in Tables 7 and 8. The required minimum values must be determined by test, and for some starters they may be slightly higher.

UL has standardized maximum permitted I^2t and I_p values and two points of the characteristics of MSCP. These values were selected based on the test results of two major manufacturers in the United States. These test results did not show any damage to the controller. A combination starter with MSCP can be listed even if the device can be damaged under fault conditions (pitting of contacts would not be regarded as

TABLE 7. Approximate I^2t and I_p Values at 600 V and 100 000 A Available Short-Circuit Current for Various MSCP Ratings Used with Standard Motors in Combination with Switches

FLMC, A	I^2t, A²s	I_p, A
10	3 000	3 200
20	6 000	4 000
30	10 000	4 800
40	13 000	6 300
50	25 000	7 900
70	54 000	11 600
100	140 000	15 100
130	204 000	19 100
175	268 000	21 000

TABLE 8. Approximate I^2t and I_p Values at 600 V and 100 000 A Available Short-Circuit Current for Various MSCP Ratings Used with Motors Being Protected by Quick-Trip Heaters

FLMC, A	I^2t, A²s	I_p, A
5	460	1450
10	1 800	2700
20	2 900	2900
30	4 900	3700
50	10 300	4800
70	13 000	6300
100	24 000	7900

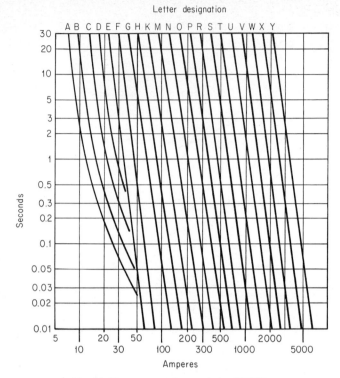

Fig. 16 Time-current characteristic of MSCPs.

damage). If any damage is permitted, the standardized values for the MSCP might still be too high for other than the presently approved equipment, and the manufacturer who has a starter of lower withstandability than the tested ones must select MSCPs which permit less let-through energy than the standardized values.

Overload relays are necessary for 3-phase systems for a combination starter with MSCP because the overload relays provide the branch-circuit-overcurrent protection. Such a combination will protect the control equipment from any running overload up to 100 000-A available short-circuit currents at rated voltage. It is required that the protector not be interchanged with any standard fuse.

The MSCP characteristics are shown in Fig. 16. They are straight lines in equal spacings in loglog scale between 0.1 and 30 s. Only the four smallest MSCP characteristics are not straight lines, because it is not necessary to make these small MSCP devices of the same costly construction as the larger ones. Each MSCP has a well-defined tolerance of ±10%. It protects a number of heater elements, usually three to four, because heater elements are selected in steps of approximately 10% and the MSCP in steps of approximately 35%. Tests are conducted with the available current which is found by a separate test as the current causing the MSCP to interrupt with its highest let-through energy. The coordination between the overload relay and the MSCP is tabulated in the heater table (Table 9). The manufacturer has tested this coordinated system, and the proper selection is therefore simplified for the user.

Each MSCP has the same barrel length, but the terminal length varies (Fig. 17). A NEMA starter cannot use MSCP sizes which have longer terminals than the determined maximum length for which the clip is built. In this way a smaller MSCP can be placed in larger combination starter units, but not in reverse. This one-way interchangeability is required by UL for safety reasons and is standardized.

TABLE 9. Heater-Selection Table, Size 0 and 1 Combination Starter

Motor current 3 O.L. relays		
Min	Max	MSCP
0.71	0.78	A
0.79	0.87	A
0.88	0.96	A
0.97	1.04	B
1.05	1.11	B
1.12	1.24	B
1.25	1.35	C
1.36	1.46	C
1.47	1.62	C
1.63	1.83	D

Motor short-circuit protectors

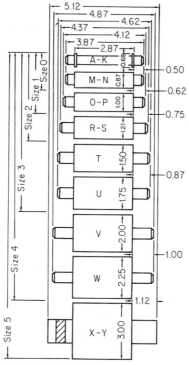

All terminals 0.56 diameter
except X-Y (1.25 x 0.50)

Fig. 17 Physical arrangement of principal MSCPs for various starter sizes.

These protectors can also be built with striker pins, which in turn trip a tripper bar when an MSCP melts. This tripper bar opens a normally closed contact in the control circuit, thus deenergizing the starter coil after melting of one MSCP. This pin is used to provide an anti-single-phase system and disconnect the load as soon as one MSCP melts (Fig. 18).

11. Comparison of Various Protective Devices. As stated earlier, the let-through energy at high faults is a good way to indicate how a protective device works under short-circuit conditions. The let-through energy at various available short-circuit currents for adjustable magnetic-trip-type circuit breakers, thermal-magnetic circuit breakers, large air breakers, and H-type fuses is not published. These fuses can be used only up to a 10 000-A available short-circuit current. For the most important fuses and the MSCP, however, the let-through energy is published. Figure 19 shows the average let-through energy in $10^3 \times (A^2s)$ depending on the FLMC if the motor is protected by RK-5 or RK-9 dual-element fuses selected for 150 and 225% FLMC. These characteristics are given in Fig. 12. Also shown in Fig. 19 are the let-through currents of RK-5 and J-type one-time fuses chosen at 300 and 400% FLMC (Fig. 10) and MSCP (Fig. 15).

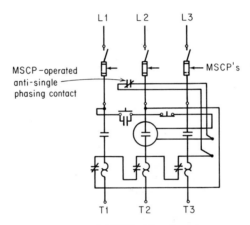

Fig. 18 Circuit with MSCP with striker pin.

An RK-9 fuse is a standard fuse similar to RK-5 and J-type with lower protection. It can be seen that the minimum let-through energy for all fuses is the let-through energy of the 15-A fuse, which is the minimum standard fuse which can be supplied. Also, the RK-9 fuse, especially on higher motor currents, has a considerably higher let-through energy than the other fuses. The RK-5 fuse at 150% has approximately the same protection as the 300% J-type fuse with the exception of small motors, but by comparing Figs. 10 and 12, the crossover point between fuse and overload relay is seen to be fairly low for the 150% dual-element fuse.

The MSCP is individually applied for various motor currents similar to a heater and has a considerably lower let-through energy. The maximum let-through energy which does not cause damage to the controller is not identical for various designs; therefore, it is not shown in Fig. 19. A figure similar to Fig. 19 could be drawn for I_p values, but the I^2t level, in combination with the controller, is the more important factor. The maximum let-through current I_p is an important factor for the withstandability of switches. Similar curves could also be plotted for less than a 100 000-A available short-circuit current and also for various fuses selected from various manufacturers. Figure 19 is based on the maximum values permitted by UL for J, RK-5 and RK-9 fuses, and MSCPs. The actual I^2t and I_p values of various manufacturers are usually somewhat lower. Actually not all I^2t and I_p values are standardized. For those not standardized, values can be interpolated. Since the actual values are usually lower than the standard value, an approximation may be permitted. The actual umbrella value is based on the I_p and I^2t values of the next larger standardized fuse.

12. Branch-Circuit Protection with Instantaneous-Trip-Type Circuit Breakers and High-Fault Circuit Protector (Ref. 16). The high-fault circuit protector (HFCP) is a fuselike device which has a characteristic similar to a motor short-circuit protector but is used in combination with instantaneous-type circuit breakers (Art. 9). The purpose of the HFCP is to combine the advantages of the instantaneous-trip-type circuit breakers for protection of motor branch circuits at low-level faults with the advantage of MSCPs on high-level faults. As with the MSCP, the HFCP must be coordinated and tested with the branch-circuit components so that the proper protective system is developed; i.e., the starter with each heater, the proper instantaneous circuit breaker, and the HFCP must be coordinated and tested by the manufacturer. The calculations based on the maximum allowable withstandability of a heater are (Art. 7)

$$I^2t = \frac{\theta C w}{R} \qquad \text{A}^2\text{s}$$

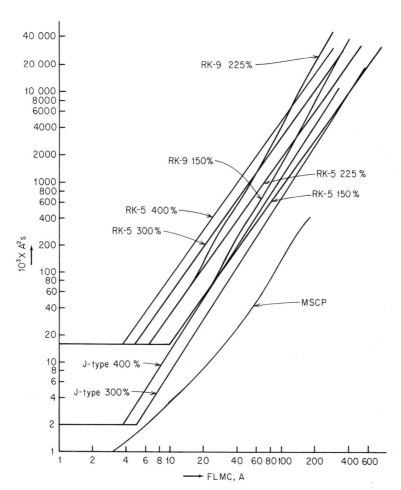

Fig. 19 Let-through energy in 10^3 A²s of RK-9, RK-5, J-type fuse and motor short-circuit protector at 100 000 A rms available short-circuit current depending on FLMC for various protection levels.

Assuming Nichrome V or a similar alloy is the heater material, which is usual, the specific heat is 0.42 Ws/°C, and the melting temperature θ is 1400°C, the withstand-ability of a heater can be calculated:

$$I^2t = 580\, \frac{w}{R} \qquad A^2s$$

The withstandability of contactors must be tested. Typical values are shown in the following table:

NEMA size	A²s
1	75 000
2	180 000
3	350 000
4	1 000 000

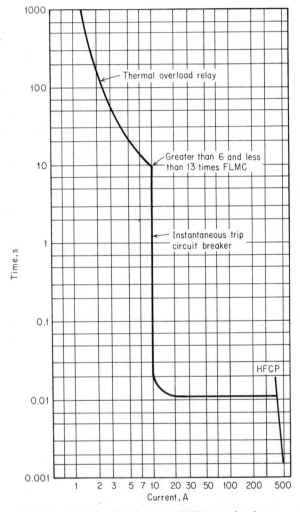

Fig. 20 Principle of combination HFCP circuit breaker.

These typical values are valid for the condition in which no damage should occur to the contactor under short-circuit conditions. They could be much higher if, for example, welding of contacts were permitted. For proper, complete protection, a protective device must be chosen which restricts the let-through energy I^2t to a value lower than the withstandability of the heater and the contactor, provided that the wire has a higher withstandability than the starter. The withstandability of wires can be calculated based on a table published by a cable manufacturer (Ref. 15). Examples are shown in the following:

Wire size	20	19	18	15	12	10	8
I^2t, A²s	3700	6000	9400	37 500	150 000	575 000	750 000

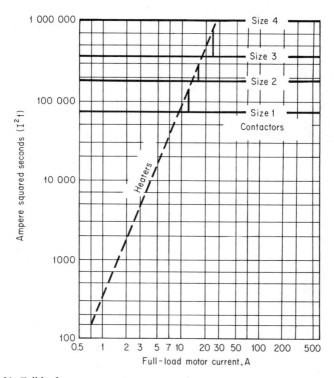

Fig. 21 Full-load motor current in amperes and approximate starter withstandability.

These I^2t values are higher than the withstandability of the appropriate starter.

Each manufacturer, in his coordination study, must develop an HFCP and the circuit-breaker curve to protect his heaters and contactors. Figure 20 shows a typical example of the coordination of the overload relay, instantaneous-trip-type circuit breaker and HFCP.

The maximum short-circuit current that the circuit breaker may interrupt without damaging any of the components can be calculated. Above this current, the branch circuit must be protected by the HFCP. The crossover point between HFCP and the circuit breaker in a time vs. current characteristic is determined by the circuit element having the lowest I^2t value in the circuit (Fig. 21). Assuming that a manufacturer has found that a certain heater has a withstandability of 10 000 A²s, and interrupting time of the circuit breaker is 16 ms, then the crossover point between the characteristics of the breaker and the HFCP must be below

$$I = \frac{10\ 000 \times 1000}{16} = 790 \text{ A}$$

To develop a coordinated system, it must be certain that at any current higher than 790 A the HFCP melts before the circuit breaker trips.

The I^2t values are in many cases too low to trip the breaker, especially since they must be conservatively determined. A large number of tests must be conducted to determine if the let-through energy of an HFCP trips the breaker. Three-phase short-circuit tests are generally not acceptable, since at least two HFCPs would melt, and therefore the trip energy may be larger than under the worst practical conditions. Single-phase tests must be conducted with one HFCP in a circuit with phase-to-neutral system voltage. In many cases, the melting time is so short that the tripping mechanism of a breaker cannot react. Various means exist to ensure tripping of a breaker. A special fused breaker can be used in conjunction with the HFCP. When the HFCP melts, a striker pin of the fuse is released which activates the tripping of the circuit breaker. Such fuse/breaker combinations are known under various trade names.

Another possibility would be to trip the starter by means of a striker pin which would be released by the blowing of the HFCP. The striker pin moves a crossbar which in turn opens a normally closed contact in the control circuit of the starter.

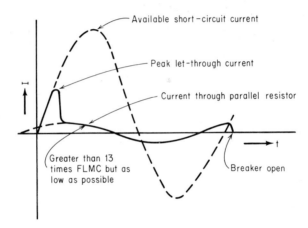

Fig. 22   Principle of interruption of an HFCP with an internal resistor.

It is most desirable, however, to be able to use standard instantaneous-trip circuit breakers which can be tripped by the let-through energy of the HFCP. This characteristic can be accomplished by a small resistor in the HFCP which is parallel to its fusible element. The purpose of the resistor is to limit the current through the circuit after the HFCP has melted to a value slightly above the maximum tripping current of the breaker for approximately 1 cycle. This I^2t value is low enough to prevent damage to any device. The breaker now trips within 1 cycle after the HFCP has interrupted the heavy fault current. The breaker itself interrupts a current which is only slightly higher than its maximum trip current (Fig. 22).

By selecting different resistors for 200- to 240-V systems and 440- to 600-V systems, the resistors are kept small and burn out after the breaker has tripped. If the voltage range is higher, the weight w at the maximum tripping current I_T is so high that the resistor will not melt fast enough.

SPECIAL PROTECTIVE DEVICES AND SYSTEMS

13. Ambient-compensated Relays.   The standard overload relay is calibrated at 40°C ambient. Most American relays trip at about 100 to 110°C. Most time-current characteristics are published for 25°C ambient temperature. The current necessary to raise the temperature in a certain time from the ambient to the tripping temperature is dependent on the ambient temperature; therefore at higher ambient temperature the overload relay will trip at a lower current than at lower ambient. The relay may even trip at FLMC in a high ambient temperature. It is possible to increase the load of the motor if the ambient temperature is lower. Heaters are selected usually under the assumption that the motor and starter have been placed in the same ambient.

If the motor and starter are in different ambients, the overload-relay tripping current is dependent on the ambient of the starter, but the permitted motor-winding temperature at the same time is dependent on the ambient of the motor. Often European relays trip on much higher temperature than 110°C. They are then less sensitive to the variation of the ambient temperature, but they do not duplicate the permitted temperature of the motor winding.

A starter with an ideally ambient-compensated overload relay trips independently of the ambient temperature of the starter. If the relay is only partially ambient-compensated, the dependence is small. The NEMA Standards express the ambient-temperature sensitivity of an overload relay as percent of change in ultimate trip current per 10°C between 40°C and a specified lower or higher ambient temperature. Ambient compensation is desirable when the ambient in which the motor is located is controlled and the ambient in which the starter is located is not controlled. There is a widespread misconception about the applications in which ambient-compensated overload relays are desirable. Such cases require special investigation. If there is an approximately constant temperature difference between the location of the motor and starter, an adjustment in the choice of the heater should be made. If the ambient in which the starter is located has approximately 10 to 15°C higher temperature than where the motor is located, the next larger heater (usually for approximately 10% more FLMC at identical ambients) should be selected. If the ambient temperature is always approximately 10 to 15°C lower, the next smaller heater is chosen.

If the difference between the temperature in which the motor is located and the ambient temperature of the thermal relay is more than 15°C, the proper heater must be selected in coordination with the manufacturer. If the ambient temperature of the trip element, bimetal or eutectic solder, at the tripping current is higher than the motor ambient temperature, then the heater must be derated to less than the rated motor current at a higher ambient temperature and derated more at a lower ambient temperature than at the standard ambient temperature (assumed to be 25°C). The influence of the ambient temperature is given by the following formula for most American relays (tripping temperature θ approximately 100°C):

$$\frac{I_s^2}{I_{st}^2} = \frac{100 - a}{75}$$

where a = ambient temperature, °C

I_s = selected heater current at ambient temperature

I_{st} = selected heater current at 25°C ambient temperature; for 40°C ambient temperature, change 75 to 60

These values are not quite correct because the influence of the starter enclosure and of the geometric configuration of the relay is neglected, but the values are sufficient to select the proper heater. When the tripping temperature θ does not equal 100°C, use the following formula:

$$\frac{I_s^2}{I_{st}^2} = \frac{\theta - a}{\theta - 25}$$

An ambient-compensated relay does not properly protect the motor if the ambient temperature changes independently of the location of the motor and the starter.

A typical application for the ambient-compensated relay is a submersible pump. The pump is in an unchanged ambient temperature, since it is in the ground and artificially cooled. The starter is either in the house or outdoors, and its ambient temperature therefore changes. Another application for an ambient-compensated overload relay may be protection of a hermetically sealed compressor. The ambient temperature of the motor winding due to the artificial cooling is always approximately constant. The starter is in different ambients. Ambient-compensated relays might also be used for a narrow panel having a great number of starters or thermal-magnetic circuit breakers which create much heat, causing these devices to influence each other. This condition would be permitted only if the cable selected is oversized, because otherwise the heat in the panel may damage the insulation without detection. This arrangement places each relay in a high ambient temperature only because the adjacent relays have a high watt loss. If, in such an arrangement, the difference between the outside temperature and the inside temperature is fairly constant, an ambient-compensated relay is still not desirable, but an adjustment in the selection of the heater might be necessary. When the ambient within the panel enclosure changes considerably because of changing loads, however, an ambient-compensated relay is desirable.

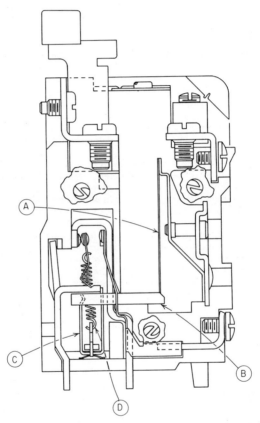

Fig. 23 Ambient-compensated overload relay with indicating contacts. (*I-T-E Imperial.*)

Until now there has been no success in building ambient-compensated eutectic-solder relays. But bimetallic relays may be built partially ambient-compensated. Figure 23 shows the relay of Fig. 2, but the ambient compensation has been added. Ambient compensation is obtained as bimetal A which moves the tripper bar B is compensated by another bimetal C expanding in the reverse direction with relation to bimetal A. The second bimetal theoretically should not be influenced by the heater, but only by the ambient temperature. Therefore the movement of the tripper bar is dependent on the difference of the work of the two bimetals and independent of the ambient temperature. Tripper bar B actuates over-the-center spring D. The pivot point of D moves with the ambient-compensating bimetal C. It is unavoidable that the heater itself influences the ambient-compensated bimetal to a small degree because the physical dimensions of the relay restrict the placement of the compensation. It is also very difficult to develop bimetallic relays in which the compensating and tripping bimetals have identical thermal and mechanical characteristics. Most designs, however, have a broad temperature range, usually given by the supplier, which could be between 40 and 160 or 170°F of adequate temperature compensation. Ambient-compensated heater-type relays usually trip faster on overloads and locked-rotor current. Attention must be given to assure that the overload relay does not cause nuisance tripping, especially after a hot start.

14. Relays with Indicating Contacts. Overload relays have a normally closed contact in the coil circuit which opens when the overload relay trips. In some cases it is desirable that the tripping of the overload relay should be indicated in some fashion, such as a warning light. Sometimes disconnecting the motor in case of an overload is not desirable; for example, in steel mills it is often less costly to burn out the motor than to interrupt the manufacturing process. In such a case the normally closed trip-

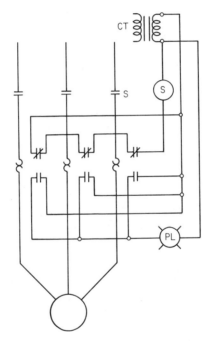

Fig. 24 Overload-relay circuit with indicating contacts.

ping contact may be omitted, and the normally open indicating contact is used to warn that the motor is in danger. Figure 24 shows the wiring of three indicating contacts. It can be seen that the indicating contacts are connected in parallel, the normally closed contacts in series. The trip indication can be dependent only on the tripping of the overload relay and must be independent of the movement of the contactor.

15. Group Motor Protection. If a conductor supplies several motors, its ampacity must be larger than or equal to the sum of all FLMC + 25% of the largest motor (NEC 430-24). The horsepower rating of a single controller for a group of motors must be equal to or larger than the sum of the horsepower ratings of all the motors of the group. The NEC, Article 430-87, allows the use of a single controller for a group of motors in cases where a number of motor drives are part of a single machine or piece of apparatus or where a group of motors are located in a single room within sight of the controller location. A number of manual starters which are protected by one circuit breaker or one fusible switch may be used for group motor protection.

Such a condition is permitted when:

1. (*a*) Each of the motors does not exceed 1 hp and the branch circuit is protected for an ampacity of not more than 20 A at 125 V or for 15 A at 600 V. (*b*) The FLMC of each motor does not exceed 6 A.

2. All motor circuits are connected to one branch if each motor controller with a specified maximum fuse or circuit breaker is approved for group installation and each fuse or time-limit circuit breaker does not exceed the permitted rating of the largest motor connected to the branch (the largest permitted rating is 400% of the largest FLMC), plus the amount of all FLMCs so that the branch circuit to the controller has sufficient cross section (NEC 430-53-06). At present, only manual starters are approved for group installation. The starters must pass the test described in UL Standards 508 in which the controller is tested with a fuse larger than 400% of its FLMC rated at 5000 A available short-circuit current. In this test, welding or complete disintegration of contacts, burnout of the current element, or other failure making the con-

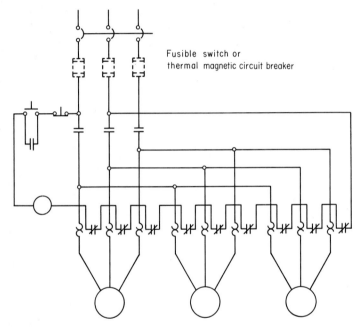

Fig. 25 One disconnect device, one contactor, and several sets of overload relays for a group of motors.

troller inoperative is acceptable, but ignition of cotton or any other manifestation of fire hazard is not acceptable. The manufacturer must mark the largest fuse or circuit breaker which can be selected to protect the controller. Test results show that, in order to pass a test with oversized fuses or oversized thermal-magnetic circuit breakers, either the heater or the relay must be completely enclosed or the enclosure of the starter must be so tight that fire in the enclosure could not ignite cotton placed outside the enclosure. The user should contact the manufacturer if he wants to know more about the exact damage level which occurs in a particular device by combining starters with overload relays with oversized fuses or oversized time-limit circuit breakers.

The combination of several sets of overload relays connected in parallel to one contactor is shown in Fig. 25. The purpose of such an arrangement is to ensure that if one motor is disconnected owing to overload, the other motors are also disconnected; for example, if the motor for circulating the coolant of a machine tool is disconnected, the main contactors of the machine tool must also be disconnected. By using only one branch-circuit-protecting device for protecting the system at short-circuit currents, considerable savings can be realized.

16. Two vs. Three Overload Relays.   As already explained, the NEC, 1968, required three overload relays if the branch-circuit-protective device is an adjustable magnetic-trip-type circuit breaker or when 3-phase motors are installed in isolated, inaccessible, or unattended locations unless the motor is protected by other approved means.   As indicated earlier, the NEC, 1971 and later, requires three overload relays for protection of a 3-phase motor.   This change was long overdue, because in nearly all cases where a branch-circuit-protecting device is a fuse or time-delay breaker, the setting or rating is so high that the actual overcurrent-protective device for the wire in a wide current range is the overload relay.   However, many installations will remain for years with two overload relays; therefore one should know that for the protection of the 3-phase motor itself, certain cases exist in which two overload relays are not sufficient.

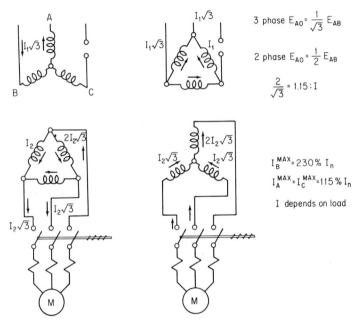

$$3 \text{ phase } E_{AO} = \frac{1}{\sqrt{3}} E_{AB}$$

$$2 \text{ phase } E_{AO} = \frac{1}{2} E_{AB}$$

$$\frac{2}{\sqrt{3}} = 1.15 : 1$$

$$I_B^{MAX} = 230\% \ I_n$$

$$I_A^{MAX} = I_C^{MAX} = 115\% \ I_n$$

$$I \text{ depends on load}$$

Fig. 26 I_1 and I_2 are primary and secondary currents of wye-delta- and delta-wye-wound transformer with one primary phase interrupted.

Such a case exists when a motor is connected to the secondary side of a transformer having a wye-connected primary and a delta-connected secondary or with the primary delta-connected and the secondary wye-connected.   As soon as one phase on the primary side is interrupted, as shown in Fig. 26, the current flowing in one phase on the secondary side of the transformer is nearly twice as high as it is in the other two phases; therefore, overload protection in two phases may not detect an overcurrent. The ratio of the current compared with a normal 3-phase load is shown.   However, if several motors (not one predominant motor) are the secondary load, the phase current becomes equalized.

Another case where three overload relays are necessary for protection is if a larger motor is connected in parallel to one small motor (Fig. 27).   If one phase is open, an overcurrent flows through the other two cables.   Since the two motors tend to keep the field magnetized, the current for the magnetization of the open phase must flow through the other two conductors.   Normally this magnetization current is not high in

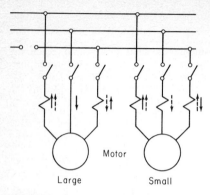 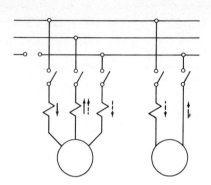

Fig. 27 Combination of small and large three-phase motors with one-phase interrupted.

Fig. 28 Single and three-phase motors with one-phase interrupted.

ratio to the motor current, but if a very small motor is connected in parallel to a large motor, the magnetization current for the third phase of the large motor flows through one phase of the small motor, and this current can overheat the small motor. This phenomenon may not be detected with two overload relays.

Finally, in a 3-phase system, if one phase is interrupted and a single-phase motor (Fig. 28) is connected in parallel to the 3-phase motor, the current flowing in the interrupted phase of the motor also flows through the 3-phase motor. This condition may cause overloading of one phase without overloading the other two phases.

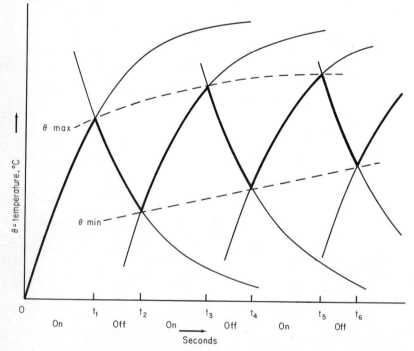

Fig. 29 Temperature rise of a relay with intermittent duty.

These examples demonstrate some of the difficulties in determining when two overload relays may have been sufficient and when not. The NEC in 1971 did, however, change the existing practice and requires three overload relays as mentioned before for all cases. Very few motor failures, however, are known to have occurred as a result of the use of only two overload relays.

17. Intermittent Duty. It is very difficult to determine the proper overload-relay setting for a motor which operates frequently or intermittently or both, because the time constant of the thermal-overload relay is much shorter than the time constant of the motor. The best protection under these conditions is obtained with inherent protectors. However, if the protection is not built in the motor, another way to protect the motor must be found. The temperature of the motor and the relay increases according to a curve shown in Fig. 29. At least in some cases the following must be considered to select the proper heater. Assuming:

$$t_s = \text{starting time, s}$$
$$t_r = \text{running time, s, after start}$$
$$t_1 = t_s + t_r = \text{on time, s}$$
$$T = \text{time constant of the overload relay, s}$$
$$t = \text{total time of one cycle, which means the running time plus the time the motor is disconnected, s}$$
$$p = \text{ratio of intermittent-duty load divided by continuous-duty load, both giving the same temperature rise}$$
$$t_1/t = \text{duty}$$
$$t_2 = \text{off time, s}$$

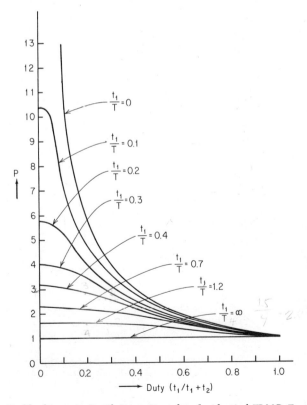

Fig. 30 Permitted load increase P with intermittent duty for identical FLMC, T = time constant, $t = t_1 + t_2$, t_1 = on-time cycle, t_2 = off-time cycle.

The following formula can be established if the temperature rises in an exponential function (Ref. 8):

$$p = \frac{1 - e^{-(t_1 + t_2)/T}}{1 - e^{-t_1/T}}$$

Figure 30 shows p as a function of t_1/T. Generally the value t_1/T is approximately 0.1 to 0.2.

The equivalent current I_e' for the time $t_1 = t_s + t_r$, i.e., for the time the motor is running, is defined by

$$I_e' = \frac{(I_R/I_N)^2 t_s + t_r}{t_s + t_r} I_N$$

where I_N = running current
$\quad I_R$ = inrush current
$\quad I_e$ = total equivalent current for the complete cycle
$\quad\quad = \dfrac{I_e'}{p}$

Figures 31 and 32 give some results from the formulas for squirrel-cage motors. It is assumed that the inrush current is five times FLMC for 40 and 25% duty. A check should be made to determine if the heater chosen is too small. If the number of operations is less than 60 per hour and the starting time is very short, a heater may be selected with the following approximate values:

Duty cycle, %	70	40	25
I_e, %	90	80	70

The current I_e determines the heater to be selected provided that no nuisance tripping occurs; otherwise, a larger heater must be selected even though it does not provide

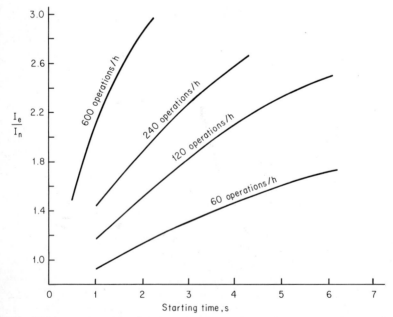

Fig. 31 Heater selection I_e/I_n for squirrel-cage motor with I_r/I_n equals 5 for 40% duty under various conditions. I_n = FLMC, I_e = selected heater current, I_r = inrush current.

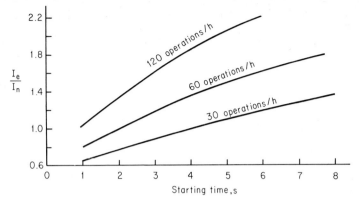

Fig. 32 Heater selection I_e/I_n for squirrel-cage motor with $I_r/I_n = 5$ for 25% duty under various conditions.

sufficient protection. The foregoing calculation is based on many simplifications (Ref. 6). In reality, the temperature does not increase in an exponential fashion, but in a very complex manner. The time intervals are not equal, but for the selection of the heater, the simplified formulas are often sufficient.

18. Inherent Motor Protectors. Inherent motor protectors are devices which are placed in the winding of the motor and respond either to the temperature of the heater directly or to the rate of rise of the motor temperature, or to a combination of temperature and current. These devices have certain advantages and disadvantages when compared with the previously described running-overload protectors. Since they are placed in the motor, they do not protect the motor branch in case of ground fault or short circuit in the wiring system or at the motor terminals. Their coordination with fuses or adjustable-magnetic-trip circuit breakers and the proper contactor is difficult, considering the variance in ratio of tripping current to full-load current of the inherent protector. The NEC permits, as stated earlier, a maximum ultimate tripping current of 170% of FLMC for motors up to 9 A, 156% of FLMC for motors up to 20 A, and 140% of FLMC for motors over 20 A. Another disadvantage is that the inherent motor protectors are placed in the motor itself and, therefore, they are part of the motor. Usually, they cannot easily be attached during the installation of the equipment. On the other hand, they have some important advantages. Since they are dependent on the motor temperature or on the rate of rise of the motor temperature, they are the only practical devices which protect against insufficient ventilation and mechanical-component failure such as improper lubrication or broken belts which may cause overheating of the motor.

Inherent protection is provided if a motor is exposed to an excessive duty cycle for which the motor was not designed. If all the figures (duty cycle, time constant, inrush current) are not available, which is usually the case, a protective device for direct response to the temperature of the motor winding is safer. If the system frequency varies, a low frequency causes lower speed of the motor. This reduction of speed can cause decreased cooling and, therefore, higher temperature without changing the current sufficiently to be detected by current-responsive devices.

Since the motor's temperature is measured, it is possible to build the motor smaller or to use a motor at a higher overload provided that the other branch-circuit components withstand the higher ultimate trip current. A difficulty with inherent protectors is that several of these protectors must be placed in one motor, since they detect a temperature rise in only one particular locality. Inherent protection is advantageous if the motor is placed in an ambient, which changes the temperature while the starter is not placed in the same location and is not subjected to the same temperature changes. Nearly all hermetically sealed motors up to 8 or 10 hp now have inherent motor protectors. Figure 33 shows a typical thermostat.*

*The writer is indebted to Albert P. White of Texas Instruments, Inc., for information about various motor protectors.

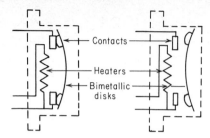

Fig. 33   Inherent protector.

If the temperature changes slowly by heat transfer, the bimetallic snap-action disk causes interruption of the current as soon as the temperature reaches a dangerous value. When the disk protects small motors, the contact may be connected directly into the motor circuit. When the disk protects larger motors, the contact may interrupt the coil curcuit of a contactor. The heat may be transferred from the heater, and at the same time, the bimetallic disk may carry the motor current. The disk may have automatic or hand reset.

Another dual-responsive sensor is shown in Fig. 34. Most of the heat transfer occurs in sensors of Figs. 24 and 25 by placing the thermostat in the winding of the motor. Another thermal-sensing device is a rod-and-tube linkage (Refs. 9 and 10). Such a rod-and-tube protector is shown in Fig. 35. The outside tube is of material having a low coefficient of thermal expansion. The differential between the two expansion coefficients results in a movement which is amplified by linkages which operate a snap-action switch. Such a device senses the rate of the current change. With slow temperature changes, the tube and rod heat at the same rate and the opening of the switch is dependent on the tube-temperature difference between tube and rod.

If the temperature changes quickly, the tube heats faster than the rod and thus the difference of the temperature of the rod and tube causes a faster opening of the switch. Because this device is extremely sensitive to heat increase due to a locked-rotor current, it is often used to protect large motors.

A bimetallic disk which is influenced by current and temperature rises may be used as a fast-acting device to protect the motor. This action may also be obtained by combining a heater element with a bimetallic disk in one protector. Such a device may be used to combine protection at locked-rotor current due to current influence and small,

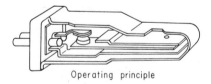

Operating principle

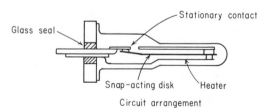

Fig. 34   Motor winding protector. (*Klixon type.*)

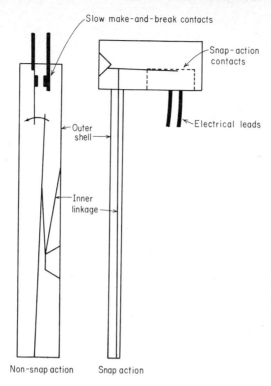

Slow make-and-break contacts

Snap-action contacts

Electrical leads

Outer shell

Inner linkage

Non-snap action   Snap action

Fig. 35 Rod-tube-linkage-type inherent motor protector.

long-time overheating due to heat transfer. The application of such a device for 3-phase motors is shown in Fig. 36.

A modern protector is a thermistor. It is very small and is made of a semiconductor, i.e., barium titanate with special impurities. The characteristic of such a device is shown in Fig. 37. The resistance of the thermistor increases suddenly at a definite temperature. This temperature can be controlled. There are other thermistors that use materials with negative temperature coefficients of resistivity. The output of a thermistor circuit is small and must be amplified to be used to energize the starter or contactor coil. Many thermistors can easily be placed in one motor, thus giving excellent protection.

19. Undervoltage Relays. If the system voltage drops under a certain value (usually 80% of the rated voltage), the motor must in some cases be disconnected from the supply line. The torque changes with the square of the voltage, and dropping the torque to $0.8^2 = 0.64$ of the normal torque may be undesirable. A special undervoltage relay is not necessary, if a magnetic starter is used, if the dropout voltage can be much less than 80%. Magnetic starters usually drop out at 40 to 50% and lower values of the rated voltage. There are very few cases in which undervoltage protection is really necessary or required by the Code.

Most undervoltage relays are magnetic devices which drop out at a predetermined voltage level. The levels are adjusted by changing the air gap or the return spring force or both. The coil has high reactance and comparatively low resistance so that the drop-out voltage does not significantly change with the temperature. Such an undervoltage relay may be combined with a time-delay action by dashpot or escapement to override small disturbances of the line voltage for a few seconds (usually ad-

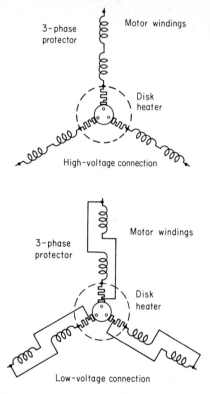

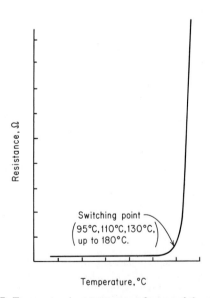

Fig. 36 High and low voltage connection of three-phase Y-connected thermally protected motor.

Fig. 37 Temperature/resistance curve for typical thermistor.

justable from 1 to 5 s). In the last few years, solid-state voltage sensors have frequently
been used instead of electromagnetic undervoltage relays. But they are at the same
time more expensive and therefore are used only if a high degree of precision is required. Generally a zener diode provides a reference voltage. If the rectified line
voltage of one phase or the average voltage of all three phases is higher than the
reference voltage, the transistor is in the off position or a unijunction transistor will
trigger. As soon as the actual line voltage falls below the reference level, either the
transistor turns on or the unijunction transistor stops triggering. This action causes
either a reed relay to drop out or a static signal, which in turn will open a coil circuit
causing a magnetic starter to drop out.

It is common to use undervoltage relays in dc applications and in combination with
wound-rotor motors. If the voltage has dropped, the wound-rotor motor or the dc
motor may not restart without having the full starting resistor in the circuit after full
voltage is reestablished. Contactors are interlocked to accomplish this goal. Once the
proper voltage is reestablished, even automatic restarting of a squirrel-cage motor
should occur only if neither personnel nor equipment can be injured.

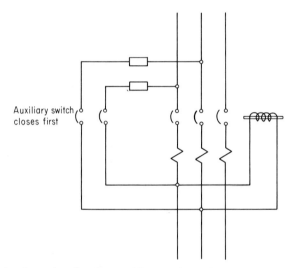

Fig. 38 Circuit breaker with undervoltage coil for several wound rotor motors. Circuit compels all
motor controllers to be in open or zero position before circuit breaker closes.

Full undervoltage protection requires three undervoltage relays between all phases
or all phases to neutral. There are some simplified conditions when one undervoltage
relay is sufficient; for example, when the load is equally distributed, the relay may be
placed between the star point and ground. As soon as this relay is energized by a
critical voltage, such as approximately half the system voltage, the main contactor disconnects. However, it is rare to find such simplified applications because equal distribution must be assured.

In many cases a solid-state voltage sensor which responds to the average voltage of a
3-phase system is sufficient. If the electrical system has several wound-rotor or dc
motors, all starters must be in the zero position before the main switch on the circuit
breaker can be connected to the supply voltage. Either a circuit breaker or a circuit
interrupter may be used with undervoltage relays.

A circuit interrupter is a circuit-breaker-type device with a dummy tripping unit, i.e.,
trip units that have a tripper bar but no current-overload- or short-circuit-sensing device. The tripper bar can be activated by an undervoltage relay causing the circuit
interrupter to trip. The circuit interrupter has a much lower interrupting capacity because it will not clear faults and overloads. Fuses must be used.

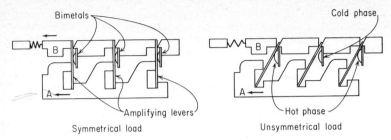

Fig. 39 Differential protection with bimetallic overload relay.

In case all motor starters use contactors for shortening the starter steps, interlocking of the contactors may be provided. If hand-operated or motor-driven controllers are used, a scheme such as that shown in Fig. 38 may be used. A two-pole auxiliary switch closes first and an undervoltage coil to the source through two resistors. If the controllers which connect the various motors to the source are in the open position (zero), the voltage is high enough so that the undervoltage magnet does not drop out and the main contact of the switch can follow.

Voltage-sensitive relays are used to remove the field of a synchronous motor when the motor is disconnected because of an overload. A field relay is usually a time-delay voltage-sensitive relay. If the field voltage of a synchronous motor drops too much, this relay is used to disconnect the motor. The field is also removed when the motor does not start properly.

20. Relays with Anti-Single-Phase Tripping Characteristics. Special bimetallic relays which trip faster if the motor is unsymmetrically loaded, especially under single-phase conditions, are used often in Europe and to a small degree in the United States (Ref. 11). The basic principle of such a device is shown in Fig. 39. The relay has two tripper bars A and B. On a normal symmetrical overload, the two tripper bars move together in parallel. However, on an unsymmetrical load, tripper bar A moves faster because of the pivoting action of the amplifying levers. Therefore, if the bimetals are strong enough to cause tripping without much restraint, the relay trips faster under single-phase conditions than on a 3-phase overload. One phase may be interrupted suddenly by the blowing of one fuse. The impedance in the three windings remains the same, and the current increases approximately 73% in the remaining phase during the single-phase operation.

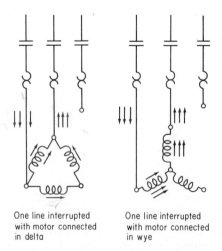

Fig. 40 Three-phase motor running after one phase is interrupted.

If the motor is connected in wye, the current through the motor winding is identical to the line current. If the motor is connected in delta, the current in each phase of the motor is only 58% $(1/\sqrt{3})$ of the line current if all three phases are connected to the line. If the motor runs on a single phase, twice as much current flows in one winding as in the other two (Fig. 40) and the temperature in the first winding increases faster than the increase due to line current. A 3-phase overload relay with a single phase should therefore trip faster than if the motor is just overloaded.

In the United States, it is common to build delta-connected motors so that they can withstand this higher current, which is ⅔ of 1.73, or 1.15 of the line current in this one phase. In the other two phases of a 3-phase motor, the current is the nominal current, i.e., 58% of the line current. A relay with anti-single-phase provision is therefore generally not necessary for the protection of motors in this country. European motors usually do not have the indicated reserve, and anti-single-phasing devices are advantageous.

With such a relay, it would be desirable that tripping occur in case of a voltage unbalance of very short duration. Even small voltage unbalances may be harmful to a motor if they prevail for a long time. A relay with an anti-single-phasing tripper-bar arrangement does not protect the motor better in such a case than a standard relay. Voltage unbalance has the following effect on motors (Ref. 17):

Voltage unbalance, %	0	2.0	3.5	5.0
Current, stator, rms %	100	101	104	107.5
Stator average loss increase, %	0	2	8	15
Rotor average loss increase, %	0	12	39	76

Note that the losses in the rotor increase much faster than in the stator because of the negative-sequence slip caused by the reversed rotation of the resulting flux.

If the motor runs on one phase, the rotor losses increase more than the stator losses due to the skin effect caused by the negative-sequence current since the rotor frequency is approximately twice as high as the system frequency. Under single-phase conditions, the rotor losses are more than seven times as high as the normal rotor losses. The stator losses increase by a factor of 3 owing to the 73% increase of the current. This factor may increase to 3½ to 4 because of the positive-sequence slip. Motors with very small reserve may not be able to withstand these increased rotor losses for the time necessary to trip a standard overload relay without the anti-single-phasing tripper-bar arrangements. This condition may fortunately not be a problem with most standard United States motors, but it may happen more often with some European motors. It is important to remember that an overload relay may not protect a 3-phase motor under single-phase conditions to the same degree as under 3-phase load.

21. Phase-Failure Relays. In some applications, it is necessary to trip immediately when single phasing occurs. Phase-failure relays must be used according to the NEC, 620-53. An example of such protection is an elevator motor, since a 3-phase motor cannot be used for inching, plug stopping, or frequent stopping when one phase is interrupted. A bimetallic relay as previously described could act too slowly after the interrupting of one phase in applications such as elevator motors where it would be hazardous to eliminate plug stopping, inching, or reversing. In such applications a phase-failure relay is required to open the circuit instantaneously and disconnect the motor when single phasing occurs.

It is not sufficient to provide three undervoltage relays for this purpose, because a running motor can maintain the 3-phase voltage after one phase is interrupted, and the interrupting of one phase would therefore not be detected quickly enough while the motor is running. The failure of one phase may, however, be detected quickly using the current unbalances in the motor branch circuit. These unbalances occur not only when one phase is interrupted on the line immediately leading to the motor, but also on the primary side of a wye-delta or delta-wye transformer.

A phase-failure relay, relay circuit, and coil arrangement are shown in Figs. 41 and 42. The current coils E_1-T_1, E_2-T_2, and E_3-T_3 are primary coils of three transformers having U-shaped magnets. One secondary coil PTR is wound around all three legs of

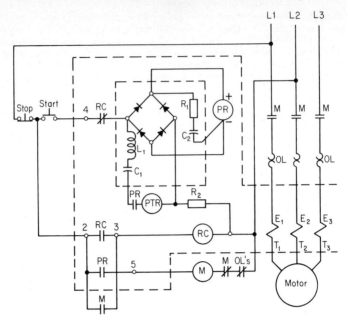

Fig. 41   Diagram of phase-failure relay connected to full voltage starter.   (*Allen-Bradley Company.*)

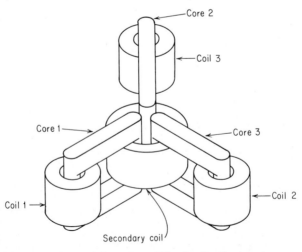

Fig. 42   Phase-failure relay.   (*Allen-Bradley Company.*)

the three U-shaped transformer magnets. The cores are built of a material which has square loop characteristics. The lowest motor current is still sufficient in the three phases to saturate the core. At fairly balanced current, if all three phases are connected, the output is maximum. If one phase leading to the motor is disconnected, the other two phases are in phase opposition, and the output is zero. If one primary phase of a wye-delta or a delta-wye transformer in a motor circuit is disconnected, the current distribution as explained above is unequal, and the output voltage of the coil *PTR* is about one-third of the symmetrically loaded value. This output voltage is not enough to keep the *PTR* relay closed, and phase failure is immediately indicated.

22. Phase-Reversal Relays. In cases where the phase reversal or an incorrect phase sequencing create a hazardous condition, for example, in elevators, cranes, or hoists, a phase-reversal relay must be used. Assuming after the repair of an elevator that the motor is connected in the wrong phase sequence, protection must be provided to ensure that the motor does not run in the wrong direction. Several systems have been developed for phase-reversal relays (Fig. 43). Relay 2 is attracted owing to the voltage

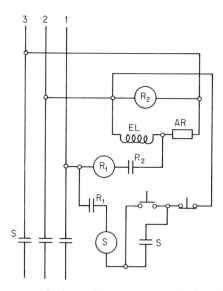

Fig. 43 Phase-reversal relay used in combination with phase-failure relay.

between phases 2 and 3. Relay 1 is directly connected between phase 1 and phase 2 through the coil *EL* and to phase 3 through the resistance *AR*. If the phase sequence is correct, the two currents through the coil and the resistance are added geometrically, but if the phase sequence changes, relay 1 drops out, disconnecting the starter.

Other systems have been developed for the same purpose, such as disk motors which are driven by magnetic systems. If the disk turns in one direction, it keeps the contact in the coil circuit closed. If the direction of the disk changes, the contact opens and the main contactor drops out.

It is also possible to develop phase-reversal or phase-failure relays with solid-state devices. The output of such a device can be a relay or static switch, i.e., a triac. A sensor causes the relay to drop out or to change the state of the static output device when the voltage conditions are not normal, as is always the case when one phase is reversed or missing. It is also possible to sense an unbalance of voltage or current by solid-state means.

23. Ground-Fault, Short-Circuit, and Field Protection for Large Motors. In case a fault occurs in a system in which large motors are used (Ref. 12), a sudden voltage dip

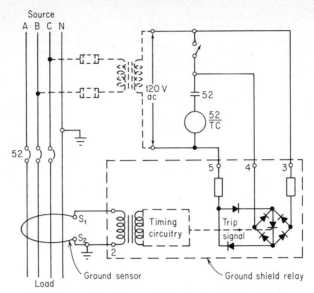

Note: Any connection from neutral to ground must be located on
source side of all ground sensors.

Fig. 44 Typical connection for ground-fault relay.   (*I-T-E Imperial.*)

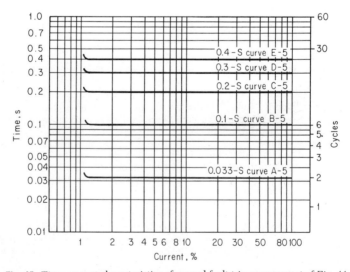

Fig. 45 Time-current characteristics of ground-fault trip arrangement of Fig. 44.

causes a slip which may cause a large synchronous motor to be unable to swing back into synchronism. This condition may occur if the duration of the fault is 10 to 30 cycles. On a 3-phase system, protection against undervoltage conditions could be obtained by using three instantaneous undervoltage relays which almost drop out when the potential drops below a specified amount. If the motor allows a short time of voltage fluctuations, the three undervoltage relays may have a very small time delay. If a ground fault occurs, two phases may remain unaffected while the third phase may have nearly zero voltage. Sometimes additional time is permitted for clearing the ground fault before the motor loses speed.

A ground-fault relay consists of a ground shield used as a sensor and a solid-state ground relay (Fig. 44). This shield is a current-balancing transformer having a low burden. If no ground fault occurs, the sum of the currents in the sensor is zero. Since an unbalance occurs because a part of the power is transmitted through the earth, the secondary side of the balancing transformer has an output. If an unbalance occurs, this output can be amplified and fed into a relay through a timing circuit. While the arrangement provides enough sensitivity, nuisance tripping is prevented by the timing circuit. The usual trip time is 0.1 s (Fig. 34), and the most usual trip setting for motor circuits is between 5 and 10 A. Such a ground-fault relay can be used to trip the shunt trip of a circuit breaker (Ref. 13). For the protection of large synchronous motors a field relay is often necessary. Such a relay is normally undercurrent relay, time-delayed or not. It operates if the field current becomes too low. It is usually in series with the exciting field. Time-delayed relays do not act if the field is low for only a short period.

24. Solid-State Overload Relays. Solid-state overload relays were, during recent years, developed for protection of large motors, mainly high voltage motors. They combine most of the necessary elements for protection. They may or may not be combined with resistor-type temperature detectors (RTD).

The running overload characteristics of motors vary. Solid-state overload devices have different selectable time-current characteristics. They may, for example, have 4 curves between 1 and 5 min tripping time at 200% ultimate trip current. They have memory of previous loads to insure proper protection in applications of repetitive operations. Due to the variations of the current-time characteristics, they may copy intermittent duty of motors. The first set of curves concerns smaller overloads up to 2 times ultimate trip current. The motor is protected by another set of characteristics with various tripping times between 200% and 1000% ultimate trip current. They can be adjusted, for example, in 6 curves between 5 and 30 s at 6 times ultimate trip current. All adjustments are performed at the secondary side of current transformers. Due to the small burden of solid-state devices, the transformer transfers linearly in a very wide current range.

The principle of the solid-state relay curves is based on the fact that a simple thermal device changes its temperature by an exponential function where T is the time constant.

$$\theta = \theta_{max}(1 - e^{-t/T})$$

If the body cools then

$$\theta = \theta_{max}e^{-t/T}$$

An electrical analog to these curves is the discharge curve of a capacitor through a resistor to ground potential. Ec is the voltage across the capacitor. This is an analog to the temperature rise. RC is the time constant, where R = the resistance and C the capacitance in the discharging current. RC must be identical to T of a thermal circuit. IR is the analog to the maximum temperature rise.

A motor is in reality a complex thermal mechanism and could be regarded as composed of n expressions

$$\theta = \theta_{max} - A_1e^{-t/T_1} - A_2e^{-t/T_2} - A_ne^{-t/T_n}$$

It is possible to develop electrical equivalent characteristics but we can also combine different adjustable single time constant analog characteristics, one set of curves for stalled rotor protection and severe overload problems. By having the choice of four curves for lower overcurrents and six curves for high overcurrent, all with memories, a fairly proper overcurrent capability characteristic of the motor can be copied.

The third curve is the fault-protective curve. This curve is adjustable for pickup currents from 4 to 16 times ultimate trip current in four steps. At the pickup current the fault-protection curve crosses the selected stalled rotor protection characteristic and causes very fast tripping. The current-time characteristic is inverse time from approximately 0.1 to 0.01 s tripping time (not counting clearing time of the contactor).

The ultimate trip current can always be adjusted at the secondary side of the current transformer, usually from 2 to 5 A.

If the solid-state relay is combined with inherent temperature-sensing devices, the inherent detector (RTD) may be placed on various spots in the machine, especially those spots which probably become the hottest (rotor or stator sensitive machines). The resistance changes with the temperature, and an electrical circuit can be developed acting as an analog to the temperature rise with the voltage drop across the RTD. $I \times R$ must be proportional to the steady-state temperature difference between the conducting copper and the temperature sensor. I/C must be proportional to the rate of temperature rise of the stator conductor or rotor bar under stalled rotor conditions (depending where the RTD measures the temperature), $R \times C$ must be equal or greater than machine rotor bar or stator copper to the RTD time constant.

The voltages V_1 between the temperature into the voltage converter and resistor of the circuit and V_2 between the resistor and capacitor C are monitored and trip the breaker. $V_2 > V_1$ by the voltage drop $I \times R$. V_1 causes tripping when the copper temperature reads 130°C (Class B insulation). Tripping due to V_2 depends on the permissible transient temperature under a stalled rotor condition. The advantage of this kind of measurement is that the hotter the motor temperature is prior to starting, the shorter will be the allowed starting time (Ref. 18).

These relays can also be combined with phase unbalance and phase failure protection. Motors having rotor bars of low conductivity may only tolerate 6% to 7% current unbalance (Ref. 18). The negative sequence pickup current can be adjusted in fractions of full load currents on the secondary side of the current transformer between 0.5 and 2 A.

The tripping time is adjustable between 0.5 to 4 s. The selected time is independent of the negative sequence current.

The solid-state relay can further be combined with ground-fault protection. The negative sequence relay also detects the ground fault. Also, phase reversal is detected by the negative sequence relay since the starting current has negative sequence.

The selection of the proper curves, pickup current, and tripping times must be made by the user. The application of solid-state relay is therefore an engineering job.

25. Adjustable Relays have been used for slow starting, using two parallel contactors and for applications which require time-overcurrent characteristics that a normal thermal-overload relay cannot provide. Two such devices are the Inducto-Therm overload relay and the magnetic-type overload relay. The Inducto-Therm relay operates on a eutectic-alloy ratchet principle (Fig. 46). The overcurrent increases the induced current in the copper heater tube proportionally to the overload current. This heating finally melts the eutectic solder, which causes opening of the contact in the same manner as a heater of resistive material trips a thermal relay. The Inducto-Therm relay may be reset manually as soon as the copper tubing is cool enough that the eutectic solder has solidified. The tripping value can be adjusted over a wide range by changing the position of the iron core in the coil, thus changing the magnetic flux. The advantages of this relay over other thermal relays are the wide adjustability and the special characteristics which may be obtained. These characteristics are suitable for protecting motors with long starting times.

While the tripping current of a directly heated bimetal can be adjusted in a limited range, a method commonly used in Europe, this arrangement has distinct disadvantages. One disadvantage is that the adjustment becomes very inaccurate and another is that the bimetal tripping temperature changes over a wide range, depending on the distance which the bimetal moves to trip the relay. The directly heated bimetal trips fast, but this characteristic is often not desirable.

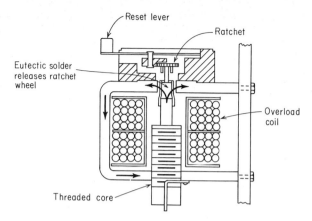

Fig. 46 Inducto-Therm overload relay. (*Allen-Bradley Company.*)

Because the applicability of the Inducto-Therm relay is determined by the tripping characteristic and not the ratio of FLMC to locked-rotor current, an Inducto-Therm relay should not be used if the locked-rotor current is less than four times FLMC.

26. Magnetic Relays. A magnetic relay (Fig. 47) can be used without a current transformer for larger motors. It consists of a coil through which the motor current flows and a movable iron core. The magnetic force at FLMC is not sufficient to move the iron core. An overload current, however, causes the iron core to overcome the restraining action of the piston in the dashpot. The relay can be adjusted to trip in the current ratio 1/3. This adjustment can be obtained by screwing the core downward or upward. The trip time of the relay can be adjusted by changing the opening of the valve in the piston. Time and current are inversely related. A wide range of characteristics is possible. The characteristics also depend on the viscosity of the fluid used in the dashpot.

Another typical magnetic relay (Fig. 48) has a coil wound around a hermetically sealed, nonmagnetic tube. The tube is filled with a silicone oil and contains a spring-biased movable iron core. At normal current, the flux is not strong enough to move the iron core toward the yoke. As the magnetic field increases, the force also increases with the square of the current, attracting the iron in the tube and pulling the iron toward the armature. With the movement of iron toward the armature, the magnetic field is increased by the reduction of the reluctance of the magnetic circuit.

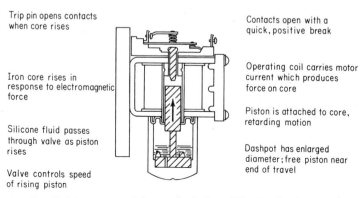

Fig. 47 Magnetic time-delay overload relay. (*Allen-Bradley Company.*)

The rate in which the iron core moves is controlled by the silicone oil, thus creating a time-inverse characteristic. When the flux reaches a particular magnitude, the armature is attracted and the relay trips. On high overloads or short circuits, the coil produces enough flux to attract the armature regardless of the position of the iron core. As explained in Art. 13, the thermal bimetallic relay with compensated bimetallic components is only partially influenced by the ambient temperature because the compensating bimetal is usually placed in the same housing as the tripping bimetal.

Magnetic relays have less dependence on the ambient temperature than most ambient-compensated thermal relays. Their only dependence is the change of the viscosity of the liquid with the ambient. If unusual precision of the tripping characteristic is required, for example, running-overload protection for thyristors in combination with short-circuit protection by fuses with extremely small let-through energy and steep time-current characteristics, magnetic relays are preferable.

Magnetic relays which trip instantaneously protect motors against excessive torque, which may mechanically damage the motor. They are necessary for protection of commutators of ac and dc commutator motors. They can be adjusted and may be used to disconnect the motor if the current increases to 2½ to 3 times FLMC. The adjustment can be obtained by changing the air gap or the spring force or both.

27. Overload Relay Bypassed during Starting. It is also possible to protect motors for large centrifuges or compressors without a release valve or other long-starting equip-

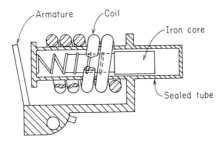

Fig. 48 Typical magnetic overload relay. (*Heinemann Electric Company.*)

ment by using two contactors in parallel and having the overload relay bypassed by the second contactor during the starting time. A properly selected adjustable magnetic circuit breaker or MSCP, which cuts the overload-relay curve of three overload relays at a very steep slope, is probably the best branch-circuit protection. Proper selection of the wires for such extreme starting conditions needs special investigation. The wire size for such motor applications must usually be larger than for standard motors.

Note that this starting arrangement may be good for only one start with the motor at ambient temperature. Before a successive start is attempted, the motor must be allowed to cool for a long period to avoid motor damage. Since the motor must have enough thermal capacity to withstand one or possibly more consecutive starting operations under such severe starting conditions, it is recommended that the motor manufacturer be consulted before this arrangement is applied. It may be possible to protect the motor during starting with a properly calculated second overload relay (Art. 17) or by inherent motor protectors (Art. 18).

28. Maintenance of Motor Protectors. Most overload relays need very little maintenance because they operate rarely (Ref. 14). Essentially, each overload relay should be checked for proper heater size, corrosion, dirt, and proper connections. A common error is incorrect initial wiring resulting in normally closed contacts of the overload relay being connected in the control circuit without the coil. This wiring error can cause welding of the tripping contacts, which prevents the overload relay from tripping out. If the electrician has corrected this mistake by proper wiring, a relay contact may remain welded, and the overload relay has lost its value. Therefore the motor circuit must be checked before usage whenever possible, for example, by tripping by hand. If an overload relay is exposed to a short circuit and the heater should burn out, it might be advisable to replace the whole overload relay unless the supplier has assured that a heater burnout would not cause a change of calibration of the relay. A bimetallic relay should be checked after heater burnout to be sure that no heater material was splashed on the bimetal. A eutectic-alloy-type relay should be checked for heater material in the ratchet. Magnetic-time-delay overload relays should be checked for freedom of movement, and relays filled with silicone oil should be checked for proper oil level. It is clear that the terminals must be tightened from time to time.

If an overload relay trips or a fuse blows repeatedly, check for accidental ground, excessive motor current, improper heater size, or internal shorts in the motor. Relays should be checked and cleaned whenever dirty. Generally thermal-overload relays should be replaced and not repaired if malfunction occurs. They are not expensive enough to be repaired, and after being repaired, they may not remain properly calibrated. If relays are repaired, they should be reused only after the calibration is checked.

If magnetic devices fail to drop out at the proper drop-out voltage, the coil voltage rating should be checked; or if the core surface is gummed or rusty, the surface should be cleaned. If the magnetic relay does not pick up at the proper voltage, the coil rating should be checked.

REFERENCES

1. National Electrical Code, 1968.
2. National Electrical Code, 1971 Preprint.
3. Underwriters' Laboratories Standards 508 and 547.
4. "Industrial Control and Systems Standards," NEMA Publication ICS-1970.
5. "Motors and Generators," NEMA Standards Publication.
6. Frank W. Kussy, "Elektrische Niederspannungsschaltgerate und Antriebe" (Electrical Distribution and Control Equipment and Systems), Technischer Verlag Herbert Cram, Berlin, 1969.
7. Robert E. Walters, Overcurrent Protection within the Motor Branch Circuit at Low Fault Currents, *IEEE Trans. Gen. Appl.*, vol. IGA-3, no. 6, November/December 1967.
8. Oelschlager, Die Berechnung von Widerstanden und Degl. für Aussetz. Betrieb (Calculation of Resistors, Motors, etc., for Intermittent Duty), *Elektrotech. Z.*, 1900, p. 1058.
9. Frank Yeaple, "Which Heat Protector for the Motor," *Prod. Eng.*, Sept. 12, 1960.
10. Jon Campbell, "Motor Protection" (Design Guide), *Machine Design*, Aug. 13, 1964.
11. K. Lehman and F. Steinbach, Des Phasenausfallschutz, ein Randproblem des Motorschutzes (Opening of One Phase, A Side Problem of Motor Protection), *BBC-Nachrichten*, June 1963.
12. H. J. Sutton, "Protective Relaying for Large Motors," AIEE Conference Paper CP-62-1092, Mar. 28, 1962.
13. "Ground Fault Application Protection Guide," *Bull.* 18.1-4A, I-T-E Imperial, 1973.
14. Robert J. Lawrie, How to Maintain Motor Controls, *Elec. Construction Maintenance*, April 1964.
15. "Manual of Technical Information," 2d ed., p. 164, Rome Cable Corp., 1957.
16. Frank W. Kussy, B. Di Marco, and K. W. Swain, "Total Motor Branch Circuit Protection with Instantaneous Trip Type Circuit Breakers and High Fault Circuit Protectors," IGA Annual Meeting, Philadelphia, 1972.
17. J. Linders, Effects of Power Supply Variations on AC Motor Characteristics, *IEEE Trans. Ind. Appl.*, 1972, p. 583.
18. Boothman, Elgar, Rehder, Woodall, "Thermal Tracking—A Rational Approach to Motor Protection," IEEE PES paper, p. 1335, made available for printing Nov. 12, 1973, presented at the IEEE meeting in New York City, Jan. 27–Feb. 2, 1974.

14

Metering and Instrumentation

L. A. FUNKE[*]

[*] Instrument Product Planner, Relay—Instrument Division, Westinghouse Electric Corporation.

FOREWORD

Certain measurements and recorded data may be vital for the safe and efficient operation of systems, processes, machines, and equipment. The instruments and metering equipment installed on switchgear, switchboards, and control equipment may be selected to provide these measurements and data.

This section was written to aid in specifying, applying, and maintaining instruments and metering equipment. Information is provided on their functions, selection, circuit arrangements, checkout during installation, and routine maintenance.

INSTRUMENT TYPES

1. Functional Characteristics. The safe, reliable, and economical operation of electrical machinery and distribution equipment requires accurate measurement of the electrical forces involved. Adequate instrumentation on switchgear and control equipment will enable an operator to load his equipment properly, to anticipate problems, and to solve them intelligently before they become reality.

Instruments, like control equipment, are built at several quality levels to match the criticality of the apparatus with which they are to be used. The instruments on a portable engine-generator, for example, ordinarily need not be as high quality as those on a large tubine-generator. In this section no further distinction will be made between instruments on the basis of their design level; so first some guidelines are given with respect to quality.

Electrical instruments are usually designed to conform to Standard C 39.1 of the American National Standards Institute. This standard recognizes two general classifications, "switchboard type" and "panel type," each with several variants (Fig. 1). The standard sets up the minimum requirements only. Table 1 is made up to be representative of the commercial practice of the several manufacturers who specialize in instruments for the industrial market.

This table is highly generalized. There are great differences between available brands, which affect quality. Plastic covers with integral windows can be the extremely abrasion-, impact-, chemical-, and heat-resistant polycarbonate; or an intermediate acrylic; or a styrene, which ranks quite low in these characteristics. In security (the ability to remain electrically intact under extremely adverse conditions) there is again a wide difference. An ac ammeter of one brand can be melted open with overload currents which will not damage an ammeter of another brand. It is therefore advisable to know the capability of the instrument that is applied.

In addition to the instruments listed, there are edgewise types within the panel classification. The 3- to 5-in sizes are sometimes used in industrial applications. Edgewise ac meters are usually uncompensated rectifier types whose limitations in ac measurements are covered in Art. 8. Dc edgewise models are comparable with conventional panel instruments in performance.

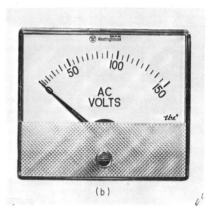

Fig. 1 Typical electrical indicating instruments. (*a*) Switchboard type. (*b*) Panel type.

Falling approximately into the performance class of switchboard instruments are several brands of 6-in edgewise instruments. These have either a moving-iron or a compensated rectifier mechanism; so they are capable of satisfactory ac measurement.

Measurement of electrical power also involves metering and telemetering (Art. 46).

2. Instrument Mechanisms. It is helpful in using instruments to have some familiarity with the operating principles involved. Any number of mechanisms have come and gone over the years, but the ones which have survived in the more common modern instruments are:

1. Permanent-magnet, moving-coil
2. Repulsion or attraction iron-vane
3. Dynamometer
4. Rotating iron-vane

a. Permanent-Magnet, Moving-Coil. In 1882 d'Arsonval made the first instrument of this type. He suspended an electromagnet in the field of a permanent magnet using a pair of fine wires which also conducted the current to the coil. As the coil was energized, it rotated against the twisting torque of the suspension wires. Edward Weston by 1888 had developed the d'Arsonval idea into a commercial instrument. He

TABLE 1. Characteristics of Electrical Instruments

	Panel type				Switchboard type				
Accuracy, %	±2				±1				
Size, in	2½	3½	4½	5½	4½	6	4½	4½	8¾
Deflection, °	90	90	90	90	100	100	180	250	250
Scale length, in	2	3	4	5	4	5	5	7	14
Readability, ft*	2–3	3–5	4–7	5–9	7	9	10	14	25
Influences, % †	1–2				0.5–1				
Magnetic shielding	dc, yes; ac, no				dc, yes; ac, yes				
Case	Plastic				Steel				
Cover	Plastic or metal				Plastic or metal				
Window	Plastic or glass				Plastic or glass				
Suspension	Usually pivot-and-jewel				Usually taut-band				
Shock resistance	Fair				Excellent				
Life expectancy	Moderate				Long				
Overload capability	Fair				Excellent				
Dust- and watertight	Fair				Good to excellent				
Security	Fair				Excellent				
Scales	Flat				Antiparallax or flat				

* Lower figure with tubular pointer, higher figure with broad-lance pointer.
† Added error for extreme operating conditions.

used an almost closed magnetic loop for permanence. He suspended the moving coil on jeweled bearings borrowed from watchmakers. He used their bronze hairsprings for restoring torque and as conductors for the current to the coil. The basic principle is used in today's permanent-magnet, moving-coil instruments for the measurement of direct current. Many modern instruments use a fine metal band to replace the pivot-and-jewel bearings and the hairspring conductors. This is known as taut-band suspension. Figure 2 illustrates both types of suspensions. Figure 3 illustrates the basic mechanism.

Current, in flowing through the coil, generates a magnetic field at right angles to the plane of the coil. The magnetic field of the coil produces a force tending to align the coil with the field of the permanent magnet, generating a torque. The torque is proportional to the current in the coil. Rotation of the coil is opposed by the hairsprings, which permit deflection proportional to torque. Thus the pointer, connected to the shaft of the coil, rotates proportionally to the current in the coil.

Modern permanent-magnet, moving-coil instruments seldom use the horseshoe magnet as shown. Instead the permanent magnet may take the form of a magnetized ring or a soft iron ring with a magnetized slug; or the permanent magnet can be the core inside the coil with only a soft iron return ring around the coil. Some designs, using specially shaped magnetic parts, rotate 250° from zero to full scale. Each of these constructions has its reasons for being used, but all operate on the basic principle first used by d'Arsonval.

Permanent-magnet, moving-coil instruments are used for dc measurement. A dc voltmeter is such an instrument mechanism in series with a resistor to limit current. A dc ammeter is the same device in parallel with a shunt which permits only a small part of the load current to flow through the instrument coil.

Today we have transducers to convert every conceivable mechanical or ac electrical measurand into a dc signal. The permanent-magnet, moving-coil instrument is used as the readout device for such signals. The first, and probably still the most used, of such converting devices is the rectifier which, when combined with a dc instrument, permits it to read ac voltage or current. Rectifier instruments are used universally in the popular volt-ohm-milliammeter (VOM) test sets. Their limitations in ac measurement will be covered in Art. 8.

b. Repulsion or Attraction Iron-Vane. The commercial use of alternating current followed that of direct current by about 10 years. The development of suitable instruments moved along in the same time schedule, but with several false starts. There

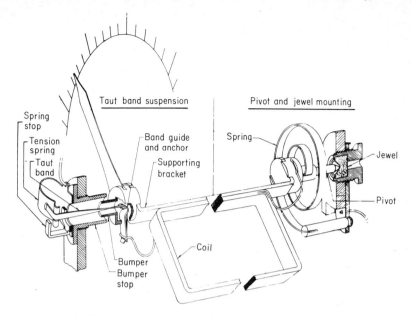

Fig. 2   Comparison of taut-band and pivot-and-jewel bearing systems.

were solenoid-plunger and induction-disk instruments for the early ac machines, but these were generally unreliable. A development by the Nalder brothers in Great Britain finally proved to be the answer. They found that if two flat pieces of soft iron were placed in a large solenoid, one piece fixed and the other free to move about a hinged edge, these pieces would separate like the covers of a book when alternating or direct current was passed through the coil. The iron pieces tended to be repelled from each other because the alternating magnetic flux induced instantaneous like poles at adjacent edges of the iron (Fig. 4).

Repulsion iron-vane instruments built by George Westinghouse in 1894 used gravity as a restoring force. Edward Weston, at the same time, also became involved with ac measurements and adapted the hairspring to the instrument, which proved more practical. It is used today as one of several configurations of the repulsion iron-vane

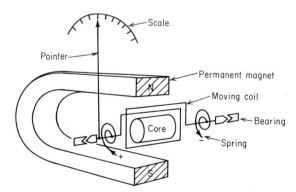

Fig. 3   Permanent-magnet, moving-coil (d'Arsonval) dc instrument mechanism of an early type using a horseshoe magnet.

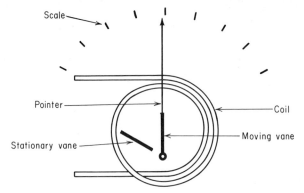

Fig. 4 Repulsion, iron-vane ac instrument mechanism of the radial-vane type. Schematic front view with torque springs and bearings omitted for clarity. Pointer arc 100°.

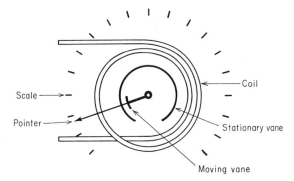

Fig. 5 Iron-vane ac instrument mechanism of the concentric-vane type. Schematic front view with torque springs and bearings omitted for clarity. Pointer arc 180 or 250°. "Rolled-out" views of the iron vanes shown in Figs. 6, 7, and 8 for the several types of vane configurations used.

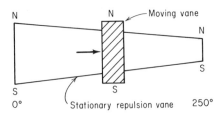

Fig. 6 Repulsion, iron-vane, circular-scale ac instrument. "Rolled-out" view of the iron vanes.

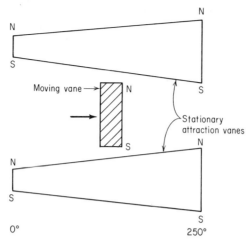

Fig. 7 Attraction, iron-vane, circular-scale ac instrument. "Rolled-out" view of iron vanes.

mechanism. Generally it is referred to as the radial-vane construction. It is used on all 90° deflection ac instruments. It is used for dc measurement only in certain highly special cases. The soft iron vanes have a tendency to retain some magnetism, which introduces a residual error on dc unless the average of two reversed readings is used.

The scale on a radial-vane instrument tends to compress at the low and high ends, and it is limited to 90 or 100° of pointer rotation. To overcome these limitations, the concentric-vane mechanism was developed (Figs. 5 and 6). Here again, both the stationary and moving vanes are surrounded by a solenoid which carries the current to be measured. The alternating current induces instantaneous like magnetic poles simultaneously on adjacent parts of both vanes, causing them to repel each other. The moving vane (Fig. 6) will travel as far as possible to the right so that like polarities on the stationary vane are as far away from its pole ends as possible. This device is called the concentric repulsion iron-vane mechanism.

The opposite principle works as well. If two stationary vanes are located as in Fig. 7, the coil will induce opposite magnetic poles on adjacent parts of the stationary and

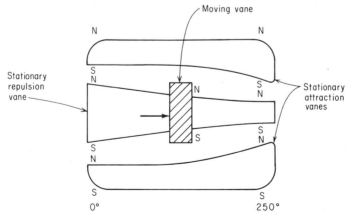

Fig. 8 Repulsion-attraction, iron-vane circular-scale ac instrument. "Rolled-out" view of iron vanes.

moving vanes. The moving vane will travel to the right until its pole ends are as close as possible to the opposite polarities on the stationary vanes. This device is called the concentric attraction iron-vane mechanism.

Either the repulsion or the attraction iron-vane mechanism will work well for up to 180° of rotation; after that they become somewhat unstable because of the length of the air gaps at either end. For 250° instruments, a combination of the two is used (Fig. 8). The vanes are shaped with curving edges to linearize more nearly the instrument scale. This mechanism is the concentric repulsion-attraction, iron-vane design used in commercial 250° ac ammeters and voltmeters.

c. Dynamometer. The third basic instrument is the electrodynamometer. This type existed in the earlier years as a laboratory curiosity and was first developed by Siemens in 1884 as a wattmeter. Edward Weston started working on a dynamometer voltmeter in 1890, but it was not until 1910 that electrodynamometer instruments, spring-restrained, with pivot-and-jewel bearings, took a practical form. But, once developed, they remain largely unchanged today.

To understand how this mechanism works, refer to the permanent-magnet, moving-coil instrument (Fig. 3). The permanent magnet could be replaced with a piece of

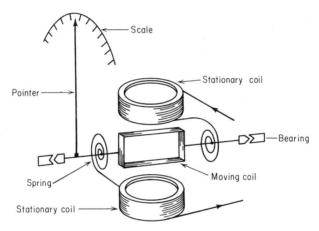

Fig. 9 Dynamometer 100° scale ac instrument connected as a milliammeter.

soft iron having a coil of wire around its legs to make it an electromagnet whenever current flowed in the coil.

With the moving coil and the magnet coil connected in series across a source of direct current, this instrument would function like the original permanent-magnet instrument. The deflection would not be linear, however, because now the field flux is a variable along with the moving-coil flux so the torque developed will be as the square of the measured current. This instrument is the iron-core dynamometer used to measure dc watts. The moving coil is connected to the voltage and the stationary coil to the current circuit.

The most common of the dynamometers eliminates the iron entirely and replaces it with two flat coils where the iron-magnet pole faces had been. The mechanism can now be used on ac because there is no iron and because both the stationary and the moving coils will change their instantaneous polarities simultaneously (Fig. 9).

The moving coil of the dynamometer can carry currents of only a few milliamperes maximum, but the stationary coils can be built as high as 100 A. For measuring voltage, coils of about 10 mA are wound on both elements. These are connected in series through a resistor across a voltage source. For small currents, as in a milliammeter, the series resistor is eliminated. For heavy currents the stationary coil is wound for the rated current, and the moving coil is shunted so it carries only a fraction of the load.

The dynamometer develops a torque proportional to the square of a current common to both the coils. If the current doubles, the flux due to the stationary coil doubles, and the flux due to the moving coil doubles. The result is four times as much deflection. With the two coils being electrically separate, it is possible to put a current winding on the stationary coil and a voltage winding on the moving coil. This connection makes the device into a multiplying instrument with the capability of measuring watts. The torque at any instant is proportional to the product of the instantaneous values of current and voltage. This value is $EI \cos \theta$, or true power.

The dynamometer meter has numerous variants adapting it for use in frequency, phase-angle, power-factor, and ratio-meter applications.

d. Rotating Iron-Vane. In the first days of the electrical industry, when machines were being paralleled, there were only synchronizing lights. Then in 1901, Dr. Paul

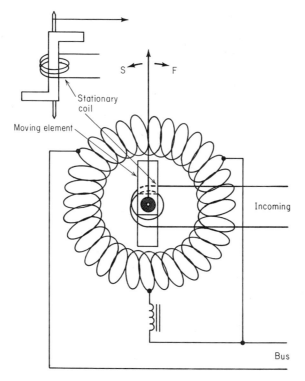

Fig. 10 Rotating iron-vane or polarized iron-vane ac instrument mechanism connected as a synchroscope.

Lincoln developed the crossed-coil synchroscope, and at the same time Dr. Frank Conrad developed the rotating iron-vane type. The Lincoln meter enjoyed popularity for a number of years, but more recently the rotating iron-vane type has taken over.

The rotating iron-vane instrument has a stator exactly like that of a small polyphase induction motor, either 2-phase or 3-phase, connected to a voltage source through a capacitor-reactor network. The network causes phase shifts which set up a rotating magnetic field in the stator.

The rotor consists of two magnetic lobes 180° apart mounted one at each edge of the stator on a soft iron shaft. The shaft passes through a coil. When the coil is energized, the lobes become opposite magnetic poles (Fig. 10).

If the stator were energized with ac and the vanes with dc, the instrument would run as a synchronous motor. The field of the vanes would lock onto the rotating field of the stator. If both the stator and the coil were energized from the same ac source, the instrument would stand still. The vane would seek that point of rest at which the maximum flux in the field and in the rotor occur at the same instant. If it were connected to two different sources of ac, the vane would still seek that position where instantaneous fluxes reach maximum simultaneously, but the instrument would then rotate at whatever speed represented the frequency difference of the systems. This characteristic permits the instrument to be used as a synchroscope to bring two machines together so that corresponding phases are exactly in synchronism.

Actually the synchroscope measures the phase angle between two voltages. By substituting a current coil on the rotor it is possible to measure the angle between a current and a voltage.

With such an arrangement the instrument becomes a phase-angle meter which, when fitted with a cosine function scale, becomes a power-factor meter.

3. Definitions. A review of the terminology used in the instrument field, but not defined in the text, is given for additional background.

a. Accuracy is the limit, expressed as a percent of full-scale value, that errors will not exceed under reference conditions. An instrument of 1% accuracy class, 150 V full scale, will have a permissible error of 1.5 V at any level of reading. This error amounts to 2% of reading at 75 V and 4% of reading at 37.5 V.

b. Influence is allowable additional error for deviation from reference conditions. There are allowable influences for such factors as temperature, position, magnetic field, frequency, and waveform. These errors are all cumulative.

c. Full-scale value is the value of the electrical quantity required to deflect the instrument across its entire scale length. In the case of a 0-50 V voltmeter the full-scale value is 50 V. In the case of a 50-0-50 V center-zero instrument it is 100 V.

d. End-scale value is the value of the electrical quantity required to deflect an instrument from zero to the end of its scale. In the above examples the end-scale value is 50 V.

e. Self-contained instrument has all the necessary equipment to perform a given measurement built into the instrument case. Thus a frequency meter may operate from a separately mounted transducer, or the transducer may be incorporated into the instrument case for a self-contained design.

f. Reference conditions are those conditions at which the instrument was designed to deliver its rated accuracy. For panel-mounted instruments these conditions are:

Temperature, °C.................. 25
Position............................. Vertical
Relative humidity, %............ 30–60
Stray field.......................... Earth only
Vibration........................... Negligible
Waveform.......................... Pure sine wave
Frequency, Hz..................... 60 ± ½

g. Rated current voltage is that value used for design purposes, usually 5 A for current coils and 120 V for potential coils, except for voltmeters, which are usually rated 150 V.

h. Overload capacity is the value of current or voltage expressed as a percent of rated value which an instrument can sustain for a stated time without damage. Momentary overloads are generally construed to be of 1 s duration. Generally, however, capabilities in the area of 2 to 10 s are of more concern. These must be obtained from the instrument manufacturers. There are no standards.

4. Checkout for Service, All Types. Recommendations are given in this article for checking specific types of instruments before placing them in service. In addition there are a number of points which apply to all instruments.

a. Receiving Check. If the instruments are received mounted on a panel, they should be inspected for obvious damage only. The pointers should be on zero and, if not, should be readily adjusted to zero as described below. Pointers should be straight and parallel with the dial. If not, the instrument should be removed and tested. If instruments are received in the original factory boxes, they should first be checked against the description in the bill of material. Then each should be inspected for obvious damage, for balance, and the pointer should be checked as recommended above. Then the instrument should be twisted sharply in the hand to drive the pointer upscale; it should return normally. The dial should be inspected for pointer tracks indicative of severe handling in transportation. If any point is questionable, the instrument should be tested. Whether or not all instruments are tested before installation is an economic judgment which varies from case to case.

b. Balance. An instrument can, but should not, become off balance during shipment. To check, hold the instrument with the dial horizontal and set the pointer to zero. Then hold the instrument with the dial vertical and rotate it so the pointer is horizontal in the 9 o'clock position. The pointer should still be touching the zero calibration line. For 90 and 100° deflection instruments check also with the pointer vertical at the 12 o'clock position and in the normal operating position at zero. The pointer should continue to touch the zero calibration line. In the case of 180° circular-scale instruments the checks are at 9, 12, and 3 o'clock and normal position. For 250° instruments the 6 o'clock position is added. The above does not apply to synchroscopes or to phase-angle and power-factor meters of the rotating iron-vane type. The synchroscope is deliberately thrown off balance so that it will seek the 3 o'clock position when disconnected. Other rotating iron-vane instruments cannot be checked for balance when deenergized. If any instruments are found to be out of balance, they should be sent to an authorized service shop for correction.

c. Zero Setting. There is a screwdriver adjuster on the front of every standard commercial instrument. Turning this device will move the pointer of a deenergized instrument slightly above or below the zero mark on the dial. To adjust, turn so the pointer is below zero. Then bring it up so the pointer splits the zero mark. Many experienced users recommend that the adjuster then be backed off a "hair." This procedure makes it unlikely that any shift in the cover, especially on panel instruments, will affect the zero adjustment.

Mechanically suppressed and some electrically suppressed instruments do not have a free zero. That is, when the instrument is deenergized, the pointer will run off scale, usually against a stop. On such instruments the pointer is adjusted to the first division on the low end of the scale by energizing the instrument to the low end-scale value. The current or voltage is checked with a portable standard, 0.5% accuracy class. The "zero" adjuster is then turned as in the instructions for zero setting until the pointer splits the first scale division.

5. Routine Maintenance, All Types. Instruments require very little attention, but periodic inspection is good preventive maintenance.

a. Cleaning. The outside surfaces of an instrument may be cleaned with a moist chamois or tissue using a mild detergent—no solvents, window sprays, or cleaning solutions containing acetone, benzene, carbon tetrachloride, or similar solvents. If there is dirt inside the instrument, it should be cleaned by a qualified instrument repairman. He should replace or install gaskets or cover cement.

b. Inspection. Instruments should be checked for loose glass or cover, zero setting, and freedom from friction if the instrument can be activated. They should also be checked for dial discoloration or smoky film on the window. These may indicate overheating. Connections should be checked, especially on dc ammeters with separate shunts.

c. Testing. Instruments should be field-tested periodically; once every 2 years is usually enough for industrial service, but experience will govern in extreme cases. Each instrument should be checked against a portable standard of 0.5% accuracy class. Either a dynamometer or an iron-vane instrument should be used for ac voltage or current tests; otherwise waveform errors may cause discrepancies in readings which are not in fact indicative of calibration errors.

d. Antistatic Treatment. All plastic instrument covers are treated with an antistatic agent during manufacture. This material dissipates in time. It should be replaced if the instrument shows a tendency to hang upscale when wiped with a dry cloth. Normally any static charge will bleed off in a few seconds. If charges persist, the covers should be sprayed with an antistatic compound (available from most manufacturers) and wiped with a soft cloth. For best results both the inside and outside surfaces should be treated.

6. Diagrams.   Caution: The connection diagrams as shown in this section are typical. The physical location of terminals is not necessarily as shown for any one manufacturer's instruments.   Resistors may be required in potential circuits depending upon the particular voltage coil ratings.   Numerous variations are possible.   These diagrams cover only the most commonly used instrument for any one function.   The diagrams published by the instrument manufacturer should always be followed.   Grounding connections from metal instrument cases are usually made to the current transformer and potential transformer common secondary ground.   These connections are not shown on the diagrams.

VOLTAGE MEASUREMENT

7. Voltmeters for ac measurement are usually made up with iron-vane mechanisms, but there are some with dc mechanisms fed through rectifiers.   The iron-vane types respond to rms values, and the rectifier types are average-responding.   The rectifier types draw less power from the circuit being measured, however.

8. Voltmeter Selection.   Electrical-power calculations are made in terms of rms values yielding true power as measured by the heating effect of a current (Fig. 11). Rectifier-type instruments respond to the lower average value of the ac wave; so the actual dial is marked by increasing each reading 11%.   This adjustment results in a dial calibrated in rms for a pure sine wave only.   Industrial power circuits rarely have a pure sine wave.   Modern plant equipment with its phase-controlled rectifiers, its resonant circuits, and its capacitors all tend to distort the voltage wave.   This distortion is especially noticeable on those power circuits in which there is no delta-wye transformer between the instrument and the load.

The errors in rectifier instruments cannot be ignored.   Figure 12 shows the relationship between error and third-harmonic content of the measured wave.   It also shows the performance of those rectifier instruments, usually switchboard types, which incorporate an rms compensating network.   The third-harmonic content is a reference standard for the performance of instruments on nonsinusoidal waveforms.   Industrial circuits often run in the 5 to 10% area, and 20% is not uncommon.   A distinction must therefore be drawn between rectifier instruments "rms-calibrated" and those "rms-compensated."   The safest approach is to use iron-vane instruments for ac power measurements unless there is a requirement for some special feature, such as low burden, available only in rectifier types.   In these cases if the waveform is uncertain, it is best to specify rms compensation such that the error due to waveform will not exceed 1% with 15% third-harmonic content.

Voltmeters may be supplied with electrically expanded or suppressed scales (Fig. 13).   There are some of these instruments using inductors and shaped iron vanes, but in recent years the trend has been toward rms-compensated rectifier instruments using zener-diode circuitry for scale shaping.   These instruments can be read much more closely because of the expanded scale.   The switchboard types usually are rated ±0.3% of midscale value while panel types run something less than double that error.   The exact accuracy depends on the actual amount of suppression or expansion.

Mechanical suppressions are also possible, but they are impractical because of the

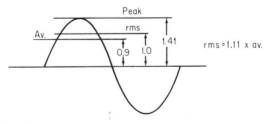

Fig. 11 Ac waveform characteristics.   Instantaneous voltages are plotted against time or phase angle.

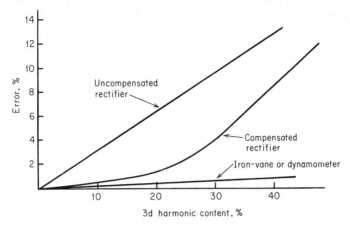

Fig. 12   Performance of various ac instrument types when reading rms values on imperfect wave-forms.

difficulty of checking zero after the instrument is in service. The limit of mechanical suppression on an ac instrument is about 35%, which affords little increased readability.

Electrically suppressed or expanded instruments should never be ordered with volt-age ranges different from catalog listings. Completely new network designs are needed for each range. The conventional voltmeters, however, can be supplied in any voltage up to 600 or 800 V with only a change in internal resistors. Scales are marked in primary voltages whenever potential transformers are used.

9. Voltmeter Connections.   Potential transformers are used to step down bus volt-age to 120 V nominal to energize an instrument with a 150 V end scale. The potential transformer capacity is rarely a problem, as typical moving-iron instruments will pre-sent a burden of less than 2.5 VA and rectifier types about 0.1 VA.

The normal transformer configuration for 3-phase, 3-wire systems is shown in Fig. 14 and 4-wire systems in Fig. 15.

In the case of Fig. 14 the voltages 1-2, 2-3, and 3-1 are measured across the corre-sponding leads from the transformer secondary. A single voltmeter may be used with a suitable instrument switch, or three voltmeters may be left permanently connected in the circuit. Transformer secondaries are usually based on 120 V nominal. The instruments used at 120 V have a normal full-scale rating of 150 V. The scales are marked with a full-scale voltage equal to 150 times the potential transformer ratio.

Figure 15 shows a set of three transformers in a conventional wye hookup. Again the secondaries are 120 V nominal; so a voltmeter connected N-1, N-2, or N-3 should be rated 150 V full scale and scaled at potential transformer ratio times 150. This volt-meter would read line-to-neutral voltage. It is customary, however, to refer to the sys-tem voltage in terms of line-to-line voltage and to scale instruments to correspond. In this case voltmeters rated 250 V at end scale would be used and connected 1-2, 2-3, and

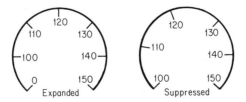

Fig. 13   Typical dial distribution for expanded-scale and for suppressed-zero, circular-scale ac instruments.

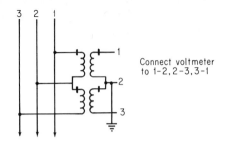

Fig. 14   Voltmeter connections for 3-phase, 3-wire circuit.

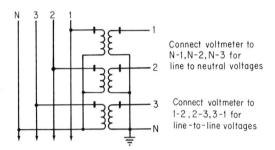

Fig. 15   Voltmeter connections for 3-phase, 4-wire circuit.

3-1. The dial would be marked at a full-scale value equal to the potential transformer ratio times 250 V.

If the voltages are assumed to be balanced, a single voltmeter connected line-to-neutral may be marked in terms of line-to-line voltage. In this case the voltmeter will have a full-scale rating of 143 V, but a scale with a full-scale value equal to 250 times the transformer ratio.

Where the line-to-neutral voltages are not needed for relays or for other instruments, it is customary to connect two potential transformers line-to-line as in Fig. 14 even though the system is wye-connected. In this case the voltmeters are applied as for 3-wire systems.

10. Voltmeter Checkout for Service.   There is little that can go wrong in the connection of a voltmeter. If it does not read at all, look for a blown potential transformer fuse or an open circuit in the meter. If it reads high or low by a factor of 1.73, it is probably connected incorrectly to the potential transformer on a wye system. Either error can also be caused by incorrect polarity of a primary or a secondary on the transformer. There will be no polarity error if both the primary and the secondary are reversed, however.

CURRENT MEASUREMENT

11. Ammeters.   An ac iron-vane ammeter has essentially the same mechanism as an ac voltmeter. The difference is that it has fewer turns of heavier wire in its coil so that load current can be passed directly through it. Rectifier-type ac ammeters differ from rectifier voltmeters in that they use an input transformer, except for some of the low milliampere ranges. Having the same mechanisms, the performance is the same with respect to average and rms measurements. Being right at the source of most current waveform distortion, the ammeters on industrial-plant feeder circuits are particularly susceptible to waveform error. Those instruments in circuits to conventional large motors are less so.

Ammeters are likely to be read anywhere on the scale; so an attempt is made to distribute the scale uniformly. This distribution is accomplished reasonably well except at the lowest 5 or 10% of the scale. Where an ammeter must be read accurately at both low and high values it should be specified for a dual range (2/1 ratio is the only practical combination), or a multiratio auxiliary current transformer may be used. Electrical or mechanical suppression is possible but not practical.

12. Ammeter Selection. Most manufacturers offer ac ammeters in self-contained ratings up to 50 or 75 A. This means that the actual load currents may be passed directly through the terminals of the instrument. Industrial circuits usually run to the hundreds or thousands of amperes; so current transformers are used to step down the current to something the ammeter can handle. The standard current for transformer circuits is 5 A *maximum*. (Note the difference between this value and the nominal value used on voltmeters.) Here, when the current transformer is loaded to rating, the associated instrument will be at full scale rather than 80% scale, as on a voltmeter. Therefore, if a 25% overload is to be read on a nominally 400-A circuit, the current transformer should be rated 500/5 and the ammeter 5 A. The scale would be 0-500 A. Sometimes, because of the protective devices on a circuit, the higher-rated transformer cannot be used. For such cases most manufacturers of switchboard instruments will supply a moving-iron ammeter rated 6.25 A with a 0-500-A scale. It is somewhat more difficult to get switchboard rectifier ammeters or any panel types in nonstandard ac ratings, particularly in small quantities.

Ammeters are often called upon to carry heavy momentary overloads during fault conditions and somewhat less severe overloads for a longer time during motor-acceleration periods and the like. Compliance with ANSI C39.1 standards is no guarantee of competence in this area. Perhaps the best criterion is experience. A brand and model should be selected which have been used in industrial power service over the years. However, if capability for infrequent overloads is not specified, a good requirement is 35 times full load for 2 s and 10 times full load for 10 s without mechanical or electrical damage. It is unnecessary to be concerned about transients greater than 35 times because, in any reasonably good design, the saturation of the iron vanes or the input transformer, combined with saturation of the current transformer, will have established a limit well below this level. For cyclic overloads, where heat dissipation may become a factor, it is best to discuss the specific problem with an instrument manufacturer. On continuous overloads ANSI C39.1 will allow one-half rated accuracy as a permanent error after 6 h at 120% load. Manufacturers who are experienced in instruments for industrial service work more closely to 200% as a continuous overload limit.

13. Ammeter Connections. Although ammeters on power circuits are usually used with current transformers to reduce measured currents to manageable levels, the transformers also serve to insulate the measuring circuits from the high bus voltage.

Figure 16 shows the conventional current-transformer configuration for 3-phase, 3-wire circuits. The current in line 1 is measured 0-1, line 3 is 0-3. Line 2 is 1-3 because, in any delta 3-phase system, the vectorial sum of currents 1, 2, and 3 is zero. Any switch used to transfer an ammeter between these points must short-circuit any current transformer not in the measuring circuit, and must short-circuit both transformers at

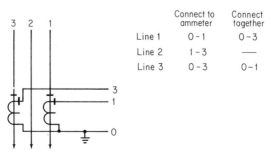

	Connect to ammeter	Connect together
Line 1	0-1	0-3
Line 2	1-3	—
Line 3	0-3	0-1

Fig. 16   Ammeter connections for 3-phase, 3-wire circuit.

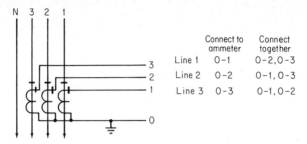

	Connect to ammeter	Connect together
Line 1	0-1	0-2, 0-3
Line 2	0-2	0-1, 0-3
Line 3	0-3	0-1, 0-2

Fig. 17   Ammeter connections for 3-phase, 4-wire circuit.

any intermediate position or in the off position.   Specialized instrument switches are made for this purpose.

The 3-phase, 4-wire system requires three transformers, one in each line, as shown in Fig. 17.   Current in line 1 is measured across 0-1, line 2 across 0-2, and line 3 across 0-3.   Now the vectorial addition of currents need not add up to zero, but rather will represent the current in the system neutral.

14. Ammeter Checkout for Service.   Ammeters are even more trouble-free than voltmeters.   If there is no reading at all, look for an unopened shorting switch on a current transformer.   If the current in line 2 of a 3-wire system is too high, look for a reversed polarity on a current transformer.   If the reversed connection exists on the primary side, the secondary side may be reversed instead of correcting the primary. *Under no circumstances should any of the current circuits be opened while the equipment is energized.*   Dangerously high voltages would exist across any break in the circuit.

POWER MEASUREMENT

15. Wattmeter and Varmeter Functions.   Electrical power has two components, that which does useful work, and that which merely circulates.   The useful power is measured in watts, the product of voltage and in-phase current.   The circulating power is measured in reactive voltamperes (vars), the product of voltage and that component of current which is 90° out of phase with the voltage.   When added vectorially, the resultant is the apparent power, or voltamperes, of the circuit.   The vector angle between the apparent and the true power is the power-factor angle θ (Fig. 18).   It is the same as the angle by which the circuit current leads or lags the voltage.   The cosine of this angle is equal to power factor or the ratio between the true power (watts) and the apparent power (voltamperes) (Art. 20).

It is apparent that any instrument for measuring watts must be able to evaluate the cosine of the phase angle θ and multiply it by the product of voltage and current.   A dynamometer mechanism is capable of making such measurements.   The voltage is impressed on the moving coil and the current on the stationary coil.   The instrument develops torques proportional to the instantaneous products of current and voltage

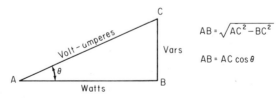

$$AB = \sqrt{AC^2 - BC^2}$$

$$AB = AC \cos\theta$$

Fig. 18   Elementary vector diagram.

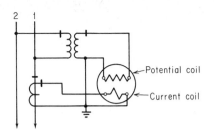

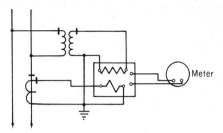

Fig. 19 Wattmeter connections for single-phase, 2-wire circuit.

Fig. 20 Watt transducer connections for single-phase, 2-wire circuit.

which is $EI \cos \theta$. Two dynamometer elements are mounted on a single shaft for 3-phase, 3-wire measurements, and with a special connection, usually referred to as $2\frac{1}{2}$ elements, on 3-phase, 4-wire.

Actually the dynamometer mechanism is used only in 100° instruments. The dynamometer does not lend itself to a construction using a taut-band suspension for the moving element. To utilize the circular-scale taut-band instrument, with its decided advantages in readability, long life, ruggedness, and stability, all major switchboard-instrument manufacturers have standardized on a combination of taut-band dc instrument with a built-in transducer to convert ac watts to a signal which the instrument can use. Panel-instrument manufacturers use a separately mounted transducer. The actual connections for a transducer are the same as those for a dynamometer to the circuit being measured.

There are numerous connection schemes which will work with wattmeters. The arrangements covered here are those most commonly used in industrial switchgear and control. Diagrams show current and potential transformers. For those low values of current and voltage not requiring transformers the instrument coils are connected directly into the circuit to replace the transformer primaries in the diagrams. The diagrams show wattmeters only. Watt transducer diagrams are the same as the corresponding wattmeter diagrams, but a separate indicating instrument is connected to the additional terminals on a transducer.

16. Types and Connections for Wattmeters. There are a number of connection arrangements for wattmeters.

a. Single-phase, two-wire measurement uses a single-element wattmeter connection (Fig. 19), or a transducer circuit (Fig. 20).

b. Single-phase, three-wire measurement uses the same two-element wattmeter used for 3-phase, 3-wire service. The connections are exactly the same with the neutral being the L_2 leg (Fig. 21).

It is somewhat less expensive to do single-phase 3-wire metering with a single-element wattmeter and a 3-wire current transformer having two primary windings. In

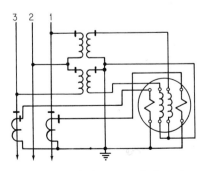

Fig. 21 Wattmeter connections, 2-element, on 3-phase, 3-wire circuit.

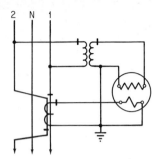

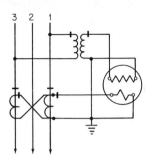

Fig. 22 Wattmeter connections for single-phase, 3-wire service using a double-primary current tramsformer.

Fig. 23 Wattmeter connections, single-element on balanced 3-phase, 3-wire circuit.

this case the wattmeter potential coil is 240 V, or a 2/1 potential transformer may be used (Fig. 22).

c. Three-Phase, Three-Wire. The conventional connections for a two-element wattmeter used on a 3-phase, 3-wire circuit provide accurate readings regardless of power factor or current or voltage unbalance (Fig. 21).

If currents and voltages are balanced, as they would be on a motor load, it is possible to use a single-element, single-phase wattmeter (Fig. 23).

d. Three-Phase, Four-Wire. Theoretically a 3-element wattmeter is required to measure 3-phase, 4-wire power. Such an instrument is impractical to build; so a modification of a 2-element meter is used. In this design, popularly known as 2½-element, there are three current coils (Fig. 24). One coil is on each element and the third is split between the two elements. The 2½-element wattmeter is correct only under balanced-voltage conditions. However, electrical-power systems or equipment will not tolerate a degree of unbalance which would cause errors outside the accuracy rating of the instrument. There is therefore no practical need for a 3-element device. The error due to 15% voltage unbalance is in the order of 0.2% of reading, which is considered negligible in any commercial power measurement.

17. Types and Connections for Varmeters. Varmeters are exactly the same as watt-meters except that a single-phase reactive compensator or a polyphase phase-shifting transformer is used in the potential circuit. These connect to the potential transformers. They have output voltages which lag the input voltages by 90°. Figures 25 to 27 show the connections to the phase-shifting devices as they are inserted into diagrams of Figs. 19, 22, and 24 for equivalent wattmeters.

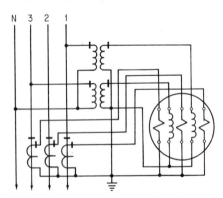

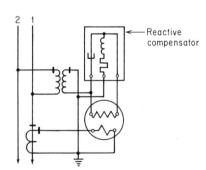

Fig. 24 Wattmeter connections, 2½-element to 3-phase, 4-wire circuit.

Fig. 25 Varmeter connections for single-phase 2-wire circuit using an external reactive compensator.

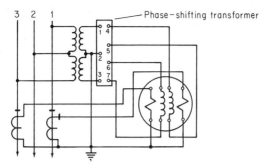

Fig. 26 Varmeter connections, 2-element, on 3-phase, 3-wire circuit, using external phase-shifting transformers.

It is possible to switch the phase shifters in and out of the circuit to make a single instrument serve as both a wattmeter and varmeter.

More recent developments in varmeters and var transducers are the self-contained types which do not require external phase shifters. These connect exactly like the wattmeters in Figs. 19, 22, and 24. It is not possible to operate these self-contained instruments as combination watt- and varmeters.

18. Meter Selection. Wattmeters as well as varmeters are rated nominally 5 A and 120 V, but they are not ordinarily calibrated for full scale at 5 times 120 or 600 W or vars, as one might expect. They are calibrated for that value of power which, when multiplied by the current transformer and the potential transformer ratios, will give the desired end-scale value on the instrument.

Wattmeters are calibrated on single-phase with all current coils in series and potential coils in parallel. Thus each wattmeter has a specified single-phase calibrating watts which will cause deflection to full scale. This value is usually marked on the instrument dial. The basic relationship is

$$W_{cal} = \frac{W_{fs}}{r_{ct} \times r_{pt} \times K} \tag{1}$$

where W_{cal} = single-phase calibrating watts
W_{fs} = full-scale watts on instrument
r_{ct} = current-transformer ratio
r_{pt} = potential-transformer ratio
K = 1 for 1-element, 2 for 2-element, 4 for 2½-element

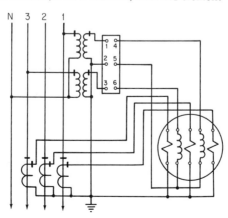

Fig. 27 Varmeter connections, 2½-element, to 3-phase, 4-wire circuit using external phase-shifting transformer.

Each wattmeter also has a self-contained rating W_{sc} which is the secondary watts which the instrument measures at full scale.

$$W_{sc} = \frac{W_{fs}}{r_{ct} \times r_{pt}}$$ (2)

The self-contained rating is the most convenient term to use in designating wattmeter calibration.

A wattmeter has a limited range of current capacity. One rated 5.0 A nominal, for example, may have a capability of being calibrated between 4.0 minimum and 6.25 A maximum. Therefore most manufacturers include in their catalogs a tabulation of preferred scales giving the minimum, nominal, and maximum wattmeter scales for any combination of current and potential transformers. It is possible to get scales other than those listed as preferred, but these will usually require special current coils.

Wattmeters can be made zero-center (10MW-0-10MW) or offset-zero (5MW-0-15 MW) for those situations where power flow may be in either direction. The instruments are, by nature, direction-sensitive; so the modification is simply a mechanical change of the zero point.

Varmeters are left-zero when there is only one generator on a system. Here vars can flow only in one direction. The usual application, however, has var flow in either direction; so most varmeters have either center-zero or offset-zero scales.

Wattmeters and varmeters on the same circuit usually have their current coils in series, with both measuring the same current. Thus the wattmeter and varmeter scales must be the same, if left-zero. If the scales are center or offset-zero, the sum of the end-scale values on the varmeter must be equal to the sum of the end-scale values on the wattmeter. Some deviations are possible, but these values must be checked out with the instrument manufacturers before being specified.

There are many other possible connection schemes for wattmeters and varmeters, but they are beyond the scope of this section. If the transformer configuration is different from the classical ones shown here, a sketch should be sent to an instrument manufacturer for evaluation. Verbal descriptions of unusual transformer and instrument hookups should not be trusted; they cause more errors than any other factor in the application of watt- and varmeters.

19. Meter Checkout for Service. Because there are more possible ways of connecting wattmeters than any other instruments, there is a greater possibility of error in calibrating or installing them. The following check should be made:

1. If a wattmeter fails to register or gives any reading differing from that calculated as correct, check all current-transformer short-circuiting devices. Then check out the connections for configuration and polarity.

2. If the instrument is a 2-element type and it fails to register at all, try shorting out one current coil. If it now reads half the estimated load, the polarity on the shorted coil is incorrect and should be reversed. If, however, the instrument is driven to the peg in the downscale direction, the chances are the nonshorted coil is reversed. The check should be repeated on the other current coil, and if half scale is now registered, the current leads should be reversed. It is possible that the current coils were correctly connected and that the voltage coils were in fact reversed but that the reversal of current leads compensated for the error. This possibility should be investigated. The equipment should not be left in service with two wrongs making a right, as this could cause some other difficulty.

3. A similar procedure will check out a 2½-element wattmeter except that each current coil provides one-third rather than one-half of the torque.

4. If these checks do not find the trouble, it is best to make actual current and voltage measurements at the instrument using a portable voltmeter and ammeter. The secondary voltages for the 2-element wattmeter must be 120 V open-delta with the common grounded, and voltages for the 2½-element must be 120/208 V wye with the center grounded. Current transformers must be shorted out while the test ammeter is being installed. It is desirable to load the current circuit with a 5-A phantom load, but any current at 2 A or more will be satisfactory.

If the connections and the current and voltage measurements are found to be correct, the instrument itself becomes suspect. The possibility of an incorrect calibration should be checked. Equation (1) will yield a value for single-phase test watts. This value should check with the information on the instrument dial. If the calibration information seems in order, an instrument technician should make a single-phase calibration check. Ordinarily it is best to send the instrument to a calibration laboratory for this work.

5. Varmeters are checked the same as wattmeters except that there are now more connections to verify because of the added phase shifter.

POWER-FACTOR MEASUREMENT

20. Power factor is generally defined as the ratio between true power (watts) and apparent power (voltamperes). In single-phase circuits this ratio can be expressed as the cosine of the phase angle between the voltage and the current. For single-phase:

$$PF = \frac{W}{VA} = \frac{kW}{kVA} = \cos \theta \tag{3}$$

Where the waveform is not perfectly sinusoidal, the relationship to $\cos \theta$ does not hold, and only the kW/kVA relationship is valid.

In polyphase circuits the kW/kVA relationship remains valid, but the kVA value must be determined by the vectorial addition of the active power and the reactive power from the several phases. In Fig. 28 the polyphase power factor is the ratio AB/AC where AB is the arithmetic sum of the wattmeter readings on each of three phases, CB is the arithmetic sum of the varmeter readings on each of three phases, and AC is the square root of $[(CB)^2 + (AB)^2]$. If the circuit has a polyphase varmeter and a polyphase wattmeter, the power factor may be determined more readily as

$$PF = \frac{W}{(W^2 + vars^2)^{1/2}} \tag{4}$$

Obviously a simple electromechanical instrument cannot perform these calculations, so in practice a circuit is used which gives an approximate direct indication using the voltage and the current in a single-phase circuit or a current and the resultant of two voltages in a polyphase circuit. The instruments are accurate only above 40% of rated current, if current and voltage are in balance, and with sinusoidal waveforms. For these reasons it is recommended that power-factor meters be generally avoided and that varmeters be used instead. An operator usually has to relate power factor to the flow of reactive power anyhow; so he might better read vars directly.

Power factor may be read from a chart (Fig. 29) rather than making the kW/kVA calculations. A power factor reading of 0.99 on a power-factor meter gives the operator a sense of being near unity. However, he has, as read from the 0.99-PF line, a considerable amount of reactive power. Under the 0.99-PF condition, a power-factor meter will hardly have moved from center, while a varmeter will have moved upscale almost 15% when the wattmeter on the same circuit reads 100%.

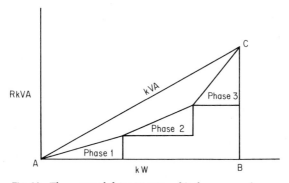

Fig. 28 The vectorial determination of 3-phase power factor.

21. PF-Meter Types. Power-factor meters are made in the crossed-coil electrodynamometer type, usually with a 100° scale arc, to cover the range 0.5 lag-1-0.5 lead or 0 lag-1-0 lead or even an offset scale such as 0.3 lag-1-0.7 lead.

With the popularity of circular-scale instruments, there has been a standardization on the rotating iron-vane instrument. This device measures phase angle directly with 360° rotation of the pointer representing the 360° vector rotation of an alternating current.

The instrument can be scaled in two quadrants as 0 lag-1-0 lead, or in four quadrants to show all possible directions of power flow and reactive exchange.

A third type uses a power-factor transducer with a dc output proportional to phase angle. The usual transducer measures the time between the instants that the voltage waves and the current waves cross zero. The readout instrument is a dc type with a scale being a cosine function of the phase angle.

22. PF-Meter Selection. Where the power flow is always in one direction, the power-factor meter need read in only two quadrants. In most circuits a range of 0.5 lag-1.0-0.5 lead is adequate. Where there will almost always be a good power factor, as with automatically switched capacitors, for example, a range such as 0.1 lag-1.0-0.8 lead

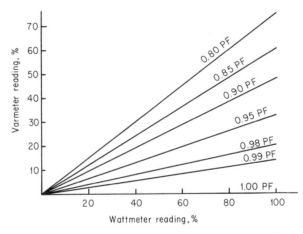

Fig. 29 Typical chart determining power factor from relative wattmeter and varmeter readings.

will permit close readings. These partial scales are available only on dynamometer and transducer types. The rotating iron-vane type, being a direct-reading phase-angle meter, can only be made with the two quadrants scaled 0 lag-1-0 lead.

Four-quadrant power-factor meters are used wherever power flow will be both in and out, such as with a tie-line panel where locally generated power is sometimes fed back into the power system. Scales are often marked in percent, 0-100-0 instead of 0-1-0.

23. PF-Meter Connections. Single-phase power-factor meters utilize the current in one line and the voltage across both lines. The dynamometer type requires a resistor/reactor network as a separate auxiliary (Figs. 30 and 31).

Polyphase meters for 3-wire systems invariably connect line-to-line for two voltages. They utilize the current in the common line. Those for 4-wire service may, if they have 208-V coils, or series resistors for 208 V, be connected in the same way. Some models, however, use line-to-neutral voltages and 120-V coils. It is important to know which type is used to avoid damage by overheating, or inaccuracy due to insufficient torque.

The polyphase connection diagrams in Figs. 32 to 34 apply to either dynamometer or rotating-vane instruments. Internally there is a difference between instruments, as the

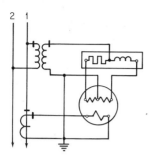

Fig. 30 Power-factor-meter connections, single-phase, dynamometer type, with external compensator.

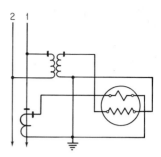

Fig. 31 Power-factor-meter connections, single-phase, rotating-vane type.

dynamometer uses two coils in a V connection and the moving-iron type uses three coils in a Y configuration.

A single-phase power-factor transducer (Fig. 35) connects exactly like a single-phase power-factor meter. The polyphase transducer connection is different (Figs. 36 and 37). It uses a current coil in one line and a single potential across the remaining two lines. For 3-phase, 4-wire service the transducer voltage coil is rated 208 V. As a practical matter a single-phase transducer would be as useful on a 4-wire application as the polyphase device. In either case the circuits must be balanced for a meaningful indication.

24. PF-Meter Checkout for Service. The following steps are recommended for checkout:

1. Verify voltage at the instrument.

2. Current circuit load must be at least 40% of rating.

3. Calculate power factor from ammeter, voltmeter, and wattmeter readings [Eq. (3) or (4), or Fig. 29].

4. If reading checks with calculations, the meter and connections may be assumed correct.

5. If the pointer of a 100 or 180° meter "pegs" hard, it indicates a reversed polarity on the current circuit. The four-quadrant meter will indicate 180° from actual power factor under the same conditions.

6. If the pointer indicates 0.5 when the power factor is known to be 1.0 or if the pointer on a moving-vane type is either 60 or 120° from the correct position, the probability is that the current transformer is in the wrong leg. If so, it is not necessary to

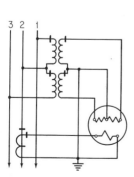

Fig. 32 Power-factor-meter connections, 3-phase, 3-wire type.

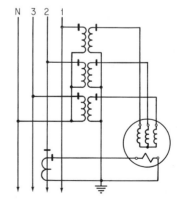

Fig. 33 Power-factor-meter connections, 3-phase, 4-wire rotating-vane type using line-to-line voltages.

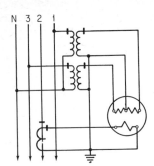

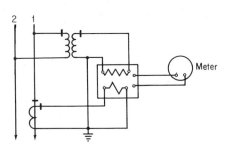

Fig. 34 Power-factor-meter connections, 3-phase, 4-wire for line-to-neutral voltages.

Fig. 35 Power-factor-meter connections, single-phase, transducer type.

change the transformer. Rotate the voltage connections until the instrument reads correctly. It may be necessary to recheck item 5.

7. If the instrument shows "lead" when it should show "lag," or vice versa, rotate two potential leads, then recheck items 6 and 5.

8. When a single-phase power-factor meter reads in the wrong direction, reverse the noncommon connections at the resistor/reactor.

9. If these checks do not produce a correct indication, the instrument should be removed and sent to a qualified instrument-repair shop for further checks.

FREQUENCY MEASUREMENT

25. Frequency-Meter Types. Both mechanical and electrical frequency meters are in general use today. The mechanical type uses a series of tuned bars of magnetic material arranged in a parallel array. The bars are tuned to vibrate at uniformly different frequencies across the range of the instrument. The assembly is placed within the field of a solenoid energized from the circuit being measured. Each bar begins to vibrate about approximately 2% below its tuned frequency; it reaches maximum amplitude at the tuned frequency, and then gradually reduces to zero amplitude at 2% above frequency. This range gives the instrument a display pattern (Fig. 38). The meter shown is a simple seven-reed model. The accuracy increases with the number of reeds and the narrowness of range.

Electrical indicating frequency meters have been made in numerous circuits with practically all types of moving elements. Most modern instruments are combinations of a frequency transducer and a dc mechanism. Some are self-contained and some use

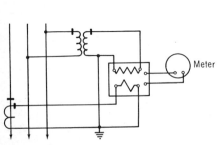

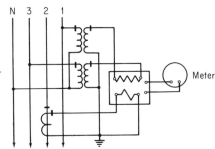

Fig. 36 Power-factor-meter connections, 3-phase, 3-wire, transducer type.

Fig. 37 Power-factor-meter connections, 3-phase, 4-wire, transducer type, 208 V potential.

separately mounted transducers. In switchboard-type instruments, and in some panel designs, the transducer approach permits the use of taut-band mechanisms with the advantages discussed earlier (Art. 15).

Frequency meters have a span equally divided above and below the center frequency. The 55-60-65-Hz range is probably the most popular for industrial use, but where frequencies are critical, commercial meters may be as narrow as 59-60-61 Hz spread across a 250° dial with an accuracy of ±0.04 Hz. Even closer accuracy is available on special instruments.

Three types of transducer outputs are used. One is linear with respect to frequency, with zero output at the low end of the scale. This type indicates deenergization by the pointer going below scale on the low end. Another type is linear in the opposite direction; it indicates off-scale on the high side when deenergized. Some of the above types do not go off scale but indicate at the end of the scale when not energized.

The third type uses a nulling-type tuned circuit within the span of the instrument. Usually the null point is marked with a green line about 30% downscale from center frequency. If the instrument is deenergized, it will rest at the green line. This position is also the zero-set point of the instrument.

26. Frequency-Meter Selection. The vibrating-reed frequency meters are relatively low in cost. They are used mostly on smaller standby or portable generators.

For larger sets, especially those supplying frequency-sensitive equipment, the electrical indicating meters are generally used. The span relates to the accuracy required in terms of absolute frequency. The indicating meter is particularly useful in studying the behavior of the governor on the generator prime mover. It will indicate speed droop, response, or hunting on an isolated machine.

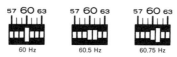

Fig. 38   Reading the vibrating-reed frequency meter.

27. Frequency-Meter Connections. A frequency meter connects to any one of the potential transformers on a system.

28. Frequency-Meter Checkout for Service. The following checks may be made:

1. If the instrument has a zero-input mark, adjust the pointer to that line.

2. The best check for frequency indication is to connect the meter to a large electric utility system. Even when they vary frequency to correct clocks, they do not get beyond the error limits of the broad-range meters. Most utilities will verify their frequency at any given time on request.

3. If the utility voltage is not available, a precision portable frequency meter, or a digital counter and stopwatch may be used. The engine governor should be blocked and the generator load should be either zero or absolutely steady during this check.

4. Transducer-type meters are usually field-adjustable to maximum accuracy at any point on the scale.

5. Because of the variety of circuits, no general instructions can be given for adjustment of a frequency meter. The manufacturer's data should be followed.

SYNCHROSCOPES

29. Synchroscope Function. Whenever an ac generator is to be paralleled with another, or connected into an operating system, it is necessary that the incoming machine be in synchronism with the energized bus before the generator circuit is closed. The instrument used to indicate synchronism is the synchroscope. It compares the instantaneous voltages on one phase of the incoming machine with the instantaneous voltages on the corresponding phase of the bus. When the two voltage waves vary together, they are in synchronism.

30. Synchroscope Connections. Synchroscope connections are shown in Fig. 39. It is absolutely essential that the phase rotation of the machine and the bus be identical and that the potential transformers be across the corresponding phases. Whether line-to-line or line-to-ground connections are used is immaterial, but they must be the same for machine and bus. Failure to make certain of proper connection could seriously damage a generator.

In using the synchroscope, it is first necessary that the bus be within 2 Hz of rated frequency and that the incoming machine be no more than 1½ Hz below or 3 Hz above the bus. The voltage on the generator should be matched to that of the bus. The speed is then changed up or down to bring it slightly below the bus frequency as indicated by a slow rotation of the synchroscope in the "slow" direction. Then speed is increased very gradually until the pointer stops at the 12 o'clock position. The generator circuit may then be connected to the bus. With some machines it is not possible to control the speed precisely; so the breaker is closed while the pointer is creeping toward 12 o'clock — some few degrees ahead, to allow for circuit-breaker closing time.

As soon as a synchroscope has served its purpose, it should be disconnected from the line on both sides. Otherwise, under certain conditions with only the stator energized, the instrument may rotate continuously and wear itself out.

31. Synchroscope Checkout for Service. The following checks should be made prior to placing the synchroscope in service.

1. Verify phase rotation and connections.

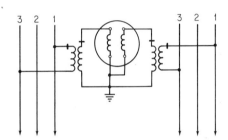

Fig. 39 Synchroscope connections.

2. It is a wise precaution to check the pointer position for accuracy. Connect both coils to the same voltage source. The pointer should pull up sharply to 12 o'clock from any other position.

3. When first used, observe the behavior of the instrument. It should begin to rotate when the incoming frequency is about 58½ Hz (with the bus at 60 Hz) and should stop rotating at approximately 63 Hz. When it is outside these limits or is deenergized, it should rest away from the 12 o'clock position. Synchroscopes are usually made slightly off balance so that they rest toward the 3 o'clock point.

4. If the instrument rotates in the wrong direction, it has improper internal connections, or it is improperly connected to the external impedance box sometimes used. If the problem is internal, the synchroscope should be sent to a qualified repair shop.

5. If the instrument does not rotate when there is a frequency difference, check for correct voltage across each coil.

GROUND DETECTION

32. Ground-Detector Function. Ungrounded systems are usually equipped with a ground detector to indicate the presence of grounds before they can cause faults. The simplest of these detectors consists of three lamps which will glow equally under normal conditions. One lamp will grow dim when a ground occurs on the phase to which it is connected. Lamps are not capable of detecting the high-resistance grounds which usually precede the dead grounds.

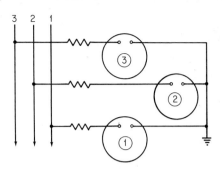

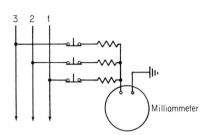

Fig. 40 Ground-detector connections for 3-phase, 3-wire circuits (single instrument).

Fig. 41 Ground-detector connections for 3-phase, 3-wire (three instruments).

33. Ground-Detector Connections.   For the greatest sensitivity to partial grounds on 3-phase, 4-wire systems the schemes shown in Figs. 40 and 41 are commonly used. The ground detector in Fig. 40 uses a 50 mA ac milliammeter with a scale marked in circuit volts.   The resistors are so selected that when a ground occurs, the instrument will read the voltage from the ungrounded line to the grounded one.   In case of a dead ground, this reading is the full line-to-line voltage.   In case of a high-resistance ground, it might be only a slight deflection.   With no ground, the instrument reads zero.   When the switches are opened momentarily in succession, the grounded line will be identified by that switch which causes the highest indication.

A second scheme (Fig. 41) uses three instruments, again 50 mA full scale, with resistors selected to read line-to-line voltages when one line is grounded.   Under normal conditions all three instruments will read approximately 86% of line-to-line voltage. When a ground occurs, the instrument on the grounded line will read less than the others.   The difference in readings is an indication of the resistance of the ground.   A dead ground is indicated when one reads zero and the other two read line voltage. Standard ac voltmeters may be used in this scheme, omitting the resistors, but the sensitivity to high-resistance grounds will be reduced.

Manufacturers of ground detectors will supply their instruments complete with the necessary resistors for any circuit voltage up to 600 V.   Where circuit voltages exceed 600 V, potential transformers are used ahead of the instruments.

Similar ground-detecting schemes are available for single-phase and dc circuits.

TEMPERATURE MEASUREMENT

34. Thermometer Function.   Electrical thermometers are often incorporated into control equipment to read the temperatures of bearings or windings on the electrical apparatus.   There are many types of electrical thermometers, some involving thermocouples, some thermistors, but those most used on heavy industrial apparatus are of the resistance type.   These use several resistance coils, called resistance temperature detectors (RTD), embedded in windings or immersed in oil sumps.   These connect through selector switches to the instrument which, by means of a resistance bridge circuit, is able to measure the resistance of the RTD.   The detectors are designed for a known resistance at a specified temperature.

Resistance changes are proportional to temperature changes; so the instrument can be calibrated directly in degrees of temperature.   Popular temperature detectors are 10 Ω of copper, 100 Ω of nickel, or 120 Ω of nickel, all measured at 25°C.   There are many other materials and resistance values being used for various reasons.   When selecting an RTD type, it is important to decide what kind of protective relaying and alarm signaling will be operated from the RTDs along with the thermometers.   Not all protective devices are available with a choice of detector resistance materials or ohmic values.

35. Thermometer Connections.   Connection diagrams are so varied that they are not shown here.

RECORDERS

36. Recorder Function. Recording instruments draw a permanent chart record of any electrical quantity or any mechanical quantity which can be converted to a proportional electrical signal. They consist of an electrical measuring mechanism which moves a pen or a stylus across a graduated chart driven by a clock mechanism. The pen or stylus may draw a line on the chart by any of several means.

1. Ink on a standard paper chart.
2. Hot stylus on a heat-sensitive paper.
3. Impact on a pressure-sensitive paper.
4. Scratching a coated film or paper.
5. Passing a current through conducting sensitized paper.

Recorders which accept an electrical signal directly into the measuring mechanism are called direct-acting recorders. The signal may be the dc output of a transducer having ac input.

37. Recorder Types. Ac voltage or current is usually measured by an iron-vane or dynamometer mechanism, but occasionally by one using a rectified permanent-magnet, moving-coil mechanism.

Ac wattmeters will use either a self-contained dynamometer mechanism or a permanent-magnet, moving-coil dc mechanism fed from a transducer.

Most mechanical quantities are measured as direct current from a transducer. Here, too, the mechanism is a permanent-magnet, moving-coil type.

These direct-acting recorders all get the energy to move the pen directly from the measured signal. This required energy limits their use to those signal sources which can stand a substantial burden. They can never be used with transducers which have a filtered millivolt output.

For use on the low signal levels there is a broad choice of potentiometric recorders. These are null-balancing servo-type devices which take their operating energy from a separate source. They respond to minute electrical signals.

38. Recorder Selection. The most common recorder for daily or weekly records is the circular-chart type. Charts make one revolution in either 1 or 7 days. These instruments ordinarily are direct-acting type for current or voltage and potentiometric type for all other measurements.

Both the direct-acting and potentiometric recorders are made in strip-chart types as well. The length of charts is fixed; so the running time of a chart is variable with the chart speed. Charts run 100 to 150 ft in length and chart speeds from $\frac{1}{2}$ to 12 in/h. Very high chart speeds of several inches per minute are supplied for special analytical work (Fig. 42).

39. Recorder Connections. Recorders connect to the measured circuits exactly as do their counterparts in indicating instruments.

40. Recorder Checkout for Service. Basic circuit troubleshooting is similar to that for indicating instruments, but recorders are so complex and so varied that no attempt is made here to cover placing equipment in service. The manufacturer's manual should be consulted.

METERING ELECTRICAL ENERGY

41. Energy-Measurement Function. The total energy flowing through an electrical circuit is measured quantitatively in kilowatthours (kWh). The instantaneous rate of power flow is in kilowatts (kW). The average rate of power flow over period of time (usually 15 min) is referred to as kilowatt demand. Each of these measurements is of interest in industrial power distribution.

42. Watthour Meters. The total energy is of obvious interest for billing purposes and for departmental cost accounting. Measurement of kWh is one of the most highly developed of electrical measurements. The induction-disk watthour meter is the only device used for this purpose in industrial switchgear. It consists of one, two, or three induction-disk "motor" elements on a single shaft. Each "motor" element has a voltage and a current coil so arranged that the torque produced is proportional to the product of voltage, current, and power factor (watts). The torque is opposed by an arrangement of permanent magnets which generate eddy currents in the induction disks. The net result is a rotation of the shaft at a rate proportional to the rate of kilowatt flow. The shaft is geared to a dial.

43. Watthour-Meter Connections. Watthour meters are connected exactly as are the indicating wattmeters in Figs. 18 through 23.

The flow of reactive power may be measured on a standard watthour meter connected through a phase-shifting transformer. The connections are the same as for a varmeter (Figs. 24, 25).

44. Demand Meters. The sizing of and therefore the investment in generation, transmission, or distribution equipment depend upon the loads to which it will be subjected. Most electrical equipment has substantial short-time overload capability, so it is realistic to factor duration of load into equipment sizing. This consideration is reflected into power billing through a demand charge. Most frequently the charge is

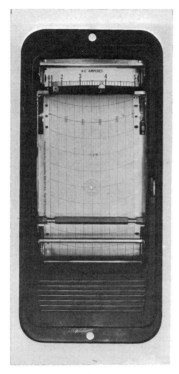

Fig. 42 Strip-chart recorder.

based upon the maximum average kW or kVA in a 15-min period. Occasionally a 30-min period is used.

Indicating demand meters are usually combined with watthour meters. The demand mechanism may be driven from the watthour-meter element to set a pointer at the highest kW for a timed interval, or it may use an independent thermal element to perform the same function. Reactive kVA demand indicators are invariably thermal except in the case of recorders.

Recording kW demand meters use a watthour-meter mechanism complete with a kWh register. The meter mechanism also drives a pen across a strip chart for a specific time interval (15 or 30 min), then trips and repeats as the chart advances. Thus there is a graphic record of average kW demand.

Recording kVA demand meters use a kWh and a kVAR-hour mechanism to drive an integrating mechanism which moves the recording pen. Otherwise it is similar to the kW demand recorder.

Connections to demand meters are the same as to corresponding watthour or varhour meters and their indicating counterparts (Figs. 18 through 25).

45. Watthour- and Demand-Meter Checkout for Service. The following checks may be made:

1. The rotation of any induction-disk meter element is from left to right as viewed from the front. Each element may be checked individually by shorting out the current coil. Reversed rotation indicates an incorrect polarity on either the voltage or current circuit. Either circuit may be reversed to correct the situation, but it is best to trace out the error and correct it. Also diagrams should be checked and corrected so that future troubleshooting will be facilitated.

2. From the r/min of the meter disk it is possible to calculate the rate of power flow:

$$kW = \frac{60 \times rpm \times K_h}{1000} \tag{5}$$

where K_h is the watthour-meter constant. The value thus determined may be checked against a wattmeter or other instruments which permit the calculation of kW. The K_h constant is marked on the meter nameplate. The same calculations apply to kvar meters.

3. Incorrect registration may be caused by shorted or reversed current circuits, or open or reversed potential circuits. The diagnosis of troubles parallels that under Art. 19.

. 4. If the difficulty is suspected to be in the meter itself, the meter should be sent to a qualified repair shop. Any attempt to repair or adjust the device in the field may result in great errors.

TELEMETERING

46. Telemetering Function. It is often necessary to locate an indicating instrument at some distance from the point where measurements are taken. When a transmission channel between the sending and the receiving point carries other than the basic measurand, the system is called telemetering. Usually such systems use transducers to convert the measurand into a proportional direct current or voltage. The dc signal is usually, but not necessarily, amplified before transmission. In the more sophisticated systems which feed into data systems or computers, the signal from the transducer is converted to digital form before transmission. Only systems using a dc output transducer with or without an integral amplifier are considered here.

47. Telemetering Types. Nonamplified transducers having sufficient output for direct telemetering are generally available for alternating current and voltage and, with limitations, for watts and vars. Generally the transducer outputs are 1 or 3 mA into a single resistance up to 5000 or 10 000 Ω. The receiver, therefore, may be a standard 1- or 3-mA instrument or recorder. The loop (channel plus receiver) resistance of the circuit must be equal to, or less than the maximum rated load resistance of the transducer. The transducer must be adjusted to the proper load resistance prior to being placed in service and must be readjusted if the circuit resistance changes.

A much more flexible, but somewhat more costly, system uses amplified transducers having an output of 1 mA into any load resistance up to a maximum, usually 10 000 Ω. These devices are known as constant-current output transducers. They are available for current, voltage, watts, vars, frequency, speed, power factor (phase angle), as well as in dc amplifier form for use with pressure, flow, and position transducers.

The readout is a standard 1-mA instrument with an appropriate scale. It is simply connected to the transducer through the transmission channel. No adjustment is necessary. If the channel is exposed to lightning or other surges, consult with the manufacturer of the transducer for recommendations on protection.

The amplified transducers are made in an optional form with the output adjustable from near zero to 10% over rating for the standard input. This permits scaling for parallel operation or totalizing and also allows a single receiver to be used for various inputs selectively. For example, if a remote receiver is to indicate readings from transducers fed by 1000/5, 800/5, and 600/5 A current transformers, the transducer outputs would be adjusted to 1 mA, 0.8 mA, and 0.6 mA, respectively.

TABLE 2. Loss Data for Voltage Coils of AC Instruments

Type	VA	W	vars	PF
Voltmeters:				
Switchboard, circular-scale, iron-vane	2.4	2.4	0	1.0
Switchboard, 100° scale, iron-vane	1.9	1.9	0	1.0
Switchboard, all types, rectifier	0.1	0.1	0	1.0
Panel, iron-vane	1.2	1.2	0	1.0
Transducer type	0.6	0.6	0	1.0
Recorder, direct-acting, strip-chart	14.4	14.2	1.6	0.99
Recorder, direct-acting, circular-chart	10.7	10.3	3.1	0.96
Wattmeters, polyphase:				
Switchboard, circular-scale, transducer	1.2	1.2	0	1.0
Switchboard, 100° scale, dynamometer type	3.6	3.6	0	1.0
Panel, transducer type	1.0	1.0	0	1.0
Recorder, direct-acting	6.0	6.0	0	1.0
Power-factor meters, polyphase:				
Switchboard, circular-scale	2.2	2.2	0	1.0
Switchboard, 100° scale, dynamometer	3.7	3.7	0	1.0
Panel, transducer type	1.0	1.0	0	1.0
Frequency meters:				
Switchboard, all types	3.5	2.5	2.5	0.7
Panel, transducer type	4.0	4.0	0	1.0
Synchroscopes:				
Switchboard, all types:				
Running	4.5	2.0	4.1	0.45
Incoming	4.6	3.7	2.8	0.80

48. Telemetering Connections. Telemetering transducers are connected exactly the same as the corresponding local-reading transducer and instrument combinations (Fig. 20). Note that all transducers are connected the same as their indicating-instrument counterparts.

LOSS DATA

Every electrical coil has an energy requirement. In the case of dc coils it is simply the product of the current through the coil and the voltage across it. However, with ac instruments, the calculations are more complex, as they involve the vectorial addition of

TABLE 3. Loss Data for Current Coils of AC Instruments

Type	Impedance	Resistance	Reactance	VA	PF
Ammeters:					
Switchboard, circular-scale	0.015	0.012	0.009	0.375	0.79
Switchboard, 100° scale	0.018	0.017	0.006	0.44	0.94
Switchboard rectifier	0.002	0.0016	0.0012	0.05	0.80
Panel, iron-vane	0.01	0.01	0.002	0.26	0.99
Transducer	0.064	0.011	0.007	1.6	0.17
Recorder, strip-chart				3.5	0.10
Recorder, circular				4.0	0.22
Wattmeters, polyphase:					
Switchboard, circular-scale	0.08	0.016	0.079	2.0	0.20
Switchboard, 100° scale	0.078	0.077	0.012	1.9	0.99
Panel, transducer type	0.08	0.046	0.007	2.0	0.20
Recorder, strip-chart				5.9	0.17
Power-factor meter, polyphase:					
Switchboard, circular-scale	0.174	0.14	0.104	4.35	0.82
Switchboard, 100° scale	0.076	0.075	0.012	1.9	0.99
Panel, transducer type	0.072	0.011	0.075	1.8	0.15

resistance and inductive loads. The "loss" of the instrument coil becomes the "burden" on the transformer (Sec. 15).

Typical loss data are shown in Tables 2 and 3. The figures shown are nominal. They will vary quite broadly from manufacturer to manufacturer, so these tables should be used for rough-estimating purposes only.

Instrument Transformers

RAYMOND C. THOMAS*

FOREWORD

Key components in switchgear and large-motor control are the instrument transformers. Their selection and application determine the success of not only instrumentation and metering, but relaying, system coordination, and special control functions.

This section provides background information to aid in applying these transformers in each of their uses.

---

* President, Acutran Instrument and Specialty Transformers; Registered Professional Engineer (Pa.); Member IEEE.

COMMON APPLICATIONS

1. Function. Instrument transformers have two primary functions. First, they serve to reduce the large number of rated primary-circuit voltages and currents encountered in modern power circuits to a common secondary base, thus permitting standardization of meters, relays, or other measuring devices. These secondary values are usually 120 V and 5 A. Second, they provide insulation between the primary and secondary circuits, thus protecting meters and control equipment from dangerously high voltages and providing safety for personnel using such devices.

Instrument transformers are divided into two general types: voltage (potential) transformers for use in voltage measurement, and current transformers for use in current measurement.

With respect to insulation, they are further classified as dry types, encapsulated or molded types, and liquid-immersed types.

2. Current Transformer Design. The current transformer is designed to have its primary winding connected in series with the circuit to be measured (Fig. 1). The primary winding is usually the high-voltage winding and is insulated accordingly.

The ratios of the primary current to secondary current are usually those which will produce 5 A at the secondary terminals (Ref. 1).

With respect to mechanical construction, current transformers are classified as follows:

1. The wound primary-type transformer has the primary and secondary windings completely insulated and permanently assembled on the core. The primary winding is usually a multiturn winding but may be a single turn encircling the core.

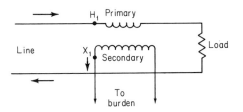

Fig. 1 Current transformer with single-ratio primary.

2. The bar-type transformer has the primary and secondary windings completely insulated and permanently assembled on the core. It differs from the wound primary type because its primary winding consists merely of a bar-type conductor mounted in the core window.

3. The window-type transformer has only a secondary winding completely insulated and permanently assembled on the core. There is no primary winding. The transformer is generally insulated for 600 volts. For higher voltages the insulation is usually supplied with the cable or bus bar to be placed through the window.

4. The bushing-type transformer, like the window type, has only a secondary winding completely insulated and permanently assembled on the core, but has no primary winding or insulation for a primary winding. The primary conductor is usually a component part of other apparatus, such as a power transformer or circuit-breaker bushing.

Current transformers can be further classified with respect to electrical connections:

1. The single-ratio transformer has a single primary and secondary winding (Fig. 1).

2. The three-wire transformer has a primary winding that consists of two equal sections fully insulated from each other and from the secondary winding and ground. This transformer is commonly used for measurement of total power in a conventional three-wire, single-phase service (Fig. 2). Three-wire transformers are used on low-voltage services only since it is difficult to provide the necessary insulation between the two primary windings.

3. The series-parallel primary transformer has primary winding divided into two or more identical sections and so arranged that they may be connected in series or in parallel so as to produce the same ampere-turns in the secondary winding (Fig. 3).

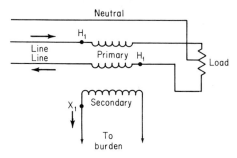

Fig. 2 Single-phase, three-wire current transformer.

4. The tapped secondary transformers are used in applications where it is necessary to have available two values of secondary current from the same secondary winding of the current transformer. This double rating is accomplished by placing a tap or taps in the secondary winding. It should be noted that only one can be used at a time and that the terminals not in use should be left open. When operated on the tap position, the accuracy is poorer than when operated on the full winding (Fig. 4).

5. The multiple-secondaries transformers consist of a single primary winding encircling two or more secondary windings on separate magnetic circuits (Fig. 5). These transformers are generally used where space is limited and the burdens must be kept separate, e.g., metering and relaying burdens. Unlike the tapped secondary transformers, the secondary winding not in use must be kept short-circuited.

3. Current Transformer Burdens. The burden of a current transformer is the load connected to the secondary winding. The term "burden" rather than "load" is used to distinguish this load from the load on the primary circuit which is being metered or controlled.

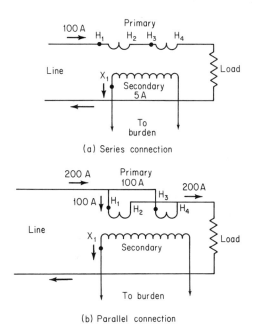

Fig. 3 Current transformer with series-parallel primary.

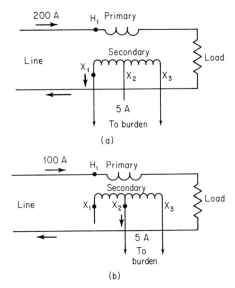

Fig. 4 Current transformer with tapped secondary.

Burden is expressed as total ohms impedance at a certain power factor. This power factor is not the power factor of the system or the load being metered or controlled. For example, the burden may be given as 2 Ω impedance at 50% PF. Burden may also be expressed in terms of volt-amperes and power factor, such as 50 VA at 50% PF.

Standard burdens for current transformers are shown in Table 1 and are based on standard 5-A secondary and 60 Hz (Ref. 1).

For current transformers having nonstandard secondary current ratings, the corresponding nonstandard burdens may be derived from the standard burdens shown in Table 1 in proportion to the square of the ratio of the 5-A current to the nonstandard current. For example, for a current transformer having a 1-A secondary current rating, the burden would be 25 times the values of ohms resistance and impedance given in Table 1.

4. Current Transformer Open-Circuit Voltage. The general rule for all current transformers is that the secondary circuit should never be opened or left open when current is flowing in the primary. If the secondary circuit is open, all the primary current is used in magnetizing the core, which results in extremely high flux with accompanying saturation of the core. Abnormally high voltages are thereby induced in the secondary winding. The amount of the secondary voltage depends on the individual design and type of core steel used. Because of the saturation of the core, the induced voltage will be extremely peaked, though the rms value may be relatively low. Un-

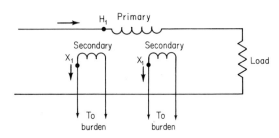

Fig. 5 Current transformer with double secondary.

TABLE 1. ANSI Standard Burdens for Current Transformers with 5-A Secondaries

Burden designation	Impedance, Ω	Volt-amperes	Power factor
	Metering burdens		
B-0.1	0.1	2.5	0.9
B-0.2	0.2	5.0	0.9
B-0.5	0.5	12.5	0.9
B-0.9	0.9	22.5	0.9
B-1.8	1.8	45.0	0.9
	Relaying burdens		
B-1	1.0	25.0	0.5
B-2	2.0	50.0	0.5
B-4	4.0	100.0	0.5
B-8	8.0	200.0	0.5

less proper measuring equipment is available, this peak voltage cannot easily be determined in the field.

Another reason why a current transformer should not be left open-circuited is that after the core becomes saturated, a magnetic bias may remain on the core after the voltage is removed. In the event that this bias should be present, the current transformer should be demagnetized to restore its degree of accuracy.

5. Voltage (Potential) Transformer Design. These transformers are designed for connection line-to-line or line-to-neutral in the same manner as a voltmeter. They are used to operate voltmeters, potential coils of wattmeters and watt-hour meters, or control devices.

The ratios of primary to secondary voltage are usually those which will produce either 120 or 115 V at the secondary terminals; however, some ratios are such as to produce approximately $\sqrt{3} \times 120$ or 115 V (Ref. 1).

Normally voltage transformers are designed for operation at 60 Hz, although many designs are suitable for use on 50-Hz circuits. Operating at frequencies less than 50 Hz will result in serious error. Also, when operation is at 10% above rated voltage, saturation of the core may result, causing serious error and excessive heating.

6. Voltage (Potential) Transformer Burdens. The burden of a voltage (potential) transformer is the load connected to its secondary terminals. It is expressed in volt-amperes, which is the secondary voltage multiplied by the current in amperes through the instruments which are connected in parallel in the secondary circuit.

To facilitate comparison of transformers on a uniform basis, standard burdens have been derived as shown in Table 2 (Ref. 1).

TABLE 2. ANSI Standard Burdens for Voltage (Potential) Transformers

Burden	Volt-amperes at 120 V	Burden power factor
W	12.5	0.10
X	25.0	0.70
Y	75.0	0.85
Z	200.0	0.85
ZZ	400.0	0.85
M	35.0	0.20

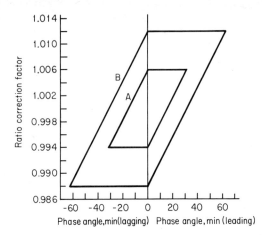

Fig. 6 Standard accuracy limits for Class 1.2 current transformers. A, parallelogram for 100% of rated current; B, parallelogram for 10% of rated current.

In addition to the standard burdens for accuracy designation, voltage (potential) transformers are usually given a thermal burden rating. This burden is the total volt-amperes which the transformers will carry at rated voltage and frequency without causing the specified temperature limitations to be exceeded. This thermal burden rating is not related to accuracy.

7. Accuracy. A fundamental basis for instrument transformer selection is accuracy. Although instrument transformers are very efficient in operation, certain losses, which cannot be entirely eliminated, do exist. Because of the resistance of the winding, there is a power loss equal to the square of the current multiplied by the ohmic-resistance (I^2R). This loss is dissipated in heat and causes a temperature rise in the windings. The result is that the power input of the primary winding is greater than the power output of the secondary winding. This difference causes the secondary voltage or current to be less than the ratio of primary to secondary turns (marked ratio) would indicate. This difference is termed *ratio error*.

In addition, a small portion of the primary current is used to magnetize the iron core. This exciting or excitation current supplies the iron losses of the core. Because of these

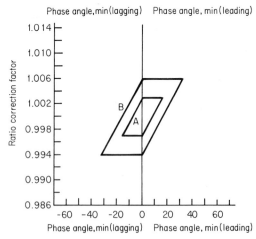

Fig. 7 Standard accuracy limits for Class 0.6 current transformers. A, parallelogram for 100% of rated current; B, parallelogram for 10% ot rated current.

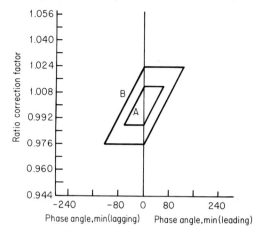

Fig. 8 Standard accuracy limits for Class 0.3 current transformers. A, parallelogram for 100% of rated current; B, parallelogram for 10% of rated current.

losses, the secondary voltage or current is not exactly 180° out of phase with the primary current or voltage. The departure from the 180° normal difference with the primary is termed *phase angle error*. In a well-designed transformer this error is measured in minutes.

The phase angle error is of little importance in measurement of power circuits of unity power factor or in circuits where only indicating instruments or control devices are used. With circuits of lower power factor where integrating instruments are used, the phase angle errors have an increasing effect upon the accuracy of power measurement.

The standard metering accuracy classes, as given in Ref. 1, are 0.3, 0.6, and 1.2, which means that the error produced by the instrument transformer shall not exceed plus or minus 0.3, 0.6, or 1.2% in the metered or measured circuit.

When an instrument transformer is used with integrating instruments, the ratio correction factor (RCF) and phase angle error must be combined into a transformer correction factor (TCF). Limits of transformer correction factor (TCF) are best defined by parallelograms. The parallelograms for current transformers are shown in Figs. 6, 7,

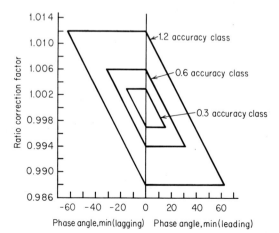

Fig. 9 Standard accuracy limits for voltage (potential) transformers.

and 8 and for voltage (potential) transformers in Fig. 9. For a specified accuracy class the combination of ratio error and phase angle error must fall within that parallelogram.

By means of this ANSI system, the accuracy of an instrument transformer may be described by listing the best accuracy class which it meets for each burden. For example, a current transformer may be rated 0.3B0.1, 0.3B0.2, 0.3B0.5, 0.3B1.0, and 0.3B2.0, or 0.3B0.1, 0.3B0.2, and 0.3B0.5, or 0.3B0.1, 0.3B0.2, and 0.6B0.5. A voltage transformer may be rated 0.3W, 0.3X, 0.3Y, and 0.3Z, or 0.3W, 0.3X, 0.6Y, and 1.2Z, or 0.3W, 0.3X.

The omission of any burden indicates that the error for those burdens exceeds the 1.2 class.

Relay accuracy classes for current transformers are designated by two symbols, such as C200 or T200, which describe the transformer as follows: C classification (formerly H classification) describes a current transformer in which the leakage flux in the core has a negligible effect on the ratio and indicates that the ratio error can be calculated with a reasonable degree of accuracy for relaying or control applications from a secondary excitation curve (Fig. 10). T classification describes a current transformer in which the leakage flux in the core has an appreciable effect on the ratio, and indicates that the ratio errors should be obtained by test rather than calculation.

The second symbol is the secondary terminal voltage which the transformer will deliver to a standard burden at 20 times the rated secondary current without exceeding a 10% ratio correction. For example, a relay accuracy class C200 means that the ratio can be calculated and that the ratio correction will not exceed 10% at any current from 1 to 20 times rated secondary current with a standard 2.0-Ω burden connected to the secondary terminals (2.0 Ω times 5 A times 20 equals 200 V).

It should be noted that tapped secondary or multiratio current transformers having a relay accuracy classification means that the relay accuracy applies only to the full winding and does not apply to the tap ratios unless so indicated.

8. Instrument Transformer Polarity. One primary and one secondary terminal of instrument transformers are usually marked to indicate the relative instantaneous direction of primary and secondary currents. These markings may consist of white dots or the letters H_1 and X_1. These markings are so arranged that, when current enters the

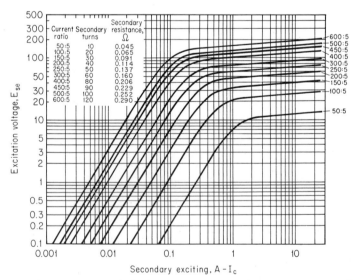

Fig. 10 Typical excitation curves for bushing-type current transformer. Data on secondary turns based on normal ratio of 600:5 and 60 Hz.

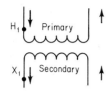

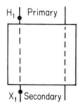

(a) Schematic diagram

(b) Simplified diagram

Fig. 11   Polarity diagram.

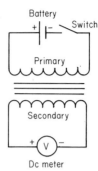

Fig. 12   Polarity check by inductive kick.

marked primary terminal, current is leaving the marked secondary terminal at the same instant (Fig. 11).

When the polarity of an instrument transformer is unknown because of the loss of polarity markers or for some other reason, it may be determined by one of several methods. The most common method used in the field, requiring a minimum of preparation and equipment to check polarity, is the inductive kick. This method may be used for both current and voltage transformers and makes use of a battery and a dc voltmeter. The voltmeter is connected across the low-voltage winding, and its instantaneous deflection is noted when the battery circuit is closed across the high-voltage winding. If the voltmeter kicks up scale, it indicates that the secondary or low-voltage lead connected to the positive (+) side of the meter should be marked and the primary or high-voltage lead connected to the positive (+) side of the battery should be marked. Should the voltmeter kick down scale, then reverse either the leads to the voltmeter or the battery until the voltmeter kicks up scale (Fig. 12).

9. Mechanical and Thermal Characteristics. In addition to considering accuracy requirements in choosing an instrument transformer, mechanical and thermal characteristics must be taken into account.

For short periods of time, during faults on the power system, current transformers may be subjected to extremely large primary currents. Therefore, in addition to being able to perform satisfactorily under normal operating conditions, the transformer should be able to withstand these fault currents. The ability to withstand such conditions is expressed by the mechanical and thermal rating of the transformer.

Disruptive forces, causing mechanical failure, are proportioned to the square of the number of primary turns. Therefore, as ampere turns are increased to obtain a higher degree of accuracy, as in the wound-type current transformer, the mechanical strength is lowered. In bar-type or window-type current transformers which have only one electrical primary turn, the mechanical strength is practically unlimited, depending upon the bracing of the conductor or cable and the spacing between the return conductor.

The thermal rating is dependent upon the size and number of turns of the conductor in the primary and secondary windings and the duration of the fault current. Increasing the size of the primary and secondary conductors increases the thermal rating, and increasing the number of turns decreases the thermal rating. If large fault currents or faults of long duration are anticipated, accuracy may have to be sacrificed.

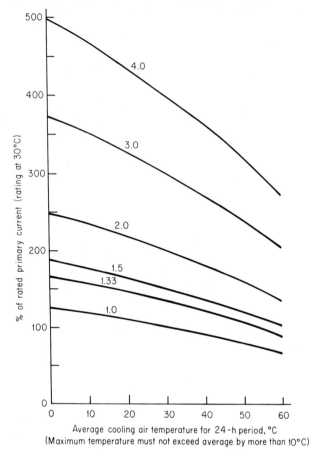

Fig. 13 Current limitations for current transformer with relation to cooling air temperature. Curve designations are continuous current-rating factors.

Since voltage (potential) transformers are connected either line-to-line or line-to-neutral, they are not subject to the same mechanical and thermal problems as the current transformer. However, they are subject to short-circuit faults in the secondary circuit. To protect the transformer from these conditions, fuses in the secondary circuit are recommended, although they may tend to increase the burden on the transformer.

It is a common practice to fuse the primary of voltage (potential) transformers used in switchgear applications. These fuses are generally of the current-limiting type and serve to protect the system from faults in the transformer and/or its secondary circuit. These primary fuses have little effect upon the accuracy of the transformer.

SPECIAL CURRENT TRANSFORMER APPLICATIONS

In special applications of current transformers it is advisable to consult the transformer manufacturer to ensure the desired results. The characteristics of the current transformer must be thoroughly understood before using it in a circuit arrangement for which it was not specifically designed.

10. Operation at Other Than 30°C. Current transformers are normally designed to operate in a 30°C ambient air temperature with a 55°C temperature rise and are rated accordingly. However, many switchgear and control applications require the current transformers to operate in ambients other than 30°C. For such applications these current transformers may be operated in accordance with the curves shown in Fig. 13. For example, a current transformer having a continuous rating factor (RF) of 2.0 at 30°C can be used at 150% of rated current at an ambient temperature of 55°C. Likewise, a current transformer with a RF of 1.0 could be operated at only 75% of its rated current at an ambient temperature of 55°C.

11. Overload Tripping. Current transformers are frequently used in trip circuits for circuit breakers and motor starters. The factors that must be considered are normal burden, saturation under short-circuit conditions, and the resulting change in ratio. This change in ratio will affect the trip time of the overload relay. Care must be used in the selection of current transformers for tripping so that proper protection will be provided under normal overloads as well as under short-circuit conditions. A ratio correction curve for the transformer will show the change in ratio resulting from various amounts of overload (Fig. 14).

Because of the changing ratios overload tripping transformers are not simultaneously used for accurate metering or instrumentation.

In the case of large low-voltage motor starters, the saturation of current transformers in the thermal-overload heater circuits can be beneficial. Under normal overloads the relay will operate to protect the motor. Under severe short-circuit conditions, the saturation, resulting from greatly increased burden, can be coordinated to prevent the heating element of the relay from burning out before system short-circuit protective devices operate.

In using a current transformer in tripping circuits, care must be taken to prevent the transformer secondary from being opened under any circumstances when load current is flowing in the primary. Dangerously high voltages which can cause personal injury or equipment damage will result.

12. Control Voltages from Current Transformers. In some special control circuits a voltage proportional to the current in a high-current circuit is required for relaying or comparator circuits. The voltage may be compared to a voltage that is responsive to some other circuit function. The differential or sum of the voltages can then be used in relaying or in regulator circuits.

To obtain a voltage proportional to current, a fixed or adjustable resistor is placed across the secondary circuit of the current transformer. A curve for the excitation voltage of the transformer may be obtained from the manufacturer (Fig. 10).

The resistor is sized in accordance with the circuit burden to obtain a voltage that will correspond with the primary current. The amount of voltage on the secondary may be

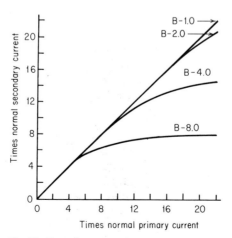

Fig. 14 Typical overcurrent ratio correction curve.

adjusted by setting the value of the resistor to correspond to a given primary current. Again care must be taken to prevent opening the secondary of the current transformer with the primary loaded. The resistor is therefore permanently connected across the secondary with taps or slide for adjusting voltage.

CARE IN INSTALLATION

Often with control or switchgear equipment current transformers may be shipped as separate equipment or may be mounted in the switchgear with secondary circuits to be connected for remote metering or relaying in the field. In these cases the current transformer is usually shipped with a jumper across the secondary terminals. This jumper is intended to prevent injury or equipment damage if the primary is connected and energized before the secondary connections are properly made. The jumper should be removed only after the secondary circuit has been carefully connected and checked.

REFERENCE

1. ANSI C57.13 Standard Requirements for Instrument Transformers.

16

Enclosures

KENNETH L. PAAPE *

* Director of Industrial Control Development, Allen-Bradley Company, Milwaukee, Wis.; Registered Professional Engineer (Wis.); Senior Member, IEEE; Chairman of NEMA Subcommittee on Enclosures and Service Conditions.

FOREWORD

When planning for or specifying switchgear or control equipment, there are many considerations involving standards and codes that determine the type of enclosure required for a given installation. This section provides an understanding of the types of enclosures and their application in accordance with the standards and electrical codes.

DEFINITIONS PERTAINING TO ENCLOSURES

The following definitions are based on Ref. 1:

Corrosion-Resistant. Corrosion-resistant means so constructed, protected, or treated that corrosion will not exceed specified limits under specified test conditions.

Cover. A cover is a formed sheet, casting, or molding without hinges, used to close the principal opening to the equipment cavity of an enclosure.

Design Test. A design test is one which demonstrates compliance of a product design with applicable standards; it is not intended to be a production test.

Door. A door is a formed sheet, casting, or molding with hinges, used to close the principal opening to the equipment cavity of an enclosure.

Driptight. Driptight means so constructed or protected as to exclude falling dirt or drops of liquid under specified test conditions.

Dusttight. Dusttight means so constructed as to meet the requirements of a specified dusttightness test.

Explosionproof. Explosionproof is an often misapplied term used in lieu of "hazardous location" to identify enclosures. See Table 1 for explanation and applications.

Hazardous Locations (Sec. 8).

Indoor. Indoor means suitable for installation within a building which protects the apparatus from exposure to the weather.

Oil-Resistant Gaskets. Oil-resistant gaskets are those made of material which is resistant to oil or oil fumes.

Oiltight. Oiltight means so constructed as to exclude oils, coolants, and similar liquids under specified test conditions.

Outdoor. Outdoor means suitable for installation where exposed to the weather.

Proof (Used as a Suffix). Proof (used as a suffix) means so constructed, protected, or treated that successful operation of the apparatus is not interfered with when subjected to the specified material or condition.

Rainproof. Rainproof means so constructed, protected, or treated as to prevent rain under specified test conditions from interfering with successful operation of the apparatus.

Raintight. Raintight means so constructed or protected as to exclude rain under specified test conditions.

Resistant. Resistant means so constructed, protected, or treated that the apparatus will not be damaged when subjected to the specified material or conditions for a specified time.

Sleetproof. Sleetproof means so constructed or protected that the accumulation of sleet (ice) under specified test conditions will not interfere with the successful operation of the apparatus including external operating mechanism(s).

Sleet-Resistant. Sleet-resistant means so constructed that the accumulation and melting of sleet (ice) under specified test conditions will not damage the apparatus.

Tight (Used as a Suffix). Tight (used as a suffix) means so constructed that the enclosure will exclude the specified material under specified conditions.

Ventilated. Ventilated means so constructed as to provide for the circulation of external air through the enclosure to remove heat, fumes, or vapors.

Watertight. Watertight means so constructed as to exclude water applied in the form of a hose stream under specified test conditions.

TABLE 1. Enclosures for Indoor Hazardous Locations (Ref. 1)

			Enclosure type No.								
Provides protection against atmospheres containing	Class †	Group †	7A or 8A	7B or 8B	7C or 8C	7D or 8D	9E *	9F *	9G *	10	12 ‡
Acetylene	I	A	Yes								
Hydrogen, manufactured gas	I	B	Yes	Yes							
Ethyl ether, ethylene, cyclopropane	I	C	Yes	Yes	Yes						
Gasoline, hexane, naphtha, benzine, butane, propane, alcohol, acetone, benzol, natural gas, lacquer solvent											
Metal dust	I	D	Yes	Yes	Yes	Yes					
Carbon black, coal, dust, coke dust	II	E	-----	-----	-----	-----	Yes				
Flour, starch, grain dust	II	F	-----	-----	-----	-----	-----	Yes			
Fibers, flyings †	III	---	-----	-----	-----	-----	Yes	Yes	Yes	-----	Yes
Methane with or without coal dust	Bureau of Mines		-----	-----	-----	-----	-----	-----	-----	Yes	

If the installation is outdoors and/or additional protection is required as covered by Tables 2 and 3, a combination-type enclosure is required. As defined by the National Electrical Code, the term "explosionproof" applies only to Type 7 and 10 enclosures which, when properly installed and maintained, are designed to contain an internal explosion without causing external hazard. The term should not be applied to Type 8 enclosures which are designed to prevent an explosion through the use of oil-immersed equipment or to Type 9 enclosures which are designed to prevent an explosion by excluding explosive amounts of hazardous dust.

* These enclosures may be ventilated, except that Type 9 will not be dusttight if ventilated.

† As described in Sec. 8.

‡ Nonventilated version only.

ENCLOSURE TYPES AND REQUIREMENTS

Enclosure types and requirements are based on Ref. 1.

1. Purposes of Enclosures.   Enclosures should be designed to protect:
1. Personnel against accidental contact with enclosed electrical devices
2. Internal devices against specified external conditions

These internal (enclosed) devices are assumed to operate in air unless otherwise specified.

Enclosures cannot protect devices against conditions such as condensation, icing, corrosion, or contamination which originate within the enclosure or enter via the conduit or unsealed openings (Sec. 8).

2. Enclosure-Type Designations.   Enclosures are designated by a type number indicating the external conditions for which they are suitable.   These type designations originated decades ago in the Industrial Control Section of the National Electrical Manufacturers Association (NEMA) for industrial-control applications but have been adopted by other segments of the electrical industry (Tables 1, 2, 3).   Enclosures which meet the requirements for more than one type of external condition are often designated by a combination of type numbers, the smaller number being given first.   Enclosure-type designations are applied to a group of controls such as control centers and switchboards as well as to individual controllers.

3. Enclosure Applications.   Tables 1, 2, and 3 may be used as a guide in selecting enclosures for specific applications.   It is quite possible that, because of local code requirements or unusual operating conditions, a different or modified type of enclosure may be required, in which case the manufacturer should be consulted.   Special care should be used in selecting enclosures for hazardous locations, and the advice of competent authorities should be solicited.

4. Nonventilated and Ventilated Enclosures.   Enclosures for switchgear and control are usually nonventilated, except that Type 1, 2, 3R, 9, and 12 enclosures may be either nonventilated or ventilated.   Type 9 and 12 ventilated enclosures are not dusttight.   Ventilated enclosures are designated by the same type numbers as nonventilated enclosures, followed by the appropriate suffix from Table 4.

5. Type 1, General-Purpose — Indoor.   Nonventilated Type 1 enclosures are intended for use indoors, primarily to prevent accidental contact of personnel with the enclosed equipment in areas where unusual service conditions do not exist (Sec. 8). In addition, they provide protection against falling dirt.   Enclosures which are intended to be flush-mounted in building walls should have provision to align the device with the flush plate and to compensate for the thickness of the wall.

Where the enclosure of floor-mounted control apparatus is within 6 in of the floor and exposed live parts within the enclosure are not less than 6 in above the lower edge, no covering is required across the bottom of the enclosure.

When completely and properly installed, general-purpose enclosures prevent the entrance of a rod having a diameter of 0.125 in, except that, if the distance between the

TABLE 2. Enclosures for Indoor Nonhazardous Locations (Ref. 1)

Provides protection against	Enclosure type No.								
	1*	2*	4	4X	5	6	11	12⁺	13
Accidental contact with enclosed equipment......	Yes	Yes	Yes	Yes	Yes	Yes	Yes	Yes	Yes
Falling dirt	Yes	Yes	Yes	Yes	Yes	Yes	Yes	Yes	Yes
Falling liquids and light splashing....................	-----	Yes	Yes	Yes	Yes	Yes	Yes	Yes	Yes
Dust, lint, fibers, and flyings............................	-----	-----	Yes	Yes	Yes	Yes	-----	Yes	Yes
Hosedown and splashing water.........................	-----	-----	Yes	Yes	-----	Yes	-----	-----	-----
Oil and coolant seepage	-----	-----	-----	-----	-----	-----	-----	Yes	Yes
Oil and coolant spraying and splashing	-----	-----	-----	-----	-----	-----	-----	-----	Yes
Corrosive agents..	-----	-----	-----	Yes	-----	-----	Yes	-----	-----
Occasional submersion.....................................	-----	-----	-----	-----	-----	Yes	-----	-----	-----

* These enclosures may be ventilated, except that Type 12 will not be dusttight if ventilated.

TABLE 3. Enclosures for Outdoor Nonhazardous Locations (Ref. 1)

Provides protection against	Type 3	Type 3R*	Type 3S	Type 4	Type 4X	Type 6
Accidental contact with enclosed equipment	Yes	Yes	Yes	Yes	Yes	Yes
Rain, snow, and sleet †	Yes	Yes	Yes	Yes	Yes	Yes
Sleet, while external operating mechanisms remain operable ‡	-----	-----	Yes	-----	-----	-----
Windblown dust	Yes	-----	Yes	Yes	Yes	Yes
Hosedown	-----	-----	-----	Yes	Yes	Yes
Corrosive agents	-----	-----	-----	-----	Yes	-----
Occasional submersion	-----	-----	-----	-----	-----	Yes

 * These enclosures may be ventilated.
 † External operating mechanisms are not required to be operable when the enclosure is ice-covered.
 ‡ External operating mechanisms are operable when the enclosure is ice-covered (sleetproof).

opening and nearest live parts is greater than 4 in, the opening may permit the entry of a rod having a diameter greater than 0.125 in but not greater than 0.500 in.

Ventilated Type 1 enclosures have the same provisions as nonventilated enclosures, except that they have ventilating openings. These openings are controlled to prevent the entrance of a rod having a diameter of 0.500 in, except that, if the distance between the opening and the uninsulated live metal parts is greater than 4 in, the opening may permit the entry of a rod having a diameter greater than 0.500 in but not greater than 0.750 in.

The design tests for Type 1 nonventilated and ventilated enclosures are the following:
1. The rod-entry test described in Art. 22
2. The rust-resistance test described in Art. 29

6. Type 2, Dripproof—Indoor. Type 2 enclosures are intended for use indoors to protect the enclosed equipment against falling noncorrosive liquids and falling dirt. They have provision for drainage. If provision is made for the entrance of conduit at the top, it consists of a conduit hub or the equivalent (Fig. 1). When completely and properly installed, these enclosures prevent the entrance of dripping liquid at a higher level than the lowest live part within the enclosure.

Where the enclosure of floor-mounted control apparatus is within 6 in of the floor and exposed live parts within the enclosure are not less than 6 in above the lower edge, no covering is required across the bottom of the enclosure.

When completely and properly installed, dripproof enclosures prevent the entrance of a rod having a diameter of 0.125 in, except at drain holes. Drain holes do not permit the entry of a rod having a diameter greater than 0.250 in. A distance greater than 4 in

TABLE 4. Types of Ventilation Used for Enclosures (Ref. 1)

Suffix	Method
CV	Convection ventilation (by natural convection)
FV	Forced ventilation without air filter (positive pressure in enclosure)
EFV	Evacuation forced ventilation without filter (negative pressure in enclosure)
FVF	Forced ventilation with inlet-air filter (positive pressure in enclosure)
EFVF	Evacuation forced ventilation with inlet-air filter (negative pressure in enclosure)
FVFF	Forced ventilation with inlet-air filter and outlet-air filter (positive pressure in enclosure)
EFVFF	Evacuation forced ventilation with inlet-air filter and outlet-air filter (negative pressure in enclosure)

"Positive pressure" means greater than atmospheric pressure; "negative pressure" means less than atmospheric pressure.

The entrance of dust into ventilated enclosures can be minimized by the addition of a filter system. However, when considerable air contamination exists and forced ventilation is required, filters fail rapidly, and it may be necessary to bring in air from a remote location.

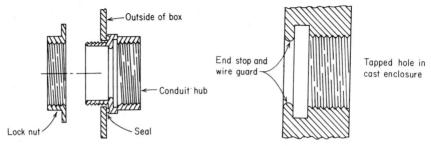

Fig. 1   Conduit hub or equivalent provision.

is maintained between nearest live parts and drain holes which are larger than 0.125 in in diameter, or barriers are installed between drain holes and live parts to prevent the contact of live parts with a straight rigid rod.

Ventilated Type 2 enclosures have the same provisions as nonventilated enclosures, except that they have ventilating openings. These openings are controlled to prevent the entrance of a rod having a diameter of 0.500 in, except that if the distance between the opening and the uninsulated live part is greater than 4 in, the opening may permit the entry of a rod having a diameter greater than 0.500 in but not greater than 0.750 in.

The design tests for Type 2 nonventilated and ventilated enclosures are the following:

1. The rod-entry test described in Art. 22
2. The drip test described in Art. 23
3. The rust-resistance test described in Art. 29

7. Type 3, Dusttight, Raintight, and Sleet- (Ice-) Resistant — Outdoor. Type 3 enclosures are nonventilated and intended for use outdoors to protect the enclosed equipment against windblown dust and water. They are not sleet- (ice-) proof. They have conduit hubs or equivalent provision for watertight connection at the conduit entrance (Fig. 1), mounting means external to the equipment cavity, and provision for locking.

The design tests for Type 3 enclosures are the following:

1. The rain test described in Art. 24
2. The dust test described in Arts. 25 and 26
3. The external-icing test described in Art. 27
4. The rust-resistance test described in Art. 29

8. Type 3R, Rainproof and Sleet- (Ice-) Resistant — Outdoor. Nonventilated Type 3R enclosures are intended for use outdoors to protect the enclosed equipment against rain. They are not dust-, snow-, or sleet- (ice-) proof. They have a conduit hub or equivalent provision for watertight connection at the conduit entrance when the conduit enters at a level higher than the lowest live part, provision for locking, and provision for drainage. When completely and properly installed, these enclosures prevent the entrance of rain at a level higher than the lowest live part.

When completely and properly installed, these enclosures also prevent the entrance of a rod having a diameter of 0.125 in, except at drain holes. Drain holes do not permit the entry of a rod having a diameter greater than 0.250 in. A distance greater than 4 in is maintained between nearest live parts and drain holes which are larger than 0.125 in in diameter, or barriers are installed between drain holes and live parts to prevent the contact of live parts with a straight rigid rod.

Ventilated Type 3R enclosures have the same provisions as nonventilated enclosures, except that they have ventilating openings. These openings are controlled to prevent the entrance of a rod having a diameter of 0.500 in, except that, if the distance between the opening and an uninsulated live part is greater than 4 in, the opening may permit the entry of a rod having a diameter greater than 0.500 in but not greater than 0.750 in.

The design tests for Type 3R nonventilated and ventilated enclosures are the following:

1. The rod-entry test described in Art. 22

2. The rain test described in Art. 24
3. The external-icing test described in Art. 27
4. The rust-resistance test described in Art. 29

9. Type 3S, Dusttight, Raintight, and Sleetproof (Iceproof)—Outdoor. Type 3S enclosures are nonventilated and intended for use outdoors to protect the enclosed equipment against windblown dust and water and to provide for its operation when the enclosure is covered by external ice or sleet. These enclosures do not protect the enclosed equipment against malfunction resulting from internal icing. These enclosures have conduit hubs or equivalent provision for watertight connection at the conduit entrance, mounting means external to the equipment cavity, and provision for locking. In addition, they have sleetproof (iceproof) operating mechanisms, the ability to support the additional weight of the ice, and the ability to withstand removal of the ice by a hand tool to permit access to the enclosure interior. Auxiliary means may be provided to break the ice and permit operation of external mechanisms.

The design tests for Type 3S enclosures are the following:
1. The rain test described in Art. 24
2. The dust test described in Arts. 25 and 26
3. The external-icing test described in Art. 27
4. The rust-resistance test described in Art. 29

10. Type 4, Watertight, Dusttight, and Sleet-Resistant—Indoor and Outdoor. Type 4 enclosures are nonventilated and intended for use indoors and outdoors to protect the enclosed equipment against rain, splashing water, seepage of water, falling or hose-directed water, and severe external condensation. They are not sleetproof (iceproof). They have conduit hubs or equivalent provisions for watertight connection at the conduit entrance (Fig. 1) and mounting means external to the equipment cavity.

The design tests for Type 4 enclosures are the following:
1. The external-icing test described in Art. 27
2. The hosedown test described in Art. 28
3. The rust-resistance test described in Art. 29

11. Type 4X, Watertight, Dusttight, Sleet- and Corrosion-Resistant—Indoor and Outdoor. Type 4X enclosures are nonventilated, have the same provisions as Type 4 enclosures, and in addition, are corrosion-resistant. If application conditions are more severe than those represented by the design test for corrosion resistance, the manufacturer should be consulted.

The design tests for Type 4X enclosures are the following:
1. The external-icing test described in Art. 27
2. The hosedown test described in Art. 28
3. The corrosion-resistance test described in Art. 30

12. Type 5, Dusttight—Indoor. Type 5 enclosures are nonventilated and are intended for use indoors to protect the enclosed equipment against fibers and flyings, lint, dust, and dirt. They have conduit hubs or equivalent provision for dusttight connection at the conduit entrance (Fig. 1) and mounting means external to the equipment cavity. Manufacturers of control apparatus have superseded the Type 5 with the Type 12. However, other related industries still carry the Type 5 designation.

The design tests for Type 5 enclosures are the following:
1. The dust test described in Arts. 25 and 26
2. The rust-resistance test described in Art. 29

13. Type 6, Submersible, Watertight, Dusttight, and Sleet- (Ice-) Resistant—Indoor and Outdoor. Type 6 enclosures are nonventilated and intended for use indoors or outdoors where occasional submersion is encountered. They protect the enclosed equipment against a static head of water of 6 ft for 30 min, dust, splashing or external condensation of noncorrosive liquids, falling or hose-directed water, lint, and seepage. They are not sleet- (ice-) proof. They have conduit hubs or equivalent provision for watertight connection at the conduit entrance (Fig. 1) and mounting means external to the equipment cavity.

The design tests for Type 6 enclosures are the following:
1. The submersion test described in Art. 31
2. The external-icing test described in Art. 27
3. The rust-resistance test described in Art. 29

14. Type 7, Class I, Group A, B, C, or D Hazardous Locations, Air-Break—Indoor. Type 7 enclosures are nonventilated and intended for use indoors, in the atmospheres and locations defined as Class I and Group A, B, C, or D in the National Electrical Code. The letters A, B, C, or D which indicate the gas or vapor atmosphere in the hazardous location appear as a suffix to the designation Type 7 to give the complete designation and correspond to Class I, Group A, B, C, or D, respectively, as defined in the National Electrical Code. These enclosures are designed in accordance with the requirements of Underwriters Laboratories, Inc., "Industrial Control Equipment for Use in Hazardous Locations," UL 698, and are marked to show the class and group letter designations.

The enclosure with its enclosed equipment is evaluated in accordance with Underwriters Laboratories, Inc., Publication UL 698, in effect at the time of manufacture. This evaluation includes proving the ability of a sample enclosure which is typical of the design to withstand the pressures of an internal explosion and not ignite an explosive mixture outside the enclosure, and the ability of the enclosed equipment to interrupt in a flammable atmosphere. The customary method of testing the enclosure consists of mounting an automotive-type spark plug with its electrodes inside the enclosure and its terminals outside the enclosure. A pressure-sensitive transducer is also mounted within or connected to the enclosure. Appropriate vapor for the group (A, B, C, or D) is mixed with air in various proportions and ignited by the spark plug until the mixture creating the highest pressure is determined. The enclosure is then placed in a lightweight wooden box having a large plastic window. The air within the enclosure and within the wooden box is replaced with an appropriate air-gas mixture and ignited by the spark plug. The mixture which results in the worst condition for propagation of a flame through a joint is usually not the same as the mixture which creates the highest pressure. The flame paths of the enclosure are considered adequate if the explosion within the enclosure does not result in sparks or flame outside the enclosure as viewed through the plastic window and does not ignite the outer flammable mixture.

Adequate strength of the enclosure is proved by a hydrostatic-pressure test of another sample enclosure. This sample enclosure must withstand without permanent deformation a static pressure of twice the maximum measured explosion pressure, and three times the maximum measured explosion pressure without rupture if the enclosure is fabricated from sheet steel. The sample enclosure must withstand four times the maximum measured explosion pressure without rupture or permanent deformation if the enclosure is made of cast metal. The increased safety factor is to provide for unknown porosity and stresses in cast enclosures.

When the flammable vapor-air mixture is ignited by arcing at the contacts of the enclosed apparatus, the high-intensity arc thus produced sometimes creates a more severe explosion than that produced with a spark plug.

15. Type 8, Class I, Group A, B, C, or D Hazardous Locations, Oil-Immersed—Indoor. Type 8 enclosures are nonventilated and intended for use indoors, in the atmospheres and locations defined as Class I and Group A, B, C, or D in the National Electrical Code. The letters A, B, C, or D which indicate the gas or vapor atmospheres in the hazardous location appear as a suffix to the designation Type 8 to give the complete designation and correspond to Class I, Group A, B, C, or D, respectively, as defined in the National Electrical Code. These enclosures are designed in accordance with the requirements of Underwriters Laboratories, Inc., Publication UL 698, and are marked to show the class and group letter designations.

The enclosure with its enclosed equipment is evaluated in accordance with Underwriters Laboratories, Inc., Publication UL 698, in effect at time of manufacture. This evaluation includes demonstrating that the oil-immersed equipment can operate at rated voltage and most severe current conditions in the presence of flammable gas-air mixtures without igniting these mixtures.

16. Type 9, Class II, Group E, F, or G Hazardous Locations, Air-Break—Indoor. Nonventilated Type 9 enclosures are intended for use indoors in the atmospheres defined as Class II and Group E, F, or G in the National Electrical Code. The letters E, F, or G which indicate the dust atmospheres in the hazardous location appear as a suffix to the designation Type 9 to give the complete designation and correspond to Class II, Group E, F, or G, respectively, as defined in the National Electrical Code. These enclosures prevent the ingress of hazardous dust. If gaskets are used, they are mechanically attached and of a noncombustible, nondeteriorating, verminproof material. These enclosures are designed in accordance with the requirements of Underwriters Laboratories, Inc., Publication UL 698, and are marked to show the class and group letter designations.

Type 9 ventilated enclosures are the same as nonventilated enclosures, except that ventilation is provided by forced air from a source outside the hazardous area to produce positive pressure within the enclosure.

The design test for Type 9 enclosures is the dust test described in Arts. 25 and 26, and in addition the enclosure with its enclosed equipment is evaluated in accordance with Underwriters Laboratories, Inc., Publication UL 698, in effect at time of manufacture. This evaluation includes a review of dimensional requirements for shaft openings, and joints, gasket material, and temperature rise under a blanket of dust.

17. Type 10, Bureau of Mines. Type 10 enclosures are nonventilated and designed to meet the requirements of the U.S. Bureau of Mines which relate to atmospheres containing mixtures of methane and air, with or without coal dust present, and which are contained in the latest revision of Schedule 2 of the U.S. Bureau of Mines, U.S. Department of the Interior, Washington, D.C., *Information Circular* 8227, and *Bulletin* 541, also U.S. Bureau of Mines.

The design tests for this type of enclosure are those described in Schedule 2. They are similar in principle to those for Types 7 and 9.

18. Type 11, Corrosion-Resistant and Dripproof, Oil-Immersed – Indoor. Type 11 enclosures are nonventilated, corrosion-resistant, and are intended for use indoors to protect the enclosed equipment against dripping, seepage, and external condensation of corrosive liquids. In addition, they protect the enclosed equipment against the corrosive effects of fumes and gases by providing for immersion of the equipment in oil. If application conditions are more severe than those represented by the design test for corrosion resistance, the manufacturer should be consulted. They have conduit hubs or equivalent provision for watertight connection at the conduit entrance (Fig. 1) and mounting means external to the equipment cavity.

The design tests for Type 11 enclosures are the following:

1. The drip test described in Art. 23
2. The corrosion-resistance test described in Art. 30

19. Type 12, Industrial Use, Dusttight and Driptight – Indoor. Nonventilated Type 12 enclosures are intended for use indoors to protect the enclosed equipment against fibers, flyings, lint, dust and dirt, and light splashing, seepage, dripping, and external condensation of noncorrosive liquids. There are no holes through the enclosure and no conduit knockouts or conduit openings, except that oiltight or dusttight mechanisms may be mounted through holes in the enclosure when provided with oil-resistant gaskets. Doors or covers are provided with oil-resistant gaskets. In addition, enclosures for combination controllers have doors which swing horizontally and require a tool to open.

When intended for wall mounting, Type 12 enclosures have mounting means external to the equipment cavity, captive closing hardware, and provision for locking.

When intended for floor mounting, Type 12 enclosures have closed bottoms, captive closing hardware, and provision for locking.

Type 12 ventilated enclosures have the same provisions as nonventilated enclosures, except that they contain both dusttight and nondusttight sections or compartments. Only the nondusttight sections or compartments are ventilated and are not subject to the dust test.

The design tests for Type 12 enclosures are the following:

1. The drip test described in Art. 23 for nonventilated and ventilated enclosures
2. The dust test described in Arts. 25 and 26 for nonventilated enclosures and the nonventilated sections of ventilated enclosures
3. The rust-resistance test described in Art. 29

20. Type 13, Oiltight and Dusttight – Indoor. Type 13 enclosures are nonventilated and intended for use indoors primarily to house control-circuit devices (see Sec. 3) such as limit switches, foot switches, push buttons, selector switches, and pilot lights and to protect these devices against lint and dust, seepage, external condensation, and spraying of water, oil, or coolant. Type 13 enclosures have oil-resistant gaskets and, when intended for wall or machine mounting, have mounting means external to the equipment cavity. They have no conduit knockouts or unsealed openings providing access into the equipment cavity. All conduit openings have provision for oiltight conduit entry.

To ensure against entry of liquid via the conduit, the conduit and the wiring in it must be tightly sealed at the point of entry to the enclosure by an oil- and coolant-resistant compound or equivalent means. The compatibility of the gasket with other sealing material and liquids to which it is exposed may have to be determined by tests. Neoprene is the material most often used for gasketing Type 13 enclosure, although it reacts unfavorably with coolants having a phosphate ester base for flame-retardant qualities.

The design tests for Type 13 enclosures are the following:

1. The oil test described in Art. 32
2. The rust-resistance test described in Art. 29

DESIGN TESTS FOR ENCLOSURES

21. Design-Test Concept. The following design tests are used to demonstrate conformance with NEMA Standard Part ICS 1-110 (Ref. 1). These design tests do not necessarily duplicate environmental conditions and are not contemplated in normal production. To ensure realistic testing, the enclosure and its enclosed equipment are mounted as intended for use in service.

22. Rod-Entry Test. This test is made by attempting to insert the end portion of a straight rod of the specified diameter into the equipment cavity of the enclosure. The enclosure conforms to the NEMA Standard if the rod cannot enter the enclosure. Openings in the bottom of the enclosure need not be tested when the enclosure is installed within 6 in of the floor.

23. Drip Test (Dripproof and Driptight). The enclosure is mounted beneath a drip pan which extends beyond all exposed sides of the enclosure to produce both splashing and dripping. The bottom of the drip pan is equipped with uniformly distributed spouts, one for each 20-in^2 pan area. Each spout has a drip rate of approximately 20 drops per minute of water. The enclosure is subjected to continuously dripping water for 30 min.

The apparatus is considered dripproof if there is no significant accumulation of water within the enclosure and no water has entered the enclosure at a level higher than the lowest live part. The enclosure is considered driptight if no water has entered the enclosure.

24. Rain Test (Rainproof and Raintight). The complete enclosure is mounted with the conduit connected but without using pipe-thread sealing compound. The rigid conduit is threaded into the opening in the enclosure and tightened with the following torque:

Nominal size of conduit, in	Torque, lbf-in
3/4 and smaller	800
1, 1¼, and 1½	1000
2 and larger	1600

A continuous water spray, using as many nozzles as required, is applied against the entire top and all exposed sides of the enclosure for 1 h at a rate of at least 18 in/h at an operating pressure of 5 lbf/in^2. The rate is determined by measuring the rise of water in a straight-sided pan which has been placed horizontally and completely within the area covered by the falling water. A rain test which is performed in accordance with Underwriters Laboratories, Inc., "Rain Tests of Electrical Equipment," Research Bulletin 23, September 1941, is considered to be equivalent to this test.

The enclosure is considered rainproof if there is no significant accumulation of water within the enclosure and no water has entered the apparatus cavity at a level higher than the lowest live part. The enclosure is considered raintight if no water has entered the enclosure.

25. Dust Test (Dusttight) — Dust-Blast Method. The enclosure is subjected to a blast of compressed air mixed with dry Type 1 general-purpose portland cement using a suction-type sandblast gun which is equipped with a $\frac{3}{16}$-in-diameter air jet and a $\frac{3}{8}$-in-diameter nozzle. The air is at a pressure of 90 to 100 lbf/in^2.

The cement is supplied by a suction feed. Not less than 4 lb of cement per linear ft of test length * of the test specimen is applied at a rate of 5 lb/min. The nozzle is held from 12 to 15 in away from the enclosure, and the blast of air and cement is directed at all points of potential dust entry, such as seams, joints, and external operating mechanisms. A conduit may be installed to equalize the internal and external pressures.

The enclosure is considered dusttight if no cement dust has entered it.

Type 1 general-purpose portland cement has been chosen because it is readily available and has a controlled maximum particle size. The analysis of a typical sample is as follows:

Particle size		% content
Mesh	In	
Coarser than 200	Larger than 0.0029	3
200	0.0029	8
325	0.0017	7
400	0.0015 or smaller	82

26. Dust Test (Dusttight) — Atomized-Water Method (Alternate to Dust-Blast Method). The enclosure is subjected to a spray of atomized water, using a nozzle which produces a round pattern 3 to 4 in. in diameter when measured 12 in from the nozzle. The air pressure is 30 lbf/in^2. The water is supplied by a suction feed with a siphon height of 4 to 8 in. Not less than 5 oz/ft of test length * of the test specimen is applied at a rate of 3 gal/h.

The nozzle is held from 12 to 15 in away from the enclosure, and the spray of water is directed at all points of potential dust entry, such as seams, joints, and external operating mechanisms. A conduit may be installed to equalize the internal and external pressures but should not serve as a drain. Metal-to-metal joints with a ground-contact surface meeting the requirements for Type 9 enclosures are not tested. Ungasketed shaft openings which have a length of path not less than $\frac{1}{2}$ in are not tested if the total (diametric) clearance between the shaft and the opening is not more than 0.005 in/$\frac{1}{2}$ in length of path. These joints and openings are protected by suitable external means during the atomized-water test.

The enclosure is considered dusttight if no water has entered the enclosure.

27. External-Icing Test (Sleet-Resistant and Sleetproof). The enclosure is mounted in a room which can be cooled to 20°F. A metal test bar 1 in. in diameter and 2 ft long is mounted in a horizontal position in a location where it receives the same general water spray as the enclosure under test. Provision is made for spraying the entire enclosure from above with water at an angle of approximately 45° from the vertical. The water is between 32 and 37°F. (As a guide, spraying facilities which provide between 1 and 2 gal/h/ft² area to be sprayed have been found effective.) The room temperature is lowered to 35°F. The spray of water is started and continued for at least 1 h, maintaining the room temperature between 33 and 37°F. At the end of this time, the room temperature is lowered to between 20 and 27°F without discontinuing the water spray. (The rate of change in the room temperature is not critical and may be whatever is obtainable with the cooling means employed.) The water spray is controlled so as to cause ice to build up on the bar at a rate of approximately $\frac{1}{4}$ in/h and is continued until $\frac{3}{4}$ in of ice has formed on the top surface of the bar. The spray is then discontinued, but the room temperature is maintained between 20 and 27°F for 3 h to ensure that all parts of the enclosure and ice coatings have been equalized to a constant temperature.

---

* Test length is the summation of the height plus the width plus the depth.

The enclosure and its external mechanisms are considered sleetproof if, while ice-laden, they can be manually operated by one person without any damage to the enclosure, the enclosed equipment, or the mechanism. When an auxiliary mechanism is provided to break the ice, it is included and utilized in the test. A separate test is required for each maintained position of each external operator. It should be possible to gain access to the enclosure interior using an appropriate hand tool without causing functional damage to the enclosure.

The enclosure and its external mechanisms are considered sleet-resistant if they are found to be undamaged after the ice has melted. Enclosures having no external cavities which trap water and sheet-metal enclosures constructed in accordance with Art. 39 are considered sleet-resistant without testing.

28. Hosedown Test (Watertight). The enclosure and its external mechanisms are subjected to a stream of water from a hose which has a 1-in nozzle and which delivers at least 65 gal/min. The water is directed at the enclosure from all angles from a distance of 10 to 12 ft for a total period of 5 min. A conduit may be installed to equalize internal and external pressures but should not serve as a drain.

The enclosure is considered watertight and dusttight if no water has entered the enclosure.

29. Rust-Resistance Test (Rust-Resistant). The enclosure or representative parts of enclosures are subjected to salt spray (fog) and are considered rust-resistant if there is no rust at the completion of the following test:

The test apparatus consists of a fog chamber, a salt-solution reservoir, a supply of suitably conditioned compressed air, atomizing nozzles, support for the enclosure, provision for heating the chamber, and a means of control. The test apparatus does not permit drops of solution which accumulate on the ceiling or cover of the chamber to fall on the enclosure being tested, nor does it permit drops of solution which fall from the enclosure to be returned to the solution reservoir for respraying. The test apparatus is constructed of materials which will not affect the corrosiveness of the fog.

The salt solution is prepared by dissolving 5 ± 1 parts by weight of salt in 95 parts of either distilled water or water containing not more than 200 ppm of total solids. The salt is sodium chloride which is substantially free of nickel and copper and which contains, when dry, not more than 0.1% of sodium iodide and not more than 0.3% of total impurities.

The compressed-air supply to the nozzle(s) for atomizing the salt solution must be free of oil and dirt and maintained between 10 and 25 lbf/in^2.

The temperature of the salt-spray chamber is maintained at 95°F, plus 2°F or minus 3°F. The nozzle(s) are directed or baffled so that none of the spray can impinge directly on the enclosure being tested.

The test is run continuously for 24 h; i.e., the chamber is closed and the spray operated continuously except for the short interruptions necessary to inspect, rearrange, or remove the test specimens, to check and replenish the solution in the reservoir, and to make necessary recordings.

At the end of the test, the specimens are removed from the chamber and washed in clean running water which is not warmer than 100°F to remove salt deposits from their surface, and then dried immediately. Corrosion products are removed by light brushing if required to observe corrosion of the underlying stratum.

The enclosure is considered rust-resistant if there is no rust except at those points where protection is impractical, such as machined mating surfaces of cast enclosures and sliding surfaces of hinges and shafts.

30. Corrosion-Resistance Test (Corrosion-Resistant). The corrosion-resistance test is the same as the rust-resistance test described above except that the exposure time is 200 h.

An enclosure is considered corrosion-resistant if it does not show pitting, cracking, or other deterioration more severe than that resulting from a similar test on passivated AISI Type 304 stainless steel after 200 h of exposure.

31. Submersion Test (Submersible). The complete enclosure is mounted in a tank with the conduit connected using pipe-thread sealing compound. The conduit is tightened with the following torque:

Nominal size of conduit, in	Torque, lbf/in
³/₄ and smaller	800
1, 1¹/₄, and 1¹/₂	1000
2 and larger	1600

The tank is filled with water so that the highest point of the enclosure is 6 ft below the surface of the water. After 30 min, the enclosure is removed from the tank, the excess water is removed from the surface of the enclosure, and the enclosure is opened.

The enclosure is considered submersible if no water has entered the enclosure.

32. Oil Test (Oiltight).   The enclosure is subjected to a stream of test liquid through a nozzle which has a ³/₈-in-diameter opening and which delivers at least 2 gal/min. The stream is directed upon the enclosure from all angles from a distance of 12 to 18 in for 30 min. If the enclosure houses an externally operated device, the device is operated at a rate of approximately thirty operations per minute for the duration of the test. A conduit may be installed to equalize internal and external pressures but may not serve as a drain. The test liquid should be water with the addition of a wetting agent which is approximately 0.1% by volume (or approximately 1.0% by weight if nonliquid) of the test solution.

The enclosure is considered oiltight, dusttight, and driptight if no liquid has entered the enclosure.

ENCLOSURE CONSTRUCTION

33. Size and Configuration.   In general, enclosures are made as small as possible while maintaining required clearances (see Art. 34) between live parts and enclosure walls or openings to conserve enclosure material and user wall or floor space. When several devices are housed in a wall-mounted enclosure, a long and narrow arrangement is more advantageous than a square or short and wide arrangement. Since conduit entries are usually made top and/or bottom, more long and narrow enclosures can be mounted on a given wall than short and wide enclosures without complicating conduit runs. Long and narrow floor-mounted enclosures likewise require less floor space than short and wide enclosures but are more prone to tipping. In determining the proportions of a floor-mounted enclosure, the minimum height is first established as that which is sufficiently high to place external operating mechanisms at convenient heights. (Some local codes require that the center of a disconnect-device operating handle, when in the on position, be not more than 78 in from the floor.)

The total height of an enclosure including base sills, lifting angles, eye bolts, etc., should generally be less than 96 in. Next the enclosure depth is established as that necessary to clear the deepest device enclosed and/or to provide sufficient depth for the entry of large conduit or a distribution bus, or to promote stability. As a guide, a control-center section 90 in high is not considered free-standing with a base less than 15 by 20 in.

34. Electrical Clearance.   In arranging devices and exposed live parts, such as bus bars, safety considerations require minimum clearances between live parts of opposite polarity (parts with a voltage potential between them) and between the enclosure and live parts. These minimum spacings, called electrical clearances, have been established by industry practice and Underwriters' Laboratories' standards, depending upon the control of transients, the voltage rating involved, the energy available in the event of a fault, and the nature of the service. For most industrial-control equipment installed in industrial locations and serviced by trained maintenance personnel, the industrial-control spacings shown in Tables 5 and 6 are adequate. For switchgear and other distribution equipment, such as transfer panels, irrigation or oil-pump controls, or the bus-bar system of a control center, greater clearances are required as shown in Table 7.

35. Provisions for Conduit.   Except for Type 12 enclosures, provisions for line and load wiring conduit are normally provided in the top and bottom end walls of a wall-mounted enclosure. If the equipment requires control wires for remote operation, an additional conduit entrance is provided for in the bottom end wall. Where the enclosure configuration prevents bottom entry as, for example, with some tank-type enclosures, equivalent provisions for conduit entry are made in the top or side end walls. Conduit hubs or the equivalent tapped openings into a casting (Fig. 1) are used when the conduit connection must be sealed against ingress of dust or liquids. For all other wall-mounted enclosures, except Type 12, conduit entry is via knockouts. The size of conduit to be provided for is determined by the number and the size of wires to be connected to the equipment (see Table 8). Wire size is determined by the lug or terminal capacity of the enclosed equipment. Line and load connections are assumed to be via separate conduits. Knockout dimensions are shown in Table 9. Except for cast enclosures or enclosures for marine applications no knockouts or other provisions for conduit are normally made in floor-mounted enclosures. Enclosure specifications for marine applications often require a gasketed removable brass plate of a certain size be furnished in lieu of conduit provisions so that the installer can easily attach conduit fittings.

36. Wiring Space. The minimum space considered adequate for wiring is shown in Table 10. This wiring space is the distance between the end of the lug or pressure wire connector or terminal screw (whichever is least) to be used by the installer and the wall of the enclosure toward which the wire will be directed.

37. Enclosure Materials (Ref. 1). All enclosures must be made of materials which

TABLE 5. Clearance and Creepage Distances for Use Where Transient Voltages Are Not Controlled or Known (Ref. 1)

		Min clearance and creepage distance, in					
		To other than enclosure walls				To walls of metal enclosure ‡	
		Clearance through		Creepage along surface in		Clearance through	Creepage along surface
Apparatus *	Voltage rating, V (rms or dc) †	air	oil	air	oil	air or oil	
General industrial control (see	0–50	0.125	0.125	0.125	0.125	0.500	0.500
Table 7 for those items of in-	51–150	0.125	0.125	0.250	0.250	0.500	0.500
dustrial control considered to	151–300	0.250	0.250	0.375	0.375	0.500	0.500
be distribution equipment)	301–600	0.375	0.375	0.500	0.500	0.500	0.500
	601–1000	0.550	0.450	0.850	0.625	0.800	1.000
	1001–1500	0.700	0.600	1.200	0.700	1.200	1.650
	1501–2500	1.000	0.750	2.000	1.000	2.000	3.000
	2501–5000	2.000	1.500	3.500	2.000	3.000	4.000
Manual motor controllers rated 1 hp or less	51–300	0.063	0.063	0.125	0.125	0.250	0.250
Control devices, not manually	51–150	0.125	0.125	0.250	0.250	0.250	0.250
operated, rated 1 hp or less or 2000 VA or less	151–300	0.250	0.250	0.250	0.250	0.250	0.250
General industrial control where	0–50	0.063	0.063	0.125	0.125	0.500	0.500
the power is limited to a short	51–150	0.125	0.125	0.188	0.188	0.500	0.500
circuit of 10 kVA or less §	151–300	0.188	0.188	0.250	0.250	0.500	0.500
	301–600	0.250	0.250	0.375	0.375	0.500	0.500
	601–1500	0.500	0.375	0.875	0.625	0.500	0.875
	1501–2500	0.875	0.563	1.375	0.750	0.875	1.375

* Clearance and creepage distances are between nonarcing uninsulated live parts and (1) nonarcing uninsulated parts of different potential, (2) metal parts (other than enclosure walls) which may be grounded when the equipment is installed, (3) exposed ungrounded metal parts, and (4) the walls of a metal enclosure, including fittings for conduit or armored cable.

† For grounded power systems, such as 3-phase, 4-wire systems, the clearance and creepage distances to ground are governed by the voltage to ground.

‡ For this requirement, a metal piece attached to the enclosure is considered to be a part of the enclosure if deformation of the enclosure is likely to reduce clearance and creepage distances between the metal piece and uninsulated live parts. See Table 9 for dimensions of conduit bushings.

§ Maximum short-circuit voltamperes is determined as the product of the open-circuit voltage and the short-circuit amperes available at the supply terminals when protective devices are bypassed. Apparatus using these spacings must include the limited power supply.

TABLE 6. Clearance and Creepage Distances for Use Where Transient Voltages Are Controlled and Known (Ref. 1)

Application [a]	Instantaneous peak working voltage, V [b]	To other than enclosure walls		To walls of metal enclosure [c]	
		Clearance through air	Creepage along surface in air	Clearance through air	Creepage along surface
For general use where random	0–50	0.030	0.030	0.500 [f]	0.250
surge and peak transient	51–225	0.075	0.100	0.500	0.500
voltages are limited to 150%	226–450	0.150	0.200	0.500	0.500
of the instantaneous peak	451–900	0.300	0.400	0.500	0.500
working voltages	901–2100	0.625	1.000	1.000	1.500
	2101–3500	1.000	1.600	2.000	3.000
For use where the power is	0–30	0.030	0.030 [e]	0.500 [f]	0.250
limited to a short circuit of	31–50	0.030	0.030	0.500 [f]	0.250
10 kVA or less [d] and peak	51–225	0.060	0.060	0.500	0.500
transient voltages are limited	226–450	0.100	0.100	0.500	0.500
to 150% of the instantaneous	451–900	0.200	0.200	0.500	0.500
peak working voltages	901–2100	0.450	0.450	0.500	1.000
For use where the power is	0–36	0.012	0.012	0.500 [f]	0.250
limited to a short circuit of	37–72	0.016	0.016	0.500 [f]	0.250
500 VA or less [d] and less than	73–100	0.030	0.030	0.500 [f]	0.250
100-V breakdown	101–225	0.045	0.045	0.500	0.500
	226–450	0.060	0.060	0.500	0.500
	451–900	0.100	0.100	0.500	0.500

[a] Clearance and creepage distances are between nonarcing uninsulated live parts and (1) nonarcing uninsulated parts of different potential, (2) metal parts (other than enclosure walls) which may be grounded when the equipment is installed, (3) exposed ungrounded metal parts, and (4) the walls of a metal enclosure, including fittings for conduit or armored cable.

[b] For grounded power systems, such as 3-phase, 4-wire systems, the clearance and creepage distances to ground are governed by the voltage to ground.

[c] For this requirement, a metal piece attached to the enclosure is considered to be a part of the enclosure if deformation of the enclosure is likely to reduce clearance and creepage distances between the metal piece and uninsulated live parts. See Table 9 for dimensions of conduit bushings.

[d] Maximum short-circuit voltamperes is determined as the product of the open-circuit voltage and the short-circuit amperes available at the supply terminals when protective devices are bypassed. Apparatus using spacings based on a limited power supply must include that power supply.

[e] Smaller creepage distances are permissible for this voltage range if the live parts of different potential are rigidly attached to an insulating surface and provided suitable precautions are applied for environmental protection.

[f] Where deflection of the enclosure wall cannot reduce the through-air spacing to the enclosure wall, the spacing through air may be 0.250 in.

TABLE 7. Clearance and Creepage Distances for Switchgear and Such Industrial-Control Apparatus Considered to Be Distribution Equipment (Ref. 4)

Voltage rating, (rms or dc) [*],[†]	Between live parts of different potential		Between live parts and grounded metal parts, including enclosure, or exposed ungrounded metal parts	
	Clearance through air	Creepage along surface	Clearance through air	Creepage along surface
0–125	0.500	0.750	0.500	0.500
125–250	0.750	1.250	0.500	0.500
251–600	1.000	2.000	1.000	1.000
601–1500	1.000	2.000	1.200	1.650
1501–2500	1.000	2.000	2.000	3.000
2501–5000	2.000	3.500	3.000	4.000

[*] Clearance and creepage distances are between nonarcing uninsulated live parts and (1) nonarcing uninsulated parts of different potential, (2) metal parts (other than enclosure walls) which may be grounded when the equipment is installed, (3) exposed ungrounded metal parts, and (4) the walls of a metal enclosure including fittings for conduit or armored cable.

[†] For grounded power systems, such as 3-phase, 4-wire systems, the clearance and creepage distances to ground are governed by the voltage to ground.

TABLE 8. Minimum-Conduit-Size Requirements (Ref. 2)

Max wire size, AWG or MCM	Nominal trade size (in) for this number of conductors					
	1	2	3	4	5	6
16 or 14	1/2	1/2	1/2	1/2	1/2	1/2
12	1/2	1/2	1/2	1/2	3/4	3/4
10	1/2	1/2	1/2	1/2	3/4	3/4
8	1/2	3/4	3/4	3/4	1	1
6	1/2	3/4	1	1	1 1/4	1 1/4
4	1/2	1	1	1 1/4	1 1/4	1 1/2
3	1/2	1	1 1/4	1 1/4	1 1/2	1 1/2
2	1/2	1	1 1/4	1 1/4	1 1/2	2
1	3/4	1 1/4	1 1/4	1 1/2	2	2
0	3/4	1 1/4	1 1/2	2	2	2 1/2
00	3/4	1 1/2	1 1/2	2	2	2 1/2
000	3/4	1 1/2	2	2	2 1/2	2 1/2
0000	1	2	2	2 1/2	2 1/2	3
250 MCM	1	2	2 1/2	2 1/2	3	3
300 MCM	1	2	2 1/2	3	3	3 1/2
350 MCM	1 1/4	2 1/2	2 1/2	3	3 1/2	3 1/2
400 MCM	1 1/4	2 1/2	3	3	3 1/2	4
500 MCM	1 1/4	3	3	3 1/2	4	4
600 MCM	1 1/2	3	3	3 1/2	4	4 1/2
700 MCM	1 1/2	3	3 1/2	4	4 1/2	5
750 MCM	1 1/2	3	3 1/2	4	4 1/2	5

Unless otherwise stated, incoming-line wires, outgoing-load wires, and remote-control wires are assumed to run in separate conduits. Industry practice for wall-mounted enclosures is to provide for 1/2 and 3/4 in, or 3/4 in only conduit for control wires, usually in the bottom end wall. Many manufacturers make alternate provision (conduit hub or knockout) for one trade size larger than the minimum shown above.

will not support combustion in air. Completed enclosures should be capable of passing the rust-resistance test (see Art. 29).

38. Cast-Metal Enclosures (Ref. 1). Cast metal, other than die-cast metal, should be at least 1/8 in thick at every point and of greater thickness at reinforcing ribs and door edges and at least 1/4 in thick at threaded holes for conduit. Die-cast metal should be not less than 3/32 in thick for an area which is greater than 24 in^2 or in which any dimension is greater than 6 in and not less than 1/16 in thick for an area which is 24 in^2 or less or in which no dimension is greater than 6 in and shall be at least 1/4 in thick at threaded holes for conduit.

Cast-metal enclosures are used for Type 1, 4, 6, 7, 8, 9, 10, 11, and 13 applications. Experience has shown that gaskets must be added for Types 4, 6, 11, and 13.

39. Sheet-Metal Enclosures. The minimum thickness required for sheet-metal enclosures varies with the size of the wall panels which constitute the enclosure. Based on the use of solid sheets without any openings other than those required for operating handles or shafts, or for ventilation, the thickness should be not less than that given in Table 11 and not less than 0.032 in if steel, and not less than 0.045 in if non-ferrous metal at points where conduit is connected. Table 11 is based on panel (any surface) widths which result in substantially uniform deflection for a given load and is appropriate for most industrial-control and switchgear applications. If, however, a sheet-steel enclosure is for a control center or switchgear apparatus which is to be listed by UL, the metal thickness must be that specified by Table 12. This table is based on panel area and many years of success in the distribution sector of the electrical industry.

Sheet steel for an enclosure intended for outdoor use is required to be not less than 0.035 in. in thickness if zinc-coated and not less than 0.031 in. in thickness if uncoated.

Sheet copper, brass, or aluminum for an enclosure intended for outdoor use is required to be not less than 0.029 in. in thickness.

TABLE 9. Knockout Diameters and Conduit-Bushing Dimensions (Ref. 1)

Nominal size of conduit, in	Knockout diam, in			Min diam at flange, in	Min overlap over max knockout, in	Height, in
	Nominal *	Min	Max			
1	2 †	3 †	4 †	5 ‡	6	7
½	0.875	0.859	0.906	1.01	0.014	0.375
¾	1.109	1.094	1.141	1.27	0.013	0.422
1	1.375	1.359	1.406	1.54	0.014	0.516
1¼	1.734	1.719	1.766	1.92	0.016	0.562
1½	1.984	1.969	2.016	2.18	0.015	0.594
2	2.469	2.453	2.500	2.68	0.018	0.625
2½	2.969	2.953	3.000	3.20	0.021	0.750
3	3.594	3.578	3.625	3.83	0.021	0.812
3½	4.125	4.094	4.156	4.40	0.024	0.937
4	4.641	4.609	4.672	4.94	0.026	1.000
4½	5.180	5.149	5.211	5.51	0.028	1.062
5	5.719	5.688	5.750	6.05	0.031	1.187
6	6.813	6.781	6.844	7.20	0.040	1.250

* It is desirable that the diameter of the knockout be held as close as possible to the nominal diameter given in col. 2.

† These diameters apply to single or concentric types only and exclude any projection of breakout ears or tabs.

‡ These diameters at the flange of the conduit bushing are minimum in order to obtain the overlap shown in col. 6 with maximum knockout diameter.

TABLE 10. Wire-Bending Space for Field Connections (Ref. 1)

Max wire size, AWG or MCM	Min bending space in, wires per pole *							
	1	2	3	4	5	6	7	8
14–8	Not specified							
6	1½							
4–3	2							
2	2½							
1	3							
000	3½	5	7					
000–0000	4	6	8					
250 MCM	4½	6	8	10				
300–350 MCM	5	8(7)	10(8)	12(9)				
400–500 MCM	6	8(7)	10(8)	12(9)	14	16	18	20
600–700 MCM	8	10(8)	12(9)	14(10)	16	18	20	22
750–900 MCM	8	12(10)	14(11)	16(12)	18	20	22	24
1000–1250 MCM	10							
1500–2000 MCM	12							

* Values in parentheses may be used when the depth of the wiring space is 12 in or more and where only a single right-angle bend is required within 12 in of the terminal.

TABLE 11a. Thickness of Sheet-Steel Enclosures

Maximum Dimensions of Any Surface—Carbon or Stainless Steel Sheet

| Without supporting frame * | | With supporting frame or equivalent reinforcing * | | |
Max width,*† in	Max length, in	Max width,† in	Max length, in	Min thickness, in
4.0	Not limited	6.25	Not limited	0.020
4.75	5.75	6.75	8.25	0.020
6	Not limited	9.5	Not limited	0.026
7	8.75	10	12.5	0.026
8	Not limited	12	Not limited	0.032
9	11.5	13	16	0.032
12.5	Not limited	19.5	Not limited	0.042
14	18	21	25	0.042
18	Not limited	27	Not limited	0.053
20	25	29	36	0.053
22	Not limited	33	Not limited	0.060
25	31	35	43	0.060
25	Not limited	39	Not limited	0.067
29	36	41	51	0.067
33	Not limited	51	Not limited	0.080
38	47	54	66	0.080
42	Not limited	64	Not limited	0.093
47	59	68	84	0.093
52	Not limited	80	Not limited	0.108
60	74	84	103	0.108
63	Not limited	97	Not limited	0.123
73	90	103	127	0.123

TABLE 11b. Thickness of Nonferrous Sheet-Metal Enclosures

Maximum Dimensions of Any Surface—Aluminum, Copper, or Brass Sheet

| Without supporting frame | | With supporting frame or equivalent reinforcing * | | |
Max width,*† in	Max length, in	Max width,† in	Max length, in	Min thickness, in
3.0	Not limited	7	Not limited	0.023
3.5	4	8.5	9.5	0.023
4.0	Not limited	10	Not limited	0.029
5.0	6	10.5	13.5	0.029
6	Not limited	14	Not limited	0.036
6.5	8	15	18	0.036
8	Not limited	19	Not limited	0.045
9.5	11.5	21	25	0.045
12	Not limited	28	Not limited	0.058
14	16	30	37	0.058
18	Not limited	42	Not limited	0.075
20	25	45	55	0.075
25	Not limited	60	Not limited	0.095
29	36	64	78	0.095
37	Not limited	87	Not limited	0.122
42	53	93	114	0.122
52	Not limited	123	Not limited	0.153
60	74	130	160	0.153

Notes appear on p. 16–19.

Notes to Tables 11*a* and 11*b*

Dimensions are for a substantially uniform deflection based on the relationship

$$Y = R \frac{Pa^2}{Et^3}$$

where Y = deflection
P = load, concentrated at the center of the plate
R = factor depending upon width-length ratio, Poisson's ratio, and other factors
a = width or smaller dimension of rectangular plate
E = modulus of elasticity
t = thickness of plate

This relationship is true for small deflections of a plate, but for large deflections and thinner plates it is necessary to consider the stresses set up as a result of elongation of the plate along its central plane or axis in addition to the stresses due to bending. These additional stresses are called "diaphragm stresses" and have the effect of greatly increasing the stiffness of thin plates as the deflection increases, so that the maximum permissible dimensions become conservative for greater loads.

* A supporting frame is a structure of angle or channel or a folded rigid section of sheet metal which is rigidly attached to and has essentially the same outside dimensions as the enclosure surface and which has sufficient torsional rigidity to resist the bending moments which may be applied via the enclosure surface when it is deflected. Construction that is considered to have "equivalent reinforcing" may be accomplished by a multiplicity of designs that will produce a structure which is as rigid as one built with a frame of angles or channels. Construction considered to be "without supporting frame" includes (1) a piece with single formed flanges (formed edges), (2) a single sheet which is corrugated, and (3) an enclosure surface loosely attached to the frame, as, for example, with spring clips.

† The width is the smaller dimension of a rectangular sheet-metal piece which is part of an enclosure and supported at all four edges. Where a surface is supported at fewer than four edges, the width is to be considered the maximum dimension between supporting edges even though it may be the long dimension of the sheet. Supported at the edge means fastened to, or otherwise having a relatively solid support, stiffening member, or adequate reinforcing by forming so as to prevent or minimize the deflection of the sheet metal. Adjacent surfaces may have supports in common and be made of a single sheet.

40. Ventilating Openings (Ref. 1). When openings are screened, the wire gage of the screen should be not less than No. 16 AWG. When the screen-mesh openings are greater than $\frac{1}{2}$ in^2 in area, the wire gage should be not less than No. 12 AWG.

Except as noted in the following paragraph, sheet metal employed in expanded metal mesh and perforated sheet metal should be not less than 0.042 in thick if steel and not less than 0.058 in thick if nonferrous. When the mesh openings or perforations are greater than $\frac{1}{2}$ in^2 area, the metal should be not less than 0.080 in thick if steel and not less than 0.122 in thick if nonferrous.

In a small device where the indentation of a guard or enclosure will not affect the

TABLE 12. Thickness of Sheet-Steel Enclosures for Switchboards and Control Centers to Be Listed by Underwriters Laboratories, Inc. (Ref. 4)

Max size of largest surface		Min acceptable thickness of sheet metal		
Length or width,* in	Area, in^2	Zinc-coated steel, in	Uncoated steel, in	Aluminum, in
24	360	0.056(16 †)	0.053(16 †)	0.075(12 †)
40	1000	0.070(14 †)	0.067(14 †)	0.095(10 †)
Over 40	Over 1000	0.097(12 †)	0.093(12 †)	0.122(8 †)

* The length and/or width of a panel may be greater than shown in this column, provided the area limitations are not exceeded, if (1) the panel is flanged the full length of both the longer sides, or (2) the panel is flanged the full length of one of the longer sides, and the unflanged side is appropriately secured to a rigid switchboard member.

† The figures in parentheses are the Galvanized Sheet Gage (GSG) numbers for zinc-coated steel, the Manufacturer's Standard Gage (MSG) numbers for uncoated steel, and the American Wire Gage (AWG) numbers for aluminum that provide the required minimum thickness of metal.

clearance between uninsulated movable current-carrying parts and ground metal, 0.020-in-thick expanded metal mesh may be employed, provided that either:

1. The exposed mesh on any one side of the surface of the device, which is so protected, has an area which is not greater than 72 in^2 and has no dimension which is greater than 12 in.

2. The width of an opening so protected is not greater than $3\frac{1}{2}$ in.

41. Operating Mechanisms Mounted on Enclosures (Ref. 1).   External operating mechanisms, such as those for disconnects, pilot devices, and resetting, when mounted on or through the enclosure, should be capable of passing the test or tests for the enclosure.

42. Legends.   Legends and instruction information appearing on the exterior of the enclosure should be applied in such a manner as to have a degree of permanence and legibility commensurate with the environment and application for which the enclosure is intended.

43. Enclosure-Door Requirements (Ref. 3).   Doors (as opposed to covers) should be provided on all enclosures which contain either fuses or apparatus which requires access to the enclosure interior during normal operation, as, for example, to reset overload relays.   Since fuses are known to explode under certain fault conditions, doors and latches must be securely attached.   Door hardware should be designed to keep the door closed and any fire hazard contained in the event of a high-energy short circuit. Experience has shown that any gasket suitable for industrial applications is effective in containing the hot gases created by a fault if the door itself is not displaced.   Doors of sheet-metal enclosures without gaskets must provide a controlled flame path (Fig. 2). A controlled flame path is a minimum length of path between two metal surfaces a maximum distance apart.   For most wall-mounted enclosures a $\frac{1}{2}$-in-long flame path (overlap between enclosure box and door) not more than $\frac{1}{8}$ in wide (clearance between box flange and door) is adequate.   Fabrication methods for larger enclosures made of heavier sheet metal usually require door clearances greater than the $\frac{1}{8}$ in, and the length of flame path is increased accordingly.   For example, the door of a floor-mounted enclosure made of 0.093-in-thick sheet steel which clears the box flange by $\frac{1}{4}$ in should have a $1\frac{1}{4}$-in-long flame path (overlap of box flange by door flange).

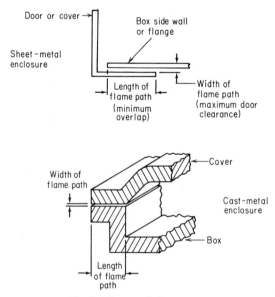

Fig. 2   Flame-path dimensions.

44. Enclosure Grounding. Metal enclosures in fixed installations are normally grounded by connection via armored cable, conduit, or metal raceway (Ref. 2). Nonmetallic enclosures and enclosures used as pendants require a terminal or terminals to bond the apparatus to a grounding conductor included in the cable or nonmetallic conduit. The fundamental requirement is a continuous low-impedance circuit to ground for the baseplate on which the enclosed apparatus is mounted and for all metallic parts which project through the enclosure. A test to demonstrate an adequate ground circuit consists of wiring a 120-V, 60-W light bulb with the ground circuit as one conductor. If the bulb glows to full brilliance with 120 V supplied, the ground circuit is adequate. If a grounding conductor is furnished by the apparatus manufacturer, it should be green with or without a yellow stripe (Ref. 3). If a screw is furnished for the installer to use in making a grounding connection, a screw with a slotted hexagonal green-colored head is preferred. Control centers, switchboards, and service-entrance equipment require grounding provisions as shown in Table 13 (Sec. 30, Art. 20).

45. Class I Hazardous-Location Requirements (Ref. 5). Enclosures for use in Class I hazardous locations are designed on the premise that at some time during their life the hazardous gases enter the enclosure and mix with air in such proportions that as the enclosed devices operate they ignite these mixtures which then explode within the enclosure. These enclosures must be strong enough to withstand the force of such explosions without rupturing and have sufficient capability along their flame paths (Fig. 2) to cool the hot gases of the explosion below the ignition temperature of a similar explosive mixture outside the enclosure. See Underwriters Laboratories, Inc., Publication UL 698, "Industrial Control Equipment for Use in Hazardous Locations."

a. Joints between box and cover must be metal-to-metal types and sufficiently long and narrow to provide an effective flame path. Gasketed joints or gasketed shaft openings are not acceptable because of their susceptibility to damage and deterioration. Threaded shafts should be considered as a way of meeting flame-path requirements in a shorter span if only partial rotation is required, to operate a reset mechanism, for example.

b. Holes in the enclosure for attaching nameplates, legend plates, hinges, closing hardware, etc., should be bottomed or closed by welding or peening the fastener to prevent accidental opening. Bolt holes which are part of a flame path may not have a diameter which is more than $1/32$ in greater than the nominal diameter of the bolt and must be at least $1/2$ in from the inside of the enclosure at their closest points. The distance from the joint to the underside of the bolt head should equal at least one-half of the required flame path. Drains, breathers, plugs, etc., must be attached in such a way as to preclude their removal from outside the enclosure.

c. Conduit provisions should be for threaded rigid conduit engaging at least five full

TABLE 13. Grounding Conductor for Grounded Control Centers, Switchboards, and Service-Entrance Equipment (Ref. 2)

Size, AWG or MCM of largest incoming conductor or equivalent for multiple conductors		Min size, AWG, of grounding lug to be provided	
Copper	Aluminum	For copper wire	For aluminum wire *
2 or smaller	0 or smaller	8	6
1 or 0	00 or 000	6	4
00 or 000	0000 or 250 MCM	4	2
0000 through 350 MCM	251 through 500 MCM	2	0
351 through 600 MCM	501 through 900 MCM	0	000
601 through 1100 MCM	901 through 1750 MCM	00	0000
Over 1100 MCM	Over 1750 MCM	000	250 MCM

When largest incoming conductor is bus bar, convert bus bar cross-sectional area to MCM and apply table. (Width in mils × thickness in mils × 0.001273 = MCM.)

* The National Electrical Code does not permit aluminum grounding conductors to be used in direct contact with masonry or earth or where they may be subjected to corrosive conditions as, for example, outdoors within 18 in of the earth.

threads, and so located that there is sufficient room in which to rotate conduit seal-off fittings.

Enclosures for use in hazardous locations should include a warning legend calling attention to the necessity of disconnecting incoming power to the enclosed equipment before opening the enclosure. This restriction on the user is one of the reasons why circuit breakers may be preferred over fuses in hazardous locations. The enclosure should also be marked for the class and group(s) for which the enclosure is designed.

46. Class II Hazardous-Location Requirements (Ref. 5). Enclosures for use in Class II hazardous locations are designed on the basis of excluding the hazardous dust. Joints between box and cover or door should be either metal-to-metal at least $\frac{3}{16}$ in wide or gasketed at least $\frac{3}{8}$ in wide. Gaskets, if used, must be made a noncombustible, nondeteriorating, vermin-proof material and mechanically attached. Adhesive may be used in addition to but not in lieu of the mechanical means. The gasket material is usually some form of asbestos or treated plant fiber. The total clearance within a shaft opening (the difference in diameters between a smooth shaft and bushing) should not be more than 0.005 in for a $\frac{1}{2}$-in-long engagement but may be increased 0.001 in for each $\frac{1}{16}$-in increase in engagement length. Threaded shafts should be considered as a

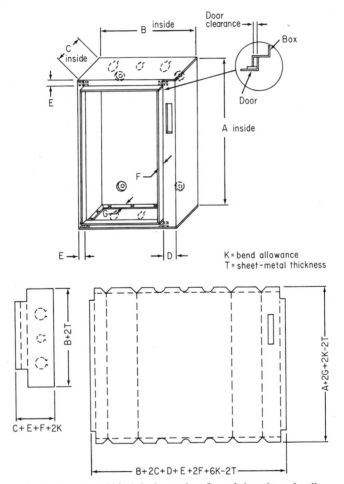

Fig. 3   Sheet-metal blank for box with unflanged sheet for end wall.

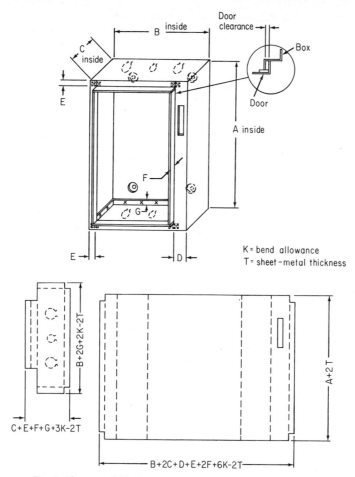

Fig. 4   Sheet-metal blank for box with flanged sheet for end wall.

way of providing an adequate dust barrier in a shorter span if only partial rotation is required, to operate a selector switch, for example.

Conduit provisions should be for threaded rigid conduit engaging at least three full threads or ¼ in, whichever is greater.

Enclosures for use in hazardous locations should include a warning legend calling attention to the necessity of disconnecting incoming power to the enclosed equipment before opening the enclosure. This restriction on the user is one of the reasons why circuit breakers may be preferred over fuses in hazardous locations. The enclosure should also be marked for the class and group(s) for which the enclosure is designed.

47. Typical Sheet-Metal Construction. Most sheet-metal wall-mounted enclosures with doors are fabricated with a single piece formed to produce a back wall, two side-walls, and a portion of the front wall (Figs. 3 and 4). The chief difference between these two types of construction is the location of the flanges which are used to resistance-weld the end walls to the back and sidewalls. In Fig. 3 the flanges are part of the back and sidewalls and the portion of the end wall to be attached is flat. A tight joint can be made in spite of slight variations in the accuracy of the forms. Conversely, in Fig. 4, where the flanges are part of the end wall, the forming of both pieces must be

TABLE 14. Sheet-Metal Bend Allowances

Metal thickness, in	0.015	0.031	0.047	0.062	0.093	0.125	Inside corner 0.156
0.015	+004	−004	−010	−016	−028	−041	−053
0.020	+007	−002	−008	−014	−026	−039	−051
0.031	+014	+007	−002	−007	−020	−032	−045
0.036	+016	+010	+002	−005	−018	−031	−043
0.040	+018	+013	+005	−003	−016	−029	−040
0.050	+023	+020	+013	+004	−010	−024	−036
0.062	+030	+026	+021	+014	−004	−016	−030
0.070	+034	+031	+026	+020	+002	−012	−024
0.078	+038	+035	+031	+025	+009	−008	−020
0.093	+046	+044	+040	+035	+021	+003	−012
0.109	+054	+052	+049	+045	+033	+015	−004
0.125	+062	+061	+058	+054	+043	+028	+011
0.140	+069	+068	+065	+062	+052	+039	+022
0.156	+077	+075	+074	+071	+062	+049	+034
0.187	+093	+092	+091	+088	+081	+070	+057
0.218	+108	+107	+106	+104	+098	+088	+076
0.250	+124	+123	+122	+121	+116	+107	+097

closely controlled to ensure a good fit. The construction of Fig. 3, however, requires that the flanges be notched to clear conduit bushings, or the knockouts must be located sufficiently far from the back wall to permit conduit bushings to clear the rear flanges if they are not notched. The greater the distance from wall to knockout, the larger the conduit offset required. Adding mounting feet or embossed mounting holes to the back wall of the enclosure (to provide clearance for the projecting screws used to mount

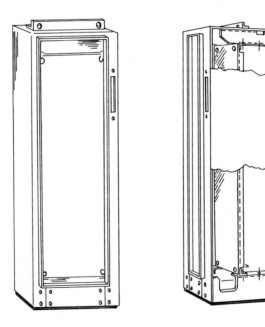

Fig. 5 Supporting-frame construction.

radius, in

0.187	0.218	0.250	0.312	0.375	0.500	0.625	0.750	1.000
−065	−077	−090	−114	−138	−187	−236	−284	−382
−063	−075	−088	−112	−136	−185	−234	−282	−380
−057	−069	−082	−106	−130	−179	−228	−276	−374
−055	−067	−080	−104	−128	−177	−226	−274	−372
−053	−065	−078	−102	−126	−175	−224	−272	−370
−048	−060	−073	−097	−121	−170	−219	−267	−365
−042	−054	−067	−091	−115	−164	−213	−261	−359
−038	−050	−063	−087	−111	−160	−209	−257	−355
−034	−046	−059	−083	−107	−156	−205	−253	−351
−023	−037	−051	−075	−099	−148	−197	−245	−343
−015	−027	−041	−067	−091	−140	−189	−237	−335
−007	−019	−031	−059	−083	−132	−181	−229	−327
+004	−012	−024	−049	−076	−125	−174	−222	−320
+018	−002	−016	−039	−068	−117	−166	−214	−312
+042	+025	+005	−023	−047	−101	−150	−198	−296
+064	+048	+030	−008	−032	−082	−135	−183	−281
+085	+072	+055	+020	−016	−062	−119	−167	−265

Read K in thousandths (0.000) of an inch from chart for a given metal thickness and inside corner radius. Interpolate as necessary.

equipment inside the box) also increases the amount the conduit must be bent to move away from the wall and enter the knockout.

The overall dimensions of cabinets constructed as shown in Figs. 3 and 4 are determined by the required inside dimensions and the formulas shown in the referenced figures. The bend allowance K in these formulas is a plus or minus figure as shown in Table 14 depending upon the thickness of the sheet metal and the edge radius of the tool used to form the cabinet parts.

If the equipment to be housed in a Type 1 enclosure consists of a single or at most only a few devices, holes in the back wall of the enclosure can be the mounting means, if there is clearance for the projecting screws. For all other applications of sheet-metal enclosures, a panel is provided on which the apparatus is mounted. In the case of wall-mounted enclosures, electricians often remove this panel to facilitate attaching conduit and pulling wires. This panel may be mounted with screws, on studs, on brackets, or on a frame as shown in Fig. 5.

Doors for sheet-metal enclosures are fabricated from blanks notched as shown in Fig. 6. The four joints which result after the door is formed may be arc-welded for increased rigidity or improved appearance.

Enclosures are manufactured from sheet metal and assembled by resistance weld-

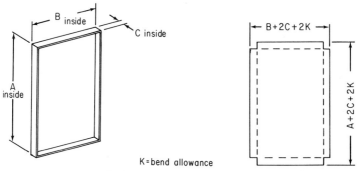

Fig. 6   Sheet-metal blank for door.

ing to meet the requirements of Types 1, 2, 3, 3R, 3S, 5, and 12. Sheet-metal enclosures with arc-welded seams are used for Type 4, 4X, 6, 7, 8, 9, 11, and 13 applications. Experience has shown that gaskets should be added for Types 2, 3, 3S, 4, 4X, 5, 6, 9, 11, 12, and 13.

48. Molded Enclosures. Enclosures for Type 1, 4X, 6, and 13 applications are sometimes made from molded nonmetallic parts and require special provisions for grounding (Art. 44).

REFERENCES

1. "Industrial Control and Systems," National Electrical Manufacturers' Association, NEMA Standard Publication ICS-1970, New York, 1970, Reaffirmed 1975.
2. National Fire Protection Association (NFPA), National Electrical Code 1971, art. 250, Grounding, and chap. 9; *Pub.* NFPA 70-1975, Boston, 1974.
3. "Industrial Control Equipment," Underwriters Laboratories, Inc., *Pub.* UL 508, Chicago, Ill., 1970.
4. "Dead-Front Switchboards," Underwriters Laboratories, Inc., *Pub.* UL 891, Chicago, Ill., 1966.
5. "Industrial Control Equipment for Use in Hazardous Locations, Class I, Groups A, B, C, and D; Class II, Groups E, F, and G," Underwriters Laboratories, Inc., *Pub.* UL 698, Chicago, Ill., 1973.

17

Ac Switchgear

MAX E. EVERETT *

* Senior Engineer, Switchgear Division, Allis-Chalmers Corporation (Retired); Registered Professional Engineer (Wis.).

FOREWORD

Selection of the proper switchgear for a particular job requires a knowledge of the considerations which have been covered in other sections of this handbook, such as the functions of switchgear, system-voltage considerations, and protection and coordination. The loads that are to be supplied by the switchgear, as well as the characteristics of these loads and therefore the protection they require, must also be known.

The engineer responsible for making the selection writes a specification for the switchgear to be purchased. He then sends copies of the specification to various manufacturers, asking for quotations. The selection is then made, based not only on price but on the reputation of the manufacturer, his willingness and ability to comply with the requirements of the specifications, and in some cases his ability to meet desired delivery schedules. Also, in a plant which already contains switchgear of one manufacturer, it may be advantageous to purchase the new switchgear from the same manufacturer, because of interchangeability of circuit breakers and other parts. This section describes available switchgear elements to assist in writing specifications and selecting a supplier.

When writing specifications for switchgear, the engineer should specify that the equipment must meet the requirements of the NEMA, ANSI, and IEEE standards. The National Electrical Code does not include standards for switchgear, and at present no switchgear or major components such as large air circuit breakers are sold with the UL label.

Switchgear specifications should describe in detail the equipment required but should leave details of design to the manufacturer. For instance, the engineer may specify that the main bus should be rated 3000 A, but the size and configuration of the bus should be left to the manufacturer. Reliable manufacturers can furnish certified test data to show that prototypes of their switchgear units have been tested and that, at rated current, the temperature rise will not exceed that allowed by the standards.

LOW-VOLTAGE METAL-ENCLOSED SWITCHGEAR

1. **Structures.** There are two basic types of low-voltage switchgear structures, those for use outdoors and those for use indoors. Indoor switchgear is, of course, considerably less expensive, but other factors must be considered. Generally for new construction, the switchgear will be located indoors if the loads are indoors and outdoors when the loads are located outdoors. If possible the low-voltage gear will be located at the transformer and directly connected to it, to form an articulated secondary-unit substation. However, it may be advantageous to locate the transformer outdoors with bus duct or cables connecting the transformer to the low-voltage switchgear. In some cases the bus duct or cables may serve the added purpose of reducing the short-circuit current available at the switchgear, but usually this is not a consideration.

The size of the switchgear will also be a factor in determining its location. The size, of course, will depend on the number of circuit breakers and the size of the breakers. Metal-enclosed, low-voltage switchgear is shown in Sec. 1, Fig. 12.

Switchgear is not available in explosion-proof enclosures. Therefore, to serve a building which may contain explosive atmospheres, the switchgear may be located outdoors. This arrangement may also be made to serve a building in which corrosive or extremely dusty atmospheres are prevalent. Outdoor-type switchgear may be used indoors in some cases where there is excessive dust, since this type of enclosure is also dusttight.

For corrosion problems, special paints may be required on the steel structure. If possible, switchgear should be located where corrosive atmospheres will not be encountered. If this arrangement is not possible, the choice between aluminum and copper bus may be made on the basis of the chemicals in the atmosphere. Special plating may be required on the bus, on contact surfaces, and on other conducting parts which cannot be painted. A chemist should be consulted to determine the best method of protecting the various metals in the switchgear.

a. *Indoor Switchgear.* If indoor switchgear is selected, it will consist of a front section, which will contain the circuit breakers, meters, relays, and controls; a bus section; and a cable-entrance section. Each circuit breaker must be isolated from all other equipment. Vents are provided in the circuit-breaker compartments, for cooling and to allow escape of ionized gases which are formed when the circuit breaker opens to interrupt fault currents. Other sections of the switchgear must also be ventilated to allow circulation of air for cooling.

Barriers between the bus compartment and the cable compartment are not required, but they may be furnished. Some engineers feel that they should be provided for safety reasons, to permit connecting or disconnecting the cables without danger of contacting the hot bus. The argument is certainly valid if frequent changes in the cable connections are anticipated. However, if barriers are provided between the bus and the cable compartments, all cable terminations should be insulated after installation, so that there will also be no danger of workmen contacting these hot terminals when making changes in the cable connections. It is assumed, of course, that the breaker feeding the cables being worked on will be open.

In double-ended substations the standards require a barrier between the two sections of the switchgear. A horizontal barrier will be placed between the upper and lower terminals of the tie breaker with a vertical barrier above the horizontal barrier on one side, and below it on the other, in such a way as to isolate the two buses and prevent an arcing fault on one end of the substation from traveling to the other end and taking out both power sources. In double-ended substations which do not have a tie breaker (spot network), the two incoming lines should be isolated in the same way. Some codes require a barrier in all cases to isolate the incoming line and prevent an arcing fault from taking out the service. However, an arcing fault will travel away from the source of power, so that is is rarely possible for such a fault on the main bus to travel toward the source and take out the service.

When drawout breakers are supplied, a hoist for removing the breakers is desirable. With indoor gear, this may not be required, since overhead cranes or other lifting facilities may be available in the building. With outdoor gear, a hoist is almost a necessity, especially when larger breakers are involved.

b. Outdoor switchgear is essentially the same as indoor switchgear with a structure built around it for weatherproofing. A walk-in aisle is provided in front of the circuit breakers to protect workmen and equipment from the weather during maintenance operations. Heaters must be provided to keep the inside temperature above the outside temperature to prvent condensation. Lights are provided in the aisle space, and a 120-V receptacle is provided for portable power tools.

c. Painting. Indoor switchgear is usually painted a light gray, ANSI 61, using lacquer or enamel. Outdoor switchgear usually is painted the same color inside, but the outside is painted a dark gray, ANSI 24. Also, outdoor switchgear is mounted on sill channels and the structure is undercoated with a heavy coat of an asphalt material to prevent rusting. Some manufacturers may use other colors as their standard paint. As previously stated, special paints, such as epoxy paints, may be required in certain corrosive atmospheres. Unless necessary, special paints, even special colors, should not be specified, as they add appreciably to the cost of the equipment.

2. Bus Bars. Copper is a better material for bus bars than aluminum, except in the few cases where corrosive atmospheres may have an adverse effect on copper. Copper has a higher conductivity than aluminum, it is more easily plated, and bolted joints can be made more easily. Also the melting point of aluminum is lower than that of copper, so that more damage will be done to aluminum buses in case of an arcing fault. However, copper is more expensive than aluminum, and most switchgear manufacturers now use aluminum for bus bars, unless copper is specified, in which case they charge a premium price. The engineer specifying the switchgear must decide whether the advantages of copper warrant paying the premium. Aluminum has proved to be satisfactory in many cases. If aluminum is used, joints may be welded, in which case it is difficult to make changes in the field. Where bolted joints are necessary, as at shipping splits, aluminum may be welded to copper. Copper joints must be silver-plated. Aluminum bolted joints may be silver-plated or tin-plated. If aluminum bus is specified, bolted joints should be made with Belleville washers to minimize the tendency to cold-flow and to maintain the tight clamp if there is some cold flow or some stretching of the bolt over a period of time.

As stated previously, details of bus design should be left to the manufacturer. Temperature rise in the bus must be kept at or below the rise permitted by the standards (ANSI C37.20–4.4.2). Many factors such as bus size, bus configuration, ventilation, current density, skin effect, bus location, bus material, and the condition of the surface of the bus have an effect on the temperature rise. For this reason, it makes little sense to specify a maximum current density when all these other factors affect temperature rise. A current density of 1000 A/in^2 copper is frequently considered the maximum permissible. But a No. 14 wire carrying 15 A has a current density of over 4000 A/in^2, and all codes permit 15 A in a No. 14 wire.

Each vertical section of low-voltage switchgear will contain from one to four circuit breakers. This requires branches and taps off the main bus, so that covering the bus with insulation is not practical. Therefore, bare buses are used, supported on insulators. The insulating material used should be flame-retardant, track-resistant, and nonhygroscopic. It must also have high impact strength to be able to withstand the stresses caused by magnetic forces when faults occur. Glass polyester is the best material available for this purpose and is used by most manufacturers.

The insulators should be assembled to the bus in such a way that there are no continuous horizontal surfaces between bus bars. Such surfaces may collect dust, which may form a high-resistance path between the bars which could develop into a fault. Dust and moisture may also cling to vertical surfaces. For this reason a creepage distance of 2 in should be maintained between phases, and from any phase to ground. A 1-in clearance in air is adequate for 600 V; however, the standards do not specify a minimum clearance. They do require that the bus withstand a high-potential test of 2200 V between phases, and from each phase to ground, for a period of 1 min.

Supports for the bus bars should be placed at frequent intervals so that the bars will not be deformed when a fault occurs. The magnetic forces on the bus depend on the amount of current and the configuration of the bus. When writing specifications, one should state that the bus should be braced to withstand the forces created by the maximum current available on the bus. Note that this current is the maximum available from the transformer, plus any current which may be contributed by motors on the load side of the feeder breakers.

3. Low-Voltage Circuit Breakers. When specifying circuit breakers, the required frame size and the desired trip rating must first of all be determined. Fused breakers can sometimes be used to reduce the frame size, thereby decreasing the cost. The choice must also be made between drawout and fixed breakers and between manually and electrically operated breakers.

a. Selecting Frame Size and Trip Rating. The proper frame size of the breaker will depend on the short-circuit current available, the continuous current, and the system voltage. Short-circuit ratings for the various standard frame sizes are given in Table 1. Some manufacturers make breakers in other than standard frame sizes. For information on the interrupting rating of these breakers consult the manufacturer.

When a transformer is supplying the switchgear, the fault current available at the transformer terminals is

$$I_f = \frac{100 \times 1000 \times T}{\sqrt{3}E(Z + 100T/P)}$$

where E = system voltage
I_f = fault current available (rms symmetrical) at the transformer terminals
T = kVA rating of transformer
Z = transformer impedance, %
P = short-circuit capacity of the primary system, kVA

If the switchgear is supplied by a generator, the formula is the same, except that T becomes the kVA rating of the generator, Z is the generator impedance, and P equals 1.

For main breakers the above formula gives the maximum short-circuit current available at the breaker. The actual current available may be less, owing to the impedance of the bus or cable between the transformer terminals and the main breaker. When the load on the switchgear is a motor load, the feeder breakers may have to interrupt the fault current available from the transformer or generator, plus any contribution from the motors. Motor contribution will depend on whether the motors are the induction or synchronous type. For most purposes it is customary to assume that 75% of the

TABLE 1. Ratings for Low-Voltage Ac Circuit Breakers

						Direct-acting trip devices			
							Min pickup setting with		
		Symmetrical interrupting rating with		Symmetrical close and latch and 30-cycle				Delayed trip	
AC volts	Frame size, A	Instantane- ous trip, A	Delayed trip, A	short-time rating, A	Max trip rating, A	Instan- taneous · trip, A	Min band, A	Inter- mediate band, A	Max band, A
208	225	25 000	14 000	14 000	225	30	100	125	150
and	600	42 000	22 000	22 000	600	150	175	200	250
240	1600	65 000	50 000	50 000	1600	600	350	400	500
	2000	85 000	55 000	55 000	2000	600	350	400	500
	3000	85 000	65 000	65 000	3000	2000	2000	2000	2000
	4000	130 000	85 000	85 000	4000	4000	4000	4000	4000
480	225	22 000	14 000	14 000	225	20	100	125	150
	600	30 000	22 000	22 000	600	100	175	200	250
	1600	50 000	50 000	50 000	1600	400	350	400	500
	2000	65 000	55 000	55 000	2000	400	350	400	500
	3000	65 000	65 000	65 000	3000	2000	2000	2000	2000
	4000	85 000	85 000	85 000	4000	4000	4000	4000	4000
600	225	14 000	14 000	14 000	225	15	100	125	150
	600	22 000	22 000	22 000	600	40	175	200	250
	1600	42 000	42 000	42 000	1600	200	350	400	500
	2000	55 000	55 000	55 000	2000	200	350	400	400
	3000	65 000	65 000	65 000	3000	2000	2000	2000	2000
	4000	85 000	85 000	85 000	4000	4000	4000	4000	4000

motors are induction and 25% are synchronous. With this ratio, the motor contribution will be four times the motor current, or

$$I_{mc} = \frac{4000 \times M}{3E}$$

where I_{mc} = motor contribution to fault current
M = sum of motor loads, kVA
E = system voltage

For the proper frame ratings for the main and feeder breakers with various sizes of transformer, see Tables 2 to 5.

The pickup or trip rating of main breakers should be approximately 125% of the continuous-current rating of the power source. However, when transformers are fan-cooled, the trip-current rating may be based on the self-cooled rating of the transformer in order to avoid using a larger frame size. For instance, a 1000-kVA transformer is rated 1204 A at 480 V. The main breaker pickup would be 1600 A if the transformer is self-cooled. However, a 1000-kVA dry-type transformer has a continuous-current rating of 1504 A when fan-cooled. To use a main breaker with a pickup of 125% of the transformer rating would require a 2000-A frame. If it is expected that the loads may exceed 1600 A for any period of time (other than for very short periods, such as motor-starting currents), a 2000-A frame should be used. (Standard transformers have a 125% overload capacity for 2 h.) However, if the load is not expected to exceed 1600 A, the 1600-A frame may be used.

Pickup settings for feeder breakers, of course, depend on the anticipated load, and the anticipated load must be calculated for each feeder. Generally, the continuous current through the breaker should not be more than 80% of the pickup setting of the breaker.

When circuit breakers are used as motor starters, the pickup or continuous-current setting may be selected from Table 6.

b. Drawout and Fixed Breakers. Low-voltage switchgear may be obtained with either fixed or drawout-type circuit breakers. Switchgear with fixed breakers is less

TABLE 2. Breaker Ratings for 208-V Transformers

Transformer rating, 3-phase kVA and % IZ	Max short-circuit kVA available primary system	Normal load, continuous A	Short-circuit current, symmetrical A			Recommended ac breaker ampere ratings *		
			Transformer alone	50% motor load	Combined	M	F	S
300, 5%	50 000	834	14 900	1700	16 600	1600	600	600
	100 000		15 700		17 400	1600	600	600
	150 000		16 000		17 700	1600	600	600
	250 000		16 300		18 000	1600	600	600
	500 000		16 500		18 200	1600	600	600
	Unlimited		16 700		18 400	1600	600	600
500, 5%	50 000	1388	23 100	2800	25 900	1600	600	1600
	100 000		25 200		28 000	1600	600	1600
	150 000		26 000		28 800	1600	600	1600
	250 000		26 700		29 500	1600	600	1600
	500 000		27 200		30 000	1600	600	1600
	Unlimited		27 800		30 600	1600	600	1600
750, 5.75%	50 000	2080	28 700	4200	32 900	3000	600	1600
	100 000		32 000		36 200	3000	600	1600
	150 000		33 300		37 500	3000	600	1600
	250 000		34 400		38 600	3000	600	1600
	500 000		35 200		39 400	3000	600	1600
	Unlimited		36 200		40 400	3000	600	1600
1000, 5.75%	50 000	2780	35 900	5600	41 500	3000	600	1600
	100 000		41 200		46 800	3000	1600	3000
	150 000		43 300		48 900	3000	1600	3000
	250 000		45 200		50 800	3000	1600	3000
	500 000		46 700		52 300	3000	1600	3000
	Unlimited		48 300		53 900	3000	1600	3000

For Tables 2 through 5:
Fully rated system:
M = main breaker (optional) with or without instantaneous-trip element
F = feeder breaker with instantaneous-trip element
Selective system:
M = main breaker (optional) without instantaneous-trip element to give selective tripping with feeder breakers F and S
F = feeder breaker with instantaneous-trip element not required to coordinate with additional protective devices nearer the load
S = feeder breaker without instantaneous-trip element which must coordinate with additional protective devices nearer the load
* Breakers are adequate for use with standard liquid-filled transformers only. For dry-type transformers use next larger beaker. See Table 1 for trip ratings.

expensive and may be considered if shutdown of the equipment does not present serious problems. However, removal of a circuit breaker of the fixed type should not be attempted unless the main bus is deenergized. Therefore, if the fixed type is used, a shutdown may be necessary in case of trouble with any of the circuit breakers, and it will definitely be required periodically for routine maintenance. For this reason the extra cost of the drawout-type circuit breaker is usually justified.

When drawout-type circuit breakers are used, a method is provided for moving the breaker from the fully withdrawn position to a test position, and to a fully connected position. For safety of the operator, it should be possible to move the breaker from one position to another with the breaker compartment door closed and only when the breaker is tripped. Also, it should not be possible to close the breaker when it is in an intermediate position between the test and the connected position.

When the breaker is in the fully connected position, it is connected to the source and to the load-side terminals, and the frame is connected to the ground bus. This is the position for normal operation. When in the test position, the breaker is disconnected from the source and from the load, but the secondary circuits for control of electrically operated breakers, and for monitoring, are connected. In this position, all secondary circuits may be tested and the breaker may be closed and tripped, either manually or electrically, without affecting the primary circuits. In the withdrawn position, all circuits to the breaker are disconnected. The frame of the breaker should be connected to ground in both the test and the fully connected position. It may or may not be connected in the withdrawn position.

c. Manual and Electrically Operated Breakers. Circuit breakers may be obtained either electrically or manually operated, except that circuit breakers with frame sizes above 1600 A must be electrically operated to comply with the standards. Breakers must be electrically operated if they are to be operated from a remote location or if they are to be used in any automatic control scheme. Manually operated breakers whose closing speed is dependent on the speed of the operator should not be used as motor starters. Manual stored-energy-operated breakers may be used as motor starters, except that if frequent starting and stopping of the motor is required, a motor controller should be used.

d. Fused Breakers. It is preferable to specify circuit breakers which have an interrupting rating equal to, or greater than, the maximum short-circuit current available at the breaker. However, it is often possible to save money by using breakers with a lower interrupting rating, in series with a fuse which is capable of interrupting the full

TABLE 3. Breaker Ratings for 240-V Transformers

Transformer rating, 3-phase kVA and % IZ	Max short-circuit kVA available primary system	Normal load, con-tinuous A	Short-circuit current, symmetrical A			Recommended ac breaker ampere ratings *		
			Transformer alone	100% motor load	Combined	M	F	S
300, 5%	50 000	722	12 900	2900	15 800	1600	600	600
	100 000		13 600		16 500	1600	600	600
	150 000		13 900		16 800	1600	600	600
	250 000		14 100		17 000	1600	600	600
	500 000		14 300		17 200	1600	600	600
	Unlimited		14 400		17 300	1600	600	600
500, 5%	50 000	1203	20 000	4800	24 800	1600	600	1600
	100 000		21 900		26 700	1600	600	1600
	150 000		22 500		27 300	1600	600	1600
	250 000		23 100		27 900	1600	600	1600
	500 000		23 600		28 400	1600	600	1600
	Unlimited		24 100		28 900	1600	600	1600
750, 5.75%	50 000	1804	24 900	7200	32 100	3000	600	1600
	100 000		27 800		35 000	3000	600	1600
	150 000		28 900		36 100	3000	600	1600
	250 000		28 900		37 000	3000	600	1600
	500 000		30 600		37 800	3000	600	1600
	Unlimited		31 400		38 600	3000	600	1600
1000, 5.75%	50 000	2406	31 000	9600	40 600	3000	600	1600
	100 000		35 600		45 200	3000	1600	3000
	150 000		37 500		47 100	3000	1600	3000
	250 000		39 100		48 700	3000	1600	3000
	500 000		40 400		50 000	3000	1600	3000
	Unlimited		41 800		51 400	3000	1600	3000
1500, 5.75%	50 000	3609	41 200	14 400	55 600	4000	1600	3000
	100 000		49 800		63 200	4000	1600 *	3000 *
	150 000		53 500		67 900	4000	3000	4000
	250 000		56 800		71 200	4000	3000	4000
	500 000		59 600		74 000	4000	3000	4000
	Unlimited		62 800		77 200	4000	3000	4000

See footnotes to Table 2.

* Breakers are adequate for use with standard liquid-filled transformers only. For dry-type transformers use next larger breaker. See Table 1 for trip ratings.

fault current. These fuses must be of the current-limiting type. Figure 1 is the trip curve of a circuit breaker with 1600-A pickup and the curve of a 3000-A current-limiting fuse. Note that at currents below about 39 000 A, the circuit breaker will trip before the fuse blows. At higher currents the fuse blows, and clears the fault, before the breaker trips. Since the breaker has an interrupting rating of 42 000 A, it will never be called on to interrupt more than its rated interrupting capacity. Therefore the combination may be used on a system having a capability of up to 200 000 A, which is the interrupting capacity of the fuse. It should be noted, however, that the breaker must be able to carry momentarily the let-through current of the fuse. Therefore, a fuse must be

TABLE 4. Breaker Ratings for 480-V Transformers

Transformer rating, 3-phase kVA and % IZ	Max short-circuit kVA available primary system	Normal load, continuous A	Short-circuit current, symmetrical A			Recommended ac breaker ampere ratings *		
			Transformer alone	100% motor load	Combined	M	F	S
300, 5%	50 000	361	6 400	1400	7 800	600	600	600
	100 000		6 800		8 200	600	600	600
	150 000		6 900		8 300	600	600	600
	250 000		7 000		8 400	600	600	600
	500 000		7 100		8 500	600	600	600
	Unlimited		7 200		8 600	600	600	600
500, 5%	50 000	601	10 000	2400	12 400	1600	600	600
	100 000		10 900		13 300	1600	600	600
	150 000		11 300		13 700	1600	600	600
	250 000		11 600		14 000	1600	600	600
	500 000		11 800		14 200	1600	600	600
	Unlimited		12 000		14 400	1600	600	600
750, 5.75%	50 000	902	12 400	3600	16 000	1600	600	600
	100 000		13 900		17 500	1600	600	600
	150 000		14 400		18 000	1600	600	600
	250 000		14 900		18 500	1600	600	600
	500 000		15 300		18 900	1600	600	600
	Unlimited		15 700		19 300	1600	600	600
1000, 5.75%	50 000	1203	15 500	4800	20 300	1600	600	600
	100 000		17 800		22 600	1600	600	1600
	150 000		18 700		23 500	1600	600	1600
	250 000		19 600		24 400	1600	600	1600
	500 000		20 200		25 000	1600	600	1600
	Unlimited		20 900		25 700	1600	600	1600
1500, 5.75%	50 000	1804	20 600	7200	27 800	3000	600	1600
	100 000		24 900		32 100	3000	1600	1600
	150 000		26 700		33 900	3000	1600	1600
	250 000		28 400		35 600	3000	1600	1600
	500 000		29 800		37 000	3000	1600	1600
	Unlimited		31 400		38 600	3000	1600	1600
2000, 5.75%	50 000	2405	24 700	9600	34 300	3000	1600	1600
	100 000		31 000		40 600	3000	1600	1600 *
	150 000		34 000		43 600	3000	1600	3000
	250 000		36 700		46 300	3000	1600	3000
	500 000		39 100		48 700	3000	1600 *	3000
	Unlimited		41 800		51 400	3000	3000	3000
2500, 5.75%	50 000	3008	28 000	12 000	40 000	4000	1600	1600 *
	100 000		36 500		48 500	4000	1600 *	3000
	150 000		40 500		52 500	4000	3000	3000
	250 000		44 600		56 600	4000	3000	3000
	500 000		48 100		60 100	4000	3000	3000
	Unlimited		52 300		64 300	4000	3000	3000

See footnotes to Table 2

* Breakers are adequate for use with standard liquid-filled transformers only. For dry-type transformers use next larger breaker. See Table 1 for trip ratings.

selected with a let-through current lower than the momentary rating of the breaker.

Fused breakers may also be used to protect downstream equipment which has a momentary or interrupting rating lower than the available fault current. For example, a 480-V system supplied by a 1500-kVA transformer with standard impedance may be subject to fault currents as high as 37 000 A symmetrical. The let-through current of an 800-A fuse with 37 000 A available is about 40 000 A peak, or the equivalent of 18 000 A symmetrical. (The asymmetry factor for fuses is 1.6. Peak current is $\sqrt{2}$ times rms current; therfore, peak current is 1.6×1.414 or 2.3 times the symmetrical current. $40\,000/2.3 = 18\,000$.) Therefore an integrally fused circuit breaker with an 800-A

TABLE 5. Breaker Ratings for 600-V Transformers

Transformer rating, 3-phase kVA and % IZ	Max short-circuit kVA available primary system	Normal load, continuous A	Short-circuit current, symmetrical A			Recommended ac breaker ampere ratings *		
			Transformer alone	100% motor load	Combined	M	F	S
300, 5%	50 000	289	5 200	1200	6 300	600	600	600
	100 000		5 500		6 700	600	600	600
	150 000		5 600		6 800	600	600	600
	250 000		5 600		6 800	600	600	600
	500 000		5 700		6 900	600	600	600
	Unlimited		5 800		7 000	600	600	600
500, 5%	50 000	481	8 000	1900	9 900	600	600	600
	100 000		8 700		10 600	600	600	600
	150 000		9 000		10 900	600	600	600
	250 000		9 800		11 200	600	600	600
	500 000		9 400		11 300	600	600	600
	Unlimited		9 600		11 500	600	600	600
750, 5.75%	50 000	722	10 000	2900	12 900	1600	600	600
	100 000		11 100		14 000	1600	600	600
	150 000		11 600		14 500	1600	600	600
	250 000		11 900		14 800	1600	600	600
	500 000		12 200		15 100	1600	600	600
	Unlimited		12 600		15 500	1600	600	600
1000, 5.75%	50 000	962	12 400	3900	16 300	1600	600	600
	100 000		14 300		18 200	1600	600	600
	150 000		15 000		18 900	1600	600	600
	250 000		15 600		19 500	1600	600	600
	500 000		16 200		20 100	1600	600	600
	Unlimited		16 700		20 600	1600	1600	600
1500, 5.75%	50 000	1444	16 500	5800	22 300	1600	1600	1600
	100 000		20 000		25 800	1600	1600	1600
	150 000		21 400		27 200	1600	1600	1600
	250 000		22 700		28 500	1600	1600	1600
	500 000		23 900		29 700	1600	1600	1600
	Unlimited		25 100		30 900	1600	1600	1600
2000, 5.75%	50 000	1924	19 700	7800	27 500	3000	1600	1600
	100 000		24 800		32 600	3000	1600	1600
	150 000		27 200		35 000	3000	1600	1600
	250 000		29 400		37 200	3000	1600	1600
	500 000		31 300		39 100	3000	1600	1600
	Unlimited		33 500		41 300	3000	1600 *	1600 *
2500 5.75%	50 000	2404	22 400	9600	32 000	3000	1600	1600
	100 000		29 200		38 800	3000	1600	1600
	150 000		32 400		42 000	3000	1600 *	1600 *
	150 000		35 600		45 200	3000	3000	3000
	500 000		38 500		48 100	3000	3000	3000
	Unlimited		41 800		51 400	3000	3000	3000

See footnotes to Table 2.

* Breakers are adequate for use with standard liquid-filled transformers only. For dry-type transformers use next larger breaker. See Table 1 for trip ratings.

TABLE 6. Application of Air Circuit Breakers to Full-Voltage Motor-Starting and Running Duty of 3-Phase, 60-Hz, 40°C Rise Motors *

Continuous-current rating of circuit breaker, A	Motor full-load current, A		Max permissible motor locked-rotor current at 60 Hz, A	Hp rating of 3-phase ac motors								
				Induction motors			100% PF synchronous motors			80% PF synchronous motors		
	Min	Max		220 V	440 V	550 V	220 V	440 V	550 V	220 V	440 V	550 V
15	9	13	120	3	7.5	7.5, 10						
20	13	17	160	5	10	15			25			
30	19	26	240	7.5	15, 20	20, 25		25	30			25, 30
40	26	35	320	10	25	30		30	40		25	40
50	32	44	400	15	30	40		40	50		30	
70	45	61	560	20	40	50, 60	25	50	60	25	40	50
90	58	78	720	25, 30	50, 60	75	30	60	75	30	50	60
100	64	87	800				40	75	100	40	60	80
125	80	109	1 000	40	75	100	50	100	125	50	75	100
150	96	131	1 200	50	100	125	60		150	60	100	
175	112	152	1 400	60		150		125		75	125	125
200	128	174	1 600		125		75	150	200		150	150
225	144	196	1 800	75	150	200						
250	160	218	2 000				100	200	250	100	200	200
300	192	261	2 400	100	200	250			300	125	250	
350	224	304	2 800		250	300	125	250	350		300	250
400	256	348	3 200	125		350		300	400			300
500	320	435	4 000	150	300, 350	400, 450		350, 400	450, 500		350	350, 400
600	384	522	4 800	200	400	500		450	600		400, 500	450
800	512	696	6 400	250	450, 500	600, 700		500, 600	700, 800		600	500, 600
1000	640	870	8 000	300, 350	600, 700	800, 900		700, 800	900, 1000		700	700, 800
1200	768	1044	9 600	400	800	1000		900			800–1000	900
1600	1023	1392	12 800	450, 500	900, 1000			1000				1000
2000	1280	1740	16 000	600, 700								
2500	1600	2180	20 000	800, 900								
3000	1920	2610	24 000	1000								
4000	2560	3480	32 000									

Locked-rotor currents are based upon motors having NEC code letters *a* through *j*, inclusive. If the locked-rotor current exceeds this value, select the circuit breaker having the next higher continuous-current rating, provided there is a calibration point on the breaker which does not exceed 140% of the motor full-load current.
* From NEMA Standard for Low-Voltage Power Circuit Breakers, Publication SG3-1958.

current-limiting fuse could be used to feed a bus duct with 20 000-A momentary rating, or a panelboard with 20 000-A interrupting-capacity breakers, even though the short-circuit capability of the system is 37 000 A.

Other factors must be considered in selecting the size of fuse to be used with a breaker for a particular application. These are:

1. Thermal coordination. Fuses generate some heat which complicates the problem of cooling the breaker. Because of this, a fuse with a high rating relative to the frame

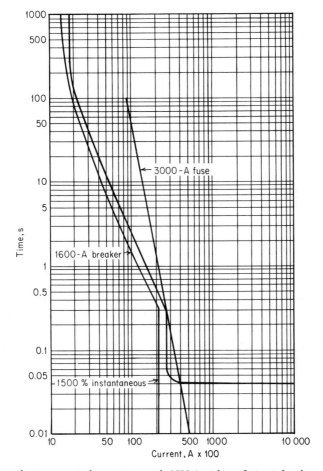

Fig. 1 Typical time-current characteristic with 1600-A pickup of circuit breaker as compared with that of a 3000-A current-limiting fuse.

size of the breaker should be used. Check with the breaker manufacturer for rules to follow to prevent excessive heating.

2. Selectivity between the fuse and the breaker trip device. The melt time of a fuse should be at least twice the total clearing time of the breaker at the current level where the trip device transfers to instantaneous pickup. In some cases the use of a breaker with long-time, short-time, and instantaneous trip will assist in meeting this requirement.

3. Selectivity with upstream devices, such as main breakers and relays, which may dictate the use of small fuses.

4. Selectivity between two fuses. If two fuses are in series in a system, each sensing the same fault current, the fuse close to the source should have at least double the continuous-current rating of the fuse near the load.

5. Reliability, which dictates the use of the largest possible fuse, to avoid false shutdowns and unnecessary fuse blowing.

6. Economics, which indicates the use of the smallest and least costly fuse to replace, yet large enough to avoid unnecessary blowing and prevent overheating of the breaker.

7. Inrush, which requires that the fuse and trip-device settings be high enough to ride through motor-starting currents.

8. Growth, which requires a fuse large enough to handle any planned increased loading on the circuit.

e. Auxiliary Devices. Circuit breakers may be equipped with various auxiliary devices. All electrically operated breakers are equipped with auxiliary switches, some of which are used in the breaker control circuits. Switches which are open when the breaker is open are designated as *a* switches, those which are closed when the breaker is open are called *b* switches. An *a* switch is used in series with the shunt trip coil to open that circuit after the breaker has tripped. It is common practice also to use this switch to operate a red indicating light which, when lighted, indicates that the breaker is closed and that the trip-coil circuit is intact. A *b* switch is used to operate a green light which, when lighted, indicates that the breaker is tripped. Other auxiliary switches may be used in the control circuit. If additional auxiliary switches are required for external circuits, they may be furnished when specified, on either electrically or manually operated breakers.

Undervoltage devices may be supplied on all air circuit breakers. These devices will trip the breaker when the voltage on the source to which they are connected falls below a certain value. They are usually self-resetting so that the breaker may be reclosed as soon as voltage is restored, and may be obtained to trip the breaker instantly on loss of voltage, or after a time delay.

Low-voltage breakers may also be obtained with a bell-alarm device. This device is merely a switch which opens or closes when the breaker trips because of overcurrent. It does not operate when the breaker trips for any other reason. Once it has operated, it must be reset manually, or when specified, it may be equipped to be reset electrically. This device, when used on an electrically operated breaker, may be used as an electrical lockout, so that after tripping on overcurrent the breaker cannot be reclosed until the bell-alarm device has been reset.

Breakers, either electrically operated or manually operated, may be equipped with a mechanical lockout device which prevents reclosing the breaker, after tripping due to overcurrent, until the device has been manually reset.

4. Protective Devices. Low-voltage circuit breakers may be obtained without overcurrent trip devices (nonautomatic), with magnetic trip devices, or with static trip devices. If supplied without overcurrent trip devices they may be used as a switch, or they may be supplied with a shunt trip device which is operated through a relay to give the desired protection. With this arrangement a source of power must be available for the shunt trip device. Batteries are the most reliable source of power, but control power may be taken from the source which is feeding the switchgear, usually through a control power transformer. However, it is necessary to make sure that such power is available when needed. A short circuit, for instance, even when on the load side of the breaker in question, may reduce the control voltage below that required to trip the breaker, so that when the contacts of an overcurrent relay close, the breaker will not trip. Also, an undervoltage relay could not trip the breaker when its contacts closed because the voltage would be too low to operate the shunt trip. In such cases a capacitor trip device may be used. This device takes power from the source. The power is rectified, and charges a capacitor. Then, if control voltage falls below that required to trip the breaker, the energy stored in the capacitor can be discharged through the shunt trip device to trip the breaker.

If it is desired to use relays for overcurrent tripping, the expense of the capacitor trip can be avoided by obtaining a breaker with instantaneous trip only. Short circuits which cause the voltage to drop below that required for the shunt trip will cause the breaker to trip instantaneously by means of the overcurrent trip on the breaker which takes its energy from the fault current. Overloads, involving fault currents of lower value, will not cause a serious drop in the applied voltage and the shunt trip device may be used, operated by the overcurrent relay.

The magnetic-type overcurrent device was the standard type used by all manufacturers for many years, and it is still available from most manufacturers. The device utilizes magnetic forces created by the current through the breaker to trigger the release mechanism which allows the breaker to trip. Various mechanical, hydraulic, or pneumatic devices are used to delay the tripping, when desired. The resulting characteristic curves have broad bands and have shapes which make coordination difficult with upstream relays and fuses, and with downstream plastic-case breakers or fuses. Also, since this is a mechanical device which is seldom operated, it is not as reliable as such a device should be.

Most manufacturers today can offer static trip devices on their low-voltage large air circuit breakers. These devices are more reliable, and they have characteristic curves with narrow bands and with shapes designed to "nest" well with curves of fuses, relays, and plastic-case breakers. These static trip devices are fed by sensors, which are special current transformers which monitor the current in each phase. The device receives power from these sensors to trip the breaker when the current in the sensor is greater than the pickup current for which the device has been set. Timing in accordance with the characteristic curves is also accomplished by the electronic circuitry.

With the static trip device, changes of pickup setting can be made easily within the range of the sensor by changing the setting of variable resistors. Changing the range of pickup settings can be done easily by changing the sensors.

Another advantage of static trip devices is that they can be easily tested and calibrated in the field. With magnetic devices it is necessary to pass high-magnitude currents through the breaker to be sure that the overcurrent device is operating properly. With the static device low-magnitude currents (5 A or less) can be used for testing and calibrating. Test sets are available from the manufacturer which operate from a 115-V source, for field testing of these breakers.

Low-voltage distribution systems may be solidly grounded, grounded through a resister, or ungrounded. The solidly grounded system is used most frequently. When it is used, some form of ground-fault protection should be provided. This protection is recommended in order to prevent arcing faults, which may be of relatively low current magnitude, from destroying much equipment before they can be cleared.

To illustrate the need for such protection, consider a 1000-kVA substation with 480/277-V secondary and a 1600-A main secondary breaker. The transformer neutral is solidly grounded. Maximum fault current is approximately 20 000 A. It has been established that in case of an arcing fault to ground the arc may be maintained if the fault current is as low as 20% of the available fault current, or in this case 4000 A. The voltage across the arc may be 100 V or higher. If such an arc is established on the main bus of a substation, it may take 30 s for a main breaker which is not equipped with a ground-fault protective device to clear the fault, depending on the setting of the breaker. Since the fault is only 2½ times the pickup setting of the breaker, it would never be set to trip instantaneously at this value. But the power in the arc is 40 kW, and in 30 s the total energy is 1200 kWs. This can be very destructive. A ground-fault protective device would trip the breaker instantly in case of such a fault, thereby keeping the damage to a minimum. It should be noted that if a phase-to-phase arcing fault is established, it will almost invariably jump to ground immediately, so that ground-fault protection will protect against any arcing fault.

In the above example we considered a fault on the main bus which would be cleared by the main breaker. Feeder breakers may also be provided with ground-fault tripping to protect against arcing faults in the motor-control centers or panelboards, or in the cables on the downstream side of the breaker.

When breakers are equipped with static overcurrent trip devices, the ground-fault protection usually can be obtained as an optional feature. If magnetic trip breakers are used, or in cases where ground-fault tripping is not obtainable as a feature of the static trip device, a separate ground-fault relay may be used. A number of such relays are available, most of them solid-state, but an electromechanical-type overcurrent relay may be used. All types require a current transformer as a ground-current sensor. This may be located on the ground connection, or where this is not possible, a current transformer which encircles all conductors, including the neutral conductor, may be used.

For solidly grounded systems, ground-fault tripping should be specified for the main breaker. Ground-fault tripping for the feeder breakers should also be specified for a selective system (Art. 8) and should be carefully considered for a fully rated system.

5. Metering. The metering to be specified with low-voltage switchgear depends entirely on the needs for the information which metering can supply. Usually the minimum requirement is for a voltmeter on the incoming power source and an ammeter to indicate total current. These instruments are of the single-phase type. They are usually supplied with switches to permit reading current and voltage of all three phases. If the voltage is greater than 240 V, potential transformers are required, since standards do not permit voltages on the panels which exceed 250 V to ground. Current transformers are required for all ammeters.

With a 3-wire system, only two potential transformers and two current transformers are required. The potential transformers are connected open delta so that all three phase-to-phase voltages can be read. The current transformers are usually in phases one and three. The switch is arranged so that when set on phase two, the current from phases one and three will both pass through the meter. In a 3-wire system, of course, the vector sum of the currents in phases one and three is equal to the current in phase two.

In a 4-wire system some current may flow in the neutral; therefore, three current transformers are required. If a voltmeter is the only instrument requiring potential input, only two potential transformers are required. However, if wattmeters or watthour meters are specified, three potential transformers are also required. Actually the $2\frac{1}{2}$-element watthour meter requires only two potential transformers connected open wye. But the third transformer is required to give a voltage reading on the voltmeter for the third phase.

When watthour meters are required, the two-element meter should be used on 3-wire systems. On 4-wire systems the $2\frac{1}{2}$-element meter is satisfactory. Three-element meters are available, but they are more expensive, take more space on the panel, and do nothing which the $2\frac{1}{2}$-element meter does not do. It is not true, as some believe, that the three-element meter is more accurate than the $2\frac{1}{2}$-element. Watthour meters may be supplied with demand registers if desired. When they are specified, the desired' time interval, usually 15 or 30 min, should always be specified.

Power-factor meters can be supplied, when desired, for either 3- or 4-wire systems. A 3-phase power-factor meter on a 4-wire system does not give an accurate reading, since it monitors the current in only one phase. However, the indication is usually satisfactory for most purposes. If accurate readings are required, three single-phase meters should be used.

Frequently it is desired to meter each feeder, with either an ammeter, a watthour meter, or both. If they are used, additional switchgear units may be required to provide the necessary panel space. Ammeters and voltmeters may be supplied with either 1 or 2% accuracy. For most purposes the 2% instrument is satisfactory, although often the 1% instrument can be obtained at no additional cost. Also, 1% instruments may be obtained with 180 or 250° scales. The 250° instrument is easier to read and may often be obtained at the same price.

Test switches or test blocks may be specified to permit plugging in portable instruments. This may be desirable to check the accuracy of the panel instruments or to obtain more accurate readings. If meter readings are to be taken only at infrequent intervals, it may be advisable, in order to reduce the cost, to omit the panel instruments and specify only test blocks, so that portable instruments can be plugged in when readings are taken.

Usually meters required for the incoming power will be mounted on a panel above the main breaker. Potential transformers and control power transformers will be mounted inside this compartment. Current transformers may be mounted in the rear, or if a main breaker is provided, they may be in the breaker compartment. Ammeters and ammeter switches for the feeders can often be mounted on the breaker panel, if the meters are the panel type having 2% accuracy. If watthour meters are required for the feeders, they must be mounted on a separate panel. This may result in an increase in the size of the switchgear.

Current transformers for the feeder breakers also can often be mounted in the breaker compartment, rather than in the bus compartment. With this arrangement, they can be changed in the field without the necessity of entering the bus compartment.

6. Wiring. Wire used for metering and control circuits should not be smaller than No. 14 AWG. It should be insulated with a material rated not less than 90°C. Crimp-type terminals should be used, and terminals should be insulated. A detailed wiring diagram is required, showing the relative location of terminals on various devices and on terminal blocks. When troubleshooting or making changes, wires can be identified by referring to this wiring diagram. The use of wire markers to identify each wire can be specified, but this will increase the cost of the switchgear considerably, since applying the markers is time-consuming.

SELECTION OF SYSTEM TYPE

When specifying low-voltage switchgear, a choice must be made between a fully rated system and a selective system. The cascade system, once considered acceptable, is no longer recommended by the standards.

7. Fully Rated System. In a fully rated system all breakers are capable of interrupting the maximum fault current available, and all breakers have dual magnetic or dual static trip devices; that is, they have long-time delay, and instantaneous elements or long-time, short-time, and instantaneous elements. If a fault occurs on a feeder of sufficient magnitude to trip the main breaker instantaneously, both the main breaker and the feeder breaker will trip, and power will be interrupted to all feeders. This arrangement may not be objectionable, since in many applications, if one part of the operation is stopped, the entire process must be interrupted. The fully rated system is less expensive than the selective system, since "dual" trip devices cost less than "selective" trip devices. Also, in a fully rated system breakers may be applied where the maximum fault current available is equal to or less than the interrupting rating of the breaker. Breakers with selective trip devices may be applied only where the maximum fault current available is equal to or less than the short-time rating of the breaker. A larger frame size may therefore be required for a selective system. The larger frame may increase the breaker cost considerably.

8. Selective System. A selective system should be specified where a fault on one feeder should not interrupt service to other feeders. With this arrangement the main breaker will not have instantaneous trip but will have only long-time and short-time characteristics. The feeder breaker may have instantaneous trip which will clear the fault before the short-time element on the main breaker operates. Where feeder breakers are feeding motor-control centers or panelboards containing plastic-case breakers, the feeder breakers may also be selective with long-time and short-time but no instantaneous trip, so that a fault on the downstream side of the plastic-case breaker will be cleared by that breaker before the feeder breaker in the switchgear trips. In this case the timing of the feeder breaker and the main breaker are coordinated, so that the feeder breaker will trip before the main for all faults on the downstream side of the feeder (see Sec. 12).

In addition to being more expensive than the fully rated system, the selective system has another disadvantage. When a fault occurs which cannot be cleared instantaneously, as on the main bus of the switchgear, the delay in clearing the fault will result in greater damage to the faulted equipment. This delay even occurs when ground-fault protection is provided, since in the selective system even the ground-fault tripping must be delayed to give the downstream breaker the opportunity to clear the fault.

A variation of the selective system is called "zone-selective." This system takes advantage of the impedance of the cables or bus duct on the load side of the feeder breaker when that breaker is feeding panelboards or control centers with plastic-case breakers. The feeder breaker has long-time, short-time, and instantaneous trips. The instantaneous trip is set high enough so that it will not trip on a fault beyond the panelboard or control center. The proper setting can be determined by calculating the fault current available at the panelboard, and setting the instantaneous above that value. This arrangement will usually be higher than fifteen times the pickup setting of the breaker so special high-set instantaneous devices are required. These devices may not be available from all manufacturers.

Use of the zone-selective system permits use of breakers on systems where the fault current available is equal to or less than the interrupting rating of the breaker, whereas as previously stated, breakers in the selective system must have a momentary rating equal to or greater than the available fault current.

9. Cascade System. The cascade system is discussed here only because in the past it has been considered an acceptable system and many such systems are in use. With this system, all breakers must have instantaneous trips and the instantaneous trip on the main breaker must be set no higher than 80% of the interrupting rating of the feeder breaker. Feeder breakers may be applied where the available fault current is up to 200% of the interrupting rating of the breaker. However, feeder breakers with interrupting ratings lower than the available fault current must be power-operated, and the control switch must be located where the operator will not be standing in front of the breaker when he closes it. Also, if a breaker operates to clear a fault in excess of its interrupting rating, it must be inspected before it is again put in operation, and it may require repair or complete replacement.

10. Low-Voltage Switchgear Layout. The size of a group of switchgear will vary somewhat with the manufacturer. When outdoor switchgear is used, space is usually not critical so that the product of any manufacturer can be used. However, the pad should not be poured until certified drawings have been received from the manufacturer, to be sure that it will be the proper size and shape.

When switchgear is to be located indoors, the room should be designed to allow ample space for the switchgear and to provide for adequate ventilation. To estimate the space required, catalogs from manufacturers usually give information which can be used to plan the layout. Breakers with 225- or 600-A frame sizes will be in units 18 or 20 in wide. Usually they can be stacked four high. Frames rated 1600 A will be in units 24 to 27 in wide and can be stacked three or four high. Frame breakers rated 3000 A require units about 30 in wide. Usually only one 3000-A breaker is permissible in a unit, although in some cases two may be furnished. Also, with one 3000-A breaker, one or two smaller breakers may be supplied in the same unit.

Breakers with frame sizes of 4000 A require units 30 to 38 in wide, depending on the manufacturer. Usually only one of these breakers can be furnished in a unit.

Fused breakers in 225-, 600-, and 1600-A frame sizes are supplied in units of the same width as unfused breakers of the same frame size. When fuses are used in series with 3000- and 4000-A breakers, they are usually in a separate compartment directly above or directly below the breaker. However, at least one manufacturer can offer a 4000-A breaker with integrally mounted fuses and no increase in space requirements.

Depth of indoor switchgear will be from 54 to 60 in, depending on the manufacturer. However, this depth may not allow enough space for outgoing cables, so in some cases it may be necessary to add a section on the rear of one or more units to provide sufficient space for the cables. Height of the indoor switchgear will be approximately 90 in.

Depth of outdoor switchgear will be approximately 94 in. Height of outdoor gear will be about 112 in. All dimensions given here are to be used only for preliminary estimating. In all cases, ample space should be allowed, so that the choice of a supplier is not limited to one who can fit his equipment into the available space.

11. Selecting Switchgear Location. Several factors must be considered when selecting the site for the switchgear. For outdoor switchgear the primary consideration is proximity to the load, although the switchgear should, of course, be located where it will not be subject to floods. If possible the switchgear should be located where it will be shaded during the hottest part of the day. Also, corrosive or explosive atmospheres should be avoided when possible.

Indoor switchgear must, of course, be protected from the weather. A dry location is preferred, but the switchgear may be located in a room in which condensation takes place, if it is supplied with heaters and is purchased with dripproof construction. Adequate ventilation is essential, so that the ambient temperature outside the switchgear will not exceed 40°C. Corrosive atmospheres should be avoided, but if this cannot be done, the switchgear should be ordered with special provisions for protecting against corrosion. Indoor switchgear must not be located in a room where explosive atmospheres are present.

Indoor switchgear should be located in a room which is free of dust, if possible. Dust may build up on insulators to form a conducting path which could cause a fault with resultant severe damage to the equipment. Dust may also cause excessive wear of breakers and may cause failure of any component with moving parts.

Switchgear should be located on a floor or pad which is level and smooth. It may be mounted on sill channels, and anchored to the floor, depending on the manufacturer's recommendations.

12. Switchgear Maintenance. To ensure trouble-free and reliable operation, switchgear should have some care, and a regular maintenance program should be established. Frequency of maintenance operations will depend on the location and the frequency of operation of the circuit breakers. In extremely dirty atmospheres, it may be advisable to clean and inspect the equipment every 2 or 3 months. Under normal conditions an annual inspection of the equipment is satisfactory, although if the circuit breakers are operated frequently it may be advisable to inspect and service them every 6 months.

Routine maintenance of the switchgear structure should involve cleaning of the structure and especially the insulating supports for the bus. All electrical connections should be checked to make sure that bolts and terminal screws are tight. Relays should be tested in accordance with the manufacturer's instructions. Megohmmeter tests may be made to check the insulation of both the primary circuits and the control circuits. Records of these readings should be kept. A low megohmmeter reading may not indicate trouble, but a reading which has been falling steadily over a period of time indicates that trouble is developing and corrective action should be taken. Such testing is often referred to as "Megger" testing and the readings as Megger readings.

Care of the circuit breakers involves periodic inspection, cleaning, and lubrication. Frequency of servicing will depend on conditions of the atmosphere and frequency of operation. Breakers which are operated several times daily should be serviced more frequently. However, breakers which are not required to be operated for long periods of time should be opened and closed several times every 2 or 3 weeks to burnish the contacts and make sure that all moving parts are operating freely. When servicing, make sure that all bolts are tight. When possible, check the overcurrent trip devices to be sure they are operating properly. With static trip devices, a function test of the overcurrent trip device can be made with very little equipment. Devices are available from manufacturers which can also be used to check calibration and timing. These devices are relatively inexpensive. Equipment for testing magnetically operated trip devices is much more expensive. For this reason, testing of these magnetic trip devices often is not practical. However, they should be inspected to see that parts move freely. Where dashpots are used for timing, they should be inspected to be sure that the oil is clean and that it flows freely through the orifice.

13. Low-Voltage Metal-enclosed Switchgear Specifications. The following is a guide which can be used in writing specifications for low-voltage metal-enclosed switchgear. Items, phrases, or sentences in parentheses are optional. Where optional items appear in groups, select the one desired. Also fill in blanks as required.

The equipment outlined in this specification will consist of 600-V metal-enclosed switchgear with necessary compartments, bus work (drawout), air circuit breakers, and miscellaneous equipment for the application.

The switchgear equipment will comply with all applicable standards of ANSI, IEEE, and NEMA.

a. Service. The switchgear sections will be (indoor) (outdoor), rated 600 V. This equipment will operate on a service voltage of _____ V, _____ Hz, _____ phase, _____ wire (grounded) (ungrounded).

b. Metal-enclosed Assembly. The assembly will consist of welded-steel breaker compartments with hinged front doors, welded-steel framework, and sheet-steel enclosure. The enclosure and welded components will be chemically cleaned, bonderized or hot-phosphate-treated, and rinsed, and given one primer coat of ANSI No. 61 indoor light-gray paint. After complete assembly, all exterior surfaces are given an additional coat of paint (ANSI No. 61 light-gray indoor) (ANSI No. 24 dark-gray outdoor.) (Provisions will be made for racking the breakers to the "connected," "test," and "disconnected" position with the door closed. An interlock will prevent racking the breaker when it is closed.) A coordinated glass-polyester insulation system will be supplied.

c. Circuit Breakers. Circuit breakers will be the low-voltage power-circuit-breaker type, 600-V class, 3-pole, single-throw (drawout mounted), electrically and mechanically trip-free with stored-energy operator. Each will have arc quenchers, main and arcing contact structure, an overcurrent trip device, a contact-position indicator (open-closed), stored-energy mechanical indicator (charged-discharged), (primary disconnecting devices, and a mechanical interlock to prevent making or breaking contact on the primary disconnects). Electrically operated breakers will also have a shunt trip device, a motor to store the energy, a spring-release coil (secondary disconnecting devices), and a four-stage auxiliary switch.

d. Bus. The main bus will be suitably rated for the application. A ground bus will extend the length of the lineup. (A _____ capacity neutral bus will be carried with the phase bus in addition to a ground bus.) All bus will be (copper) (aluminum).

e. Weatherproof Housing (Option). Outdoor weatherproof construction will be used. Front and rear doors will be gasketed and hinged, front doors will have provision for padlocking, and rear doors will be bolted. An aisle, approximately 38 in wide and accessible from any one of the front doors, will be provided at the front of the switchgear lineup to facilitate inspection and testing of the circuit breakers and associated equipment while protected from the weather. One hand-operated crane, mounted above the switchgear aisleway, will be provided to facilitate removal and handling of the circuit-breaker elements.

The following equipment will be furnished within the outdoor weatherproof switchgear: light sockets for interior illumination of the aisle, convenience receptacles as a source of 120-V ac power for electric tools, necessary space heaters to prevent condensation of moisture, a switch for all the space heaters, and a switch for the lamps.

f. Detailed Specifications. The group of switchgear will include the following:

1 set of _____-A, 3-phase, 3-wire, main bus

1 ground bus

____ (neutral bus, _____ A)

____ (main) (feeder) air circuit breakers, _____-A symmetrical interrupting capacity at _____ V (manually) (electrically) operated with (dual static) (selective static) (dual magnetic) (selective magnetic) trip device rated _____ A (add breaker modifications as desired)

____ key interlocks for main breakers, for interlocking with primary interrupter switches

____ (main) (feeder) circuit-breaker control switch(es), each complete with one red and one green indicating light

____ (tie) (feeder) air circuit breakers, _____ A symmetrical interrupting capacity at _____ V, _____ (manually) (electrically) operated with (dual static) (selective static) (dual magnetic) (selective magnetic) trip device rated _____ A (add breaker modifications as desired)

____ (tie) (feeder) circuit-breaker control switches, each complete with one red and one green indicating light

____ feeder air circuit breaker, _____ A symmetrical interrupting capacity at _____ V, (manually) (electrically) operated, with (dual static) (selective static) (dual magnetic) trip device rated _____ A (add breaker modifications as desired)

____ provision for the future addition of a (tie) (feeder) air-circuit-breaker element, _____ A symmetrical interrupting capacity at _____ V (manually) (electrically) operated (add breaker modifications as desired)

____ provision for the future addition of a feeder air-circuit-breaker element, _____ A symmetrical interrupting capacity at _____ V, (manually) (electrically) operated (add breaker modifications as desired)

____ feeder circuit-breaker control switches, each complete with one red and one green indicating light

____ sets of necessary suitable clamp-type terminals (one set per feeder) for purchasers outgoing feeder cables (or specify special terminals as desired)

_____ sets of space heaters (standard on outdoor)

_____ control power transformers, dry-type, suitable kVA capacity, single phase, _____-120/240-V ratio, complete with primary current-limiting fuses, to supply auxiliary power from the low-voltage bus for space heaters, lights, convenience outlets, transformer fans, and electrically operated breakers

_____ current transformers, _____-A ratio, for main bus metering

_____ potential transformers, dry-type, _____-120-V ratio, complete with primary current-limiting fuses

_____ potential transformers, dry-type, _____-120-V ratio, complete with primary current-limiting fuses for ground detector (lights) (voltmeters) (_____)

_____ current transformers, _____ ampere ratio, for feeder metering

_____ current transformers, _____ ampere ratio, for feeder metering

_____ voltmeters, single-phase, indicating, _____-V scale

_____ voltmeter transfer switches, 3-phase

_____ ammeter, single-phase, indicating, for (main bus) (feeder) metering, _____-A scale

_____ ammeter transfer switches, 3-phase

_____ wattmeters, 3-phase, indicating, for (main bus) (feeder) metering, _____ MW scale

_____ power-factor meters, indicating, for (main bus) (feeder) metering

_____ varmeter, 3-phase, indicating, for (main bus) (feeder) metering _____ Mvar scale

_____ watthour meters, _____-element, (with _____ minute demand attachment), for (main bus) (feeder) metering

_____ sets of three ground-detector lights

_____ sets of three ground-detector voltmeters, indicating

_____ potential test blocks

_____ current test blocks

_____ sets of nameplates, as required

   g. Breaker Modifications.   (Add the following to each breaker item as required):
Overcurrent bell-alarm device (manual reset) (electrical reset)
Overcurrent lockout device (electrical) (mechanical)
Shunt trip device (standard on electrically operated breakers)
Instantaneous undervoltage device
Time-delay undervoltage device
Auxiliary switch with _____ contacts

   h. Accessories.   A set of standard low-voltage switchgear accessories will be furnished, including, but not necessarily limited to:

_____ crank for manual operation of the breaker drawout mechanism

_____ lifting yoke for each type of breaker element

_____ maintenance closing device for electrically operated breakers

_____ test plug, less cable, for drawout watthour meters and/or test blocks

13.8-kV METAL-CLAD SWITCHGEAR

   14. Switchgear Structures.   Metal-clad switchgear may be obtained for use indoors, or in weatherproof structures for outdoor use. Generally, if the loads are located indoors the switchgear will be indoors, and outdoor loads will be served by outdoor switchgear. However, outdoor switchgear may be used to serve indoor loads when indoor space is limited, when corrosive or explosive atmospheres are present inside the building, or when the indoor atmosphere is exceedingly dusty. Outdoor switchgear is, of course, more expensive than indoor switchgear and therefore should be used only when the added cost can be justified or when indoor space is not available.

   15. Indoor Switchgear.   Metal-clad switchgear structures differ from the standard low-voltage switchgear structures in several respects. By definition of metal-clad switchgear the circuit breakers must be of the removable type. Circuit breakers must be enclosed, that is, barriered from other components, as in low-voltage switchgear, but buses, potential transformers, and control power transformers and cable terminals must also be enclosed in separate metal compartments. All metal barriers must be grounded. Shutters must be provided which close automatically when a breaker is withdrawn, to prevent operators from contacting the primary disconnecting device, or the main bus, which may be energized. Also, interlocks must be provided to prevent moving the breaker into or out of the connected position when it is closed, and to prevent closing the breaker when in an intermediate position. Instruments and relays may be mounted on the door through which the breaker is inserted into the cubicle, but when this is done, a barrier must be provided between the instruments and the breaker.

Circuit breakers may be moved into the connected position in metal-clad switchgear by either the "horizontal-drawout" or the "vertical-lift" method. When the horizontal-drawout method is used, the breaker is moved horizontally into position in the cubicle; then a racking mechanism is provided to force the breaker into the operate position where the primary disconnects are fully engaged. The secondary disconnects for control of the breaker must also be in contact. A test position is also provided where the secondary disconnects are in contact but the primary contacts are separated by a safe distance.

Vertical-lift breakers are moved into the cubicle beneath the stationary primary disconnects, then raised with either a manually or electrically operated hoist until the primary and secondary contacts are fully engaged. No test position is provided, but a "plug jumper" is furnished, so that when the breaker is outside the cubicle, it can be connected to the control circuits in the switchgear for test of the electrical operation of the breaker.

Most manufacturers use the horizontal-drawout method of inserting circuit breakers. From the user's standpoint there is no particular reason for choosing either method, and the selection of the switchgear should not be based on the method of breaker insertion.

16. Outdoor Switchgear. When metal-clad switchgear is to be installed outdoors, it is purchased with a weatherproof housing; otherwise it is the same as the indoor switchgear. The weatherproof housing may be provided without an aisle, or with an aisle space in which maintenance men can withdraw the breakers, and test and repair breakers or other equipment. Lights and convenience outlets are provided in the aisle. Heaters must be provided in all outdoor units to prevent condensation inside the switchgear.

Outdoor switchgear may be obtained with a "standard" aisle which has a single line of breaker and auxiliary units with an enclosed aisle space at the front. It may also be obtained with a "common-aisle" space, which has two lines of breaker and auxiliary units facing each other, with an enclosed aisle between the two lines of units.

Switchgear without an aisle may be selected for reasons of economy, but in most cases the switchgear with an aisle space would be recommended. In case of trouble, in inclement weather, it may be practically impossible for workmen to inspect and repair the equipment if no aisle space is provided.

The choice between standard-aisle and common-aisle switchgear will depend on the number of units involved and the size and shape of the space available for installing the switchgear. When a relatively large number of units are to be supplied in a lineup, the common-aisle construction will be less expensive.

With either common-aisle or standard-aisle construction, a door will be provided at each end of the aisle. For the safety of the workmen, the doors must be supplied with "panic bars" which permit the doors to be opened from the inside by merely pushing on the bar, even though the doors have been padlocked on the outside.

17. Painting. Standards call for indoor switchgear to be painted light gray, ANSI 61. The interior of outdoor switchgear will also be painted ANSI 61, but the exterior will be dark gray, ANSI 24. Either lacquer or enamel may be used. Other types of paints, such as epoxy paints, may be specified where corrosive atmospheres are present and special protection is required. Other colors may also be specified, but special paints add appreciably to the cost of the switchgear and should not be specified unless the additional cost can be justified.

18. Bus Bars. Bus bars for metal-clad switchgear may be copper or aluminum. Although copper is the better material for the purpose, it has become so expensive that aluminum is usually furnished as standard. Copper is available only at a premium price.

Aluminum bus-bar joints may be welded or bolted. When bolted joints are used, the bus should be silver- or tin-plated. Belleville or similar washers should be used to give a large compression area, minimize cold flow, and prevent loosening of the joint.

When aluminum bus joints are welded, joints at shipping splits may be made the same as above, or the aluminum may be brazed to a copper bar which can be bolted to a similarly prepared copper bar in the adjoining section. Similar terminals may be prepared for connecting incoming and outgoing cables or bus duct.

When copper bus is used, joints are bolted. The copper should be silver-plated at the joints. Belleville-type washers are not required.

In metal-clad switchgear all bus and connections must be covered with an insulating material. This insulation should withstand the rated line-to-line voltage of the switchgear when applied between the bus and a conducting covering on the outside of the insulation. It should be pointed out, however, that the outside of the insulation is not necessarily at ground potential and that it is therefore not safe for a workman to touch the insulation when the bus is energized.

Insulation used on bus bars may be tape but is usually a tubing which slides over the bus bar before it is bolted in place. The inside of the tubing usually has a conducting coating to prevent corona in the air space between the bus bar and the tube. The material used for this tubing should be flame-retardant but need not be nontracking, since this insulation does not extend from phase to phase, or phase to ground.

Bus-bar joints may also be insulated with tape. Most manufacturers use some type of boot which can be quickly assembled to the bus to cover the joint. Various methods are used to prevent corona inside the insulation. A metal cover or a semiconducting putty may be applied to the joint before the insulation is applied. When boots are used, they may be filled with a suitable compound to exclude air and provide a uniform potential gradient within the boot.

The insulated bus must be supported on insulators which should be flame-retardant and track-resistant. Both glass polyester and porcelain are widely used for this purpose.

In addition to the main bus, a ground bus must be provided, extending the length of the switchgear. This bus should be bolted to each switchgear unit, and when the switchgear is installed, it should be connected to the station ground.

19. Circuit Breakers. Circuit breakers used in metal-clad switchgear have a voltage rating, an interrupting rating, and a continuous-current rating. Standard voltage ratings for metal-clad switchgear are 4.16, 7.2, 13.8, and 34.5 kV. For other system voltages the next higher rating is used.

Except for switchgear rated 34.5 kV, all metal-clad switchgear presently available will be supplied with oilless-type circuit breakers. Breakers for use in 34.5-kV switchgear may have the arc interruption in an oil chamber, an SF_6 gas-filled chamber, a vacuum chamber, or an air blast chamber. Options are now available for 4.16-, 7.2-, and 13.8-kV switchgear with breakers having arc interruption in a vacuum in place of the air-type circuit breakers that have been in use for many years.

Preferred ratings of breakers available for use in metal-clad switchgear are given in Table 7. To obtain the interrupting rating at voltages other than the nominal voltages given, follow the instructions in the table footnote. To find the required interrupting rating at a particular location, check with the utility supplying the power. They will advise the MVA available from their lines. For feeder breakers the fault current available from motors must be added to that available from the power-company lines. Synchronous-motor contribution to the fault, in MVA, will be 4.8 times the motor rating, in MVA. Induction-motor contribution to a fault is 3.6 times the motor rating, in MVA.

Selection of the continuous-current rating of the breaker will depend on the load to be carried. Breakers are available with continuous-current ratings of 1200, 2000, and 3000 A.

Circuit breakers used in metal-clad switchgear are power-operated. Both stored-energy closing and solenoid-closing breakers are available, although the stored-energy operator has become the standard for most manufacturers. It has the advantage that it requires less power to operate the breaker so that smaller batteries, or a smaller control-power transformer may be used. However, one should consider that with a solenoid operator there is only a very remote chance that two breakers will close simultaneously, but with the stored-energy operator, there is a possibility that two or more of the motors which charge the closing springs will be running simultaneously. Therefore, the control-power source should be large enough to operate more than one breaker.

TABLE 7. Schedule of Preferred Ratings for Indoor Oilless Circuit Breakers (Symmetrical-Current Basis of Rating)

Line No.	Nominal voltage class, [a] kV, rms	Nominal 3-phase MVA class [a]	Rated max voltage, [b] kV, rms	Rated voltage range factor K [c]	Low frequency kV, rms	Impulse [d] kV, crest	Rated continuous current at 60 Hz, [e] A, rms	Rated short-circuit current (at rated max kV), [f,g] kA, rms	Rated interrupting time, [h] cycles	Rated permissible tripping delay Y, s	Rated max voltage divided by K, kV, rms	Max symmetrical interrupting capability [i]	3-s short-time current-carrying capability [j]	Closing and latching capability 1.6 K times rated short-circuit current [j,k]
	1	2	3	4	5	6	7	8	9	10	11	12	13	14
												K times rated short-circuit current		
1	4.16	75	4.76	1.36	19	60	1200	8.8	5	2	3.5	12	12	19
2	4.16	150	4.76	1.36	19	60	1200	18	5	2	3.5	24	24	39
3	4.16	250	4.76	1.24	19	60	1200	29	5	2	3.85	36	36	58
4	4.16	250	4.76	1.24	19	60	2000	29	5	2	3.85	36	36	58
5	4.16	350	4.76	1.19	19	60	1200	41	5	2	4.0	49	49	78
6	4.16	350	4.76	1.19	19	60	3000	41	5	2	4.0	49	49	78
7	7.2	250	8.25	1.79	36	95	1200	17	5	2	4.6	30	30	49
8	7.2	500	8.25	1.25	36	95	1200	33	5	2	6.6	41	41	66
9	7.2	500	8.25	1.25	36	95	2000	33	5	2	6.6	41	41	66
10	13.8	250	15	2.27	36	95	1200	9.3	5	2	6.6	21	21	34
11	13.8	500	15	1.30	36	95	1200	18	5	2	11.5	23	23	37
12	13.8	500	15	1.30	36	95	2000	18	5	2	11.5	23	23	37
13	13.8	750	15	1.30	36	95	1200	28	5	2	11.5	36	36	58
14	13.8	750	15	1.30	36	95	2000	28	5	2	11.5	36	36	58
15	13.8	1000	15	1.30	36	95	1200	37	5	2	11.5	48	48	77
16	13.8	1000	15	1.30	36	95	3000	37	5	2	11.5	48	48	77

Notes to Table 7:

These ratings were prepared by the EEI-AEIC-NEMA Joint Committee on Power Circuit Breakers.

For service conditions, definitions, and interpretation of ratings, tests, and qualifying terms, see American National Standards C37.03-1964, C37.04-1964, C37.04a-1964, C37.04b-1970, C37.09-1964, and C37.09a-1970.

The interrupting ratings are for 60-Hz systems. Applications on 25-Hz systems should receive special consideration.

Current values have been rounded off to the nearest kiloampere except below 10 kA, where two significant figures are used.

[a] For reference only. Figures in Col. 2 must not be used for evaluation of circuit breaker in any specific application. Actual application must be based on rated short-circuit current at rated maximum voltage and in accordance with Notes f and g.

[b] The voltage rating is based on American National Standard Voltage Ratings for Electric Power Systems and Equipment (60-Hz), C84.1-1970, where applicable and is the maximum voltage for which the circuit breaker is designed and the upper limit for operation.

[c] The rated voltage-range factor K is the ratio of rated maximum voltage to the lower limit of the range of operating voltage in which the required symmetrical and asymmetrical current-interrupting capabilities vary in inverse proportion to the operating voltage.

[d] 1.2×50 μs positive and negative wave. All impulse values are phase-to-phase and phase-to-ground and across the open contacts.

[e] The 25-Hz continuous current ratings in amperes are given herewith following the respective 60-Hz rating: 1200–1400, 2000–2250, 3000–3500.

[f] To obtain the required symmetrical current-interrupting capability of a circuit breaker at an operating voltage between $1/K$ times rated maximum voltage and rated maximum voltage, the following formula shall be used:

Required symmetrical current-interrupting capability

$$= \text{rated short-circuit current} \left(\frac{\text{rated max voltage}}{\text{Operating voltage}} \right)$$

For operating voltages below $1/K$ times rated maximum voltage, the required symmetrical current-interrupting capability of the circuit breaker shall be equal to K times rated short-circuit current.

[g] With the limitation stated in 04-4.5 of American National Standard C37.04-1964, all values apply for polyphase and line-to-line faults. For single phase-to-ground faults, the specific conditions stated in 04-4.5.2.3 of American National Standard C37.04-1961 apply.

[h] The ratings in this column are on a 60-Hz basis and are the maximum time interval to be expected during a breaker-opening operation between the instant of energizing the trip circuit and interruption of the main circuit on the primary arcing contacts under certain specified conditions. The values may be exceeded under certain conditions as specified in 04-4.8 of American National Standard C37.04-1964.

[i] Current values in this column are not to be exceeded even for operating voltages below $1/K$ times rated maximum voltage. For voltages between rated maximum voltage and $1/K$ times rated maximum voltage, follow Note f above.

[j] Current values in this column are independent of operating voltage up to and including rated maximum voltage.

[k] If currents are to be expressed in peak amperes, multiply values in this column by a factor of 1.69, which is a ratio of 2.7/1.6.

Solenoid-operated closing mechanisms are simpler and require less maintenance than the stored-energy mechanisms. For this reason many engineers still specify the solenoid-type operator.

Generally batteries are preferred as a power source for operating the circuit breakers in metal-clad switchgear. Power may be obtained from a control-power transformer, but when this is done, capacitor trip devices must be used for overcurrent tripping to be sure that the breaker will trip in case of a short circuit which reduces the system voltage, and consequently the control voltage, below the valve required to trip the breaker. Larger industrial plants will have a station battery which will supply control power for all the switchgear in the plant. Batteries are usually 125 V, although some installations have 250-V station batteries (see Sec. 6, Pt. 3).

If no station battery is available, the control-power source must be purchased with the switchgear. If there are only a few circuit breakers in the switchgear, a control-power transformer may be specified with capacitor trip devices for each breaker. But capacitor trip devices are expensive, and if a large number of breakers are involved batteries will be less expensive. Of course, a battery charger must also be provided and included in the cost. For outdoor switchgear the battery and charger may be located inside the weatherproof structure.

Usually, it is preferable to have circuit breakers close and trip from the same power source; however, at times the cost of the power supply can be reduced by using a control-power transformer to close the breaker, and a battery to trip the breaker. When this is done, a 48-V or even a 24-V battery may be used for tripping. With stored-energy-operated breakers, a 48-V battery may be used for closing, that is, to operate the solenoid which releases the energy in the springs to close the breaker. The spring-charging motor, however, will be operated from the control-power transformer.

All circuit breakers supplied with metal-clad switchgear will be provided with the auxiliary switches required for the operation of the breaker, plus at least one spare *a* and one spare *b* switch. An *a* switch is closed when the breaker is closed; a *b* switch is closed when the breaker is open. If more auxiliary switches are required, they should be specified. The extra switches may be provided on the breaker, or they may be mounted in the cubicle.

20. Relaying and Metering. Relays and meters to be specified with each breaker will depend on the use of the breaker. The various uses for these circuit breakers are enumerated below, together with metering and relaying equipment which will be required, and that which may be specified as optional. Three line diagrams, showing meter and relay connection, are shown in Sec. 1, Figs. 1 through 7. Typical panel arrangements are also shown. Symbols used are listed below.

SYMBOLS *

A	Ammeter
Adc	Ammeter, dc
Asw	Ammeter switch, to transfer indicating ammeter from one phase to other phases without opening current-transformer circuit
CL	Current-limiting
CO	Changeover switch, voltage regulator, manual-automatic operation
CS	Control switch
FM	Frequency meter
G	Governor control switch, to raise or lower load, and for synchronizing
L	Synchronizing lights, to check synchroscope
M	Governor motor
N	Neutral
RTD	Resistance temperature detector $(10\ \Omega)$
S	Synchroscope, to check synchronism of generator or line
Ssw	Synchronizing switch, connects generator and bus voltages to synchroscope and interlocks closing of main generator breaker
TI	temperature indicator, used with 10-Ω resistance temperature detector
Vi	Voltmeter, voltage of incoming machine
V. Adj	Voltage-adjusting rheostat, to preset voltage level
VAR	Varmeter, to measure reactive kVA
Vdc	Voltmeter, dc, voltage of field
VR	Voltage regulator, automatically maintains preset voltage with cross-current compensation for parallel operation
Vr	Voltmeter, voltage of bus, running
Vsw	Voltmeter switch, to transfer voltmeter from one phase to other phases or off
W	Wattmeter, to measure load kW
WHM	Watthour meter, totalizes watthours (integrating meter)
40	Field failure relay, detects loss of field on the generator
41	Field breaker, with field discharge resistor contact
46	Negative-sequence overcurrent relay, protects machine from unbalanced current

* A more complete list of device function numbers is given in Sec. 1.

47/27	Reverse phase sequence and undervoltage relay, protects motor from starting in reverse or with low voltage
49	Overload thermal relay
50	Instantaneous overcurrent relay attachment
50N	Instantaneous overcurrent ground relay
51	Overcurrent phase relay
51N	Overcurrent ground relay
51V	Overcurrent relay with voltage restraint, desensitizes overcurrent relay for remote feeder faults. Sensitizes relay for faults on or near bus
52, 42, 6	Circuit breaker
59	Overvoltage relay
64F	Field ground detector relay, alarm-initiating
64G	Generator neutral ground relay, protects generator from line-to-ground fault
86	Hand reset auxiliary lockout relay, must be reset to start generator. Trips generator breaker and field breaker. Shuts down prime mover
87G	Generator differential relay, protects generator from internal phase-to-phase and phase-to-ground faults

INCOMING LINE (see Sec. 1, Fig. 1)

BASIC

1 Ammeter, ac, indicating (A)
1 Ammeter transfer switch, 3-phase (Asw)
1 Circuit-breaker control switch with indicating lights (CS-52)
2 Relays, time and instantaneous, with built-in test facilities (50/51)
1 Relay, time and instantaneous, with built-in test facilities (50/51N)
3 Current transformers, suitable ratio

OPTIONAL

3 Directional overcurrent relays (67) *
1 Directional overcurrent ground relay (67N).* Instead of three, overcurrent relays (50/51 and 50/51N)
1 Wattmeter, polyphase, indicating (W)
1 Varmeter, reactive voltampere/indication (VAR)
2 Voltmeters, running and incoming (V)
1 Watthour meter, polyphase, power in or out
2 Watthour meters, polyphase, power in and out
1 Synchroscope
1 Synchronizing switch, with keyed handle
1 Potential transformer, line synchronizing
2 Potential transformer-, meter-relay, synchronizing, open delta
1 Auxiliary potential transformer, polarization with wye-wye potentials (addition)
1 Potential transformer, for wye-wye potentials

FEEDER (see Sec. 1, Fig. 2)
BASIC

1 Ammeter (A)
1 Ammeter transfer switch (Asw)
1 Control switch, pistol grip with lamps (CS-52)
2 Overcurrent relays, time and tap as required (51)
1 Overcurrent relay, time and tap as required (51N)
3 Current transformers, ratio as required (CTs)

OPTIONAL

2 Instantaneous attachments for overcurrent relays (phase) (50)
1 Instantaneous attachment for overcurrent relay (ground) (50N)
1 Wattmeter, polyphase (W)
1 Varmeter, polyphase (VAR)
1 Watthour meter, polyphase (WHM)
1 Set of potential transformers with current-limiting fuses, if required

* Use with two incoming lines.

BUS TIE (see Sec. 1, Fig. 3)

BASIC

1   Control switch for circuit breaker, complete with one red and one green position-indicating lamps (CS52)

OPTIONAL

1   Ammeter (A)
1   Ammeter switch for phase-to-phase transfer (Asw)
1   Wattmeter to show power flow and direction
3   Differential current relays, bus (87B)
1   Trip and lockout auxiliary relay, trips and locks all connected breakers (86)
1   Supervision light, 86 trip circuit (L)
3   Current transformers, metering
6   Current transformers, for differential relays for both buses

INDUCTION MOTOR (3-*Phase, Full-Voltage Start*) (see Sec. 1, Fig. 4)

BASIC

1   Ammeter indicating, ac (A)
1   Control switch, with position-indicating lamps (CS)
1   Undervoltage relay to prevent starting or running on low voltage, adjustable voltage setting (27)
2   Thermal-overload protective relays with instantaneous overcurrent trip attachments, or two elements mounted in one case (49/50)
1   Overcurrent ground relay with instantaneous trip attachment (50/51G)

OPTIONAL

1   Ammeter switch, for transferring ammeter to each phase separately (Asw)
1   Omit undervoltage relay (27)
1   Add undervoltage and phase-sequence relay to protect against starting and running motor with low voltage and/or accidental phase-sequence reversal (47/27)
3   Differential current-protective relays for protecting motor against phase-to-ground internal faults, standard speed
1   Lockout relay hand reset for use in conjunction with differential relay to shut down and prevent restarting of motor without resetting (86)
6   Current transformers, three mounted in metal-clad switchgear and three mounted in three neutral leads brought out of motor by purchaser for differential relays
1   Thermal relay for use with 10-Ω embedded detector coil (RTD) provided by machine manufacturer (49)
1   Set of three-phase surge capacitors and three lightning arresters to be mounted at and connected to motor terminals by purchaser
2   Potential transformers, removable type for device 27 or 27/47 if others on bus are not provided

SYNCHRONOUS MOTOR (*Full-Voltage Start*) (see Sec. 1, Fig. 5)

BASIC

1   Ammeter, indicating, load current (A)
1   Varmeter, indicating, reactive voltamperes (VAR)
1   Ammeter, indicating, dc field of motor (A dc)
1   Control switch, with position-indication lamps (CS)
1   Undervoltage, relay, single-phase (27)
2   Thermal-overload relay with instantaneous overcurrent attachments or two elements, mounted in one case (49/50)
1   Overcurrent ground relay with instantaneous trip attachments (50/51G)
3   Current transformers, ratio as required
1   Set field application relays and necessary devices for automatic closing of field switch including locked-rotor protection, field failure, incomplete sequence protection, start to running transition timer
1   Field discharge switch, electrically operated (41)
1   Field discharge resistor, mounted above

OPTIONAL

1 Ammeter switch, for transferring ammeter to all phases (Asw)
1 Power-factor meter, in lieu of varmeter
1 Undervoltage and phase sequence, 3-phase relay. (Omit device 27 basic) to prevent starting or running in the event of accidential reverse phasing or low voltage (47/27)
1 Current balance relay, 3-phase, protects motor against phase unbalancing and single-phase starting and running (46)
1 Thermal relay, for use with resistance temperature detectors (RTD) 10 Ω, located at machine and provided by purchaser, device (49)
3 Differential current relays, percentage type, standard speed, device (87)
1 Auxiliary shutdown and lockout relay, hand reset (86)
2 Potential transformers, if required, relays and meters
6 Current transformers for differential relays, three mounted by purchaser and in external wye leads
1 Wattmeter, indicating, motor load (W)
1 Surge capacitor, three phase and three lightning arresters. Mounted near and connected to motor terminals.
1 Watthour meter, 3-phase, with built-in test facilities (WHM)
1 Temperature indicator for use with RTDs in machine and furnished by purchaser
1 Temperature-indicator transfer switch, for transferring indicator between RTDs
2 Potential transformers, for metering and relaying

GENERATOR AND EXCITER (see Sec. 1, Fig. 6)

BASIC

4 Current transformers, three metering, one voltage regulation (CTs)
1 Field equipment auxiliary compartment
1 Field circuit breaker, electrically operated, 250 V, 600 A (41)
3 Potential transformers with primary current-limiting fuses, two for metering and relaying and one for voltage regulator (PTs)
1 Ac ammeter, indicating with phase-transfer switch (A), (Asw)
1 Dc ammeter, indicating for field, complete with shunt (Adc)
1 Varmeter, reactive voltampere indication (VAR)
1 Wattmeter, indicating power (W)
1 Control switch, hand-operated, for main circuit breaker (CS-52), complete with one set of one red and one green indicating lights
1 Control switch, governor (G)
1 Control switch, remote-field rheostat or exciter-field rheostat or provisions for mounting manually operated exciter-field rheostat
1 Provision for mounting generator field discharge resistor
3 Overcurrent relays with voltage restraint (51V)
1 Space for voltage regulator and wiring
1 Instrument switch, voltmeter transfer (V Sw)
1 Synchronizing switch, with removable keyed handle

OPTIONAL

1 Set of three surge-capacitor and lightning arresters for remote mounting (by customer) at generator terminals
1 Voltmeter, indicating, field volts (V dc)
1 Temperature indicator, for use with 10-Ω resistance temperature detectors (T)
1 Temperature-indicator transfer switch, six position and test
1 Voltmeter, incoming, ac complete with transfer switch with keyed removable handle (V-I)
6 Current transformers, three mounted in switchgear and three mounted by purchaser at neutral of generator
1 Field failure relay, reactance type (40)
1 Negative-sequence current relay (46)
1 Neutral ground relay (64G)
3 Differential current relays (87)
1 Differential auxiliary relay, hand-reset lockout (86)
1 Field ground-detector relay (64F)

GENERATOR NEUTRAL (see Sec. 1, Fig. 7)

BASIC

1   Circuit-breaker control switch with position-indicating lamps (CS)

OPTIONAL

6   Current transformers, three for differential relays and three for metering and other protective relays if required

21. Wiring. The secondary wiring in metal-clad switchgear must be enclosed in metal channels or in conduit to isolate it from the primary circuits. (An exception to this rule may be made for short lengths of wire where connections are made to instrument transformers.) The wire should not be smaller than No. 14 AWG if stranded wire is used, or No. 12 AWG if solid. Flexible wire must be used for wiring across a hinge to a panel.

The insulation on the wire must meet the requirements for type TA, TBS, or SIS as described in the National Electrical Code. It must withstand a high-potential test of 1500 V to ground for a period of 1 min.

22. Switchgear Layouts. In metal-clad switchgear, breakers are not stacked as in low-voltage switchgear. Each breaker unit will contain only one breaker. For preliminary plans of floor space required for the switchgear the following figures can be used. Floors or pads should not be poured, however, until certified drawings of the equipment have been received from the supplier.

Indoor 5-kV metal-clad switchgear units for 1200- or 2000-A breakers are 26 in wide. Height will vary from 72 to 92 in, depending on the supplier. Depth will vary from 56 to 80 in. An aisle space is required on the breaker drawout side wide enough to permit removal of the breaker. This may be from 28 to 50 in. An aisle space is also required on the other side of the switchgear for use in maintenance. A 36-in-wide aisle is usually recommended.

Indoor 5-kV switchgear units for 3000-A breakers will be 36 in wide, 64 to 94 in deep, and about 90 in high. Aisle space on the drawout side should be 38 to 50 in wide; the aisle space on the other side can be the same as for the smaller breakers.

Indoor 15-kV metal-clad switchgear units for breakers of any rating are 36 in wide. Height will be approximately 90 in, and depth will vary from 81 to 105 in. Aisle space on the drawout side must be from 49 to 66 in wide, depending on the supplier. Usually a 36-in aisle space on the other side is sufficient, but some manufacturers may recommend more.

Generally auxiliary units are the same size as breaker units, except for special requirements, such as power-company metering units. When such cubicles are required, they should be constructed to the power-company specifications, and drawings of the units should be approved by the power company before manufacture.

Outdoor switchgear units will be the same width as indoor units of the same rating. They will be somewhat higher, since a sloping roof is usually supplied. The depth will depend on the width of the aisle (if any) and whether standard-aisle or common-aisle type of construction is used. Aisle spaces are usually from 70 to 110 in wide.

23. Installation and Maintenance. When metal-clad switchgear is received, it should be inspected as soon as possible to be sure it has not been damaged in shipment. If the equipment is to be stored, it should be in a dry location. Outdoor switchgear will be supplied with space heaters in each unit. These heaters should be energized while the equipment is being stored. If indoor switchgear is to be stored outdoors, or in an unheated room, temporary heaters should be provided in the switchgear to prevent condensation.

When the equipment is installed, outdoor switchgear should be located as near the loads as is practical. If possible, it should be kept out of direct sunlight, especially during the hottest part of the day.

Indoor switchgear should be installed in a clean, dry room if possible, with adequate ventilation. If it must be installed in a damp room, it should be purchased with space heaters and with dripproof type of construction. If a dusty atmosphere is unavoidable, frequent cleaning should be planned.

Conduit should extend above the floor in accordance with the supplier's recommendations. Cables should be pulled in and connected, giving careful consideration to phase rotation. In accordance with standards, cable terminations will be phases 1, 2, and 3 from left to right (when facing the front of the switchgear), from top to bottom or from front to rear.

All cable terminations should be insulated after cables have been connected. Usually the supplier furnishes insulating material for this purpose. The manufacturer's instructions should be followed for insulating cable terminations.

After installation, the switchgear should be carefully inspected before the bus is energized. The manufacturer's instructions regarding lubrication should be followed. With breakers in the test position, operation of each of the breakers should be checked. All relays should be set following instructions in relay instruction books. Relay trip settings should be coordinated with other relays and with the power company when connecting to power-company lines.

All protective relays should be tested to be sure that they will trip the breakers as intended. This procedure should be done preferably by passing enough current through the relay (or applying sufficient voltage to the relay) to operate the relay contacts. If this is not practical, at least the contacts should be closed manually to be sure that the relay trips the breaker.

Megohmmeter tests may be made on the primary, and on the control circuits, to be sure all circuits are properly insulated. High-potential tests may also be made for this purpose. High-potential tests on the bus are made by applying 14.3 kV on 5- and 7.5-kV switchgear, or 23 kV on 15-kV switchgear, for a period of 1 min, from all bus bars to ground. High-potential tests on secondary circuits are made by applying 1200 V for 1 min from the circuit under test to ground.

After the equipment is tested and inspected, the breakers are racked into the connected position, the bus is energized and the load picked up. One breaker is closed at a time, with as little load on the feeder as possible. Meters are checked as load is increased, and any indication of overheating is watched for carefully.

After the equipment is in service, periodic inspections should be made to be sure that equipment will continue to function properly. Frequency of inspection will depend on conditions under which the switchgear operates and the frequency of operation of the breakers. If dust and dirt are present, the equipment should be inspected frequently. At each inspection, the equipment should be cleaned, taking special care to remove dirt from all insulators. At the same time, checks should be made of contacts on all instrument and control switches, all secondary connections to terminal blocks and other devices, and bus bars for any indication of overheating. Bolts should be tightened in joints if overheating is due to loose joints. Any blown fuses or burned-out indicating lights should be replaced. All circuit breakers should be serviced in accordance with manufacturer's instructions. Where space heaters are provided, all heaters and thermostats should be checked and any which are not operating properly replaced.

24. Typical Metal-clad Switchgear Specifications. The following sample specification may be used as a guide in writing specifications for metal-clad switchgear. Select the proper word or phrase from each group of words or phrases in parentheses. Omit paragraphs, such as "Weatherproof Housing," when not required.

The equipment outlined in this specification will consist of (indoor) (outdoor) metal-clad switchgear with drawout air circuit breakers. The complete switchgear sections will be of coordinated design so that shipping groups are easily connected together in the field into a continuous lineup. Necessary standard connecting materials will be furnished.

The general arrangement and single-line diagram of the switchgear will be as shown on attached sketch Nos. _____.

The switchgear equipment will comply with all applicable standards of ANSI, IEEE, and NEMA.

a. Service. The switchgear sections will be rated _____ kV, with a maximum design voltage of _____ kV. This equipment will operate on a service voltage of _____ V, _____ phase, _____ wire, _____ Hz.

b. Insulation Levels. The assembled switchgear structures will be designed for the following insulation levels:

Standard insulation tests, 60 Hz _____ kV

Standard full-wave impulse (withstand) tests _____ kV

c. *Dimensions.*   Approximate dimensions of the switchgear are shown on the sketch included with this specification.   The circuit breakers will be removable (from) (opposite from) the control-panel side.   An aisle space of _____ in is recommended to permit withdrawal of the circuit-breaker element.

d. *Circuit Breakers.*   The circuit breakers will be rated _____ kV, _____ kVA interrupting capacity at _____ V _____ A momentary, with a continuous-current rating as given in the detailed specifications, and with other characteristics and ratings listed on the attached circuit-breaker data sheet.

The circuit breakers will be 3-pole, single-throw, mechanically and electrically trip-free, with position indicator, operation counter, auxiliary switches, primary and secondary disconnecting devices, and a mechanical interlock to prevent making or breaking load current on the primary disconnects.

(Select one of the following circuit-breaker operating methods.)

1. The circuit breakers will be arranged for (125) (250)-V dc close and trip.   The (125) (250)-V dc source (furnished by the purchaser) (obtained from a battery located in the vicinity of the switchgear, furnished by the purchaser).

2. The circuit breakers will be arranged for 240-V ac solenoid-rectifier close and capacitor trip.   The 240-V ac source will be (obtained from a control-power transformer in the switchgear) (furnished by the purchaser).

3. The circuit breakers will be arranged for 240-V ac solenoid-rectifier close and (24) (48)-V dc trip.   The 240-V ac source will be (obtained from a control-power transformer in the switchgear) (furnished by the purchaser) and the (24) (48)-V dc source will be obtained (from a battery in the switchgear) (from a battery located in the vicinity of the switchgear, furnished by the purchaser).

4. The circuit breakers will be equipped with a (48) (125) (250)-V dc, (120) (240)-V ac stored-energy-closing mechanism and (a 240-V ac capacitor trip device) (_____-V ac) (_____-V dc trip coil).   The (_____-V dc) (_____-V ac) closing and tripping power source will be obtained from (a battery provided in the switchgear) (a control-power transformer in the switchgear) (a separate supply provided by the purchaser).

5. _____

e. *Meters and Relays.*   All indicating meters will be semi-flush-mounted switchboard type.   Watthour meters will be semi-flush-mounted, drawout type, with built-in test facilities.   Overcurrent relays and the like will be semi-flush-mounted, drawout type, with built-in test facilities.

f. *Basic Structure.*   Each circuit-breaker compartment will contain an electrically operated drawout mechanism with automatic shutters and safety interlocks, and will also include the following:

1. Hinged front panel
2. Primary and secondary disconnecting devices
3. Control-circuit cutout device
4. A set of _____ A, 3-phase, 3-wire, insulated main bus and connections
5. Ground bus
6. Necessary terminal blocks, small wiring and control buses, where required
7. Cable supports, where required
8. Engraved nameplate or card holder, as required
9. _____

g. *Compartment.*   (The) (each) auxiliary compartment will include the following:
1. Hinged front panel
2. A set of _____-A, 3-phase, 3-wire, insulated main bus and connections
3. Ground bus
4. Necessary terminal blocks, small wiring and control buses, where required
5. Cable supports, where required
6. Engraved nameplate, or card holder, as required
7. _____

h. *Weatherproof Housing, Standard-Aisle Design.*   The switchgear will consist of indoor-type circuit breaker and auxiliary compartments located in a weatherproof steel housing having an operating aisle space of sufficient size to permit withdrawal of the circuit breakers for inspection, test, or maintenance.   An access door will be located at each end of the aisle, each with provision for padlocking on the outside, but also arranged so that the door can be opened from the inside regardless of whether or not it has been locked on the outside. The aisle space will have adequate lighting which will be controlled by means of a switch at each access door. Included in the switchgear will be the following items:

1. Space heaters in each compartment as required to prevent condensation
2. One thermostat in each compartment, to control space heaters
3. One lamp receptacle for each compartment, operated from switch at either access door
4. Two utility duplex receptacles, one at each access door, for electric tools, extension cords, etc.

The complete enclosure will rest on sill channels running parallel to the length of the switchgear. The ends of the enclosure will be sealed with covers to prevent entrance of debris. The underside of the enclosure and sill channels will be undercoated with asphaltic material to prevent rusting.

The switchgear will be shipped in convenient groups for easy erection in the field, and shipping groups will not exceed (_____) ft long.

i. Weatherproof Housing, Common-Aisle Design. The switchgear will consist of two lineups of indoor-type circuit breaker and auxiliary compartments located in a weatherproof steel housing having a common operating aisle space of sufficient size to permit withdrawal of the circuit breakers for inspection, test, or maintenance. An access door will be located at each end of the aisle, each with provision for padlocking on the outside, but also arranged so that the door can be opened from the inside regardless of whether or not it has been locked on the outside. The aisle space will have adequate lighting which will be controlled by means of a switch at each access door.

Included in the switchgear will be the following items:

1. Space heaters in each compartment as required to prevent condensation
2. One automatic thermostat in each compartment, to control space heaters
3. One lamp receptacle for each compartment, operated from switch at either access door
4. Two utility duplex receptacles, one at each access door, for electric tools, extension cords, etc.

The complete enclosure will rest on six sill channels running parallel to the length of the switchgear. The ends of the enclosure will be sealed with covers to prevent entrance of debris. The underside of the enclosure and sill channels will be undercoated with asphaltic material to prevent rusting.

The switchgear will be shipped in convenient groups for easy erection in the field, and shipping groups will not exceed (_____) ft long.

j. Weatherproofing, Conventional Design (without Aisle). The switchgear will be designed for outdoor service, and weatherproof construction will be used. Each compartment will be equipped with hinged front and rear doors, each with provision for padlocking.

The following equipment will be furnished within each unit of the outdoor switchgear:

1. Two lamp receptacles for interior illumination, one in front and one in the rear
2. One utility duplex receptacle, for electric tools, etc.
3. Space heaters, thermostatically controlled as required to prevent condensation

A switch for all the space heaters and a switch for all the lights will be conveniently located on one compartment.

k. Incoming Line Unit or Feeder Unit
Units No. _____ : _____

(This) (each) unit will contain the following:

_____ air circuit breaker, rated _____ A
_____ current transformers _____ A ratio, _____ kV, for _____
_____ current transformers _____ A ratio, _____ kV, for _____
_____ potential transformers, drawout mounted, _____ 120-V ratio, complete with primary current-limiting fuses
_____ sets of suitable cable-terminating facilities, lugs or pothead, as required
_____ potheads, _____ conductor, _____ kV, _____

____ sets of clamp-type terminals, for _____

____ roof bushings, rated _____ A, _____ kV _____

____ provision for connection to a bus duct rated _____ A, _____ kV, at the (top) (bottom) of the unit

Mounted on the (inner) hinged front panel:

____ circuit-breaker control switch, complete with one red and one green (and one _____) _____ indicating light

____ voltmeter, single-phase, (indicating) (recording) _____ V scale

____ voltmeter transfer switch, 3-phase

____ ammeter, single-phase, (indicating) (recording) _____ ampere scale

____ ammeter transfer switch, 3-phase

____ watthour meter, _____-element, drawout type (with _____-min demand attachment)

____ relays (instantaneous) and (time) overcurrent, drawout type, device No. (50) (51) (50/51)

____ ground relay (instantaneous) and (time) overcurrent, drawout type, device No. (50N) (51N) (51/51N)

____ directional overcurrent relays, drawout type, device No. 67

____ differential relays, drawout type, device No. 87

____ auxiliary relay, _____ device No. _____

____ reclosing relay, with (one) (three) reclosures, (automatic) (hand) reset, device No. 79

____ reclosing relay cutout switch

l. Auxiliary Unit

Units No. _____ : _____

(This) (each) unit will contain:

____ current transformers _____ A ratio, _____ kV, for _____

____ potential transformers, drawout mounted, _____ 120-V ratio, complete with primary current-limiting fuses

____ control-power transformer, stationary mounted, _____ kVA, _____ phase, 60 Hz, _____ - _____ V ratio, complete with drawout mounted primary current-limiting fuses, and secondary molded-case circuit breaker interlocked with the primary fuses

____ lightning arresters, single-pole, _____ type, _____ kV

____ battery (24) (48) V, _____ type, complete with rack and standard accessories

____ trickle charger, static type (without voltage regulation) (with automatic charge control) complete with ammeter, voltmeter, and rheostat

____ sets of suitable cable-terminating facilities, lugs or pothead, as required

____ pothead(s) _____ conductor, _____ kV, _____

____ set(s) of clamp-type terminals, for _____

____ roof bushings, rated _____ A, _____ kV

____ provision for connection to a bus duct rated _____ A, _____ kV, at the (top) (bottom) of the unit

Mounted on the (inner) hinged front panel:

____ voltmeter, single-phase, (indicating) (recording) _____ V scale

____ voltmeter transfer switch, 3-phase

____ ammeter, single-phase (indicating) (recording) _____-A scale

____ ammeter transfer switch, 3-phase

____ watthour meter, _____-element, drawout type, (with _____-min demand attachment)

m. Tie Breaker

Units No. _____ : _____

(This) (each) unit will contain the following:

____ circuit breaker, rated _____ A

____ current transformers, _____ A ratio, _____ kV, for _____

____ (provision for connection to a tie bus duct rated _____ A, _____ kV, at the top of the unit)

Mounted on the (inner) hinged front panel:

____ circuit-breaker control switch, complete with one red and one green and one _____ indicating light

_____ ammeter, single-phase, (indicating) (recording) _____-A scale
_____ ammeter transfer switch, 3-phase
_____ relays (instantaneous) and (time) overcurrent, drawout type, device No. (50) (51) (50/51)
_____ ground relay, (instantaneous) and (time) overcurrent, drawout type, device No. (50N) (51N) (50/51N)

  n. Accessories. The following accessories will be supplied:
1 Test device, consisting of plugging units and cable, for testing breaker and control circuits while breaker is outside and adjacent to unit
1 Maintenance closing tools
1 Racking-in tool
1 Fifth wheel or other device to facilitate moving breaker when outside unit
1 Arc chute-lifting device
1 Test plug, less cable, for drawout relays and watthour meters
1 Set of spray-on touch-up paint

  o. Special Accessories
1 Inspection and test cabinet, indoor type, with necessary control switches
1 Ground and test device, _____
1 Spare-type _____ circuit-breaker element, rated _____ A, _____ kV

13.8-kV METAL-ENCLOSED INTERRUPTER

25. Structures. Metal-enclosed interrupter structures are designed to house high-voltage (4.16- to 34.5-kV) interrupter switches. These switches, in addition to a continuous-current rating, have an interrupting rating which may or may not be equal to the continuous-current rating of the switch. Other items such as high-voltage fuses, instrument transformers, instruments, and lightning arresters may also be mounted in these structures.

Although enclosed interrupters are most commonly used as the power-input section of a substation, they may be used in a bank as a power-distribution system, providing control and protection of feeders to various loads.

  a. Indoor and Outdoor Structures. Structures for indoor and outdoor equipment are usually about the same, except that the outdoor structures have a sloping roof. Doors and bolted plates are gasketed to make the cubicles weatherproof. When the switches are used as the input section of a substation, the structure is arranged to bolt directly to the transformer. A transition unit may or may not be required for this purpose.

These structures usually have a welded, angle-iron frame to which side plates, front and rear doors or bolted plates, roof, and (for outdoor units only) floor plates are fastened. Interior barriers, such as those between units, must be made of No. 11 USS gage steel. Exterior plates may be No. 14 USS gate steel. However, expulsion-type fuses, frequently used in these cubicles, may cause a momentary high pressure inside the unit when they blow. The structure must be capable of withstanding this pressure without being permanently distorted. For this reason, cubicles housing fuses of this type should be designed in accordance with standards established by the manufacturer of the fuse. This problem does not apply to structures housing fuses of the current-limiting type.

Bus bars in structures of this type are bare, unless insulation is required, because of insufficient clearance between phases, or from phase to ground. Cables may be used as buses in this type of switchgear. Cables may be insulated, but if properly supported and spaced from other phases and from ground, the insulation need not be rated for the full voltage rating of the switchgear.

When fused switches are supplied, interlocks must be provided to prevent access to the fuses unless they are deenergized, and to prevent energizing the fuses when the access door is not closed and locked. This interlocking may be made with a mechanical arrangement, or by the use of key interlocks. Key interlocks may also be used to prevent improper operation of the switch. For instance, when a switch is feeding a substation, a key interlock may be used between the switch and the transformer secondary breaker to prevent operation of the switch, unless the secondary breaker is open. This precaution is necessary if the interrupting rating of the switch is less than the full-load rating of the transformer. Even if the switch can interrupt the rated transformer load, it is advisable to prevent operation under load, since less maintenance work will be required.

Structures of this type, as with other types of switchgear, should be phosphate-treated, or bonderized, before painting. After this treatment the interior must be given at least one coat of paint, and the exterior at least two coats. Standard colors are ANSI 61 light gray for the interiors of all units and for the exterior indoor units, and ANSI 24 dark gray for the exterior of outdoor units. Either enamel or lacquer may be used. If the equipment will be subjected to corrosive fumes, a special paint such as an epoxy paint may be specified. Since special paints or special colors add appreciably to the cost, they should not be specified unless the additional cost can be justified.

b. Structures for Drawout-Type Switches. For most applications using interrupter switches, the switch is bolted into the cubicle. However, drawout-type switches are available, and they may be specified if it is desired to minimize outage time for mainte-nance, simplify maintenance procedures, or in the case of outdoor switchgear, assure that maintenance operations can be carried out in any weather.

When drawout, fused switches are supplied, the fuses are located on the drawout carriage. Drawout switches and structures must of course be supplied with primary and secondary (when required) disconnects, as well as a mechanism for racking the switch in and out of the connected position. Interlocks must be provided to prevent racking the switch in or out when the switch is closed. However, fuses are made in-accessible unless the carriage is withdrawn from the cubicle; therefore, no interlock is required to prevent access to the fuses when the switch is closed.

26. Switches. All switches in equipment of this class must have, in addition to the main switch blade, an interrupter which is designed to extinguish the arc which results when the circuit is opened, and to protect the main blade from being damaged by the arc.

a. Standard interrupter switches may be rated 600 or 1200 A and 4.8 or 13.8 kV. They are gang-operated; that is, all three blades open and close simultaneously. The interrupter is frequently an auxiliary blade which opens in an arc chute after the main blade has opened. The arc chute is made of a special plastic material which evolves a gas in the presence of the arc. The gas given off extinguishes the arc rapidly. Other types of interrupters are used which confine the arc in an enclosure, elongate it rapidly, and deionize the gases, causing rapid extinction. All such interrupters meet the re-quirements of the standards and are equally satisfactory.

Switches should be selected which are stored-energy-operated; that is, they are opened, and closed, by springs, so that the speed of opening and closing is not depend-ent on the speed of the operator. With this type of operating mechanism the switch, when closed, will always have full contact (if properly adjusted) and cannot be left with the main blade open and the arcing contact closed. Also, with a stored-energy mecha-nism, the springs must be charged after the switch is closed, before it can be reopened. This assures a time delay, so the switch cannot be opened before the protective device has operated, if the switch is closed on a fault.

b. Selector Switches. Frequently it is desired to feed a load, such as a substation, from either of two sources. Some selector switches are available, which are three-position switches with a single blade for each phase which can be moved from "line 1" position to an open position, then to a "line 2" position. Provision is made, of course, for arc interruption in either of the line positions.

Another method of providing a choice of sources is to use a disconnect-type (a switch which cannot be used to interrupt current) two-position (line 1, line 2) switch in series with a standard interrupter switch. The switches are interlocked so that the discon-nect switch cannot be operated unless the interrupter switch is open. The disconnect switch, of course, need not have a stored-energy operator. Both switches are mounted in the same cubicle.

If power-company lines are to be feeding the selector switch, the utility should be consulted before either of the above types of switches are ordered. Some power com-panies require barriers between their lines when more than one line enters a cubicle, and barriers between the lines cannot be provided with either of the selector switches described above.

When selection of sources is required and incoming lines must be isolated from each other, the duplex arrangement is supplied. In this arrangement, two load-break switches are used, each in its own cubicle. Each incoming line is connected to the jaw of a switch, and the blades of the two switches are connected together and connected to the load, through a set of fuses mounted in one of the cubicles when fuses are required. A barrier is provided between the two cubicles so that one incoming line is isolated from the other. If required, key interlocks can be furnished so that only one switch can be closed at a time, so that the two sources cannot be paralleled.

c. Electrically Operated Switches. Switches may be obtained which are electrically operated. These may be specified when remote operation is desired. The springs are charged by an electric motor. Closing the control switch energizes a solenoid which releases the springs to close or open the switch. The motor then starts automatically to recharge the springs. When electrically operated switches are specified, a source of control power must be provided, either from an existing source or from a control-power transformer furnished with the equipment.

27. Fuses. When required, fuses may be supplied in switch cubicles to protect the load and the cables which feed it. Since these switches are most frequently used to feed substation transformers, the selection of fuses for this purpose is covered here.

a. Selecting the Fuse Rating. Transformers are fused to protect them against damage which might result from a short circuit on the secondary. Overload protection cannot be provided with fuses. If overload protection is desired, a thermal sensor of some type should be used.

For the protection of transformers, a fuse must be selected which will blow and clear the fault before the transformer is damaged by the short-circuit current. It must not melt as a result of the inrush current when the transformer is first energized.

Fuse manufacturers publish time-current curves showing the characteristics of their fuses. Two such curves must be considered: the minimum-melting-time curve and the total-clearing-time curve.

The inrush current to a transformer may be considered to be twelve times its rated current for 0.1 s. A fuse must be selected which has a minimum-melting-time curve that is above this inrush point. The inrush current may be less than twelve times, and if it is desired to use a fuse with a lower rating, the manufacturer of the transformer should determine what the inrush current will be.

Transformers should withstand a short circuit on the secondary for a period of time which depends on the impedance of the transformer. These time periods are given in Table 8. A fuse must be selected which has a total-clearing-time curve which falls below this "withstand" point (Fig. 2).

The minimum-melting-time and the total-clearing-time curves for a 125-A fuse are shown. The transformer inrush point for a 1000-kVA transformer (twelve times rated current for 0.1 s) is below the minimum-melt-time curve, as required. Two withstand points are shown, one for a delta-delta- and one for a delta-wye-connected transformer. The point for the delta-delta-connected transformer is 21 000 A and 3.75 s. This value is obtained from the table by interpolation, considering a transformer impedance of 5.75%, which is standard for load-center transformers of this rating.

The withstand point for a transformer with a wye-connected secondary is at a current value which is 58% of the current for a delta-connected secondary. This lower value must be used because the current in the fuse (line current) is 1.73 times the phase current. In case of a line-to-neutral fault on the secondary the fuse must clear the fault

TABLE 8. Transformer Withstand Time

Impedance, %	Max short-circuit current (times rated current)	Time, s
4 and below	25	2
5	20	3
6	16.6	4
7 or higher	Short-circuit current	5

Interpolate to determine intermediate values.

in less than 3.75 s with 21 000 A in the secondary, or 21 000 divided by the transformer ratio in the primary winding (phase current). The line current then is 1/1.73 or 0.58 times the phase current, when a phase-to-phase fault exists.

As shown, the 125-A fuse is satisfactory for either a delta-delta or a delta-wye transformer. Wye-delta transformers are seldom used, but if one were used, the withstand point would be the same as for the delta-wye connection.

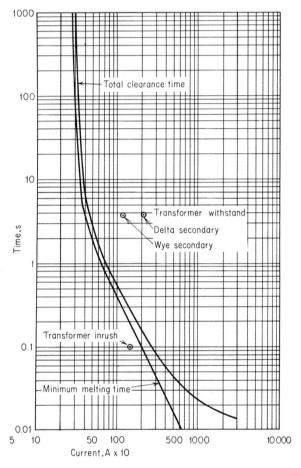

Fig. 2   Time-current characteristic of transformer primary fuse.

The curves of other fuses might fall between the two points, so that more than one fuse might be satisfactory. In this case the choice may be made based on a coordination study, or it may be made arbitrarily. In no case, however, should a fuse be used which has a rating lower than the rated current of the transformer. Most fuses will not blow on less than about twice their rated current, but if they carry current continuously in excess of their rating, they may be damaged to the extent that their characteristics may change and they may fail to interrupt their rated fault currents.

Usually, it is not necessary to plot curves in order to determine the proper fuse rating for the protection of a transformer. Fuse manufacturers publish tables with their recommendations for fuses to protect transformers of various ratings. In most cases, these recommendations can be followed. However, for coordination with other protective devices in the system, it may be desirable to use fuses other than those recommended in the tables. In this case, one can check the curves as outlined above.

In addition to selection of the current rating of the fuse, a choice must be made between current-limiting and expulsion-type fuses.

b. Expulsion- vs. Current-Limiting-Type Fuses. The expulsion type of fuse is often less expensive than the current-limiting fuse of the same voltage and current rating. It is always true, however, that replacement cost of the expulsion-type fuse will be less than that of the current-limiting fuse. With the expulsion-type fuse, only the fuse element needs replacing when the fuse has blown; with the current-limiting fuse the entire unit must be replaced.

Generally speaking, the expulsion-type fuse has a lower interrupting rating than the current-limiting fuse of the same voltage rating. Whatever fuse is chosen, it must, of course, have an interrupting rating equal to or greater than the available fault current at the point of application.

When expulsion fuses blow because of a short circuit, there may be a very loud report, similar to that of an explosion. Current-limiting fuses are sealed and if properly applied will not generate any sound when they blow. The current-limiting fuse may be preferred in some cases for this reason.

Current-limiting fuses, when interrupting high fault currents, will do so in the first quarter cycle, before the current has reached its peak in that initial cycle. They are called "current-limiting" because the fuse opens the circuit before the current reaches the peak value it would reach if no fuse, or an expulsion fuse, were in the circuit. Because the current is interrupted so rapidly, a voltage surge is generated which may damage lightning arresters on the line (upstream) side of the fuse. This characteristic is a problem only if the voltage rating of the arrester is less than the voltage rating of the fuse, if the arrester is located relatively close to the fuse, and if the available fault current is high relative to the interrupting rating of the fuse (Art. 20). It should be pointed out that no such problem exists with expulsion-type fuses. Where current-limiting fuses are used, the problem can usually be avoided by selecting arresters with a voltage rating as high as or higher than the voltage rating of the fuse. If this arrangement is not desirable, the fuse manufacturer should be consulted for the maximum permissible available fault current when arresters of low voltage ratings are used.

28. Surge Protection. Lightning arresters are frequently installed in switch cubicles to protect transformers which the switch is feeding, to protect the switch itself, to protect incoming lines, or to protect other equipment which the switch is feeding.

In general lightning arresters should be located as close as possible to the equipment they are to protect, and the length of the conductor from the arrester to the protected equipment should be kept to a minimum. Thus connecting the arrester directly to the transformer terminals seems wise when arresters are mounted in the switch cubicle for the protection of the transformer. However, when this is done, the incoming line is left unprotected when the switch is open. For this reason, it is common practice to connect the arrester to the incoming line ahead of the switch, since the distance from this point to the transformer is usually short. If the switch is connected to the transformer (or other load) by a run of cable or bus duct, the arresters should be located at the transformer. If the line and the line terminals of the switch require protection, an additional set of arresters may be advisable.

Obviously, if arresters are to protect equipment, the basic insulation level (BIL) rating of the equipment must be higher than the sparkover voltage of the arrester. This voltage condition will usually exist if the rated voltage of the arrester is not significantly higher than the system voltage, that is, if an arrester is chosen with a standard voltage rating higher than but as close as possible to the system phase-to-phase voltage. Arresters may be applied on an effectively grounded system when the rated voltage of the arrester is equal to or greater than the phase-to-neutral voltage. An effectively grounded system is one where the ratio of zero-sequence reactance to positive-sequence reactance X_0/X_1 is between 0 and +3, and the ratio of zero-sequence resistance to positive-sequence reactance R_0/X_1 is between 0 and +1, under all operating conditions and for any amount of connected capacity. Unless one can be sure that this condition exists, the arresters should be specified with a voltage rating equal to or greater than the line-to-line voltage.

29. Installation and Maintenance. General rules for installation of metal-enclosed interrupters are the same as for the installation of other types of switchgear. The cubicles should be set on a level floor to avoid distortion of the frame. Indoor switches should be in a dry, well-ventilated room, free of dust or corrosives. Adequate aisle space should be provided for operation of the switch and for maintenance operations. If expulsion-type fuses are used, the equipment should not be located in a building where the loud report of a fuse blowing might startle people working in the vicinity.

Maintenance of switches is relatively simple. Periodic inspection is advised, the frequency depending on frequency of operation and on the environment. Insulators should be kept clean. Arcing contacts will require replacing at intervals depending on the current being interrupted and the number of operations. Bearings should be lubricated as instructed by the manufacturer. After expulsion-type fuses have blown, any insulators located below the fuses should be inspected and cleaned if necessary. Other maintenance procedures may be required as the manufacturer instructs.

30. Metal-enclosed Interrupter Specifications. The following specification may be used as a guide in writing a specification for metal-enclosed interrupter-switch assemblies. The proper word or phrase should be selected from each group of words or phrases in parentheses and the blanks filled in as required.

The (incoming-line section) (switch lineup) will consist of an (indoor) (outdoor) metal-enclosed cubicle(s), including the following:

a. Single air interrupter (drawout) switch, 3-pole, 2-position (open-closed). The interrupter switch will be stored-energy-closed and stored-energy-opened (manually) (electrically) operated, with the operator located on the front of the unit. Each operator will have indicating targets to show position of switch blades (open-closed) and condition of charging springs (charged-discharged). A window will be located on the front panel for visual inspection of switch blades.

b. Duplex selector (drawout) switch, consisting of two 3-pole, 2-position air interrupter switches. The two switches will provide three positions (line 1-open-line 2). The switches will be key-interlocked to prevent both from being closed at the same time. Each interrupter switch will be stored-energy-closed and stored-energy-opened, (manually) (electrically) operated, with the operator located on the front of the unit. Each operator will have indicating targets to show position of switch blades (open-closed) and condition of charging springs (charged-discharged). A window will be located on the front panel for visual inspection of switch blades.

c. Selector switch, consisting of one 3-pole, 2-position (open-closed) 600-A air interrupter switch and one 3-pole, 2-position (line 1-line 2) disconnect switch, both mounted in a single cubicle. The interrupter switch will be stored-energy-closed and stored-energy-opened (manually) (electrically) operated, with the operator located on the front of the unit. Each operator will have indicating targets to show position of switch blades (open-closed) and condition of charging springs (charged-discharged). The interrupter switch is to be connected in series with the disconnect switch. The disconnect-switch operating handle is to be interlocked with the interrupter switch to prevent operation when the interrupter switch is closed. The disconnect-switch operating handle will have indicating targets (line 1-line 2) to show position of switch blades. Windows will be provided for visual inspection of switch blades.

(The) (each) interrupter switch will be rated:

System voltage _____ kV
Voltage class (4.16) (13.8) kV
Impulse level (60) (95) kV
Continuous current (600) (1200) A
Interrupting rating (600) (1200) A
Momentary rating (40 000) (60 000) A
Fault closing _____ A

The following will also be included:

_____ set(s) of three (expulsion) (current-limiting) fuses, type _____, with an interrupting rating of _____ rms A, _____ equivalent kVA at _____ V.

The fuses will be mounted in the lower compartment of the unit between the switch and outgoing connections. The fuse-access panel will be hinged and interlocked with the operating mechanism to prevent opening when the switch is closed or the operator springs charged, and to prevent operating the switch when the panel is open.

____ incoming (loop-feed) line(s) will enter through the (top) (bottom) of the unit and will terminate at suitable clamp-type (cable lugs) (potheads) (roof bushings).

____ the incoming cable will be _____-conductor _____ MCM _____ kV, _____ in outer diameter.

____ provision for direct connection to the adjacent _____ kVA transformer section. (Details of transformer termination should be supplied.)

____ space heater, thermostatically controlled (standard with each outdoor unit, optional in indoor units). The space heater will be energized by a (115) (230)-V ac, single-phase supply furnished (by the purchaser) (by the supplier).

____ lightning arresters, rated _____ kV, single-pole, (station) (intermediate) (distribution) class, mounted within the unit.

____ key interlock so that operation is possible only when (the associated transformer secondary breaker is open) (all the feeder breakers are open).

____ an auxiliary unit will be provided with a hinged front panel, on which will be mounted the following:

____ voltmeter, single-phase, indicating, semi-flush-mounted _____ scale

____ voltmeter transfer switch, 3-phase

____ ammeter, single-phase, indicating, semi-flush-mounted, _____ scale

____ ammeter transfer switch, 3-phase

____ watthour meter, _____-element, drawout type, semi-flush-mounted, (with demand attachment), (15) (30)-min demand interval

Mounted within the unit will be the following optional equipment:

____ current transformer, _____ secondary, _____/5-A ratio, stationary mounted, for primary metering

____ potential transformer, fused, _____ Hz, _____ V ratio, stationary mounted, for primary metering

18

Dc Low-Voltage Switchgear

GUY W. CHAMPNEY *

FOREWORD

Low-voltage ac power circuit breakers are used in large quantities (several thousand per year) and have been widely publicized as to ratings, dimensions, coordination, limitations, etc. Their application has become familiar to most users of switchgear.

In contrast, dc breakers are used in much smaller quantities (in hundreds rather than thousands). Their applications are often rather special and few published data are available. Hence application engineers and ultimate users of dc switchgear often need more information about how to choose and apply these breakers. The purpose of this section is to review the general characteristics of dc breakers and provide background data on their application in dc switchgear.

* Senior Design Engineer, Low Voltage Switchgear, Westinghouse Electric Corporation; Senior Member, IEEE (Retired).

GENERAL

1. Interruption of direct current is distinctly different from the interruption of alternating current and is generally more difficult at comparable voltages and currents. An ac breaker usually interrupts at or very near normal current zero. Ideally, an ac interrupter does not need to develop much arc voltage drop (during the current part of the cycle) but must develop the ability to withstand voltage very quickly after the instant of current zero. A dc interrupter, however, must develop an arc voltage greater than the applied circuit voltage *while current is flowing* to force the current to zero. Because the dc interrupter usually must absorb much more energy than an ac interrupter for comparable voltages and currents, a given contact structure and arc chute will have less voltage and current-interrupting capacity on direct than on alternating current.

CLASSIFICATION OF Dc CIRCUIT BREAKERS

Dc circuit breakers may be classified by speed, voltage, continuous current, and special applications. However, other classifications related to NEMA Standard (Ref. 1) are also used. General-purpose breakers are not specifically designed to be fast and usually do not limit the maximum fault current except in slowly rising current. General-purpose breakers are often identical in design to standard ac breakers, but they are given different ratings when they are applied on dc. Usually with breakers designed for ac service but used on dc, the continuous current-carrying ability will be higher owing to the absence of skin effect, eddy currents, and hysteresis. However, the maximum current the breaker can interrupt will be less and the maximum circuit voltage, on which it may be applied, will be much less.

Current-limiting breakers are, by standards, put in classes: high-speed and semi-high-speed. Both types, when properly applied, act quickly enough to stop the rise of fault current before it reaches the ultimate or E/R value.

Voltage classes recognized by NEMA are 250, 275 for mining, 750, 850, 1500, and 3000.

Frame-size current ratings range from 225 to 12 000 A. Note, however, that all classifications do not cover the complete ranges of speed, current, and voltage classes (Tables 1 to 3).

2. General-Purpose Breakers. For many applications on lower-voltage and lower-powered circuits, the maximum possible fault current will be so low that all the equipment on the circuit can withstand the maximum current for about 0.20 s (200 ms), and a so-called general-purpose or slow-speed breaker may be used. These breakers are not fast enough to do much current limiting except when the current rises very slowly, and they must therefore be so chosen that they can interrupt the maximum current with maximum voltage on the circuit. On older-type breakers which have no arc chutes or which have simple barrier-type arc chutes, the arc energy is largely dissipated in the wide-open free air space always provided above and/or around these breaker types. These breakers are not sensitive to arc energy. When inductance and current are high, energy ($\frac{1}{2} Li^2$) will be high and a larger, more persistent ball of fire (ionized gas) will be thrown high above the breaker, but the interruption will be completed. When these breakers are applied on higher-powered circuits, it is essential that adequate clear space be provided above the breaker. Before these breakers are applied, the general outline drawings which show the gas-clearance space required should be studied, or the manufacturer's switchgear engineers should be consulted.

Other general-purpose dc breaker types, which have arc chutes that constrict the arc in a relatively small space, have upper limits on the amount of arc energy they can absorb. Therefore, these types have an interrupting current rating which is dependent upon the circuit inductance (Table 1a). Note that when the circuit inductance exceeds a specified value for each frame size, the maximum amperes that each particular frame size is required to interrupt are reduced as the inductance is increased. The reduction is to maintain an approximately constant circuit stored energy (Ref. 1). Table 1a relates particularly to the application of general-purpose-type breakers on 250 V. For higher voltages (500-600-750) older, slow-speed, general-purpose types may sometimes be used. The general trend is to use semi-high-speed breakers or high-speed breakers on circuits of 500 V or higher. Since higher-speed breakers limit the short-circuit

TABLE 1a. Ratings for General-Purpose Dc Circuit Breakers (Ref. 1)

Line No.	Frame size, A	Rated voltage, V	Rated max voltage, V	Rated max short-circuit current, A *	Max inductance for full interrupting rating, μH †	Circuit stored-energy factor W, kWs †	Continuous-current rating range of direct-acting trip ratings, dual overcurrent trip or instantaneous overcurrent trip, A ‡
	1	2	3	4	5	6	7
1	225	250	300	15 000	267	30	15–225
2	600	250	300	25 000	160	50	40–600
3	1600	250	300	50 000	80	100	200–1600
4	3000	250	300	75 000	50	140	2000–3000
5	4000	250	300	100 000	32	160	4000
6	5000	250	300	100 000	32	160	5000
7	6000	250	300	100 000	32	160	6000

The above values apply to one pole of the circuit breaker.
* For the basis of short-circuit current ratings (interrupting ratings), see Ref. 1.
A circuit breaker whose coils have a continuous-current rating lower than those listed for the breaker under a particular interrupting rating shall be given an interrupting rating corresponding to the greatest interrupting rating under which the coil rating is listed.
† If expected inductance to point of fault exceeds the value given in Col. 5, obtain the reduced interrupting rating from the formula

$$I = 10^4 \sqrt{\frac{20W}{L}}$$

where W = the value in Col. 6 and L = actual inductance, μH.
‡ Standard coil ratings are 15, 20, 30, 40, 50, 70, 90, 100, 125, 150, 175, 200, 225, 250, 300, 350, 400, 500, 600, 800, 1000, 1200, 1600, 2000, 2500, 3000, 4000, 5000, and 6000 A.   For standard settings, see ANSI C37.17-1962.

TABLE 1b. Test-Circuit Values, General-Purpose Dc Circuit Breakers (Ref. 1)

Line No.	Frame size, A	Test No.	Rated max voltage, V	Test circuit Current, A	Test circuit Resistance, Ω	Test circuit Inductance, μH	Circuit stored-energy factor W, kWs
1	225	a	300	15 000	0.020	267	30
		b	300	7 000	0.043	1200	30
2	600	a	300	25 000	0.012	160	50
		b	300	9 000	0.033	1200	50
3	1600	a	300	50 000	0.006	80	100
		b	300	13 000	0.023	1200	100
4	3000	a	300	75 000	0.004	50	140
		b	300	15 000	0.020	1200	140
5	4000	a	300	100 000	0.003	32	160
		b	300	17 000	0.018	1200	160
6	5000	a	300	100 000	0.003	32	160
		b	300	17 000	0.018	1200	160
7	6000	a	300	100 000	0.003	32	160
		b	300	17 000	0.018	1200	160

TABLE 2. Ratings and Test-Circuit Values for Semi-High-Speed Dc Power Circuit Breakers (Ref. 1)

Line No.	Frame size, A *	Test No.	Rated voltage, V	Rated max voltage, V	Current E/R value, A	Rate of rise, A/μs	Resist-ance, Ω	Induct-ance, μH
					Test circuit			
1	2 000–10 000	a	750	800	160 000	5.0	0.005	160
		b				1.7	0.005	470
2	2 000–10 000	a	850	1000	160 000	5.0	0.006	200
		b				1.7	0.006	590
3	2 000–10 000	a	1500	1600	160 000	5.0	0.01	320
		b				1.7	0.01	940
4	2 000–10 000	a	3000	3200	160 000	5.0	0.02	640
		b				1.7	0.02	1880

For the basis of short-circuit current ratings (interrupting ratings), see Ref. 1.
* Frame sizes are 2000, 4000, 6000, 8000, and 10 000 A.

current, they are chosen because the higher voltage usually results in higher rates of rise and such high ultimate currents that some current limitation is desirable even though not absolutely necessary.

3. Current-limiting Breakers. On higher-powered circuits, especially at the higher voltages, the value of the fault current may increase to the point that damage may result to the equipment or the breaker may be unable to interrupt the fault current. It is uneconomical to operate under these conditions, and it becomes necessary to slow down the rise of fault currents and interrupt them.

The manner in which the current builds up in a dc circuit depends on the supply circuit. If the supply circuit is true dc such as is produced by a battery or a dc generator, the current will build up along the well-known exponential curve that is characteristic of a resistance-inductance circuit. If the supply circuit is ac, as with power rectifiers, the current can build up with a peak occurring during the first cycle of the ac supply frequency and then drop to some sustained value. The early peak is caused by the offset current in the ac phase currents. This early peak means that, except for high-speed dc breakers which part contacts in a short enough time to prevent the peak from being reached, a dc breaker on such a circuit can be subjected to the mechanical stresses of currents that can be appreciably higher than the current it will have to interrupt. If the breaker mechanical and interrupting ratings are at the same current level, the first-cycle peak current will determine the application limit, not the smaller sustained current that must be interrupted.

TABLE 3. Ratings and Test-Circuit Values for High-Speed Dc Circuit Breakers (Ref. 1)

Line No.	Frame size, A *	Test No.	Rated voltage, V	Rated max voltage, V	Current E/R value, A	Rate of rise, A/μs	Resist-ance, Ω	Induct-ance, μH
					Test circuit			
1	1 200–12 000	a	750	800	200 000	15.0	0.004	53
		b				5.0	0.004	160
2	1 200–12 000	a	850	1000	200 000	15.0	0.005	67
		b				5.0	0.005	200
3	1 200–12 000	a	1500	1600	200 000	15.0	0.008	107
		b				5.0	0.008	320
4	1 200–12 000	a	3000	3200	200 000	15.0	0.016	215
		b				5.0	0.016	640

For the basis of short-circuit current ratings (interrupting ratings), see Ref. 1.
* Frame sizes are 1200, 2000, 3000, 4000, 6000, 8000, 10 000, and 12 000 A.

A simple rough estimate of the first-cycle peak can be made. A short circuit on the dc terminals of the rectifier can be considered as an extension of a 3-phase short circuit on the ac side. A maximum asymmetry factor which gives the highest current in the ac supply, determined by multiplying the factor by the steady-state ac fault current, can be calculated for a 3-phase short circuit as a function of the X/R ratio of the circuit to the terminals of the rectifier and then the maximum asymmetry factor can be found from the curve in Fig. 1. The steady-state fault current is calculated from the impedance to the terminals of the rectifier. This value is then multiplied by the maximum asymmetry factor to approximate the first-cycle peak.

4. Semi-High-Speed Breakers. Present industry standards define the interrupting rating of semi-high-speed and high-speed breakers in terms of the rate of rise of fault current (Ref. 1) (Table 3). For semi-high-speed breakers, the fault current is to be

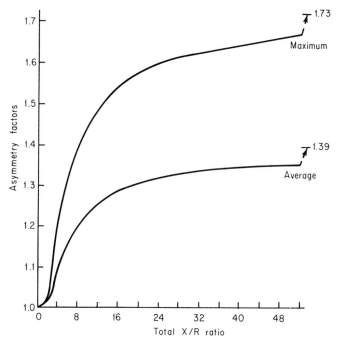

Fig. 1 Asymmetry factors are a function of X/R ratio and are used to approximate the first-cycle peak.

limited within 0.03 s after the beginning of the fault currents having a rate of rise between 1.7 and 5.0 A/μs.

A study of the circuit values specified for testing the ability of a dc breaker to meet the standards established for semi-high-speed breakers is helpful in understanding what is involved. The standards presume the true dc type of supply described earlier. A test circuit resistance of 0.005 Ω is specified for both the 1.7 and 5 A/μs rate of rise di/dt. Using the maximum voltage rating of 800 V, the sustained or E/R value of current becomes 160 000 A. In a true dc circuit the initial di/dt is simply E/L; so with an E of 800 V the L must be 160 μH for a 5.0 di/dt and 470 μH for a 1.7 di/dt. The equation for the exponential buildup of current is $I = E/R[1 - e^{-(R/L)t}]$. Using 800 for E, 0.005 for R, 0.000 16 for L, and 0.03 for t, the current after 0.03 s, for the di/dt of 5, would rise to slightly less than 100 000 A if there is no current-limiting action. Changing only L to 0.000 47, after 0.03 s the current would rise to only 43 800 A with the 1.7 di/dt if there is no current-limiting action.

For circuits in which the initial di/dt does not exceed 5 A/μs, the semi-high-speed breakers are available. Figure 2 shows the most severe circuit on which the semi-high-speed breaker is required to limit within 0.03 s. At lower rates of rise or with smaller ultimate currents the breaker will limit the fault current to proportionately less than shown in Fig. 2. At di/dt of less than 1.7, the tripping speed may get slower and the current limit may not be reached in 0.030 s, although the maximum current will be smaller.

On a true dc circuit that has a rate of rise exceeding 5 A/μs, a semi-high-speed breaker may still be applied if the sustained or E/R current does not exceed the mechanical strength and/or electrical-interruption ability of the breaker. In pure dc circuits the action of the semi-high-speed breaker will prevent the current from attaining the 160 000-A E/R current theoretically possible with the specified test-circuit parameters.

On an ac-supplied dc circuit, the first-cycle peak must be given consideration. For this type of circuit the current can get higher than the current that is interrupted, so the breaker and all other series components must be able to withstand the higher current. This type of circuit gives an appreciably higher di/dt for the same kW rating. Here, as in the case of the true dc circuit described above, the breaker must not be subjected to any current higher than its momentary rating.

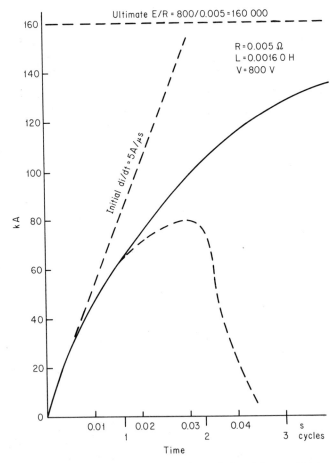

Fig. 2 Predicted performance for NEMA SG3.14 standard test for semi-high-speed circuit breakers.

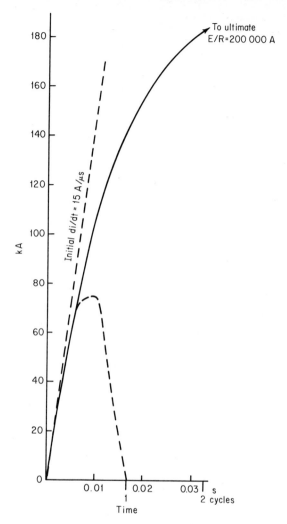

Fig. 3   Predicted performance for NEMA SG 3.14 standard test for high-speed circuit breakers.

5. High-Speed Breakers.   Present industry standards define the interrupting rating of high-speed breakers in terms of the rate of rise of fault current (Ref. 1) (Table 4). The fault current is to be limited within 0.01 s after the beginning of the fault for currents having an initial rate of rise between 5 and 15 A/μs (Fig. 3).

The circuit values specified for testing the ability of a dc breaker to meet the standards established for high-speed breakers give a potential sustained fault current of 200 000 A.   For the 5 A/μs minimum rate of rise the test-circuit inductance is 160 μH for the 750-V service voltage (800 V maximum); for the 15 A/μs maximum rate of rise the test-circuit inductance is 53 μH.   Using these figures in the exponential rate-of-rise formula (Art. 4), the currents after 0.01 s are 106 000 and 44 400 A.

The short time in which a dc high-speed breaker must act to be current-limiting necessitates the use of mechanisms with a small amount of mass.   It usually requires a mechanism that has a holding-coil-type latch rather than a mechanical-type latch which would require time to disengage and accelerate.

TABLE 4. Breaker-Application Characteristics

	Auxiliary power supplies				Main mill drives					Transportation power supplies				
	1	2	3	4	1	5	2	3	4	1	5	2	3	4
Breaker characteristics:														
Current.................. A	B	C	C	A	D	B	C	C	A	D	B	C	C
Voltage .... E	F	F	F	E	G	F	F	F	E	G	F	F	F
Interrupting............. H	I	J	K	H	L	I	J	K	H	L	I	J	K
Breaker attachments:														
Closing mechanism.. M	N	N	N	M	O	O	O	O	M	O	O	O	O
Shunt trip............... P	Q	Q	Q	P	Q	Q	Q	Q	P	Q	Q	Q	Q
Undervoltage release R	R	R	R	R	S	R	R	R	R	S	R	R	R
Overcurrent............. T	U	V	W	T	X	U	V	W	T	X	U	V	W
Reverse current	Y	...	Z	...	AI	Y	...	Z	...	AI	Y	...	Z
Auxiliaries.............. B1	B1	B1	B1	B1	B1	B1	B1	B1	B1	B1	B1	B1	B1

Numbered notes indicate breaker use:
1 = ac breakers
2 = machine breakers
3 = feeder breakers
4 = tie breakers
5 = anode breakers
Lettered notes indicate breaker characteristics:
A: The minimum current rating of the breaker shall be the 2-h rating of the equipment. For auxiliary power supplies and main mill drives, this current shall be 125% of the full-load continuous rating. For transportation power supplies this current shall be 150% of the full-load continuous rating. The 1-min or 10-s rating, etc., shall be disregarded when the application is made.
B: The minimum current rating of the machine or cathode breaker shall be 125% of the continuous unit output for auxiliary supplies and mill drives and 150% for transportation units.
C: The minimum current of the utilization equipment. For a tie breaker, it shall be the maximum continuous current that may be transferred between buses.
D: Anode breakers are rated on the dc output current of the machine. The ratings shall be the same as for the cathode breaker. For a 6-phase single-way rectifier the actual anode current is 0.289 times the dc current.
E: The breaker shall be suitable for the maximum system voltage that may be applied to the breaker.
F: The breaker shall be suitable for the maximum voltage that may be applied to the breaker. In addition, the maximum arc voltage that the breaker may develop should be specified for those applications where possible high arc voltages may damage the equipment. An example is the application of breakers on thyristor power supplies.
G: Anode breakers are rated on the dc voltage of the rectifier. This should be stated when anode breakers are ordered.
H: The breaker should be specified on the maximum system kVA available at the breaker terminals for medium- and high-voltage breakers. For low-voltage breakers, the maximum current available.
I: The breaker should be specified on the maximum current from the source or, in the case of multiple units, the maximum reverse current that is available for a fault within the source. With rectifiers the first ½-cycle peak should be considered when semi-high-speed breakers are used to ensure that the current is not above the mechanical capability of the breaker.
J: The breaker shall be suitable for interrupting the maximum output current of the source or sources.
K: The breaker shall be suitable for interrupting the maximum current that may be transferred between buses.
L: Anode breakers shall be suitable for interrupting the maximum arc-back current.
M: The closing mechanism may be manual, electrical, mechanical, or pneumatic depending on the breaker type.
N: The closing mechanism may be manual, electrical, or mechanical depending on the breaker type.
O: The closing mechanism may be electrical or mechanical.
P: All breakers shall be equipped with a shunt trip mechanism.
Q: All breakers shall be equipped with a shunt trip mechanism. Optional when undervoltage

release is supplied. High-speed breakers have an inherent undervoltage release and do not have shunt trips.

R: Optional.

S: Optional except for cathode breakers with reverse-current trips. On cathode breakers, the coil of the reverse-current-trip mechanism and the undervoltage release shall be in series except when high-speed breakers are used.

T: Medium- and high-voltage breakers shall detect overcurrents with CTs and relays, with time delay, and shall trip the breaker through the shunt trip. Low-voltage breakers shall detect over-currents through a series-overcurrent trip attachment, with time delay or may be equipped to trip by means of the shunt trip and relays, similar to the medium- and high-voltage breakers.

U: Machine breakers, unless there are no feeders, shall trip from the ac side through the use of relays with time delay, and the shunt trip. They shall be equipped with reverse-current trip mechanisms or relays to detect internal machine faults. No time delays shall be used. When there are no feeders, they become equivalent to feeders and should have instantaneous trips (see Note V).

V: Feeder breaker should have instantaneous trips only. Time-delay trips should almost never be used, as most utilization equipment is not capable of interrupting short-circuit currents. The small contactors used to control motors, etc., can interrupt normal overloads, but their ability to interrupt short circuits is limited. As an example, 100-A contactors are limited to about 1000 A (10 times). Higher currents may destroy or damage them. The opening time of these contactors is about 7 to 10 cycles (on a 60-Hz basis), whereas the opening time of a breaker used for this service is 2 to 5 cycles. Therefore, a breaker, without any intentional time delay, can open fast enough to protect the auxiliary equipment.

W: Tie breakers may be applied three ways:

1. Tie adjacent buses together.
2. Tie remote buses together.
3. Tie remote buses together and also be used as a feeder.

In Case 1, they may or may not be equipped with overcurrent trips. If they are so equipped, they should have time delay to permit the feeders to trip before the tie breaker trips so that the whole system is not deenergized.

In Case 2, they should be equipped with time-delay overcurrent trip to protect the tie cables in case of a fault on the cable and also permit a feeder breaker to clear if the fault is on a feeder from the remote bus.

In Case 3, the tie breaker should have instantaneous trips only to protect the utilization equipment (see Note V).

X: Anode breakers have reverse-current trips only. All faults except arc-backs should be cleared by the ac or cathode breakers.

Y: Machine or cathode breakers should have reverse-current trips (except on Ward-Leonard systems or equivalent). For rectifiers, the reverse-current trip can be a part of the breaker. However, for cumulatively compounded MG sets or rotary converters, relays should be used because of the extreme sensitivity required. This arrangement is important to prevent "runaway" in case of loss of ac voltage.

Z: Tie breakers are not equipped with reverse-current trips unless it is desired to feed power in one direction only.

A1: Anode breakers have an inherent reverse-current trip.

B1: All breakers should be equipped with the auxiliary equipment that is required for the proper application and safety requirements for the specific application. These consist of auxiliary switches, cutoff switches, secondary contacts, mechanical interlocks, key interlocks, etc.

In general, for any circuit in which the rate of rise of fault current is in the range of 5 to 15 A/μs, a high-speed breaker would be required; but each application must be checked to be sure that a semi-high-speed breaker could not be used. Regardless of the rate of rise, if the first-cycle peak or the sustained current is within the capabilities of a semi-high-speed breaker, a semi-high-speed breaker could be used. There are, however, cases where the high-speed breakers must be used because the utilization equipment may not be able to withstand the maximum let-through currents of a semi-high-speed breaker.

In a true dc circuit it is unlikely that the initial rate of rise will ever exceed about 10 A/μs. Higher rates of rise will occur only on ac-supplied dc circuits with a first-cycle peak. When a high-speed breaker is subjected to fault currents with rates of rise which fall below the minimum of the prescribed range, the current will be limited to proportionally less than for the higher rates of rise, and the breaker may be somewhat slower.

CIRCUIT STUDIES FOR NEW APPLICATIONS

When a dc circuit breaker is being considered for a new or unusual application, the following circuit factors must be reviewed: voltage, resistance, inductance, and circuit arrangement. In addition to the circuit constants, parallel sources of power and feedback from driven apparatus must be considered. As with ac-circuit-breaker application, a complete circuit or system diagram is essential. This diagram should include the electrical characteristics of all components which may affect any possible switching or fault-interrupting conditions to be imposed on the breaker. Not only rotating machine voltages but also superimposed ac voltages from rectifier circuits, battery voltages, and counter emf from electrochemical cells. In higher-powered circuits, all resistance, inductances, and parallel sources should be carefully estimated. For simplicity, it is often convenient or even necessary to reduce a complex multiunit circuit to an equivalent simple series circuit consisting of source voltage or voltages, resistance, inductance, and the breaker in question.

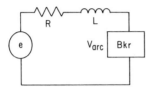

e = total instantaneous voltage (may include several sources and sometimes ac superimposed on the dc)

R = total circuit resistance

L = total circuit inductance

When a fault occurs, but before the breaker begins to open, the initial rate-of-rise equation is

$$\frac{di}{dt} = \frac{e}{L}$$

In high-powered circuits, it is important to get a good estimate of this initial di/dt. After the breaker starts to open, arc voltage begins to develop and affect the current according to the general relation

$$\frac{di}{dt} = \frac{e - iR - V_{arc}}{L}$$

From this equation, it is apparent that the voltage, resistance, and inductance must be known at least approximately before the performance of a breaker in a new circuit can be predicted.

6. Arc-Energy Absorption. Those breaker types which have arc chutes that constrict the arc in a relatively small space dissipate much of the arc energy in heating the materials of the arc chute. Therefore breakers of this design have very definite limits as to how much arc energy they can absorb. When an arc chute becomes too hot, the breaker arc-voltage drop becomes too small to force the current to zero. The interrupting time will prolong to ultimate failure. The amount of energy that a specific arc chute can absorb and still be expected to interrupt the fault current successfully is expressed in watt-seconds (Ws).

The importance of knowing the circuit inductance, at least approximately, can hardly be overstressed. In addition to setting the initial rate of rise $di/dt = e/L$, inductance affects the energy initially stored in the circuit ($\frac{1}{2} Li^2$). This energy plus energy from the source during the interrupting interval must all be dissipated in the arc chute.

7. Accessory Devices. For a circuit breaker to perform properly its primary protective functions of interrupting faults or overloads in the main circuit and in some cases provide routine switching of the main circuit, various accessory and secondary

---

* May include several sources and sometimes ac superimposed on the dc.

control-circuit devices are usually required to be mounted on the breaker. Not all devices listed below are applicable to all breaker types. Furthermore, the number of devices that can be mounted simultaneously, on a single breaker, may be limited. These devices include:

1. Series-overcurrent trip devices
   a. Dual long delay and instantaneous (sometimes called ITL trip)*
   b. Long delay and short delay *
   c. Instantaneous only
2. Reverse-current trips
3. Shunt trips
4. Undervoltage trips
   a. Instantaneous
   b. Delayed *
5. Impulse trips
   a. Bucking bar
   b. Electronic trips
6. External relays; operating through shunt trip or on undervoltage release
7. Manual trips
8. In addition to the above trip devices, the following auxiliaries may be supplied:
   a. Automatic trip-indication switch
   b. Lockout attachment
   c. Operation counter
   d. Auxiliary switches
   e. Position indicators
   f. Key interlocks
9. Usually most high-speed breakers do not have mechanical latches to hold them closed but use some form of magnetic latch with a holding coil. They are also usually tripped by an impulse system (see item 5).
10. The controls for general-purpose and semi-high-speed breakers are similar to the controls used for low-voltage ac breakers. They may be equipped for only manual closing and tripping, in addition to a series-overcurrent trip, or they may have electrically operated closing devices and some form of electrically operated tripping devices.

The controls for high-speed breakers are special as is true with all special application breakers. They may or may not be equipped with any of the accessories listed above. When applying these breakers, consult the manufacturer of the breaker for analysis of the application. The following application information will also be helpful.

APPLICATION

Breaker applications can be divided into two general categories: the type of service and the breakers used for each service.

8. Auxiliary Power Supplies. The most commonly used service is for auxiliary power in factories and mills. Usually, power for this service is supplied at a nominal voltage of 250 V dc. In the past, the source of this dc was MG sets or rotary converters. From about 1935 to 1960, the multianode rectifier and later the single-anode rectifier, either Ignitron or Excitron rectifier, was used. Today, silicon-diode rectifiers are almost universally used for this type of service.

The output of the converting apparatus is fed to the distribution bus through the machine breaker. In the case of the rectifier this breaker is called the cathode breaker. Feeder breakers connect the utilization apparatus to the distribution bus. When two sources of power are available, the distribution buses are usually connected by a tie breaker. In some installations the two buses may be connected without a tie breaker. In this case, the installation becomes equivalent to a single source of larger capacity. Also, in this case, it is most economical to connect the power sources to opposite ends of the distribution bus, as under this condition, the bus will need to have sufficient ca-

---

* These devices are very seldom used on dc breakers except for tie feeders or very special applications.

pacity to carry output of only the larger source. The standard output rating of conversion equipment today is 100% continuously, 125% for 2 h, and 200% for 1 min. The machine breaker should be rated to carry the 125% for 2 h. For example, for a 1000-kW, 250-V, dc source, the full-load current will be 4000 A with 5000 A for 2 h. The breakers must be rated 5000 A. The breaker must also be able to interrupt the short-circuit output current of the source. For most conversion equipment, the maximum short-circuit output current is about ten to twelve times the continuous rated current, so that in this example the breaker must be able to interrupt 50 000 A. Typical applications are given in Figs. 4 through 8.

9. Main Mill Drives. The power supplies for dc main mill drives may be MG sets, variable-voltage rectifiers, or thyristors (SCRs). The most common configuration, today,

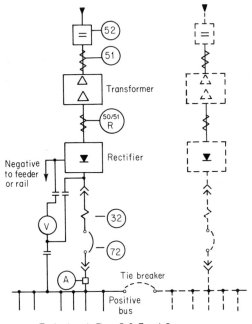

To feeders in Figs. 5, 6, 7 and 8.

Fig. 4 Single-line diagram of single- or double-ended substation used to supply feeders.

is to have an individual power supply for each motor. Some of the smaller mills may have a common supply that feeds all the main drive motors.

Motors for the smaller mills, such as merchant mills, bar mills, and rod mills, range from 500 to 2500 hp for each mill stand or group of stands. A few mills have had multiple motors on a group of stands up to 5000 hp from one power supply. On cold-strip mills and the finishing stands on hot-strip mills, the motor will range from 2500 to 12 000 hp per stand. In almost all cases, there is an individual power supply for each stand. The roughing stands are usually synchronously driven.

The dc voltage for these drives is almost always in the 600- to 700-V range, although a few have been designed for higher or lower voltages.

10. Machine Breakers with Thyristors (SCRs). When thyristors are used as the main power source, a thyristor section is supplied for the mill for the rolling operation and also a second section, called the inverter section, to slow the mill down to threading speed. The rolling operation will require a thyristor to supply the full capacity continuously. The inverter-section duty cycle is such that breakers of much reduced continuous-current rating may be used. The interrupting duty on the rolling section

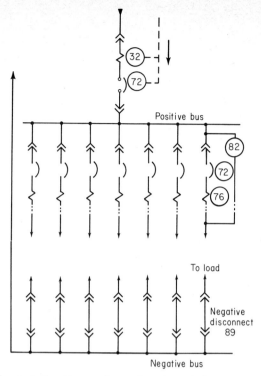

Fig. 5 Feeder single-line diagram of general service with drawout circuit breakers.

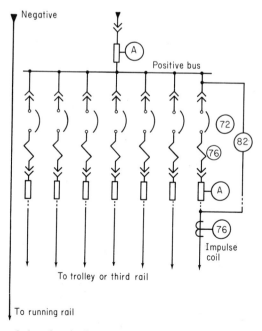

Fig. 6 Feeder single-line diagram for transportation service with drawout circuit breakers.

requires the application of high-speed breakers to protect the thyristors. The duty on the inverter section is even more severe. If an inverter fault occurs, the breaker has imposed on it not only the dc voltage but the ac voltage as well. Special very high-speed breakers have been developed for this application. Special arc chutes have also been developed for these breakers as all thyristors have a maximum peak reverse voltage, PRV, that can be applied across them without damaging them. For the normal mill operating voltage, the thyristors will have a maximum PRV of about 2400 V. The arc voltage is limited to about 2200 V. Mill drives are all special, and the manufacturer should be consulted for the proper breaker application.

11. Traction Service. The power supplies for traction service have been MG sets, rotary converters, mercury-arc rectifiers, or silicon rectifiers. The silicon rectifiers are used almost exclusively on all new installations. The most common operating voltage is 600 V, although some of the latest heavy-duty installations, for high-speed rapid-transit service, operate at 1000 V. Breakers are supplied for the cathode. This arrangement connects the output of the rectifier to the bus. Track breakers (feeder breakers) which connect the power source to the track feeders and "gap" breakers which isolate the track between stations are also used.

Rectifiers for this service have a 2-h overload rating of 150% instead of 125%, which is standard for most other applications. They may also have overload ratings of 300% for 1 or 5 min and 400 or 450% for a very short time, such as 15 s. The breakers are applied on the basis of 150% of the converter rating.

It is common practice, to ensure continuity of service, to install more than one converter in a station. Multiple units are also used to supply the necessary capacity.

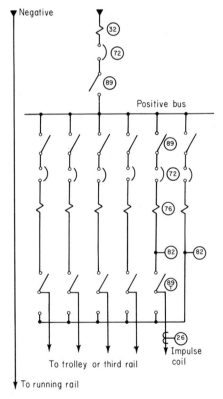

Fig. 7 Single-line diagram of unit substation for transportation service using fixed circuit breakers.

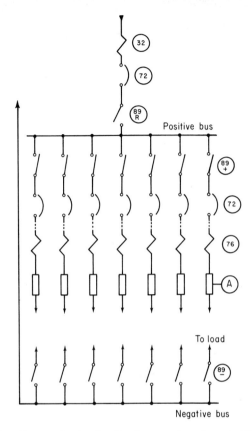

Fig. 8 Single-line diagram of unit substation for general service using fixed circuit breakers.

Great care must be taken to ensure that the interrupting capacity of the breakers is not exceeded. If the maximum current cannot exceed 150 000 A (taking into consideration the ac offset in the first cycle), semi-high-speed breakers may be used. If the system parameters will permit maximum currents in excess of the above, high-speed breakers must be used. Voltages above 750 V (up to 3000 V) require the use of the high-speed breakers.

The machine breakers should be equipped with reverse-current trips, and the feeder or "gap" breakers should have instantaneous, nonpolarized trips. These breakers may also have undervoltage release and/or shunt trips (Figs. 4 to 7).

REFERENCE

1. NEMA Standard SG3, Part 14.

19

Bus Duct
T. F. BRANDT, JR.
J. B. CATALDO

FOREWORD

Bus duct is a general term given to a prefabricated structure for the distribution of electrical energy. It is available for both distribution and utilization voltages in applications for power-station generators as well as small industrial plants. This section gives construction and application information for both classes of bus duct.

Part 1

Distribution-Voltage Bus Duct

T. F. BRANDT, JR.*

1. General Types. Distribution of electric energy at the distribution voltages is done with three different kinds of metal-enclosed bus structures. The first is non-segregated-phase bus where the 3-phase conductors are enclosed in a single housing (Fig. 1a). This bus is structurally similar to the low-voltage bus described in Part 2. It is, of course, insulated for the higher voltage levels, typically, 2.3, 4.16, or 13.2 kV. In certain cases where a higher degree of security is desired in an effort to provide continuous electric power, a metal barrier is interposed between the individual phase conductors. This form of bus, called segregated-phase bus, provides security from phase-to-phase short circuits, as all short circuits must be phase-to-ground (Fig. 1b). Installations requiring a high degree of integrity have generally employed isolated-phase construction, wherein each of the phase conductors is totally enclosed in its own grounded metal enclosure (Fig. 1c). In the isolated-phase construction there is no possibility of a phase-to-phase fault occurring. For this reason, it has found particular applications in the leads from generators of high-capacity generating stations.

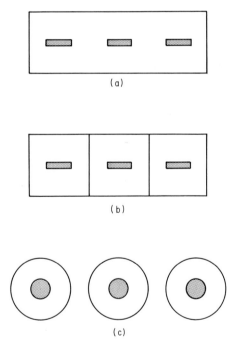

(a)

(b)

(c)

Fig. 1 (a) Nonsegregated-phase bus cross section. (b) Segregated-phase bus cross section. (c) Isolated-phase bus cross section.

* Manager, Switchgear Engineering, I-T-E Power Equipment, I-T-E Imperial Corp.; Registered Professional Engineer (Pa.); Member IEEE.

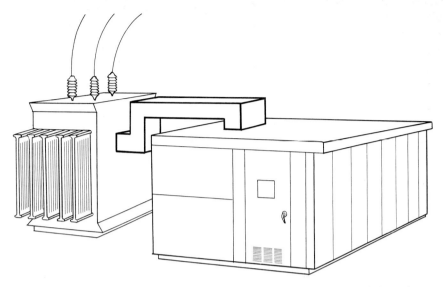

Fig. 2 Nonsegregated-phase bus duct connects transformer (left) to top of metal-clad switchgear.

2. **Nonsegregated-phase bus** has particular application in distribution in conjunction with power switchgear. It is used principally to connect transformers to switchgear lineups or to interconnect switchgear lineups (Figs. 2 to 4). Application consideration must include the continuous current to be carried as well as the operating voltage. Requirements for distribution-voltage bus are found in Ref. 1. The temperature rise of the conductor, including joints, is limited to 65°C over an assumed 40°C ambient. In certain protected indoor applications it is appropriate to use ventilated enclosures. For outdoor application or less desirable indoor environments, the enclosure must be total.

Since the enclosure and bus are relatively rigid, expansion joints must be included in long straight runs. A well-designed joint will have flexible connectors separating the ends of the bars with enough expansion space for the amount of bar in the run. Furthermore, the cover must be able to move. Since the housing is normally much cooler than the bars, the housing expansion is much less. There results a differential

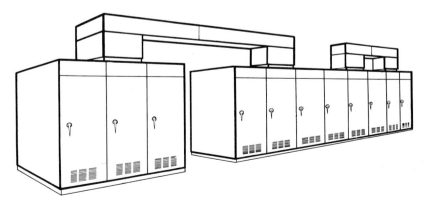

Fig. 3 Nonsegregated-phase bus duct interconnects sections of metal-clad switchgear.

expansion between bus and enclosure which must be accommodated in this joint. Outdoor equipment adds a further complexity, for the joint must not leak. A similar provision must be made where the bus duct enters a building or connects to a transformer.

It is extremely important that the structure of the duct have sufficient mechanical integrity to withstand the forces of short-circuit currents. Typically such a duct is rated for 60, 70, or 80 kA momentary current.

The conductors of nonsegregated duct, used with metal-clad switchgear, are insulated to minimize the opportunity for phase-to-phase faults and to control arc motoring. It must be recognized that, with parallel conductors extending for considerable distances, an arc established between the conductors will be driven to the end remote from the source by the self-induced electromagnetic field. Thus, while faults are extremely rare, when they do occur, the insulated conductors localize the problem.

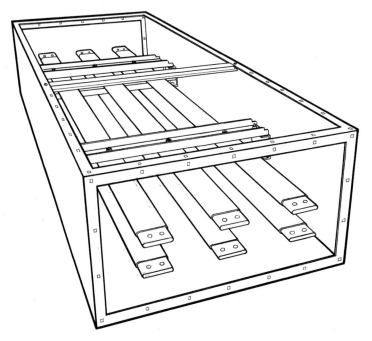

Fig. 4   Construction of 5-kV 3000-A nonsegregated bus structure.

3. **Segregated-phase bus** provides a higher degree of integrity in that the ground plane interposed between phase conductors precludes phase-to-phase faults. This type of structure, of course, is more expensive than the nonsegregated bus, although it does provide increased security for the system.

4. **Isolated-phase bus** was introduced in the late 1930s when generating-station designers found themselves faced with the requirements for increasing generator capacity and greater security. The matter was really brought into focus when certain existing generating stations were being modernized with large generators. The conductors required for the larger currents involved simply could not be fitted into the existing building without recourse to the isolated-phase-type structure. The bus has grown in popularity from its early beginnings as segmented coaxial conductors, totally self-cooled, to large, continuously bonded, forced-air-cooled installations of more than 30 000 A continuous current. Security is of prime importance in generator leads, since there are normally no short-circuit protective devices in the generator leads.

Therefore, if any difficulties develop at generating voltage, the only means to clear the fault is to deenergize the field of the generator and remove the steam from the turbine, a time-consuming procedure. If a fault does exist, under these conditions, considerable damage will be done before the generation can be reduced to a safe level. A second concern is that of system instability which could result if a large generator was suddenly removed from the operating system. The use of isolated-phase bus has eliminated this cause of concern.

Application of isolated-phase bus involves determination of the required continuous duty from the generator, the operating voltage, the insulation level, and certain other thermal characteristics of the structure. The larger bus used today is generally forced-air-cooled so the determination of the appropriate size of the conductor and housing must be based upon a consideration of the continuous current of the generator and the size of the cooling plant required to maintain the operating temperature of the bus. A typical construction of isolated-phase bus is shown in Fig. 5. Security of the sys-

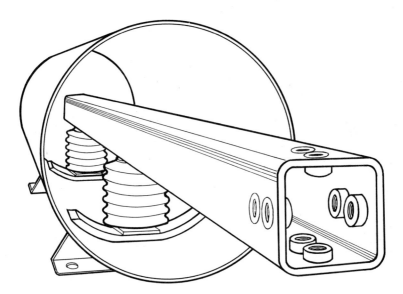

Fig. 5 Construction of isolated-phase bus structure.

tem also requires that there be sufficient time to reduce loading in an orderly fashion if the cooling plant ceases to operate. Depending upon the size of the unit, the size of the system, and the particular operating procedures, a period of $\frac{1}{4}$ to 1 h may be needed to remove the generator from the system. In this transient period the bus temperatures may rise above the values considered satisfactory for continuous duty but must not exceed damaging values. The exact numbers involved will vary with the specific design being considered and should be obtained from the manufacturer.

The forced cooling of isolated-phase bus is done with an enclosed system in which air is dispatched down the center phase and returned through the two outer phases (Fig. 6). The header where the air is exchanged between the phases must have suitable deionizing baffles to intercept any arc that may have been established and prevent its communication to the other phases. The baffle must be constructed to provide minimum head loss for the cooling air and labyrinthian path to deionize any incident plasma.

Modern high-capacity bus is designed with the enclosure continuously bonded, providing a concentric conductor around the phase conductor. This arrangement

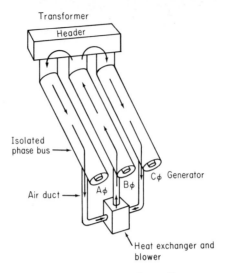

Fig. 6   Isolated-phase bus cooling-airflow system.

provides two significant operating features derived from the absence of external magnetic field characteristic of coaxial structures. First, there is little or no difficulty from the heating by induction of supporting steel in the building. Second, forces between phases during short circuit are practically negligible.

Since isolated-phase bus is used as part of a generating station, proper economic evaluation of the design proposal includes consideration of initial cost and operating cost. The initial cost is determined from the price of the duct, the cooling equipment, field assembly, and installation. The operating cost is determined from the cost of the energy consumed in bus losses, energy to operate the auxiliaries, cost of money, and other considerations peculiar to the individual installation. Once the size of conductor, enclosure, length of run, and phase spacing are determined, the projected losses can be calculated by the procedures described in Ref. 2. Aluminum is generally used for both enclosure and conductor. Energy consumed by auxiliaries can be determined simply from considering the size of the heat-exchanger motors and circulating-fan motors. The capacity of the fan is derived from the volume of airflow and pressure drop required to maintain the temperature. Calculation of this group of figures is quite involved. Most manufacturers use digital computers to do the work. In this way several alternative designs can be evaluated quickly, enabling the user to realize the benefits of an optional design.

Part 2

Utilization-Voltage Bus Duct

J. B. CATALDO*

5. General Types. Bus duct for utilization voltages of 600 V or less is designed in a variety of types and for many applications. Its essential purpose is to transmit relatively large amounts of power from a power source to utilization points where individ-

* Vice President—Engineering, I-T-E Distribution and Controls, I-T-E Imperial Corp.; Member IEEE.

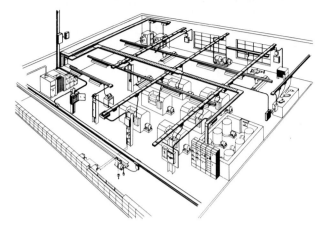

Fig. 7 Busway versatility in typical plant (from Ref. 23).

ual loads can be conveniently plugged in or tapped. Prefabricated in the factory in nominal 10-ft lengths, bus duct of all varieties is shipped to the site and installed in runs of varying lengths. Fittings and other accessories such as transformer and switchboard connections, elbows, tees, plugs, and trolleys are available to complete the installation.

Types of bus duct include feeder, plug-in, short-run, weatherproofed, current-limiting, and trolley. Feeder duct is designed for the efficient transmission of power from one point to another with as low a voltage drop as possible. In addition to its installation in industrial plants in essentially horizontal runs, feeder-type duct is also employed in vertical-riser applications in high-rise buildings for the transmission of power from a lower to an upper floor.

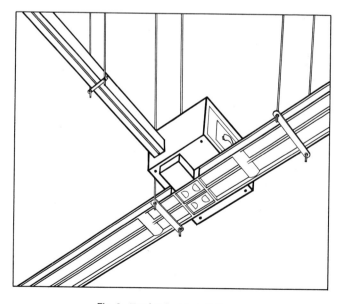

Fig. 8 Feeder-duct installation.

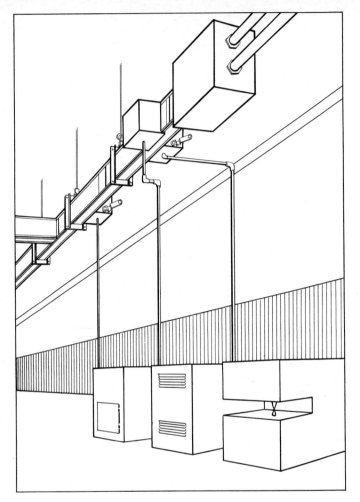

Fig. 9 Plug-in-duct installation.

Plug-in bus duct differs from feeder duct in that openings are provided along the duct length for power taps.   This type of duct was essentially designed for industrial applications where outlet points are required for lines of machinery that could be regrouped as production requirements changed.   The availability of frequent plug-in openings has made this type of duct suitable also for riser applications where power can be made available at each floor.   Feeder and plug-in duct are available for ac and dc applications in current ranges between 225 and 6500 A with full neutral and ground conductors. Plugs can be obtained in both circuit-breaker and switch types in current ranges from 30 through 600 A.   Figure 7 illustrates in one composite drawing the use of feeder and plug-in duct (Figs. 8 to 10).

Bus duct is also available for outdoor application for both short and long runs. Transmission of power is often required from outdoor transformers to indoor switchboards through an intervening building wall.   Because of the wide variability of dimensions among transformers and switchboards, this type of duct is not greatly standardized and the runs are short—usually less than 30 ft.

Long runs of duct are sometimes required outdoors for the transmission of power between buildings. Because in such applications it is essential that efficiency be maintained, a weatherproofed type of indoor feeder is available (Fig. 11).

Another duct that is used on occasion is the current-limiting type. It is sometimes considered necessary to reduce the available short-circuit current of a system to a value that is within the capabilities of a desired protective device. For this purpose, the bus bars are arranged in a configuration so as to interpose a desired amount of impedance in the run between the incoming transformer bank and the switchboard housing the protective device. Care must be taken in such applications to ensure proper voltage regulation and maximum temperature rises.

Bus duct is also available for mobile and continuous power takeoffs. Applications for a mobile system include power tools where the trolleys facilitate the supply of electricity to the tool on a moving or flexible production line. Bus duct of some mobile types is available in prefabricated lengths in ranges to 150 A at 600 V ac or 250 V dc with trolleys of various types, depending on the job application. Other types are fabricated at the installation site in ratings of several hundred amperes.

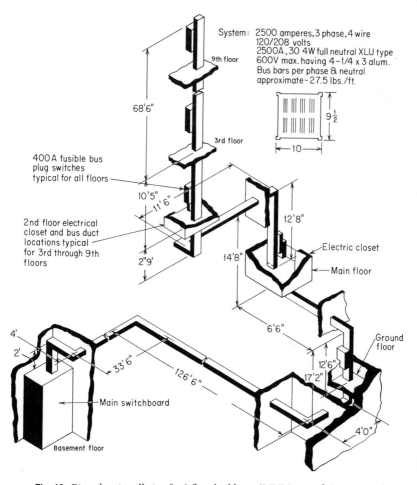

Fig. 10 Riser-duct installation for 9-floor building. (*I-T-E Imperial Corporation.*)

Fig. 11 Weatherproofed-feeder duct. (*I-T-E Imperial Corporation.*)

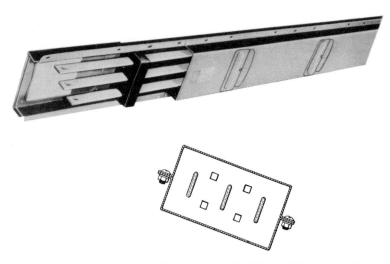

Fig. 12 Plug-in duct with bare bus bars. (*I-T-E Imperial Corporation.*)

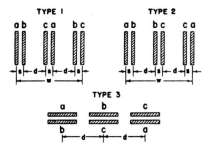

Fig. 13   Paired-phase bus-bar configurations.

6. Feeder and Plug-in Duct.   Bus duct is essentially composed of conductors, spaced or insulated from each, and supported within a metallic enclosure.   The required operating and application characteristics often dictate the bus-bar configuration and construction elements such as insulation and enclosure design.

a. Bus-Bar Configurations.   One of the older but simpler forms of duct (Fig. 12) employs bare bus bars spaced apart periodically by an insulator and enclosed in a simple two-piece steel enclosure.   This early form was primarily designed for plug-in applications and hence required sufficient spacings for plug-in fingers with bare bus bars.   Where voltage-drop requirements are not particularly stringent, a duct with this form of bus-bar spacing having line-to-line voltage-drop values of almost 3 V/100 ft with distributed loading is entirely satisfactory.   Duct of this type is available in ratings to 1000 A at 600 V ac.

As industrial loads increased, a demand for the more efficient transmission of power produced low-reactance duct that reduced the line-to-line voltage drop to about 1½ to 2 V/100 ft.   Two general types of designs are available, both utilizing at least two bus bars per phase in close proximity to reduce reactance.   Paired-phase types are shown in Fig. 13, and an interlaced configuration is illustrated in Fig. 14.   The close spacings require that the bus bars be insulated.   Both types of configurations are available in feeder and plug-in types.

With further increase in loads and a demand for compactness and better use of conductor materials, a stacked configuration has emerged in the last several years (Fig. 15).   The duct employs insulated bus bars in contact with each other and with the enclosure.   A low reactance is maintained equivalent to the paired-phase and interlaced configuration.   In addition, the enclosure acts as an additional cooling medium for more efficient use of the conductor material.   The configuration also results in a very compact duct ideally suited to riser applications in high-rise buildings.

b. Voltage Drop.   The geometrical arrangement and construction elements of bus duct provide the essential operating characteristics of voltage drop and temperature rise.   In ac applications, the problem is one of arranging the bus bars so that currents will flow with as uniform a density as possible for minimum ac resistance.   In addition, the generated heat must be dissipated to minimize temperature rise.

Fig. 14   Interlaced bus-bar configuration.

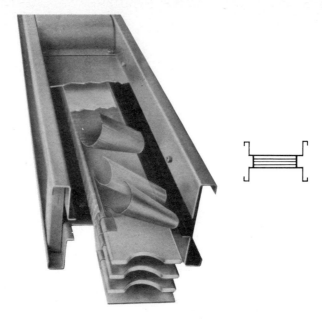

Fig. 15 Stacked bus-bar configuration. (*I-T-E Imperial Corporation.*)

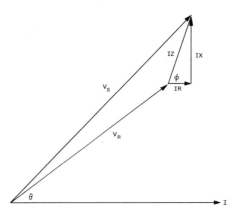

Fig. 16 Vector diagram for voltage-drop determination. *IR*, duct resistance-voltage drop; *IX*, duct reactance-voltage drop; *IZ*, duct impedance-voltage drop; ϕ, duct impedance angle; θ, load power factor; V_S, voltage at line end of duct; V_R, voltage at load end of duct.

In considering line-to-line voltage drops for proper regulation, it is important to determine if the loads will be distributed or lumped. Manufacturers' data will generally indicate the maximum value in line-to-line volts per 100 ft for all application conditions. In the configuration of Fig. 12, a drop of about 3 V/100 ft for a distributed load is about average, while a drop of 1½ to 2 V/100 ft is average for various manufacturers for the configurations shown in Figs. 13 to 15. Table 1 shows the voltage drop for plug-in or feeder duct. For many systems, this information is sufficient. However, there are particular instances where a more detailed study is necessary because of varying loads and unusual duct lengths. In such instances the actual impedance values of the duct must be considered and can be obtained from the manufacturer.

While the voltage drop in dc bus arrangements is an arithmetical calculation, employing current and resistance values, the determination of voltage drop in ac bus is more complicated and involves the vectorial sum of voltage drops as a function of duct and load power factors. The vector diagram in Fig. 16 illustrates this summation for a single-phase circuit. Voltage drops can be obtained from a manufacturer as a function

TABLE 1. Voltage Drop—Plug-in or Feeder Duct
3-Phase, Line-to-Line Drop per 100 Ft, at Rated Current with Distributed Loading

Duct capacity, A	Voltage drop by % load power factor, V *										
	100	90	80	70	60	50	40	30	20	10	0
Aluminum Conductors											
225	1.33	1.53	1.62	1.64	1.62	1.58	1.52	1.44	1.35	1.25	1.17
400	1.28	1.84	1.99	2.03	2.04	2.03	2.02	1.99	1.92	1.82	1.70
600	1.26	2.10	2.40	2.48	2.53	2.53	2.52	2.49	2.43	2.37	2.25
800	1.85	2.05	2.00	1.92	1.80	1.68	1.53	1.38	1.21	1.04	.88
1000	1.85	2.02	1.98	1.89	1.77	1.65	1.50	1.35	1.19	1.08	.87
1200	1.68	1.87	1.91	1.85	1.76	1.65	1.53	1.40	1.25	1.09	.94
1350	1.65	1.83	1.82	1.76	1.67	1.57	1.44	1.32	1.18	1.03	.89
1600	1.72	1.91	1.87	1.80	1.71	1.58	1.46	1.30	1.16	1.00	.84
2000	1.72	1.92	1.88	1.79	1.69	1.58	1.46	1.31	1.17	1.01	.86
2500	1.60	1.78	1.75	1.71	1.61	1.53	1.41	1.28	1.13	.99	.83
3000	1.62	1.80	1.77	1.69	1.61	1.48	1.35	1.23	1.08	.95	.78
4000	1.52	1.76	1.70	1.60	1.54	1.40	1.26	1.18	1.02	.88	.76
5000	1.55	1.75	1.75	1.73	1.67	1.55	1.45	1.30	1.15	.98	.80
Copper Conductors											
225	1.74	2.10	2.08	2.07	1.99	1.87	1.78	1.64	1.50	1.33	1.12
400	1.00	1.70	1.91	2.07	2.12	2.17	2.18	2.17	2.12	2.08	1.99
600	1.20	2.25	2.55	2.55	2.77	2.86	2.85	2.85	2.78	2.70	2.49
800	1.64	1.90	1.91	1.88	1.81	1.72	1.61	1.50	1.37	1.23	1.08
1000	1.37	1.62	1.66	1.64	1.60	1.53	1.46	1.37	1.27	1.16	1.04
1200	1.65	1.94	1.98	1.97	1.92	1.84	1.75	1.64	1.52	1.39	1.25
1350	1.42	1.75	1.78	1.78	1.74	1.68	1.61	1.51	1.40	1.28	1.16
1600	1.29	1.62	1.69	1.69	1.66	1.61	1.54	1.46	1.37	1.26	1.18
2000	1.40	1.79	1.88	1.89	1.84	1.79	1.71	1.62	1.52	1.40	1.29
2500	1.23	1.53	1.60	1.56	1.53	1.49	1.41	1.34	1.24	1.12	1.02
3000	1.23	1.55	1.62	1.62	1.56	1.51	1.46	1.35	1.26	1.17	1.03
4000	1.20	1.46	1.56	1.58	1.56	1.46	1.40	1.34	1.20	1.10	.98
5000	1.20	1.52	1.55	1.55	1.52	1.48	1.42	1.30	1.20	1.07	.97
6000	1.20	1.53	1.56	1.59	1.56	1.53	1.47	1.38	1.26	1.17	1.02
6500	1.11	1.46	1.56	1.59	1.56	1.49	1.40	1.30	1.24	1.04	.94

* Values for nonventilated duct 225 to 600 A and ventilated edgewise mounted duct 800 A and higher.

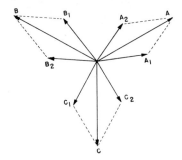

Fig. 17   Bar and phase-current distribution for balanced loads.

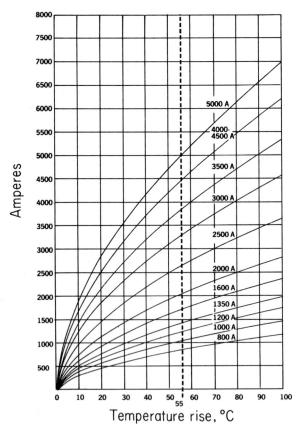

Fig. 18   Example of temperature-rise curves for aluminum bus bars, ventilated case, bars mounted edgewise.

of duct rating and load power factor. For applications such as feeder duct where the load can be considered lumped, the values shown in Table 1 are multiplied by 2.

Manufacturers can also supply the impedance characteristics of bus duct in terms of ac resistance, reactance, and impedance values which can be used by the system designer to calculate the resultant voltage drop for various load conditions and length of run.

The voltage drops of bus duct are usually unbalanced, depending on the bus-bar configurations. Where long runs of high-amperage duct are encountered, the voltage unbalance can be objectionable. The paired-phase bus-bar arrangement of Fig. 13 has an almost even balance among the phase-voltage drops owing to the almost complete neutralization of magnetic fields. Figure 17 shows the currents in each phase and each bar in which essentially equal and opposite currents result in each bus pair. Regardless of the bus-bar configuration, however, unbalanced voltage drops can be eliminated by interphase ties periodically installed along a bus-duct run.

c. Temperature Rise. The current-carrying ability of bus duct is a function of many factors including bus-bar configurations, type of enclosure (enclosed or ventilated), proximity of bus bars to the enclosure, mounting position, etc. In accordance with UL Standard for Busways and Associated Fittings (Ref. 21) and NEMA Standard on Busways (Ref. 20), the ampere rating of a given duct design is based on a permissible maximum temperature rise of 55°C above an ambient of 40°C at any point in the duct system, including the joint. All commercially available bus duct employs bus-bar sizes based on such tests. Manufacturers of bus duct also make available curves which give temperature rises at other than full load. This information is particularly important where high ambient temperatures are likely to be encountered and the duct must be derated in accordance with maximum safe operating temperatures (Fig. 18).

d. Conductor Materials. Copper and aluminum are the two major bus-bar materials used today. Because of its progressively increasing costs, copper has been largely supplanted by aluminum in recent years. While commercial-type aluminum bus bars have a conductivity of only 55% as against 100% for commercial-grade copper, the lighter weight and relatively lower cost have been responsible for this trend (Table 2).

Essential in the choice of conductor material for bus bars is its treatment in making safe and effective joint connections. It is axiomatic that a bus-duct run is a series of joint connections with intervening bus bars, and that the effectiveness of the whole

TABLE 2. Physical Properties of Copper and Aluminum

Property	Copper	Aluminum 55 EC-grade
Specific gravity	8.93	2.70
Weight:		
$\quad$ lb/in³	0.322	0.0977
$\quad$ lb/ft³	558	169
$\quad$ lb/ft/in²	3.875	1.172
Resistivity (at 20°C):		
$\quad$ $\mu\Omega$/ft/in² cross section	8.23	14.81
$\quad$ Ω/cmil ft	10.48	18.86
Linear expansion, temp. coefficient:		
$\quad$ Per °C	0.000 016 8	0.000 023
$\quad$ Per °F	0.000 009 3	0.000 013
Resistance, temp. coefficient (at 20°C):		
$\quad$ Per °C	0.003 93	0.003 63
$\quad$ Per °F	0.002 18	0.002 02
Ratio: density of aluminum to copper	0.304	
Specific heat, Btu/lb/°F or g-cal/g/°C	0.092	0.230
Thermal conductivity,		
$\quad$ W/in/in²/°C	9.85	5.30
All data are for 20°C (68°F).		

is greatly dependent on the efficiency of the joints. When a joint is first made, it has a certain resistance and, therefore, a certain voltage drop. These characteristics are affected in part by the bolting methods and pressures designed for the joint and the effectiveness of field installations. Another important consideration is the condition of the contact surfaces. If such surfaces are susceptible to the formation of non-conducting oxides and other products of chemical action, the resistance and the voltage drop increase. This condition in turn produces an increase in temperature which, then, accelerates the chemical effect to create a still higher resistance, until burning and arcing occur.

The compounds formed by the atmosphere in both copper and aluminum are non-conducting. One of the ways for improving the surface-contact areas is to apply a silver coating, since the compounds formed by silver are conducting and will tolerate higher operating temperatures. The 55°C maximum temperature rise in the UL Standard is based on a silvered joint area, whereas a 30°C maximum is imposed for bare copper bars.

The application of a silver coating to copper bars is fairly simple. However, the tough insulating oxide film formed on aluminum bars presents a more complicated problem. There are several methods used by various manufacturers but, essentially, they consist of chemically removing the oxide and replacing it with a substratum material such as zincate or stannate. A buffer metallic layer, such as copper or bronze, is then added to provide a base for the final coating. The choice of the final layer, usually tin or silver, is a function of its intended use. In plug-in duct, the inherent expansion of bus bars due to periodic heating and cooling causes a large relative movement at the plug-in points between bar and finger. The burnishing action of silver makes it ideal as a finish for the maintenance of low-resistance and friction-free motion. In feeder duct, the essentially stationary joints make it possible to use either silver or tin as a final coating.

When properly designed, fabricated, and installed, bus duct employing copper or aluminum bus bars is comparable in every respect. Considering joint connections, both types of bus-bar material can be intermixed provided that contact-surface preparations are equivalent and that bolt methods and pressures are properly applied.

e. Short-Circuit Considerations. In practice, bus duct can be subjected to two kinds of short-circuit faults: bolted and arcing. The bolted type results from a direct line-to-line or line-to-ground connection and can initiate the flow of the maximum currents possible. The arcing fault results in variable current depending on the length and other characteristics of the arc and can result in a current 25 to 50% of the bolted fault.

Bolted short-circuit ratings for bus duct are based on its ability to withstand the stresses created by the short-circuit current without a decrease in electrical spacings

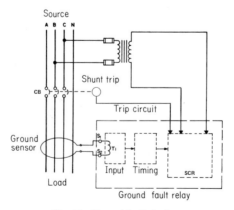

Fig. 19 Ground-fault device.

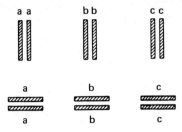

Fig. 20   Bus-bar configuration for high-impedance duct.

or other damage that would not permit the duct to be reused without repair. NEMA Standard Publication BU-1-1971 provides details for applicable tests.

The relatively short spacings between bus bars in modern bus duct create high forces with high fault currents. Manufacturers must make the bus-duct structure sufficiently rugged to withstand such forces. While such ratings are based on a 3-cycle current, improved safety factors can be obtained by the selection of proper protective devices. Molded-case circuit breakers have total interruption times of between ½ to 1 cycle and, when properly applied within their interrupting ratings, provide excellent protection. Where available short-circuit currents are very high, current-limiting fuses limit both available peak current and time (less than ½ cycle) to provide possibly the best protection. The shorter time permitted by both these protective devices results in materially reduced deflection of bus bars, and increases short-circuit safety factors.

Because of their high variability, arcing faults become difficult to test for and to protect against. The only safe rule is to rely on the best protective device that can be applied to meet the load conditions. A safe rule is to employ a protective device with the lowest instantaneous-trip setting consistent with in-rush-load requirements. With higher currents, current-limiting fuses or circuit breakers in combination are preferable.

In recent years, ground-fault devices have become available which can be used effectively to protect against arcing faults. Such devices (Fig. 19) include a ring which surrounds the phase and neutral conductors of a system and senses in a zero-sequence manner. Hence any current not flowing back into a system, such as from a line-to-ground arc, is sensed and can cause the protective device to open when a set value is exceeded. Commercial types have ground-current ranges from 30 to 2000 A.

7. Current-limiting Duct. It is sometimes necessary to reduce the available short-circuit current from a supply network to a value that is equal to the interrupting capacity of a particular protective device, or to the short-circuit ratings of a switchboard. The use of current-limiting busway may be an economical way of solving a protection problem.

The principle of this type of duct is to connect the phase bars so as purposely to introduce greater reactance (Fig. 20). When connected in this manner, voltage drops can

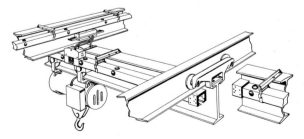

Fig. 21   Trolley-duct application.   (*I-T-E Imperial Corporation.*)

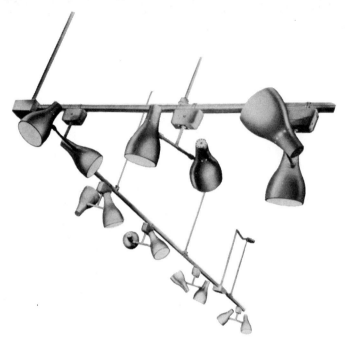

Fig. 22 Lighting duct. (*I-T-E Imperial Corporation.*)

increase by a factor of 4 to 5 over the low-impedance configuration. The voltage-regulation calculations for a particular installation must, however, be carefully checked. For this purpose, manufacturers of current-limiting duct provide voltage-drop tables as a function of load power factor similar to Table 1. In addition, imped-ance values are also available for maximum line-to-line voltage-drop calculations.

The current densities in bus bars are affected in a manner almost opposite to that of low-reactance duct. Where the current distribution is almost uniform in the paired-phase arrangement, owing to minimum skin and proximity effects, the high-impedance arrangement creates high skin effects and current crowding at bus-bar edges. A high ac resistance results, giving higher and less uniform temperatures. Hence the current rating and temperature rise of duct to be used for current-limiting purposes must also be carefully checked.

8. Trolley Duct. This type of duct has been designed for applications requiring the availability of power on a continuous or mobile basis. Where power is required for portable tools for operators who move along a production line, or for moving loads such as cranes, trolleys become an ideal method for such supply. They also become a convenient method for the continuous supply of power at nonregular points, as for lighting arrangements.

Trolley duct is available in 2- and 3-pole forms, in many ratings of current, voltages, and types, depending on application (Fig. 21). Such trolleys are designed in various types, with and without rollers, with protective devices, and with hangers for tool support.

A smaller form of trolley duct is available in ratings to 50 A, 250 V ac. This form (Fig. 22) is ideally suited for the supply of power to lighting fixtures, permitting great flexibility in the rearrangement of lighting. Trolleys as well as fixed takeoff devices are also commercially available for this form of duct.

REFERENCES

1. "Switchgear Assemblies Including Metal-enclosed Bus," ANSI C37.20-1969, American National Standards Institute, Inc., New York.
2. "Calculating Losses in Isolated Phase Bus," ANSI C37.23-1970, American National Standards Institute, Inc., New York.
3. R. H. Albright, A. Conangla, A. C. Bates, and J. B. Owens, "Isolated-Phase Metal-enclosed Conductors for Large Electric Generators," AIEE Conference Paper CP 62-242.
4. C. H. Asperen, Mechanical Forces on Bus Bars under Short-Circuit Conditions, *AIEE Trans.*, November 1922.
5. G. E. Buchanan and J. S. Banas, "The High Current Generator Bus," AIEE Conference Paper CP 62-282.
6. G. E. Buchanan, Laboratory and Field Test Experience with High-Capacity Isolated-Phase Busses, *AIEE Trans.*, October 1959.
7. A. Conangla and H. F. White, Isolated Phase Bus Enclosure Loss Factors, *IEEE Trans. Power Apparatus Systems*, vol. PAS-87, pp. 1622–1628, July 1968.
8. H. B. Dwight, "Electrical Coils and Conductors," McGraw-Hill Book Company, New York, 1945.
9. R. C. Elgar, R. H. Rehder, and N. Swerdlow, Measured Losses in Isolated-Phase Bus and Comparison with Calculated Values, *IEEE Trans. Power Apparatus Systems*, vol. PAS-87, pp. 1724–1730, August 1968.
10. S. C. Killian and K. Boyajian, Isolated-Phase Telescoping Bus Duct, *AIEE Trans.* 56-989.
11. S. C. Killian, Induced Currents in High Capacity Bus Enclosures, *AIEE Trans.*, vol. 69, pp. 166–171, 1950.
12. A. B. Niemoller, Isolated-Phase Bus Enclosure Currents, *IEEE Trans. Power Apparatus Systems*, vol. PAS-87, pp. 1714–1718, August 1968.
13. J. G. Noest and A. A. Milusich, A History of the Development and Application of Isolated-Phase, Metal-Clad Bus Equipment, *IEEE Trans.*, vol. 68, CP 132, December 1967.
14. J. B. Owens, Metal-enclosed High-Voltage Switching Station, *IEEE Trans.*, vol. 2110A, Nov. 4, 1958.
15. A. H. Powell and N. Swerdlow, Single Insulators Isolated-Phase Aluminum Bus, *AIEE Trans.*, vol. 77, part III, pp. 808–813, 1958.
16. O. R. Schurig and H. P. Kuehni, Temperature Rise and Losses in Solid Structural Steel Exposed to the Magnetic Fields from A. C. Conductors, *AIEE Trans.*, pp. 184–202, 1926.
17. O. R. Schurig and M. K. Sayre, Mechanical Stresses in Bus Bar Supports during Short-Circuit, *AIEE Trans.*, vol. 44, pp. 217–233, 1925.
18. W. F. Skeats and N. Swerdlow, Minimizing the Magnetic Field Surrounding Isolated-Phase Bus by Electrically Continuous Enclosures, *AIEE Trans.*, 62-171.
19. N. Swerdlow and M. A. Buchta, Practical Solutions of Inductive Heating Problems from High Current Busses, *AIEE Power Apparatus Systems*, no. 46, pp. 1736–1746, February 1960; also *AIEE Trans.*, vol. 78, part III-A, pp. 825–931.
20. "Busways," NEMA Standard Publication BU-1-1971.
21. "Busways and Associated Fittings," Underwriters' Laboratories Standard 857.
22. Derio Dalasta and Richard R. Conrad, A New Ground Fault Protective System for Electrical Distribution Circuits, *IEEE Trans. Ind. Gen. Appl.*, vol. IGA-3, no. 3, May/June 1967.
23. Electrical Power Distribution for Industrial Plants, *IEEE Pub.* 141, August 1969.
24. "Electrical Buses and Bus Structures," *I-T-E Imperial Corporation Bull.* 2603-2A.
25. J. B. Cataldo and N. Shackman, "Short Circuit Protection of Busway Systems with Current Limiting Fuses," *IEEE Trans. Paper* 56-46.

20

Substations

S. E. McDOWELL *

* Senior Staff Engineer, Switchgear Division, Allis-Chalmers Corporation; Registered Professional Engineer (Wis. and Tex.); Member, NSPE.

FOREWORD

Because substation equipment for industrial power-distribution systems is so versatile and is available in so many physical and electrical arrangements, considerable study and planning must be given to its selection and location to assure the most practical choice for the application.

This section is written to aid in specifying and applying substation equipment for industrial use.

INTRODUCTION

1. Review of Steps Considered in Planning an Industrial Distribution System. In the planning of an industrial power-distribution system, there are six basic steps to be considered in detail:

1. Select the distribution voltage based on:
   a. Plant demand (present and future)
   b. Standard equipment ratings
   c. Nature and physical location of loads
   d. Economy
2. Arrange loads in logical groupings:
   a. Tentatively designate individual "large" motors
   b. Group small loads in economical substation sizes to fit plant layout
   c. Divide substation loads into feeder groups
3. Consider special load requirements:
   a. Starting requirements of "large" motors
   b. Operation of arc furnaces, welders, and other cyclic loads
   c. Critical or vital loads
4. Select plant distribution system from the following:
   a. Radial (old style)
   b. Radial (modern)
   c. Secondary selective
   d. Primary selective
   e. Looped primary
   f. Secondary network
   g. Combinations of above
5. Select system components for:
   a. Main-supply substation
   b. Secondary distribution system
   c. Distribution/utilization circuits
6. Comparison of various alternates, taking into account:
   a. Initial cost
   b. Safety
   c. System reliability
   d. Operation and maintenance—safety, simplicity, and economics
   e. Flexibility for future expansion

Most of the above points are covered in other sections of the book and will not be detailed here. This section covers step 4 and touches partially upon steps 5 and 6.

A review of Fig. 1 will indicate that while load-center systems do not comprise the entire electrical-distribution system of an industrial plant, they do provide the important heart of any well-designed, efficient, and economical system. They are the basis upon which the fundamental principles of a modern distribution system rest.

These principles apply equally well to both primary and secondary unit substations, but this section primarily covers load-center systems serving loads at 600 V or below, since these are predominant in the typical industrial plant.

2. Load-Center Systems—Concepts and Advantages. The basic concept of load-center systems is to distribute power at highest economical voltage level (2.4 to 13.8 kV generally) to areas of concentrated electrical load (load centers) where voltage is transformed to the lower utilization voltage and delivered to utilization equipment via relatively short secondary feeders.

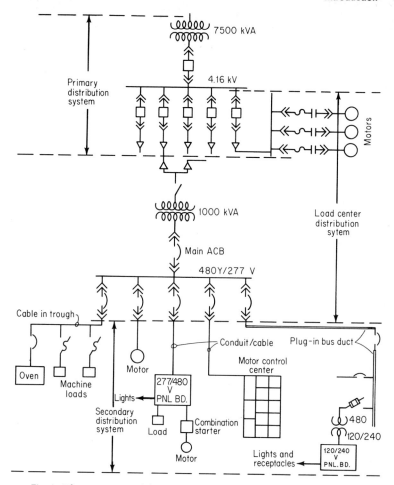

Fig. 1   Three segments of distribution—primary, load-center, and secondary.

Where total plant load is such that it exceeds the most economically sized substation (750 to 1500 kVA), the load is served by two or more smaller substations rather than by one large substation at the point of primary service entrance; again, the secondary feeders are relatively short.

In the old-style radial system, the total load was served by a single substation at the point of primary service entrance. This arrangement resulted in long and heavy secondary feeder cables. Figure 2 shows the difference between the old-style radial and the modern load-center systems.

Many advantages accrue to the owner from the use of the load-center concept.

Initial costs are lower. Carrying power to load centers via smaller high-voltage cables rather than large multiple low-voltage cables results in a substantial saving in cable costs. Figure 3 is a graphical illustration of this saving.

While transformer costs are somewhat higher for the load-center system, this cost is offset by the fact that smaller substation sizes are utilized. Less costly, lower-interrupting-capacity breakers can therefore be utilized for protection of secondary feeders (Sec. 17, Tables 2 to 5). Table 1 lists various transformer sizes and secondary-feeder application data for four basic secondary voltages.

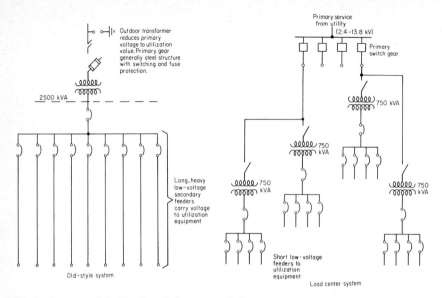

Fig. 2 Comparison of old-style radial system with long, expensive secondary cables, left, with load-center system with short secondary-cable runs, right.

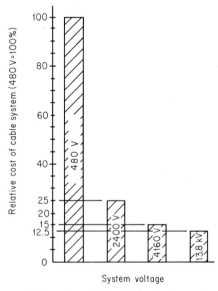

Fig. 3 Comparison of cable costs to distribute 1000 kVA at different voltage levels.

TABLE 1. Standard kVA/Voltage Ratings of Primary-Unit-Substation Transformers

Rated high voltage, V	Rated low voltage, V			
	2400, 2400Y/1385, 2520, 4160Y/2400, 4360Y/2520	4800, 5040, 8320Y/4800, 8720Y/5040	6900, 7200, 7560, 12 470Y/7200, 13 090Y/7560, 13 200Y/7620, 13 800Y/7970	12 000, 12 600, 13 200, 13 800, 14 400
6900, 7200	1000–3750			
12 000, 13 200, 13 800	1000–7500			
22 900	1000–7500	1000–10 000	1000–10 000	
26 400, 34 400	1000–7500	1000–10 000	1000–10 000	1000–10 000
43 800, 67 000	1500–7500	1500–10 000	1500–10 000	1500–10 000
115 000	3750–7500	3750–10 000	3750–10 000	3750–10 000
138 000	3750–7500	3750–10 000	3750–10 000	3750–10 000

All connections delta unless noted.
kVA ratings separated by dash indicate that all the following intervening ratings are included: 1000–1500–2000–2500–3750–5000–7500–10 000.
Askarel is applicable in units rated 34.4 kV high voltage and below.

Primary switchgear for the load-center system can be made to add much greater selectivity to the overall system as compared with the old-style system at a nominal increase in cost. Figure 4 gives total system-cost comparison based on use of two primary feeders for the load-center system with one primary feeder for the old-style system. Total substation capacities were assumed to be 3000 kVA for the load-center system and 2500 kVA for the old-style system serving a single-floor plant with an area of 250 000 ft.²

Better voltage regulation for utilization equipment results from a reduction of voltage drops inherent in long low-voltage cable systems which were prevalent in the old-style systems. This means a more efficient system and reduced costs due to the lower power losses found in a load-center system. Figure 5 shows voltage profiles typical of the old-style and load-center systems.

Load-center systems also offer increased safety to personnel and equipment since all live parts are isolated and completely metal-enclosed.

Because the load-center concept divides the total plant load into a small number of widely dispersed substations, the continuity of service is increased over the old-style system. In addition many different circuit arrangements such as secondary selective and primary selective are available in the load-center system. These arrangements can greatly increase service continuity at a minimum extra cost; these systems will be described later.

The load-center system offers greater flexibility, since small units may be added whenever and wherever they are needed to take care of increased plant load. In the old-style system, it was general practice to put in larger substations than actually required in anticipation of future load growth; here, additional secondary feeders were added up to the capacity of the original substation and beyond that point a larger substation was added, again in anticipation of future load growth, which meant wasted investment if plant expansion did not come about (Fig. 6).

If future plant expansions require greater continuity of service for a particular manufacturing process, a load-center system initially installed as a simple radial system can easily and economically be converted to a secondary-selective or other desired system. This conversion cannot be made with the old-style system.

In the old-style system, addition of transformer capacity would automatically increase the short-circuit currents on all previously installed feeder breakers. It is conceivable that this increased short-circuit capability may exceed the breaker ratings and thus preclude the proposed expansion.

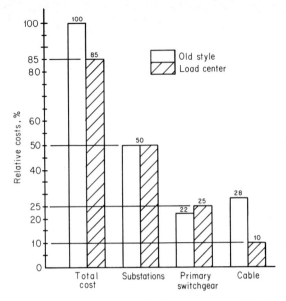

Fig. 4 Cost comparison between old-style and load-center systems of Fig. 2.   Total cost of old-style system assumed to be 100%.

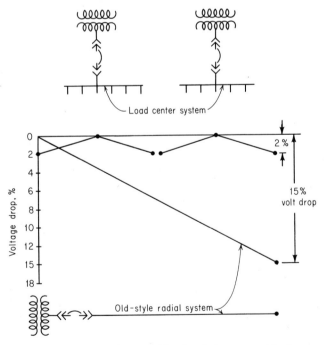

Fig. 5 Comparison between voltage drops in old-style radial system and load-center distribution system.

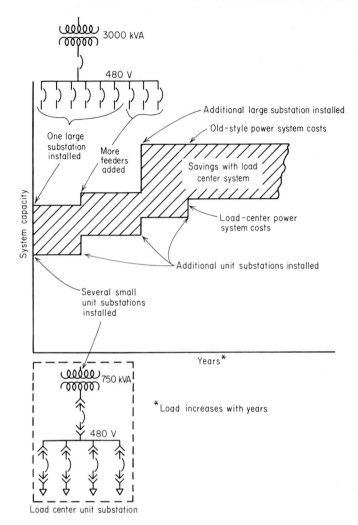

Fig. 6 Savings of load-center distribution system with load increases compared with old-style radial system (Ref. 1).

Load-center substations are of coordinated factory design with basic subassemblies completely factory-assembled, for connection of two or three sections in the field. This arrangement materially reduces and simplifies the engineering, planning, and purchasing work involved by the industrial-plant personnel.   Field handling and installation is simplified and less costly.

Substations can be located practically anywhere in or around the industrial plant— indoors, outdoors, balconies, rooftops, basements, or manufacturing floor area.   They occupy less space than old-style substations, since they are compact, functionally designed units with total flexibility in the arrangements of component sections.

Load-center substations also have far greater salvage value than the old-style substations.   If manufacturing processes change, or if it is desired to change the distribution system or plant layout, load-center substations can be easily and economically relocated to fit the new plans.

We can now summarize all advantages of the load-center system compared with the old-style system as follows:
1. Lower initial cost
2. Lower operating costs
3. Reduced power losses
4. Improved voltage regulation—greater efficiency
5. Increased safety
6. Increased flexibility
7. Simplified engineering, planning, and purchasing
8. Simplified and lower-cost installation
9. Increased reliability and continuity of service
10. Space saving (can be located anywhere), improved appearance
11. Higher salvage value
12. Easier to expand

3. Basic Definitions are as follows:

a. Unit substation is a substation consisting primarily of one or more transformers which are mechanically and electrically connected to and coordinated in design with one or more switchgear or motor-control assemblies or combinations thereof.

b. Articulated unit substation is a unit substation which consists of:

1. An incoming-line section which provides for the connection of one or more incoming high-voltage circuits, each of which may or may not be provided with an interrupting device as a component of the incoming section
2. A transformer section which includes one or more transformers
3. An outgoing section which provides for the connection of one or more outgoing feeders, each of which shall be provided with an interrupting device as a component of the outgoing section

Articulated unit substations are manufactured as subassemblies, intended for connection in the field.

c. Primary unit substation is a unit substation of which the low-voltage section has a voltage rating of 1000 V or above.

d. Secondary unit substation is a unit substation of which the low-voltage section has a voltage rating of less than 1000 V.

PRIMARY UNIT SUBSTATIONS

4. Application of Primary Unit Substations.   Primary unit substations are used predominantly in electric utility systems, municipal pumping stations, and transportation systems but are also widely used as main power-supply substations for industrial plants (Figs. 1 and 7).

Primary unit substations are also used within industrial plants as part of the load-center system where concentrated loads such as large motors (generally 250 hp and above), ovens, and furnaces are more economically served at voltages ranging from 2400 to 13 800 V.

The load-center principle of transmitting power to concentrated load centers at high voltages and then stepping down to utilization voltage applies equally to primary or secondary unit substations.

5. Ratings of Primary Unit Substations.   Table 1 gives the standardized ratings for primary unit substations as listed in NEMA Standards.   Larger size ratings are available.   Actual rating limitations, however, are based on economics and continuous current and interrupting ratings of available secondary switchgear.   Generally, the economics and flexibility of two or more smaller substations located close to centers of concentrated loads (load-center concept) rather than a single larger substation are limiting factors.

Section 17 covers the selection and application of switchgear ratings available today.

6. Components of Primary Unit Substations.   The articulated primary unit substation (Art. 3b) consists of three basic components: incoming-line section, transformer section, and outgoing section.

a. Incoming-Line Section.   The incoming-line section of a primary unit substation is available in several arrangements, one of which is best suited for each particular application and for each primary service voltage; they are as follows:

1. A cable-terminal chamber can be mounted on the side of the transformer tank to enclose the high-voltage wall bushings and provide for termination of primary cables generally entering from below. Air-filled chambers with or without potheads are normally used for voltages up to 15 kV while oil-filled chambers with potheads are available for voltages up to and including 69 kV.

2. Oil-filled switches combined with the above terminal chambers are available to provide a means for disconnecting the transformer without interrupting service to other loads served from the same primary line. These switches are only able to interrupt transformer magnetizing current and should therefore be interlocked with the secondary breakers to prevent operation under load. Oil switches are available up to and including 69 kV.

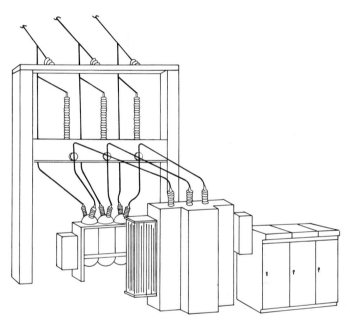

Fig. 7 Typical primary unit substation serving industrial plant.

3. Air-filled load-interrupter switches are available for primary voltages up to and including 34.5 kV and can be combined with current-limiting or power fuses to provide short-circuit protection as well as a disconnect means.

4. Metal-clad switchgear incoming-line arrangement using drawout circuit breakers provides disconnect means and short-circuit protection. This arrangement is the ultimate in primary protection, and while it costs more, it also provides for more flexibility in system design, operation, and maintenance. Metal-clad switchgear is available for primary voltages of up to 34.5 kV.

5. The terminal structure available in steel or aluminum design and for all primary voltages has high-voltage cover bushings on the transformer with disconnect and short-circuit-protection devices being mounted on the structure.

Figure 8 compares the cost of these five incoming-line arrangements.

b. Transformer Section. The transformer section of primary unit substations generally utilizes oil as the insulating medium, since these substations are usually installed in outdoor locations. Transformers are available, however, with askarel liquid (nonflammable) for indoor installations, but the use of askarel is limited to a maximum high-voltage rating of 34.4 kV. Dry-type transformers are also available, but their use is extremely small.

Standard ratings are listed in Table 1, and Table 2 gives other standard characteristics of these units.

The bushings utilized on these transformers are selected to have an insulation level equal to or greater than the insulation levels of the windings to which they are connected.

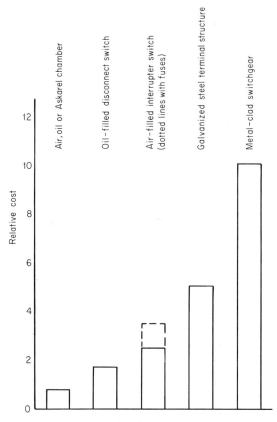

Fig. 8 Cost comparison for various incoming-line arrangements in primary unit substation for 15-kV outdoor service.

Primary-unit-substation transformers are available with automatic load-tap-changing (LTC) equipment where voltage-regulation requirements necessitate. The standard LTC equipment is provided to regulate voltage in a plus or minus 10% range from nominal rating.

c. Outgoing Section. The outgoing section of primary unit substations is available in the following options: metal-clad switchgear, fused or unfused air interrupter switches and motor-control assemblies.

Metal-clad switchgear (Secs. 1 and 17) is designed to meet the requirements of ANSI Standards C37.20 and C37.03 through C37.11 as applicable.

Air interrupter switches are designed to meet the requirements of ANSI Standards

TABLE 2. Standard Design Characteristics of Primary-Unit-Substation Transformers

Insulation Class, Basic Impulse Level and Tap Voltages (High-Voltage Only)

Rated voltage, V	Insulation class, kV	BIL, kV	Standard high-voltage taps, V at rated kVA
2400, 2520	2.5	45	
4160Y, 4360Y, 4800,			
5040	5.0	60	
6900	8.7	70	7245/7070/6730/6555
7200	8.7	70	7560/7380/7020/6840
7560, 8320Y, 8720Y	8.7	70	
12 000	15.0	95	12 600/12 300/11 700/11 400
13 200	15.0	95	13 860/13 530/12 870/12 540
13 800	15.0	95	14 400/14 100/13 500/13 200
12 600, 14 400, 12 470Y,			
13 090Y, 13 200Y, 13 800Y	15.0	95	
22 900	25.0	150	24 100/23 500/22 300/21 700
26 400	34.5	200	27 800/27 100/25 700/25 000
34 400	34.5	200	36 200/35 300/33 500/32 600
43 800	46.0	250	46 200/45 000/42 600/41 400
67 000	69.0	350	70 600/68 800/65 200/63 400
115 000	92.0	450	120 750/117 875/112 125/109 250
138 000	115.0	550	144 900/141 450/134 550/131 100

Self-cooled/Forced-cooled Rating and Impedance Voltage

Self-cooled kVA	Forced-cooled kVA	% Impedance voltage at rated high voltage					
		6900–22 900 V	26 400–34 400 V	43 800 V	67 000 V	115 000 V	138 000 V
1 000	1 150	5.5	6.0	6.5	7.0	7.5	8.0
1 500	1 725	5.5	6.0	6.5	7.0	7.5	8.0
2 000	2 300	5.5	6.0	6.5	7.0	7.5	8.0
2 500	3 125	5.5	6.0	6.5	7.0	7.5	8.0
3 750	4 687	5.5	6.0	6.5	7.0	7.5	8.0
5 000	6 250	5.5	6.0	6.5	7.0	7.5	8.0
7 500	9 375	5.5	6.0	6.5	7.0	7.5	8.0
10 000	12 500	5.5	6.0	6.5	7.0	7.5	8.0

Above impedance without LTC; with LTC, add 0.5%.

TABLE 3. Maximum Ratings of Switches and Fuses for 5- and 15-kV Service

System voltage	Continuous and load-break rating, A	Current-limiting fuse		Expulsion-type power fuse	
		Continuous A	Interrupting MVA	Continuous A	Interrupting MVA
2 400	1200/1200	450	210	1200	270
4 160	1200/1200	450	360	1200	270
4 800	1200/1200	200	310	1200	270
7 200	1200/1200	200	620	1200	600
12 000	1200/1200	200	650	1200	600
13 200	1200/1200	200	715	1200	600
13 800	1200/1200	200	750	1200	600

Manufacturer's catalog data should be consulted for complete listing of available fuse ratings.

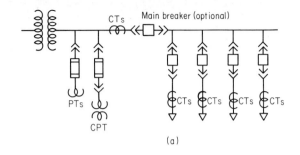

(a)

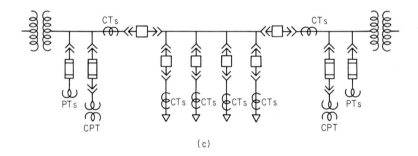

(b)

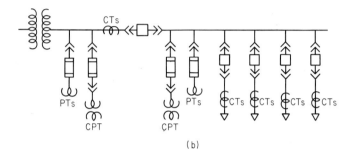

(c)

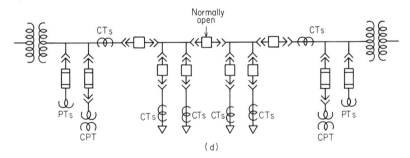

(d)

Fig. 9 (a) Radial-type primary unit substation. (b) Typical primary (distributed)-network-type primary unit substation. (c) Spot-network-type primary unit substation. (d) Low-voltage (secondary) selective primary unit substation.

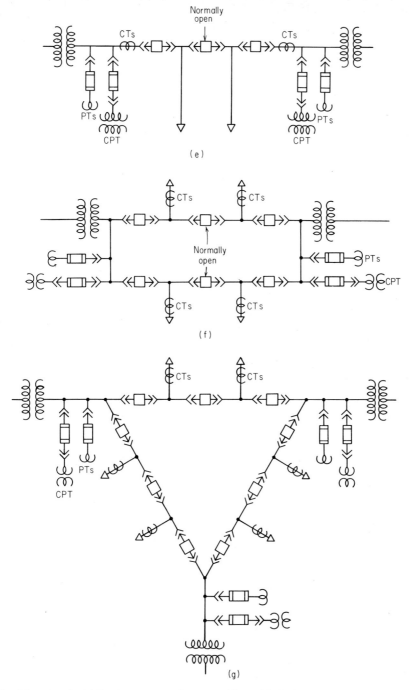

Fig. 9 (*continued*)   (*e*) Duplex primary substation.   (*f*) Double-duplex primary unit substation. (*g*) Multiplex primary unit substation.

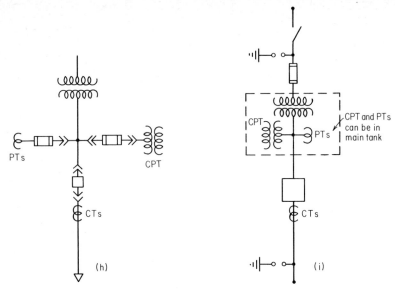

Fig. 9 (*continued*) (*h*) Integral (single-circuit) primary unit substation. (*i*) Mobile substation.

C37.20 1965-20-6.4 and C37.30 as applicable. Fuses are designed to meet ANSI Standard C37.46. Table 3 lists available switch and fuse ratings.

Motor-control assemblies where utilized are designed to meet the requirements of NEMA Standard IC-1 and are limited in use to secondary voltages of up to 5-kV class.

The choice of option for the outgoing section should be based not only on economics but also on the application involved. Where transient fault protection, overload protection, and ground-fault protection are required, the choice must be metal-clad switchgear with power circuit breakers and protective relays. However, if the application is basically an indoor type with incoming and outgoing feeders underground, the normal requirement will be for interruption of a fault which is permanent in nature—this protection can be handled adequately by a fused interrupter switch which is less expensive than metal-clad switchgear. All aspects of the particular application must be fully investigated before a final choice is made.

7. Types of Primary Unit Substations. There are seven types of primary unit substations:

a. Radial. A radial unit substation is one which has a single transformer section and an outgoing section for connection of one or more outgoing feeders (Fig. 9a). It may or may not be provided with a main secondary interrupting device (Art. 17).

b. Primary-network substation, also known as distributed network (Fig. 9b), consists of a single transformer section whose outgoing side is connected to a main bus through a power circuit breaker equipped with relays arranged to trip the circuit breaker on reverse power flow, and to reclose the breaker when voltage of proper value, phase angle, and phase sequence is restored at the secondary terminals of the transformer. The main bus is provided with two or more multiple feeders and is tied through these feeders to other distributed network substations.

c. Spot-network substation (Fig. 9c) is equipped with two transformer sections, each of which is connected on the secondary side to a common main bus through a main secondary circuit breaker. Each breaker is provided with relays to trip the circuit breaker on reverse power flow and to reclose the breaker when voltage of proper value, phase angle, and phase sequence is restored at the secondary terminals of the transformer. The common main bus is provided with two or more outgoing feeders.

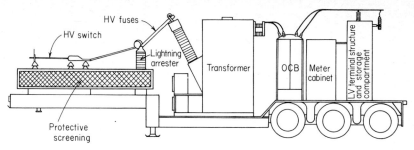

Fig. 10 Highway-type mobile substation. Primary protection provided by fuses while secondary protection is provided by oil circuit breakers with meter/relay cabinet.

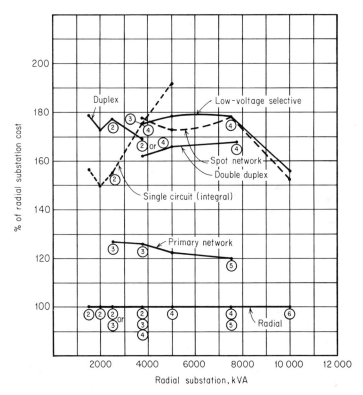

Fig. 11 Comparative costs of radial, primary-network, spot-network, low-voltage-selective, duplex, double-duplex, and integral types of unit substations based on 34.5-kV subtransmission and 4.16-kV feeders. Circled numbers indicate feeder positions of Table 6.

   d. *Low-voltage-selective* or secondary-selective substation (Fig. 9d) consists of two transformer sections each of which is connected on the secondary side to a separate bus through a suitable switching and protective device. The two separate buses are connected through a normally open switching and protective device. Each separate bus contains one or more outgoing feeders.

   e. *Duplex* substation (Fig. 9e) has two transformer sections whose outgoing sides are connected to an outgoing feeder. The two feeders are connected on the feeder side through a normally open switching and protective device.

   The duplex-substation principle can be expanded to serve four outgoing feeders with an arrangement referred to as the double-duplex substation (Fig. 9f) or to serve six or more feeders by what is referred to as a multiplex arrangement (Fig. 9g).

   The multiplex station contains an even number of feeders greater than four, and is fed by half as many transformers as feeders served.

   f. *Integral* substation, also known as single-circuit type (Fig. 9h) consists of a single transformer section whose outgoing side is connected to a single radial feeder. It differs from other primary substations in that it is constructed in one piece rather than individual pieces for incoming transformer and outgoing sections.

   g. *Mobile* substation (Fig. 9i) consists of the same basic elements as other primary substations but has these components mounted on some form of readily transportable device which is usually a rubber-tired highway trailer. These substations require special design equipment owing to the weight and size restrictions imposed by state highway authorities. Figure 10 illustrates a typical mobile substation.

   Tables 4 and 5 show operation and application comparisons of the various primary unit substations described above, and Fig. 11 shows comparative costs based on equivalent transformer ratings and feeder positions of Table 6.

TABLE 4. Substation-System-Operation Comparison Table (Ref. 3)

Type of substation	Subtransmission line outage	Transformer fault	Breaker bypass facilities	Voltage regulation
	Radial Subtransmission System			
Radial	Outage until line restored	Out of service until arrival of mobile unit or field cutover	Mobile unit or field cutover	Bus, single-level *
Primary network.................	No outage	No outage	None required	Bus, single-level *
	Loop Subtransmission System			
Radial	Outage until primary switched to good line	Out of service until arrival of mobile unit or field cutover	Mobile unit or field cutover	Bus, single-level *
Duplex..............................	90-cycles outage on half of feeders	90-cycles outage on half of feeders	Power circuit breaker	Individual feeder (grou feeder for double-du
Spot network	No outage	No outage	Mobile unit or field cutover	Bus, single-level°
Low-voltage-selective	90-cycles outage on half of feeders	90-cycles outage on half of feeders	Mobile unit or field cutover	Bus, group feeder or dual-level°
	Multiple Subtransmission System			
Radial	Outage until primary switched to good line	Out of service until arrival of mobile unit or field cutover	Mobile unit or field cutover	Bus, single-level *
Duplex..............................	90-cycles outage on half of feeders	90-cycles outage on half of feeders	Power circuit breaker	Individual feeder (gro feeder for double-d
Spot network	No outage	No outage	Mobile unit or field cutover	Bus, single-level°
Low-voltage-selective	90-cycles outage on half of feeders	90-cycles outage on half of feeders	Mobile unit or field cutover	Bus, group feeder or dual-level°

* If required, supplementary feeder regulators can be used or regulator-reactor-transfer bus can be included at the sub
to give individual feeder regulation.

TABLE 5. Substation-Application Comparison (Ref. 3)

Substation type	Load-area characteristics		Particular field of application
	Load densities	Type of area	
Radial with radial subtransmission	Light to medium	Rural, suburban, residential, small industrial	Areas where economy is important, and service continuity is not critical
Single circuit integral	Light to medium	Rural, suburban, residential, small industrial	Areas where radial service is satisfactory and economics favor small unit size. Also for temporary or seasonal loads, industrial loads, and new residential developments
Radial with multiple or loop subtransmission	Light, medium, heavy	Suburban, residential, small industrial	Similar to simple radial except that higher degree of service continuity is required
Primary network	Medium to heavy with fairly uniform load density	Suburban, residential, commercial, small industrial	Areas where high degree of service continuity at moderate cost is required. Limited somewhat in application by load-area geography
Duplex	Medium	High-class residential, commercial, small industrial	Areas where very high degree of service continuity is required, where individual feeder regulation is desirable, and/or where small (2-feeder) substation size is economical
Double duplex	Medium	High-class residential, commercial, small industrial	Similar to duplex except larger substation size and group instead of individual feeder regulation
Spot network	Heavy	Commercial, urban downtown, industrial	Areas where service continuity is very important and where there is a heavy concentration of load in a small area
Low-voltage-selective	Medium to heavy	High-class, residential, urban commercial, industrial	Similar to double-duplex and radial with multiple or loop subtransmission but in larger sizes

SECONDARY UNIT SUBSTATIONS

8. Application of Secondary Unit Substations. Secondary unit substations form the heart of all industrial-plant electrical distribution systems. They are used to step down the primary voltage to the utilization voltage at various load centers throughout the plant.

Many factors must be considered when selecting and locating substations. Most important of these are load groupings by kVA, voltage rating, service facilities, safety, ambient conditions, continuity of service, aesthetic considerations, lighting requirements, space available, outdoor vs. indoor location, and plans for future expansion.

In general, substations and equipment operating on system voltages over 15 kV are located outdoors. This limitation is mainly due to building costs and to a certain extent to personnel safety considerations. At voltages 15 kV and below, location can be indoors, outdoors, or various indoor/outdoor combinations depending upon space

TABLE 6. Summary of Equivalent Transformer Ratings and Number of Feeder Breaker Positions for Substations Rated 34.5 to 4.16 kV (Ref. 3)

Simple radial		Radial with loop or multiple subtransmission		Primary network		Duplex	
Transformer kVA rating	Feeder position	Transformer kVA rating	Feeder position	Transformer kVA rating	Feeder position	Transformer kVA rating	Feeder position
1 500	2	1 500	2	....	....	2-transformer 1000/1150	2
2 000	2	2 000	2	....	....	2-transformer 1500	2
2 500	2,3	2 500	2	2500/3125	3	2-transformer 2000	2
3 750	2,3,4	3 750	3	3750/4687	3	2-transformer 2500/3125	2
5 000	4	5 000	4	5000/6250	4	....	....
7 500	4,5,6	7 500	5	7500/9375	5	....	....
10 000	4,6	10 000	6	....	....	....	....
15 000	8	15 000	8	....	....	....	....

availability and atmospheric conditions. Highly corrosive atmospheres generally dictate indoor locations.

Indoor equipment is less expensive and usually easier to maintain than comparably rated outdoor equipment, but unless it can be located in portions of the plant used to house other equipment, the additional building costs may outweigh these advantages.

It is important that the plant engineer consider all facets of the total system and plant-process requirements in selecting substation designs to be utilized. Additional time spent in the planning stage will preclude costly alternatives at a future date.

9. Ratings of Secondary Unit Substations. Standardized ratings for secondary unit substations are listed in NEMA Standards (Table 7). This is not to say that other sizes are not available; actual limitations are based on economics in addition to continuous-current and interrupting ratings of available secondary switchgear or control equipment.

In general, these ratings will meet nearly 100% of all load-center system requirements when all factors of design and economics are considered.

10. Components of Secondary Unit Substations. An articulated secondary unit substation (Art. 3) consists of three basic components: incoming-line section, transformer section, and outgoing section.

a. Incoming-line section of a secondary unit substation is available in several arrangements, one of which is best suited for each particular application. These arrangements are the same as those covered in Art. 6a for primary unit substations plus the addition of fused or unfused oil cutouts.

Oil cutouts are generally mounted in an air-filled terminal chamber with provisions for a terminating 3-phase incoming feeder. They are limited in use to a maximum transformer rating of 300 kVA at 2400 V and 500 kVA at 4160 V with fuses, or 750 kVA at 2400 V and 1000 kVA at 4160 V as unfused disconnects. Oil cutouts are available for indoor or outdoor use.

It should be noted that the incoming-line section of secondary unit substations will seldom be rated at voltages above 15 kV because these substations are generally located within the plant distribution system and are served by the plant primary voltage, which is usually 15 kV or below. In smaller plants, where load requirements and feeder lengths will economically allow a main distribution voltage of 480 V or below,

Double duplex		Spot network		Low-voltage-selective		Single circuit	
Transformer kVA rating	Feeder position	Transformer kVA rating	Feeder position	Transformer kVA rating	Feeder position	Transformer kVA rating	Feeder position
....	....	....	....	....	....	2-transformer 750	2
....	....	....	....	....	....	2-transformer 1000	2
....	....	....	....	....	....	2-transformer 1500	2
2-transformer 2500/3125	4	2-transformer 2500/3125	4	2-transformer 2500/3125	4	3-transformer 1500	3
2-transformer 3750	4	2-transformer 3750	4	2-transformer 3750	4	4-transformer 1500	4
2-transformer 5000/6250	4	2-transformer 5000/6250	4,6	2-transformer 5000/6250	4,6		
....	....	2-transformer 7500	6	2-transformer 7500	6		
....	....	2-transformer 10 000/12 500	8	2-transformer 10 000/12 500	8		

it is possible that the incoming voltage to the secondary unit substation will be in the 25-, 34.5-, or 69-kV class.

b. Transformer section for secondary unit substations is available in four basic designs:
1. Oil-immersed for outdoor use (or indoor vault)
2. Askarel (nonflammable)-immersed for indoor or outdoor use
3. Open ventilated dry for indoor use
4. Sealed dry (gas-filled) for indoor or outdoor use

Standard ratings are listed in Table 7, and Table 8 gives other standard characteristics of these units.

TABLE 7. Standard kVA/Voltage Ratings of Secondary-Unit-Substation Transformers

	Rated low voltage, V					
	Liquid-filled		Ventilated dry type		Sealed (gas-filled) dry type	
Rated high voltage, V *	208Y/120 240	480Y/277 480	208Y/120 240	480Y/120 480	208Y/120 240	480Y/120 480
2400	112.5–1000	112.5–1500	112.5–1000	112.5–1500	750–1000	750–1500
4160, 4800	112.5–1000	112.5–1500	112.5–1000	112.5–1500	750–1000	750–1500
6900, 7200	112.5–1000	112.5–2500	300–1000	300–2500	750–1000	750–2500
12 000, 12 470, 13 200, 13 800	112.5–1000	112.5–2500	300–1000	300–2500	750–1000	750–2500
22 900	500–1000	500–2500				
34 400	750–1000	750–2500				

kVA ratings separated by dash indicate that all the following ratings are included: 112.5–150–225–300–500–750–1000–1500–2000–2500.

* All connections delta with 2 to 2½% taps above and below rated.

TABLE 8. Standard Design Characteristics of Secondary-Unit-Substation Transformers

Insulation Class and Dielectric-Test Levels

Rated voltage, V	Insulation class, kV	Liquid-immersed		Ventilated dry type		Sealed (gas-filled) dry type	
		Low-frequency test, kV	BIL, kV	Low-frequency test, kV	BIL, kV	Low-frequency test, kV	BIL, kV
208Y/120, 240, 480Y/277, 480	1.2	10	30	4	10	10	30
2400	2.5	15	45	10	20	15	45
4160, 4800	5.0	19	60	12	25	19	60
6900, 7200	8.7	26	75	19	35	26	75
12 000, 12 470, 13 200, 13 800	15.0	34	95	31	50	34	95
22 900	25.0						
34 400	34.5						

Self-cooled/Forced-cooled Rating and Impedance Voltage

Rated high voltage, kV	Self-cooled kVA	Forced-cooled kVA			% impedance voltage
		Liquid-filled	Ventilated dry type	Sealed dry type	
All voltages	112.5	NA	NA	NA	Not less than
2400 through 13 800	150	NA	NA	NA	2.0
	225	NA	NA	NA	
	300	NA	NA	NA	Not less than
	500	NA	NA	NA	4.5
	750	862	1000	NA	
	1000	1150	1333	NA	5.75
	1500	1725	2000	NA	
	2000	2300	2667	NA	
	2500	3125	3333	NA	7.5

As in the case of primary unit substations, the bushings utilized for secondary-unit-substation transformers are selected to have an insulation level equal to or greater than the insulation levels of the windings to which they are connected.

c. Outgoing section of secondary unit substations is available in the following options:

1. Metal-enclosed switchgear with stationary or drawout low-voltage power circuit breakers.

2. Metal-enclosed switchgear or switchboard construction with molded-case circuit breakers.

3. Metal-enclosed switchboard construction with switch and fuse combinations.

4. Motor-control-center assemblies.

5. Combinations of the above devices.

Secondary sections using molded-case circuit breakers or fused switches are designed to meet the applicable portions of Underwriters Laboratories, Inc., Standards 891, 489, 98, 198, and 512, as well as NEMA Standard AB1-1964 (Fig. 12).

Motor-control centers utilized as secondary equipment are designed to meet the applicable portions of NEMA Standard IC1-1965 (Fig. 13).

The selection of the best secondary equipment for each particular application must take into account all functional requirements, economic factors, and safety considerations, since at first glance, low-voltage power circuit breakers, fused switches, and molded-case breakers may appear to do the same job; each, however, has its own area of application.

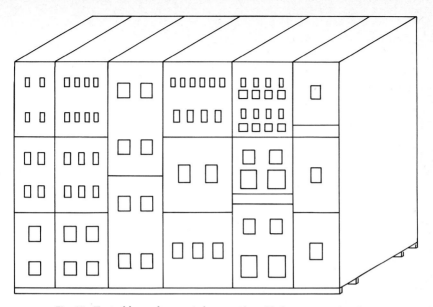

Fig. 12 Typical low-voltage switchgear with molded-case circuit breakers.

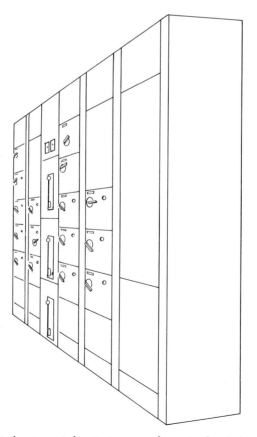

Fig. 13 Typical motor-control-center-type secondary-unit-substation outgoing section.

From a purely economic standpoint, fused switches generally present a lower initial cost followed by molded-case circuit breakers; power circuit breakers present the highest initial cost. To evaluate fully a selection based on economics, replacement costs of fuses should be considered. This will include the cost of stocking a sufficient quantity of fuses for emergencies as well as the man-hour cost of securing and replacing blown fuses. For circuit breakers, one must evaluate periodic maintenance costs.

From a safety standpoint, all three devices are completely safe if properly applied and maintained throughout their life. All can be hazardous if maintenance is neglected or operating personnel become careless. Circuit breakers will deteriorate in time if not properly maintained, while fuses are basically fail-safe devices. Fused switches can pose safety problems if a larger-size fuse or solid connection is substituted after a fuse blows. They are also more hazardous than circuit breakers during replacement of fuses, unless drawout fused switches are used. There is also the possibility of an operator's opening a switch under overload conditions before a fuse can blow, but at a time when the current flowing in the circuit is higher than the switch can safely interrupt.

From an application standpoint, fused switches are completely adequate for switching and protection of circuits subject to infrequent faults and switching duty as long as they are applied within their voltage and current ratings and interrupting ability. For circuits subjected to frequent faults and switching or where downtime is expensive and sustained loss of power results in loss of material in process, circuit breakers should be used, since they can be reclosed quickly.

Current-limiting fuses are available with interrupting ratings up to 200 000 A, which gives them a distinct advantage in applications which present high short-circuit currents (Table 9). They also have the advantage of high speed of operation and current-limiting action which affords definite protection against the damage to or possible destruction of the equipment it is protecting. These fuses are combined with switches or circuit breakers in many applications. In these arrangements the switch or breaker must be capable of withstanding the peak let-through current of the fuse. When combined with circuit breakers, the devices are coordinated so that the circuit breaker provides overload protection and short-circuit protection within its capability, while the fuse provides short-circuit protection over and above the circuit-breaker rating (Fig. 14).

The following summary comparison of fuses vs. circuit breakers is taken from IEEE 141, "Electric Power Distribution for Industrial Plants," August 1969:

Advantages of fuses are:
1. Fuse-interrupting ratings are greater than those of available circuit breakers.
2. Less expensive.
3. Simple mechanically.
4. Less space requirement.

TABLE 9. Typical 600-V Switch and Fuse Ratings

Device	Continuous-current rating, A	Fuse data*	
		UL class	Interrupting rating, A rms
Disconnect (safety) switch	30–1200		
Bolted pressure switch	1000–4000		
One-time fuse	1–600	K-5	50 000
Renewable fuse	1–600	H	10 000
Dual-element fuse	1/10–600	K-5	100 000
Current-limiting fuse.................	1–600	J	200 000
Current-limiting fuse.................	400–6000	L	200 000

* Fuse data based on Shawmut fuses (The Chase-Shawmut Co.).

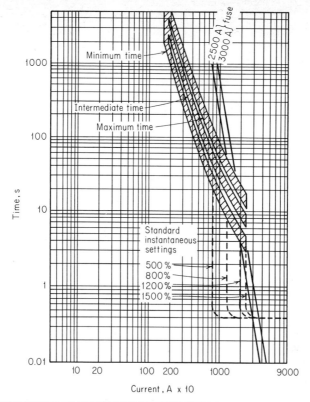

Fig. 14 Typical 600-V, fused electrically operated circuit breaker rated 200 000-A symmetrical interrupting. Curves show coordination between breaker trip setting and current-limiting fuse for 1600-A frame breaker.

5. Current-limiting fuses will limit damage to protected equipment by the let-through current of short circuits.

6. Fuses are easier to coordinate, since their melting and clearing characteristics are more consistent.

7. Fuses are fail-safe devices, while deterioration of a breaker may cause unsafe operation.

Disadvantages of fuses:

1. Fuse time-current blowing characteristics are less accurate than relay-controlled circuit-breaker tripping characteristics, with the result that coordinated circuit protection is less reliable. In some cases it is impossible to provide adequate overcurrent protection to specific equipment such as transformers, regulators, and motors.

2. Failure of one fuse in a 3-phase circuit may result in single-phase operation of ac motors with consequent possible damage to the motor, as well as to the process operation driven by the motor.

3. Fuses are not capable of remote operation.

4. Fuses are more hazardous to personnel than circuit breakers during replacement of fuse elements, unless drawout fused switches are used.

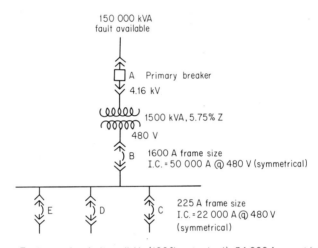

Total secondary fault available (100% motor load) = 34 000 A symmetrical

Fig. 15   Typical distribution system in which low-voltage circuit breakers are applied in cascade.

5. Fuses must be replaced after blowing, with the possibility that an incorrectly rated fuse element may be reinstalled.

6. In a polyphase circuit, the opening of a fuse, in response to a fault condition, may severely reduce the magnitude of current continuing to flow to the fault and make impossible the operation of remaining fuses in the circuit, resulting in an inability to interrupt current flow to the fault.

7. It is necessary to stock replacement fuses.

Section 17 covers application of power circuit breakers together with details of selective tripping applications, fully rated applications, and cascade applications, but it is considered important here to review the cascade application as it applies to power and molded-case circuit breakers.

The cascade system offers the advantage of lower cost but presents the disadvantage of possibly shutting down a much greater portion of the plant than the faulted feeder. Because the cascade breaker (breaker C in Fig. 15) is applied at a point where the available fault current exceeds its rating, it costs less. It must, however, rely on breaker B to interrupt faults exceeding its rating, and all circuits fed by breaker B are shut down.

Frame sizes, A	Rated continuous current, A	1-pole ac				2- and 3-pole, ac		2- and 3-pole, ac — 600-V ac rated circuit breakers						1-, 2-, and 3-pole, dc	
		120 V 120/240 V		277 V		240 V 120/240 V		240 V		480 V		600 V		125 V 125/250 V	250 V
		Sym-metrical	Asym-metrical	Sym-metrical	Asym-metrical	Sym-metrical	Asym-metrical	Sym-metrical	Asym-metrical	Sym-metrical	Asym-metrical	Sym-metrical	Asym-metrical		
100	0–100	5 000	5 000	......	......	5 000	5 000	......	......	......	......	......	......	5000	......
100	0–100	7500	7500	10 000	10 000	7 500	7 500	......	......	......	......	......	......	5000	10 000
100	0–100	......	......	......	......	......	......	18 000	20 000	14 000	15 000	14 000	15 000	......	10 000
100	0–100	......	......	......	......	......	......	65 000	75 000	25 000	30 000	18 000	20 000	......	100 000
200	125–220	......	......	......	......	10 000	10 000	100 000	......	100 000	......	100 000	......	......	100 000
225	125–225	......	......	......	......	10 000	10 000	......	......	......	......	......	......	......	......
225	70–225	......	......	......	......	......	......	22 000	25 000	18 000	20 000	14 000	15 000	......	10 000
225	70–225	......	......	......	......	......	......	25 000	30 000	22 000	25 000	22 000	25 000	......	10 000
225	70–225	......	......	......	......	......	......	65 000	75 000	35 000	40 000	25 000	30 000	......	20 000
225	70–225	......	......	......	......	......	......	100 000	......	100 000	......	100 000	......	......	100 000
400	200–400	......	......	......	......	......	......	35 000	40 000	25 000	30 000	22 000	25 000	......	20 000
400	200–400	......	......	......	......	......	......	65 000	75 000	35 000	40 000	25 000	30 000	......	20 000
400	200–400	......	......	......	......	......	......	100 000	......	100 000	......	100 000	......	......	100 000
600	300–600	......	......	......	......	......	......	42 000	50 000	30 000	35 000	22 000	25 000	......	20 000
600	300–600	......	......	......	......	......	......	100 000	......	100 000	......	100 000	25 000	......	100 000
800	300–800	......	......	......	......	......	......	42 000	50 000	30 000	35 000	22 000	25 000	......	20 000
800	300–800	......	......	......	......	......	......	65 000	75 000	35 000	40 000	25 000	30 000	......	20 000
800	600–800	......	......	......	......	......	......	100 000	......	100 000	......	100 000	......	......	100 000
1000	600–1000	......	......	......	......	......	......	42 000	50 000	30 000	35 000	22 000	25 000	......	20 000
1200	700–1200	......	......	......	......	......	......	42 000	50 000	30 000	35 000	22 000	25 000	......	20 000

For further information see NEMA Standard AB-1-1964.

An individual breaker will not necessarily carry all the voltage ratings listed for its frame size, nor will a given frame size necessarily have a full complement of rated continuous currents.

Molded-case breakers are generally equipped with instantaneous overcurrent trips, and this should be considered in their use in selective systems.

Breakers listed above with interrupting ratings of 100 000 A (symmetrical) are integrally fused breakers. Unfused breakers may be used above their interrupting rating when properly coordinated with current-limiting fuses, according to the recommendations of the manufacturer.

Rules governing cascading of power circuit breakers are:

1. Number of cascade steps shall not exceed two.

2. Breakers used in cascade must be recommended for such application by the manufacturer.

3. Breakers in cascade must have an interrupting rating of at least one-half the short-circuit current available at the point of application.

4. If breaker C in Fig. 15 is cascaded, breaker B must be fully rated.

5. It is recommended that all breakers be inspected after a short-circuit interruption and particularly a cascaded breaker.

6. Cascaded breakers should be electrically operated from a remote location to protect operating personnel.

7. All breakers in a cascaded system should have instantaneous trip with instantaneous trip of breaker B in Fig. 15 set at 80% of interrupting rating of cascaded breaker C to assure proper backup protection.

8. Molded-case breakers *should not* be cascaded under any circumstances but must be applied only where available short-circuit current at point of use is within the breaker's interrupting rating (Table 10).

Motor-control-center application is discussed in Sec. 23. The starters consist of a switching device (contactor) and an overload-protective device which is generally a series overcurrent relay. Generally, the starter is combined with a molded-case circuit breaker or fused switch which provide for short-circuit protection and are used as a disconnecting means. These combination starters in motor-control-center construction (Fig. 13) are used for control and protection of plant low-voltage motors and are preferred over use of power circuit breakers, since contactors are designed for the repetitive-duty requirements of this application. Electrically operated power circuit breakers are sometimes used for motor-starting duty, but their use is generally restricted to applications where motor horsepower ratings are large and the starting and stopping is infrequent.

11. Types of secondary unit substations include radial, secondary-selective, distributed-network, and spot-network.

a. Radial substation as depicted in Fig. 16a has a single step-down transformer and an outgoing section which provides for the connection of one or more outgoing feeders. It may or may not be equipped with a secondary main breaker.

b. Secondary-selective substation (Fig. 16b) has two step-down transformers, each connected to an incoming high-voltage line. The outgoing side of each transformer is connected to a separate bus through a suitable switching and protective device. Each bus provides for the connection of one or more outgoing feeders, and the buses are connected by a normally open switching device.

c. Distributed-network substation of Fig. 16c has a step-down transformer with the outgoing side connected to a main bus through a directional protective device which trips on reverse power flow to the transformer and recloses upon restoration of voltage of correct value, phase angle, and phase sequence. The main bus provides for one or more outgoing feeders, as well as one or more tie connections to similar unit substations.

d. Spot-network substation (Fig. 16d) has two step-down transformers each connected to an incoming high-voltage line. The outgoing side of each transformer is connected to a common main bus through a directional protective device which trips on reverse power flow to the transformer and recloses upon restoration of voltage of correct value, phase angle, and phase sequence. The main bus provides for one or more outgoing feeders.

12. Selection of Substation Ratings (Ref. 2). Assuming that the system circuit arrangement and substation voltage ratings have been decided, the selection of the kVA rating of the substation is largely a matter of economics.

Three major system components affect the overall system cost. These are primary cable, substations, and secondary cable. These factors work contrary to one another; so the most economical system as affected by substation kVA rating can be obtained only by looking at all three at once.

As the number of substations increases in a given area, the length of primary feeder cable required to serve these substations increases. Conversely, as the number of

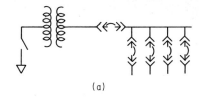

(a)

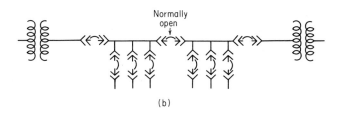

(b)

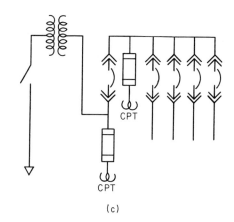

(c)

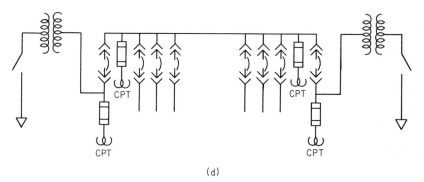

(d)

Fig. 16 (*a*) Typical radial secondary unit substation. (*b*) Typical secondary-selective secondary substation. (*c*) Typical distributed-network secondary substation. (*d*) Typical spot-network secondary unit substation.

substations in a given area increases, the amount of secondary feeder cable required decreases. The substation cost per kVA varies depending upon the kVA rating of the substation (Fig. 17).

The curves indicate that there is a very definite minimum system cost as a function of substation size for different voltage levels. There are also other factors which have an influence on substation rating. The most important of these are:

1. Higher primary voltages may require substations with larger kVA ratings so that a greater kVA per primary feeder can be handled without unduly complicating the substation overcurrent-protection problem. For example, when the primary voltage is 13.8 kV, it is desirable to have a loading of 4000 to 7500 kVA per primary feeder.

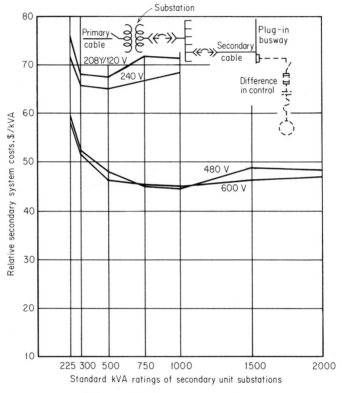

Fig. 17 Approximate installed cost of 208-, 240-, 480-, and 600-V radial secondary systems (Ref. 2).

With total primary-feeder loadings of this magnitude and using standard 500-kVA substations with typical diversity factors, individual primary overcurrent protection would be required (National Electrical Code, 1975 Edition, Article 450-3). If 750-, 1000-, or 1500-kVA substations were selected, primary feeder loadings of this same order of magnitude could be obtained without individual overcurrent protection. This point is of particular importance in systems with primary voltages of 15 kV and above.

2. Large spot loads can sometimes justify larger unit substations. For example, a single furnace, a single large oven, or a single large welder may justify substations of the order of 2000 or 2500 kVA at 480 V secondary since there is no secondary distribution. With a very few secondary breakers, there is therefore nothing to be saved

in that part of the system to offset the lower cost per kVA of the larger substation. Where these larger spot loads are encountered, it is well to consider using two smaller substations rather than one larger substation, particularly if the number of feeder breakers for the secondary exceeds three or four. For general factory areas, two 1000-kVA unit substations will nearly always provide a lower overall system cost than one 2000-kVA substation. Many times a system using two 750-kVA substations will be less expensive overall than a system using one 1500-kVA substation per load area.

3. Space available for unit substations sometimes dictates larger substation kVA ratings. For example, one 1500-kVA substation instead of two 750-kVA substations may be necessary because the space can be found for the one larger unit, whereas it could not be found for the two smaller units. Conversely, other areas may allow space for two smaller units, but not for the larger one. Other local factors may affect the kVA size of substations in a particular plant. In general, however, the points listed on the preceding pages are the major factors involved in the selection of substation sizes for most industrial plants.

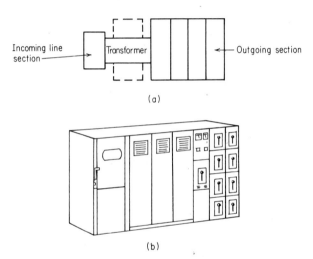

(a)

(b)

Fig. 18 (a) Plan view of typical single-ended secondary unit substation with all components in line, indoor or outdoor installation. (b) Typical single-ended load-center substation with primary load-break switch.

13. Physical Arrangements and Location. Many physical arrangements are available for secondary unit substations. This flexibility enables the plant engineer to locate these substations wherever economical space is available near the various plant load centers.

Three basic locations can be considered in the plant layout design. These locations often affect system costs as well as efficiency of distribution. They are:

a. Building Roof. This location should be considered where primary distribution is overhead. It also provides a location away from high ambient temperatures or contaminated atmospheres found in some industrial plants. It is also extremely desirable for use where plant load areas change frequently. Since it is a remote location, it usually dictates the need for more expensive electrically operated feeder breakers.

b. Balconies. Many plants can be designed with expanded metal or sheet-steel flooring on steel support beams with floor placed high enough off the main floor so that space below can be utilized for washrooms or shop offices. This keeps substations out of normal flow of traffic and, in some plants such as food processing, out of the range of splashing from equipment-hosing operation.

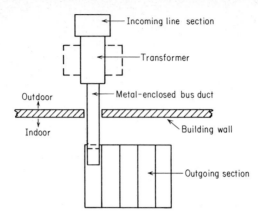

Fig. 19   Plan view of typical single-ended secondary unit substation indoor/outdoor arrangement with transformer in back or front of outgoing section.

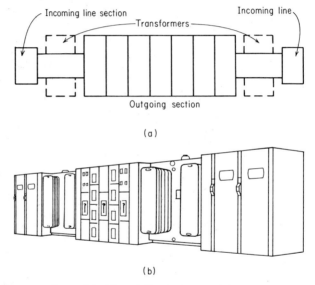

Fig. 20   (*a*) Plan view of typical double-ended secondary unit substation with all components in line, indoor or outdoor installation.   (*b*) Typical double-ended 480-V secondary unit substation.

c. Main Floor Level.   This location can be indoors or outdoors adjacent to the plant.   The indoor areas should be low-value areas which are not or cannot be used for main process or manufacturing lines.   To take advantage of load-center principles, the locations considered must be such that secondary feeders are kept as short as possible.

The physical arrangement of substation components is extremely flexible.   The primary and transformer sections can be located outdoors with secondary equipment indoors, double-ended arrangements can be completely in-line, across the aisle, or can have transformers behind the secondary equipment or above it.   This flexibility allows the plant engineer to arrange equipment to fit the space configuration available.   Figures 18 through 22 show some of the most commonly used physical arrangements.

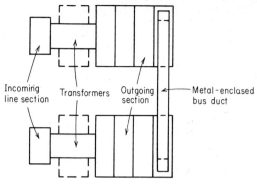

Fig. 21   Plan view of typical double-ended secondary unit substation made up of two single-ended substations connected across a common aisle by bus duct.

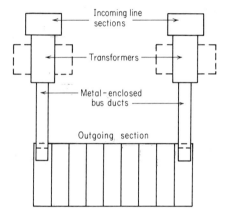

Fig. 22   Plan view of typical double-ended secondary substation with transformers in back or front of outgoing section.   Substation can be indoor, outdoor, or combination type.

SELECTION OF CIRCUIT ARRANGEMENTS

14. Background for Circuit Selection.   The power-systems engineer has a choice of many different circuit arrangements.   Although the various arrangements may be compared on several bases, the choice usually resolves itself into the selection of an arrangement which will provide the required degree of reliability at the minimum cost.   In arriving at a satisfactory compromise between cost and service reliability, the following fundamental considerations should be kept in mind.

The cost of providing electric power to a particular area in an industrial plant includes fixed investment charges and capitalized losses in addition to the cost of the power itself.   A great deal more can be done to control investment charges and losses than can be done to control the power cost.   Thus, the principal concern when attempting to reduce overall costs is with the initial investment.

In controlling initial investment, far more can be accomplished by proper selection of circuit arrangement than by economizing on equipment details.   When cost reductions are necessary, they should never be made at the sacrifice of safety and performance by using inferior apparatus.   Reductions should be obtained by using a less expensive distribution system with some sacrifice in reserve capacity and reliability.

The degree to which extra expenditures should be added to the plant distribution system to increase its service reliability depends upon the characteristics of the manufacturing process (Ref. 1) and the reliability of the power source (Ref. 2). Both these factors must be considered in arriving at a solution to this problem.

a. Characteristics of Manufacturing Process. In many manufacturing plants, a short outage of power can be tolerated. In other plants, short outages of power result in the spoiling of considerable material in process or cause shutdowns over a large area. In the former cases, where short shutdowns can be tolerated, it is questionable if anything but the simplest system can be justified on an economic basis. In the latter cases, it might be possible to justify considerable expense for increasing the reliability of the distribution system.

b. Power-Source Reliability. Service reliability at the point of utilization depends upon the reliability of the power source. Therefore, every effort should be made to obtain a power source having a degree of reliability commensurate with plant requirements.

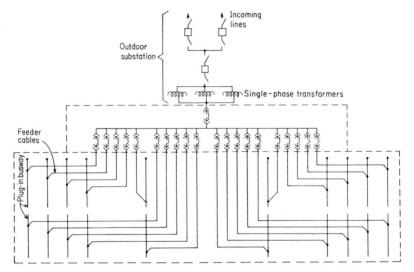

Fig. 23 Old-style distribution system for a medium-sized plant using one large substation and long secondary feeders (Ref. 2).

If the source circuits are cable, and particularly underground cable, the probability of faults in them will be much less than in overhead, open-wire circuits. However, while the frequency of occurrence is low, as much as 24 h may be required to locate and repair the fault.

Faults in overhead open-wire circuits may be either transient or permanent in nature. In the case of transient faults, service is generally restored immediately by reclosing of the line breaker. Outages due to permanent faults may require about 8 h to locate and repair.

It is important to keep in mind that service-reliability figures are highly variable on various parts of a system and on various systems. Historical records of similar installations under similar conditions will provide a reasonable guide as to what to expect (Ref. 24).

15. Types of Circuit Arrangements. Many types of circuit arrangements are possible, although the most commonly used power-system arrangements can usually be classified as one of the following basic types: radial, secondary-selective, primary-selective, looped-primary, and secondary-network. A brief discussion of each of these basic circuit arrangements follows.

   a. Radial. In the radial arrangement, there is only one primary feeder and one transformer through which a given secondary bus is served. Earlier types of radial systems usually consisted of an outdoor transformer supplying the load through several low-voltage secondary feeders (Figs. 23 and 24).

   The radial-type circuit arrangement of the load-center system is the least expensive in the majority of installations, since there is no duplication of equipment.

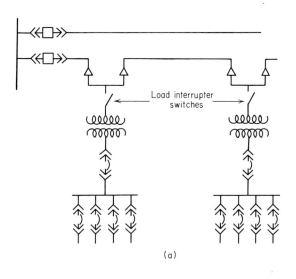

(a)

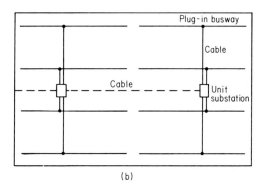

(b)

Fig. 24 Typical radial-circuit load-center distribution system (Ref. 2).

   If sufficient substation capacity is used, the radial arrangement will adequately care for practically any diversity that will be encountered by shifting of load. With adequate, properly installed equipment, the system is safe, simple, easy to operate, and easy to expand by merely extending a medium-voltage feeder or adding a new feeder and substation. Good voltage regulation is provided because of the short secondary feeders.

   It must be recognized that should a primary cable or transformer fail, service is lost to the area supplied by the faulty equipment until it is repaired. Furthermore, maintenance on primary feeders or transformers requires the complete deenergization of the area served by the equipment being maintained. Many engineers feel that the deenergization during maintenance is far more of a handicap than forced outages. Forced outages are so rare that temporary connections can be resorted to, if necessary, to keep essential loads in operation.

b. Secondary-Selective System. Two of the several possible arrangements of the secondary-selective system are shown in Fig. 25. This system utilizes two primary-feeder circuits and two transformers to supply each load-center area. This arrangement

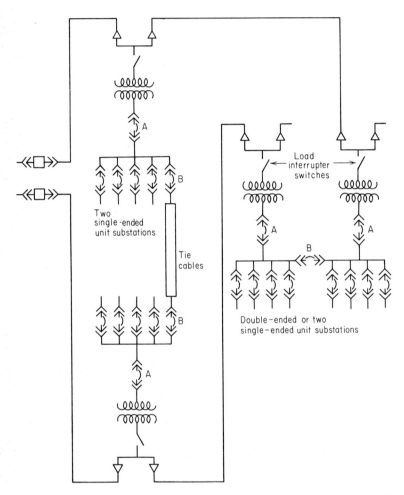

Fig. 25 Typical secondary-selective arrangement of load-center distribution system (Ref. 2).

provides duplicate paths of supply from the source to each secondary bus and makes it possible to provide power at all secondary buses when a transformer or primary-feeder circuit is out of service.

The secondary-selective arrangement may be achieved through a tie between two single-transformer substations or through the use of double-ended substations. The tie breaker is normally open, and the system operates as two parallel radial systems entirely independent of each other beyond the power-supply point. The tie breaker is normally interlocked with the two transformer breakers so that it cannot be closed unless one of the transformer breakers is open. This practice minimizes the short-circuit duty imposed on the low-voltage feeder-circuit breakers.

Since the load is normally divided equally between the two bus sections, half of the load in the area is dropped in the case of a primary-feeder fault. Service can be quickly restored to the interrupted loads by opening the transformer breakers associated with the faulted feeder and closing the bus tie breakers at all load centers. If the system is to carry the entire plant load with one primary-feeder circuit out of service, each primary feeder must be capable of carrying the entire load, and sufficient reserve transformer capacity must be provided. The reserve transformer capacity installed will usually be based on the magnitude of essential loads.

c. Primary-Selective System. This arrangement differs from the radial and secondary-selective systems in that two primary feeders are brought to each substation transformer (Fig. 26). Half of the transformers are normally connected to each of the two feeders. The system is designed so that when one primary feeder is out of service, the remaining feeder has sufficient capacity to carry the entire load.

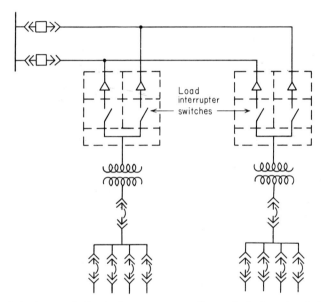

Fig. 26 Typical primary-selective load-center system with two single-throw interlocked primary interrupter switches (Ref. 2).

With this arrangement, as with the secondary-selective arrangement, service to half the load is interrupted when a fault occurs on a primary-feeder circuit. To restore service quickly to all loads following the loss of one feeder, the transformers normally supplied from the faulted feeder can be switched to the good feeder. In cases where the fault may be in a transformer, preferred operating procedure would be to open the circuit breaker on the energized feeder and switch one transformer from the other feeder to the good feeder. When this switching has been accomplished, the good feeder would be energized. This procedure would be followed until all transformers had been switched or until the feeder breaker tripped because the last transformer connected to it was faulted.

If switching is to be done while one primary-feeder circuit is energized, the safest way to make the transfer from one feeder to the other is with adequate power circuit breakers. However, the cost of such an arrangement is relatively high, and it is normally not used unless an automatic-transfer scheme is desired. Usual practice is to use two load-interrupter switches properly interlocked, to accomplish the transfer from one feeder to the other.

The primary-selective system provides a higher degree of service reliability than the radial system and about the same as the secondary-selective system, depending on the amount of reserve transformer capacity installed in the secondary-selective system. The flexibility of this arrangement to handle shifting or growing loads is the same as the radial system, assuming the same reserve transformer capacity in both systems.

   d. Looped-Primary System.   The systems considered thus far have radial primary feeders. In cases where the centers of load are relatively far apart, the use of looped

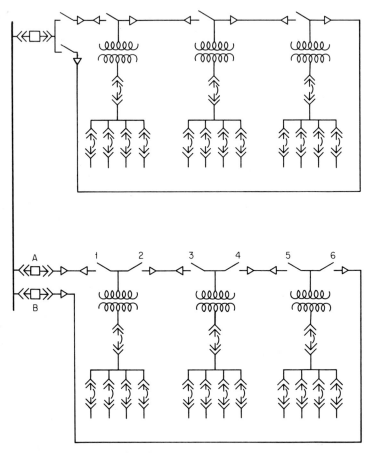

Fig. 27 Typical looped-primary load-center system with sectionalizing switches for supplying load-center substations (Ref. 2).

primary feeders may offer some advantages. Figure 27 shows two forms of the looped-primary arrangement.

   The looped-primary arrangement at the top of Fig. 27 has a single primary feeder breaker and one sectionalizing load-interrupter switch at each transformer, while the lower loop arrangement utilizes two primary-feeder breakers and has two sectionalizing switches at each transformer. When a transformer or primary-feeder fault occurs in either loop arrangement, the primary-feeder breaker or breakers will open and interrupt service to all loads served from that loop. The fault can be located by opening all load-interrupter switches and closing them one at a time in sequence. It is strongly recommended that the primary-feeder breaker or breakers be open before any switch is operated. This practice will eliminate the possibility of closing the switch on a fault. Although modern load-interrupter switches have fault-closing ratings, deliberate closure on faults as with any switching device should be avoided.

When the fault has been located, it can be isolated from the system by leaving the appropriate load-interrupter switch or switches in the open position. In the upper loop arrangement of Fig. 27, one transformer and its loads must be out of service for a transformer fault or a fault in the loop between transformers until the fault is eliminated. In the lower loop arrangement, the only time that loads are subjected to an extended outage is in the case of a transformer fault. The two switches at each transformer make it possible to isolate any fault location on the loop and still provide service to all transformers.

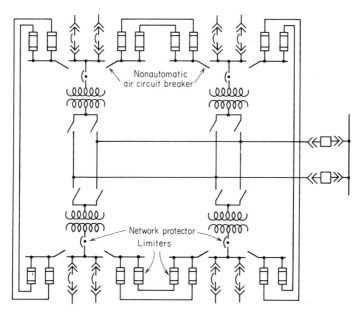

Fig. 28 Typical primary-selective secondary-network arrangement of load-center distribution system.

The upper loop arrangement will cost little more than the radial arrangement and offers the advantage of providing service to all loads except those served from one transformer when either a transformer or primary-feeder fault exists on the system. The lower arrangement costs less than the primary-selective arrangement and provides service to all loads when a primary-feeder fault has been isolated, like the primary-selective arrangement. A transformer fault will cause an extended outage to the associated loads in either the looped-primary or primary-selective arrangement.

The main disadvantage of the looped-primary arrangement is that a primary-feeder or transformer fault will cause an interruption of service to all loads.

e. Secondary-network system may take the form of either a distributed network or a spot network, and either form may utilize a primary-selective-feeder arrangement.

The primary-selective secondary-network arrangement is the form of network system used most frequently in industrial plants (Fig. 28).

This system differs from the radial, secondary-selective, primary-selective, and looped-primary-system arrangements in several important aspects. The major difference is that a primary-feeder or transformer fault will not cause even a momentary interruption of power to any of the loads. The transformer secondaries are interconnected and operated in parallel, and two or more primary-feeder circuits are used to supply the system. In effect, there are several parallel paths of power supply from any load on the secondary back to the power source. In many industrial secondary-network systems plug-in busway is used for the tie circuits between transformer buses.

If two primary feeders are used to supply the primary-selective network system, half the transformers will normally be connected to each feeder with adjacent transformers on different feeders. In the event of a primary-feeder fault, the fault is isolated from the system by the automatic tripping of the primary-feeder circuit breaker and all the network protectors associated with the faulted circuit. Following these tripping operations, the entire load is supplied over the remaining feeder and half of the network transformers. All transformers can be restored to service by manually switching the deenergized units to the remaining feeder. When the faulted circuit is repaired and the appropriate transformers have been reconnected to the feeder, the network protectors associated with those transformers will close automatically when the feeder breaker is closed.

The network protector is the device that makes it possible to operate the two primary-feeder circuits in parallel. The network protector consists basically of an electrically operated air circuit breaker that is controlled by a directional power relay and by a phasing-voltage relay. When a primary-feeder fault occurs, there is a flow of power from the secondary to the fault through all the network protectors associated with the faulted feeder. The directional power relay in each of these network protectors operates and trips the appropriate network protectors to isolate the fault from the secondary. In the meantime, the primary-feeder breaker has tripped to isolate the fault completely from the rest of the system. When the fault is eliminated and voltage is restored on the feeder by closing the feeder breaker, the network relays on all associated network protectors will cause the protectors to close automatically if conditions are such that power will flow from the primary to the secondary.

In addition to providing a high degree of continuity of service to the loads, the network system with its interconnected secondaries inherently offers flexibility to meet shifting or growing loads. In the radial, secondary-selective, primary-selective, and looped-primary arrangements, the magnitude and characteristics of the loads to be supplied from a given transformer secondary bus are governed by the rating of the particular transformer. In the network system the tie circuits between transformers make it possible for adjacent transformers to share load and thereby permit loads on some buses that are in excess of the kVA rating of the transformer at those buses. The amount of power that can be transferred between transformer buses will be determined by the tie-circuit impedance, the transformer impedance, and the characteristics of the load.

f. Spot Network. The secondary-network arrangement described above is designed to serve loads that are reasonably distributed within an area. However, if there are concentrations of critical loads that are widely separated and there is very little load in the areas between them, the spot-network arrangement is more economical than the conventional-network arrangement. The spot-network arrangement operates on the same principle as the conventional network. The main difference is that the transformers are all connected to the same bus (Fig. 29).

The array of possible circuit arrangements can be broken down into three broad types—radial, selective-radial (primary and/or secondary selective), and network. Actual costs of these various systems vary widely, but using an approximation of mean values, the following system-cost comparison can be made:

Radial system.................................. 100%
Selective-radial............................... 130%
Network... 160%

16. Use of Transformer Secondary Main Breaker.   It might be appropriate here to consider the principal reasons for using a transformer secondary main breaker.   They are:

1. It provides a simple and rapid means for deenergizing the entire load on a transformer.

2. In the event of trouble or a fault in the transformer or primary-feeder cable, it is sometimes desirable to disconnect the transformer from the secondary bus and supply this bus from another source to maintain service to important loads.   A main breaker allows this change to be accomplished quickly, easily, and in complete safety to operating personnel.

3. It provides bus-fault protection and backup protection for feeder breakers.   Also, less sensitive primary protection can be used if a main secondary breaker is used.

4. It simplifies a key-interlocking scheme when interlocking is required between primary switches and secondary protective devices.

5. It makes coordination with primary protective devices easier to obtain.

6. It provides added flexibility for expansion.   Many of the substation types covered

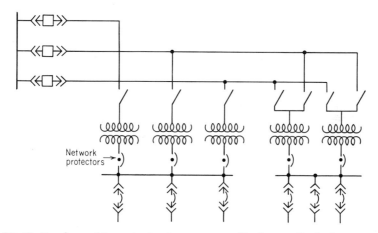

Fig. 29   Two forms of the spot-network arrangement of load-center distribution system.

in Arts. 6 and 10 as well as the overall circuit arrangements covered here require a transformer secondary main breaker.   Including a main breaker in a radial substation at the outset enables future expansion into a more elaborate scheme without extensive and costly field modifications.

The continuous-current rating of transformer secondary main circuit breakers should be approximately 25% greater than that of the transformer to enable them to carry short-time loads above transformer nameplate rating.   These short-time loads are frequently encountered because of short duty cycles or low ambient temperature conditions.

Because consideration must be given to present or future forced-air-cooled transformer ratings of the transformer in selecting the continuous-current rating of the main breaker, the breaker rating is selected to be 125% of this maximum rating.

17. Interlocking.   Interlocking of various equipment within an electrical distribution system is important for many reasons.

Many switching devices used to disconnect transformers from the primary power-supply system have the ability to open or close only the magnetizing current of the transformer and must therefore be interlocked with secondary load-interrupter switching devices to prevent opening or closing of the primary device until all secondary loads are disconnected.   This provides safety to operating personnel and protects the equipment from severe damage or total destruction.

In many cases primary switching devices such as oil-filled or air interrupter switches are used on transformers which have full-load currents less than the interrupting rating of the switch, but even here precautions must be taken, since there is no real assurance that the current flowing in the circuit will not be in excess of the switch rating at the time the switch is opened.   Currents in excess of full load are caused by such things as temporary or prolonged overloads, transformer or motor-starting currents, motor-accelerating currents, circulating currents in tie circuits, or faults in the system. Since this possibility does exist, it is recommended that all primary switching devices except for electrically operated power circuit breakers be key-interlocked with secondary interrupters to assure that all load is removed from the transformer before the primary switches are operated.

Interlocking is also utilized to assure proper system operating procedures.   Where two or more incoming-line circuits are provided, interlocks are used to prevent parallel operation of the incoming circuits.   Where certain operating steps must be followed in a particular sequence, interlocking assures that this sequence is followed.

SUBSTATION TRANSFORMERS

18. Bases for Transformer Selection.   The selection of a substation transformer should start with several basic considerations (Table 11).   First is the kVA rating, which is a direct function of initial and future planned loads.   Next is the voltage rating and voltage ratio, voltage taps, and insulation level.   The insulation medium, type of cooling, and impedance are also chosen to complete transformer selection.   At times, however, sound level is an important consideration.

Standard kVA ratings of primary substation transformers are given in Table 1, and other standard characteristics of these transformers are shown in Table 2.   For secondary substations, these data are shown in Tables 7 and 8.

19. Insulating Medium.   In nearly all cases, industrial-plant primary substations are located outdoors and have oil as the insulating medium.   For indoor applications, a nonflammable liquid (askarel) is used.   If oil is used indoors, however, the transformer must be located in a vault because of the fire hazard which oil presents.   In a few instances primary-substation transformers may be of the dry type, although this would be limited in kVA and voltage rating owing to increased size requirements of dry-type design.

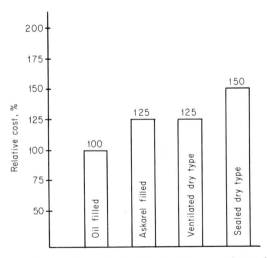

Fig. 30   Relative cost of secondary-unit-substation transformers utilizing different insulating mediums.

For secondary-substation transformers, the insulating medium may be selected from oil, askarel, open (ventilated) dry type or sealed dry type, depending upon location and ambient conditions. As in the primary substation, oil-filled transformers must be located outdoors or in indoor vaults; askarel units are suitable for either indoor or outdoor application as are sealed dry units. Open dry-type units can be used only in clean, dry indoor locations.

Figure 30 shows the relative costs of these four insulating mediums. While oil-filled transformers are the least expensive, they are generally limited to outdoor applications. Askarel units have one disadvantage in that toxic chlorine gas is generated should arcing occur in the tank and they must be equipped with a pressure-relief vent leading to an outdoor flue. Both liquid-filled units require periodic maintenance of liquids to maintain the insulation integrity.

Ventilated dry-type units eliminate the liquid-maintenance problem and are of lighter weight than liquid-filled units but are somewhat larger and present other

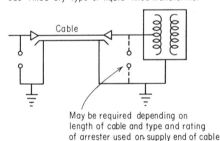

Fig. 31 Recommended placement of lightning arresters for surge protection of unit-substation transformers (Ref. 7).

problems of their own. Dirt accumulates within the housing, which requires periodic removal of side plates for cleaning. They must also be located in relatively dry locations to prevent moisture-absorption problems.

The newest and most expensive transformer is the sealed dry type. It utilizes inert gas at positive pressure as the insulating medium and offers absolute sealing against air, moisture, and other contaminants. There is little maintenance required and little need for a special location when installing.

Note that both ventilated and sealed dry-type transformers have lower basic impulse levels. Caution must therefore be used in applying these transformers where exposure to lightning or other voltage surges is present. It is recommended that lightning arresters be used for all dry-type transformers to limit voltage surges to values below the surge rating of the transformer. Figure 31 shows typical surge-protection application.

Table 11 shows application data for selection of secondary-unit-substation transformers.

20. Voltage Ratio and Taps. Voltage ratio and taps are selected to provide the desired voltage at the secondary terminals. These voltage considerations should take into account the variations which may occur in the supply voltage as well as total voltage drop through the plant distribution system. The primary voltage is generally selected to equal the nominal supply-voltage value with two $2\frac{1}{2}\%$ taps above and below the rated voltage to adjust for variations in supply voltage. Note that these taps are for deenergized operation only and are not intended for adjusting output voltage as the load varies.

The taps above rated voltage increase the number of turns in the primary winding. Therefore, with a fixed primary voltage, selection of a higher tap will result in a lower secondary voltage output, and while this does compensate for transformer and distribution-system voltage drop, it is not an intended means for secondary regulation.

TABLE 11. Selection of Transformer Section of Load-Center Unit Substations (Indoor Service) * **(Ref. 2)**

Application	Liquid-type askarel	Dry type	
		Ventilated	Sealed
Exposure to lightning:			
Where transformers are connected to circuits exposed to lightning and the usual protection is provided †	Yes	No	No
Where study has determined that the amount of lightning exposure is negligible, or the possible resultant voltage stresses can be adequately taken care of by lightning protection	Yes	Yes	Yes
Atmospheric conditions:			
Where atmospheric conditions are clean, such as in plants producing aircraft, instruments, precision parts and certain types of machine shops, assembly plants, food-processing plants, and in clean, dry vaults	Yes	Yes	Yes
Where dirt conditions are severe, such as foundries, steel mills, flour mills, cement mills, or other similar dusty or dirty locations	Yes	No	Yes
Where moisture conditions are severe, such as in geographical locations of high humidity and where the transformers may be subjected to partial or total submersion	Yes	No	Yes
Where units are subjected to acid, oil, or corrosive vapors	Yes	No	Yes
Where units are subjected to oil or flammable vapors and location is classified as hazardous or semihazardous (see Article 500 of National Electrical Code for guidance)	Yes	No	Yes
Where transformers are located overhead on platforms or in roof trusses	No ‡	Yes	Yes
Future application:			
Where possible rearrangement may cause transformers to be moved outdoors at a later date	Yes	No	Yes

* In a condensed guide such as this, it is possible to cover only the more usual types of installations. If should therefore be recognized that there are many special applications or combinations of factors that may affect the decision as to the type of transformer section to be used. In such cases sound engineering judgment coupled with the economics of the situation should be used in reaching a decision.

† The basic insulation level of dry-type transformers is only about half that of liquid-filled transformers, an important factor in case of exposure to lightning or switching surges.

‡ From an engineering standpoint, a dry-type or an askarel transformer section is equally satisfactory for overhead locations, but since the ventilated dry-type transformer section is generally the lighter construction, a less expensive supporting structure may be provided.

Keeping precise voltage at loads is becoming a major consideration in many plants, especially for computers and other electronic loads. Primary substation transformers can be provided with load-tap-changing (LTC) equipment where automatic voltage regulation is required. This equipment generally provides plus or minus 10% regulating range and automatically compensates for voltage variations which result from changing loads. Since equipment costs are a factor, the decision to use LTC depends upon voltage limits of supply and plant voltage-regulation requirements. The control can be set to hold voltage constant or to raise it slightly as load increases, to compensate for voltage drop in the plant's distribution system. LTC in individual secondary substations is uneconomical and impractical.

21. Forced Cooling. Fans may be used to increase the transformer capacity above the self-cooled rating. This increase is related to transformer size as provided for by standards. The fans are automatically controlled by relays sensitive either to top-oil temperature or to winding temperature. Today's transformers for primary and secondary unit substations are designed for adding fans at a future date. This capability assures adequate internal leads, bushings, and tap-changer capacity for the increased rating.

For complete information on application and loading of transformers, see NEMA and ANSI Standards listed in references at the end of this section.

TRANSFER SCHEMES

Momentary power interruptions can result in plant shutdowns and material losses measured in thousands of dollars where continuous-process lines are involved.

While power interruptions cannot be totally eliminated by transfer schemes, their duration and effect can be minimized by providing an alternate power source and some method of load transfer to the alternate source.

Electric utilities are very conscious of the need for service continuity to their customers and through many years of effort have eliminated many factors which contributed to outages. In addition, power interruptions due to foreseeable equipment failures have been almost completely eliminated by advance planning.

Despite efforts made by utilities, outages do occur. Industrial power users must therefore do all that is possible to eliminate effects from outages.

There are many ways to design transfer schemes; some are simple, and some are quite complicated and expensive. Each application presents its own problems and will dictate the economics involved in arriving at the best individual scheme to use.

22. Manual Transfer Schemes. Manual transfer schemes will generally suffice for plants which do not have critical process lines or other requirements which would dictate an absolute minimum loss of power continuity. The simplest scheme is one in which all transfer would be completely manual. The operator closes and trips all breakers manually in a specific sequence.

As a slightly more sophisticated manual scheme, an arrangement such as that shown in Fig. 32 could be used. This scheme would find application where transfer for normal maintenance is to be accomplished while keeping both buses energized. The tie breaker is closed manually, and as soon as it closes, the previously selected main breaker is tripped. This scheme momentarily parallels the two transformers and would result in a short-circuit current duty of two transformers rather than one, should a feeder fault take place at the instant transfer is being made.

Also, if one bus is deenergized owing to a primary system fault, the transfer must be accomplished by first opening the main breaker feeding the deenergized bus. If this is not done, both secondary buses may be deenergized when the tie breaker is closed, since the dead bus would be directly connected to the primary-feeder fault.

23. Automatic Transfer Schemes. There are two automatic transfer schemes. The first (Fig. 33) is used where the two transformer secondaries could be momentarily but not continuously paralleled and where load is such that the transfer can be made at high speed. A 100% lighting load is a typical example. In this scheme, the two incoming-line breakers are normally closed and the bus tie is normally open. If a voltage failure occurs on either incoming line, that breaker is tripped by undervoltage relays (device 27) and the bus tie breaker will close automatically to maintain voltage on both buses. When voltage of proper value, phase angle, and phase rotation is restored on the faulted line, the breaker on that line will close and the tie breaker will trip, thus returning the system to its normal operating condition. No intentional time delay is involved in this scheme, and since the operation is accomplished through use of breaker auxiliary switches, it is relatively inexpensive.

Where motor loads are involved, an automatic transfer scheme becomes more complicated because when motor-supply voltage fails, motors continue to generate a voltage during the time required to complete a transfer. This generated voltage presents a problem when trying to synchronize the two buses across the tie breaker with no control over either voltage.

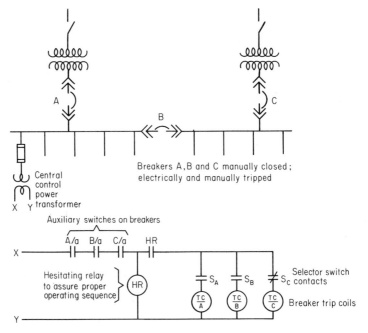

Fig. 32 Typical scheme for transfer of load from one secondary bus to the other in a secondary-selective system to permit removing one transformer or primary feeder without dropping the secondary service (Ref. 1).

If throwover is held back until no residual motor voltage exists, there will be a total interruption of power and the tie breaker will close on a dead bus. Restarting the motors on a dead bus would result in a high inrush current which could possibly trip out the second supply breaker.

Since the purpose of the automatic transfer scheme is to keep both buses energized, it is necessary that the transfer be made while the motors are still running, but at a point where the residual voltage will not be of sufficient magnitude to damage the motors.

To eliminate the problem of high inrush on automatic throwover, a time delay may be introduced to allow residual voltage to decay to a safe value.

Another scheme uses relays to measure the level of residual voltage to determine

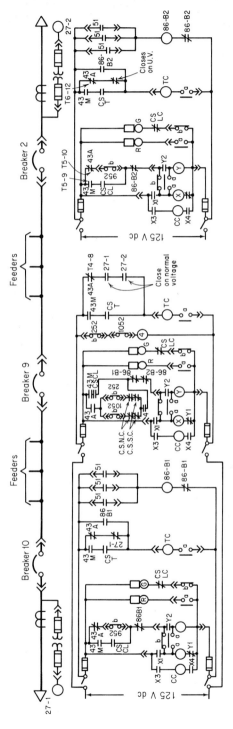

Fig. 33 Automatic transfer of a lighting load is accomplished by control scheme which closes the tie breaker if an undervoltage causes either incoming breaker to open (Ref. 25).

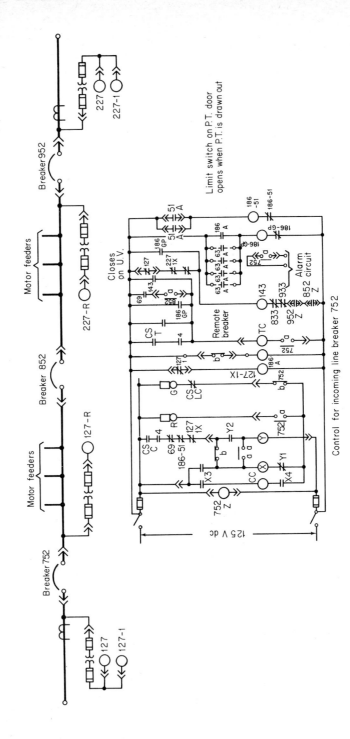

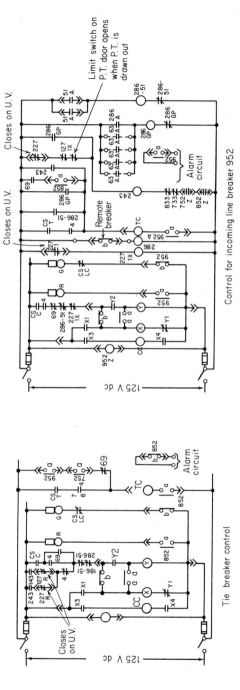

Fig. 34 Automatic transfer of motor loads is accomplished without timing devices. Device numbers and functions are labeled according to NASI Standards (Sec. 1, Table 3 and Sec. 17, Art. 20).

the exact time for transfer. This scheme eliminates the need for precisely setting timing relays (Fig. 34).

Used for a motor load, this automatic transfer scheme utilizes specially calibrated instantaneous undervoltage relays to hold back closing of bus-tie breaker 852 until the residual voltage decays to a safe value. In a normal operating procedure, with normal voltage on relays 127 and 227, incoming-line breakers 752 and 952 are closed and bus-tie breaker 852 is open. If undervoltage occurs on either of the two incoming lines and there is at least 75% voltage on the other line, as determined by relays 127-1 and 227-1, an automatic throw-over will be initiated.

Throwover is accomplished by picking up relays 143 and 243 through the normally closed contacts of breaker cell switches 833 and 733 or 933 and control-voltage relays 852Z and 952Z or 752Z. Relay 143 or 243 then trips the incoming-line breaker and closes the bus-tie breaker. The bus-tie breaker will not close unless residual voltage on the deenergized bus has decayed to 25% of normal, as determined by relays 127-R and 227-R. Also, the bus-tie breaker will not close if either of the two incoming-line breakers has tripped because of an overcurrent condition, as checked by the normally closed contacts of overcurrent lockout relays 186-51 and 286-51.

This scheme holds back a load transfer until the residual voltage has reached a predetermined value. It will be noted that no synchronous or pneumatic timers are needed and that the time delay is not necessarily a constant value. If the voltage decays to a safe value in 20 cycles, a transfer operation will take place in that time. This method eliminates any assumed time settings and assures a transfer in an absolute minimum interval (Ref. 25).

Many schemes are available, and it is the duty of the system-design engineer to select the one which gives the desired continuity of service at the desired initial and long-range cost.

REFERENCES

1. D. Beeman, "Industrial Power Systems Handbook," McGraw-Hill Book Company, New York, 1955.
2. "Electric Power Distribution for Industrial Plants," IEEE Publication 141, August 1969.
3. H. M. Bankus, Distribution Substations in System Planning, *Elec. South*, Atlanta, Ga., February–June 1955.
4. S. E. McDowell, Breakers or Switches for Intermediate Voltage Service? *Allis-Chalmers Elec. Rev.*, Milwaukee, First Quarter, 1961.
5. H. A. Wright, Industrial Control or Switchgear? *Allis-Chalmers Elec. Rev.*, Milwaukee, Second Quarter, 1956.
6. "Primary Unit Substations," NEMA Standards Publications 201, 1970.
7. "Secondary Unit Substations," NEMA Standards Publications 210, 1970.
8. National Electrical Code, National Fire Protection Association, Pamphlet 70, 1968; ANSI C1-1968.
9. "High Voltage Power Circuit Breakers," NEMA Standards Publication SG4-1975.
10. "Low Voltage Power Circuit Breakers," NEMA Standards Publication SG3-1975.
11. "Molded Case Circuit Breakers," NEMA Standards Publication AB1-1969.
12. "Power Switching Equipment," NEMA Standards Publication SG6-1974.
13. "Power Switchgear Assemblies," NEMA Standards Publication SG5-1971.
14. "High Voltage Power Fuses," NEMA Standards Publication SG2-1975.
15. "Industrial Control," NEMA Standards Publication ICS-1970.
16. "Transformers, Regulators, and Reactors," NEMA Standards Publication TR1-1974.
17. "Power Switchgear," ANSI Standards Publication Series C37.
18. "Transformers, Regulators and Reactors," ANSI Standards Publication Series C57.
19. "Safety Requirements for Dead Front Switchboards," UL Standard 891, Underwriters Laboratories, Inc.
20. "Safety Requirements for Branch Circuit and Service Circuit Breakers," UL Standard 489, Underwriters Laboratories, Inc.
21. "Enclosed Switches," UL Standard 98, Underwriters Laboratories, Inc.
22. "Fuses and Fuse Holders," UL Standards 198 and 512, Underwriters' Laboratories, Inc.
23. G. J. Meinders, Fused Low Voltage Power Circuit Breakers, *Allis-Chalmers Eng. Rev.*, vol. 29, no. 2, 1964.
24. W. H. Dickinson, "Report on Reliability of Electrical Equipment in Industrial Plants," AIEE Technical Paper 62-61.
25. W. E. Schwartzburg, Automatic Load Transfer, *Allis-Chalmers Elec. Rev.*, First Quarter, 1955.

21

Distribution Panelboards and Switchboards

W. B. PERKINS*

*Manager, Commercial and Industrial Product Marketing Section, Distribution Equipment Division, Square D Company.

FOREWORD

Careful planning and consideration of the industrial-plant layout calls for the most effective use of panelboards and switchboards. Their application and maintenance are of major importance to plant safety and continuity of operation. This section cites the standards and good practice necessary for proper application and maintenance of this equipment.

GENERAL

The National Electrical Code defines a panelboard as "a single panel or group of panel units designed for assembly in the form of a single panel; including buses, automatic overcurrent devices, and with or without switches for the control of light, heat or power circuits; designed to be placed in a cabinet or cutout box placed in or against a wall or partition and accessible only from the front."

A switchboard is similarly defined as "a large single panel, frame, or assembly of panels, on which are mounted, on the face or back or both, switches, overcurrent and other protective devices, buses and usually instruments. Switchboards are generally accessible from the rear as well as from the front and are not intended to be installed in cabinets."

In general, both types of equipment provide the same function. They are convenient methods of grouping together overcurrent devices fed from a common source to provide circuit protection to a number of feeder or branch circuits. As the fundamentals of application are essentially the same, what follows with few exceptions might be applied equally well to either category. It will deal, however, primarily with panelboards and small switchboards. Larger switchboards may include more specialized or larger pieces of equipment for which specific data should be obtained from individual manufacturers.

1. Arrangements. Panelboards and distribution switchboards are normally manufactured with:

1. Plug fuses only
2. Switches and fuses (cartridge or plug)
3. Circuit breakers

Panelboards with cartridge fuses must include a switch ahead of the fusible element. All three types of equipment are for use on circuits of 600 V ac and below.

Some panelboards provide only a set of lugs for wiring to the panel bus. Others may incorporate a main fusible switch or main circuit breaker as a disconnect as well as overcurrent protection for the entire panelboard and termination for incoming cable.

Other specialized panelboards offered by some manufacturers include a version of either a fusible or circuit-breaker panelboard provided with motor-starter units. These panelboards provide many of the same features as a motor-control center, allowing centralized control and protection of a number of motor loads.

2. Standards. Both Underwriters Laboratories and the National Electrical Manufacturers Association have standards for construction, testing, and performance of panelboards and switchboards. The applicable regulations most frequently referred to are:
Panelboards:
NEMA Standards Publication PB-1
Standards for Panelboards, UL 67
Standards for Cabinets and Boxes, UL 50
"Service Entrance," UL Publication 869
Switchboards:
NEMA Standards Publication PB-2
Standards for Safety, Dead-Front Switchboards, UL 891

CODE REQUIREMENTS

The references to the National Electrical Code contained in this section are based on the 1975 edition but are not intended as a complete or comprehensive list of all the code requirements that may affect panelboard and switchboard installations. The most recent edition of the National Electrical Code, as well as any state or local codes that apply, should be consulted to determine the compliance with regulations of any existing or contemplated installation. Numerous paragraphs under the general heading of "Wiring Methods," "Branch Circuits," "Services," "Grounding," and other sections of the Code, as well as "Switchboards and Panelboards," may all be pertinent.

Article 384 of the National Electrical Code applies to panelboards and switchboards. Two sections of this article are of particular interest. They establish two types of panelboards and provide rules peculiar to these types.

3. Lighting and Appliance Branch-Circuit Panelboards. Paragraph 384-14 states in part, "a lighting and appliance branch circuit (L&abc) panelboard is one having more than 10% of its overcurrent devices rated 30 A or less, for which neutral connections are provided." Any panelboard that contains more than 10% of its overcurrent devices rated at 30 A or less with neutral connections provided for these circuits automatically falls under the special rules applicable to lighting and appliance branch-circuit panelboards. The main bus of a panelboard falling into this category is sized by the manufacturer on the basis of an average loading of 10 A per branch circuit for the number of branch locations provided (NEMA PB-1-2.07). Panelboards not conforming to the requirements of an L&abc panelboard—i.e., not having a large number of single-pole lighting circuits—are generally referred to as power panelboards or as power and feeder distribution panelboards. In practice, power panelboards often contain some single-pole lighting branch circuits as well as 2-pole or 3-pole motor branch circuits.

Section 384-15 states that no L&abc panelboard may contain more than 42 overcurrent devices other than those provided for the main—a 3-pole circuit breaker being considered for the purposes of Section 384-15 as three overcurrent devices and a 2-pole breaker as two overcurrent devices.

EXAMPLE: Analysis of the loads to be served in a small industrial plant indicates the need for the following circuits. The circuit-breaker panelboard feeding these loads is to be connected to a 3-phase, 208Y/120-V system.

Twenty 15-A single-pole lighting circuits

Five 30-A 3-phase motor circuits

Two 50-A 3-phase motor circuits

One 100-A 3-phase motor circuit

The lighting circuits will be single-phase circuits, containing one single-pole circuit breaker and a neutral connection per circuit. Obviously, more than 10% of the devices enclosed in this panelboard will be of this kind; therefore, the panelboard, whatever its type, must be considered a lighting and appliance branch-circuit panelboard. The 3-phase motor circuits will be protected by 3-pole branch breakers—a total of eight 3-pole circuit breakers in this case. Therefore, there will be $20 + (3 \times 8) = 44$ overcurrent devices installed in this panelboard. This arrangement is not in accordance with NEC Section 384-15, and some other means must be found to provide the service required.

Alternatives would include:

1. Install two smaller panelboards instead of the one originally contemplated.

2. Subfeed from the original panelboard to a second much smaller one accommodating only two or three of the motor circuit breakers.

3. Recombine some of the lighting load so that the area previously fed by eight 15-A circuits would be fed instead by six 20-A circuits, eliminating two overcurrent devices. Six 20-A branches provide the same single-pole circuit capacity (120 A) as the eight 15-A branches previously employed.

It is quite probable that the latter method would be the more economical solution, if it is feasible.

4. Main Overcurrent Devices. Section 384-16 requires that lighting and appliance branch-circuit panelboards shall be individually protected on the supply side by not more than two sets of overcurrent devices. This arrangement allows for use of a single main breaker or set of fuses, or split-bus construction with two breakers or sets of fuses as may be needed in special metering or service-entrance equipment (Figs. 1 and 2).

There are two exceptions to this requirement for a main overcurrent device in an L&abc panelboard:

1. If the overcurrent device protecting the feeder that supplies the panelboard is the

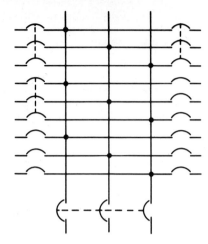

Fig. 1 Single-main set of overcurrent circuit-breaker devices.

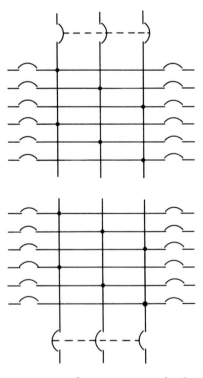

Fig. 2 Two-main set of overcurrent circuit-breaker devices.

same rating as the panelboard bus or is rated less than the panelboard bus, this overcurrent device would provide the protection required by Section 384-16.

2. Should the L&abc panelboard be used as residential service-entrance equipment, it is not required to have main overcurrent protection, provided that 15- or 20-A branch circuits are fed from bus that is protected by a main overcurrent-protective device, i.e., split bus with main only on the distribution section.

Section 230-71 requires service-entrance equipment containing not more than six subdivisions of the service. A subdivision is defined as those circuits which can practically be disconnected with one motion of the hand. Main overcurrent protection is used where more than six subdivisions would otherwise be required.

When used as service-entrance equipment, the neutral must be bonded to the panelboard enclosure and to the service ground by means of a main bonding jumper. See Paragraphs 384-3(c) and 384-27.

5. **Branch Taps.** Panelboards may be fed by taps from a main feeder provided that the restrictions of Section 240-21, Exception 2 or 3, are met.

The panelboard, if not an L&abc panelboard, may be of the main-lugs type if the length of the tap does not exceed 10 ft and the cables constituting the tap have an ampacity not less than the sum of the computed loads on the circuits supplied by the tap conductors and not less than the ampere rating of the panelboard.

If the panelboard has main overcurrent protection, with a main breaker or fusible switch sized to protect the tap conductors, the length of the tap may be increased to 25 ft provided its size is at least one-third that of the main feeder.

6. **Fluorescent Lighting.** Paragraph 210-6 limits the use of 277-V fluorescent lighting and sets up criteria which are commonly met by the use of distribution panelboards. Such lighting must:

1. Be at least 8 ft above floor level
2. Be in a commercial-industrial installation with trained maintenance personnel in attendance
3. Be in a permanently installed fixture
4. Have no switch at the lamp itself

Requirement 4 dictates that some form of panelboard be supplied to provide switching function as well as overcurrent protection for the lighting circuits. Circuit-breaker panelboards or fusible panelboards employing cartridge fuses in a fusible switch are commonly used. On circuit-breaker panelboards note must be given to the voltage rating of the circuit breaker, and only those types approved for use on systems with 277 V or more to ground can be used safely.

7. **Continuous Load.** Section 384-16(c) of the National Electrical Code limits the load on any overcurrent device in a panelboard to 80% of its rating where the normal operation of the load will continue for 3 h or more unless the device is specifically approved for continuous duty at 100% of rating.

In connection with the National Electrical Code regulations, Section 220-2 for calculation of branch-circuit loads and Sections 210-22 and 210-23 for maximum loading of branch circuits should be checked.

BRANCH-CIRCUIT RATINGS

All elements of a panelboard generate some amount of heat during normal operation. Overcurrent devices, including thermal-magnetic circuit breakers, time-lag magnetic circuit breakers, fusible switches, panelboard bus, and both incoming and line-side wiring contribute to the total heat load on the panelboard. With the exception of magnetic-only circuit breakers, the overcurrent devices employed in the panelboard are also heat-sensitive, relying in one way or another on the sensing of abnormal heat generation to open under overload conditions. In sizing branch circuits, it is imperative, therefore, that loading be kept to a reasonable figure to avoid undue heat buildup resulting in nuisance tripping or blowing of circuit-protective devices. The most common cause of panelboard malfunction is excessive operating temperature due to overloaded conductors.

Code requirements limit individual circuit breakers or fused switches for branch circuits to loads of not over 80% of their rating. To achieve longer component life and more reliable operation, many authorities consider limitations of 70% of nominal rating better practice for thermal-magnetic circuit breakers and fused switches employed in panelboards. In the case of 40°C calibrated thermal-magnetic circuit breakers or ambient-compensated circuit breakers an 80% loading factor is probably acceptable, but the 70% factor for all other circuit-protective devices is recommended unless the device is specifically approved for continuous duty at 100% of rating.

8. Load Calculation. The sizing of individual branch circuits is a subject too broad for the scope of this section. Ample literature can be found elsewhere to provide guidance in load calculation. Briefly, however, any analysis of branch-circuit loads should include consideration of:

1. Demand factor—a measure of that part of the total connected load which will actually be called for at any given time and is usually averaged over a short period. Peak load divided by total connected load.

2. Load factor—a measure of the distribution of load in time over a brief period, for instance, over 1 day. Average load divided by peak load.

3. Growth factor—a measure of the load that may be expected in the future as compared with present total connected load. Future total connected load divided by present total connected load.

Lighting loads will be determined from a consideration of the square footage to be lighted, the intensity of light desired in foot candles as dictated by the use of space and the type of lighting to be applied, incandescent, fluorescent, or vapor. Motor loads require data on the number, horsepower, and design characteristics of the motors employed. Lighting installations are not frequently expanded and require little, if any, growth factor, whereas with motor-load installations, a growth factor of 50% is not unusual. Lighting loads are generally all on at once, however, requiring high demand factors as compared with motor loads. The load factor depends on the use of a facility and defies generalization.

9. Main Ratings. In general, the sizing of main bus or main breakers of a distribution panelboard or switchboard is more mechanical and easier than the proper sizing of branch circuits. Once branch-circuit sizes are known, code regulations, standard construction, and good engineering practice will lead easily to a choice of mains. Certain NEMA and UL regulations will influence the choice—for instance, NEMA Standard PB-1 (PB-1-2.07) dictates that manufacturers size the main bus of lighting and appliance branch-circuit panelboards by assuming no less than an average of 10 A for each overcurrent-protective device, except if the panelboard is designed for a specific installation where the known loading is less than 10 A per branch. A single-phase L&abc panelboard having room for 20 circuits will therefore normally be offered only in main ratings of 100 A per phase and larger. The following general formula may be used to calculate main size:

$$M = 1.25(\Sigma I_L + I_{m1} + I_{m2} + \cdots + 1.25\, I_{m0})G$$

where M = rating of main lugs or overcurrent device
ΣI_L = sum of all lighting load currents, but not branch-circuit breaker ratings, after applying any applicable demand factor
I_{m1} = full-load running current of motors 1, 2, 3, . . . after applying applicable load factors to each individual motor
I_{m0} = full-load running current of largest motor
G = growth factor as determined from judgment of anticipated growth of load area

If specific load information is lacking, an approximation of proper main size can be obtained from the following rule of thumb based on experience with properly engineered installations. Note that figures used are branch-circuit overcurrent-device ratings and not connected loads.

M for 100% motor load = sum of branch-circuit overcurrent-device ratings ×0.6
M for 50% motor/50% lighting = sum of branch-circuit overcurrent device ratings ×0.7
M for 100% lighting load = sum of branch-circuit overcurrent-device ratings ×0.8

10. Voltage and AIC Ratings. Voltage ratings are specified to equal or exceed system voltages, and are relatively straightforward. Care must be exercised not to exceed the voltage rating of main or branch-circuit breakers or fuses. For instance, circuit breakers or fuses rated 120/240 V are suitable for systems having no more than 120 V phase-to-ground and 240 V phase-to-phase and must not be applied on systems having 240 V phase-to-ground. A similar condition exists for circuit breakers rated at 277/480. Circuit breakers or fuses having a single voltage rating such as 600 V can be safely used on systems having up to that voltage phase-to-phase or phase-to-ground. Circuit breakers should never be applied on systems having any voltage in excess of the breaker rating.

Not so well understood is the short-circuit interrupting-capacity rating applied to circuit protective devices. Ampere-interrupting capacity (AIC) is commonly given in symmetrical amperes, and is a measure of the ability of the device to open safely on a system with the specified short-circuit amperes available at the line side of the device in the event of a direct short on its load side (Ref. 4). This interrupting capacity varies with the system voltage (Secs. 4, 12, and 13).

In general, the factors affecting these calculations will be available primary power, transformer size and impedance, secondary-distribution-system impedance, including cable, switchboard, and panelboard bus, etc., and motor-feedback contributions. Two means of achieving the proper match between the device interrupting capability and the system available fault current are possible:

1. Use circuit breakers of sufficiently high rating.
2. Reduce the available current by the use of current-limiting fuses or current-limiting circuit breakers located ahead of panelboard or by increasing the length of cable from the distribution transformer to the panelboard to introduce current-limiting impedance. Note that molded-case circuit breakers should not be cascaded to achieve adequate short-circuit interrupting ability unless the combination is specifically approved for that purpose. Small molded-case circuit breakers tend to open more quickly than larger frame sizes of molded-case breakers and therefore attempt to perform the job of interrupting fault current prior to the operation of the larger breakers with which they might be cascaded.

To meet the requirements of higher short-circuit current-level systems, most panelboard manufacturers have available various types of circuit breakers with higher interrupting capability which may be substituted for standard circuit breakers in a given panelboard. Generally the use of these higher-capacity breakers will prove less expensive than the addition of current-limiting fuses, either in a fusible main switch in the panelboard or ahead of it as an individual fusible safety switch or switchboard-mounting fusible switch. The higher-capacity industrial circuit breakers are commonly available with ratings of 65 000 A rms symmetrical at 240 V and up to 35 000 AIC at 480 V. Current-limiting circuit breakers may have ratings of 100 000 AIC or even higher at 480 V.

While proper operation of an individual circuit breaker at a given fault level may be indicated by its rating, there is still the possibility that when mounted in a panelboard in close proximity to other circuit breakers and live panelboard bus it may not function properly at its rated maximum. Panelboard bus distortion, due to magnetic forces and the ionized conductive gas vented by the circuit breaker as it functions, makes proper operation of the circuit breaker in the panelboard less certain than when it is isolated under test conditions for determinimg its rating. For this reason, Underwriters Laboratories in 1973 established additional test requirements for panelboards that allow listing the entire panelboard with a specific short-circuit rating. The additional tests are optional at the manufacturer's discretion. A short-circuit rating on a UL-listed panelboard indicates that the entire integral piece of equipment—panelboard enclosure, interior assembly, and breakers—has been tested together and is capable of interrupting fault currents when used on systems with up to the rated short-circuit current available. Panelboards not carrying a rating may not be suitable for use on systems with over 5000 A available in the case of circuit-breaker equipment, or 10 000 A available in the case of fusible equipment.

11. Load Balance. Care must be exercised in laying out the panelboard that the total connected load is balanced between the various phase conductors of the panel. If this procedure is not followed, excessive current can be carried through one conductor of

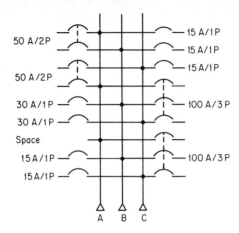

Fig. 3 Balanced loading with phases *A*, *B*, and *C* carrying 310 A.

the panelboard bus, causing it to overheat. Single-phase loads on a 3-phase panelboard should be divided equally between the *A*-, *B*-. and *C*-phase conductors (Fig. 3).

On single-phase, 3-wire systems the phase-to-phase and phase-to-ground loads should be distributed so as not to overload either phase conductor or the neutral.

APPLICATION AND INSTALLATION

12. Phase Identification. NEMA Standards require that phase orientation of bus bars in panelboards and switchboards be such that identification reads *ABC* from left to right, from front to rear, or from top to bottom. Most panelboards built today are of the distributed-phase bus type that fulfills this requirement (Fig. 4).

Some panelboard types, where the connector to the panelboard main bus structure is part of the overcurrent device rather than part of the panelboard, allow installation of branch-circuit devices so that any phase relationship desired can be obtained.

The circuit breakers may be installed in accordance with NEMA Standards, or if for

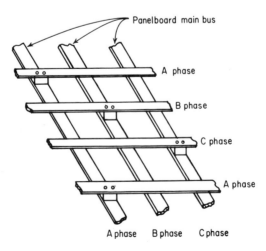

Fig. 4 Phase orientation determined by connectors.

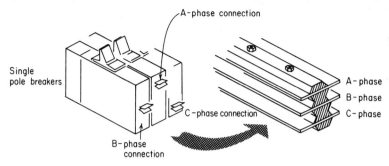

Fig. 5 Phase orientation determined by position of branch-circuit devices.

any reason deviation from the standard is desired, they can be installed in some other arrangement (Fig. 5).

13. Panelboard and Switchboard Location (Refs. 8, 9, 10). Standard panelboards and cabinets are intended to be installed in a clean, dry, noncorrosive location, where the normal ambient temperature does not exceed 100°F. Where there is excessive moisture in the atmosphere, a weatherproof-type panelboard and cabinet should be installed (Sec. 16). In dusty locations or locations exposed to flying particles, a dusttight or weatherproof construction should be used. If it is impossible to locate the panelboard in other than a high-ambient-temperature location, the panelboard current-carrying parts, overcurrent devices, wires, and cables should have a current-carrying rating at least double the actual capacity of the branch-circuit and main-bus-bar loads.

Panelboards and switchboards function best when the length of the cable run to the loads they serve is kept to a minimum. This procedure generally requires that the load grouping be relatively concentrated in a physical area that is not excessively large. If the loads are widely scattered over a broad area, other means of power distribution (such as plug-in busway) may prove to be more advantageous (Sec. 19). Short load-side wiring not only serves to reduce installation cost, but helps reduce voltage drop as well as minimize cost in circuit rearrangement. A 1% voltage drop on load-side wiring is an acceptable target for most installations. One commonly used criterion is that the panelboard should be so placed that no branch circuit must run more than 100 ft to the nearest point of power utilization.

In most commercial installations, the panel can be surface- or flush-mounted on a wall and be near enough to the loads to satisfy the conditions above. In industrial installations, however, it is frequently impossible to find a suitable wall in a largely open manufacturing area. Column-width panelboards designed to fit between the flanges of 8- or 10-in wide-flange structural-steel beams are specifically designed to meet the needs of such installations and are offered by most manufacturers. Generally the circuit breakers are mounted in one row, as opposed to double-row mounting usually employed with flush- or surface-mounted standard-width panelboards. A pull box containing the neutral connections for all branch circuits is mounted above the panelboard with gutter connecting the two components. This arrangement avoids the necessity of running neutral cables into the panelboard gutter (Fig. 6).

14. Panelboard Installation (Refs. 8, 9, 10). The cabinet, consisting of a box and removable front, should be installed, leveled up, and securely fastened to the mounting surface, utilizing all of the mounting holes provided in the panelboard cabinet. It is important that the cabinet be secured to a flat and even surface, or otherwise adjusted to keep the back of the panelboard true and plumb. Conduits can then be installed in the walls of the box through conduit openings or knockouts provided for that purpose. After conduits are installed, all unused openings should be closed to keep dust and moisture out of the panelboard cabinet. After wires are pulled in the conduits and before the panelboard interior is installed, the cabinet should be inspected to see that none of the corrosion-prevention finish has been scratched off during installation. If it is scratched off, paint or other protective coating should be applied to keep the cabinet from corroding in the bare places.

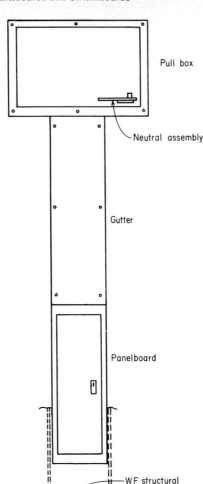

Fig. 6 Column-width panelboard with pull box.

Until installation, the panelboard interior should be kept in a clean, dry, and normal-temperature location. Panelboards are packed at the factory to withstand shipment and reasonable handling on the job. Care should be exercised to avoid damage to the panelboard while unpacking and handling prior to installation.

After unpacking, and before installation in the cabinet, the panelboard should be carefully inspected to see that all connection and mounting screws are tight. Connections may be loosened during shipment or nandling after leaving the factory. The interior should then be installed and secured in place in the cabinet by the mounting means provided. Panelboards and cabinets are provided with adjustable mounting means to permit alignment of the panelboard interior and cover even though the cabinet may not be exactly flush.

The panelboard interior should then be connected, making sure that the wires are tightly secured in the terminals provided. After connecting, the wires should be neatly arranged in the gutters. It must be remembered that the wires in the gutters of the panelboard generate heat within the panelboard enclosure, and accordingly, unnecessary or excessive amounts of wiring in the panelboard gutters are a source of panelboard heating difficulties.

After the panelboard interior is connected, the cabinet should be carefully cleaned of cut ends of wire and foreign substances, and the door and trim immediately installed to protect the panelboard and wiring. If the building is not completed before the panelboard front is installed, heavy cardboard or other material should be used to protect the panelboard and its wiring from damage, dirt, and defacing during building construction. Before electric power is supplied to the panelboard, all wiring should be checked and tested for grounds, short circuits, etc.

CARE OF EQUIPMENT

15. Maintenance. All overcurrent devices installed in panelboards or switchboards are heat-producing elements. While it is true that overheating is the most common cause of panelboard faulty operation, it is equally true that undue alarm at elevated operating temperatures is common. Panelboards and switchboards normally operate at relatively high temperatures. Switchboards may be ventilated to dissipate excessive heat, but panelboards must rely on radiation from the enclosure to maintain proper operating temperatures. A thermometer should be used rather than the sense of touch in testing for temperature, although a rough guide recommended in Refs. 9 and 10 is to feel the surface of the deadfronts, boxes, trims, and circuit breakers with the palm of the hand. If contact cannot be maintained for at least 3 s, chances are the equipment is running too hot and further temperature readings are necessary. The primary effect of cumulative heating due to grouping devices within an enclosure is a tendency for individual circuit breakers to trip out and fuses to blow on circuit loads less than their current ratings. If this difficulty is encountered, the panelboard must be checked under load, as the heat contributed by adjacent branch circuits may be the determining factor.

If excessive operating temperature is found, the load-side wiring should be checked for excessive amounts of wire. The practice of leaving loops of wire on the load side of branches contributes additional I^2R losses in the wire for which the panelboard was not designed. All load-side wiring should be clipped as short as possible.

Joints and lugs should be checked for tightness to reduce possible heating points. Joints may loosen during transportation or installation, or by vibration transmitted through building or conduit, or by creep of conductor material caused by alternately heating and cooling as the load is turned on and off. It is good practice to check joints for tightness at the time of installation and again 3 months later. Thereafter, yearly maintenance inspections are recommended.

NEMA PB 1.1-1975 recommends the following torque for accessible bolted connections, where specific manufacturer's recommendations are not available:

Bolt diam.	Tightening torque
#8 ($^5/_{32}$ in)	10–15 lbf-in
#10 ($^3/_{16}$ in)	15–20 lbf-in
$^1/_4$ in	5–7 lbf-ft
$^5/_{16}$ in	10–12 lbf-ft
$^3/_8$ in	18–20 lbf-ft
$^1/_2$ in	40–50 lbf-ft

Follow manufacturer's recommendations when tightening mechanical lug wire binding screws.

16. Circuit Breakers. Circuit breakers are usually sealed units and require no internal inspection or maintenence. The sealed case should not be opened. Sealing is necessary to prevent loss of calibration, which can easily occur if proper precautions are not taken on reassembly of the case and cover. If improper circuit-breaker operation is discovered, the entire breaker should be replaced as a unit. Some circuit breakers have means for mechanically tripping the latch mechanism. This arrangement not only assures proper operation, but exercises the trip latch mechanism and helps prevent binding of these parts. Such breakers may be mechanically tripped as part of a yearly procedure. When a breaker is first installed or has not been operated for a long time, the breaker should be switched from on to off a few times. This will clean the contacts and help prevent overheating.

17. Switches. Switches mounted in a panelboard or switchboard require periodic inspection and maintenance to see that they are kept clean, lubricated, free from dirt and moisture, and operating at a reasonable temperature. Joints, slides, bearings, and moving parts should be lubricated with any good light lubricating grease if needed. However, conducting joints should not be lubricated. Sliding electrical contacts and blade hinge joints such as found on knifeblade switches may be abrasively cleaned, wiped clean and free of abrasive particles, and then coated with a light covering of high-grade petroleum jelly or high-temperature grease. This coating may be applied with a wiping action of an emery cloth to clean the contact surfaces.

18. Fuses. Fuse heat amounts to roughly three-quarters of the total heat in a properly operating switch. Replaceable link fuses are a major cause of switch heating problems. If this type of fuse is to be used, extreme care should be exercised that the fuse links are properly and securely installed. Internal fuse link bolts should be checked and tightened periodically. Usually one-time fuses require less maintenance and result in lower operating temperatures than replaceable link fuses. Some dual-element fuses operate nearly one-third lower in temperature than the zinc-link type. Often this lower temperature will eliminate heating problems. The fuse manufacturer's service recommendations should be followed for maintenance, cleaning, and replacement of links.

19. Heat Damage. Any spring-type current-carrying part or fastener such as a lock washer or dished pressure washer which must inherently maintain spring tension to assure proper operation should be carefully inspected for signs of overheating. If severe oxidation or discoloration indicates it has been overheated, it should be replaced, as the spring tension is probably reduced by the heating. This care applies particularly to fuse clips and blade jaws.

20. Short Circuits. Normally the overcurrent-protective device on the circuit will prevent any electrical damage except at the actual point of short circuit. However, the high mechanical stress as generated by heavy short-circuit currents may do considerable damage to the conductors and insulation of a panelboard or switchboard. A thorough inspection of the entire device after any large fault current must be made to ensure that no mechanical damage to either conductors or insulation has taken place. As a final check, a 500- or 1000-V megohmmeter should be used to test for concealed damage to insulation. While a 1-MΩ reading is normally considered safe for any individual circuit breaker, a panelboard or switchboard with multiple paths to ground may exhibit considerably less resistance and still be operating properly.

21. Arcing Damage. Some organic insulating materials carbonize when subjected to the heat of an electrical arc. When carbonized, they themselves become a conductor and lose their insulating qualities. Following an arcing fault, careful inspection of the apparatus for any signs of carbon tracking on the insulation must be made. Any insulation found to be carbon-tracked must be replaced before the reliability of the apparatus can be restored.

22. Fluorescent-Lighting Problems. One application in particular deserves mention in connection with service problems. Fluorescent lighting presents conditions peculiar to that type of load and should be fully understood to avoid operational difficulty. Both 480Y/277-V and 208Y/120-V systems with fluorescent-lighting loads produce third-harmonic voltages at a frequency of 180 Hz. The fundamental 60-Hz current in the three phases of the system adds to zero, but the third-harmonic currents are additive (Fig. 7). The sum of the 180-Hz component of current in all phases must be carried by the neutral. Tests have indicated that the magnitude of this neutral current may be as much as 90% of the fundamental frequency current in any one phase. This current results in heating in the panelboard and under severe conditions may cause nuisance tripping of the branch breakers. As the fluorescent-lighting loads are usually

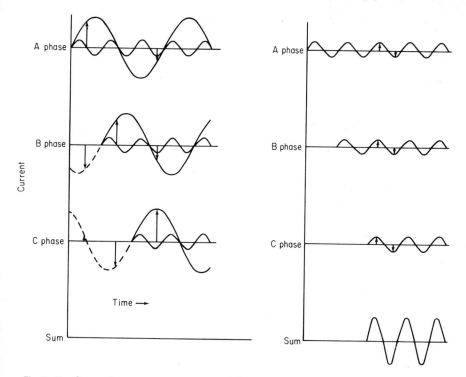

Fig. 7 Fundamental sums to zero at any point (left). Third harmonic sums to value of three times phase contribution (right).

continuous in nature and of a high demand factor, this type of installation should be loaded to no more than 70% of the individual branch ratings. Panelboards for fluorescent-lighting loads must be provided with 100% capacity neutral because of the large neutral currents and to allow for the possible unbalance in the phase currents resulting in additional neutral current (NEC Section 220-22).

23. Housekeeping (Ref. 8).   A panelboard is a major link in the low-voltage distribution system. It can be the point of convergence of a large number of branch circuits providing light and power to a large area of the plant or building it serves. As such, an outage at the panelboard can cause widespread disruption and work stoppage. Where downtime of equipment or disruption of office routine is expensive, panelboard failure can be costly indeed. Considering the importance of the distribution panelboard, it should be a major point of attention in a preventive-maintenance program but all too often is the forgotten element in the secondary distribution system. Preventive-maintenance procedures will often prevent costly failures.

Panelboard interiors should be kept clean and free of accumulations of dust and debris. Knockouts in the enclosure which have not been used for wiring the panelboard should be left intact or if they have been opened should be closed to prevent the entrance of dirt, vermin, or rodents. The latter have been the known cause of panelboard failure. In areas of atmospheric contamination, chemical fumes, excessive humidity, or high ambient temperature, insulation should be carefully inspected for any signs of deterioration. Any discoloration, cracking, flaking, or other noticeable change of any insulation should be investigated. Plating on bus bars, lugs, and connectors can also be adversely affected by some atmospheric conditions. Some discoloration may take place, however, under normal conditions without affecting the serviceability of the equipment.

It is also important to watch for condensation in conduits, particularly those which enter in the top of the cabinet. Condensation of moisture in the conduit can cause moisture to drop down on bus bars or other current-conducting parts, tending to create leakage currents or arc flashovers. If the presence of moisture is detected in the cabinet and is due to such condensation, it is recommended that the conduits be sealed where they enter the cabinet and a means be provided to drain them to keep any condensation from coming in contact with electrical parts of the panelboard or switchboard.

REFERENCES

1. D. Beeman, "Industrial Power Systems Handbook," McGraw-Hill Book Company, New York, 1955.
2. Charles I. Hubert, "Preventive Maintenance of Electrical Equipment," McGraw-Hill Book Company, New York, 1955.
3. Joseph F. McPartland, "Electrical Systems for Power and Light," McGraw-Hill Book Company, New York, 1964.
4. Russell O. Ohlson, Procedure for Determining Maximum Short Circuit Values in Electrical Distribution Systems, *Trans. IEEE*, vol. IGA-3, no. 2, March/April 1967.
5. "Electric Power Distribution for Industrial Plants," IEEE 141, The Institute of Electrical and Electronics Engineers, New York, 1969.
6. "Electric Systems for Commercial Buildings," IEEE 241, The Institute of Electrical and Electronics Engineers, New York, 1964.
7. Editors of *Power*, "Industrial Electrical Systems, Energy Systems Engineering," McGraw-Hill Book Company, New York, 1967.
8. "NEMA Instructions for the Installation, Operation and Care of Panelboards," National Electrical Manufacturers Association, 1942.
9. "NEMA Instructions for Safe Installation, Operation and Maintenance of Panelboards," Pub. PB 1.1-1975, National Electrical Manufacturers Association.
10. "NEMA Instructions for Safe Handling, Installation, Operation and Maintenance of Switchboards," Pub. PB 2.1-1975, National Electrical Manufacturers Association.

22

Polyphase-Motor Control

T. F. BELLINGER*

*Consulting Engineer, Controls Division, Allis-Chalmers Corporation; Registered Professional Engineer (Wis.).

FOREWORD

Ac polyphase-motor controls generally serve in a wide range of applications in which larger, more costly machines are controlled and protected. They differ considerably from the general-purpose controls covered in Sec. 2 or in Sec. 23 because the general-purpose starters are built in high-volume production to meet industry needs for economical control of an almost infinite number of different applications for smaller motors.

When applying larger ac polyphase motors, greater care must be given to selecting the proper control to meet the load and power-system requirements. This section is written to aid in this selection.

GENERAL

1. Definition and Ratings. Ac polyphase motor-control equipment is a general terminology which covers a broad complement of voltage and horsepower ratings and innumerable combinations of equipment arrangements and operational functions. All such equipment is designed and produced in accordance with NEMA Standards Publication "Industrial Controls and Systems."

Both motor starter and motor controller terms are used interchangeably to describe equipment units; "motor starter" usually implies simpler start-stop low-voltage functional units. Large low-voltage equipment generally implies ratings above 100 hp commencing with NEMA size 5 up through NEMA size 9 (Sec. 2, Table 1). Equipment with ratings of 100 hp and below is considered to be in the smaller-equipment class suitable for plug-in construction in motor-control-center arrangements (Secs. 2 and 23).

High-voltage motor-control equipment usually commences in the area of 100 hp and extends up to several thousand horsepower with application limits provided in Table 1. Low-voltage equipment would normally be considered below 100 hp because of equipment size and cost. High-voltage installations are considered primarily because of higher efficiencies in power transmission and distribution, improved power-system stabilities, direct utilization of generated or service power, maximum horsepower ratings, etc.

Motor-starting equipment for polyphase motors is basically identifed by current National Electrical Manufacturers Association (NEMA) industrial control standards. The equipment is divided into two distinct voltage classes, low-voltage and high-voltage. Low-voltage is defined as equipment for application on service voltages below 600 V. High-voltage motor-control equipment ranges from 2200 to 4800 V. The most common applications are 2300 and 4160 V. Above 4800 V, metal-enclosed switchgear equipment must be employed for polyphase motor operation. Metal-enclosed switchgear and metal-clad switchgear can be employed for motor-starting service throughout the two motor-control-equipment voltage ranges, but switchgear equipment has definite duty-cycle limitations and usually is considerably more costly (Sec. 1). In general, the service voltage for which motor-control equipment is specifically designed and applied can operate continuously at a plus 10% value and a minus 15% value.

TABLE 1. Horsepower and Current Ratings of Class E Controllers and Line Contactors

Size of controller and contactor	Current ratings,		2200−2400 V, 3-phase			4000−4800 V, 3-phase		
				Synchronous motors			Synchronous motors	
	Continuous	Service limit*	Induction motors	80% power factor	100% power factor	Induction motors	80% power factor	100% power factor
H2	180	207	700	700	900	1250	1250	1500
H3	360	414	1500	1500	1750	2500	2500	3000

*1.15 times the continuous-current rating of the controller.
From "Industrial Control and Systems," NEMA Standard 1970.

2. Purpose and Functions. Polyphase motor-control equipment can provide isolation and short-circuit protection, motor switching, operating sequence, and motor protection. Although the primary application of such equipment is for polyphase motors, it can also be applied with power transformers, resistance banks, induction furnaces, lamp banks, capacitor banks, etc. Each particular type of load will have specific application factors which may affect the controller rating and performance. NEMA industrial control standards, National Electrical Code, and manufacturers' application data should be consulted for specific details. These paragraphs will be primarily directed toward motor-control applications.

3. Isolation and Fault Protection. System fault protection in combination with motor-control equipment is provided by fuses or circuit breakers. Such devices when furnished will normally also provide or be an integral part of a disconnect means for isolating the controller motor circuit from the feed power system.

Extreme caution must be exercised in selecting and coordinating correct fault protection. Such protective devices not only should be compatible with the power system but also must be coordinated with controller overcurrent capabilities to provide for optimum protection and without nuisance tripping during motor starting and operation (Sec. 13).

Proper application data must be used when selecting fuses or circuit breakers for loads other than motor applications. Fault protection is provided in high-voltage motor controllers by high-interrupting-capacity, current-limiting fuses. Such fuses are especially designed for motor service with capability of carrying high starting currents during prolonged acceleration without fuse deterioration or nuisance blowing. These fuses normally are without actual current ratings but are provided with application designations with typical melting time, total arc clearing time, and current-limiting characteristics (Figs. 1 to 3).

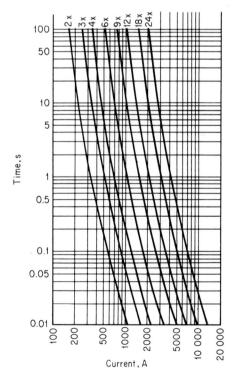

Fig. 1 Typical minimum melting curve for high-voltage current-limiting motor-controller fuses rated 80 000 A asymmetrical.

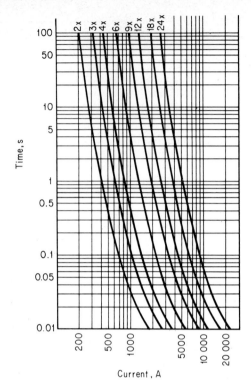

Fig. 2 Typical total fault-clearing-time characteristics of high-voltage current-limiting motor-controller fuses rated 80 000 A asymmetrical.

Controller fuses are available in various sizes up to motor full-load currents approaching 600 A. These fuses will clear faults on systems delivering up to 80 000 A asymmetrical in less than 1 cycle. The peak let-through values of fault current are substantially lower than available peaks (Fig. 3).

High-voltage motor-controller fuses must be carefully coordinated with controller overload protection to obtain intersection of corresponding protective characteristics and to provide protection for the fuses from needless blowing on starting currents. Adequate coordination requires values of motor full-load current, service factor, full-voltage locked-rotor current, accelerating time, and operating duty cycle. The fuses must be selected to permit two successive starts of the motor with the motor and fuses being initially at room ambient. Motor-controller fuses are suitable for application with transformers and other types of ac polyphase loads where overload protection is provided other than through the high-voltage fuses.

4. Motor Switching. Motor switching consists of energizing and deenergizing the motor stators to accomplish starting and stopping operations. In motor-control equipment, switching operations are accomplished with contactors. Contactors are power-utilization electromechanical switches which are magnetically closed and magnetically held. They are multiple-pole switches capable of repetitive operations with mechanical life expectancy that may approach a million or more operations. High-voltage contactors may have increased interrupting ratings up to 50 000 kVA at specified service voltages. Contactor switches are normally the air type, but oil-immersed and vacuum types are often employed in high-voltage motor-control equipment (Table 2).

Contactors should not be confused with circuit breakers, although they may be used interchangeably for specific switching applications. In principle, circuit breakers are

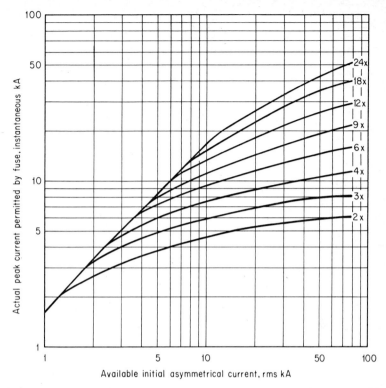

Fig. 3 Typical peak let-through current-characteristic curves for high-voltage current-limiting motor-controller fuses rated 80 000 A asymmetrical.

rugged, massive mechanisms unsuited for a repetitive duty cycle; they have a limited operational life and are designed for short-circuit power interruptions.

Some special-purpose contactors may also be available with mechanical or magnetic latches to maintain them in the operating position. In this case, the contactors operate initially closed magnetically and are maintained closed by means of the latching mechanism. Opening or tripping must then be initiated to release the latch mechanically to

TABLE 2. Horsepower Ratings of High-Voltage Ac Contactors

	hp					
	2200 – 2300 V			4000 – 4600 V		
	Synchronous			Synchronous		
8-h open rating, A	100% power factor	80% power factor	Induction	100% power factor	80% power factor	Induction
100	450	350	350			
200	900	700	700	1500	1250	1250
400	1750	1500	1500	3000	2500	2500

From "Industrial Control and Systems," NEMA Standard 1970.

open the contactor. Such contactors still retain contactor ratings and are used primarily where a maximum degree of load continuity is required. Typical loads are feeder transformers, capacitor banks, and critical motor loads. Latched contactors require special consideration in application and operation and generally are nonstandard in motor-control equipment.

5. Operating Sequence. Ac motor-control equipment is capable of starting and operating in automatic sequence which may incorporate external personnel safety functions, equipment safety functions, or load-control functions in addition to inherent functions built into the controller unit. Where remote contacts must be connected into the controller circuit, the impedance for remote connections must be kept within safe limits. Excessive impedances can prevent initial operation or dropout of motor-control equipment.

Remote sequencing or interlocking may include pressure switches, level switches, position switches, temperature switches, feeds, clutches, load valves, pilot stations, and numerous others. Inherent sequencing will normally be a factor in increment-type starting of squirrel-cage motors and for most synchronous and wound-rotor controllers. Squirel-cage motors starting on full voltage normally do not incorporate any inherent sequencing in their associated motor starters or controllers.

The starting and operation of ac polyphase motors is normally initiated through the use of pilot or control contacts or through the use of manually operated push buttons or switches. Motors are usually started by closing the initiating contact which maintains an uninterrupted control power circuit during motor operation. Interruption of this control power by initiating protective-device contacts causes the motors to stop. Any sensing or manual contact which is maintained open and closed may be used directly for starting and stopping of motors. It is called 2-wire control and provides low-voltage release operation (Fig. 4).

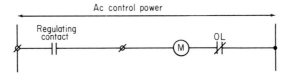

Fig. 4   Two-wire control circuit for controlling motors with maintained function contact provides "low-voltage release."

Another operating circuit is the 3-wire control circuit which is functional only with momentary-start and momentary-stop push-button contacts. After motor starting is initiated, an auxiliary interlock or seal contact on the motor line switch closes to bypass the start push-button contact. Opening the momentary-stop push button or any of the protective contacts deenergizes the control circuit to stop the motor. Such a control circuit is referred to as 3-wire control and provides low-voltage protection (Fig. 5). The 3-wire control circuits are applicable only for manual operation.

A third control circuit is a basic 3-wire control but provides a feature of the 2-wire circuits. This circuit is referred to as "time delay under voltage" but is more properly known as time-delay loss-of-voltage circuit (Fig. 6). The motor starts and stops nor-

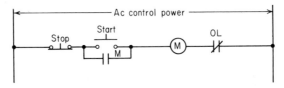

Fig. 5   Three-wire control circuit for controlling motors with momentary-start and -stop push buttons; provides "low-voltage protection."

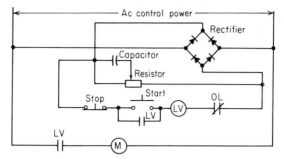

Fig. 6 Time-delay loss-of-voltage protection circuit for controlling motors provides automatic time-delay restarting on power loss.

mally as for a 3-wire circuit. However, should a voltage failure occur during motor operation, the seal contact will remain closed for a short set delay. Should the voltage be restored before the delay expires, the motor will automatically restart as with a 2-wire control circuit. Should the seal contact open before voltage is restored, the motor must again be restarted manually with the start push button. Extreme care should be exercised in using these time-delay restarting circuits.

Several motors attempting to restart automatically and simultaneously may overburden the power system. This type of control circuit should be applied to critical squirrel-cage motors, and time delays can be staggered with most critical drives having the longest delays. Generally 2-wire control circuits or time-delay loss-of-voltage circuits are recommended for use only with squirrel-cage motors. These 2-wire controls usually are considered hazardous for synchronous motors or other types of motors with very limited starting winding capabilities.

6. Motor Protection. Standard motor-control equipment provides for thermal-type overload protection and for voltage protection in the form of "low-voltage release" or "low-voltage protection." Two-phase overload protection is adequate for the majority of applications, but 3-phase protection is recommended as a standard practice. Running-overload relays normally may not protect for stall, especially with the relays initially at room ambient. The factors which affect the selection of proper thermal-overload protection are motor nameplate current, service factor, accelerating time, frequency of starting, and coordination with the fault protector (fuses or circuit breaker).

Increment-step controllers, synchronous controllers, wound-rotor controllers, and other types of more complicated units with additional protective functions as dictated by the operating hazards will be described below. Numerous special protective functions can be obtained on ac industrial-control equipment. Although these are most commonly associated with high-voltage controllers, many protective functions can be obtained on large low-voltage units where warranted. Some of the more common special protective functions that can be obtained are inverse overcurrents, calibrated undervoltage, instantaneous or inverse time, current open phase, current balance, voltage balance, phase-sequence reversal, underfrequency, current differential, zero sequence or ground fault, embedded-detector stator temperature and bearing temperature, controller internal condensation, stall protection, frequent starting, and negative sequence. Refer to Sec. 12 for detailed information regarding motor protection.

POLYPHASE SQUIRREL-CAGE INDUCTION-MOTOR STARTING

7. Polyphase squirrel-cage induction motors account for the large majority of all ac drives. Such motors are self-starting, requiring only that power be connected to the stator terminals to start and operate. They operate at normal speed approaching but below their base synchronous speeds. The operating speeds are nonadjustable with speeds determined by the connected loads. The starting and running speed-torque characteristics vary considerably with the specific motor design corresponding to drive requirements. The kVA requirements for starting and accelerating squirrel-cage motors are several times greater than for normal operation.

Squirrel-cage induction motors are normally started across the line (full-voltage starting) to permit the application of starting equipment with least complexity and investment. Such starting provides maximum starting torques and minimum accelerating times. Figure 7 illustrates current and power-factor characteristics vs. motor speeds. Locked-rotor currents of some motors may approach ten times their full-load operating currents.

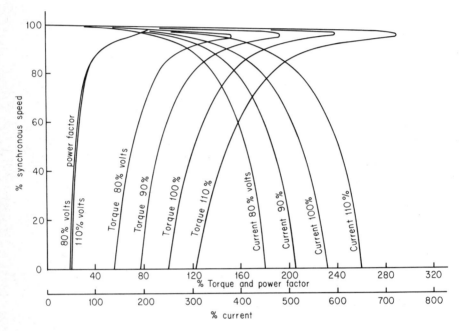

Fig. 7 Full-voltage starting characteristics for a large, 4000-V ac polyphase squirrel-cage induction motor, also typical for synchronous motors with starting windings.

8. High-Voltage Motor Control.   High-voltage motor-control equipment is basically of floor-mounted, free-standing, metal-enclosed, front-access construction. Controllers equipped with air-break contactors are available for tier mounting in a common structure with common vertical bus similar to low-voltage motor-control-center construction. Individual controller structures can be set side by side to form group assemblies and be provided with common horizontal power bus. An industry standard for air-break contactor controllers permits alignment of top-mounted bus structures of controllers of various manufacturers. Standard bus ratings are available at 1000, 1200, and 2000 A with the 1000-A rating being standard. Ratings and required control functions may limit or restrict tier mounting of individual controllers. The individual motor controllers are equipped with high-interrupting-capacity current-limiting fuses normally having an 80 000-A asymmetrical interrupting rating and line-isolating switch externally operated.

High-voltage motor-control equipment with oil-immersed contactors are normally not tier-stacked. Current-limiting fuses and an isolating switch are provided the same as for air-break controllers. Controllers may be grouped together and equipped with common horizontal power bus. Oil-immersed controllers normally will not bus-align with high-voltage air controllers. Controllers with vacuum contactors can be grouped and bus connected like air contactor units and can also be tier-mounted.

9. Motor-Control-Equipment Enclosures. Ac motor-control equipment is available in various types of sheet-steel enclosures designed for personnel protection and to protect internal equipment from harmful environments (Sec. 16). Both low-voltage and high-voltage control equipment are standard with NEMA Type I general-purpose enclosures. Free-standing, floor-mounted enclosures are normally available with NEMA I with gaskets, NEMA III, NEMA IV, and NEMA XII enclosure types. Walk-in arrangements for outdoor locations are generally available. High-voltage oil-immersed controllers can be obtained for Class 1, Group D, Division II, semihazardous locations. Limited special units are available for Class 1, Group D, Division I, hazardous locations (Sec. 8).

10. Full-Voltage (Across-the-Line) Squirrel-Cage-Motor Control. Certain factors must be considered for starting polyphase squirrel-cage motors on full voltage because of the high starting currents which prevail throughout motor acceleration. These factors are construction and condition of drive motors, type of loads being driven, and the power systems to which the motors are connected. Most new motors today are so constructed that they are suitable for across-the-line starting. The primary concern is the bracing of the coil ends extending from the stator iron. Older motors which have developed embrittled insulation may not be capable of full-voltage-starting shocks. Most driven equipment is capable of full-voltage starting and acceleration.

If squirrel-cage motors are to be started on full voltage, the power systems must be capable of supplying the high starting burdens. Generator or substation capacities and distribution circuits must be capable of supplying starting burdens without excessive voltage drop at controller and motor terminals. Excessive voltage drop may sufficiently reduce torque capabilities of motors, preventing successful acceleration to operating speeds. Torque capabilities vary as a square of the per unit values of terminal voltages. Motor-voltage guarantees are normally plus and minus 10% of nameplate, while controllers have a guarantee of plus 10% and minus 15% of nameplate. Motors and control equipment can both be seriously damaged by operation outside these limits. Low line voltage is the most common and most serious of all operating hazards for ac electrical-equipment installations.

The control equipment for full-voltage starting of either low-voltage or high-voltage ac polyphase squirrel-cage motors is similar in performance although vastly different in construction and size. The basic controller functions consist of line isolation or disconnect, short-circuit protection, line switch or contactor, and running-overload protection.

The line-isolating switch may be a separate component or in combination with the short-circuit protector. Short-circuit protection may be provided by circuit breakers or fuses on low-voltage controllers but normally will be in the form of high-interrupting capacity current-limiting fuses in high-voltage controllers. Magnetically operated contactors powered from one phase of the stator power source are used to connect and disconnect the motor stators.

Running-overload protection is normally provided with current-sensing bimetal thermal relays. This protection has been supplied in 2 or 3 phases; however, 3-phase protection is now standard. Controllers are completely metal-enclosed with compartment doors interlocked with line-isolating switches to provide personnel safety. Figure 8 is a full-voltage squirrel-cage motor controller.

With disconnect switch closed, operating the start push button energizes the operating coil of the contactor M, which connects line power to motor stator terminals. The opening of the stop push button or the overload-relay contacts interrupts the seal current to the operating coil, which in turn interrupts stator power. When the line contacts close, locked-rotor current instantly flows and proportionally high starting current continues as the motor accelerates to operating speed. Jogging the start button for rotor positioning places a severe burden on both controller and motor. The 3-wire push-button circuit shown may be replaced with a 2-wire control contact. Adequate control voltage for the contactor operating coil must be maintained throughout motor starting and operation. The contactor dropout point on low voltage is indefinite, and serious damage can develop should a sustained low-voltage condition exist.

11. Reduced-kVA Squirrel-Cage-Motor Control. Squirrel-cage ac polyphase motors should be started across the line on full voltage provided the motors, driven loads, and power systems are capable of withstanding starting impacts. If any of these three condi-

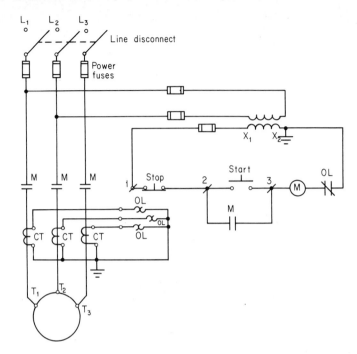

Fig. 8 Across-the-line motor controller, providing full-voltage starting.

tions cannot be met, starting methods should be considered which cushion motor start-ing by reducing starting kVA. This reduction is accomplished by various methods, which in effect reduce the values of line starting currents with line voltage theoretically maintained constant. The lowering of line currents below full-voltage starting currents is accomplished through starting elements which reduce the voltage at motor terminals or through reconnection of stator windings on full line voltage. Reduced-kVA starting methods are commonly referred to as "reduced-voltage starting," erroneously where stator windings are reconnected on line voltage.

Reduced-kVA starting employs line or primary reactors, primary resistors, neutral re-actors, autotransformers, part-winding stator connections, and wye-delta stator connec-tions. Each method basically reduces starting currents below normal full-voltage start-ing currents (Table 3). The substantial reduction of motor-starting torques associated with reduced-kVA starting requires serious consideration in equipment applications. Where motors are actually started on reduced terminal voltages, the motors must be capable of accelerating their loads to near full speed on the selected value of reduced voltage or the soft-starting advantage will be lost when transfer to full voltage occurs (Fig. 9).

Various methods of reduced-kVA motor starting involve the following application fac-tors:

1. Reduced motor breakaway and accelerating torques
2. Substantially longer accelerating times
3. Maximum motor speeds obtainable on starting connections
4. Compliance with starting limitations
5. Equipment duty cycle
6. Higher cost of starting equipment
7. More complex starting equipment

a. Primary-Reactor and Primary-Resistor Motor Control. Primary (line) reactors or resistors are commonly applied in motor-control equipment to reduce starting kVA.

TABLE 3. General Comparison of Starting Conditions among Three Types of Reduced Voltage Starters

Characteristic	Autotransformer type	Primary-resistor type	Reactor type
Starting line current at same motor terminal voltage	Least	More than autotransformer type	
Starting power factor	Low	High	Low
Quantity of power drawn from the line during starting	Less	More than autotransformer type	
Torque	Torque increases slightly with speed	Torque increases greatly with speed	
Torque efficiency	The autotransformer-type starter provides the highest motor-starting torque of any reduced-voltage starter for each ampere drawn from the line during the starting period	Low	Low
Smoothness of acceleration	Smooth	Smoother. As the motor gains speed the current decreases. Voltage drop across resistor or reactor decreases and motor terminal voltage increases. Transfer is made with little change in motor terminal voltage	
Cost	Equal	Lower in small sizes, otherwise equal	Equal
Ease of control	Equal	Equal	Equal
Maintenance	Equal	Equal	Equal
Safety	Equal	Equal	Equal
Reliability	Equal	Equal	Equal
Line disturbance	Varies with line conditions and type of load		

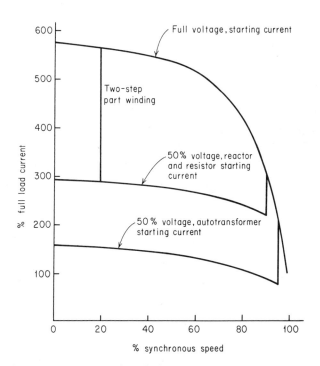

Fig. 9 Typical starting-current curves for polyphase motors with reduced-kVA starting. Transfer from starting voltage to line voltage is accomplished without circuit interruption.

Such starting elements are inserted between the line and motor terminals to lower the motor terminal voltages. Starting current flowing through series reactors produces reverse emf while series resistors produce *IR* drops which in either case subtract proportionally from applied line voltage. The values of starting voltages can be varied by the selection of starting-voltage taps. Starting elements are designed from values of motor locked-rotor currents, motor-starting power factors, duty cycles, and starting-kVA limitations. Primary-reactor control equipment is commonly associated with high-voltage applications, and primary-resistor controllers are commonly associated with smaller-horsepower low-voltage applications. Primary reactors for general application are NEMA medium-duty providing 50, 65, and 80% starting-voltage taps. Primary resistors for general applications are NEMA Class 114 providing 65 and 80% starting-voltage taps. Primary-reactor and primary-resistor starting methods provide rising motor terminal-voltage characteristics with increasing motor speeds and are without power interruption during transition from starting connections to normal operating connections. Motor torques on these starting voltages are basically proportional to the square of the per unit voltage tap. Starting performance with primary resistors or primary reactors is similar except for the lower power efficiency of starting resistors. Table 4 gives reactor and autotransformer duty-cycle ratings.

Primary-reactor or primary-resistor reduced-voltage motor-control equipment provides basically the same operation and protective functions as described for full-voltage control equipment. Some may include time-limit or overtemperature protection for the starting elements, which usually are short-time, intermittent rated. The closing of contactor 1M (Fig. 10) causes current to flow through the starting element, applying reduced voltage to motor terminals. After the motor has accelerated to adequate speed, contactor 2M closes to short out the starting elements, thus connecting full-line voltage to motor terminals. The full-voltage transition is normally controlled by a preset time delay but may also be initiated by a current or speed signal.

b. Neutral-Reactor Motor Control. Reduced-kVA starting with neutral reactors is identical in performance to that described for primary or line reactors in the preceding paragraph. Neutral reactors require special motors having both ends of each phase winding brought out to terminals (total of six terminals). The motors operate wye-connected, and the starting-reactor elements are connected in the neutral. This starting method simplifies switching connections and permits a lower level of insulation for the neutral reactor than would be required for an equivalent line reactor. The neutral-reactor starting method is generally not employed on motor-control equipment but is more likely to be encountered on high-voltage switchgear equipment above 5-kV ratings.

c. Autotransformer Motor Control. The most effective reduced-kVA, reduced-voltage method employs starting autotransformers. The same basic design considerations must be given for these starting elements as for primary reactors and primary resistor. However, autotransformer circuits provide near constant voltage on the starting connections regardless of motor speeds. Such autotransformers are generally NEMA medium-duty class with 50, 65, and 80% starting-voltage taps. Starting currents are reduced by direct transformation of line voltage to the selected lower starting voltage which is connected to motor terminals.

Theoretically, the current drawn from the power line is inversely proportional to the per unit starting voltage squared. The line must also supply magnetizing currents which may be 10 to 20% of full-load current. Autotransformers may either be 2-leg or 3-leg, with similar performance characteristics. The slight phase unbalance present with 2-leg autotransformer circuits normally has negligible effect in the starting of motors. Where very large motors are started under rigid restrictions, 3-leg autotransformers would be employed.

Autotransformer-type reduced-kVA motor-control equipment is more complex and more expensive than comparable primary-reactor equipment. Generally autotransformer-type units are not provided for high-voltage motors except where the current drawn from the power line on starting must be held at an absolute minimum. However, autotransformer-type motor-control equipment is conventional for low-voltage applications. Here power systems are usually found to be less substantial, especially for start-

TABLE 4. Duty-Cycle Rating of Polyphase Motor-Starting Autotransformers and Reactors

Rating. Heavy-duty service includes applications which require frequent starting or jogging, such as certain steel-mill drives and other special drives, including those having extremely high inertia.

Medium-duty service includes applications to motors which drive loads such as fans, pumps, compressors, and line shafts.

Heavy Duty:

On	1 min
Off	1 min
Repeat	4 times (for a total of 5 cycles)
Rest	2 h
On	1 min
Off	1 min
Repeat	4 times (for a total of 5 cycles)
Tap	Lowest tap
Load	Motor with rotor blocked or connected to equivalent inductive load

Medium duty—magnetic controllers for motors more than 200 to 3000 hp, inclusive:

On	30 s
Off	30 s
Repeat	2 times (for a total of 3 cycles)
Rest	1 h

Medium duty—manual controllers, 300 hp or less:

On	15 s
Off	3 min 45 s
Repeat	3 times (for a total of 4 cycles)
Rest	2 h
On	15 s
Off	3 min 45 s
Repeat	3 times (for a total of 4 cycles)
Tap	65%
Tap current	300% of motor full-load current
Power factor	50% or less

Medium duty—magnetic controllers for motors 200 hp or less:

On	15 s
Off	3 min 45 s
Repeat	14 times (for a total of 15 cycles)
Rest	2 h
On	15 s
Off	3 min 45 s
Repeat	14 times (for a total of 15 cycles)
Tap	65%
Tap current	300% of motor full-load current
Power factor	50% or less
On	30 s
Off	30 s
Repeat	2 times (for a total of 3 cycles)
Tap	65%
Tap current	300% of motor full-load current
Power factor	50% or less

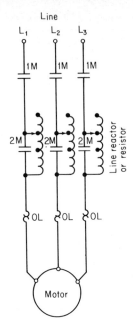

Fig. 10 Elementary power circuit for primary reactor or resistor reduced-kVA polyphase motor controllers.

ing of larger-horsepower motors. In theory, the motor starting on 50% primary reactor tap would draw half of full-voltage locked-rotor current from the power line; the same motor being started on a 50% autotransformer tap would draw only one-fourth of the normal full-voltage locked-rotor current from the power line.

The basic operation and protective functions are similar to those described in Art. 11*a*. In addition, mechanical as well as electrical interlocking may be required to prevent circuit contactors from accidentally being simultaneously closed, which would initiate a circuit fault. A conventional 2-leg autotransformer-type motor controller is shown in Fig. 11. Contactors *S* and *M* first simultaneously close to establish the starting circuit to connect reduced voltage to the motor terminals. After the motor reaches adequate speed, contactor *S* opens, providing a second reduced-voltage connection with tap-portion autotransformer coils in series-reactor connection with outside motor phases. The "run" contactor closes as the third step to connect full voltage to the motor terminals for normal operation. This arrrangement provides a closed transition between start and full-voltage operating connections which is normally initiated by preset time delay.

The motors must be capable of accelerating their connected loads to near full speed on the reduced-voltage connections in order to achieve minimum starting shocks. Transferring to full-voltage connections below optimum speeds results in line currents which may approach the severity of starting from rest on full line voltage. Open-transition start connections may be used but are considered less desirable because of high current steps at transfer. Starting autotransformers are normally intermittently rated and must be applied and operated within their rating capabilities (Table 4).

d. Wye-Delta Motor Control. Polyphase squirrel-cage induction motors that are designed for delta operation and have both ends of each phase winding brought out to terminals provide a simple method of reducing starting kVA through stator winding connections. These motors are started with their stator windings connected in wye across line voltage. The motors must be capable of accelerating their connected loads to a point approaching their normal operating speeds. At this point the motor-control

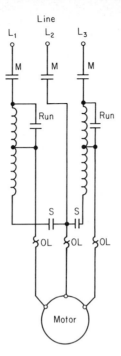

Fig. 11 Elementary power diagram for 2-leg autotransformer reduced-kVA ac polyphase motor controller.

equipment will disconnect the wye stator circuit from the line and reconnect to the line with stator windings in delta connection. The transition from wye to delta is an open transition with starting currents completely interrupted and reestablished on delta connection.

Closed transition, however, is possible by initiating power resistor circuits with currents flowing from the line during the interval of stator-winding reconnection. The transition resistors are connected first in parallel and then in series with the motor windings to simulate starting currents but produce negligible torques. The currents which flow through the resistor transition circuits approach the values of currents in the wye starting circuits and are produced only to maintain an uninterrupted flow of line current. This feature is desirable to maintain advantages on certain power systems on which step regulators are employed.

Delta motors which start on wye produce only one-third of normal breakaway and starting torques while operating on the wye connections. This method is without the adjustment flexibility that is provided with the reduced-voltage starting methods. Because of low fixed breakaway and accelerating torques, wye-delta starting is limited to low-inertia drives which have provisions to be completely unloaded during start-up. Practically all wye-delta starting is employed on low-voltage installations. The closed-transition circuit is normally considered optional and is applied only where power-system stability conditions so dictate.

Wye-delta motor-control equipment supplies the same basic operational and protective functions as outlined for full-voltage motor-control equipment. Line contactors S and $1M$ (Fig. 12) close simultaneously to connect the motor windings in wye across line voltage. After the motor accelerates approaching full speed, a preset time delay interrupts contactor S and initiates the closing of contactor $2M$. With contactor $1M$ having been initially closed on start-up, the closing of contactor $2M$ connects the motor in delta across the line for normal operation.

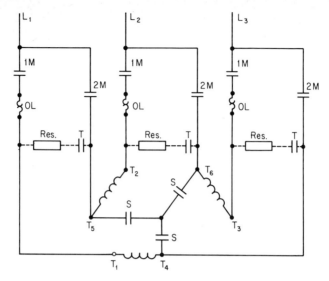

Fig. 12 Elementary power diagram for wye-delta run reduced-kVA polyphase motor controller. Transition to run is open, but closed transition can be provided with dotted resistor circuit.

Mechanical as well as electrical interlocking usually is provided between contactors *S* and *2M*. The closed-transition circuit is shown dotted. With the addition of this circuit, the transition sequence would be for contactor *T* to close to connect transition resistors in parallel with respective stator phase winding in wye connection. The closing of the *T* contactor is immediately followed by the opening of contactor *S*, which then establishes the transition resistors in series with phase windings in delta connection. The final step is the immediate closing of contactor *2M*, shorting out the transition circuit, for normal delta operating connections. Since closed-transition resistors have very short time intermittent ratings, circuits to provide their protection are recommended.

e. Part-Winding Motor-Control Equipment. Reduction of motor-starting kVA under limited starting conditions is possible through partial polyphase stator connections to the power lines, referred to as part-winding starting. This method requires that the stator windings of motors be comprised of two or more parallel winding circuits per phase with individual line terminals for each corresponding winding. Two parallel windings per phase would thus require six line terminals, permitting motors to be started with only one winding connected to the line followed by the connection of the second winding to the common power line placing both windings in parallel for normal operation.

Motors which are suitable for starting on part winding may be especially constructed for that purpose or may be of dual-voltage construction such as those available for 230/460-V operation where phase windings would be connected in parallel for the lower operating voltage. Part-winding starting can be accomplished here by first providing two independent internal stator circuits which are identical; the first would be connected to the power line and then the second in close sequence. Large motors which are inherently wound with parallel-phase conductors lend themselves for special connection box-termination for the part-winding-starting method. Three separate windings for three-step starting is usually the maximum that would be employed.

Theoretically, the normal full-voltage motor-starting current would be divided equally between the number of windings provided in the motor. A two-winding, two-step installation should result in starting kVA with one winding connected equal to one-half of that for both windings being connected simultaneously. However, the distribution of currents is usually unequal and may exceed a 2/3 ratio. The unbalance of currents is produced by unequal distribution of slot fluxes with various partial winding connections and from external differences in circuit impedances. Starting-current unbalance can be somewhat compensated through special provisions in motor-control equipment. A difference in normal running currents between the separate motor windings can also be expected.

Part-winding starting has limited applications and is primarily employed to stay within limits of motor-control equipment and to reduce the initial starting shock. As a rule, motors are not permitted to accelerate on partial winding because of the hazard in connecting successive windings to the line as motors approach operating speeds. In this area, the air-gap energy of motors theoretically can produce current surges up to twenty times full load if the remaining winding is connected at the least favorable point of phase relationship between power-line voltage and induced voltage on open motor terminals.

To minimize substantially the chance of such harmful power surges, successive part-winding steps are initiated at very low rotor speeds. Such stepping offers no reduction of accelerating currents but provides for increment stepping of the initial starting currents to their full locked-rotor values. This arrangement reduces the initial starting shocks to load equipment and permits regulating equipment on power systems to be actuating as additional starting steps are completed. In general, part-winding starting of polyphase motors should be applied with caution.

Part-winding motor-control equipment supplies operational and protective functions similar to those described for full-voltage motor-control equipment. Contactor $1M$ (Fig. 13) closes first to connect one of the motor windings to full line voltage. Following a

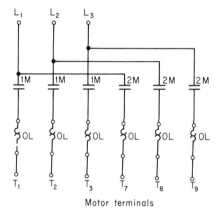

Motor terminals

Fig. 13 Elementary power diagram for two-step part winding reduced-kVA polyphase motor controller must be applied with caution.

short time delay at very low rotor speed, contactor $2M$ closes to make the second winding connection. Both motor windings are thus connected in parallel for normal acceleration and operation. Running-overload protection is provided for each of the independent motor windings because of possible unbalance of load currents.

POLYPHASE-SYNCHRONOUS-MOTOR STARTING

12. Polyphase synchronous motors have stators similar to squirrel-cage induction motors and have rotors with dc slip-ring circuits which must be energized for normal operation. They operate at constant base speeds corresponding to line frequency and number of machine poles (r/min = 120 × frequency/number of poles). Synchronous motors are employed primarily to obtain high pullout torques, constant operating speeds, or generation of leading reactive kVA for system-power-factor correction. They require conventional ac polyphase power sources for their stators and suitable dc power sources for their rotor fields. For normal operation, synchronous motors must be brought near full operating speed, at which point the dc power is connected to the rotating field through slip rings. The motors are accelerated to their synchronizing speeds by means of either built-in starting windings or external auxiliary drives. Nearly all conventional synchronous motors now manufactured have built-in rotor starting windings. Such starting windings are also referred to as squirrel-cage windings, pole-face windings, damper-bar windings, or amortisseur windings. Starting windings are actually squirrel-cage induction bars located in the faces of the dc rotor poles. They produce accelerating torque only and have relatively short time intermittent ratings (Fig. 14). As starting windings they become inoperative at synchronous speeds but serve to dampen any tendency of the rotor to oscillate in angular position with relation to the stator field. Starting characteristics vary with the motor-starting-torque capabilities.

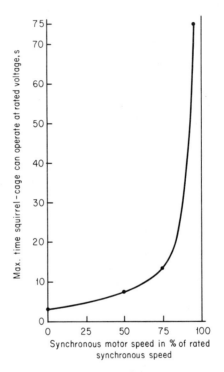

Fig. 14   Thermal-capacity characteristic curve of synchronous-motor-starting winding.

The starting of synchronous motors involves two basic switching functions. The first is the energizing of the stator to produce breakaway torque and acceleration to synchronizing speed, and the second is the energizing of the dc rotor field at the optimum speed and rotor-stator pole relationship. For motors having built-in starting windings, the same equipment considerations are required as for full-voltage or reduced-kVA starting methods for squirrel-cage induction motors. All factors relating to stator circuits are identical. If starting windings are not available, these motors must be driven by external means to synchronizing speed, at which point the proper field switching must be initiated. Because few motors are without starting windings, major emphasis is on starting equipment for conventional motors that contain self-starting cage windings.

13. Field Protection and Dc Power Application. During acceleration of synchronous motors on their squirrel-cage windings, a voltage is induced into the dc windings through transformer action producing proportionally high voltages in the rotor circuits because of the large stator-to-field turns ratio. These induced voltages would normally be far in excess of the puncture voltage levels of the 125-V or 250-V dc insulation classes of field windings. These high voltages are lowered by connecting external starting or discharge resistors across the field slip rings to reduce the open-circuit voltages to safe values. Such voltages are normally held to a maximum of 1200 to 1500 V.

The discharge resistors are a function of motor design, normally being specified by the motor manufacturer, although such resistors are usually provided as an integral part of the motor-control equipment. In addition to bleeding down the induced voltage, the resistor circuits provide a contribution to synchronizing torques at corresponding speeds. The discharge resistors also serve to protect the motor fields from overvoltage on interruption of dc field currents. Field switches for synchronous motors are equipped with two main poles for switching the dc excitation power and a normally closed discharge pole for switching the discharge resistor circuits. The switch operation must be such that there is a momentary overlap between the normally open poles and the normally closed discharge pole as the switch operates to ensure continuity through the rotor circuit at all times. An open field circuit during motor start-up or during field switching will likely cause the field insulation to be punctured.

The motor-induced field currents which flow through the discharge resistors have speed-current characteristics similar to that in Fig. 15. The circuit frequency is proportional to slip, which is equal to the line frequency at stall and zero at synchronous speeds. The current and frequency characteristics of the induced field currents provide a convenient signal for initiating the various field application and protective functions that are extremely critical for the operation and protection of synchronous motors.

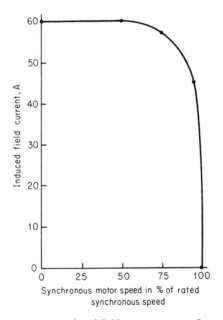

Fig. 15 Typical synchronous-motor-induced field current; current flows through discharge resistor on motor acceleration.

A frequency signal near synchronizing speeds, normally well above 95%, in combination with polarity sensing provides a signal for initiating the closing of the field contactor to energize the rotor field. The initiating signal normally senses both speed and slip phase angle to provide the optimum rotor-stator angular relationship for field application. For most synchronous-motor drives, the optimum field application point occurs as the slip current changes from negative maximum to near zero (Fig. 16). Connecting the dc field at this point provides a maximum pull-in torque and minimum disturbance. Field initiation in less favorable areas may result in serious line disturbances

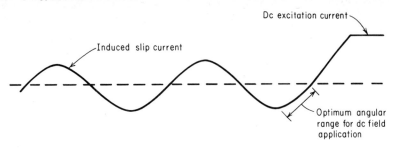

Fig. 16 Field excitation is usually applied at 97 to 99% synchronous speed, and for most motors the angular range shown is optimum.

or complete failure of motors to accelerate and lock into synchronism. The point of field application normally is not critical except for motors that are required to be synchronized under substantial loads. Random field applications like that provided by time-delay relays are normally satisfactory for motors starting unloaded (Fig. 17).

14. Protection for Synchronous-Motor Starting Windings. The most critical protection for synchronous motors is that for the starting windings. These windings are short-time-rated for starting duty only and are most vulnerable at locked-rotor conditions. Optimum protection provides for stall protection while still permitting slip protection in accordance with the characteristic curve of Fig. 14. Squirrel-cage protection is required on motor start-up to ensure that proper sequential synchronizing occurs and that motors will operate continuously in synchronism. The protection must detect and operate for any condition of prolonged subsynchronous operation beyond the thermal capability of the starting windings. Squirrel-cage protection is normally provided through a protective system which integrates subsynchronous speeds with time.

Once synchronous motors have been initially and successfully synchronized, loss of synchronism or pullout may be detected by the presence of induced slip current or voltage superimposed on the dc excitation source. Frequently, the same system used for field application can also be employed for pullout protection where field sensing is used for both functions. Stator power factor may also be used for pullout detection and protection. Such systems respond to various conditions of lagging power factor and remain inoperative under normal operating conditions in which power factors are leading or unity.

Pullout protection which is responsive to stator power factor must be bypassed during motor acceleration and field application to prevent tripping during initial synchronizing of motors. Provisions can be made for immediate shutdown or resynchronizing of synchronous motors following pullout. Because of the vulnerability of synchronous motors, the best practice is to provide for immediate shutdown on pullout except for those installations where positive protection against all combinations of operating hazards is assured.

15. Synchronous-Motor Operating Hazards. Some of the major operating hazards for synchronous motors are operating abuse, low line voltage, low excitation current, and excessive shaft load. The pullout-torque capabilities of synchronous motors are a function of stator and field power. Supplementary protection for synchronous motors is provided through the use of field voltage and current relays and with stator frequency relays. Extreme care must be exercised, especially with large synchronous motors, if attempts are made to reconnect motors to the line after momentary power interruption. Reconnection of line voltage which is substantially out of phase with open-circuit motor terminal voltage can result in extremely high current and torque surges capable of creating system disturbances and mechanical damage. Systems incorporating reclosing circuit breakers and resynchronizing after motor pullouts fall within this danger area.

Additional hazards for synchronous motors are jogging, too frequent starting, stalling, and excessive accelerating times. Any of these conditions are serious hazards to starting windings. Even the best protective system may not protect under such extreme operating conditions. Although protective systems usually follow the heating curves of the

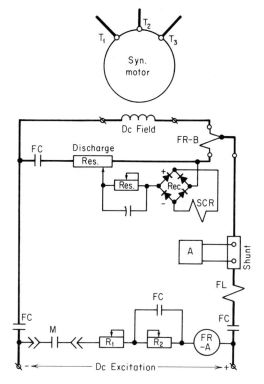

A	Ammeter
FC	Field contactor
FL	Field loss delay
FR	Field application relay
M	Main contactor
Res.	Resistor
R_1, R_2	Resistors
Rec.	Rectifier
SCR	Squirrel cage protective relay

Fig. 17 Field portion of synchronous-motor controller circuit includes field discharge, excitation switching, sensing for field application, starting winding protection, and low excitation protection.

motor windings adequately, it is nearly impossible for them to follow motor cooling because of the great difference in masses between protective devices and motors. Human judgment is still important in protecting synchronous motors.

16. Synchronous-Motor Field Excitation. The dc power for the excitation of synchronous-motor fields may be obtained from plant buses, direct-driven dc generators, individual MG sets, or rectifiers. Provisions are normally required for field-current adjustments. Usually, field currents are set to optimum values only after the motor fields have reached maximum operating temperatures. On cold start-ups, the field currents may be initially 20 to 40% high but will decrease to normal as operating temperatures are reached. The field currents are usually maintained as initially set during motor operation except for those applications in which reactive kVA or power factor is being controlled or regulated.

The rectifier source is the only dc power that can be provided as an integral part of the motor-control equipment. Such rectifiers are normally of the silicon type with transformer taps provided for field-current adjustments. The rectifiers are often powered from the stator line. Adjustment taps are usually in 4 to 5% increments in sufficient number to adjust down to approximately 70% of normal excitation requirements. Full-range continuous adjustment with or without automatic voltage regulation can be provided through the use of silicon-controlled-rectifier supplies.

Dropping resistors or rheostats are normally used with other types of excitation sources for field-current adjustments. Where excitation power is taken from fixed-voltage supplies, such as a plant bus, dropping resistors or rheostats must be used in series with motor fields. These resistors or rheostats usually represent large values of power dissipation and are located externally to motor-control equipment. Series-motor field resistors are available as an integral part of motor controllers but are mounted externally under protective vented covers.

Field currents supplied by individual MG sets or direct-driven exciter generators are set with generator field rheostats. These rheostats may also be obtained as integral parts of controllers but are usually of sufficiently low wattage to be internally mounted with an external operating means. Problems occasionally are experienced with difficulty in voltage buildup of direct-driven exciter generators. Such generators are normally self-excited, and adequate excitation voltages may not be available when the synchronous motors have reached their synchronizing speeds. This problem is minimized through various field forcing means, such as shorting out the field rheostats until normal output of voltage is available.

The trend is toward rectifiers as field-excitation sources. Rectifiers are more efficient and require less maintenance than comparable dc generators. Generator sources provide some operational advantage under marginal pull-in conditions and tend to hold up field currents during momentary voltage dips because of rotating inertias.

17. Brushless-Synchronous-Motor Control. Limited ratings of synchronous motors are available without field slip rings and are referred to as "brushless synchronous motors." Such motors have rotors equipped with rectifier and sensing circuits providing self-contained dc supplies for the rotor fields. The ac power to the rotors is supplied from alternator armatures or through rotary transformers which are shaft-driven. Rotor field currents are adjusted by proportional values in the stationary excitation windings. The rotor circuits contain means for static switching of discharge resistor and excitation power at preset speed–slip angle points. Thus brushless motors provide self-contained excitation sources and automatic field application. However, vital protective function for overload, stall, pullout, field loss, prolonged acceleration, etc., must be provided by motor-control equipment.

Since field circuits are not available for protective signals, only stator power quantities are available to initiate such protective functions. These require monitoring of stator currents, power factors, accelerating times, or motor speeds. The motor-control equipment used in conjunction with brushless synchronous motors is identical to full-voltage and reduced-kVA control equipment for squirrel-cage induction motors except for the addition of the required special protective functions.

WOUND-ROTOR INDUCTION-MOTOR CONTROL EQUIPMENT

18. Polyphase wound-rotor induction motors have stators identical to squirrel-cage induction motors but have rotors with windings brought out to three slip rings. They also have similar speed-torque characteristics but provide the added flexibility of control of starting currents, starting torques, or operational speeds. Desired control operation is obtained through external connections of appropriate values of resistance into the rotor slip-ring circuits. At full-load, full-speed operation, the slip rings of wound-rotor motors must be externally shorted. Operation at reduced-speed points results in reduced efficiency because of resistor power losses.

Wound-rotor induction motors can operate with an infinite number of different speed, torque, or current characteristics. Different performance criteria are obtained by the selection of the external resistance which is connected into the rotor slip-ring circuit. Each different value of external resistance provides a new speed-torque characteristic curve (Fig. 18). Portions of different characteristic curves can be successively sequenced in order to provide controlled acceleration to a desired operating speed.

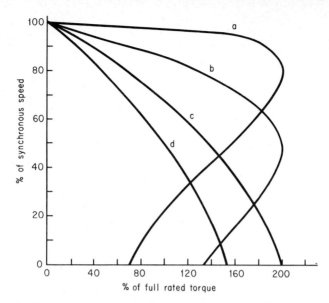

Fig. 18   Family of speed-torque curves for a large polyphase wound-rotor induction motor showing effects of external rotor resistance. Curve *a* is with 0 external resistance, while curves *b*, *c*, and *d* are with increasing values of external rotor resistance.

Primary current, secondary current, and torque are directly related. The control of speed is a function of load, and for a given value of secondary resistance, the wound-rotor motor will operate at a speed at which the resulting developed motor torque equals the load torque. Although secondary resistance can be calculated exactly for a specified operating speed, that speed will be obtained only with the existence of the corresponding load. Calculated values of secondary resistance can be connected so that constant torque such as 100% can be developed at any speed from full-load rated speed to stall with primary and secondary currents remaining theoretically constant. Continuous operation at reduced speeds is practical down to approximately 50%. Below this point, speed may vary considerably with slight changes in load and operation may be unstable. Operating efficiency at reduced speeds may also influence the minimum continuous-operating-speed point. For example, reduced-speed operating at 100% torque creates losses in the resistor rotor circuit equal to nameplate horsepower times slip. Slip is equal to 1 minus per unit synchronous speed. The overall motor efficiency is the product of the rotor efficiency times the stator efficiency.

Wound-rotor motors can be employed for controlled motor starting and acceleration, for continuous-speed regulation, or for a combination of both. This control is accomplished with the proper motor-control equipment and corresponding secondary-resistance design.

19. Secondary-Resistance Considerations.   Design factors for secondary resistance to provide controlled operation of wound-rotor motors involve similar considerations whether for starting-duty or for continuous-speed-duty operation. The theory actually involves the internal design parameters of the individual motors which normally are available only to the motor designer. Practical designs, however, are possible. These designs make use of basic assumptions and approximations which negligibly affect the final results and are adequate for most applications. Secondary-resistance designs which require consideration of internal-motor-design parameters are necessary for applications requiring precise regulation or coordination, but such approaches are not covered here.

A simple understanding of the operation of wound-rotor induction motors can best be obtained by an examination of performance on the speed-torque characteristic curve with resistance shorted out in the area from 0 torque 100% speed to a torque point slightly above 100% where the saturation effect can be neglected and the performance can be considered as a straight-line function (Fig. 18). With this assumption, a value of resistance can readily be calculated that will provide, for example, 100% torque at 0 speed where all the stator input appears as power loss in the rotor resistance. The rotor reactance can justifiably be neglected in this case because the real or resistance component of impedance will be many times larger, as required to obtain 100% torque at 0 speed. With such a condition of unity slip, the rotor frequency is equal to the stator frequency and stator and rotor currents are equal to the rated nameplate values. External rotor resistance is normally wye-connected to simplify resistance-design considerations.

Per unit analysis for resistance design makes use of the value of total rotor resistance, designated R_0, that will produce 100% motor torque at 0 speed. Required are the values of rotor or secondary current I_2 and rotor voltage E_2. I_2 is a value of shorted slip-ring current with motor operating at full rated nameplate capacity. E_2 is the value of voltage across open-circuit slip rings at stall. The voltage across each leg of the rotor wye-connected resistance is $E_2/\sqrt{3}$. The value of total secondary resistance per leg becomes $R_0 = E_2/\sqrt{3} \times I_2$. The total value of secondary resistance required to operate at 100% torque at a desired speed can readily be calculated by merely multiplying $R_0 \times$ per unit slip.

For example, at 80% synchronous speed, the corresponding slip is 0.2 and the total secondary resistance value, R_s, to operate at this speed and produce 100% torque is $0.2 \times R_0$. The value of R_0 is the total secondary resistance value per leg in ohms and includes the internal resistance, R_I, of the rotor winding. To obtain the value of the external resistance R_x, the internal winding resistance R_I must be subtracted; thus $R_x = R_0$ (per unit slip) $- R_I$. The value of R_I can often be neglected for calculating resistance for starting duty only. It can be obtained from an existing motor by carefully measuring the value of resistance from slip ring to slip ring and dividing by 2. A value of secondary resistance required to operate at a desired speed point with other than 100% torque is obtained by multiplying the 0 speed resistance R_0 by per unit slip and dividing by per unit torque. Total secondary resistance required to operate at 50% synchronous speed and 25% torque would equal $0.5 R_0/0.25 = 2 R_0$. The value of external resistance, R_x, would then be obtained by subtracting internal resistance from $2 R_0$. It is important to remember that the actual operating speed at which a wound-rotor induction motor will operate with a given value of rotor resistance is determined only by actual shaft load. Design of secondary resistance for wound-rotor motors involves consideration of motor speed-torque curves, load speed-torque curves, equipment inertias, operating duty cycle, and operation parameters.

The per unit method given for calculating secondary resistance is simplified but does contain some error because the saturation of the motor has been ignored. Methods for exact calculation are available but require internal motor data from the motor manufacturer.

Table 5 lists recommended starting-duty resistor classes for various standard applications. Motor-control equipment normally provides for the acceleration of wound-rotor motors by inserting full resistance on initial start-up followed by successive steps of shorting out resistance increments with the final slip ring shorted as the motor accelerates to full speed (Fig. 19). Table 6 shows the standard number of accelerating steps. Steps may be added or subtracted to meet the requirements of individual drives.

The secondary resistors must be designed for continuous duty if motor operation at reduced speeds is required. Unless the resistors are designed for exact motor loads, the desired operating speeds will not be obtained. Usually, continuous operation below 50% speed is avoided because of low efficiency and relatively wide speed changes with slight changes in loads. Starting-duty and continuous-duty resistors c.n be combined into common assemblies.

Secondary resistors for wound-rotor motors may be an integral part of the motor-control equipment, supplied as functional assemblies for separate mounting or as loose modules for site mounting and interconnection. Secondary resistors must be installed with adequate cooling provisions. The resistors may either be convection- or forced-air-cooled. Resistor modules are usually of the noncorrosive, nonbreakable type that may

TABLE 5. Duty-Cycle Ratings of Wound-Rotor-Motor Secondary Resistance and Similar Applications

Approx % of full-load current on first point starting from rest	Class numbers						
	5 s on 75 s off	10 s on 70 s off	15 s on 75 s off	15 s on 45 s off	15 s on 30 s off	15 s on 15 s off	Continuous duty
25	111	131	141	151	161	171	91
50	112	132	142	152	162	172	92
70	113	133	143	153	163	163	93
100	114	134	144	154	164	164	94
150	115	135	145	155	165	175	95
200 or over	116	136	146	156	166	176	96

From "Industrial Control and Systems," NEMA Standard 1970.

When an armature shunt resistor is added, the class number shall include the suffix AS. For example, Class 155-AS is a resistor which includes an armature shunt and which will allow an initial inrush of 150% with the armature shunt open.

When a dynamic braking resistor is added, the class number shall include the suffix DB. For example, Class 155-DB.

operate in the area of 500°C. Secondary resistors must be protected from operation in excess of their duty-cycle capabilities.

20. Wound-Rotor-Motor Control Equipment. The stator control equipment for wound-rotor induction motors is identical to that for squirrel-cage induction motors. Full-voltage starting equipment is used regardless of the rotor-control system employed whether for starting duty only or in combination with speed-regulation duty. Reduced-voltage starting is not normally employed because it defeats the operational advantages provided by wound-rotor motors. Secondary- or rotor-controller units may be separate from stator control but are often combined into functional-integral assemblies. Secondary controllers usually incorporate the associated resistors designed both for specific motor data and application requirements.

a. Automatic secondary control equipment which employs magnetically operated contactors is by far the most common type and is usually furnished with stator control. The secondary circuit provides for motor starting with all slip-ring resistance connected and a selected number of secondary contactors close in succession, shorting out resistance to accelerate motors to full speed. The contactors may be either 2- or

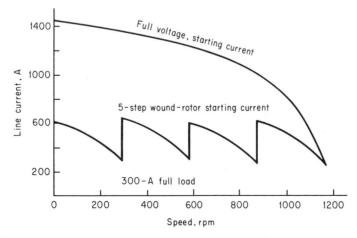

Fig. 19 Sawtooth resistor stepped acceleration curve for a large wound-rotor induction motor.

TABLE 6. Accelerating Steps for Wound-Rotor Controllers Employing Magnetic Secondary Contactors

hp rating	*Min No. of accelerating contactors*
15	1
75	2
150	3
300	4
600	5
1200	6

Values are for wound-rotor induction-motor controllers for nonplugging reversing and nonreversing service.

When a controller is used for reversing plugging service, one plugging contactor is added to the number of contactors required for acceleration.

3-pole as the application dictates, providing balanced starting points with acceleration stepping initiated through functions of speed, currents, time delays, or combination of these functions. Time-delay stepping is the method most commonly used. For starting duty, intermediate stepping contactors are intermittently rated with only the final shorting contactor having a full continuous rating. Where continuous-duty speed points are required, corresponding contactors must have a continuous rating for their respective load currents. A complete wound-rotor controller with a contactor-type secondary is shown in Fig. 20.

b. Manual drum switches may be used for stepping secondary resistance for starting duty as well as speed-regulation duty. Here, the manual drum switch and secondary resistors are independently mounted and are separate from the stator controller. The drum switches must be set to insert all external resistance on initial start and be manually advanced through successive contact steps to complete motor acceleration, with the last step shorting out the slip rings for normal operation. Initial accelerating points usually alternate between balanced and unbalanced rotor connections.

Drum switches are available for acceleration and continuous-speed regulation. Drum switches and resistors must be compatible and suitably rated for speed-regulation duty where required. Speed-regulation operating points have balanced rotor connections. Drum switches are available with various control contacts to be used in conjunction with the stator controller to ensure that drum switches are returned to the initial position before motor starting. The use of secondary drum switches places responsibility for proper stepping and overall equipment protection in the hands of the equipment operators. Manual drum switches are normally limited to about 700 hp maximum. Motor-operated drum switches for higher horsepower ratings were once available but are now almost obsolete. Secondary drum switches require resistor banks that are individually designed for the particular application and drum-switch circuitry. For small motors, usually 25 hp or less, manual secondary control can be provided in the form of ganged faceplate-type rheostats which combine both switching and resistance in a common assembly. Such rheostat applications are uncommon, however.

Secondary control schemes are available which incorporate control reactances and conventional resistor stepping into functional systems for the purpose of obtaining stepless regulation of desired operating characteristics. Such operating systems are employed for conveyors, hoist, elevators, etc., where smooth acceleration under close torque limits is required. The reactors employed may be fixed, incorporating variable-frequency regulation, or may be of the saturable-core type with feedback control.

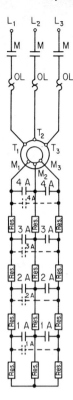

Fig. 20 Wound-rotor induction motor controller; resistor stepping is accomplished with sequenced shorting contactors.

Reactor type regulation is accomplished usually at the expense of considerably higher starting currents because of the high reactive components introduced. Reactors are incorporated primarily for starting service and are shorted out during normal operation. A design of reactor regulating systems is involved and requires complete motor internal stator and rotor winding constants. A simplified wound-rotor-motor secondary regulating system using a nonsaturable reactor in conjunction with secondary resistance is shown in Fig. 21. Control reactors may also be located in stator circuits but normally are employed in secondary circuits because of greater flexibility in the presence of slip frequency.

Stepless regulation of wound-rotor motors is also available through the use of liquid rheostats. Electrodes are raised and lowered into conducting electrolyte which forms a secondary resistance. Such equipment includes electrode-positioning mechanisms and heat exchangers for cooling electrolyte solutions. Such control systems are complex and costly and normally can be justified only where precise regulation of large motor ratings is required.

SPECIAL MOTOR OPERATING FUNCTIONS

Several special operating functions can be provided by standard motor-control equipment. The most common of these are reversing, plugging, dynamic braking, and inching.

21. Reversing. Reversing the direction of rotation of a polyphase motor is accomplished by reversing any two stator leads. This function is accomplished in the

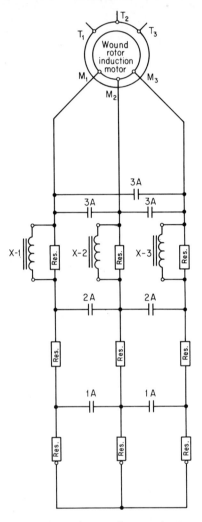

Fig. 21 Reactor-resistor secondary circuit for wound-rotor induction motor; provides for smooth torque-regulated acceleration with few steps.

motor-control equipment by providing an additional line contactor which reverses phase rotation from the first or forward line contactor. The forward and reverse contactors are mechanically and electrically interlocked to prevent accidental simultaneous closing of both switches. All protective functions are common to both forward and reverse operations. Often, the control circuit for reversing is set up for jog operation only. Speed switches or time-delay settings may be employed to prevent the reversal of motor rotation until rotors have reached standstill.

Reversing can be initiated through the various types of reduced-kVA starting as well as for full-voltage starting as described for squirrel-cage induction motors. Reversing can be provided for any type of ac polyphase motor, for such operation does not affect rotor control functions as required for wound-rotor and synchronous motors. Motor-control equipment may be provided with protective circuits to prevent motor operation in a reverse direction (Fig. 22).

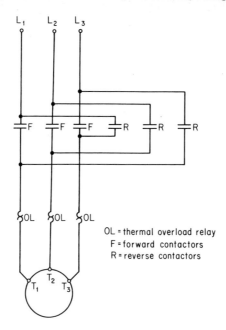

Fig. 22 Elementary power circuit for reversing ac polyphase motor controller.

22. Plugging. Plugging is a reverse operation of motors when being stopped to bring them more quickly to rest. Such operation is employed to provide personnel safety and to eliminate unusually long coasting times. The plugging operation is performed automatically by the motor controller which initiates the closing of the reversing contactor following the opening of the forward contactor. The motor-developed torque drives the motor to rest, at which point a speed switch disconnects the reversing contactor to prevent motor acceleration in the reverse direction.

The forward and reverse contactors are electrically and mechanically interlocked to prevent simultaneous closure. Where large power quantities are involved, the forward contactors may still be carrying interruption arcs if the closure of the reversing contactors is too fast. To eliminate this possibility, forward load monitoring is often employed to lock out the reversing circuit until the forward power interruption has been completed. The average plugging torque is somewhat below the motor stall torque and is produced at a stator current that is a few percent in excess of locked-rotor current. Plugging can be accomplished with reduced-kVA circuits for slowdown of high-inertia drives to minimize the line current during plugging operations. Plugging is normally employed only with squirrel-cage induction or wound-rotor induction motors. Wound-rotor induction motors may be plugged using different values of secondary resistance to meet the desired torque or stator current limitations. Synchronous motors are normally not plugged but are dynamically braked.

23. Stator Dynamic Braking. A method is available for braking squirrel-cage induction or wound-rotor induction motors by applying a proportionally low-voltage dc power to a single stator phase winding to produce braking torque while motors are being brought to rest. Similar to plugging, the braking contactor is closed to apply the dc braking power as soon as the forward or running contactor opens. The braking contactor must have a rating equal to the ac line voltage in order to isolate the dc power circuit adequately while the motor is operating normally. The maximum average braking torque that can be obtained will be less than that for full-voltage stator plugging (Fig. 23).

The value of dc braking voltage must be sufficient to produce approximately 150%

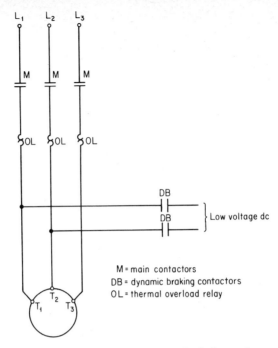

Fig. 23 Elementary power diagram for dynamic braking of polyphase induction motors. Dc power connected to one stator phase produces decelerating torque after main contactor is opened.

stator nameplate current through the phase-winding resistance. The dc power supply may be any normal source such as plant bus, MG set, or rectifiers. For small low-voltage motors, dc-charged capacitor banks may be discharged into a phase winding to produce braking torques. The advantages of dc stator braking are that a control of braking torque is possible and high plugging stator currents are eliminated.

24. Synchronous-Motor Dynamic Braking. Synchronous motors produce the highest braking torques possible for ac polyphase motors through dissipating generated power in external resistance loads. The braking contactor closes the resistor load across the motor terminals after the running line contactor opens. Full dc rotor field is maintained until after the motor comes to rest. The dynamic-braking-resistance load is specifically designed for the individual motor applications. Average braking torque can be in excess of normal running torques. The braking contactors must have voltage ratings equal to line contactors and are usually of the normally closed spring-loaded type for installations requiring personnel safety. Electrical and mechanical interlocking is normally supplied between line and braking contactors.

Dynamic-braking functions can also be provided in synchronous-motor controllers in conjunction with forward- and reverse-operation control functions. For dynamic-braking applications, the field-excitation supply must be held constant throughout the braking interval and therefore cannot be driven from the motor shaft being braked (Fig. 24).

25. Inching. Inching is a special control function that can be provided to cause synchronous motors to rotate at approximately $\frac{1}{2}$ to 1% of normal synchronous speeds. Inching is provided for ease and convenience of precise positioning of driven equipment for maintenance or inspection. Such an operation requires substantial auxiliary equipment and special provisions built into associated synchronous-motor control equipment. Inching of synchronous motors is accomplished by connecting the very-low-frequency low-voltage 3-phase power source to the motor terminals with rotor fields energized. Inching rotational speeds are obtained as rotors lock in synchronism

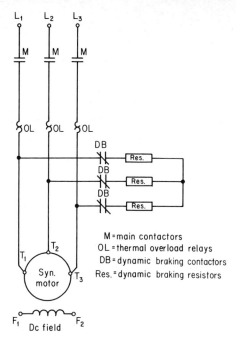

Fig. 24 Elementary diagram of power circuit for dynamic braking of synchronous polyphase motors. Dc field must be maintained during braking operation.

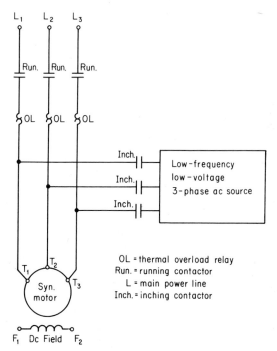

Fig. 25 Elementary diagram of power circuit for low-frequency inching of polyphase synchronous motors. Typical inching speeds are $1/2$ to 1% of normal.

with the applied low-frequency stator power supply. The low-frequency power is produced by modulating a constant dc power source through systematic switching of stepping contactors which establish series-parallel stator winding connections in progressive sequence resulting in rotational stator field. The stepping contactors are normally pulsed with a variable-speed program-type cam switch. It modulates the dc power, which is required to be approximately one-fortieth of the ac-stator voltage rating and the current capacity of approximately 125 to 175% of the ac-stator nameplate current. The power requirements may vary considerably from these values depending upon each individual installation.

A common power-supply system can be employed with several synchronous motors at the same location provided that interlocking prevents simultaneous inching operation. Inching operation is a system function and requires close coordination among power source, drive motor, load, and control equipment (Fig. 25).

REFERENCES

1. Paisley B. Harwood, "Control of Electric Motors," John Wiley & Sons, Inc., New York, 1956.
2. Gerhart W. Heumann, "Magnetic Control of Industrial Motors," John Wiley & Sons, Inc., New York, 1961.
3. Richard W. Jones, "Electric Control Systems," John Wiley & Sons, Inc., New York, 1959.
4. Robert W. Smeaton (ed.), "Motor Application and Maintenance Handbook," McGraw-Hill Book Company, New York, 1969.
5. "Industrial Controls and Systems," NEMA Standards Publication ICS-1970, New York.

23

Control Centers

R. A. GERG*

FOREWORD

In planning for new facilities or a revamp of old facilities for manufacturing or processing, careful consideration should be given to the use of motor-control centers and their advantages over separately mounted control equipment. Early planning will permit advantageous location close to the operations they control or in clean rooms if necessary. This section is written to provide convenient information in selecting and properly applying motor-control centers.

GENERAL

1. **Motor-control center** is a term that generally refers to a collection of various motor-control equipment and feeders assembled in a series of steel-clad enclosures. Some early motor-control centers were not enclosed but were an assembly of control components mounted on steel or slate panels. The modern control center is usually made up of a series of 20-in-wide by 90-in-high by 12- to 20-in-deep singular steel structures that can be bolted together to form a continuous control lineup.

Enclosures for mounting larger-sized starters may be available in 30- to 36-in widths and depths greater than 20 in. In the majority of cases they have a common horizontal 3-phase bus with a 3-phase vertical bus connected to it for the purpose of supplying

---

*Electrical Project Engineer, Reduction Systems Division, Allis-Chalmers Corporation.

Incoming line section

6- or 12- in. horizontal wiring trough available
depending on conduit entry point; 12-in. trough
shown at top of control center, 6-in. trough at
bottom

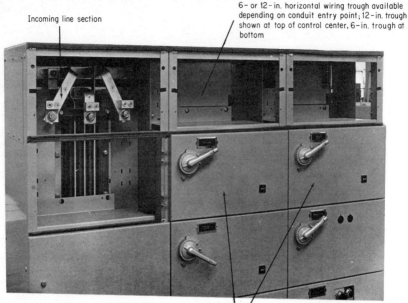

12-in.-high module 6 size 1 combination circuit breaker or fusible-type starters,
reversible or nonreversible, can be mounted in a single vertical structure, front-mounted
and/or rear-mounted, without crowding

Fig. 1 Typical control-center construction. 24- and 36-in-high modules accommodate sizes 2, 3, and 4 starters. 42-, 48-, 54-, 60-, and 72-in-high modules house sizes 5 and 6 starters. All normal forms of standard starting methods are available: nonreversing, reversing, multispeed full-voltage control, primary resistor, autotransformer, part-winding, and star-delta reduced-voltage control.

power to the various components in the vertical structures. In most instances each vertical structure has three vertical bus bars approximately 6 ft in length.

Various control components are usually assembled in 12-in modular height units or multiples thereof and are plugged onto the vertical bus by means of spring-tempered stab fingers similar in construction to fuse clips (Figs. 1 and 2). Typical size units might be 12, 18, or 24 in high. Although the 12-in-high basic module is used in this example, basic modules such as $9\frac{1}{3}$, 14, and $18\frac{2}{3}$ in high are not uncommon.

The modern trend is to install complete motor-control centers in separate air-conditioned and/or pressurized rooms, thus isolating them from the environment found in manufacturing or building areas. The separate room allows for convenient and quick maintenance without interference from surrounding equipment. Various combination starters and feeder tap units are used in manufacturing areas to control the 600-V class of motors and other electrical equipment involved in a particular portion of a manufacturing cycle or process or physical plant area.

Motor-control centers provide a compact and convenient assembly of many of the electrical components required to serve a given area in an installation. With the exception of certain accessories these control centers permit the location of control equipment in a single convenient area. Components mounted within the control center can be easily serviced without necessarily disturbing the operation of the balance of the equipment. In some cases the complete compartment can be removed for maintenance or other reasons and a new one plugged in its place. The old unit can be removed to a convenient workbench for repair or inspection without interfering with the operation of a system.

When problems of dust or other atmosphere hazards exist, the control center either should be mounted in a separate pressurized room or should be individually pressurized and preferably of gasketed construction.

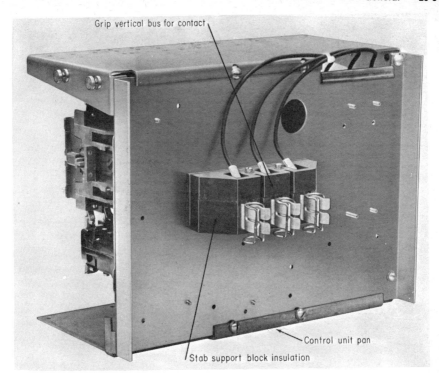

Grip vertical bus for contact

Control unit pan

Stab support block insulation

Fig. 2 Rear view of typical control unit.

2. Standards. Motor-control centers are manufactured in accordance with the requirements of the National Electrical Manufacturers Association (NEMA), Underwriters' Laboratories (UL), The National Electrical Code, specific users and consultants, and other standards. The selection of standards is usually thoroughly spelled out by the consultant or the user's engineering department. The current NEMA Standards, the basic reference in most specifications, is ICS-1975. In general, all specifications relate and tie together to provide a uniform requirement for the manufacturer's design of this class of equipment. However, some discrepancies do occur between NEMA and other established specification authorities. Perhaps the most significant deviation existing at present is the temperature-rise limitations of horizontal and vertical bus structures. NEMA Standards allow a 65°C temperature rise for both structures, and Underwriters' Laboratories allows for a 50°C maximum temperature rise above the ambient temperature. The NEMA 65°C temperature rise is generally considered a safe and allowable temperature rise provided that all joints and connections are suitably tightened and plated with tin or silver.

Complete houses with either center-aisle construction having units located on both halves of the aisle or single-aisle construction having units on one side may also be purchased. It is important that purchasing specifications are explicit as to the type of enclosure desired or the manufacturer may furnish the least expensive without the desired features. The complete house with single or double aisle allows complete maintenance of equipment regardless of surrounding conditions.

In the selection of outdoor housing for a motor-control center the consultant and the manufacturer must assure that the equipment will not be damaged or malfunction owing to the interior ambient temperature. Such outdoor houses located in desert locations have developed interior ambient temperatures so high that all the thermal-magnetic circuit breakers tripped before energy was applied to the bus or any of the units within the motor-control center.

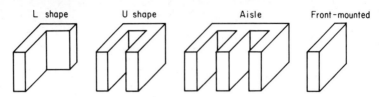

Fig. 3 Various control-center assembly combinations are available to facilitate mounting in locations with space limitations.

Proper ventilation or air conditioning may be required in areas similar to the southeast and southwest. Likewise space heaters may be required where condensation is apt to occur on interior surfaces when equipment is partially or completely deenergized.

3. Configurations.   Motor-control centers are available in a variety of configurations as shown in Fig. 3. The most common arrangement is in a straight line with front-mounted equipment. The second most common is a straight line of motor-control centers with equipment mounted front and back. The back-to-back mounting of control components in a single motor-control-center structure or lineup of structures may offer problems of high ambient temperatures within the vertical structure. This condition is particularly apt to occur in areas similar to the southern half of the United States. When back-to-back equipment is involved, the structure depth is usually 20 in. The narrower-depth structure of 12 to 14 in is limited to front mounting of equipment. As with the 20-in back-to-back construction the narrower-depth structure may result in higher ambient temperatures than the front-only 20-in-depth structure.

Many manufacturers have available as standard a separate vertical wiring trough for each structure. This vertical wiring trough with its separate door is in the majority of cases 4 to 6 in wide. Many users prefer this arrangement since it allows them, with proper safety precautions, to wire or change wiring to a unit within a vertical structure while other units are energized. This vertical wiring trough frequently takes space from the individual unit or pan and may result in crowding the components mounted within it.

Transportation and installation damage is possible when motor-control centers are moved. The horizontal main power bus is frequently installed in the top 10 or 12 in of the 90-in-high structure. Therefore, the center of gravity of the motor-control center is above the physical center, and tipping of the unit is possible unless suitable precautions are taken in transportation and installation.

4. Enclosures.   A motor-control center can be procured in a variety of enclosure styles. The NEMA 1 enclosure is by far the most common. This enclosure covers all control and bus work and will minimize the amount of dust and dirt that might obtain access to it. When a greater degree of exclusion of dust is desired, the control center can be purchased in NEMA 1 gasketed or NEMA 12 (Sec. 16). The NEMA 12 enclosure is more resistant to dust than the NEMA 1 gasketed. NEMA 3 and 4 enclosures can be purchased to exclude dust and water and are basically used for outdoor installations.

The outdoor enclosures can be purchased in three basic styles. The most economical construction is the type in which the motor-control center is completely gasketed and usually furnished with a roof deck to exclude falling moisture. This construction can successfully exclude rain and dust; however, any maintenance of equipment must be done in the environment surrounding the unit. If it is raining or dusty at the time the trouble occurs, the unit or access door has to be opened, and it too will be subjected to the rain or dust.

5. Incoming-Line Space.   A frequent problem encountered by the electrical contractor during the installation of a motor-control center is inadequate incoming-line space. Incoming-line cables may enter the center from the top or bottom, front or rear, or side of the enclosure. These cables can be single or multiple conductor per phase and in sizes up to 1000 MCM. The most frequent problem is inadequate space to bend and terminate cables (Table 1). This problem can be solved by specifying additional space,

TABLE 1. Minimum Bending Diameter, in. of Rubber Power Cable without Metallic Shielding of Armor*

Thickness of conductor insulation, $1/_{64}$ in	Conductor size	
	Up to 500 MCM	500 MCM and above
To 8	$6 \times D$	$8 \times D$
9–12	$8 \times D$	$10 \times D$
13–20	$10 \times D$	$12 \times D$
21 and above	$12 \times D$	$12 \times D$

D = outside diameter of cable, in.
*Anaconda Wire & Cable Co.

a separate cubicle, or a top hat (Fig. 4). The cost of these incoming-line arrangements will often be offset by reduced installation and maintenance costs.

A main breaker or fused disconnect is necessary to protect the bus of each individual motor-control center. This breaker or disconnect may be located in the motor-control center, in 600-V switchgear, or separately mounted. In any event it must be coordinated with all the other breakers and fuses in the total system involved. Because the types of faults encountered by this main disconnect device are usually arcing, line to line or line to ground, significant damage can be avoided by careful selection of this device (Sec. 13).

6. Components. Motor-control centers were originally designed to provide for mounting or housing motor-control combination starters having circuit breakers or fused disconnects in conjunction with motor starters. They also provide feeder taps which are circuit breakers or fused disconnects used to protect feeders and remote equipment such as heaters or motor starters remotely located from the area of the motor-control center. In many applications, however, the motor-control center not only is made up of combination motor starters and feeder taps but also includes a wider range of other equipment. In large industrial plants such as steel mills the motor-control center may include transformers for various purposes, and many structures may contain only relays or various special-processing control systems. In general, a motor-control center may well house all the electrical equipment of the 600-V classification as well as other nonelectrical control devices required in a specific process. The most common modification from the combination motor starter and feeder tap units is the inclusion

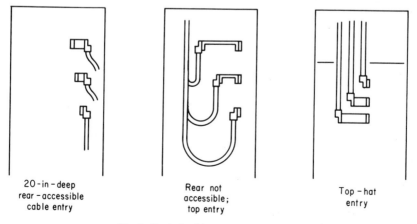

20-in-deep
rear-accessible
cable entry

Rear not
accessible;
top entry

Top-hat
entry

Fig. 4 Typical cable-entry arrangements.

Class I (NEMA ICI-22.02)   Class II (NEMA ICI-22.02)

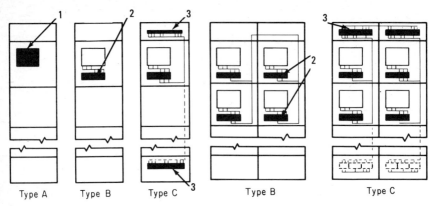

Fig. 5 Standard terminal arrangements. Dashed lines indicate alternate location of terminal board. (1) Individual control unit. (2) Control-unit terminal block. (3) Master-section terminal board.

within the motor-control center of a lighting transformer and distribution panels of the 115- and 230-V ratings.

7. Standard Type and Class. Various types of common motor-control centers can be purchased (Fig. 5). Basically they are defined as Class I Type A, Type B, or Type C and Class II, which is furnished in Type B and Type C (Table 2). The most commonly specified unit today is NEMA Type IB.

The combination starter units NEMA Types IB, IIB, IC, and IIC are equipped with terminal blocks. Depending on the manufacture, only control or power and control wiring may be terminated at these terminal blocks. These arrangements are used with NEMA starter sizes 1 to 4, inclusive. Some manufacturers use fixed terminal blocks and others have available the so-called pull-apart type. The latter type has the internal wiring of all the units terminated on the portion of the block permanently attached to the unit. All external wires coming to the unit are connected to the removable portion of the terminal block. Thus no wires need be disconnected when a unit is to be removed from a motor-control center. Only the removable portion of the terminal block needs to be disconnected from the stationary portion.

INSTALLATION REQUIREMENTS

8. Short-Circuit Capacity. In the actual utilization or specification of a motor-control center to be used for a specific application, the user, designer, or his consultant must

TABLE 2. Abbreviated Control-Center Construction Specifications

Class I
    Type A ...Control unit with circuit breaker or fusible disconnect wired to line side of starter only

    Type B ...Same as Type A but control-circuit leads are wired to fixed terminal block on control unit

    Type C ... Same as Type B but leads brought to control-unit terminal block from master-terminal boards located at top or bottom of motor-control center. No interwiring or interlocking between starters or cubicles on all types in this class

Class II
    Type B .. Same as Class I, Type B but has wiring between control units in the same or adjacent cubicles

    Type C ...Same as Class II, Type B except interwiring is made from master terminal boards at the top and bottom of the control center

first take into consideration the power system supplying the center. NEMA Standards ICS-1975, part ICS-2-322 established short-circuit ratings of the horizontal and vertical bus in symmetrical amperes available at the incoming-line terminals. These values, including any motor contributions, are 10 000, 14 000, 22 000, 30 000, and 42 000 A. Some user specifications call for nonstandard short-circuit or withstandability ratings above these values such as 65 000, 85 000, and 100,000 A. There are no industry standards in the industrial-control field covering the design and testing of motor-control-center horizontal and vertical bus at values above 42 000 A symmetrical. However, adequate design criteria are available from other fields such as switchgear. When a procurement specification is prepared for a motor-control center, the short-circuit rating must be a prime consideration, since it determines the basic selection of the type of breaker or fused disconnect to be contained within the structures of the motor-control center.

The short-circuit capacity of the system associated with a motor-control center can be determined by an analysis taking into account the short-circuit contribution of the transformer feeding the center, the impedance of the power line between the transformer and the center, and the total current contributions of all motors connected to this transformer. NEMA Standards establish design and test standards for the vertical and horizontal bracing of 22 000 and 42 000 A symmetrical. A motor-control center for a system with short-circuit capacities of 10 000 or 14 000 A would be braced for 22 000 A, whereas a motor-control center for systems with 30 000- or 42 000-A capabilities would be obtained with bus bracing of 42 000 A.

9. Coordinating Components. Once the proper horizontal and vertical bus bracing has been selected, careful consideration must be given to the suitability of the combination motor starters and feeder tap units to withstand potential fault currents. This potential fault current must be considered in the selection of all other electrical components to be mounted within the center (Fig. 6). Decisions made relative to the interrupting capability of the breakers and fuses utilized within the motor-control center have to be made on the basis of the requirements of the system, the motor-control-center, and economic considerations. There are a number of arrangements to choose from, and each has features that should be considered (Fig. 7). For systems with short-circuit capabilities of 10 000, 14 000, and 22 000 A symmetrical, the following should be considered. The combination starter designated D for 460- and 550-V systems is in most instances equipped with a circuit breaker, rated 14 000-A symmetrical interruption capacity. This type of breaker is normally supplied for systems capable of supplying a 22 000-A short circuit at the incoming terminals of the motor-control center. Current NEMA Standards require that the manufacturer of the motor-control center test prototypes of the equipment furnished to prove that they are capable of interrupting fault currents up to maximum capacity of the supplying system. When a fault occurs at the terminals of the combination motor starter, these standards require that any damage must be contained within the unit involved even though the starter or breaker may require repair or replacement. The same criteria would apply if the combination starter contained a fused disconnect in place of a breaker.

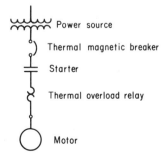

Fig. 6 Transformer limits available system fault current through breaker.

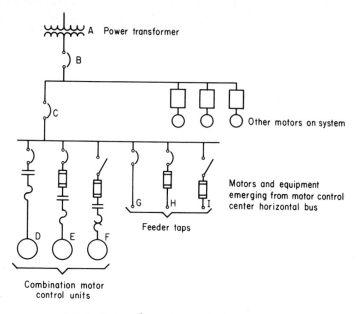

A   Power transformer

Other motors on system

Motors and equipment
emerging from motor control
center horizontal bus

Feeder taps

Combination motor
control units

Fig. 7   Variety of control-center feeder arrangements.

It should be noted that NEMA Standards specify that the combination starter be tested with its output terminals shorted with minimum-length connectors. Thus, the impedance of the breaker, the starter, thermal-overload-relay heater elements, and internal wiring are in series during this test. The combined impedance of these elements of the combination starter will reduce the magnitude of the short-circuit current. These standards are based on practical field experience, which has shown that the probability of a fault occurring between the starter terminals and the terminals of its thermal-overload relay or at the load terminals of the combination starter's circuit breaker or fused disconnect is remote. However, such faults do occur, and the user must weigh this possibility against the increased cost of the motor-control center to prevent them, the cost of downtime, and safety hazards. If the circuit breaker or fused disconnect interruption capacity is inadequate for this increased fault current, total destruction of the combination starter may occur and the fault may spread to the vertical and then to the horizontal bus structure. If it spreads to the bus, the main breaker will be required to clear it or the entire motor-control center could be destroyed. The standards cited above spell out the capabilities of the system and leave it up to the user to determine his own cost vs. risk factor.

10. Higher Interruption Capacity. All manufacturers offer, at an increase in cost, higher-interruption-capacity breakers with a 30 000- to 40 000-A rating. While the user of breakers with a 22 000-A fault-capacity system can be assured that they will not require maintenance or replacement under fault conditions, the possibility of repair or replacement of the starter of the combination motor-control unit remains unchanged. A user may choose a circuit breaker with current-limiting fuses E (Fig. 7) for his combination-starter application. This selection may result in an increase in unit cost over the cost of the higher-interruption-capacity breakers. However, the user can be assured that after fuse replacements, required only when faults exceed the current-interruption capacity of the breaker, both starter and breaker will need no repair or replacement. By proper selection of the type of fuse and disconnect F, the user can also be assured of no damage to any component in the combination motor starter when subjected to a bolted-fault condition.

A feeder tap must be selected to interrupt a bolted fault at its load terminals. Here again the user has an option of determining the class of protective equipment involved.

He may select a breaker that has the capacity to interrupt but may require repair or replacement after it has cleared a bolted fault. The selection of the proper fuse and disconnect will assure that the maximum fault is cleared with nothing more than the fuses requiring replacement (*G*, *H*, and *I*). For systems of 30 000- and 42 000-A symmetrical short-circuit capacity, the basic criteria of component selection remain the same as the considerations given for a 22 000-A system. If a user selects a combination starter for a system with this potential magnitude of short-circuit capacity, he must select the components of the combination motor-control unit and feeder taps with considerably more care than for a system of less capacity. The circuit breaker available (*D*) will have a maximum interrupting capacity at 460 V between 30 000 and 40 000 A symmetrical. The interruption capacities available are determined by the frame ampere rating chosen. The 100-A frames have less interrupting capacity than the larger ampere-rated frames.

The manufacturer supplying a combination motor controller of this class is required by industry standards to have tested these units with the criterion that when a bolted fault occurs at the terminals of the motor starter, the breaker or fused disconnect must interrupt the short circuit and any damage must be confined to the unit. However, as with equipment rated 22 000 A, the breaker, fuses, disconnect, and/or starter may require maintenance or replacement (*E*). Proper selection of fuses using the arrangement of *F* assures that no damage due to overload or fault will occur to the components of the system, other than the fuses. The fuses, however, must be replaced after a fault or overload regardless of the magnitude, whereas when circuit breakers with current-limiting fuses (*E*) are used, faults of less than maximum may result in the fault being cleared by the breaker without damage to the fuse and without the fuses requiring replacement. Feeder taps using the arrangement of *G* may be used for systems of 30 000-A symmetrical short-circuit capacity but should never be used for systems of 42 000-A symmetrical total short-circuit capacity. Systems with 42 000-A short-circuit capacity limit the selection of combination motor-starter units and feeder taps to the current-limiting equipment of *E*, *F*, *H*, and *I*. The arrangements of *D* and *G* must not be used with power systems with these capacities.

The short-circuit current available at the incoming terminals of a motor-control center can be limited by various means. The most logical method is based on selecting a logical and proper size of incoming transformer *A* to allow for a practical and economical selection of motor-control-center components. Current-limiting reactors may be selected for the incoming line to limit the short-circuit current capability. However, this type of protection has shortcomings and may result in reduced voltage at the terminals of larger motors during starting.

Current-limiting fuses can be used in incoming-line sections; however, their effectiveness is limited to motor-control-center loadings of 600 A and less. Most current-limiting fuses of higher sizes will pass significant short-circuit currents and thus are relatively ineffective in limiting fault currents within a motor-control center.

11. Thermal-Magnetic Breakers.   These breakers, when properly coordinated with a motor-starter overload relay, will provide fault protection to the motor and its power conductors (Fig. 6). Thermal-magnetic breakers have limitations in that coordination of their characteristics with those of the thermal-overload relay of lower-horsepower starters is difficult or impossible in many situations. This problem is most apparent in the analysis of high-impedance faults. Under the above circumstances the thermal capacity I^2t of the thermal-overload-relay heating element is less than the minimum reaction time of the thermal-magnetic breaker. Thus magnetic-trip-only breakers may be preferred for the protection of lower-horsepower motors. It must be recognized, however, that this type of breaker will never provide any overload protection to the motor and its conductors, only fault protection. However, as previously noted, such breakers are significantly better in the fault protection of smaller-sized motors when subjected to high-impedance faults. The combination of a magnetic-only or thermal-magnetic breaker with current-limiting fuses provides the ultimate in protection for users who prefer the convenience of reestablishing the circuit after minor faults merely by closing the handle without fuse replacement.

12. Fused Combination Starters.   Fuses are available with a wide range of interruption and performance capabilities. The so-called National Electrical Code fuse rating with a maximum of 10 000-A interruption capability usually requires application in motor circuits of four times the full-load current of the motor that it is intended to protect. Under these conditions, this fuse will provide little if any overload protection to the cables between the starter and the motor itself.

Dual-element current-limiting fuses are rated at 100 000- and 200 000-A symmetrical interruption capacity, and allow the current rating of the fuse to be approximately 20% above the full-load motor current. These fuses consist of a time-delay thermal element and a high-interruption-capacity element. There are current-limiting fuses designed for the specific purpose of minimizing fault currents. These fuses as well as the dual-element types are available in the quick-trip variety which requires application between 300 and 400% motor full-load currents as well as delayed-trip types with characteristics of a time-delay-type fuse. They are also available for systems with up to a 200 000-A short-circuit capacity.

13. Types of Faults. Three basic types of faults experienced in motor-control centers and the equipment they control or supply power to are bolted faults, arcing faults, and high-impedance faults. A bolted fault that causes the power sources to deliver their maximum short-circuit capacity is extremely rare. The arcing fault, the most common type of short circuit in this class of equipment, starts from many causes such as insulation failure and careless placement of tools or wire. The wire or small tool quickly dissipates in the fault, and a line-to-line or line-to-ground arcing fault remains which can be extremely destructive if not quickly extinguished by the protective device.

A problem has developed because the National Electrical Code permits a 15-A thermal-magnetic breaker to protect all motors of 3 hp and less at 460 V. A 20-A breaker may be selected to protect a 5-hp motor. Many heater elements, whether a separate entity or integral with a temperature-sensing element such as a bimetal or solder pot, will be destroyed before the thermal-overload relay can respond and open its contacts to disconnect the motor from the power source. A thermal-magnetic breaker in this current range will not open before the heater element disintegrates and arcing occurs across the heater element, which is not designed to be a circuit interrupter. The end result may well be a phase-to-phase arcing fault at the line terminals of a motor starter which could result in the complete destruction of the combination motor starter.

In the above case, a current above twenty times motor full-load current will cause the breaker to open and thus will protect all elements of the combination motor starter. It is not too unusual to experience a high-impedance fault of 9 A on a motor drawing 1 A full-load motor current. At 9 A most thermal-overload relays will open and protect the circuit. However, at higher currents in motors of 3 to 5 hp the fault currents may well be more than ten times full-load current, and the heater element of the thermal-overload relay may open and form an arc.

To prevent such destructive failures, it is important for the user or his consultant to specify the proper form of protection. Most manufacturers will furnish whatever type of circuit protection a customer specifies. A properly selected and adjusted magnetic-trip breaker or fusible combination starter will prevent the destruction of thermal-overload heater elements at currents above ten times full-load current and thus will prevent any damage to the motor-control center.

Proper adjustment of a magnetic-trip breaker is to set the trip unit at maximum trip current and then start the motor. The breaker should not trip at this setting. The trip-current setting is then reduced notch by notch and the motor is started in each position. When the breaker trips, the current should be set back one notch. If no tripping occurs when the motor is started again, the breaker should be sealed in this position. Care must be exercised that the motor is not overheated during this process.

14. Maintenance. Care of equipment is as important for a motor-control center as it is for any other electrical equipment. All bus structures of all motor-control centers are designed to minimize the accumulation of dust. However, it is not unusual to find a NEMA I enclosed motor-control center located in an extremely dusty area. Under such conditions frequent cleaning and removal of the dust from the bus structure and combination motor-starter units is required. In some areas and in certain manufacturing environments a motor-control center may be subjected to high humidity. The combination condensed moisture and dust may become an electrical conductor. Regardless of how carefully the equipment has been designed to prevent the occurrence of a line-to-line or phase-to-ground faults, they may occur without proper and timely cleaning. A space heater in each vertical structure will minimize condensation within the structures of a motor-control center.

Circuit breakers of the type used in motor-control centers require no maintenance. In general, if a molded-case breaker fails, it should be replaced rather than repaired. Such failures are infrequent.

Most modern starters require a minimum of maintenance. Most contactors size 1 through 4 have cadmium oxide contact tips. There is an axiom about this type of contact tip that the worse they look the better they will perform. They should never be filed, and no other method such as sandpaper should be used to clean their surfaces. They should be replaced, however, when they have worn to less than one-half of their normal thickness. It is usually not necessary to replace them then, but it is a good maintenance practice and will reduce the possibilities of unexpected failure. At the same time the contacts are being changed, it is logical to replace the contact springs.

As previously stated, the center of gravity of a motor-control center may be above its physical center. It is therefore important to take the center of gravity into account when moving a motor-control center from one location to another.

A motor-control center, as well as other types of electrical equipment, may arrive at a jobsite before the building it is to be housed in is completed or is properly heated and enclosed. It is important to keep the motor-control center free of moisture. To prevent condensation, the proper amount of heat should be applied within the motor-control center until the main bus is energized and the motors are drawing power from the motor-control center. If the motor-control center is equipped with a space heater, it is advisable to connect it on a temporary basis. If no space heaters are provided, a 200- or 300-W light bulb for each 20-in-wide structure will usually provide adequate heat and prevent damage due to excessive moisture accumulation. Frequently, during construction, extreme amounts of dirt and dust are present. Before the main bus is energized under these conditions, all parts of the motor-control center should have a thorough cleaning and vacuuming.

As a word of caution, the main and vertical buses must often be energized before wiring is completed to the other units within the motor-control center so that a few units can be connected to their loads. Unless extreme caution is exercised in pulling and connecting wires to the balance of the units, personal injury or bus faults may occur. Often the individual combination motor starters and other units are removed from the motor-control center while wires are pulled from various conduits into the motor-control center. It is equally common during this period to find these units sitting on the floor or on workbenches resting on the vertical bus stabs (Fig. 2). In this position these bus stabs are prone to damage. Even though most manufacturers may back up these fuse-clip-type stabs with sturdy springs, laying motor-control-center units in this position may result in closing, bending out of position, or opening them. Then when the unit is returned to the motor-control center these stabs will stub, bend to one side of the vertical bus, or fail to apply the required pressure to the vertical bus. Under these conditions overheating or arcing may occur.

Motor-control centers have increased in popularity significantly over the past 20 years. Their use has been accelerated by the convenience of installation of a large group of control components in a single area. The centers not only allow for convenient maintenance but result in significant savings in the installation of electrical and other control equipment. With this increased use of motor-control centers, frequently the specifier has not been thoroughly cognizant of the difference between a separate combination motor starter and a motor-control center. Usually the leads to a combination motor starter are quite small in size and of considerable length compared with the wiring to a similar unit mounted in a motor-control center. The short-circuit capability of the separately mounted combination motor starter is usually less than 10 000 A. A motor-control center which has a potential short-circuit capacity from 10 000 to 85 000 A or more requires a complete and comprehensive analysis of the components selected to be placed within the control center itself.

24

Dc-Motor Control

A. H. MYLES
T. H. BLOODWORTH

FOREWORD

Direct-current motors have been used in a variety of applications throughout the years and are still used for many industrial drives (Ref. 1). Dc motors are used where:

1. Wide-speed range with essentially stepless variation in speed settings is required.

2. Either variable or constant output torque is needed, or a combination of both, as required in most processes.

3. Fast acceleration, deceleration, or reversal of rotation is required such as for hoists, traction, propulsion, metal rolling, or processing.

4. Fine accuracy of speed control is needed, such as for tension reels.

5. Accurate speed correlation between two or more parts of a process line must be maintained.

6. High overload torque is needed at the lower part of a wide-speed range process.

7. Variable regenerative braking torque is needed.

In each application of these motors, there are a number of choices of control equipment that might be used. This section was written to aid in understanding magnetic or variable-voltage dc controls that have been in use for many years. Knowledge of these control methods will aid in applying solid-state dc control systems (Sec. 26).

Part 1

Dc Magnetic Control

A. H. MYLES°

GENERAL

Although magnetic control may not provide the smooth or stepless acceleration or deceleration of motor drives or the precise speed regulation provided by static control methods, it is considered entirely satisfactory for most control applications. In addition to economical considerations, dc magnetic controls utilize devices developed to a high degree of reliability, versatility, and ease of adjustment and maintenance.

Part 1 covers conventional magnetic control for dc motors, emphasizing the advantages and disadvantages of different acceleration methods and control circuits for the starting, stopping, reversing, and speed control of dc motors.

1. Dc-Motor Types and Characteristics. The various types of dc motors and their characteristics must be well understood to apply and care for dc magnetic motor-control equipment. There are three basic types of dc motors, and they differ in the way their magnetic fields are energized or excited. These motors are identified as shunt-, series-, or compound-wound (Ref. 1).

The most commonly used motor with magnetic starters is the shunt motor. Its excitation is provided by a field winding of many turns which are connected in parallel with the armature circuit. The field normally is energized by full line voltage to give a constant field strength. However, the field strength may be intentionally reduced by inserting series resistance. This ability to reduce the field strength provides for speed adjustment of an otherwise relatively constant speed machine. The shunt motor has the disadvantage of having limited starting and overload capabilities.

The series motor differs from the shunt motor in that its field is excited by a winding that is connected in series with the motor armature and therefore carries load current. The series field is of rather few turns, having low resistance and full-load current capacity. The series motor has good starting and overload characteristics, but its speed regulation with load is poor and overspeed at light loads may occur. The series motor is therefore used with starters only for special applications where high starting or overload torques are encountered and where the motor load never diminishes to a low value.

The compound-wound motor has both shunt and series field windings and therefore

°Chief Engineer, Heavy Industry Division, Square D Company (Retired); Fellow Member, IEEE; Life Member, AISE; Member Cleveland Engineers Society.

has better starting and overload characteristics than the shunt motor. It has a definite stable no-load speed, but speed regulation with load is a compromise between a shunt- and series-wound motor.

In addition to the three types of motor fields described, dc motors may also have one or more series field windings designated as interpole, stabilizing, or compensating. Interpole series windings are commonly used and are located on pole pieces between the main motor field poles. They improve the maintaining of the magnetic neutrals and thus provide good commutation with reduced sparking. The stabilizing series winding is applied to shunt motors designed for high speed by shunt field weakening to prevent unstable or runaway speeds that could result from the demagnetizing effect of armature reaction. The compensating series field winding is wound directly on the main field pole pieces. It tends to correct the distortion of flux in the main field pole faces caused by armature reaction and decreases the danger of flashover between brushes.

2. Starting Dc Motors. All dc motors, when rotating, are also generators, since the armature conductors are rotating in a magnetic field. The generated voltage, called counter electromotive force (CEMF), is proportional to speed and field strength and opposes the applied armature voltage. Under loaded or motoring conditions, the CEMF is always less than the applied voltage. However, under overhauling-load conditions, the CEMF may exceed the applied voltage. The armature current will depend upon the difference between applied and CEMF voltage and the resistance of the armature circuit.

With a constant-potential line supply, the CEMF of the motor is zero at the instant of start and resistance must be inserted in series with the armature to limit the current to safe starting values of usually less than 200% rated load current. This series resistance is then reduced in steps as the motor accelerates and the CEMF increases. The motor starter or controller performs the function of reducing this series resistance during acceleration. Table 1 lists the horsepower ratings and number of acceleration contactors for different controller sizes and at various line voltages (Ref. 4).

DC-MOTOR STARTERS

3. Dc Reduced-Voltage Motor Starters. Regardless of the acceleration method, the basic types of dc-motor starters are determined by the functions performed. Depending upon the service required, the following types are available as standard starters but may include various optional features:

Nonreversing, form NR, for constant-speed motors
Nonreversing, form NR, for adjustable-speed motors
Nonreversing with dynamic braking, form NRD, for constant-speed motors
Nonreversing with dynamic braking, form NRD, for adjustable speed motors
Reversing nonplugging with dynamic braking, form RNPD, for constant-speed motors
Reversing nonplugging with dynamic braking, form RNPD, for adjustable-speed motors

ABLE 1. Horsepower Ratings of Dc Magnetic Starters

Size of ontroller	8-hr open rating, A	115 V		230 V		550 V	
		hp rating	No. of accelerating contactors	hp rating	No. of accelerating contactors	hp rating	No. of accelerating contactors
1	25	3	1	5	1		
2	50	5	2	10	2	20	2
3	100	10	2	25	2	50	2
4	150	20	2	40	2	75	3
5	300	40	3	75	3	150	4
6	600	75	3	150	4	300	5
7	900	110	4	225	5	450	6
8	1350	175	4	350	5	700	6
9	2500	300	5	600	6	1200	7

Since only two or three control-circuit wires are required, the control-circuit (pilot control) device may be located remotely from the starter. This arrangement permits location of both the starter and control-circuit devices for maximum operating efficiency and installation economy. The pilot control device may provide either low-voltage protection or low-voltage release (Secs. 2 and 3).

Each type of starter may be designed using one of the various methods of acceleration.

a. Current-limit acceleration is not used extensively on modern motor starters. The control circuitry is somewhat complex as are the acceleration relays. Its principal application is for high-inertia drives requiring long starting times but where there is no danger of stalling. Current-limit acceleration has the disadvantage of not forcing the acceleration. If the motor current does not reduce to the relay operating value, for any reason, the acceleration sequence will not continue and the series resistor may be overheated.

Figure 1 shows an elementary circuit diagram of a typical motor starter for a shunt motor, with current-limit-acceleration relays. The control consists of a momentary start-stop master, a main line contactor *M*, two acceleration relays 1AR and 2AR, two acceleration contactors 1A and 2A, and two steps of resistance. The acceleration relays must be designed for fast response and the main contacts of *M*, 1A, and 2A must close before their corresponding normally open (timed) interlock contacts. This sequence allows the acceleration relays to become energized and open their normally closed contacts before 1A or 2A becomes energized. When the motor current then reduces to the set value of the relays, the corresponding relay will drop out to close 1A and 2A in a current-limit sequence, shorting out the line resistance and accelerating the motor to full speed.

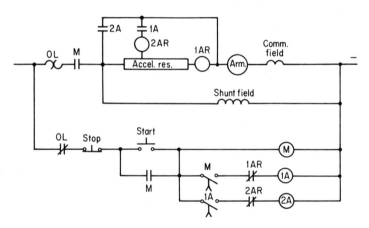

Fig. 1   Constant-speed, current-limit dc starter.

b. CEMF acceleration is similar in performance to current-limit acceleration, having the disadvantage of being nonforcing. CEMF acceleration is limited to small motor sizes and to starters having one or two steps of acceleration. The advantages are simplicity and low cost.

Figure 2 shows the elementary circuit diagram of a CEMF starter for a compound motor, with two acceleration contactors. The main line contactor *M* is controlled from a start-stop master. Acceleration contactors 1A and 2A are connected across the motor armature with sequence interlocking. The acceleration contactors have partial-voltage coils with a series resistor having a low thermal coefficient of resistance so that normal coil heating will have minimum effect on their pickup voltage value. Contactors 1A and 2A are adjusted for pickup at predetermined armature voltages. Since armature voltage increases with motor speed, 1A and 2A close in sequence to short out series resistance and accelerate the motor to full speed.

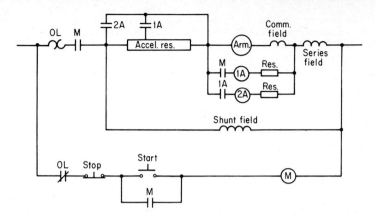

Fig. 2 Constant-speed, CEMF dc starter.

c. Definite-time acceleration motor starters are the most widely used type for accelerating constant-speed or adjustable-speed dc motors. Positive acceleration is assured, eliminating the danger that an unusually heavy load may prevent the motor from accelerating sufficiently to cause short-circuiting contactors to operate, as may be the case with current-limit or CEMF methods. Consistent starting time is provided regardless of variation in load. Starters with time-limit acceleration are universally applicable to general-purpose and machine-tool applications.

Many types of timing devices have been used to close the acceleration contactors, such as solenoid- or motor-driven timers, fluid dashpot relays, inductive time-limit contactors or relays, and pneumatic timers. Pneumatic timing elements actuated by the operation of succeeding contactors provide a popular means of acceleration. Individual time adjustment of each step of acceleration gives a flexible and straightforward starter control.

d. Time-current acceleration combines the advantages of current limit or CEMF with definite time. This method provides forcing acceleration but also provides increased time for heavier loads. Time-current acceleration therefore provides longer time per acceleration step for heavy starting loads than it does for light starting loads. For many mill and heavy-industry applications, the starting conditions vary considerably from one start to the next and time-current acceleration provides greatly reduced current and torque peaks without the danger of failing to accelerate.

Figure 3 shows the general construction of a time-current relay. The coil is con-

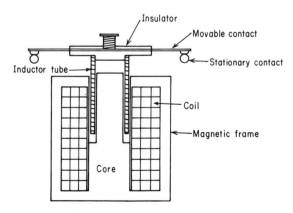

Fig. 3 Time-current acceleration relay.

nected in series with the armature in the same manner as for a current-limit relay. When the armature circuit is closed, the sudden inrush of current causes the flux in the relay magnetic circuit to increase very rapidly. This changing flux induces a current in the copper or aluminum inductor tube. This induced current is acted upon by the relay magnetic field, causing the tube to jump vertically and open its normally closed contacts. The relay tube then settles because of gravity, and its rate of descent is determined by the strength of the relay magnetic field. For heavy load currents the magnetic field is strong and the time for the relay to close its contacts is long. Light load current gives shorter relay time and thus faster acceleration of the motor.

The circuit diagram is similar to that for current-limit starters (Fig. 1). The relay contacts are normally closed and the sequence interlocking need not be of time-delay design since relay operation is very fast.

4. Nonreversing Constant-Speed and Adjustable-Speed Motor Starters. Figure 4 shows the elementary wiring diagram for a nonreversing constant-speed dc reduced-voltage motor starter with pneumatic time-limit acceleration. Pressing the momentary-start push button closes contactor M to energize the motor shunt field and apply reduced voltage to the motor armature. Contactor M makes its own holding circuit and mechanically actuates a pneumatic timing device which, after a predetermined time delay, closes its contacts to close contactor 1A, short-circuiting part of the series resistor and further accelerating the motor. When 1A closes, a similar time-delay unit closes 2A to short-circuit the remaining series resistance and the motor runs at full field, full voltage.

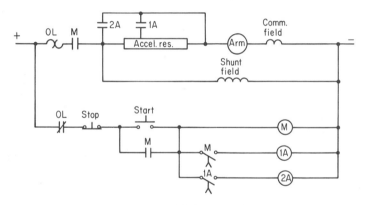

Fig. 4   Constant-speed, definite-time dc starter.

The shunt field is connected on the load side of M and the field must build up, as does the motor torque, each time the motor is started. This delay is not serious with small motors. When the motor is stopped by opening M, the shunt field is disconnected from the line but no discharge resistor is required since the field has a discharge path through the motor armature circuit.

Figure 5 shows an elementary wiring diagram for a similar nonreversing starter but providing starting for adjustable-speed motors. In addition to the equipment shown for Fig. 4, this starter is equipped with a field acceleration relay that automatically functions to provide full field to base speed of an adjustable-speed motor, and to limit the current drawn from the line during acceleration from full-field speed (base speed) to that of the field rheostat setting.

With power available the shunt field is energized at the rheostat setting. Pressing the momentary-start push button closes contactor M to apply reduced voltage to the motor armature. M makes a holding circuit for the coils of M, 1A, 2A, and FA shunt coil. The series coil of FA is additive with the FA shunt coil, and FA contacts close to short the field rheostat and apply full field to the motor. The pneumatic timing elements on M and 1A close their contacts in timed sequence to close 1A and 2A and short-circuit all

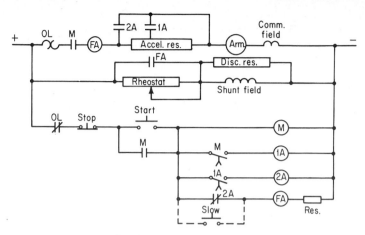

Fig. 5 Adjustable-speed, definite-time dc starter.

series resistance to apply full voltage and full field to the motor. The normally closed interlocks of 2A open to deenergize the shunt coil of FA, and this relay is then under the control of its current coil only. When the motor armature current reduces to normal, relay FA opens its contacts to insert the field rheostat in series with the shunt field. The armature current will again rise and FA will again pick up to short out the rheostat. The armature current again decreases, and this rapid-repeat operation continues until the motor accelerates to rheostat speed setting.

For applications in which it is desirable to run occasionally at base, or full-field speed, for short periods of time, a (slow) push button can be added as shown dotted (Fig. 5) to short out the 2A normally closed control interlocks, closing relay FA to short out the field rheostat. The sudden increase of field strength increases the motor CEMF, and line current may reverse. If the line current reverses, the FA series coil opposes the shunt coil, and FA may again drop out. FA can, therefore, close and open several times to provide deceleration control similar to its operation as an acceleration relay except that it is now controlling field strengthening instead of field weakening. The FA shunt coil is a partial-voltage coil with a series resistor having a low coefficient of resistance so that coil heating will not materially change its operating characteristics.

5. Reversing Constant-Speed and Adjustable-Speed Motor Starters. Acceleration methods for reversing starters are, of course, identical to those described for nonreversing starters. However, applications requiring reversal tend to require frequent starts, stops, and reversals, and stopping time becomes an important factor. Therefore the basic standard for reversing starters includes the feature of nonplugging with dynamic braking. The nonplugging feature prevents the application of reverse power to the motor armature until the motor has dynamically braked to near a stop. Dynamic braking of constant-speed motors or motors with less then 2/1 speed range is generally at full field for quick stops.

Figure 6 is the elementary wiring diagram for a reversing constant-speed dc starter with nonplugging, dynamic-braking feature. Pressing the forward or reverse push button closes the corresponding reversing contactors, and the motor starts on reduced voltage. The nonplugging relay NP picks up immediately to open its normally closed contacts and prevents closing the opposite set of reversing contactors until the motor again comes to a stop and NP opens. Interlocks on the reversing contactors close to energize the normally closed dynamic-braking contactor DB, which opens the dynamic-braking circuit. DB and 1A mechanically operate timing devices which close in timed sequence to close 1A and 2A, accelerating the motor to full voltage. Pressing the stop master opens the directional contactors. DB then closes, and the motor stops by dynamic braking.

Dynamic braking of adjustable-speed motors with 2/1 or higher speed range normally

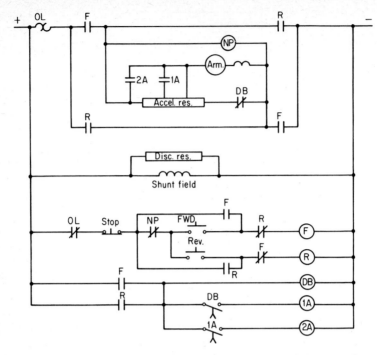

Fig. 6 Reversing, constant-speed, nonplugging, dynamic-braking, definite-time dc starter.

requires control of field strengthening during dynamic braking as well as control of field weakening when accelerating to high speed. If field-strengthening control is not provided, braking must be applied at the existing field strength to prevent excessive armature voltage and braking current. However, this arrangement does not take advantage of full available braking effort.

Two control schemes are normally employed to take advantage of available braking effort. One method changes the field from weak value to full field in a single step and at a speed which will not produce excessive voltage or current.

The other method uses a field deceleration relay and field resistor. The relay contacts are normally closed and short out the resistor. The relay has a series and shunt coil similar to the field acceleration relay and telegraphs to keep the field from strengthening too rapidly.

Figure 7 shows the elementary wiring diagram for a reversing adjustable-speed dc starter with dynamic braking on weak field at high speed and on full field at reduced speed. Starting and acceleration are similar to those for adjustable-speed nonreversing starters except that the 2A acceleration contactor is connected across the armature circuit. When 2A is closed by the timing unit, it makes its own holding circuit through a voltage-reducing resistor. 2A now becomes a CEMF contactor. Pressing the stop push button opens the directional contactors and acceleration contactor 1A, and closes the normally closed dynamic-braking contactor DB. 2A remains energized and the field acceleration relay FA cannot close. The motor slows by dynamic braking on weak field. As the motor slows down, its CEMF decreases and at near base speed 2A drops out to close FA, and braking is now at full-field strength. This control has the advantage of not being complex but has the disadvantage in that an operator can strengthen the field quickly by manual rheostat means and thus cause high armature voltage.

If danger exists in strengthening the field too rapidly by manual change of the rheostat described, protection can be provided by a field deceleration relay (Art. 11).

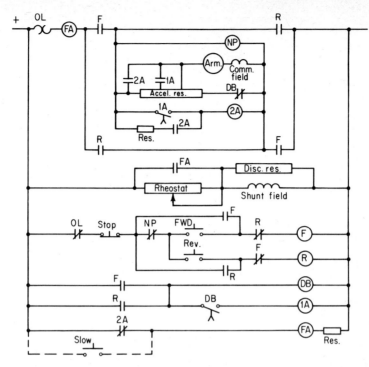

Fig. 7 Reversing, adjustable-speed, nonplugging dynamic-braking, definite-time dc starter.

DC-MOTOR CONTROLLERS

6. Dc Constant-Voltage Magnetic Controllers. Direct-current constant-voltage magnetic controllers may be used with shunt-wound, series-wound, or compound-wound dc motors. The types of controllers, in general, fall into one of the following, with special modifications available to meet unusual requirements:

Nonreversing, form NR
Nonreversing with armature shunt, form NRAS
Nonreversing with dynamic braking, form NRD
Reversing plugging, form RP
Reversing plugging with armature shunt, form RPAS
Reversing plugging with dynamic braking, form RPD
Reversing nonplugging with dynamic braking, form RNPD
Reversing dynamic lowering, form HD

Contactor horsepower ratings for dc magnetic control depend upon whether the motor is applied on a continuous or intermittent basis. Also, special-purpose applications, such as steel-mill auxiliaries, may limit the minimum size of contactors used and allow fewer acceleration contactors owing to the more rugged motor design. Table 2 lists the horsepower and minimum number of acceleration contactors considered standard for both continuous- and intermittent-duty controllers. The number of acceleration contactors is exclusive of the plugging contactor of reversing controllers (Ref. 4).

Dc magnetic controllers, in general, are applied on applications requiring frequent starts, stops, and reversals and often require control of the motor at reduced speed for indefinite time periods. The voltage-reducing power resistors, therefore, require greater current-carrying capacity than starters and should be capable of carrying full-load motor current for at least 25% of the time on (NEMA Class 150) each minute (Ref. 4).

TABLE 2. Dc Controller Horsepower Ratings and Number of Acceleration Contactors

Contactor size	Continuous-duty		Intermittent-duty		Min no. of acceleration contactors	
	Ampere rating	hp at 230 V	Ampere rating	hp at 230 V	General-purpose	Steel-mill
1	25	5	30	7.5	1	
2	50	10	67	15	2	
3	100	25	133	35	2	2
4	150	40	200	55	2	2
5	300	75	400	110	3	2
6	600	150	800	225	4	3
7	900	225	1200	330	5	3
8	1350	350	1800	500	5	4
9	2500	600	3350	1000	6	5

Dc magnetic controllers may incorporate the various methods of acceleration described for starters. However, since the service is usually severe and variable, the time-current method of acceleration gives excellent motor and drive protection.

7. Nonreversing Controllers. Figure 8 shows the elementary wiring diagram for a nonreversing controller as applied to a compound-wound mill-type motor. The shunt field is energized whenever the main-line-isolation switch is closed. Therefore, full torque is available when the armature circuit is closed. A shunt-field-discharge means is required and is usually a nonlinear-type resistor. Both inverse-time and instantaneous-trip overload relays are standard for running-overload and fault protection.

With the master switch in the off position and the main and control isolation switches closed, the undervoltage relay will close and make its own holding circuit. The master may now be moved to any desired speed point to close contactors 1M and 2M, applying reduced voltage to the motor. Acceleration relay 1AR will operate to reclose after a time delay proportional to the load. Acceleration contactor 1A will now close if the master is moved to second or third point. 2A contactor will close under control of 2AR if the master is on third point.

8. Nonreversing with Armature Shunt and/or Dynamic-Braking Controllers. If the motor is shunt- or compound-wound, dynamic braking at the master off position and armature shunt operation on a slow-speed master position can be obtained by using a normally closed contactor and resistor. If the motor is series-wound, slow-speed armature shunt control is possible, but it is more economically provided by using a normally open contactor which is closed on the slow-speed master position. Limited dynamic braking with series motors is available, but only with extra equipment.

Figure 9 shows an elementary wiring diagram which will provide four possible operating speeds and dynamic braking at the off position. Moving the master to first position closes 1M and 2M to apply reduced voltage to the armature which is shunted by the dynamic-braking resistor. This shunting resistor gives increased series field strength and a greater voltage drop across the acceleration resistor, causing a slower motor speed. When the master is moved to second point, DB opens to remove the armature shunt, giving increased speed. Moving to the third and fourth master points will close 1A and 2A under time-current delay to accelerate the motor to full speed. If higher running speeds are desired by shunt-field weakening, a field-acceleration relay FA can be added as shown in Fig. 5.

Returning the master to first position reestablishes the armature shunt for positive slowdown. At the off position, the armature circuit is disconnected from the power supply and dynamic braking becomes effective.

9. Reversing Plugging Control. The term "plugging" refers to the principle of reversing the armature connections while the motor is running in a given direction. Under this condition a countertorque is developed that retards rotation. The motor therefore slows down and stops. If power is not removed, the motor will accelerate in the op-

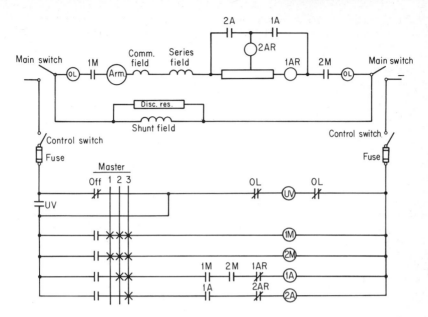

Fig. 8   Controller, dc, nonreversing.

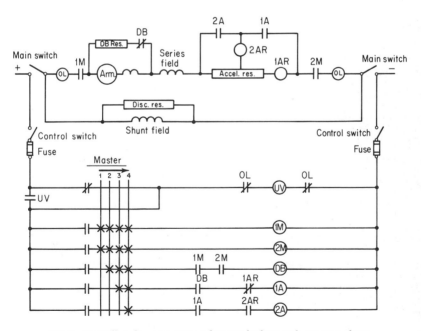

Fig. 9   Controller, dc, nonreversing dynamic braking with armature shunt.

posite direction of rotation the same as for starting from standstill. During the slow-down, or plugging operation, a plugging relay functions to keep all the resistors in series with the armature circuit and thus prevent excessive motor current. This series resistance is greater than permitted when starting from rest, since at the moment of plugging the motor CEMF adds to the line voltage, tending to force more current through the armature circuit.

Reversing plugging control is applicable to series-, shunt-, or compound-wound motors. Slow-speed armature shunting is readily applied and dynamic braking if the motor has a shunt field. Also series-generator-type dynamic braking may be obtained with a series-wound motor (Art. 10).

Figure 10 is an elementary wiring diagram of a dc reversing plugging control. Operation is the same for either direction of rotation. With power available, the main and control switches closed, and the master in the off position, the undervoltage relay will close and make its own holding circuit. If the master is now moved to first position forward, 1F, 2F, and M will close to apply reduced voltage to the motor. Moving the master to second position will close P without delay. The PR relay remains deenergized, since the CEMF of the motor is blocked by the rectifier in series with PR coil. Succeeding master positions accelerate the motor to full-load speed with time-current-relay protection.

With the motor running in the forward direction, moving the master to reverse opens all contactors and closes contactors 1R, 2R, and M. The motor CEMF is now in the same direction as the line voltage and across the PR relay in a direction to conduct and causes PR to pick up and open its normally closed contacts. P, 1A, or 2A cannot close, and the motor slows down by plugging torque. Near standstill the motor CEMF approaches zero and PR closes its contacts to allow acceleration in the reverse direction.

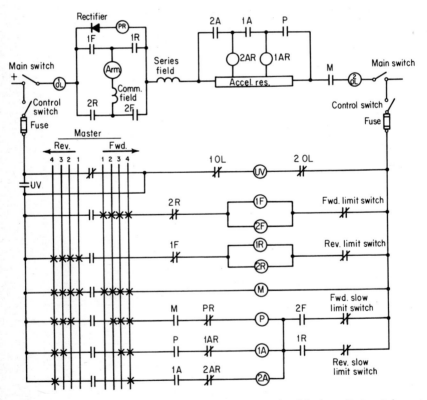

Fig. 10 Controller, dc, reversing, plugging with slow-speed and final-stop limit switches.

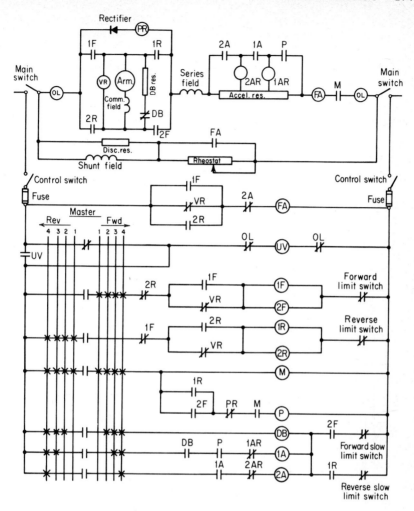

Fig. 11 Controller, dc, reversing, plugging dynamic braking.

Slowdown and final-stop limit switches are shown as they might be applied to a reversing control where mechanical load or friction will provide sufficient rate of slowdown and stop.

10. Reversing Plugging—Dynamic-Braking Controller. Figure 11 shows an elementary wiring diagram for a dc reversing plugging dynamic-braking controller for a compound-wound motor, which is also arranged to give armature-shunt slow speed on the first master position. An adjustable-speed shunt-field winding is shown, and therefore, a field-acceleration relay is included. Also included is a voltage relay *VR* which has a close differential between pickup and dropout voltage and is adjusted to drop out at approximately full-field speed.

When the motor is plugged at a speed above full-field speed, all contactors are deenergized. The *DB* contactor closes to apply dynamic braking at weak field. The *VR* relay is energized, and its normally closed contacts prevent the directional contactors from closing. The motor slows down, and at approximately full-field speed, *VR* drops out to close the corresponding directional contactors, and plugging becomes effective.

On first master position, *DB* is closed and an armature-shunt slow speed is provided. If plugging is to first point only, plugging torque is greater than for other master points and a compromise between plugging torque and armature-shunt speed may be required.

If the motor is series-wound, dynamic braking is not readily provided and plugging control with a normally open armature-shunt contactor will provide normal plugging operation and a strong armature-shunt slow-speed first position. Figure 12 shows such a circuit diagram, and it will be noted that the plugging relay *PR* removes the armature shunt even if the master is on first-position plugging.

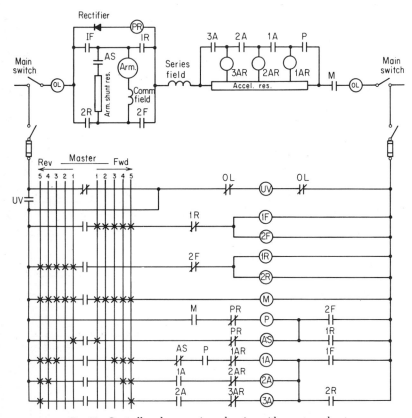

Fig. 12 Controller, dc, reversing, plugging with armature shunt.

For certain reversing applications, such as bridge drives of hot-metal cranes or outdoor cranes, emergency dynamic braking in case of power failure is essential. With additional equipment, a series motor can be connected to provide dynamic braking on power failure. The motor becomes a series generator, and the energy of the drive is absorbed in a dynamic-braking resistor (Fig. 13). The two dynamic-braking contactors 1*DB* and 2*DB* have normally closed contacts and are held open or energized during all normal reversing plugging operations and the rectifier bridge around the series field is noneffective. On power failure 1*DB* and 2*DB* close. All other contactors open. The series-field and dynamic-braking resistors are connected across the motor armature by means of the full-wave bridge rectifier so that the field current does not reverse for ei-

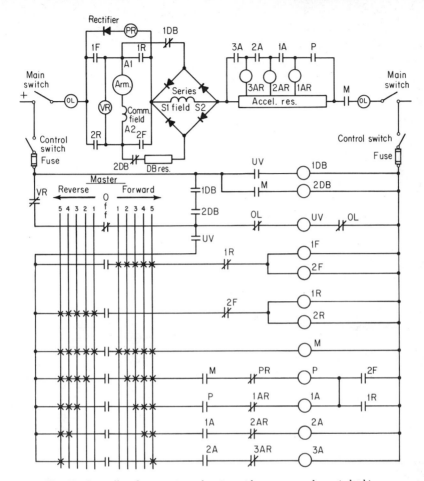

Fig. 13 Controller, dc, reversing, plugging with emergency dynamic braking.

ther direction of rotation. The armature current does reverse, and dynamic braking is obtained. The *VR* relay is used for safety interlocking and prevents removing the dynamic braking if power returns before the drive has stopped.

11. Reversing Nonplugging Dynamic-Braking Controller. For shunt-wound adjustable-speed motors of over 2/1 speed range on reversing applications, a reversing nonplugged dynamic-braking controller is normally applied. Both a field-acceleration and a field-deceleration relay are required (Fig. 14). If the motor is of general-purpose design, the field winding may not stand full excitation at standstill and a protective resistor with shorting contactor *FE* may be included. This resistor can also serve as the field-deceleration resistor. Slowdown and final-stop limit switches are shown as they might be applied with such control.

Acceleration by closure of contactors *M*, *1A*, *2A*, and *3A* and braking by the spring-closed contactor *DB* is the same as described previously. The directional contactors are controlled by a voltage relay *NP* connected across the motor armature. They are energized at start through normally closed contacts of this relay and form their own holding circuit. The relay has a full-voltage coil which picks it up as the motor acceler-

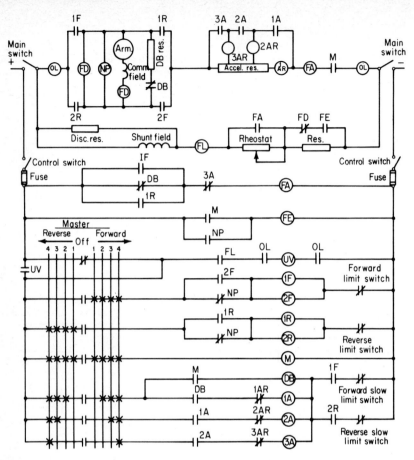

Fig. 14 Controller, dc, reversing, nonplugging, dynamic braking, adjustable speed with armature shunt.

ates to base speed. Upon reversal of the master, all contactors are deenergized, removing power and applying dynamic braking, but the nonplug relay is held in by armature voltage and keeps the directional contactors from closing until the motor has reached a slow speed. The relay drops out at about 18 to 20 V, energizing the directional contactors, which initiate normal acceleration in the new direction.

The *FE* relay is closed at start by contacts of *M* and is held in during braking by normally open contacts of the *NP* relay. Its purpose is to reduce the field heating during off periods by inserting a resistor in the field circuit.

The *FL* relay is supplied for shunt motors where loss of field would cause runaway of the motor. The coil must stand continuous full-field current and must hold in to below the weakest field current. In addition to these requirements, the selection of the *FL* coil winding and of the field-protective resistor must be such as to permit the *FL* relay to pick up on the low current allowed at a weak-field rheostat setting.

When a rheostat is used to weaken the field for speed control, it is usually important to keep full-field strength as long as there is any resistance in series with the armature to obtain the maximum torque for acceleration. Weakening the field before the series resistance is short-circuited may result in a decrease of speed even to the point of stalling if the motor is being started under load. The desired field control during acceleration is provided by the normally open relay FA having shunt and series coils. The shunt coil is energized through normally closed contacts of the last accelerator to short out the rheostat during braking and during acceleration to base speed. Acceleration on up to the rheostat speed is controlled by the series coil, which closes FA contacts whenever the current rises above 125% full load, and then reopens the contacts as the current drops to between 110 and 100% full load.

Similarly, field control during deceleration is provided by another double-coil relay FD whose shunt and series coils are connected to oppose each other during normal running so that its normally closed contacts remain closed. The series coil must be connected so as to carry armature current not only during normal running but also during braking. The shunt coil must have the proper polarity relative to the series coil (normally opposing).

In a reversing controller with dynamic braking, in which the series coil must be inside the directional contactors to be included in the dynamic-braking circuit, the polarity of FD shunt coil must be changed each time the direction of motor rotation changes. The shunt coil should also be energized during braking. A simple way to fulfill these conditions is to connect both coils inside the directional contactors as shown.

Action of FD relay while the motor is running at high speed is as follows: Assume the field is suddenly strengthened either by manual operation of the rheostat or by closure of FA relay contacts by FA shunt coil when the master is moved to a slow-speed point. The increased flux increases the CEMF above the impressed voltage so that the armature current reverses. It could reach an excessive value before the motor has time to slow down. However, FD relay is set to pick up with line voltage on its shunt-coil circuit and with 125% full-load reverse current in its series coil. The actual voltage on the shunt-coil circuit at this instant may be somewhat greater than line voltage, but the pickup of the relay will still occur close to 125% reverse current. FD contacts thus open to insert resistance in the field circuit until the reverse armature current falls below the relay dropout point of 110% FLC. FD holds the current between these limits until the motor slows down to near full-field speed.

Action of FD relay during dynamic braking is similar to the above, except that the shunt coil, instead of being on approximately line voltage, now has higher than line voltage at the start and nearly zero voltage at the end of the braking period. This arrangement gives a period when FD contacts hold open, then a fluttering period, and finally a period during which they stay closed.

12. Reversing Dynamic Lowering Hoist Control. Dc-hoist applications requiring control of overhauling loads in the lowering direction universally require a series-wound motor. Its characteristics of high available torque and series-generator-type braking are well suited for hoists (Ref. 5). In the hoisting direction, it is connected as a series motor and in the lowering direction it is connected as a shunt motor. A series-wound magnetic holding brake and a power-type overhoist-limit stop are readily applicable. Dynamic braking is effective when entering the overhoist-limit stop or when lowering an overhauling load whether or not line power is available.

The power supply may be from a 230-V constant-potential bus or from a rectifier power supply. If the rectifier supply is 230-V 3-phase ac, the dc voltage is over 300 V, and increased speed and horsepower result (Ref. 3).

Figure 15 shows an elementary wiring diagram of a dc dynamic-lowering hoist control. The H and 1L contactors are the hoist and lower positive line contactors, respectively; the 1A, 2A, 3A, and 4A contactors are the acceleration contactors for cutting out series resistance in the hoisting direction and for control of the field strength in the lowering direction. Contactors 2L and 3L are used for increasing the voltage across the armature in the lowering direction. The spring-closing contactor DB is the lowering dynamic-braking contactor connected in series with the dynamic-braking resistor R7-R10.

If the master switch is moved from the off point to the fifth-point hoist, H and M close instantly followed by 1A without time delay; contactor DB opens, and contactors 2A,

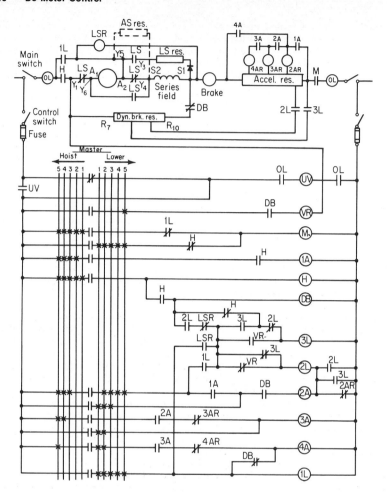

Fig. 15 Controller, dc, dynamic-lowering hoist.

3A, and 4A close in timed sequence under the control of the time-current relays 2AR, 3AR, and 4AR, respectively.

For some applications, very slow first-point hoisting speeds are desirable to take up slack cable, to hoist an empty hook, or for very light loads at a slow speed. A motor shunt connection obtained by having contactor DB closed on first-point hoist will provide such results. In this circuit, additional protective equipment should be added to the control to assure setting of the brake when the power limit switch trips if the master switch is on first-point hoist.

Many schemes have been used to provide setting of the brake with a motor-shunt connection. An extra collector bar can be used to assure brake setting, or a CEMF relay LSR can be connected to open the undervoltage relay. A CEMF relay, in addition, can provide protection in lowering out of the limit switch by picking up to stop the hoist drive if excessive speed is reached before the limit switch resets.

In lowering, the motor is connected for operation as a shunt motor. The performance-characteristics curves for a 230-V constant-potential power supply are shown in Fig. 16. The accompanying table is a point-by-point description of operation.

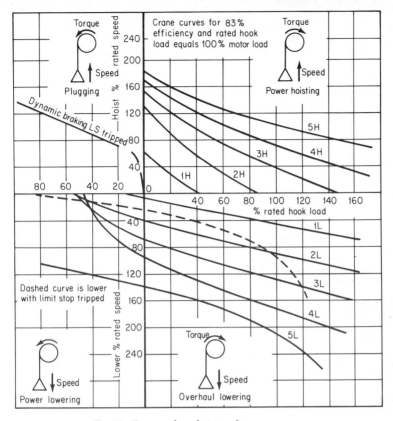

Fig. 16   Four-quadrant hoist-performance curves.

Lower—Acceleration (Moving the Master Slowly)

Point	Operation of contactors	Purpose
1	*M, 1L, 4A. 3A,* and *2A* close, *DB* remain closed	Provides strong current for brake release. Motor is connected as a shunt motor. Dynamic braking is provided in case load is overhauling
2	*2L* closes before *DB* opens	To reduce but maintain dynamic braking, voltage to the armature is increased
3	*3A* and *4A* open	To weaken field and reduce dynamic braking
4	*2A* opens	To weaken field and reduce dynamic braking
5	If the CEMF of motor is about 50%, the CEMF relay *VR* picks up to permit *3L* to close. *2L* opens, after *3L* closes	CEMF relay prevents excessive current through armature. Field is further weakened and dynamic braking reduced. Voltage to armature is increased

If the master switch is moved quickly to the last point lower, $1L$, M, and $4A$ close instantly and $2L$ closes after $1L$ closes with time delay of $2AR$. After $2L$ closes, DB opens followed by opening of $4A$ to provide acceleration. First-point circuit is held for a time period to give maximum current for brake release. When the motor attains sufficient speed, the CEMF relay VR picks up to permit closure of $3L$. $2L$ opens after $3L$ closes.

Lower—Deceleration (Moving the Master Slowly)

Point	Operation of contactors	Purpose
5 to 4	CEMF relay opens; $2L$ closes. $3L$ opens after $2L$ closes. DB remains open	To maintain and strengthen dynamic braking
4 to 3	$2A$ closes	Strengthens the field and increases dynamic braking
3 to 2	$3A$ closes followed by $4A$	Strengthens the field and increases dynamic braking
2 to 1	DB closes; $2L$ opens	Maintains and increases dynamic braking
1 to off	$1L$, M, $2A$, $3A$, and $4A$ open; DB remains closed	Dynamic braking maintained and brake sets

If the master switch is moved quickly from the fifth-point lower to the off point, $3L$ opens, DB closes, and $1L$ and M open in rapid sequence.

13. Entering the Hoist Limit Switch. With the master switch on the off point or on any hoisting point, the circuit is established to permit the brake to set when the limit switch trips. The CEMF of the motor is in a direction from A_2 to A_1, and the current will be maintained in the normal direction in the field from S_2 to S_1. The current reverses in the armature to provide dynamic braking, the motor acting as a series generator with the limit stop resistor LS across its terminals.

Should the operator move the master switch to any of the lowering points at the instant of tripping the limit switch, the motor will be plugged as a series motor with line voltage in the same direction as the CEMF. The brake does not set and hook travel is stopped by plugging.

The use of antifriction bearings and better hoist design may result in relatively high light-hook-hoisting speeds requiring an increase in the limit-switch resistance (LS RES) to obtain satisfactory motor commutation. The increase of the limit-switch resistance causes an increase in the stopping distance which, for some installations, cannot be allowed. The logical solution to this problem is to limit the maximum light-hook-hoisting speed, which in turn permits the use of lower limit-switch resistance. Both cause the stopping distance to decrease. The maximum hoist speed for most installations can be limited to 225% of rated motor speed by the addition of a permanent teaser-field resistance (not shown) connected across limit-switch terminals Y_6-Y_4. The value of this resistance depends upon the crane-hoist efficiency or percent of motor torque required to hoist an empty hook. For most cranes, 800% resistance having 1/8 full-load current-capacity is satisfactory.

14. Lowering out of Hoist Limit Switch. On all points when lowering out of the power limit switch, the motor is connected as a series motor, and under this condition, with a heavy overhauling load and with the brake released, there is no retarding torque. If the motor is not to reach an excessive speed, it is important that the limit switch reset in as short a travel distance as possible.

Numerous means of limiting the motor speed when lowering with the limit switch tripped have been used. Figure 15 shows an armature shunt (shown dotted) of 40% resistance, AS, connected across terminals Y_4-Y_5 for this purpose. The relay LSR picks up to control line resistance. A blocking rectifier in series with the limit-switch resistor greatly increases the armature-shunt retarding torque. Another method has been to use the CEMF relay LSR to open the undervoltage relay and the line contactors and set the brake should the speed become excessive before the limit switch resets.

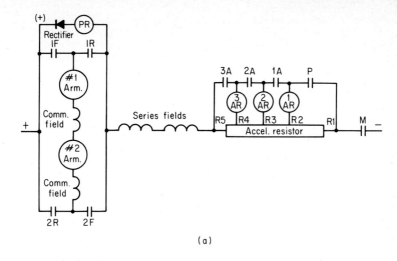

(a)

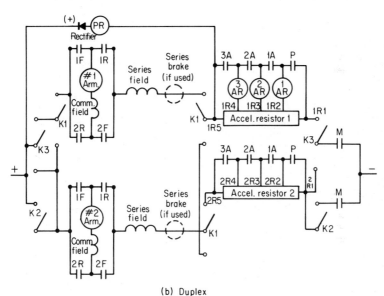

(b) Duplex

Fig. 17 (*a*) Two motors permanently in series. (*b*) Duplex control with cutout knife switches.

Oper.	Motor		Accel. res.		Kn. sws.		
mode	1	2	1	2	K1	K2	K3
A	Run	Run	In	In	Up	Up	Up
B	Run	Off	In	Out	Up	Down	Up
C	Off	Run	In	Out	Down	Down	Up
D	Off	Run	Out	In	Up	Down	Down

15. Duplex Motor Control. Owing to slight mechanical and electrical variations between different dc motors of the same rating, parallel operation of two or more motors coupled to the same drive requires special considerations (Ref. 2). If the motors are shunt-wound, a large variation in torque corresponds to a small variation in speed. Since the motors must run at the same speed, they will not equally divide the load unless special load-regulating control is employed. If the motors are series-wound, their drooping speed-torque characteristic provides for fairly equal division of load when connected for parallel operation.

For many heavy-industry applications, such as the travel drive of gantry cranes and unloaders or the bridge drive of large steel-mill cranes, multimotors provide better application of motor torque. The smaller motors also meet dimension limitations more easily. Several possible power-circuit arrangements may be used for such applications employing series-wound motors and reversing, plugging magnetic control. If the predominant requirement is full starting and acceleration torque but not the inherent high free-running speed of the series motor at full voltage, then the motors can be connected in one or more pairs with the two armatures in series for reversing and with the fields in series. The control is the same as for one motor, since rated motor current flows through the two motors. Voltage across each motor armature is one-half normal, and top speed and horsepower is one-half normal (Fig. 17a). If full-running speed and horsepower are required, then the motors must either be designed for half voltage or must be connected with armatures and fields in parallel.

For parallel operation of series motors normal practice is to control each motor separately (Fig. 17b). All devices are rated for single-motor capacity, and normal operation is achieved during acceleration, running, and plugging operation. In addition, provision is readily made to disconnect one motor and still operate under emergency conditions (Ref. 2).

The table under Fig. 17b lists four possible modes of motor operation and acceleration. The resistor use and corresponding knife switch combinations are shown. Acceleration and plugging control are provided except for mode D.

Occasionally, owing to economic or space considerations, it is desirable to operate two or more motors connected in parallel using a common acceleration resistor and common acceleration contactors. With this connection, there is no problem during acceleration or running under load. However, during plugging the motor CEMF is in the same direction as line voltage, and if the internal resistance of one motor is slightly less than that of the other, a greater current will flow in it. Its field strength is increased as is its voltage. Armature and field current continue to rise, causing a high circulating current through the two motors, and one motor will drive while the other retards the load. This difficulty during plugging operation can be overcome by crossing the motor fields during plugging (Fig. 18).

The disadvantage of a common acceleration control of parallel motors is that the acceleration devices may be of larger size than the reversing devices.

Part 2

Variable-Voltage Drives—Motor-Generator Types

THOMAS H. BLOODWORTH°

GENERAL

The static-type thyristor power supply is predominant today for variable-voltage dc drives (Sec. 26) primarily because of lower equipment and installation costs, and higher operating efficiency. Also, the maintenance of bearings, brushes, and commutators required on motor-generator sets is eliminated. However, some application considerations may dictate the use of motor-generator power supplies.

°Chief Systems Engineer, Process Electrical Systems, Allis-Chalmers Corporation (retired).

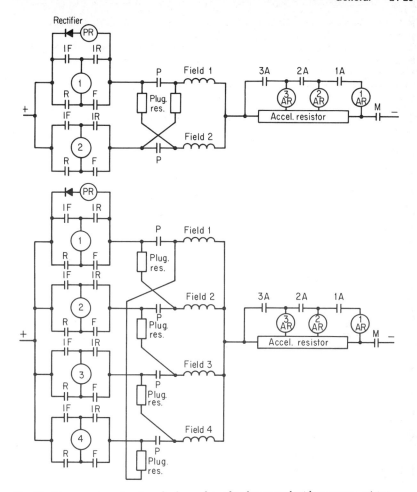

Fig. 18 Dc, reversing, plugging, duplex and quadruplex control with common resistor.

16. Application Considerations.   Several conditions should be considered.

a. Load Variations.   Load variations may cause wide fluctuations in the ac power system. Short-period load variations can be minimized by use of a flywheel motor-generator set.

A thyristor power supply reflects load variations directly into the ac system. Thyristor power-drive systems should be used only on ac power systems stiff enough to withstand the load fluctuations.

b. System Power Factor.   Thyristor power supplies have a lower power factor than systems supplied by motor-generator sets, particularly when the thyristor power supply is operating at low voltages and where considerable phase control is required. System power factor can be improved by using motor-generator sets driven by synchronous motors. Plants having large concentrations of thyristor supplies and induction-motor drives may find it advantageous to consider synchronous motor-generator sets for new dc-drive power supplies.

When a thyristor power system is applied on a weak ac transmission line, sufficient thyristor power-transformer capability must be provided to supply full dc output at minimum ac line voltage. Then, the thyristor power supply will be operating with considerable phase control—and at reduced power factor—when ac line voltage is normal. Also this arrangement increases the equipment first costs.

c. System Disturbances. Transient voltage disturbances on the ac power system are reflected directly to the dc output of the thyristor power system, until compensated for by action of the voltage-regulating system.

Harmonic-wave distortions are also introduced into the ac power system feeding the thyristor power supply. Motor-generator sets do not exhibit these undesirable effects.

d. Load Capabilities. Dc generators supply relatively ripple-free output from minimum to maximum voltage. The high ripple content in the output of thyristor power supplies operating at reduced voltages may contribute to the heating and commutation problems of dc-drive motors. In extreme cases, it may be necessary to use a series reactor to smooth out the power pulses to the motor.

The inherent ability of a dc generator to recover from the effects of high peak overloads makes this power supply more competitive with thyristor power supplies for some applications. High-intermittent-load capabilities for static power supplies usually mean furnishing thyristors capable of greater continuous output than is required for continuous duty.

e. Overhauling Loads. Whenever a load requires energy to be returned to the ac power line (regenerative braking), the cost of a thyristor power-supply unit is increased appreciably. In many such applications, motor-generator sets may become more attractive. Dynamic braking may be used with thyristor power supplies when a constant rate of deceleration or a minimum decelerating time is not required. This arrangement reduces the thyristor costs at the expense of increased overall losses. Dynamic braking discards the energy of deceleration as heat from the dynamic-braking resistors instead of returning it to the ac power line.

f. Maintenance Requirements. While the use of motor-generator variable-voltage power supplies requires less technical competence in maintenance personnel, the current use of static regulating equipment for motor-generator sets makes the same level of competence necessary as for maintaining the thyristor power supplies.

17. Basic Variable-Voltage System. The basic variable-voltage drive consists of an adjustable-voltage power supply electrically connected to a dc motor. The sophistication of the control equipment is determined by the drive requirements and the system disturbances.

a. Load Disturbances. Internal disturbances are defined as those changes in the system which change the power level of the output. The variations in loading impressed upon the shaft of the motor are an obvious example. The required recovery time from a load disturbance is the major factor in determining the system control requirements.

b. Internal Disturbances. Internal disturbances are defined as those changes in the system which effectively alter the magnitude of system parameters. The heating of a field winding and the change in speed of a generator may be cited as examples. These changes are relatively long-term disturbances.

Change in ac voltage for a thyristor dc power supply can be a very-short-time internal disturbance compared with those encountered in a motor-generator dc power supply.

MOTOR-GENERATOR DRIVES

18. Motor-Generator Variable-Voltage System. The basic motor-generator-type variable-voltage drive, formerly referred to as a Ward Leonard system, consists of an adjustable-voltage generator electrically connected to a dc motor. The generator is always separately excited. Motors are usually shunt-wound, but compound-wound motors may be used to meet some speed-torque requirements. Pusher motors on steel-mill reheat furnaces are a typical example of a compound-wound motor application for a variable-voltage drive.

Smaller motor-generator variable-voltage-drive systems may use a constant-potential dc supply for generator and motor shunt-field excitation, with direct rheostatic controls. Larger systems, because of the magnitude of excitation power to be varied, or systems requiring more sophisticated controls employ separate exciters.

19. Open-Loop Variable-Voltage Generator Regulation. Operation of motor-generator variable-voltage systems is more easily analyzed by determining the effects of load and system disturbances, first on the generator and then on the motor.

To determine the effects of these disturbances, consider the case of a separately excited dc generator with a manually operated rheostat in the exciter shunt field to adjust output voltage (Figs. 19 and 20).

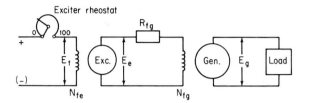

Fig. 19   Generator system with manual rheostat voltage adjustment.

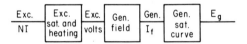

Fig. 20   Generator-system block diagram.

The exciter rheostat is adjusted to give 250 V, O-A on the no-load saturation curve O-E_0, NL (Fig. 21). This excitation requires 7.2 A in the generator shunt field. As the generator is loaded, the terminal voltage E_t would drop from the generated voltage E_o, owing to internal IR drop and the demagnetizing effects of armature reaction, to the full-load saturation curve O-B, to a value of 237 V. This change in terminal voltage with load is the "load regulation" of the generator. It is defined as the rise in generator voltage when load is reduced from full load to zero load, normally about 5% of rated voltage.

The generator shunt-field voltage for 7.2 A with a cold field O-A is 167 V. As the generator heats up, excitation drops to the "hot-field" resistance line O-D, reducing the field current to 6 A and dropping output voltage O-E. Also, the exciter would drop from 167 V cold field O-F to 125 V hot field O-G, reducing generator field current to 4.5 A. This change reduces the generator output voltage to 180 V O-H. This open-loop system regulation would be $[(250 - 180)/250] \times 100 = 28\%$. Operation of the exciter rheostat to maintain 250 V output on the generator under loaded, hot-machine conditions O-J results in 8 A I_f and 217 V on the generator shunt field O-K.

At the end of the day, the plant is shut down and all machines cool off. Next morning, the exciter output is 250 V O-L, giving 11 A O-M generator shunt-field excitation with a cold field. The generator no-load voltage O-N would be 295 V. System regulation would be $[(295 - 250)/250] \times 100 = 18\%$. Regulation is reduced under these considerations because of the effects of saturation in the exciter and generator.

20. Closed-Loop Variable-Voltage Generator Regulation.   The open-loop performance of the variable-voltage system described above would be unsatisfactory for most process operations. Some type of "feedback" system must be used. In this system a signal from the controlled variable is "fed back" to the system input in a manner to reduce output deviations. A feedback system should be highly responsive to changes in the input or reference quantity but should exhibit a high degree of indifference to other disturbances.

The feedback signal b (Fig. 22), which is a measure of the controlled variable c, is compared with the reference signal r. The difference in strength of these signals determines the point at which the control elements hold the output of the controlled variable.

The physical process represented by the summing point may take several forms. The essential characteristic is that it must reverse the sign or sense of one signal. It may be a

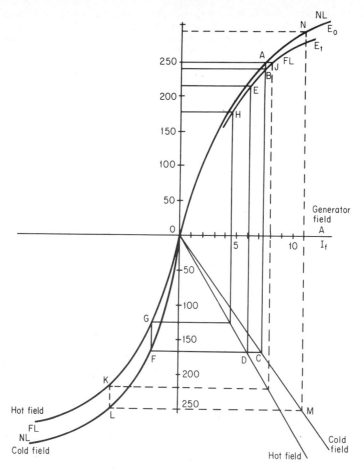

Fig. 21   Open-loop generator system.

circuit that operates to take the difference between the two quantities r and b ($e = r - b$) such as a reference and a feedback winding on a rotating regulator or a magnetic amplifier. It may also be a physical element whose nature is such that as the quantity e increases, the controlled variable c decreases. With a dc motor-speed regulator, when shunt field increases, motor speed decreases.

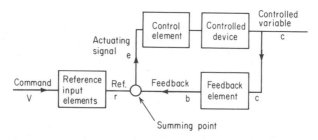

Fig. 22   Generalized block diagram of a closed-loop system.

With the exception of descriptions of main-drive and auxiliary-drive control circuits, this section will be limited to use of rotating regulators because of the ease in visualizing their operation. Only control fields such as reference, feedback, IR compensating, and load balance will be shown. Self-excitation circuitry will be omitted.

A diagram of a simplified rotating regulator is shown in Fig. 23. The summing point of Fig. 22 consists of two regulating exciter control fields which are connected so that their ampere-turns (At) oppose each other. The reference field At tends to increase the exciter voltage and, subsequently, the generator voltage. The feedback control field At tends to decrease the exciter voltage. This signal provides feedback, because the signal from the regulated quantity tends to lower the value of the regulated quantity. A positive feedback would have the control field At adding to the reference field At.

The controlled variable, which in this example is the generator terminal voltage E_t, is fed back to the feedback control field according to the factor N_c/R_c. This factor determines the amount of feedback control field At which is produced by generator terminal voltage E_t.

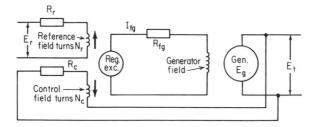

Fig. 23 Simplified diagram of a variable-voltage generator with a rotating regulator.

Figure 24 shows a block diagram of the schematic shown in Fig. 23. The terminal voltage E_t and the actuating signal (reference ampere-turns) are related by two different functions. One is the static characteristic of the generator voltage in terms of the net exciter At as determined by the components in the top portion of the loop in the block diagram (Fig. 24). The other is the relationship obtained through the bottom feedback portion of the loop.

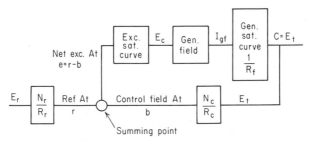

Fig. 24 Simplified block diagram of a variable-voltage generator with a rotating regulator.

Amplification required for a proportional regulator to maintain the regulated quantity within given limits may be determined graphically (Fig. 25) as follows: The generator no-load saturation curve O-E_0 and full-load saturation curves, hot O-C and cold O-M field lines, and exciter curves are the same as those shown in Fig. 21. First, determine system parameters for worst conditions, which include full load, hot field, and lowest exciter output. Draw the 250-V line to the load-saturation curve O-A. Then drop down

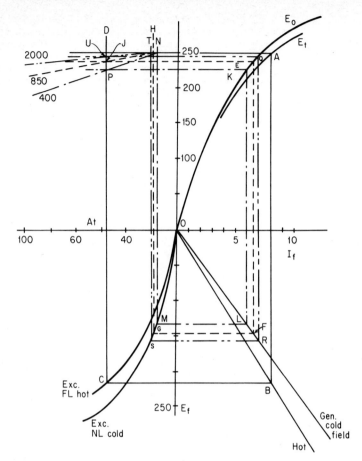

Fig. 25 Graphical determination of amplification required for a proportional regulator.

to the hot-field curve O-B. Next, go horizontally to the exciter hot-field, full-load curve O-C. Construct the vertical line C-D.

Construct a similar set of curves for desired limiting voltage, under most favorable conditions, which are cold fields in both exciter and generator. Assume a permissible variation of 5% or 250–237.5 V. Draw the 237.5-V line to E on the no-load saturation curve O-E_0 and to J on the line drawn previously. Then construct lines E-F, F-G, and G-H. Then construct a line through points H and J. The slope on this line is the feedback gain required in the regulating system. The extension of this line to intersection with the horizontal axis gives the ampere-turns required in the feedback field to give the desired regulation, 850 At for holding output voltage in the 250–237.5-V range through complete load and thermal cycles.

Increasing voltage tolerance to 10% or 250–225 V drops feedback ampere-turns to 400 At. Limiting voltage tolerance to a 2.5% range or 250–243.75 V increases feedback ampere-turn requirements to 2000 At (Fig. 25).

21. Variable-Voltage Motor Regulation. A variable-voltage dc motor may be controlled to maintain many quantities such as speed, torque, tension, or position in a predetermined relationship. Regulation may be open-loop or closed-loop, depending upon system requirements.

a. Open-Loop Motor-Control Systems. Motor-speed variations with load and temperature must be considered when determining the operating characteristics of the open-loop motor-generator drive system.

A shunt-wound motor which does not have a stabilizing winding has a tendency to rise in speed above one-half rated load, because of loss in flux from distortion in flux distribution caused by armature reaction. It is desirable to provide a stabilizing effect on the main poles to maintain the total air-gap flux. This stabilization can be accomplished by providing a small field on the main poles, connected in parallel with the commutating-pole windings. This compensating field is connected so that its ampere-turns add to those of the shunt-field winding. A resistor in series with the stabilizing winding is adjusted to give a 2 to 5% speed drop at rated motor load (Fig. 26).

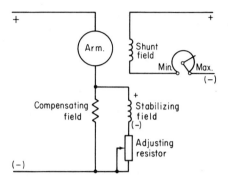

Fig. 26 Adjustable stabilization field to minimize effects of armature reaction on dc motor.

The connections to the stabilizing field must be switched on a reversing motor-generator variable-voltage system to keep the shunt and stabilizing field ampere-turns additive under both forward and reverse operation. This reversal can be most readily accomplished by using polarizing rectifiers to maintain a constant direction of current flow through the motor armature and commutating poles (Fig. 27).

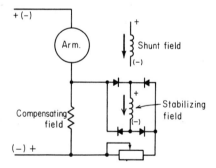

Fig. 27 Connection of stabilizing field for reversing drive motor.

The maximum field-weakening speed range of any dc motor is limited by the decrease in commutating ability. This decrease in commutating ability is caused by a factor called "divergence." The commutating field which has optimum strength at base speed has too much strength at higher speeds (Fig. 28). Commutating ability may be improved by shunting the commutating field during part of the field-weakening range. More than one step of field shunting may be required for critical drives or very wide field-weakening speed ranges (Fig. 29). Contactor S_1 is closed at one-third of the field-weakening range C (Fig. 28), shunting the commutating winding resistors R_1 and R_2 and eliminating divergence. Divergence increases as speed is increased to two-thirds of the field-weakening range D when contactor S_2 is closed, shorting resistor R_2, and divergence is again eliminated. Divergence will increase through the balance of the

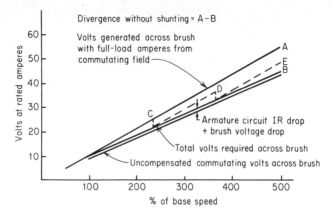

Fig. 28 Typical divergence of actual and required commutation volts.

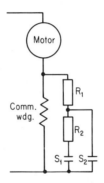

Fig. 29 Two-step commutating-field shunting.

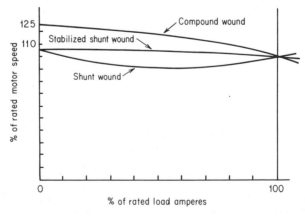

Fig. 30 Typical speed regulation for shunt and compound motors.

field-weakening range but never exceeds one-third of the value attained without commutating-field shunting B-E. For single-step commutating-field shunting, contactor S_1 is picked up at 50% of the field-weakening range and contactor S_2 is omitted.

Typical speed-regulation curves for shunt-wound, stabilized shunt-wound, and 25% compound-wound dc motors are shown in Fig. 30. These curves are based upon a constant terminal voltage. Any decrease in terminal voltage with increase in motor load will cause a corresponding decrease in motor speed, in addition to the motor-load regulation.

A simplified diagram of a variable-voltage system with a motor-operated speed-setting rheostat is shown in Fig. 31. The self-excited exciter has a cumulative series field for "flat" compounding. This series field is sized to give the same terminal voltage at full load as at no load. The exciter terminal voltage will exceed the no-load value at intermediate loadings. The exciter is normally mounted with the generator on smaller, high-speed motor-generator sets. A separate exciter set is usually more economical for installations with large, slow-speed motor-generator sets.

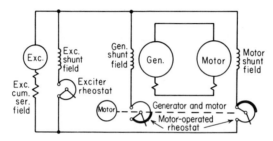

Fig. 31   Variable-voltage system with motor-operated speed-setting rheostat.

The exciter rheostat will require adjustment to maintain rated output voltage until its shunt-field temperature stabilizes. The motor-operated rheostat is employed to give a constant rate of acceleration and deceleration. Acceleration and deceleration may be at different rates, by suitably controlling the speed of the motor driving the rheostat.

The drive system in Fig. 31 utilizes a two-ring motor-operated rheostat. At start-up, the rheostat position results in minimum generator shunt-field current and full field on the motor. Rotation of the rheostat through the first half of its travel increases generator field current and terminal voltage while maintaining full motor field current. At the 50% travel position, the generator will be at full voltage and the motor will run at base speed. Further rheostat travel weakens the motor shunt field and increasing motor speed. At full rheostat travel, the generator will be at full voltage and the motor will run at top speed (weak field). Loss in speed due to a reduction in generator voltage caused by shunt-field heating will be partially offset by additional motor-field weakening caused by shunt-field heating.

When the application requires constant torque over the full speed range, separate field rheostats are provided for the generator and motor shunt fields. The motor-field rheostat is positioned to give the desired operating speed, and acceleration and deceleration is accomplishd by varying the generator voltage.

b. Closed-Loop Motor-Control Systems.   Closed-loop regulation for single-motor variable-voltage systems is best accomplished by varying the generator output as required to hold the regulated quantity within permissible limits because of the shorter response time. This arrangement results in the addition of a block for the motor-load curve, following the generator-saturation-curve block in the simplified block diagram shown in Fig. 24.

When motors are connected to a common power source, regulation must be obtained by varying motor excitation. A variable-voltage motor regulated to hold constant armature current is shown in Fig. 32. This system has many applications in industry, such as helper drives in the paper and metals industries and tensioning devices in metal-rolling

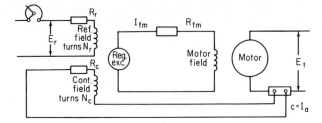

Fig. 32 Simplified diagram of a variable-voltage motor regulated to hold constant armature current.

mills and processing lines. The simplified block diagram for this system is shown in Fig. 33.

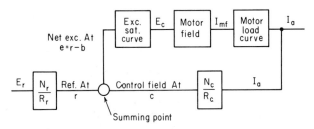

Fig. 33 Simplified block diagram of a motor regulated for constant armature current.

A summing point in the conventional sense may not be needed, as shown by the speed-watching system in Fig. 34. A dc shunt motor with a variable-voltage armature supply has its speed regulated by shunt-field control. The motor has two shunt-field windings, one of which is connected through a suitable amplifier to a tachometer on the preceding drive. A speed-reference rheostat is provided in the amplifier input to select the speed differential to be maintained between the two drives. The second field is connected, through an amplifier, to the drive tachometer.

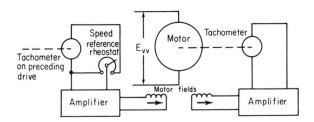

Fig. 34 Simplified diagram of a speed-matching system for a dc variable-voltage motor.

The steady-state (constant-voltage) speed characteristic S for the motor is a function of the regulating or net field current I_f, $S = f_0(I_f)$, and the overall characteristic of the tachometer amplifier is a function of motor speed S, $I_f = f_2(S)$. These two characteristics are shown in Fig. 35, and their intersection for voltage E_0 gives the operating point S_0. When the voltage is changed, both characteristics and the intersecting point change, giving speeds $S_1, S_2, \ldots, S_n$, etc. Voltage is normally varied to change system speed.

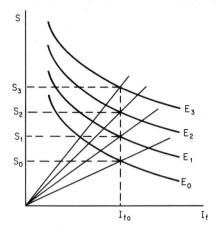

Fig. 35   Speed-matching system for a variable-voltage dc motor.

CONTROL CONSIDERATIONS

22. Protection. Variable-voltage-drive machines must be protected and utilized economically, as well as regulated to hold the controlled variable within acceptable limits. The ac-drive motors for motor-generator sets will be protected by their associated starters or switchgear, as described in other sections of this handbook.

a. Overload Protection. The armature circuits of the machines are protected against overloads by current relays (*OL*) and by current-limiting control circuitry. Current-overload relays may be classified as inverse-time, instantaneous, or rate-of-rise. They act to disconnect the machines when the relay operating point is reached. Current-limiting controls act upon the shunt fields of the motors or generators to reduce armature currents.

1. Inverse-time overload relays. The operating time of these relays varies inversely with the amount of current passing through the relay coil. They may be set very close to the anticipated maximum operating load. Some inverse-time relays also include an instantaneous element.

2. Instantaneous overload relays. These relays consist of a coil connected in series with the machine armature, surrounding a magnetic core. When the current reaches a preselected value, the magnetic core is quickly raised, opening the relay contacts, and causing the machine to be shut down.

3. Rate-of-rise tripping element. These devices are built into high-speed or semi-high-speed air circuit breakers. They open the breaker whenever current increases at an excessive ampere-per-second rate, indicating a solid short circuit or a commutator flashover. They are frequently applied on large common bus systems to interrupt fault currents before they reach their peak value.

4. Current-limit controls. A simplified diagram of a variable-voltage generator with a current-limit control is shown in Fig. 36. The generator terminal voltage is selected by operation of the three-point master switch (M_1, M_2, M_3) in the reference field of the regulating exciter. The ampere-turns of the regulating exciter feedback field increase as generator voltage increases until a balance point is reached.

The voltage drop across the generator commutating and compensating field windings is used as a measure of current. The terminal voltage of the current-limit exciter is adjusted to make the voltage drop across the halves of the voltage-dividing resistor (*A-B* and *B-C*) equal to the drop across the generator compensating and commutating field windings when current limit is to be initiated. Current flow from the current-limit exciter through the current-limit control field is prevented by the blocking rectifiers R_1, R_2. Whenever the voltage drop across the compensating and commutating fields in nor-

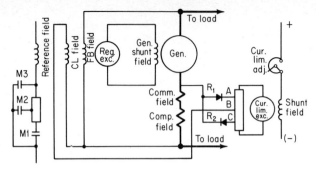

Fig. 36 Simplified diagram of a variable-voltage generator with current-limit control.

mal operation exceeds the voltage drop across the halves of the voltage-dividing resistor, current flows through the regulator current-limit field, opposing the reference field and dropping generator voltage. During regenerative braking, the action of the current-limit field is reversed, dropping generator load.

b. Overvoltage Protection.   Generator voltage may be raised to a dangerous level by malfunction of the regulating circuits, such as shorting of resistors in open-loop controls, or an open circuit in the feedback portion of a closed-loop regulator. Motor CEMF may be raised to a dangerous level by overhauling loads or during dynamic braking if the motor shunt field is suddenly strengthened when the motor is operating in the field-weakening range.

Overvoltage relays *OV* are designed to have a definite voltage pickup, adjustable by varying magnet air-gap or spring tension. When connected across motor or generator armatures during normal operation, functioning of the relays deenergizes the controls and disconnects the machines. Dynamic braking is applied to the motors. When overvoltage occurs during dynamic braking, operation of the overvoltage relay results in reduction of motor excitation and CEMF.

c. Voltage-Check Relays.   These relays (V_1, V_2, etc.) are designed to have a definite voltage dropout. They are connected across generator and motor armatures to prevent closure of armature contactors or circuit breakers while voltage is present across the armatures. The relay contacts, in control-setup circuits, close when armature voltages drop to a value suitable for start-up.

d. Low-Voltage Relays.   These relays (*LV*) monitor control-circuit voltage, opening their contacts when control voltage drops to an unsafe level.

e. Field-Loss Relays.   The coils of these relays (*FL*) are connected in series with motor shunt fields. They have a wide operating range, picking up and closing their contacts at weak-field values, and withstanding the heat generated by full-field current in the coils. If shunt-field current drops below the weak-field values, the relay opens its contacts, deenergizing control circuits and shutting down the motors.

f. Ground-Fault Protection.   Variable-voltage motor-generator systems are usually operated ungrounded. The basic theory is that a single ground will not cause a shutdown. Operation continues until a scheduled maintenance-shutdown period when the ground fault can be located and cleared.

1. Ground-detection voltmeters. Two voltmeters may be connected in series across the dc bus or machine terminals, with the junction of the two voltmeters grounded through a resistor. When a ground fault occurs, the voltmeter connected to the grounded side of the bus or machine drops to zero while the other indicates full system voltage. A single zero-center voltmeter may also be connected to the center tap of a resistor connected across the bus, with the other terminal grounded. The voltmeter will normally read zero but shows full system voltage when a ground occurs. It indicates the grounded side of the bus by direction of voltage indication.

2. Ground-detection relays. Voltage relays may be connected in the same manner as the voltmeters described above. A ground on either bus will cause a relay to operate, sounding an alarm.

g. Overheating Protection. Continuous operation of motors and generators over-loaded results in greatly reduced insulation life. Each 12°C temperature rise above the nameplate rating of the machine reduces the life of the insulation by one-half.

1. Machine-winding protection. The temperature of shunt, commutating, and compensating windings may be monitored by embedded temperature detectors. The detectors may be switched to a device indicating temperature to determine the "hot spot" or limiting highest temperature. The temperature indicator may be left on the hot-spot detector for indication, and a contact provided to sound an alarm on overtemperatures.

2. Bearing-temperature protection. Bearing temperatures for sleeve-bearing machines are monitored by temperature detectors. These devices are mounted in the bearing pedestals with the temperature-sensitive element in the babbitt bearing lining. Because of the danger of wiping sleeve bearings, operation of a bearing temperature device normally shuts down operation. Bearing-temperature detectors are available which signal two temperature levels. The lower temperature contact sounds an alarm; the second trips all breakers and stops operation.

h. Overspeed Protection. Overspeed trip attachments are provided as a safety precaution. These mechanical devices open contacts when motor speed exceeds the rated top speed by 15%, disconnecting the motor from the generator and applying dynamic braking.

i. Condensation Protection. Space heaters are usually provided on large dc generators and motors when they are to be shut down for more than a few hours. Otherwise, sudden rises in ambient temperature would result in excessive sweating on the machine metal parts. Space heaters should have sufficient capacity to raise machine metal-component temperatures to from 5 to 10°F above ambient. They are usually installed in a location where convection airflow will transmit the heat over the greatest possible internal surface of the machine.

The space heaters may be turned on automatically (with a manual lockout switch) whenever the machine is shut down. The heaters may also be turned on by a time-delay device with a preset delay after shutdown.

23. Ventilation. Open-frame generators may be self-ventilated if located in a clean atmosphere. When motors have to operate for long periods of time at rated armature current at low speeds, self-ventilation is not adequate and separate ventilation must be provided. This ventilation may be provided by a top-mounted motor-drive blower, or by a duct connecting to a central ventilating system.

Motors and generators located in dusty or corrosive atmospheres are normally ventilated by ducts connected to a central ventilating system. The system may be fed from a clean atmosphere or a recirculating system with filtered, cooled air. Vane switches may be provided in the air ducts to ensure airflow before equipment may be started, and to sound an alarm if airflow ceases during operation.

24. Generator Load Sharing. Generators connected in parallel will not share load equally, even when built at the same time, from the same design. This is due to material and manufacturing tolerances, and unequal ventilation.

a. Load Balance by Brush Shifting. Brush shifting is sometimes used to balance loading on paralleled generators. Shifting the brushes against machine rotation reduces generated voltage and drops load on the machine. Excessive brush shifting will have an adverse effect on commutation and brush life.

b. Load Balance by Differential Series Fields. Use of series field differential to the shunt field will cause the more heavily loaded machine to lower voltage and drop load. The series field should be capable of dropping the generator voltage from 20 to 30% from no load to full load. The use of differential series fields alone for load balance requires additional range in the shunt field, and additional regulator action to maintain voltage.

c. Load Balance by Cross-connected Series Fields. The use of differential and cumulative series fields for load balancing of two paralleled generators is shown in Fig. 37. Each series field should have 15 to 20% of the shunt-field ampere-turns. The cross-connected series fields provide load balancing for both forward and reverse operation, as current flow in the series field reversed when the shunt-field current is reversed.

Balanced loading is also maintained during regenerative braking as the effect of the series fields is reversed. The differential series field becomes cumulative to the shunt

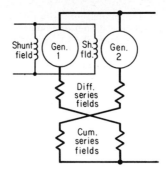

Fig. 37   Generator load balance with cross-connected series fields.

field and the generator carrying the highest load will have its CEMF increased, dropping load.

d. Load-Balance Regulator.   Theoretically, any number of generators can be load-balanced by cross connection of differential and cumulative series fields, such as diff. G_1 to cum. G_2, diff. G_2 to cum. G_3, diff. G_3 to cum. $G_n, \ldots,$ diff. G_n to cum. G_1. Practically, three cross connections are the maximum for load-balancing large dc generators by this method.   Complexity and cost of bus interconnections between machines become prohibitive beyond this number.

Figure 38 shows load balancing for four generators.   Pairs of generators are load-balanced by cross connection of series fields as shown in Fig. 37.   A load-balance regulator is provided for equal load division between pairs of generators.   The load-balancing regulating exciter has a single control field, responsive to the difference in voltage drop across the differential and cumulative series fields of the two pairs of generators.   The exciter control-field ampere-turns drop to zero when the load is balanced between the two pairs of generators.

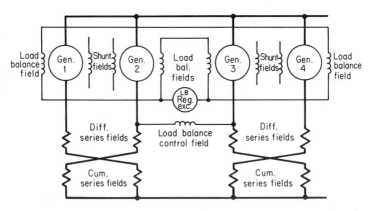

Fig. 38   Generator load balance with cross-connected series fields and load-regulating exciter.

25. Motor Load Sharing.   Motors connected in paralled encounter the same load-sharing problems as those previously described for generators.   Small motors such as table drive motors usually have high enough armature-circuit resistance to provide reasonable load balance.

a. Unidirectional Drives.   Motors for unidirectional operation can be load-balanced by the use of series field, similar to generator load-balance methods.   The differential and cumulative series fields must be interchanged, as strengthening the motor-field flux reduces loading on a motor in parallel operation.

b. Reversing Drives. The use of series fields for load balancing of dc variable-volt-age reversing motors is impractical, owing to the necessity of reversing the polarity of the series fields whenever the direction of motor rotation is changed by the reversal of generator voltage.

Figure 39 shows load balancing for two reversing motors connected in parallel to a variable-voltage generator. A load-balancing regulator is provided for equal load division between the motors. The load-balancing regulating exciter has a single con-trol field, responsive to the difference in voltage drop across the commutating and compensating windings of the two motors. The exciter control-field ampere-turns drop to zero whenever the load is balanced between the two motors.

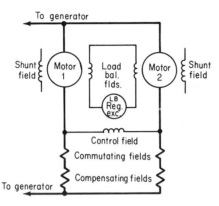

Fig. 39 Reversing dc motors with load-balancing regulator.

Additional load-balancing regulators can be provided if additional motors are con-nected in parallel with the two motors.

26. Braking. Process requirements may dictate a controlled deceleration while maintaining a fixed relationship between system components. Some drives can be disconnected and allowed to coast to a stop. Emergency conditions may require dis-connecting drives and bringing them to a standstill in the minimum possible time. Other drives may require restraint to prevent rotation after stopping.

a. Regenerative Braking. Coordinated acceleration and deceleration are most read-ily attained by use of motor-operated rheostats or similar time-rate control devices. Operating speed of the motor-operated rheostat during deceleration is determined by the net decelerating horsepower requirements of the most critical drive unit.

Horsepower required to decelerate from r/min_t to standstill in t s at a constant rate with constant motor field and armature amperes by reducing generator voltage is

$$\mathrm{hp} = \frac{Wk^2 \times r/min_t^2}{1.615 \times 10^{-6} \times t} \tag{1}$$

Load horsepower during deceleration must be subtracted from the decelerating horse-power obtained from Eq. (1) to determine the net horsepower during deceleration.

b. Dynamic Braking. Dynamic braking is applied to a dc motor by disconnecting the motor armature from its power source and connecting a resistor across the motor armature circuit. The motor shunt field remains energized, and excitation may be increased during braking for some applications.

The dynamic-braking resistor is connected across the motor armature by a spring-closed contactor or by a normally open contactor or circuit breaker for larger motors (Fig. 40). The dynamic-braking contact is closed by the opening of the motor line con-tactor or breaker.

The peak dynamic-braking current is limited by the commutating ability of the motor and the torque limitations of the driven machine. The dynamic-braking current

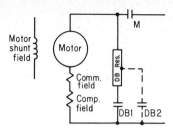

Fig. 40   Dc-motor dynamic braking.

will decay from its maximum value A-B (Fig. 41) as the motor decelerates, with consequent loss in decelerating torque. Decelerating time may be reduced by use of multiple-stage dynamic braking. If contactor DB_2 is closed at time t_1 (Fig. 40), current will again rise to the peak value and decay along curve C-D (Fig. 41).

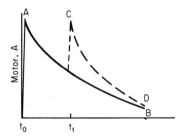

Fig. 41   Decay of dynamic-braking current.

c. Magnetic-Shoe Brakes.   Mechanical brakes may be used for deceleration as well as for preventing rotation after motors are brought to rest. The most common type is the spring-applied, solenoid-released. Small brakes may be ac- or dc-operated. Larger brakes have dc solenoids. The torque rating of the brake is usually from 25 to 50% of motor torque at the motor top speed.

TYPICAL DRIVE CONTROLS

27. Main-Drive Control Circuits.   Variable-voltage dc main drives are used where frequent rapid acceleration is required, or where precise adjustable-speed regulation or speed matching is desired.

a. Hot-Strip-Mill Drive Control.   A simplified schematic diagram of a hot-strip mill with magnetic-amplifier regulators is shown in Fig. 42. Mill speed, including acceleration and deceleration, is controlled by operation of the speed-set motor-operated rheostat, which acts on the reference winding of the three generator magnetic-amplifier voltage regulators.

Load is balanced between generators on each motor-generator set by means of cross-connected differential and cumulative series fields. Loading is balanced between pairs of generators by means of load-balance windings on the voltage regulators. A winding responsive to the loading of each of the other pairs of generators is connected cumulative to the speed (voltage) reference field, tending to increase excitation on the pair of generators carrying the lightest load.

The speed of the individual drive motors is adjusted by positioning of the motor-operated and vernier rheostats in the motor magnetic-amplifier speed regulator reference winding circuit. The motor-operated rheostat is used for setup adjustments, while the lever-operated vernier rheostat is used for minor speed adjustments during rolling.

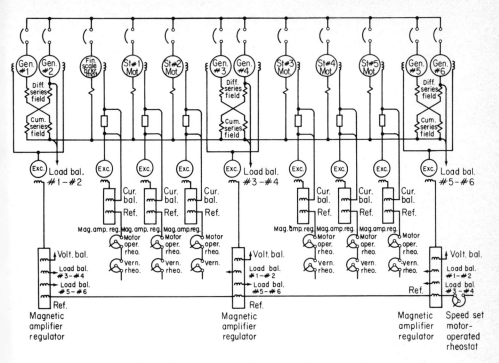

Fig. 42 Simplified schematic diagram of hot-strip-mill motors and generators.

b. Single-Stand Reversing Cold-Reduction Mill (Fig. 43).   The coil to be reduced is mounted on a feed reel (not shown) with a drag generator to provide back tension for the first pass. The strip is jogged through the mill stand and clamped on the winding reel, the wrap counter on the winding reel is set to zero, stall tension is applied, and the mill is accelerated.

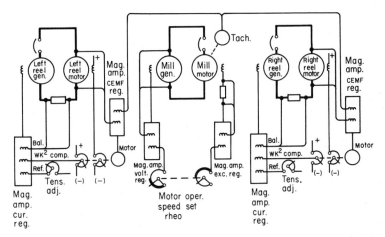

Fig. 43 Simplified schematic of a single-stand reversing cold-reduction mill with magnetic-amplifier regulators.

The rate of acceleration and deceleration is determined by the operating time of the motor-operated speed-setting rheostat. This rheosat operates on the reference winding of the generator voltage regulator to bring the generator up to rated voltage during the first half of rheostat travel. During the last half of its travel, the rheostat operates on the reference winding of the motor-excitation regulator to bring the mill motor to top speed.

When stall tension is applied, the reel generator current regulator is activated by the reference winding, and generator voltage is increased until the balance-winding ampere-turns balance the reference-winding ampere-turns. At this time the winding-reel rheostat is automatically set to the empty-mandrel (weak-field) position.

The CEMF regulator compares the reel-motor CEMF (motor terminal volts, less internal armature-circuit IR drop, compensated for by control circuitry omitted for simplification) with the mill-speed-indicating tachometer. As the coil on the winding reel builds up, armature current increases owing to the greater torque arm. The current regulator reduces generator voltage to maintain constant current. This reduction in voltage unbalances the CEMF regulator which operates the motor field rheostat to increase motor excitation. The increased torque per ampere drops armature current until the current regulator returns armature current and generator voltage to the selected value. Thus reel-motor excitation is changed as reel diameter varies, holding motor CEMF proportional to strip speed, thereby maintaining constant tension.

The motor field rheostat holds its position when the mill is shut down with a partial coil, so that correct field is maintained for acceleration and deceleration. A second ring of the motor field rheostat varies Wk^2 compensation in accordance with reel diameter, so that the current regulator is recalibrated for maintaining tension during acceleration and deceleration.

Figure 44 shows variation in Wk^2 compensation for a reel with a 16.5-in mandrel with a maximum coil diameter of 72 in and a maximum weight of 40 000 lb. Mill accelerating rate is 133 ft/min/s, normal stopping rate is 200 ft/min/s, and emergency stopping rate by regenerative braking is 250 ft/min/s.

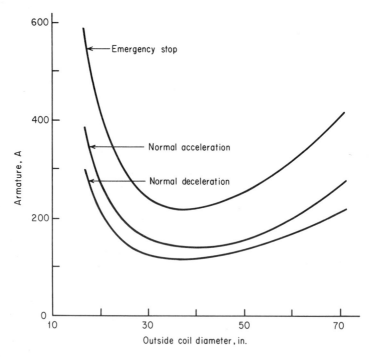

Fig. 44   Delivery reel with Wk^2 compensation.

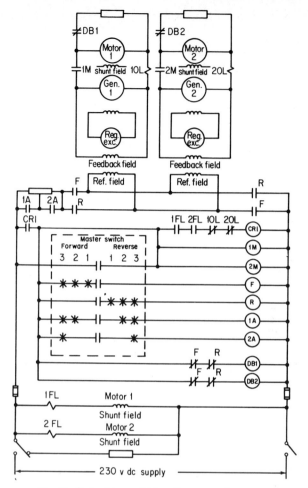

Fig. 45 Twin-drive variable-voltage control.

When the mill stand is used as a measure of strip speed, the unwinding reel regulator must be compensated for reduction taken in the mill, as the strip before reduction is not moving as fast as is indicated by the mill tachometer. This difference would result in tension higher than indicated, by an amount proportional to the reduction. A manually operated percent-reduction rheostat can be used for the recalibration. The use of separate tachometers for strip-speed indication for each reel driven by idler rolls on each side of the mill is preferred, as no recalibration is necessary.

When the coil is unwound from the feed reel, the end of the strip is clamped to the reel mandrel, the wrap counter is set to zero, the regulator is transferred to the reel motor, and stall tension is applied. Subsequent passes are made without releasing tension.

28. Auxiliary-Drive Control Circuits.   Variable-voltage auxiliaries are used where auxiliary motor speeds must be coordinated with main-drive speeds, such as mill tables, feed rolls, and coilers; where rapid acceleration and deceleration is required, such as for steel-mill screwdowns; or where rapid acceleration and deceleration with position stopping is mandatory, such as shears.

a. Twin-Drive Variable-Voltage Control (Fig. 45).   This control is suitable for operation of front and back mill tables, feed rolls, etc. With the addition of a magnetic clutch and end-of-travel limit switches, this control would also be suitable for operation of steel-mill screwdowns.

   b. Mandrel-Type Down Coiler.   Hot-strip mills may have two or three coilers. The center-driven mandrel-type down coiler has been furnished on most of the mills installed recently. Variable-voltage control is used with mill-type motors for pinch rolls, wrapper rolls, and mandrel drive. One generator is used with the pinch rolls and a second generator for the mandrel and wrapper rolls (Fig. 46).

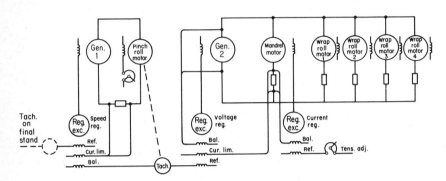

Fig. 46   Simplified schematic of a center-drive mandrel-type down coiler.

   The speed regulator controls voltage of generator 1 so that the no-load speed of the pinch-roll motor runs 6 to 8% above the speed of the strip leaving the finishing-mill final stand.   A second rotating regulator controls the voltage of generator 2 to hold the no-load speed of the empty mandrel and the wrapper rolls to 6 to 8% above the pinch-roll speed.   This speed differential between motors minimizes the chances of slack in the strip and a possible cobble.   The mandrel-motor regulator automatically changes from a speed to a tension regulator when the strip is wrapped around the mandrel and the motor is loaded.   Line resistors in the wrapper-roll motor circuits provide speed matching with the mandrel under load.

   c. Variable-Speed Runout Tables.   Runout tables transport material from a finishing stand to a succeeding operation such as coiling or cooling beds.   Individual rolls may be driven by dc or ac motors.   Dc motors are used when frequent, rapid reversal of tables is required.

   The unidirectional variable-frequency drive (Fig. 47) would normally include provisions for jogging in reverse.   The output of the finishing-stand tachometer is amplified to have sufficient power for the reference windings of the regulators for the table section generators, shears, etc.   Typical operating alternator range is from 20 to 90 Hz, giving a table-speed range of 4.5/1.   The lead speed-adjusting rheostats permit differential adjustment of speeds for the different table sections.   Table 1 following the shear normally runs above finishing-stand speed to introduce a space or gap between bars or sheets.   The final table may be slowed down to assist deceleration on the cooling beds for bars.

   .   The motors driving the alternators are operated with constant excitation, as this is a constant-torque application.   An induction motor rated 0.68/3.25 hp at 250/1205 r/min will have a rated load current of 8.7 A at low speed and 8.2 A at high speed.   The volts-per-cycle regulator will hold the alternator output to 80 V at low speed and raise it to 312 V at top speed.

   Each alternator has a circuit breaker for overload and short-circuit protection.   The table motors connected to an alternator are divided into groups by contactors.   Only one group is shown for each alternator.   Groups may consist of 10 to 15 motors.   Each motor is protected by a manually operated circuit breaker.

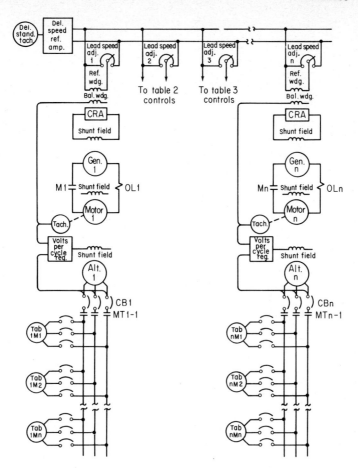

Fig. 47 Simplified schematic of variable-frequency table drives.

REFERENCES

1. Robert W. Smeaton, "Motor Application and Maintenance Handbook," McGraw-Hill Book Company, New York, 1969.
2. Charles S. Siskind, "Electrical Control Systems in Industry," McGraw-Hill Book Company, New York, 1963.
3. L. J. Srnka, M. C. Davies, and A. H. Myles, DC Magnetic Crane Hoist Control for AC Powered Cranes, *AIEE Trans.*, 60-30.
4. "Industrial Control," NEMA Standards Publication ICS-1970.
5. D. G. Fink and J. M. Carroll, "Standard Handbook for Electrical Engineers," 10th ed., sec. 20, para. 181, McGraw-Hill Book Company, New York, 1968.

25

Variable-Frequency Solid-State Ac Drives

BORIS MOKRYTZKI*

*Manager, Ac Drives, Engineering, Robicon Corporation; Senior Member, IEEE; Registered Professional Engineer (Ohio).

FOREWORD

An endless array of new ac-drive applications have developed out of the rapid growth of solid-state technology. Controls for these drives provide an economical means of gaining a wide range of control features with safety, reliability, and maintainability.

Essential parts of the drive system including the motor, inverter, power source, and load are covered in this section. Basic inverter circuits, systems, and combinations are identified and evaluated to aid in their application.

GENERAL

A variable-frequency ac drive is composed of a frequency converter operating from a power source, a control system which directs and excites the converter, and an ac motor and its load. The power source may be of fixed- or variable-frequency ac, unilateral dc, or bilateral dc capable of accepting the regenerative energy. The converter can be rotating, magnetic, or static, while the load can be either a synchronous or asynchronous machine or groups of machines, which, again, may require power flow in both directions. As in dc, the control involves either speed, torque, or position. More specifically, the present state-of-the-art ac drive employs a solid-state converter to supply a squirrel-cage or synchronous-reluctance motor.

At first glance, the ac drive is thought to be slated as an alternate dc-drive replacement, although it does have some unique applications such as in synthetic-fiber spinning. And, while it does share many external similarities with the dc system, it is unquestionably more sophisticated. This situation is complicated by the newness of an art developed literally over the past 5 years and still in a state of refinement. Fortunately, it is possible to identify basic principles, which when simplified provide an excellent means of comparing and evaluating an ac-drive system. While many new developments are still to come, these fundamental principles of operation, once understood, can provide guidance in gaging innovations.

AC MOTOR

The induction motor has traditionally been thought of as a constant-speed device, since it usually operates from a fixed-frequency power source. The disadvantages of operating such a mode are exemplified at starting where high inrush currents must be experienced to achieve reasonable starting torques. The advent of the thyristor converter as a source of variable voltage and frequency transforms the ac motor to a variable-speed device comparable with a dc motor. Much can be and has been said about the relative desirability of the ac drive which includes the ability to operate at high speeds and in adverse environments plus being light, compact, etc.

1. Induction-Motor Construction. The squirrel-cage induction motor is composed of a stator incorporating a magnetic structure and a set of polyphase windings. The rotor is composed of a set of magnetic punchings or laminations into which has been cast (or fabricated) a simple set of aluminum or copper bars and end rings to form a squirrel-cage winding. The rotor is virtually a solid, containing no electrical insulation other than that which is inherent on the surface of the laminations and conductors. As a result, it is mechanically rugged and able to operate at higher speeds and temperatures. At the same time, the operation of the machine can be efficient, as demonstrated by the following analysis.

Before any drive system can be applied effectively, the output quantities of the motor must be expressed in terms of its controllers' output quantities. These quantities, in turn, must be realized from and related to controller inputs. The operation of the familiar shunt-wound dc motor is defined by the speed and torque transfer functions

$$T = K_t I_A \qquad E_A = K_e S$$

where T = torque
S = speed
K_t = torque constant
K_e = induced-voltage constant
E_A = induced armature voltage
I_A = armature current

A similar set of relationships can be developed for the induction motor.

2. Variable-Speed Induction-Motor-Drive Analysis (Ref. 3). If excitation losses ¡and stator voltage drops are neglected, a polyphase voltage $V \sin 2\pi FT$ (F is the frequency) applied to the stator windings of a motor produces a rotating magnetic field in the stator whose angular speed is $\omega_S = \dfrac{2\pi F}{p}$ (p is the number of poles) and whose magnitude is $\phi \sim \dfrac{V}{\omega_S}$. If the speed of the rotor is ω_r, then

$$\omega_d = \omega_S - \omega_r \tag{1}$$

where ω_d = slip speed
The voltage induced in the rotor is

$$V_r = K\phi_S \omega_d \tag{2}$$

From this, a rotor current I_r is produced where

$$I_r = \frac{V_r}{Z_r} = \frac{V_r}{\sqrt{R_r^2 + (\omega_d L_r)^2}} \tag{3}$$

where Z_r = motor impedance
R_r = rotor resistance
L_r = rotor reactance

This, in turn, produces a flux

$$\phi_r \sim I_r \sim \frac{\phi_s \omega_d}{Z_r} \tag{4}$$

This flux lags ϕ_s by an angle

$$\theta = \frac{\pi}{2} + \tan^{-1} \frac{\omega_d L_r}{R_r} = \frac{\pi}{2} + \delta \tag{5}$$

The motor torque is given by

$$T \sim \phi_r\phi_s \sin \theta = \phi_r\phi_s \cos \delta \tag{6}$$

Therefore, from (1), (4), and (5),

$$T \sim \phi_r\phi_s\omega_d \frac{R_r}{Z_r^2}$$

$$= \phi_s^2 \frac{\omega_d R_r}{Z_r^2} \tag{7}$$

$$= T \sim \left(\frac{V}{\omega}\right)^2 \frac{\omega_d R_r}{R_r^2 + (\omega_d L_r)^2} \tag{8}$$

Thus independent of the excitation frequency and speed, torque is dependent only on slip ω_d and volts per cycle $\dfrac{V}{\omega_s}$. Differentiating the expression for torque with respect to slip and setting it equal to zero yields

$$\left(\frac{\omega}{V}\right)^2 \frac{1}{R_r} \frac{dT}{d\omega_d} = \frac{R_r^2 + (\omega_d L_r)^2 - 2L_r\omega_d^2}{Z^4} = 0 \tag{9}$$

$$R_r^2 = (\omega_d L_r)^2 \quad \text{or} \quad \omega_{d\max} = \frac{R_r}{L_r} \tag{10}$$

This is known as the breakdown slip associated with the breakdown point.

   3. **Synchronous Reluctance Motor.**   Operation with a synchronous reluctance motor is similar to that of a separately excited motor in which the rotor acting as a magnet spins in step with the rotating field, its torque being proportional to the excitation and the magnitude of rotor-stator flux angle (Ref. 11). However, the rotor is similar in construction to the squirrel-cage assembly, except that preferred magnetic paths are established by channeling the rotor flux with cast-aluminum magnetic barriers. The motor starts and is brought up to near synchronous speed as a squirrel-cage machine. Torque pulsations then occur, bringing the rotor and attached load into synchronism. The ability to synchronize or "pull in" is a function of the level of excitation, resulting from the motor design. The pull-in torque involves the magnitude of the accelerating impulses vs. the connected inertia, the stiffness of the output shaft, etc. Also, like a wound-rotor synchronous machine, pullout can occur if excessive load is applied. Furthermore, the starting cage tends to provide damping of oscillation in the torque angle or when a load disturbance is suddenly applied (Art. 24).

INVERTER

   While other means of producing ac, such as the cycloconverter, are available, the inverter which operates from dc is more flexible and hence more popular. A number of different types of high-power inverters available today use thyristors which act as power switches. These thyristors must be provided with a means of turnoff, almost universally employing charged capacitors in a technique known as forced commutation (Ref. 13). The ability of an inverter to provide a given amount of turnoff time is proportional to the size of the commutation capacitors and the voltage to which they are charged and inversely proportional to the current to be commutated (Ref. 1). The circuit of Fig. 1 is a 3-phase thyristor bridge with the commutation circuitry removed. The parallel opposed rectifiers carry reactive current. The thyristors of each phase conduct for alternate 180° intervals, with the sequencing of each phase displaced by 120°. A set of 3-phase output voltages are formed with the "six-step" wave shape of Fig. 2. These voltages have been used successfully with motors (Ref. 5). The motor, therefore, does not need a sine wave and can tolerate or reject a surprisingly large amount of harmonic voltage content. The harmonics, principally the 5th and 7th, produce about 20 to 30% additional losses over the sine wave in a standard motor.

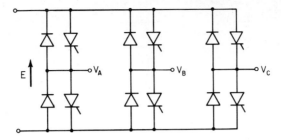

Fig. 1 Three-phase inverter.

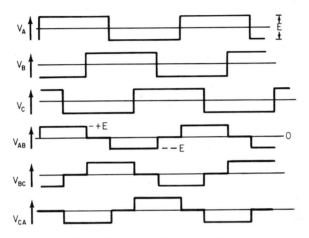

Fig. 2 Waveforms of 3-phase inverter.

4. Commutation. The thyristor is a three-terminal, four-layer semiconductor which can block current of either polarity. It can be triggered to conduct heavily in one direction if it simultaneously experiences a forward voltage V and a positive gate current I_g (Ref. 12). In this condition, it exhibits a low forward drop of approximately 1 V similar to a conventional semiconductor rectifier. It can be returned to a blocking state by bringing its anode current to zero for an interval sufficiently long to allow the carriers to deionize or recombine. This minimum interval to allow the unit to regain its blocking ability is defined as the turnoff time t_q. In ac-supplied circuits, turnoff is usually accomplished by the natural tendency of current to come to zero as the voltage polarity alternates. This action can be simulated in dc circuits, for example, by using series capacitor or resonant loads. The action is known as natural commutation or natural turnoff because of the natural tendency for the current to go periodically to zero.

To achieve turnoff in most dc circuits, current must be forced out of the conducting thyristor into an alternate and better path or circuit. This technique is defined as forced commutation and represents the almost exclusive turnoff method in ac-motor drives (Fig. 3a). The current is diverted to the closed switch, which is a more favorable path owing to the 1-V characteristic forward drop of the thyristor.

Neglecting the fact that ultimately to stop the current, the switch must be opened with an associated arc, etc., the method, in general, fails at high current when the increasing voltage drop of the switch makes the 1-V drop of the thyristor a more attractive conduction path.

This condition can be reversed by placing an active source, such as a battery, in the circuit to promote or aid current flow in the switch (Fig. 3b).

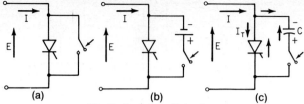

Fig. 3 Basic turnoff circuits.

Although the switch is again a better path, the solution remains impractical. A more acceptable solution is the use of a charged capacitor (Fig. 3c).

The capacitor acts as a battery for intervals as short as the several tens of microseconds necessary to achieve turnoff (Fig. 4).

When the switch in the circuit of Fig. 3c is closed, the capacitor is applied directly across the thyristor. The capacitor thereby assumes the entire load current and as a result is discharged linearly. The unit is assumed "held off" as long as the capacitor voltage exists; i.e., the thyristor is biased with a reverse polarity. The "hold-off" time to Fig. 4 can be determined as follows:

$$Q = I \times t = C \times V \qquad t \simeq \frac{CV}{I} \tag{11}$$

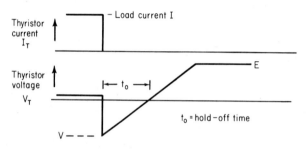

Fig. 4 Waveforms at turnoff.

This equation holds reasonably true even though many turnoff-circuit variations are present. That is, the hold-off time increases with increasing capacitor voltage and capacitor size, and decreases with current to be commutated. A thyristor will turn off if the applied hold-off time t_0 is greater than the necessary turnoff time t_q. Thus a commutation failure can occur if an unexpected current surge causes t_0 to drop below t_q. Turnoff time t_q is normally determined by test, and selected units yield values ranging nominally from 10 to 50 μs (Ref. 30). Thus, for a 300-V supply, a maximum load current of 100 A, and a turnoff time of 30 μs the minimum required capacitance is

$$C = \frac{t \times I}{V} = \frac{30 \times 10^{-6} \times 100}{300} = 10 \ \mu F \tag{12}$$

Figure 5 shows a practical turnoff circuit in which the switches of Fig. 3 have been replaced by an auxiliary thyristor A. If thyristor A is assumed on, capacitor C will charge through R to the potential E with the polarity shown. Triggering A places the capacitor voltage across A, diverting the load current I and causing turnoff. This circuit is representative of the family of circuits known as choppers or dc switches. The average load voltage is controlled by varying the on to off time of the switch as shown.

5. Inverter Circuits.   Inverters can generally be described as devices which produce ac from dc. As mentioned before, the particular type of inverter most involved in ac drives employs forced (or capacitor) commutation and hence is known as the forced

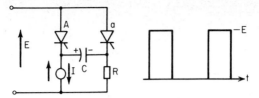

Fig. 5 Simple chopper.

commutated inverter. Before thyristors the parallel inverter was operated with mercury-arc devices (Ref. 27). One version is shown in Fig. 6a. The commutation method employed involves diverting current from a conducting thyristor to a path containing a newly fired thyristor and a charged capacitor. This operation is demonstrated by assuming thyristors A and A' are conducting, and the capacitor is charged with the polarity shown. Firing the B units causes turnoff as the current through A is diverted through the capacitor via B; similarly A' is turned off by B'.

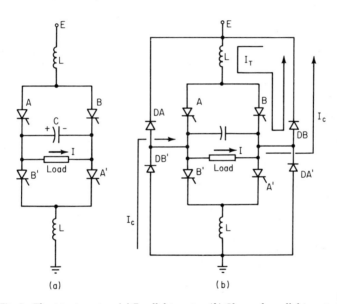

Fig. 6 Thyristor inverter. (a) Parallel inverter. (b) Clamped parallel inverter.

Inductors L serve to limit the momentary short-circuit current which tends to flow as the charged C is connected across the supply voltage E. The capacitor size can be determined as mentioned before. After a short time, turnoff is completed, and the charge on the capacitor reverses in preparation for the time when the B units must be turned off by firing the A units. Each time a new set of units are fired, a new half cycle of square wave is initiated (Fig. 7). The circuits work well for resistive loads but become awkward for reactive loads. For example, if the load in Fig. 6a is an inductor in which a current I is established as a result of a half cycle of conduction by the A unit, firing the B units extinguishes the A thyristors. The load current I continues to flow, and since the A units are off and the B units cannot accept a reverse current, the current must flow into the capacitor C. The capacitor voltage reverses as it charges to a potential higher than supply voltage E, since it ultimately must accept all the load energy. The final capaci-

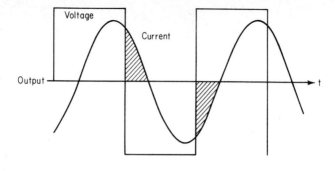

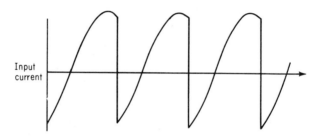

Fig. 7 Waveforms of square-wave inverter.

tors voltage V_f is determined by considering the initial load energy and the final capacitor energy.

$$1/2L \times I^2 = 1/2CV_f^2 \tag{13}$$

To keep the overshoot voltage from becoming excessive, C must be made relatively large for reasonably reactive loads. The capacitor size must therefore be optimized to fit the power factor of the load.

6. Load Clamping.   To overcome the problem of sizing capacitors, the circuit of Fig. 6b was developed (Ref. 1). It is composed of the basic parallel inverter to which has been added a set of clamping, regenerative, or feedback rectifiers DA, DB, DA', and DB'. These rectifiers do not affect the load voltage until it attempts to exceed the supply voltage E; then it acts as a clamp, limiting the load voltage to E volts. This operation is demonstrated by considering commutation with an inductive load as before.

When the A units are extinguished as before, the current persists and reverses the charge on the capacitor. However, when it attempts to cause an overshoot, the clamping circuit is activated and the load current assumes the path of I_c in Fig. 6b. That is, the load current flows through DB, the supply source E, and DB' back to the load. Rectifiers DA and DA' clamp during the opposite half cycle. The action is demonstrated by the waveforms which depict an inductor load (Fig. 7). The shaded portions of the current flow in the clamping circuit. Note that the supply system is assumed to be a battery or a rectifier with an output capacitor large enough to accept the clamping current. Such a capacitor is usually of the inexpensive electrolytic type. The use of clamping rectifiers is virtually universal in all the various types of motor drives because they make the operation of the commutating circuit (capacitor) and the level and shape load voltage almost independent of load power factor. They are the principal means by which an inverter can regenerate and are perhaps best referred to as regenerative rectifiers.

7. Trapped Energy.   One detrimental effect of using the clamping circuit of Fig. 6b is found by considering the current in inductors L after commutation. The inductor is charged with an incremental current ΔI as it experiences a high voltage (300 V) for the brief interval of turnoff.

$$e = L \frac{di}{dt} \tag{14}$$

or

$$\Delta I = \frac{e \times t}{L} = \frac{Et_0}{L} \tag{15}$$

As it attempts to reassume a quiescent current by assuming a negative current increment $-\Delta I$, and a corresponding negative or discharge voltage, it is found that unlike a conventional parallel inverter, the clamped parallel inverter limits the negative inductor voltage to the forward drop of a newly fired thyristor B and its associated clamp rectifier DB (typically 2 to 3 V = E').

$$-\Delta I = \frac{E'\Delta t}{L} \tag{16}$$

This inductor discharge with low voltage implies a long time Δt. More significantly, it implies that the energy imparted to the inductor L by the capacitor C at commutation is trapped in the form of a circulating current I_T (Ref. 1). To establish equilibrium, this energy must be dissipated in the form of watts W where

$$W = f \times 1/2 \; CE^2 \tag{17}$$

where $1/2 \; CE^2$ = capacitor energy = energy imparted to inductor
$\quad\quad f$ = switching frequency

At frequencies above 100 or 200 Hz, this added dissipation requires some semiconductor derating. At frequencies above 300 Hz, a clamping transformer is required which introduces a voltage in the trapped-energy path. While this arrangement does tend to conserve some of the trapped energy, it adds cost and complexity to the system. In addition, it adds steps to the output waveform which, while small, are affected by load power factor and have been shown to produce instabilities in motor operation. This drawback is particularly true in synchronous-reluctance-motor applications.

Figure 8a shows one phase of a more familiar form of the inverter which operates in the same manner as the circuit of Fig. 6b (Ref. 1). Figure 8b shows one phase of a different type of inverter which employs small auxiliary thyristors to turn off the main

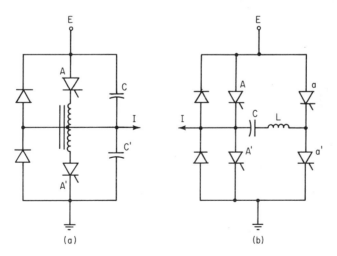

Fig. 8 Forced-commutated inverters. (a) McMurray-Bedford circuit. (b) Impulse-commutated inverter.

thyristors (Ref. 2). For example, thyristor A can be turned off by firing thyristor a and applying charged capacitor C. Whereas the other circuits covered tend to be limited in frequency range by trapped-energy effects, the circuit using auxiliary thyristors can be used at much higher frequencies with good efficiency.

8. Inverter-Circuit Design.   A comprehensive study of the design of inverters in general is beyond the scope of this book; however, a detailed analysis of one form of circuit will serve to illustrate the general approach. Using this approach as well as the references cited in this section, the detailed operation of many circuits can be predicted. Furthermore, many common characteristics can be identified among seemingly vastly different circuits (Refs. 6, 4).

a. Analysis of the Impulse Commutated Inverter (Refs. 21, 2).   The circuit of Fig. 8*b* is described basically by the waveform of Fig. 9. A conducting thyristor is turned off by a sinusoidal commutating pulse derived from an LC pulse-forming circuit. Note that for a given load current I_0, the thyristor current diminishes as the pulse current replaces it (interval I). The thyristor current is zero when the pulse and load current are equal as shown in Fig. 9. The second interval (interval II) involves a pulse current greater than the load current, the excess flowing in the parallel opposed rectifier (I_{RA}). This interval is the hold-off period when the previously conducting thyristor is prevented from con-

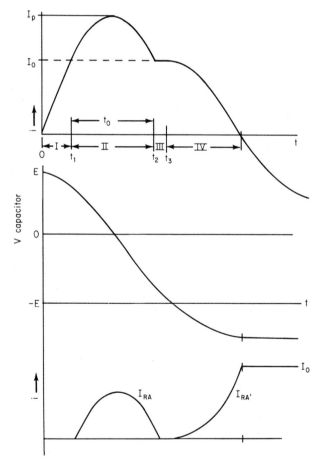

Fig. 9   Waveforms of commutation circuit.

ducting as it experiences a small negative voltage. During this period, the capacitor voltage goes through zero. This period is terminated when the pulse current falls to the level of the load current. It is sustained during interval III by the load (assumed inductive) until the terminal voltage assumes a level equal to the lower bus voltage. At this time the lower rectifier A' conducts, clamping the output to the lower bus (interval IV). Note that at this time the capacitor, which had assumed the clamped voltage, continues to charge as the current in the commutating circuit decays or relaxes. While these conditions are somewhat idealized (peripheral reactances have been neglected), they closely approximate the action of the circuit.

b. *Commutation-Circuit Analysis.* The circuit may be analyzed by studying the relationship between the pulse and load currents.

$$I(s) = \frac{V(s)}{Z(s)}$$

$$= \frac{E}{s} \frac{1}{Ls + 1/Cs}$$

$$= \frac{E/L}{s^2 + 1/LC} = \frac{E/L}{[s + j(1/LC)][s - j(1/LC)]}$$

Inverse-Laplace-transform yield:

$$i = E\sqrt{\frac{C}{L}} \sin \omega t \qquad \omega = \frac{1}{\sqrt{LC}}$$

if $I_{max} = I_p$,

$$\frac{I}{Ip} = \gamma = \frac{I}{E\sqrt{C/L}} \qquad \theta_0 = 2\cos^{-1}\gamma$$

$$t_0 = \frac{\tau\theta_0}{\pi} \qquad \tau = \pi\sqrt{LC}$$

$$t_0 = \sqrt{LC}\ 2\cos^{-1}\gamma$$

$$Ipt_0 = E\sqrt{\frac{C}{L}}\ t_0 = EC\ 2\cos^{-1}\gamma$$

$$C = \frac{Ipt_0}{E\ 2\cos^{-1}\gamma} = \frac{It_0}{\gamma E\ 2\cos^{-1}\gamma}$$

$$C\frac{E}{It_0} = \frac{1}{2\gamma\cos^{-1}\gamma}$$

Similarly,

$$\frac{Ip}{t_0} = \frac{E\sqrt{C/L}}{\sqrt{LC}\ 2\cos^{-1}\gamma} = \frac{E}{L\ 2\cos^{-1}\gamma}$$

$$L\frac{Et_0}{Ip\ 2\cos^{-1}\gamma} = \frac{Et_0}{l\ 2\cos^{-1}\gamma}$$

$$L\frac{I}{Et_0} = \frac{\gamma}{2\cos^{-1}\gamma}$$

Note that capacitor C and inductor L have been computed in a normalized form which when plotted can yield circuit values dependent on universal qualities of input voltage E, maximum current to be commutated I, and minimum turnoff (hold-off) time t_0. These values are plotted in Fig. 10. An optimum value is suggested by the minimum value of the capacitor curve near $\gamma = 0.7$. (Note other factors such as peak or maximum di/dt may

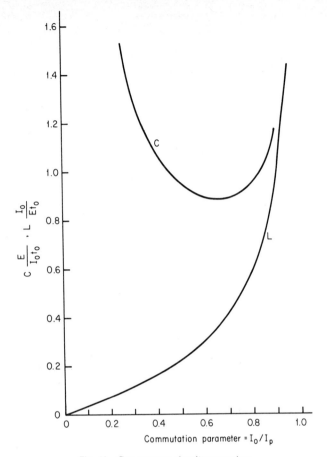

Fig. 10   Commutator-circuit parameters.

be more critical than capacitor size.) For the value

$$C\frac{E}{It_0} = 0.9$$

and

$$L\frac{I}{Et_0} = 0.43$$

for a dc input voltage of 300 V, a hold-off time of 30 μs, and a maximum load current of 300 A,

$$C = 0.9\frac{It_0}{E} = 0.9\frac{300 \times 30 \times 10^{-6}}{300} = 27\ \mu\text{F}$$

$$L = 0.43\frac{Et_0}{I} = 0.43\frac{300 \times 30}{30} = 130\ \mu\text{H}$$

The parameters for the circuit of Fig. 8a are plotted in Fig. 11 for comparison. Other circuits can and have been analyzed in the same way. Furthermore other parameters such

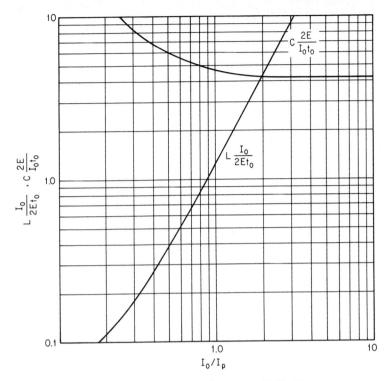

Fig. 11 Commutation parameters of McMurray-Bedford inverter.

as trapped energy and quantities such as size of suppressor components may also be plotted (Ref. 21).

9. Dc-Side Commutation. The inverters described thus far achieve commutation by affecting or diverting the current at the ac nodes and are known as ac-side commutated inverters. Another family of circuits achieve commutation by affecting currents at the dc nodes in a process known as dc-side commutation (Ref. 15). The process is principally the same as in ac-side commutation except that groups of thyristors (top or bottom or all) are normally commutated simultaneously. The advantages are usually reduction in number of commutation components (one capacitor may be used) and perhaps a better control of trapped energy. Some of the disadvantages include the need to commutate higher currents, more often, etc.

VOLTAGE CONTROL

Most inverter systems are distinguished by their method of voltage control. Each has its relative advantage and disadvantage and perhaps its place. Simplicity can be traded for performance or control-circuit complexity for power-circuit size, etc.

10. Phase-Shift Method. A simple method of voltage control (Fig. 12) provides an output of two phase-shifted inverters which are combined, usually with transformers (Ref. 9). The resultant voltage is a function of the phase angle ϕ between the inverters and is given by

$$VA + VB = V_A\, 2 \cos\left(\frac{\phi}{2}\right) \tag{18}$$

When used with a six-step waveform, the range of voltage control is small, typically less

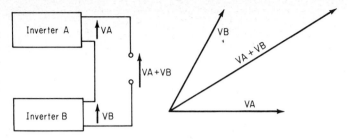

Fig. 12   Phase-shift method of voltage control.

than 2/1, because the resulting harmonics become intolerable. Tap-changing schemes on the output transformer or multiple-stepped waveforms can extend the range, but the practice is complicated and expensive.

11. Variable-Output Transformer.   Perhaps the most reliable method for achieving voltage control in motor drives over moderately wide frequency ranges is the use of a variable-output transformer (Fig. 13) (Ref. 5). Although frequency ranges of 20/1 or greater can be achieved with this method, the system has inherent limitations. These limitations result from the fact that the voltage applied to the primary of the output transformer is constant irrespective of frequency. Thus, the transformer must have a large amount of iron to support the high volt-seconds applied at low frequencies, making it bulky and expensive. Also, the response of the transformer drive is slow. The advantage of the system is its simplicity and its ability to handle high inrush loads at low voltage settings when the current reflected to the inverter output is a function of the step-down ratio. It has had wide application in the textile industry but cannot be considered a general-purpose drive.

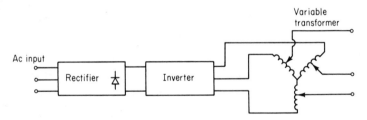

Fig. 13   Voltage control with variable-output transformer.

12. Variable Dc Voltage Input.   Figure 14 shows a six-step inverter system which uses a variable dc input to yield an adjustable output voltage (Ref. 16). The system requires an auxiliary commutating circuit to precharge the commutation capacitors, because if their voltage is allowed to fall with the input, the commutating ability of the inverter falls accordingly [Eq. (11)].

If a phase-controlled rectifier or chopper is used to form the variable dc input, the cir-

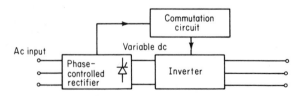

Fig. 14   Voltage control with variable dc input.

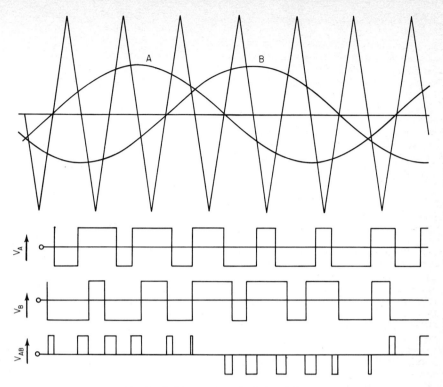

Fig. 15 Subharmonic method of modulation.

cuit is reasonably fast and can be considered a general-purpose inverter. While it is a popular circuit, it has several disadvantages. First, like the variable-output-transformer method, it requires two power controllers in cascade, one for frequency control and one for voltage control. Second, it requires an input dc filter which adds cost and time delay to the system. Third, to achieve sustained regeneration, the system requires a dual converter (phase-controlled rectifier) or a reversible chopper. All these factors are costly or limit the flexibility of the system.

13. Pulse-Width Modulated Inverter. The pulse-width modulated (PWM) inverter is an extremely effective motor controller accomplishing voltage and frequency adjustment in a single circuit (Ref. 14). Proper sequencing of the bridge in Fig. 1 produces an output characterized as composed of discrete pulses of fixed height and frequency and adjustable width (Fig. 15) (Ref. 10). The inverter voltage is controlled by the average width of the pulses, and the harmonics are controlled by modulating to a sinusoidal envelope. If a sine-wave envelope is used, the harmonics of the output waveform are principally those associated with the modulating frequency. These can be set high enough to be rejected by the motor.

The PWM inverter is a fast, linear device and its response is virtually instantaneous. As a power amplifier it is comparable with the dual converter of dc systems and its speed of response makes it applicable to virtually any feedback loop. It can be considered as a "black box" with power inputs and outputs, as well as two control inputs, one for frequency and one for voltage.

WAVE SHAPING

The ideal wave shape for exciting motors is logically a sine wave. The waveforms derived from the switching action of thyristor inverters are of a stepped or pulsed nature. When studied with Fourier analysis, the fundamental frequency as well as the harmonics of these waves can be determined (Ref. 18). Motors have been found to tolerate a surprisingly high harmonic content in their driving waveforms. However, these harmonics tend to produce added losses and increased current, and should be optimized to keep a sensible balance between losses and inverter complexity.

14. Harmonic Canceling. If a number of inverters of specific phase displacement are connected in zigzag, certain harmonics can be eliminated or canceled as the synthesized output assumes a stepped wave envelope which approximates a sine wave, as shown in Fig. 16 (Ref. 7). If each of these phases is composed by a pulse-width-modulated square wave derived from a pair of phase-shifted inverters or a single pulse-width modulated inverter, the voltage level of the fundamental can be varied while the canceled harmonics remain absent from the output. Economically, such schemes can be applied only at higher output powers, since a multiplicity of inverters are required along with output transformers with multiple secondaries for each phase. Furthermore, while the levels of the lowest-frequency uncanceled harmonics are low relative to the maximum value of fundamental, they become significant as the fundamental is diminished (1) because their magnitude relative to the fundamental becomes high and (2) because in motor drives their frequency drops with the fundamental, making them less able to be smoothed by the reactances of the machine.

15. Three-Phase Inverter. A more practical method of wave shaping, as shown before, is accomplished in the 3-phase inverter of Fig. 1 which is a simplified drawing of three single-phase sections of the type shown in Fig. 8. The particular wave shape developed (Fig. 2) is suitable for driving ac motors. The harmonics which are present in the wave, principally the 5th and 7th, produce added but acceptable motor losses. Unlike the harmonic-canceled technique, the circuit requires a minimum number of switches and no output transformers and as a result can be applied economically at low powers.

16. AVI Wave Shaping. One advantage of the adjustable-voltage-input (AVI) inverter is that it utilized a minimum number of switches or commutations (three per phase per cycle) to produce a usable waveform. The disadvantage is that at low frequencies, the commutation circuitry is being used inefficiently. At the same time, a need for voltage boost or *IR* compensation for high loads produces an overexcited condition at low loads, greatly exaggerating the harmonic effects producing high peak currents and torque pulsations (Ref. 29). By adding two pairs of extra commutations, which would be no penalty to the commutation circuit at low frequency, it is possible to reduce greatly the 5th- and 7th-harmonic content (Ref. 8). This extends the lower-frequency capability of the AVI but does add complexity in the logic needed to generate the new waveform and to make the transitions between it and the normal six-step wave. By adding more switches and some complex sequences, the waveform can be further refined, making lower and lower frequency operation feasible.

Extending the process of the preceding paragraphs is of course turning the AVI into a PWM inverter since the multiple switches is the essence of PWM. With PWM wave shaping in voltage control, the practical modulating or carrier frequencies in most cases are not a sufficiently high multiple of the desired operating frequency. Note that the maximum allowable carrier may not always be determined on a frequency basis but sometimes on a minimum transition time as dictated by the commutation and turnoff times. Thus it is often necessary to optimize waveforms to produce minimum har-

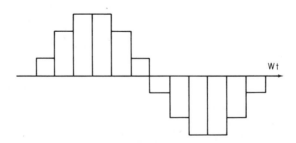

Fig. 16   Stepped waveform of harmonic cancellation.

monics for a given and sometimes low carrier ratio (Ref. 22). The minimum carrier frequency needed to produce balanced modulation is three times the operating frequency, i.e., $f = 3F$. The ratio of carrier to output or fundamental frequency is therefore 3. It has been shown that to produce stable waves containing only odd harmonics (single-carrier, 3-phase system), the carrier must be synchronized and a triple multiple of the output frequency. Furthermore, only a six-step envelope can utilize even multiples. Thus, as frequency is swept from high to low values, carrier ratios can be changed as follows: 3, 6, 9, 12, 15, etc., for six-step waves and 3, 9, 15, 21, etc., for multistep and sine waves (Ref. 21).

A convenient term for evaluating or comparing waves with respect to harmonics is the harmonic-loss factor LF which equals $\Sigma(H_n/N)^2$. The term H_n/N is a measure of the harmonic current which flows because it is simply opposed by the leakage reactances of the motor and hence is attenuated as the inverse of the frequency. (H_n is the level of the particular harmonic voltage as related to the fundamental frequency voltage.)

Note the expression assumes a constant rotor resistance which is probably not true when skin effects are considered. Thus, the answer is slightly optimistic. As a standard of comparison, the six-step may be used since experience with it is available and since motors have been sized to accommodate its losses or loss factor.

17. PWM Wave Shaping. The traces of Fig. 15, while demonstrating a method of voltage control, also describe a method of wave shaping. However, modulating to a sine-wave reference does not ensure a sine-wave envelope or effective envelope on the output when the difference between the output and carrier or modulating frequency is insufficient, that is, when there are insufficient pulses per half cycle to resolve the sine envelope. This condition usually occurs at the higher operating frequencies. In other cases, it may be more economical or practical to modulate to a stepped-wave envelope. In other instances the output of two or more inverters may be combined, thereby allowing the use of six-step envelopes with the resulting wave still being of high quality. Such a simple wave is shown in Fig. 17. The modulating

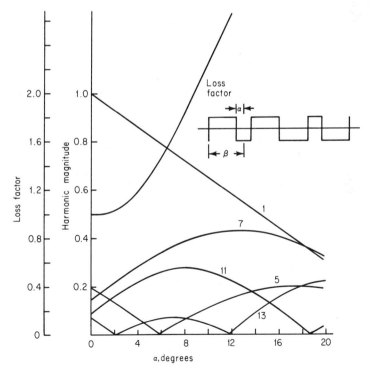

Fig. 17   Harmonic magnitude and loss factor with ratio of 3 and 6 steps.

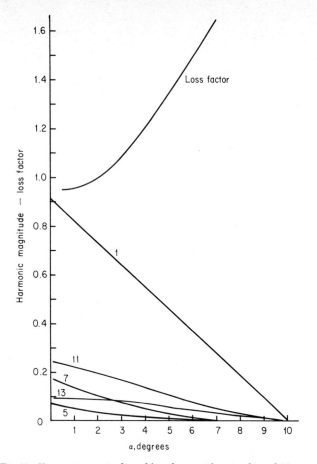

Fig. 18 Harmonic magnitude and loss factor with ratio of 9 and 12 steps.

envelope is a six-step wave and the ratio of carrier to output frequency is 3. Figure 18 shows modulation to a 12-step envelope with a ratio of 9. Note that the quality of the 12-step wave is little improved as measured by the loss factor, indicating lack of resolution. Figure 19, however, demonstrates that 12 steps with a ratio of 15 does yield a significant improvement in loss factor. The excellent effect of combining two six-step waves in zigzag, i.e., displaced by 30°, is shown in Fig. 20. The basic principle illustrated by these plots is that harmonics are identifiable and controllable through a number of techniques.

18. Harmonic Analysis. The voltage output of all inverters is of either a stepped or a pulsed nature. Since the harmonics inherent in such waves can yield significant heating in motors, a brief but general description of how these harmonics are analyzed is presented. The familiar Fourier series is a convenient means for describing the harmonic content of inverter output (Ref. 14). It has the general form:

$$f(\omega t) = \frac{a_o}{2} + \sum_{ns_1}^{\infty} (a_n \cos n\omega t + b_n \sin n\omega t)$$

The coefficient is given by

$$a_o = \frac{1}{\pi} \int_{-\pi}^{\pi} f(\omega t) \, d\omega t$$

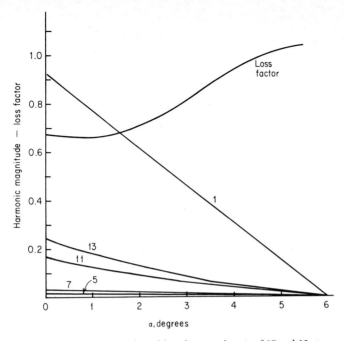

Fig. 19 Harmonic magnitude and loss factor with ratio of 15 and 12 steps.

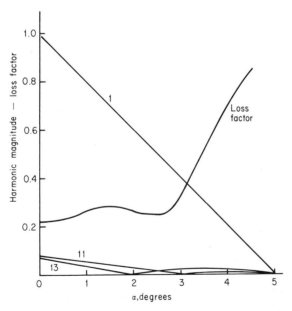

Fig. 20 Harmonic magnitude and loss factor with ratio of 9 and zigzag for 12 steps.

$$a_n = \frac{1}{\pi} \int_{-\pi}^{\pi} f(\omega t) \cos (n\omega t) \, d\omega t$$

$$b_n = \frac{1}{\pi} \int_{-\pi}^{\pi} f(\omega t) \sin (n\omega t) \, d\omega t$$

Most often it is convenient to analyze the phase-to-ground output of each phase of an inverter (V_{ao} of Fig. 2). Since most significant or usable inverter waveforms are balanced and symmetrical, some simplifying steps may be made such as neglecting the dc term a_o or neglecting triple harmonics in 3-phase inverters since these qualities cancel when the outputs of two or more phases are summed. Symmetry with respect to the two half cycles of a wave eliminates all even frequencies, i.e., $2F$, $4F$, etc. Adding symmetry about the 90° point of a wave eliminates the cosine terms. Thus a balanced symmetrical wave, made up of steps or pulses, is represented by

$$H_n = \frac{4}{n\pi} \sum_{n=odd}^{\infty} \sin n\alpha \times \sin n\beta$$

where α is half the pulse or step width and β is the angle as measured from the beginning of the wave (zero degrees) to the center of the step or pulse as shown in Fig. 17. To demonstrate the utility of the technique, the simplest PWM wave (ratio 3) can be visualized as composed of a square wave of amplitude 1 combined with a pulse centered at 90°, whose width is α and whose height is minus 2. Writing an expression using the simplified formula

$$H_n^s = \sin \alpha_1 \sin \beta_1 - 2 \sin \alpha_2 \sin \beta_2$$
$$\alpha_1 = \beta_1 = \frac{\pi}{2}$$

Therefore,

$$H_n = \frac{4}{n\pi} \sum_{n=odu}^{\infty} 1 - 2 \sin \alpha_2 \times \sin n \frac{\pi}{2}$$

These remarks are made to provide an insight into the generation of inverter waves. Highly complex waves must be studied using computers. Paradoxically a complex wave with many pulses usually yields harmonics directly caused by the modulating envelope or the carrier frequency.

INPUT CHARACTERISTICS

The prime input power source supplies energy to, and can take power from, the inverter or frequency-changer system. It is often important, if not critical, in what form or mode energy is exchanged, or what transients or conditions are imposed by the source on the converters or by the converter on the source. In most cases, an intermediary converter is placed between the source and the inverter. A rectifier fed by the ac line is a typical example. In some instances, the inverter or converter terminals are applied directly to the source, such as a battery feeding a PWM inverter.

19. Ac Line—Fixed-Voltage Rectifier. When the rectifier is considered as an ac load, it is an efficient, virtually unity-power-factor system. The resulting dc bus is suitable for energizing nonregenerating inverters with the addition of a large capacitor, usually of the electrolytic type. The capacitor is connected at the input of virtually all inverters and choppers and is often considered as a power-factor-correcting means. Its prime purposes are for carrying the input ripple currents of the inverter as well as providing a stiff source for the commutating-circuit transients by nullifying the line impedance. The stiffness of the capacitor with respect to faults makes possible extremely large transients at the time of shoot-throughs (Art. 28).

Other problems in using a fixed dc bus are (1) bus variations due to line-voltage variations and (2) difficulty in initially charging the capacitor. Standard line variations can range to ±10% of a nominal voltage. Crest riding can add an additional 4%, yielding a possible net 25% increment. This variation produces commutation difficulties at low voltage and added stress for the semiconductors at high voltage. Charging a large capacitor bank can damage fuses or weld contactor tips placed upstream from the bank. To correct this problem, additional inserted line impedance or a charging sequence using a resistor and an auxiliary contact may be required. Apart from being the simplest and least expensive power source, the fixed dc bus has the valuable quality of being able to exchange power between different inverters connected or referred to it.

20. VIT Source. The variable-input transformer-rectifier source is equivalent to the ac line-rectifier system except that the initial charging of the capacitor bank can be controlled.

21. Battery. The battery forms an excellent dc source, since it inherently has the qualities of a capacitor. Some capacitance may be required at the inverter terminals to nullify the impedance of the power leads at commutation frequencies. The battery is similar to the ac-derived dc bus except that its charge must be maintained and its operating voltage may swing beyond the 10% limits of a conventional line. It is inherently capable of being able to take large doses of regenerative energy. The specific charge/discharge characteristics are dependent on the type and construction of the battery.

22. Ac Line—Phase-controlled Rectifier. For variable-voltage inputs, the phase-controlled rectifier (PCR) is perhaps the most direct approach. Like the fixed-voltage dc bus, it must be supplied with a filter capacitor which is, however, larger than the one used for fixed-voltage buses. On the other hand, it is easily charged or initialized by the natural action of phase control. Another necessary addition is an input inductor to minimize capacitor current ripple due to the voltage ripple at the output of the PCR. The prime disadvantages of the PCR are (1) poor power factor presented to the line, (2) the inability to exchange power between independent inverters via the dc bus, and (3) the tendency to distort the line waveform.

23. Fixed Dc Bus Chopper. Another popular variable-voltage source is the combination of a dc bus and a dc chopper. Like the PCR, the chopper requires added filtering and a larger input capacitor. If the chopper is not regenerative (one-quadrant operation), it is simpler than the PCR scheme. If it is regenerative, it can tie its input to a common bus, thereby reclaiming the feature of energy exchange via the dc bus. This feature is valuable and is used extensively in fiber spinning to accommodate regenerating sections and to provide a common seat or reference for battery or capacitor backup. Like the dc bus it presents an excellent power factor to the line. A regenerating chopper can allow the use of a smaller than normal capacitor in variable-voltage applications, since power can be referred to a higher fixed-voltage bus.

24. Design of Inverter Input Filters. For a fixed dc bus, the input current can be assumed constant. The output current, neglecting harmonic ripple and assuming sine-wave inverter outputs, can be visualized as a mirror image of the PCR waveforms with a phase-angle delay equivalent to the power-factor angle. These currents are plotted in Fig. 21. As the power factor diminishes, leading or lagging the ac or ripple current, the current which must be carried by the capacitor increases. While it is certain that a motor will not present an extreme of zero power factor, it is also difficult at times to pinpoint the exact working conditions, which may vary with time. Rather than employing duty cycles, it is perhaps better to assume a worst-case design of a 0.5 PF with an ac current of 15%. Adding the harmonic current of the motor, which again can range from 10 to 50%, depending on waveform, operating point, excitation, etc., the total capacitor current can be identified as $I_c{}^2 = I_{ac}{}^2 - I_h{}^2$, where I_c is capacitor current, I_{ac} is idealized inverter current, and I_h is motor harmonic current. For inverters with low-harmonic-output currents, the capacitor current is easily predicted. For severe harmonic conditions, a direct measurement of capacitor current as the inverter is exercised through its operating conditions is necessary. The size of the required capacitor is obtained by selecting a capacitor of adequate voltage and applying the proper rating factors, such as frequency of ripple and temperature, to obtain the can rating. The number of cans required is obtained by dividing the measured or calculated ripple current by the can rating. For series operation in extreme applications involving severe duty cycles, critical lifetime requirements, shock, vibrations, etc., the capacitor manufacturer must be consulted.

For variable-voltage systems, the ripple current from the PCR must be added. It can be calculated by using the curve of Fig. 21. Since it also represents the content of a 3-phase full-wave-rectified voltage, the worst-case condition seems to occur at low voltage. For this PCR controller, the worst-case ripple factor is 29%.

The total capacitor current is the rms value of the input and inverter ripple currents. The size of the capacitor bank may be a function of other factors, such as stability or ride-through.

25. Regeneration. Regeneration is taking energy from the load and returning it to

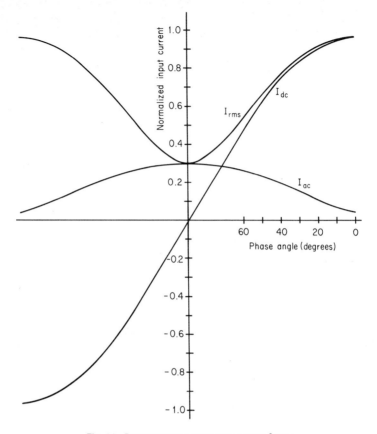

Fig. 21   Inverter input current vs. power factor.

the source and can be distinguished by the quantity of energy involved taken over a given cycle. One extreme is the small amount of energy (watt-second) reclaimed by the input capacitor during a subcycle period in feeding a reactive load. The other extreme is sustained reverse energy flow such as when maintaining tension on a material being processed. Between the two extremes are progressively greater power pulses from small transients owing to downward slewing or decelerations in the process of regulation to larger pulses for braking varying amounts of inertia. The cost of adding regeneration capability is directly related to the level of energy to be returned. In some cases, the level of energy is small enough to be absorbed by the storage and losses of the system. In other cases, such as the battery or multiple inverters on a common bus, regeneration is inherent since the net power flow is into the process. However, when these features are not available or when the energy returned exceeds the capabilities of the system, other means must be employed. The simplest form of regeneration or energy control is the dissipative snubber, a resistor which is switched into the circuit, usually across the dc bus, to absorb the energy. The switch can be a mechanical contactor, but if the speed or degree of control is critical, a solid-state switch similar to a chopper can be employed. The resistor must be sized to accept both the peak energy returned as well as the averaged or duty-cycle energy. If this average energy is excessive, the cost of the grid or the power costs represented by the energy may dictate a nondissipative regeneration system. While most power companies do not acknowledge returned power, costs may be recouped by redistributing the reclaimed energy to other points of the user's facility, by reduced demand charges, or by reduced feeder-size requirements.

Regenerating from a fixed dc bus requires a phased-back PCR operating as a synchronous inverter connected in parallel opposition with the rectifier (the term inverter is used even with a phase-control device, since the regenerative power flows from dc to ac) (Ref. 27). The level of phase-back determines the level of output voltage which is normally set at least slightly above the rectified output; thus as the energy is returned to the bus from one or more inverters, it first replaces the rectifier output current feeding the losses and loads of the inverter connected to the bus. Once that sink is satisfied, the energy blocks the rectifier and begins to charge the capacitor as the input voltage rises and eventually spills over into the regenerative PCR, which may be switched off in idle periods and activated only when necessary. Note that if the regenerative sink were not present, the capacitor voltage would rise until the system would be damaged. Limiting the voltage rise by inverter action would, however, deprive the load of braking. The advantage of a common bus is that many inverters can share a large economical regenerating apparatus. A disadvantage of the regenerating PCR is that if the line is lost during a regeneration cycle, shoot-through or commutation failure in the PCR is inevitable. Variable-voltage systems are the least adaptable to regeneration. A simple solution is the use of a regenerative chopper which refers each inverter to a common bus. For ac input systems, sustained or high-level regeneration can be obtained only by providing a phased-back PCR acting as a synchronous inverter as in the case of a fixed dc bus. Disadvantages are (1) multiple inverters require multiple regenerating apparatus, and (2) the regenerating PCR must track the rectifying PCR, yielding poor power factor and requiring more complex control.

26. Stability. If the voltage output of an inverter is not stable with load speed, speed instabilities can occur (Ref. 28). Envelope regulation or droop can cause speed instability or oscillation between two or more motors fed from a given inverter. Such instabilities can be stabilized by voltage feedback. Another type of instability occurs when the voltage waveform within a portion of a half cycle varies or is distorted with loading. This condition can occur when using a trapped-energy-return transformer. The instability takes the form of an oscillation and has been observed with synchronous reluctance motors. The best solution for instability is the use of stiff inverters such as the PWM. But even here, if the power supply is not stiff, instabilities can occur on a multimotor system.

27. Overload. As with any motor drive using a semiconductor power supply, overload limits are imposed as follows:

1. By the motor on a long-term iron-copper thermal basis
2. By the supply on a short-term 2-min thermal basis involving heat sinks and a transient thermal impedance of the semiconductors
3. By the commutation circuit in the form of a peak commutating ability (this circuit is similar to the commutation limit of a dc motor which, if exceeded, can lead to flashover)

PROTECTION

28. Fuses. Because of the stiff nature of the inverter input and the characteristic low impedance associated with the power circuit, special means must be employed to protect the semiconductors (Ref. 23). Faults are generated and characterized when one thyristor is triggered while the other thyristor of its phase fails to achieve turnoff, placing both units in conduction across the input capacitor in an action called "shoot-through." The only limiting means is the low power-circuit impedance, primarily inductance. Fusing must be applied even more diligently than in conventional PCR circuitry in which high faults and the delicate nature of the semiconductor pellet have been long recognized. Fortunately, these techniques, at their worst, are extensions of the more extreme cases of phase control, and by comparison with some cases, namely, slow, high-energy faults, are less severe.

The key to internal power-circuit protection is the fast fuse. Its construction is simple, one or more silver ribbons (notched a number of times) suspended between two copper pole pieces and encased in a sand-filled insulating tube. For normal currents, the relatively low losses of the filament are conducted to the pole pieces, which act as heat sinks. When the fast fault is applied, high levels of heat are generated, and because of the thermal delays of the package the pole pieces are thermally decoupled. The heat

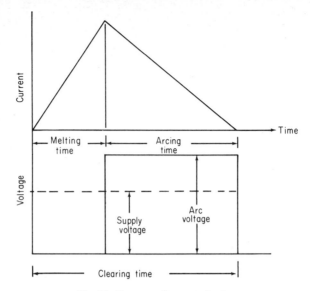

Fig. 22   Fuse waveforms at clearing.

is thus concentrated at the necked-down areas, causing localized melting and the creation of multiple and virtually simultaneously formed arcs along the length of the filament. The sand acts to quench and contain the arcs. Externally, the capacitor-fed fault can be observed as a rapid, almost linear rise of current ($di/dt = E/L$) until the cumulative heating effect i^2t causes melting, as indicated by the abrupt downward change in slope (Fig. 22). The rate of current decay is determined by the difference between the supply and the arc voltage E_{arc}:

$$\frac{di}{dt}\text{ (decay)} = \frac{E - E_{arc}}{L}$$

Note that if the voltage is insufficient, the current persists, destroying the fuse. The arc voltage is the sum of the arc voltages of the individual melting zones. Increasing the number of nicks increases the arc voltage, thus diminishing the arcing time. Total clearing time is the sum of the melting time and the arcing time. However, excessive arcing voltage may injure blocking semiconductors not involved in the fault (Ref. 26). Fuses are commonly available for dc supply voltages of 250, 500, 600, 700, and 1000 V. Special fuses and combinations of fuses can be used for special applications. If a fuse arcing voltage does not appear directly on a blocking semiconductor, a higher-voltage fuse may be substituted, with a possible penalty in extended melting time. The net result, however, may yield a reduced clearing I^2t. Insufficient or marginal arc voltage is also indicated by upward bowing of the current during arcing.

Another common malady is insufficient arcing ability. During arcing, all the link is consumed, and the arc then can bridge the pole pieces and leave the fuse body. The use of x-ray techniques to examine a spent fuse will reveal the amount and distribution of unconsumed material. Other modes of failure are caused by voids or insufficient sand and damaged links which result in normal currents causing the melting of a single neck. The insufficient arc voltage will produce prolonged arcing, possibly extending to the pole pieces.

The critical factor in the fuse is the temperature of its links. The condition or quality of its terminations, which act as sinks to the internally generated heat, plus the temperature of the connected bus and the ambient affect the melting point of the fuse. Ambient factors such as cooling air or water and thermal drops at the termination help determine the fuse current rating (Ref. 25). The current rating is determined by the conditions which create damage to the links. A fuse which has experienced a surge and appears good may be checked by measurement of its resistance on a Kelvin bridge. This resistance is extremely precise to obtain reproducibility between devices. If there is any doubt as to the fuse integrity, it should be replaced immediately.

29. Crowbars. Another form of protection is the crowbar, generally a high-powered, fast-acting switch used across the input terminals of the device to be protected. It is used to protect semiconductors which cannot coordinate with the input fuses. When a fault or shoot-through is determined, the crowbar is fired, absorbing a significant portion of the I^2t of the input fuse. A crowbar is also used where sinking massive energy is required owing to a fault in other equipment connected to the line.

30. Output Fuse Coordination. Often multiple motors are connected to a single supply. A fault in any motor should clear the affected branch fuse, allowing certain of the remainder of the loads to continue to function without achieving a current trip. The obvious key to success is the relative sizing between the main and branch fuses. Often the branch fuses must be oversized to permit line starting of a high-inertia load. Thus, as a minimum, the branch fuse must melt and arc totally clear before the main fuse approaches near enough to melting to be damaged. In other applications where service must be sustained, the total rise in current during the fault must not cause tripping. In extreme cases the fault impedance is assumed (bolted) while the limiting impedance is merely the line drop (Sec. 13).

31. Fault Calculations. The most effective way of ensuring proper fusing is to place the proposed fuse in the circuit and induce the fault, monitoring the current with a noninductive shunt. For triangular surges I^2t is given by

$$I^2t = (I_{peak})^2 \times \frac{t_{clear}}{3}$$

For other wave shapes, this figure can be obtained graphically. Melting time and peak current can be obtained using Fig. 23 and knowing the impedances of the fault path.

COMPONENTS FOR VARIABLE-SPEED DRIVES

32. Thyristors. Thyristors used in inverters and cycloconverters have basic blocking voltage and current ratings as do semiconductors in PCR systems (Ref. 30). In addition, they must be qualified with respect to:

1. Turnoff time t_q, the time required to recover blocking ability after carrying a specified current in a specified circuit.

2. The rate of reapplied, usually linearly, forward voltage dV/dt to a specified level at the end of the turnoff interval, which is not to be confused with "critical" or one-time dV/dt.

3. The repetitive rate of rise of current di/dt in a unit to a specified current level at a specified frequency.

4. Short-term surge current $(1.5 \mu s)$ for coordination with fast fuse clearings specified in I^2t as well as peak current and the di/dt of the leading edge.

5. Gate current and voltage to fire.

Note that units are specified concurrently to simulate action in real circuits. Stressing a device with respect to one characteristic can affect the other ratings. For example, for a given device, t_q increases with a reduced dV/dt.

The inverter thyristor differs in design from conventional phase-control units in that they require heavier concentrations of dropping which, while reducing turnoff time, also increase forward drop. Thus, for a given volt-ampere rating, the inverter thyristor power loss is usually larger than the phase-controlled thyristor power loss. Since the inverter unit may switch more often, or under a more severe di/dt, units must be designed to minimize turnon losses. In the past, center-fired and multiple gates have been employed (Ref. 20). Figure 24 shows conventional, as well as amplified-gate construction which allows reduced gate firing requirements. Insufficient drive can cause localized burnout near the gate region. Mechanically, the heat sinking of the inverter units is the same as that of conventional thyristors.

33. Rectifiers. Rectifiers used in inverter circuits must be rated with respect to reverse recovery, since they will experience more severe recovery transients and more

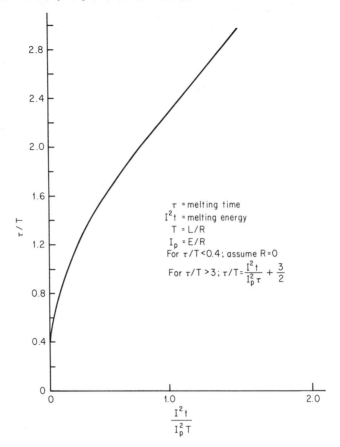

τ = melting time
I^2t = melting energy
$T = L/R$
$I_p = E/R$
For $\tau/T < 0.4$; assume $R = 0$
For $\tau/T > 3$; $\tau/T = \dfrac{I^2 t}{I_p^2 \tau} + \dfrac{3}{2}$

Fig. 23   Normalized melting time vs. melting energy for RL-limited fuse clearing.

often than conventional rectifiers (Ref. 19). A measure of suitability is "recovered charge" or the transient recovery current which flows under specified conditions. In the process of recovery some rectifiers snap off, causing severe transients. This phenomenon must be identified and controlled for specific applications.

34. Commutating Capacitors. Capacitors used to achieve turnoff in inverters operate typically on 10% pulse duty cycles at repetition rates of up to several hundred hertz at resonant frequencies of 5 to 50 kHz. The low-duty cycles employ high peak ratings, thus stressing terminations. High repetition rates and high resonant frequencies imply high internal losses. Conventional capacitors constructed with periodically inserted tabs and metallized units are normally inadequate with respect to the terminations. The extended foil capacitor is wound to allow one foil to overlap one side of the dielectric, while the other foil overlaps the alternate side. A completed roll contains the exposed edges of the foils on opposite ends. These edges not only yield the low electrical impedance simulating an infinite number of taps but also provide an extremely low thermal path to the losses generated in the core of the unit. Furthermore, when compressed and soldered, the edge forms a seat of a mechanically strong high current termination. The principal dielectric for these units is oil-impregnated paper, and they are available in sizes up to several tens of microfarads.

35. dV/dt Capacitors. Capacitors used in dV/dt suppression circuits, although small in size, typically 0.1 to $1.0\mu F$, experience much the same type of duty as the commuta-

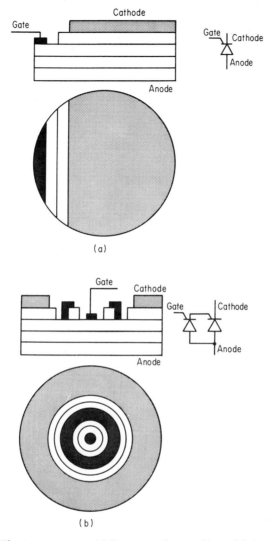

Fig. 24   Thyristor structures. (*a*) Conventional type. (*b*) Amplified-gate type.

tion capacitors. While currents and duty cycles are lower, the resonant frequencies are at least a magnitude higher. Thus these capacitors are usually also of the extended foil type.

   36. Electrolytic Capacitors.   Virtually every inverter in existence utilizes a bank of electrolytic capacitors as an input filter. These are normally of the highly developed mass-produced computer grade. They have a well-defined rating/derating format based on ambient temperature, frequency, and level of current. Their lifetime, properly applied, ranges to 10 years, providing compact, inexpensive service.

SPECIAL FUNCTIONS

   37. Ride-through.   Often in critical applications, such as synthetic-fiber spinning, it is desirable to maintain operation, or to ride through short but definite power interruptions. Three principal alternate power sources have been employed, depending on the source to be obtained and the time the load must be sustained (Sec. 7).

1. Flywheel alternator.   In this scheme an alternator connected to a flywheel typically floats on the line. Upon sensing loss of power, the affected line is isolated and power is drawn from the energy stored in the flywheel. This, while bulky, and perhaps a maintenance problem, is a sure way of supplying inverters with PCR supplies.

2. Batteries. Floating batteries on a central bus are a simple way to ensure an instant power reserve; however, inverters must be capable of operating directly off the bus or through a chopper.

3. Capacitors. A bank of electrolytic capacitors charged to a voltage higher than the bus can sustain power at a given voltage through a chopper. While the bank is easier to maintain than batteries for a power reserve of over several seconds, the size of the system becomes excessive.

38. Paralleling.   The outputs of two or more power units can be combined to produce high power outputs. Terminal-to-terminal paralleling is practical if the power units are well matched or synchronized. With modern electronics, the switching of two systems can be made virtually simultaneous. Currents can be regulated or matched within 10%. Where such matching is not practical, differentially connected paralleling transformers can be used.

Sometimes they are referred to as "spanning" reactors, since each half of the winding must be capable of carrying half the load current and supporting half the unbalance voltage. A better use for the reactors is to combine the two inverter outputs in zigzag with respect to fundamental frequency-producing waveforms with reduced harmonics.

Extending this method to PWM by displacing or "interdigitating" the carriers so that the notches of one inverter coincide with the pulses of the other yields a waveform whose effective carrier frequency is four times the carrier of either inverter. Another variation of this technique is the use of a series reactor to produce essentially the same output waveform as the spanning-reactor technique. Extending this method to multiple inverters connected by a scheme of zigzagging and interdigitation by a series of spanning reactors produces a system of highly refined performance. These combinations can be reproduced with multiple isolation transformers, but in all cases, care must be exercised at variable frequencies with a volt-second demand due to voltage boost at low frequencies which may cause saturation.

39. Transfer.   In some industries it is necessary to transfer a spare inverter onto an operating load for periodic maintenance without interrupting service. This transfer is normally accomplished by a terminal-to-terminal paralleling after manually or automatically synchronizing the controls. The original inverter is then removed. The procedure is reversed when the original inverter is returned to action after maintenance.

CONTROL CIRCUITRY

40. Simple Control.   The speed reference (Fig. 25) is fed to a time-ramp generator which provides a safe slewing rate to the speed command. Its output is applied to two distinct control channels, one for frequency and one for voltage control. The frequency channel contains a voltage-controlled oscillator, a ring counter, and a set of drivers which excite the gates of the thyristors in the power circuitry. The voltage channel is a regulator which may be a modulator to control the pulse width of the chopper, a PWM inverter, or drive the input of a phase-control circuit of a PCR. It is normally supervised by a voltage feedback loop. A voltage boost for IR compensation adjusted for maximum load is applied as a reference to the input of the voltage control. Such a control does not normally allow regeneration, but equipped with a highly accurate oscillator (0.05% stability), it has been employed in the synthetic-fiber industry.

Figure 26 shows a more flexible control scheme using controlled slip. The error between the reference and the speed as sensed by a tachometer is applied to a circuit which scales and limits it to form a slip signal. Slip, being representative of the speed difference between the rotor and stator field, measured in hertz, r/min, radians per second, etc., usually ranges at full load from 1 to several percent of base speed. Summing slip and speed, as derived from the tachometer, produces a frequency reference. Note

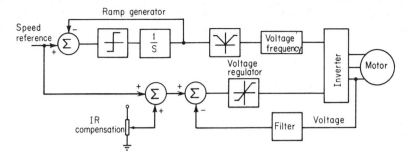

Fig. 25 Simple inverter control.

that for a negative or negative-driving reference, the error and hence the slip becomes negative. Thus the inverter frequency drops below the rotor synchronous frequency, causing braking. The frequency-signal polarity determines direction before it is rectified and applied to the voltage-controlled oscillator. The slip signal can be used to provide IR drop compensation, but it must be scaled up somewhat by a factor K since full-load boost can range to above 10% base voltage. This scheme has the unique capability of natural and instantaneous limiting. For example, the motor is at rest and the speed reference is suddenly stepped to its maximum value. The error is high but the slip, being limited, causes the inverter to supply slip frequency and threshold voltage using high but normal current and torque. This condition will persist if the motor is stalled. If the motor is free to accelerate, the frequency and the voltage rise with speed as directed by the tachometer.

Another case would be at full reference, full speed. If a shock load stalled the motor, the frequency and voltage would drop as fast as the speed, always maintaining maximum slip even at stall. Current and voltage limits could be applied by clamping slip or reducing slip. For example, in regeneration the line refuses to accept the current, causing dc voltage to rise, activating a spill-over limit, which in turn acts to reduce slip, reducing current and maintaining voltage.

These circuits are merely representative of many possibilities. They not only represent some of the complexities of controlling and coordinating two quantities instead of one as in dc, but also should demonstrate some of the potentialities of ac speed control.

41. Speed of Response. It has been found that an ac motor demonstrates a small signal-transfer characteristic similar to that of a dc machine, namely, an underdamped second-order system. However, it has also been found that its response has much less damping than is the case of the dc motor. Furthermore the response characteristics are

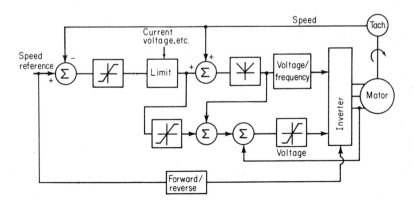

Fig. 26 Adjustable-slip control system.

subject to resonances at certain specific operating frequencies. Thus, in attempting to design a closed-loop system using an ac motor, one starts with a disadvantage. This characteristic is further complicated by the problem of using various types of inverters, which present their own unique set of problems with respect to the transfer function. On an elemental level, the slewing rate of voltage and frequency (independently) is some measure of response. Certainly the LC output filter of a variable-voltage inverter along with any peculiar dead-time responses of the phase-controlled rectifier or chopper are prominent in the voltage-transfer function of that system. By contrast, the inherent delays of the PWM system are fast enough to be considered nonexistent except for the subcycle interval. The delays of the frequency generation of each inverter are more subtle and perhaps more evasive because the input of this circuitry is discrete in the form of a string of pulses. An illustration of this characteristic might be the output of a voltage-controlled oscillator at a given frequency. After a pulse has been delivered, the control input is changed by 10%, but no effect of this change is observed on the output until the next pulse, which will be spaced at 90% of the period of the previous pulse. This delay phenomenon must be factored into the transfer function of the inverter.

Even more interesting is the hypothetical case of an AVI inverter with a six-step output, and a voltage-control means whose response is assumed instantaneous. As each pulse of the oscillator occurs, a step is formed in the six-step output of the inverter. The system frequency and voltage are fixed. Now again the speed command is raised 10% just after a pulse or step. Just as before, no frequency information (no step) is achieved in the output for 90% of the previous period. However, the output experiences an immediate rise in voltage. This rise can be transcribed into a phase shift or perhaps a frequency change. Other possibilities include transfer functions of output variation vs. voltage change with constant frequency and output variation vs. frequency change with constant voltages. Frequency has a more profound effect, since a small frequency change yields a relatively large slip.

It is thus apparent that dynamic response cannot be measured by observing the envelope of an inverter output unless only extremely slow systems are considered. Furthermore, response must be measured for extremely small subcycle intervals, with quantities such as rate of change of phase being extremely important, especially for very low frequencies. Even the concepts of slip control become inadequate, and the only significant dynamic quantities are stator-flux-vector position and its rate of change. All significant changes in motor output must be preceded by changes in these two quantities.

Measuring output response on a flux-vector basis is difficult in all but the most ideal cases applied to the most refined inverters. For example, the stator flux vector for a motor excited by a smooth sine wave rotates with a constant angular velocity while that of a six-step output changes rate abruptly as each step is experienced. For the sine-wave case, any perturbation of the output is a deviation from the constant velocity originally established in the vector.

For the six-step case, deviation must be measured as deviations from the perturbation (torque pulsations) already established by the six-step wave itself. In this technique, a transfer function can be ascribed to an inverter. This transfer function relates specifically to the effect of the inverter output on the motor. It is a direct function of the discreteness or the number of steps of inverter output. Thus the time response is affected by operating frequency. For example, a six-step wave is "fast" at high frequency, since its dead time or hesitation time is $\frac{1}{6}$ of a 60-Hz period, or about 2 ms. At low frequency it is slow, since its dead time may be as long as 60 ms at a 2-Hz operating frequency. Conversely, a 24-step wave has a dead-time response 4 times faster than the six-step wave or a delay of 15 ms in a 2-Hz frequency. Thus dead time equals 1/operating frequency times the number of steps. For a sine envelope, the time response of a PWM inverter is a function of the modulating or pulse frequency, and the slewing rate of the oscillator.

Fast response is achieved by rapid movement of the flux vector. This change may require a burst of "slip" beyond the normal slip operating range. Such techniques are still in the state of development and often represent proprietary knowledge. However, they are mentioned to provide insight into the workings of a drive system and to indicate the true potentialities of an ac drive with respect to dynamics.

42. The Adjustable Current Input (ACI) Inverter. The ACI inverter is a distinc-

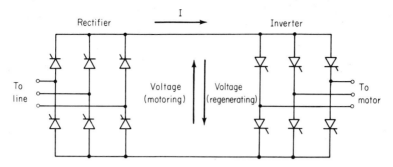

Fig. 27 Adjustable current input inverter circuit.

tive form of inverter employing a front end composed of a controlled rectifier and large filter inductor acting as an adjustable current source whose output is fed to the input nodes of the inverter (Fig. 27). The action of the inverter is to employ commutating capacitors to switch or steer this smooth and constant current from phase to phase, forming an output current waveshape similar to the voltage waveshape of Fig. 2. Thus the inverter can be thought of as determining the output frequency and ultimately the motor speed as determined by the sequencing frequency of the inverter. Similarly the level of inverter voltage output is determined by programming the level of current source output. Note that the level of programmed current is absolutely and safely limited even under direct or bolted short circuiting of the inverter terminals. When compared to voltage source inverters, the circuit is distinguished by the disadvantage of being bulky and somewhat larger in size. This size results from the presence of the large filter reactor and the need for large or oversized commutating capacitors to accommodate the switching energy of the motor leakage reactance. A unique feature is the ability to regenerate power directly back into the ac line. This characteristic is achieved by reversal of inverter terminal voltage which, if applied to a phase-controlled rectifier front end acting as a phase-controlled synchronous inverter, feeds motor power back to the line for controlled braking of the motor.

The beneficial qualities of the inverter include simplicity of construction and operation and, because of the long hold-off time provided by the large commutating capacitors, the ability to use the conventional "phase control" type of thyristor. The inverter, because of the high quality of its output waveform, is often selected, with the AVI, where maximum efficiency is absolutely essential, as in pump drive applications.

43. Cycloconverter. The cycloconverter is essentially a reversible phase-controlled rectifier whose output envelope is programmed as a sine wave and periodically reversed to form an ac wave whose frequency is lower than the input line (Ref. 27). The advantage of the system is that regeneration is achieved in a single stage. Disadvantages include a multiplicity of devices, poor input power factor, and a relatively low output frequency with a maximum of about one-third the input frequency. While this may be acceptable at high frequencies, it is normally disadvantageous at line frequencies. However, for high powers and low speeds the system can be very practical.

44. Motors and Motor Selection. The manipulation of motor designs often involves adjusting the level of stator leakage reactance for harmonic filtering since harmonic slips are high and their reactances represent the only significant impedance internal to the motor (Ref. 29). If harmonic voltages of a particular inverter are high, external reactance may have to be added (Ref. 24). The most ideal, yet perhaps unattainable, solution would be an inverter output wave with negligible harmonics. Excessive harmonics lead to excessive rotor and stator losses, necessitating oversized frames. Note that ac motors have in the past been rated at a given horsepower at a given speed. Variable-speed operation for fan-cooled motors normally requires larger frames before harmonics are considered. Harmonics often add an increment of another frame.

The rotor resistance can be varied by the size, geometry, and composition of the bars

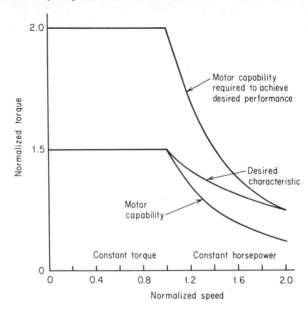

Fig. 28   Motor characteristics for constant-horsepower mode of operation.

of its cage. While having an effect on the harmonic losses, the resistance has an even more profound effect on the damping characteristics of the machine. Other electrical considerations include oversizing the motor to accept added losses or to prevent saturation due to overexcitation at low speeds. Mechanical considerations to allow higher speeds, such as the structure of the rotor and the quality of bearings, are important as a result of available higher-than-line frequency-inverter outputs. Electrically, high frequencies may require better core materials. All these factors are left to the subtle optimization of the inverter/motor-system designer (Ref. 29).

Even if the inverter produced a pure sine wave of variable frequency, some reorientation is often required to make the transition to variable-speed ac technology. Operationally, the motor performs in two distinct modes, constant-torque and constant-horsepower. The oversizing and special designs for low-speed operation have already been mentioned in the constant-torque mode. Furthermore, the breakdown torque may be increased for intermittent operation by increasing the volts/hertz ratio. The increase is in proportion to the square of that ratio. However, in the constant-horsepower mode, which is characterized by constant-voltage output as the frequency is increased, the breakdown torque is reduced by the inverse square of the frequency. To achieve a specific breakdown torque or breakdown-torque margin, the motor may have to be oversized (Fig. 28). In any event overframing in ac is inherently much less expensive than for dc.

REFERENCES

1. W. McMurray and D. P. Shattuck, A Silicon Controlled Rectifier Inverter with Improved Commutation, *AIEE Trans.*, 61-718, June 1961.
2. W. McMurray, SCR Inverter Commutated by an Auxiliary Impulse, *Proc. 1964 Intermag Conf.*, April 1964.
3. C. J. Amato, Variable Speed with Controlled Slip Induction Motor, *IEEE Conf. Record, Industrial Static Power Conversion Conference*, November 1965.
4. B. D. Bedford and R. G. Hoft, "Principles of Inverter Circuits," p. 208, John Wiley & Sons, Inc., New York, 1964.
5. A. J. Humphrey, Precise Speed Control with Inverters, *IEEE Conf. Record, Industrial Static Power Conversion Conference*, November 1965.

6. K. G. King, Variable Frequency Thyristor Inverters for Induction Motor Speed Control, *Direct Current Mag.*, February 1965.

7. A. Kernick, J. L. Roof, and T. M. Heinrich, "Static Inverter with Neutralization of Harmonics," AIEE Conference Paper 61891.

8. F. G. Turnbull, "Selected Harmonic Reduction in Static D-c–A-c Inverters," AIEE Transactions Paper 63-1011.

9. C. W. Flairty, A 50-KVA Adjustable Frequency 24-Phase Controlled Rectifier Inverter, *Direct Current Mag.*, December 1961.

10. A. Schonung and H. Stemmler, Static Frequency Changers with "Sub-Harmonic" Control in Conjunction with Reversible Variable-Speed A-c Drives, *Brown Boveri Rev.*, vol. 51 (8/9), pp. 555–577.

11. J. P. Landis, Static Inverter A-c Motor Drives, *Proc. 1963 IEEE Textile Conf.*, Charlotte, N. C.

12. Boris Mokrytzki, Solid State A-c Motor Drives, *Conf. Record, 13th Ann. IEEE Pulp & Paper Conf.*

13. A. J. Humphrey, Inverter Commutation Circuit, *1966 Conf. Record, IEEE Industry and General Applications Group.*

14. Boris Mokrytzki, Pulse Width Modulated Inverters for A-c Motor Drives, *1966 IEEE International Convention Record, Power Industry and General Applications Section.*

15. D. L. Duff, Optimum Design of Input-Commutated Inverter for A-c Motor Control, *Conference Record, 1968 IEEE Industry and General Applications Conference.*

16. R. L. Risberg, Wide Speed Range Inverter Section Motor Drive, *IEEE Conf. Record, 1969 IGA Group.*

17. D. L. Duff, Practical Considerations in the Design of Commutation Circuits for Choppers and Inverters, *IEEE Conf. Record, 1969 IGA Group Meeting.*

18. C. J. Amato, A Simple and Speedy Method for Determining the Fourier Coefficient of Power Converter Wave Form, *IEEE Conf. Record, IGA 1969 Group Meeting.*

19. Ronald C. Whigham, High Frequency Performance of Fast Recovery and Conventional Power Rectifier Diodes, *IEEE Conf. Record, IGA 1969 Group Meeting.*

20. David Cooper, Richard Williams, and Frank Durnya, Development, Characterization and Application of a New Geometry Inverter High Power Switch, *IEEE Conf. Record, IGA 1969 Group Meeting.*

21. L. Penkowski and K. Pruzinsky, Fundamentals of a Pulse Width Modulated Power Circuit, *IEEE Conf. Record, IGA 1970 Group Meeting.*

22. R. Fox and R. Adams, Modulation Techniques for Pulse Width Modulated Inverter Systems, *IEEE Conf. Record, IGA 1970 Group Meeting.*

23. E. T. Schonholzer, Fuse Protection for Power Thyristors, *IEEE IGA Conf. Record, 1970 Group Meeting.*

24. P. J. Tsivitse and E. A. Klingshirn, Optimum Voltage and Frequency for Polyphase Induction Motors Operating with Variable Frequency Power Supplies, *IEEE Conf. Record, IGA 1970 Group Meeting.*

25. Michael Goldstein, Fuse for Semi-Conductor Protection—A Special Breed, *IEEE Conf. Record, IGA 1970 Group Meeting.*

26. Phil C. Jacobs, Jr., Application of Fuses with Power Semi-Conductors in Direct Current Circuits, *IEEE Conf. Record, IGA 1970 Group Meeting.*

27. H. Rissik, "Mercury Arc Current Converters," Sir Isaac Pitman & Sons, Ltd., London, 1935.

28. V. B. Honsinger, Stability of Reluctance Motors, *Proc. IEEE Winter Meeting*, January 1972.

29. P. J. Tsivitse and F. P. Heredos, "Induction Motor Analysis and Design for Variable Frequency Operation," IEEE 1971 Machine Tool Conference, Cincinnati, Ohio.

30. "General Electric SCR Handbook," 4th ed., 1967.

31. C. F. Wagner, Parallel Inverter with Inductive Load, *IEEE Trans.*, vol. 55, September, pp. 970–980, 1936.

32. K. P. Phillips, Current Source Converters for A-c Motor Drives, *1971 IEEE/IGA Conf. Record*, p. 385.

26

Variable-Speed Solid-State Dc Drives

L. E. NICKELS*

FOREWORD

Modern solid-state variable-speed drives have made motor-generator-type drives obsolete. New drives for heavy machinery such as steel-mill main and auxiliary drives are now entirely controlled by solid-state-type equipment. These drives reduce equip-

*L. E. Nickels and Associates Converter Engineering (Formerly Chief Development Engineer, Semiconductor Products, Allis-Chalmers Corporation); Registered Professional Engineer (Wis.); Member IEEE.

ment and installation costs, save on maintenance, and provide higher operating efficiency.

Although the size range of variable-speed solid-state drives is considerable and includes both single-phase and 3-phase converters, this section provides application information on only 3-phase systems for industrial use. In these installations various arrangements of control are available, and this section will aid in evaluating the various choices of solid-state control and its protection.

GENERAL

Dc motors have been extensively used during the past decade for variable-speed applications. The applications were based mainly on the development of the thyristor (SCR) semiconductor as the means of controlling armature voltage. However, even prior to the use of thyristors, dc motors had been widely used for variable-speed applications. Dc motors are well suited for variable-speed use because of high starting torque and ease of speed control by variable-voltage armature supplies.

Dc motors were first controlled using a rheostat or a tap switch to vary armature resistance for variable-voltage speed control. This method was followed by an ac motor–dc generator controller, known as the Ward Leonard system (Sec. 24). Variable armature voltage for driving a dc motor was obtained by varying the field strength of the dc generator which was driven at constant speed by an ac induction motor. The system was reliable and flexible and had reversing as well as regenerative capabilities. In effect, it served as a performance standard for the more recently developed static drives. Later developments produced one of the first static converters known as the mercury-arc (pool) rectifier. Although this method was used extensively in Europe, it was not used too frequently in America. Recently, thyristor converters have been accepted for variable-speed drives.

Many applications exist for variable-speed drives in industries such as metal-rolling-mill main and auxiliary motors, pelletizing plants, cement plants, paper mills, machine tools, plastics, and transportation. These drives range in size from fractional horsepower to over 10 000 hp.

THE DC MOTOR

Dc motors, with both series- and shunt-connected fields, are used for variable-speed applications. Series motors are used primarily in transportation applications which require very high starting torque. However, the shunt motor is used more extensively with thyristor converters. This discussion is concerned with the shunt motor and its control.

1. Motor Construction. Conventional dc shunt motors are designed primarily for use with dc generators. The generators provide a smooth dc voltage and current which is relatively free of harmonic distortion. Appreciable derating of motors is necessary when driven by thyristor converters (Art. 9). Frequently dc reactors are furnished to ensure compatibility. Motors designed specifically for use with thyristor converters utilize one or more of the following:

1. Larger commutator insulated for higher rates and levels of voltage
2. Laminated frame to reduce eddy currents
3. Laminated interpoles

Use of such motors results in more economical drive systems because these motors are capable of operating without dc reactors and will operate with thyristor converters which have low-frequency harmonics (180 Hz). Most modern variable-speed dc drives use this type of motor.

2. Motor Characteristics. The speed of a dc motor is directly proportional to armature counter EMF and inversely proportional to field strength (Ref. 1). This characteristic is illustrated by the well-known equation

$$S = \frac{K(V - I_a R_a)}{\phi} \tag{1}$$

where S = speed
   V = terminal volts
   I_aR_a = armature current and resistance
   ϕ = field strength
$V - I_aR_a = E$ = counter EMF

With a constant field strength, speed varies directly with counter EMF. Also, if I_aR_a is held constant as it would be during current-limit control, S varies directly with V, which is also the converter output voltage ED. Typically this method of speed control is used below the motor base-speed rating.

Above base speed, $V - I_aR_a$ is constant and ϕ is decreased either manually or automatically to increase motor speed. This method, known as constant-horsepower control, is used when higher speeds at reduced power are required (Fig. 1a).

Equation (1) exemplifies the ease with which dc-motor speed can be controlled. This motor is also well suited to deliver smooth high torque near zero speed.

Armature current of a dc machine is a function of terminal voltage and counter EMF as shown in the equation

$$I_a = \frac{V - E}{R_a} \tag{2}$$

At zero speed, current amplitude equals V/R_a. It is therefore important to provide a slow buildup of V so that the counter EMF can increase with speed to limit I_a. This characteristic is known as a "soft start" and is created by an applied ramp of reference voltage in the converter controller. The theory of current feedback in a regulator is to reduce V to maintain I_a within the dc-motor rating.

Torque characteristics are expressed by the equation

$$T = KI_a\phi \tag{3}$$

Dc shunt-motor torque T is directly proportional to I_a at constant field strength. During a current-limit mode of operation, the motor torque is constant. This characteristic is largely responsible for the use of dc motors for variable-speed drives, i.e., constant torque from zero to base speed. Above base speed, if I_a is constant and ϕ decreases, T will decrease proportionately and S will increase inversely proportionately, which results in a constant-horsepower mode of operation.

$$\text{hp} = \frac{2\,TS}{33\,000} \tag{4}$$

Standard NEMA industrial motors (Ref. 2) are rated to carry 100% load continuously, 150% for 1 min. Heavy-duty mill motors may carry up to 300% for 1 min. Other applications require motors to carry overload currents within these extremes. A current-limiting function is provided within the converter controller to limit I_a within the motor-overload rating. For example, a general-purpose motor, I_a would be limited to 150%. Figure 1b shows dc-motor characteristics.

THE CONVERTER

3. **Application.** A converter is described as equipment that will change electrical energy from one form to another. This change is accomplished in two different modes of operation. The first mode is rectification in which the converter changes alternating current or voltage to direct current or voltage. The second mode is known as inversion in which the converter changes direct current or voltage to alternating current or voltage. During rectification, power flow is from ac to dc. Conversion is accomplished with as few as three or six thyristors, although multiples of these frequently are connected in parallel to increase capacity.

Although the converter and its controls are relatively simple, dc variable-speed drives provide satisfactory performance in the most demanding applications. These applications range from simple, general-purpose, nonreversing drives with a fixed shunt field

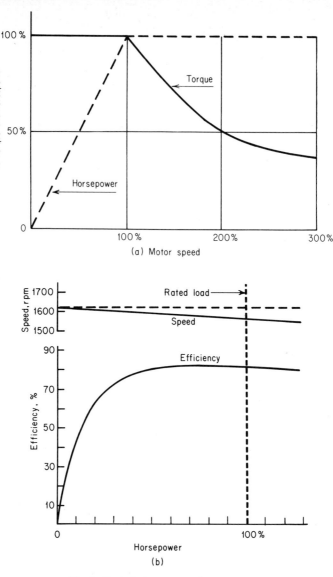

Fig. 1 Typical shunt-motor characteristics.

and a 5% voltage regulator to sophisticated steel-mill reversing drives with shunt-field controls and high-performance 0.1% speed regulators.   Generally, the latter also provide inversion capabilities for fast deceleration via power pump back into the ac line.

Solid-state converters, through proved performance, provide a high degree of reliability.   Modern forms of modular construction and packaging techniques allow quick replacement of components or assemblies to minimize downtime if a fault occurs.   The compactness and light weight of modern converters eliminate the need for elaborate foundations and require a minimum amount of floor space.   In many instances, the converters can be wall-mounted or packaged with other system assemblies to utilize floor space efficiently.

4. Classifications. Standard designations are used to describe converter characteristics (Ref. 6). An operating-quadrant diagram (Fig. 2) is used for this purpose. The designations are intended to describe the functional characteristics of converters, but not necessarily the circuits or components used. Converters are identified by "form" letters to classify functional capabilities as follows:

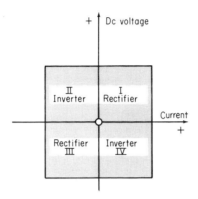

Fig. 2 Operating quadrants of thyristor converter units.

Form A. A single converter capable of unidirectional voltage and current flow. Operates as a rectifier in quadrant I only.

Form B. A double converter capable of providing a positive voltage and current or a negative voltage and current. Has no inversion capability and operates as a rectifier in quadrants I and III.

Form C. A single converter capable of bipolarity voltage and unidirectional current flow. Operates as a rectifier in quadrant I and as an inverter in quadrant IV.

Form D. A double converter capable of bipolarity voltage and bidirectional current flow. Operates as a rectifier in quadrants I and III and as an inverter in quadrants II and IV. Performance is identical to that of a Ward Leonard drive.

These forms refer to the converter capabilities only and not the motor. Motor direction and power flow may be reversed by armature switching or field reversal without the use of an elaborate Form D converter. For example, a motor can be reversed with a Form A converter and a reversing switch or a reversing field. Similar combinations may be obtained for regeneration (reverse power flow).

5. Characteristics. Converters in the rectifier mode provide a variable dc armature voltage for controlling dc-motor speed. Solid-state converters utilize thyristors, or thyristor-diode circuit combinations to accomplish this control (Refs. 4, 5, and 9). Silicon diodes are two-terminal devices that have one or more rectifying junctions (two layers) within an encapsulation, usually hermetically sealed. One terminal is known as the anode and the other as the cathode. Heavy current flow occurs through the cell from anode to cathode whenever the anode is at least 1.5 V positive with respect to the cathode. Only a small leakage current can flow when the cell polarity is reversed (Fig. 3*a*). Reverse current flow from cathode to anode is extremely low compared with the forward current flow. Once current is flowing from A to C, a diode has no inherent capability of turning off. Either a mechanical disconnect must be used to interrupt cell current, or the current must be allowed to decrease to zero through a decrease in source voltage.

Thyristors are three-terminal devices that have four layers and three junctions within an encapsulation. The three terminals are called *anode, cathode,* and *gate.* Operation

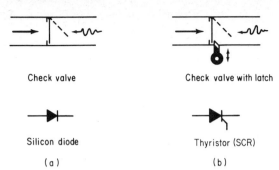

Check valve   Check valve with latch

Silicon diode   Thyristor (SCR)

(a)   (b)

Fig. 3 Semiconductor check-valve analogy.

(Fig. 3*b*) is similar to the diode; i.e., anode volts must be positive with respect to the cathode. However, no current will flow until the latch (gate) is released. Once the cell has been gated, operation is the same as that of a diode; i.e., current must reduce below the minimum holding current to turn off the device. The device cannot inherently stop the flow of forward current.

Characteristically, thyristors are ideal, efficient, static switches that are either on or off and have relatively low power losses in either mode. They have no capability of being modulated between these two modes, which would result in high power losses (such as a transistor).

6. Phase Control. Variable dc output-voltage control of a converter is obtained through a synchronized turn on (gating) of thyristors with respect to the ac source voltage. Control of the output is obtained by gating the thyristors at a variable point within the cycle. This operation is commonly referred to as phase control. The measurement of phase control is expressed as a delay angle α which is an electrical angular measurement. The delay angle indicates the point in a cycle where thyristor gating occurs with respect to the point in a cycle at which a diode assumes conduction without phase control. A diode is considered to have $\alpha = $ zero. Phase control of thyristors produces notches in the dc output waveform. Notching increases as α increases and reduces converter output voltage. When $\alpha = 0$, converter output is maximum, and when α is maximum, converter output is zero, or a maximum negative for a regenerative drive (Figs. 4 to 6). For a rather heavily loaded motor, the output voltages may be expressed by the following equations:

Fig. 4 $E_D = 1.17 \, E_s \cos \alpha$ (5)

Fig. 5 $E_D = 2.34 \, E_x \cos \alpha$ (6)

Fig. 6 $E_D = 1.17 \, E_s \, (1 + 1 \cos \alpha)$ (7)

These equations provide the average dc output voltage, which is the armature voltage of a rather heavily loaded machine. A lightly loaded motor results in a higher counter EMF which tends to follow the peak output voltages rather than the average values. The result is considerable speed regulation (no load to full load) due to the notching effect. Figures 4 to 6 illustrate the characteristics of heavily loaded motors (Art. 8).

The voltage wave shapes of 3-phase converters result in thyristor current conduction for approximately 1/3 (120°) of a cycle. Using 3 thyristors, load current is conducted for a full cycle (360°) with each thyristor sequentially conducting for 1/3 cycle. Sufficient inductance in the load is necessary to maintain a relatively flat, smooth, continuous dc current. Many of the thyristor data and current ratios published are based on continuous current flow which results in a rectangular wave of current through each thyristor. This assumption simplifies calculations of current levels, power dissipation, volt-ampere ratings, etc. With a relatively inductive circuit, 1/3-cycle conduction can occur regardless of α. The average and rms thyristor currents for 3-phase converters (1/3-cycle

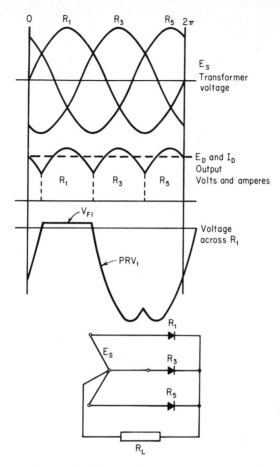

Fig. 4 Three-phase single-way circuit theory, calculations, wave shapes, and output-voltage characteristics with respect to phase control. (*See also next page.*)

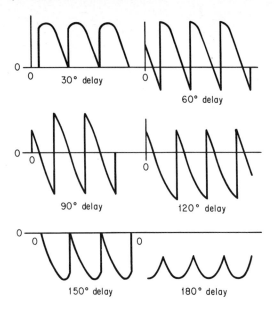

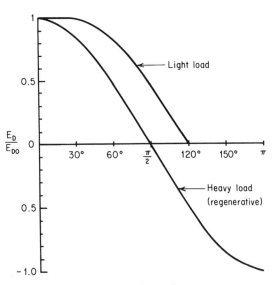

Fig. 4 (*Continued*)

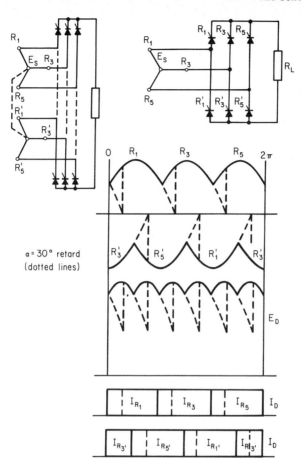

Fig. 5 Three-phase double-way circuit theory, calculations, wave shapes, and output character-istics with respect to phase control are same as in Fig. 4. (*See also next page.*)

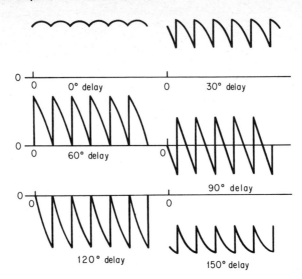

180° delay equals maximum negative voltage E_D

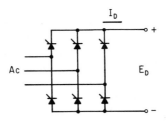

Fig. 5 *(Continued)*

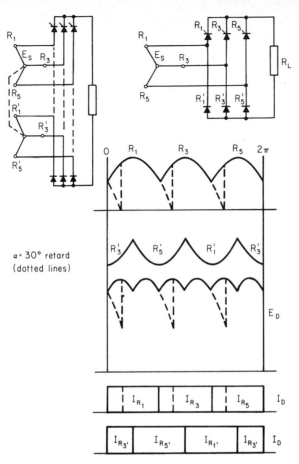

Fig. 6 Theory, calculations, wave shapes, and output-voltage characteristics for a 3-phase, double-way circuit with half-wave control. (*See also next page.*)

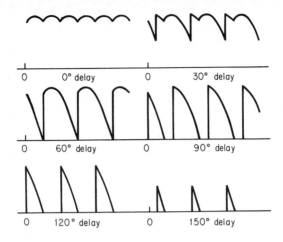

At 180° delay $E_D = 0$

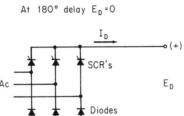

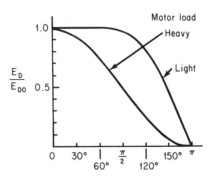

Fig. 6   *(Continued)*

conduction) are

$$I_{av} = (I_D + 120°) \div 360° \text{ or } \frac{I_D}{3} \tag{8}$$

$$I_{rms} = (I_D^2 \times 120° \div 360°)^{1/2} = \frac{I_D}{\sqrt{3}} \tag{9}$$

However, load current is frequently discontinuous; i.e., pulses of current are separated by periods of zero current. Discontinuous current usually occurs at moderate loads and/or when a large amount of phase control is required. This discontinuity increases thyristor loading and usually requires derating if continuous operation of this type is necessary. While the average current remains unchanged for large amounts of phase control, the rms current increases (Fig. 7). Square waves of current were used to simplify calculations. Similar results will occur with sinusoidal wave shapes. Diodes are usually rated on an average-current basis, whereas thyristors are rated on an rms-current basis. The rms design data provided for thyristors are essential for proper application at large delay angles.

Semiconductors are rated on the basis of not exceeding a critical junction temperature. This value is 190°C for a silicon diode or 125°C for most silicon thyristors. Operat-

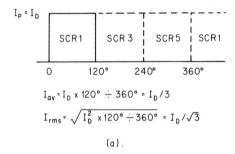

$$I_{av} = I_D \times 120° \div 360° = I_D/3$$

$$I_{rms} = \sqrt{I_D^2 \times 120° \div 360°} = I_D/\sqrt{3}$$

(a).

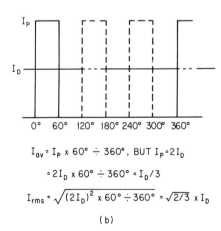

$$I_{av} = I_P \times 60° \div 360°, \text{ BUT } I_P = 2I_D$$

$$= 2I_D \times 60° \div 360° = I_D/3$$

$$I_{rms} = \sqrt{(2I_D)^2 \times 60° \div 360°} = \sqrt{2/3} \times I_D$$

(b)

Fig. 7 Examples of continuous and discontinuous current-flow modes of operation. (a) Continuous-current mode. (b) Discontinuous-current mode. Note that $I_{rms} = \sqrt{2} \times$ as large as it is in continuous-current mode.

ing in excess of these values will usually result in semiconductor failures. Proper design of the cooling system prevents the devices from reaching these levels in normal operation at rated loads (including overloads). Because these devices are small and have little thermal capacity, it is necessary to mount them on larger aluminum or copper extrusions, known as heat sinks, for cooling purposes. The thermal characteristics of semiconductors, heat sinks, and the mounting interface are expressed as thermal resistance in °C per watt. The thyristor-application maximum continuous-current rating is determined from the temperature rise of the junction and the total thermal resistance. A typical example follows:

T_A (ambient) = 40°C
T_J (max thyristor junction) = 125°C
ϕ_{JC} (thermal resistance junction to case) = 0.2°C/W
ϕ_{C-HS} (thermal resistance case to heat sink) = 0.1°C/W
ϕ_{HS-A} (thermal resistance heat sink to ambient) = 0.25°C/W
P = power, W

$$T_J - T_A = P(\phi_{JC} + \phi_{C-HS} + \phi_{HS-A})$$

$$P = \frac{T_J - T_A}{\phi_{JC} + \phi_{C-HS} + \phi_{HS-A}}$$

$$= \frac{125 - 40}{0.20 + 0.10 + 0.25} = \frac{85}{0.55} = 155 \text{ W} \tag{10}$$

Using the value of 155 W, and knowing the conduction time for the application, the average current is obtained from the thyristor average forward power-dissipation curve (manufacturer's published thyristor data). Assuming the peak forward-voltage drop of the cell is 2.0 V at this current results in 155/2 or 75.5-A average. A 3-phase circuit produces an output 3 times this amount or 226.5 A (I_D) [Eq. (8)].

Another factor affecting the application of thyristors is the class of fault protection required (Art. 18). If the thyristor is required to withstand the peak value of fault current shown in Fig. 8, it is necessary to coordinate the thyristor's peak subcycle current rating with the peak subcycle value of fault current (including offsets). Usually this value is a more demanding criterion for the application of thyristors and results in a conservative continuous-current application. Since more thyristor capacity is required for this class of protection, converter costs are higher. For a correct converter design, it is a necessity to consider continuous current, class of protection required, and the peak fault currents to which semiconductors are subjected.

7. Commutation (Ref. 5).   Thyristor turnoff is accomplished by discontinuous current or by reducing the cell current to zero. Commutation achieves this turnoff by turning on the next thyristor in its proper sequence. This operation also results in regulation due to loading.

Generally, the forward-voltage drop of rectifiers is less significant than the other factors which contribute to regulation. Of these, the commutating reactance is most significant in large power rectifiers. In circuits shown in Figs. 4 to 6, the load current is commutated from one rectifier which had the highest instantaneous voltage to a subsequent rectifier which now has the highest instantaneous voltage. In these circuits, commutation takes place every third or sixth of a cycle. During the commutating period, called the angle of overlap u, the current will be reduced to zero in the rectifier which is completing conduction during its cycle and will increase to equal the dc load current in the rectifier which is starting conduction for its cycle. This transfer cannot take place instantly (except at no load) because of the source reactance and leakage reactance of the main transformer. This total reactance is of salient importance during this period, and is known as the commutating reactance.

During commutation, two anodes will conduct current simultaneously, i.e., $I_1 + I_3 = I_D$. To do this, both anodes must have the same potential. The resultant circuit in effect during commutation becomes a short circuit. The magnitude and buildup of short-circuit current (commutation current I_c) are limited by the instantaneous voltage difference between phases and the commutating reactance (Fig. 9).

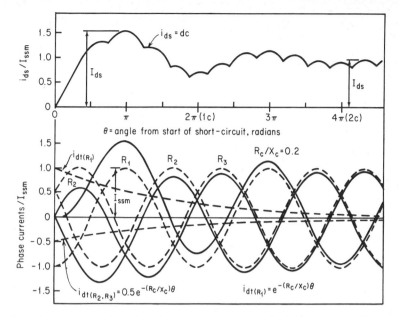

Fig. 8 Ac line currents and resultant dc current waves during dc short circuit. $\pi = 8.3$ ms on a 60-Hz basis.

The angle of overlap u may be calculated by solving the proper equation of Fig. 9. An analysis of the equation indicates that as the reactance is increased, the magnitude of commutating current I_C is decreased and as a result the angle of overlap increases (Fig. 10). The angle of overlap u corresponds with the change in voltage waveform between a and b. The output voltage E_D will follow the voltage curve E_u during the angle of overlap. The new average voltage can easily be calculated by integrating the waveform. The voltage waveform E_u is a cosine wave with its crest value occurring at point a, which is the commutating point at no load ($u = 0$).

The current of each phase will also be affected as a result of commutation. It is interesting to observe that while the dc load current I_0 is directly involved in determining the duration of commutation, it in no way affects the calculation of commutating current. The drop E_x in the dc circuit is directly proportional to load current and reactance of the transformer as a result of overlap.

Commutating reactance is measured by short circuiting that part of the secondary winding between which commutation takes place and applying to the primary a sinusoidal voltage of a value sufficient to supply rated primary current. From this value, the percent reactance may be obtained as follows:

$$X\% = \frac{I_s X_L}{E_S} \tag{11}$$

where I_s = rated current

E_S = rated secondary voltage

X_L = measured secondary reactance, Ω (Fig. 11)

As previously stated, the resistance (copper loss) is an important factor in determining regulation. The total effect may be obtained by taking the copper loss of the main transformer, the interphase transformer if used, and any filter chokes if used. The voltage

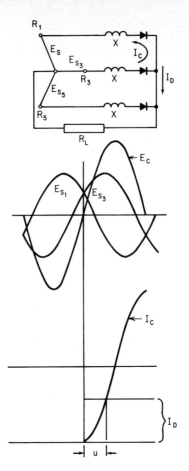

Fig. 9 Angle-of-overlap calculations.

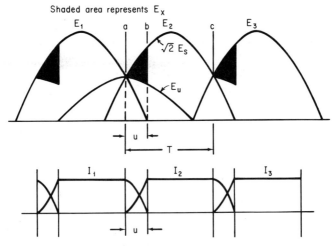

Fig. 10 Voltage and current waveforms during commutation.

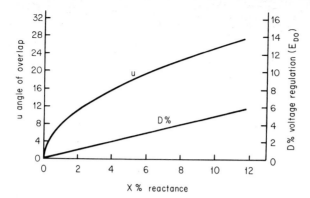

Fig. 11 Angle of overlap and percent voltage regulation as a function of percent transformer reactance with delta-delta, delta-wye connections.

drop E_r produced by these losses is

$$E_r = \frac{P_r \text{ (total)}}{I} \tag{12}$$

The general equation for direct voltage at a given load is

$$E_{DL} = E_{DO} - d - E_r - {}_sE_F \tag{13}$$

For design work the equation is more useful in the form

$$E_{DO} = \frac{E_{DL} + E_r + {}_sE_F}{1 - d\%/100} \tag{14}$$

As an example, the design will be computed for a converter rated 250 kW at 500 V_{dc} (E_{DL}). The transformer reactance is 6% and the copper loss is 2.5 kW (P_r). A 6-phase double-way circuit with a delta-wye transformer is used. The peak reverse voltage (PRV) per leg is 1.045 E_{DO} (Fig. 5). To allow a sufficient safety factor, one thyristor per leg with PRV rating of 1200 V will be used. Therefore, the forward voltage drop per leg is 1.5 V. However, the circuit is a double-way circuit which has the negative leg of the bridge in series with the positive leg of the bridge. As a result, the total drop ${}_sE_F$ will be 3 V. The voltage drop due to copper losses equals 2500/250 or 10 V. The voltage regulation due to the transformer reactance ($X\%$) is equal to $D\%$, or 3.0 from Fig. 11. The dc no-load voltage E_{DO} is

$$E_{DO} = \frac{500 + 10 + 3}{1 - 0.03} = \frac{513}{0.97} = 529 \text{ V} \tag{15}$$

$$\text{Voltage regulation} = \frac{(529 - 500) \times 100}{500} = 5.8\% \tag{16}$$

Of this percentage, the silicon-diode drop of 3 V represents 0.6%. Refer to Fig. 12 for converter efficiency.

From Fig. 5, the transformer secondary voltage is

$$E_S = \frac{E_{DO}}{2.34} = \frac{529}{2.34} = 226 \text{ V } (391 \, E_L) \tag{17}$$

The transformer secondary current I_S is equal to 0.82 I_{DC} or 410 A rms. The transformer secondary kVA is calculated to be

$$\text{Transformer kVA} = 3 \, E_S \, I_S = \frac{3 \times 226 \times 410}{1000} = 278 \text{ kVA} \tag{18}$$

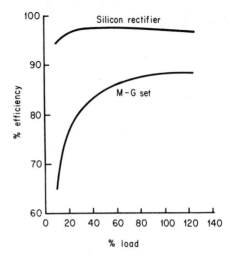

Fig. 12 Efficiency vs. load comparison.

However, if a standard 460-V source had been used,

$$\text{kVA} = \sqrt{3}\,E_L\,I_S = \sqrt{3} \times 460 \times 410 = 326 \text{ kVA}$$

Economics, or standardization, usually determines which kVA will be required.

8. Wave Shapes (Ref. 5).   Figures 4 to 6 illustrate typical circuits used for converters at this time. All these circuits can best be understood by closely examining Fig. 4, which is the basic circuit used in the other figures as well. This arrangement is known as a 3-phase single-way circuit, because the ac phase-current flow is unidirectional. A four-wire source is required. The fourth wire is used for a dc current return to the source. Conventional wye-connected transformers cannot be used as a source because of the large dc current flowing into the control. A zigzag connection is therefore used to prevent transformer saturation. The equations of Fig. 4 show that the source VA = 1.47 times the dc power output. The high ratio of input VA to output dc watts results from single-way operation which is $\sqrt{2}$ as large as double-way operation (bidirectional phase-current flow). Since many applications do not have a four-wire ac system, one with a wye-connected zigzag transformer must be provided within the converter. However, this circuit does possess at least one major advantage; i.e., it has inverter capabilities and can be operated in quadrants I and IV. The circuit is used for regenerative applications and has an economic advantage for low-horsepower drives, and also for larger-horsepower drives where a transformer is a requirement. Figure 4 also shows the phase current for a continuous-current mode of operation, and the output-voltage characteristics over the entire range. Note that from 90 to 180° delay, the output voltage is negative, which signifies inverter operation.

Figure 5 consists basically of two Fig. 4 circuits connected in series. This circuit is a 3-phase double-way circuit and is also known as a 3-phase full-wave bridge. A comparison of wave shapes and equations of this circuit with those of the previous circuit is as follows:

1. Twice as many thyristors used.
2. Twice as many rectifier legs.
3. No change in average or rms continuous current per device.
4. The dc output voltage is doubled for the same input voltage and the same PRV stress per cell.
5. Harmonic frequency is 6 × fundamental as opposed to 3 × fundamental.
6. Operates off a 3-wire ac supply.
7. Source VA is only 70% as high for the same output power.
8. Six-pulse gating circuit required.

9. Capable of operation in quadrants I and IV.
10. Transformer not required.

Output-voltage characteristics for this circuit are as shown in Fig. 5; however, E_{DO} on an absolute basis is double that of Fig. 4 [Eq. (6)]. The above factors are significant in establishing the popularity of this circuit.

Another 3-phase full-wave bridge circuit that is widely used is that shown in Fig. 6. It has features similar to those of Fig. 5, except the harmonic frequency is 3 × the fundamental, three gating pulses are required, and the circuit is not capable of regeneration (operation in quadrant I only). However, it does have an economic advantage over its all-thyristor counterpart. Figure 6 also illustrates the wave shapes at various angles of retard and the relationship of voltage vs. angle of retard. This relationship results from adding a variable 3-phase half-wave dual-polarity output to a fixed 3-phase half-wave-diode unipolarity output which eliminates the negative output voltage on the curve [Eq. (7)].

Further reduction of harmonics results from the use of a 12-phase circuit as shown in Fig. 13. Two 3-phase full-wave bridge circuits are required along with two dc reactors. The ac source voltage and gate signals of one bridge are displaced 30° with respect to the other bridge. This displacement can be obtained through the use of a transformer having a delta primary and a double (wye, delta) secondary. One bridge is connected to each secondary. Each bridge produces the usual 6-pulse dc output. However, since there is a 30° displacement between outputs, the resultant is 12 uniformly spaced pulses

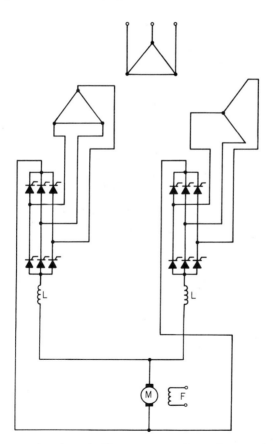

Fig. 13 Connection of 12-phase, double converter unit for supplying an adjustable speed.

and the fundamental harmonic frequency is 720 Hz. This circuit is not economical for smaller drives because of the dual-bridge and transformer-secondary requirement. However, it is frequently used for large drives where multiple bridges are required for additional capacity, and the dc-motor ripple requirements become more stringent. Dc reactors are used to reduce circulating current between the 2-phase displaced bridges, which tends to produce 120° thyristor conduction. As an added advantage, reactors also reduce fault current di/dt, which facilitates coordination with protective devices (Art. 18).

9. Ripple (Ref. 3). The preceding information and referenced figures illustrate the presence of harmonics on the dc output voltage. Harmonics consist of frequencies which are multiples of the source frequency. These multiples depend on the number of discrete thyristor commutations per cycle. For example, the harmonic fundamental multiples of Figs. 4, 5, and 6 are 3, 6, and 3 which result in a fundamental harmonic frequency of 180, 360, and 180 Hz, respectively. The amplitude of this harmonic depends on the delay angle, or its equivalent normalized output voltage, its frequency, and the commutating angle u. Higher multiples of the fundamental harmonic frequency exist but can frequently be neglected because the amplitude is relatively low and circuit effects are minimal (Figs. 14 and 15).

Ripple current is approximately equal to ripple voltage divided by the dc-load reactance (including the reactance in the dc loop). Reductions in ripple current may be obtained by using a different circuit to obtain higher-frequency harmonics or by adding reactors in the dc loop.

Ripple voltage affects the performance of dc motors as follows:
1. Temperature
2. Commutation
3. Speed regulation

Dc motors used with dc generators have sparkless, or black, commutation over a relatively wide range of field strength. The same dc motor driven by a converter with low harmonic frequencies results in a relatively narrow range of field strength to produce black commutation. This condition is an indication of the inability of the dc motor to operate satisfactorily when commutating pole flux and armature current are not matched for magnitude and phase relationship.

Typical industrial motors are rated 100% continuously and have a 1.15 service factor and a 60°C rise when supplied by a dc generator. The same motor supplied by an SCR drive having low harmonic frequencies is rated 100% continuously with unity service factor and a 70°C rise because of high ripple voltage. This derating occurs because of the thermal and commutating effects. Reductions in ripple current either by increasing harmonic frequencies or by added dc reactance can sufficiently improve the condition so the dc motor will run cooler and can have a higher service factor.

10. Power Factor (Refs. 4 and 6). The power factor of a thyristor converter is a product of distortion and displacement. The distortion factor is caused by harmonics in the ac line current which add to the reactive voltamperes. The distortion factor is unity if no harmonics exist to alter the ac-line-current sinusoidal waveform. The displacement power factor is affected by converter phase control, commutating reactance, dc loop reactance, and transformer exciting current. As the name implies, this factor is a displacement of the ac-line-current fundamental with respect to the ac line voltage (Figs. 5 and 10). Whereas a diode rectifier can have a relatively high constant power factor because phase control cannot occur, thyristor converters frequently have an extremely variable low power factor as motor speed is changed. The displacement power factor is approximately equal to the ratio of E_D/E_{DO}. Curve 1 of Fig. 16 shows this relationship, and curves 5, 7, and 9 show the effects of reduced power factor due to commutating reactance. Curve 3 tends to reflect the power factor of a typical variable-speed drive. The mode of converter operation changes from rectification to inversion where these curves intersect the abscissa.

Standard watt- and varmeters measure the fundamental of ac line currents and not the distortion of the wave. Consequently, they measure displacement power factor only, which is a ratio of total watts input to total input voltamperes.

11. Regeneration (Ref. 6). This is the ability to pump power from the dc motor into the ac supply through inverter operation of the converter. Since all converters deliver a

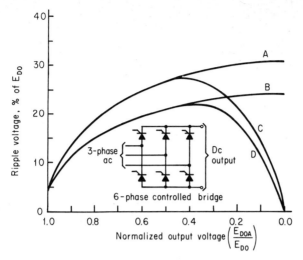

Fig. 14 Ripple voltage vs. normalized output voltage for 6-phase controlled ridge circuit. Curve *A*, rms ripple, inductive load; curve *B*, 6th harmonic, inductive load; curve *C*, rms ripple, resistive load; curve *D*, 6th harmonic, resistive load.

unidirectional dc current, it is necessary through external circuitry or the use of more than one converter to provide bidirectional current or voltage flow from the motor. This flow is essential because the counter EMF polarity of the motor does not change when the mode of motor operation is changed to regenerate. The converters shown in Figs. 4, 5, and 13 are suitable for regeneration because they can operate in quadrants I and IV. The following methods are usually used to obtain four-quadrant operation of the motor (Fig. 17):

a. Motor-field reversal requires one single Form C converter and a field-reversing switch, which may be static. The theory is to produce a reversal of counter EMF polarity with the motor rotating so power can be pumped back through the converter, operating

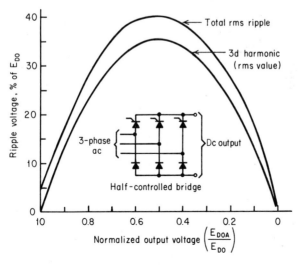

Fig. 15 Ripple voltage vs. normalized output voltage for half-controlled bridge circuit.

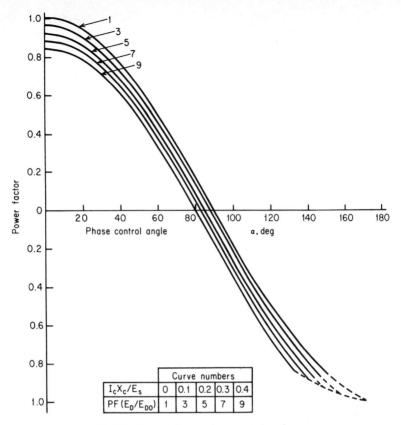

Fig. 16 Power factor vs. phase-control angle.

	Curve numbers				
I_cX_c/E_s	0	0.1	0.2	0.3	0.4
PF (E_D/E_{DO})	1	3	5	7	9

as an inverter. After regeneration has been accomplished, and with the same field polarity as used during regeneration, the motor rotation will reverse and accelerate when armature current is increased.

b. Motor-armature reversal by reversing switch requires one single Form C converter and an armature-reversing switch, which is usually mechanical. The theory is to produce a reversal of voltage polarity with the motor rotating, similar to method *a*, except through the use of an armature switch. Although the machine polarity does not change, the converter polarity does whenever switching occurs. Motor rotation is reversed similar to method *a*.

c. Static armature reversal requires two single Form C converters connected in inverse parallel, which results in a Form D converter. The theory is to provide two converter current paths, i.e., one for each direction of motor current. Regeneration is accomplished, with the motor rotating, by stopping current conduction of the converter which had been on, then turning on the converter which had been idle. This operation results in a reversal of armature current with no change in armature voltage polarity, and pump-back occurs. During regeneration, the delay angle α of the converter decreases and the mode changes from inversion to rectification; this transition occurs at zero speed. Further decreases in α result in a positive converter voltage, which reverses motor rotation and armature voltage polarity.

CONTROLS

Adequate controls are provided to ensure operation of the converter and motor within their current and voltage ratings. Additional controls are required for abnormal

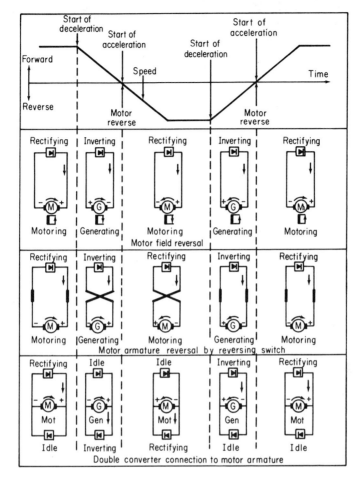

Fig. 17 Operating sequences of converters serving a reversible dc motor.

operating conditions to deenergize the drive and avert serious equipment failures. Various functions are added to a basic drive system to accomplish this purpose.

12. Protective controls are primarily auxiliary and power-control functions which react when faults occur in them. Typical examples are:

a. Single-phase protection—deenergizes drive to prevent this mode of operation.

b. Field loss—utilizes a relay to function on the loss of field current and prevent overspeed operation.

c. Field economizing—reduces field strength to prevent motor overheating at standstill.

d. Overvoltage protection—(1) senses converter output voltage at zero speed and reacts to protect the motor by preventing the dc-loop contactor from being closed if the voltage is abnormally high; (2) with motors connected in series, excessive armature voltage across either motor results in a shutdown.

e. Protective interlocks such as thermal-overload, instantaneous-overload, and over-temperature initiate a drive shutdown by applying gate blocking, reducing regulator reference to zero, and tripping the dc-loop contactor.

Additional controls are frequently required for system interlocking to achieve a programmed shutdown.

13. Speed Regulator—Voltage Feedback. Speed regulators function to accelerate a motor from standstill to a preset speed and regulate that speed within a narrow operation deviation band during steady-state conditions. This control is accomplished by applying a dc reference voltage in positive for quadrant I converters and an armature voltage negative-feedback signal into a summing junction of the regulator. The difference between these two voltages represents an error signal into the regulator and is amplified to produce a higher-voltage output signal which drives the thyristor gating circuits. With a large error signal where reference exceeds voltage feedback, the delay angle α is decreased and converter output voltage increases to accelerate the motor. Acceleration continues to occur until the preset speed is reached and the feedback signal nearly equals the reference voltage.

To prevent excessive armature current, a negative-current-feedback signal is fed into the regulator and dynamically reduces the regulator output to maintain current within rated limits during acceleration. Additionally, the application of gate drive voltage from the regulator is applied at a controlled rate to allow time for current control, and to protect the application from excessive rate-of-speed changes. Whenever the reference signal is greater than the sum of the armature current and voltage negative feedback, acceleration occurs. Deceleration occurs when the total negative feedback exceeds the reference. Reference voltage can be supplied by a rheostat excited from a dc voltage from a process reference bus or from a process controller (1 to 5 mA, 4 to 20 mA, 10 to 50 mA).

This type of regulator usually has a speed-deviation band of 3 to 5%. The tolerance is high because speed regulation is achieved indirectly from armature-voltage feedback. Since motor speed is proportional to counter EMF, not armature voltage, the regulation tolerance is high. Regulators of this type are equipped with IR compensation which utilizes a small amount of positive-current feedback to reduce the deviation-band width. This type of regulator is economical and is used where a high degree of accuracy is not required (Fig. 18).

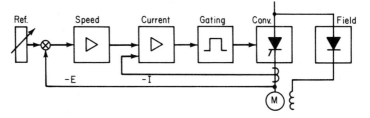

Fig. 18   Speed regulation using voltage-feedback control.

14. Speed Regulator—Speed Feedback. Operation of this regulator is similar to the previous regulator except that armature-voltage feedback is not used. Instead, negative-voltage feedback is obtained from a tachometer generator which is mechanically coupled to the dc motor. The tachometer output voltage varies directly with motor speed. Since the negative-feedback voltage signal is directly proportional to speed, the regulator will accurately control motor speed. Corrections will automatically be made for variations in armature current, and IR compensation is not necessarily required. However, a negative current feedback for current-limit control is used, as previously described.

The accuracy of this regulator is extremely good and has a deviation band (tolerance) of 0.1 to 1.0%. The degree of accuracy is a function of the tachometer, amplifier gain, reference, and thermal conditions (Fig. 19).

15. Regulator Accuracy (Ref. 14). Regulators are classified by the accuracy with which a precise motor speed can be maintained when subjected to varying conditions during steady-state operation. The limits of accuracy are defined as deviation bands. Regulator performance increases with a decrease in deviation band. The steady-state performance of a regulator is affected by the operating variables and the service deviations.

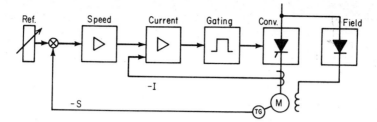

Fig. 19 Speed regulation using speed-feedback control.

a. The operating variable is motor torque for the regulating systems described. With the service conditions held constant, variations in torque of 10 to 100% of rated value are corrected by the regulator to maintain speed within a specified operating deviation band. This band is expressed as a percentage of maximum rated speed.

b. The service deviation is a combination of variables including temperature and source voltage. Temperature may vary up to 15°C within limits of the applicable service conditions, and source voltage may vary 10% (total). The regulator will maintain speed within a specified service deviation band during any 1-h interval following a 30-min warm-up period.

Two numbers are used to describe the operating and service deviation bands. The first number represents the operating band, and the second number represents the service deviation band. Standard operating and service deviation bands for variable-speed drives are 20, 10, 5, 2, 1, 0.5, 0.2, and 0.1% of rated speed. A typical variable-speed drive may have an operating deviation band of 0.5% and a service deviation band of 1%. The total deviation band is usually less than the sum of both deviation bands.

16. Shunt-Field Control. The regulators previously described are for controlling speed by varying armature voltage below motor base speed. Higher-speed control is obtained by manually or automatically controlling the field strength. Manual field control is obtained with a rheostat, and automatic control is obtained with thyristors located in the field circuit. If service deviations occur, the field strength must be adjusted to hold constant speed when a manual field control is used. A field regulator automatically maintains constant speed under these conditions (Fig. 20).

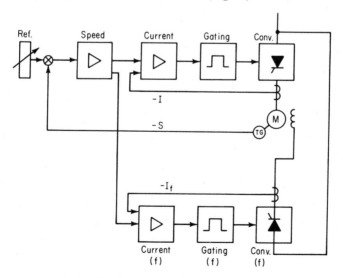

Fig. 20 Speed regulator with automatic field control.

Typically, two regulators are required for automatic control of the motor above base speeds. A voltage-feedback regulator is used to control armature voltage below base speed, and a speed-feedback regulator is used to control field current. When armature voltage is less than that required to provide base speed, full-field current is supplied and the system regulator performs as though no field control was used. However, when the armature voltage has increased to provide base speed without satisfying the desired speed set point, field current is decreased by the field regulator and motor speed increases until the speed reference has been satisfied.

17. Four-Quadrant Control (Ref. 6). Basically, the regulator requirements are the same as for a single-quadrant drive. However, because armature current must be transferred from one rectifier to another, logic is provided to accomplish the switching. This logic makes use of a bipolarity reference, a polarized current feedback, and a polarized speed or armature-voltage feedback. When the reference is positive, operation will be identical to that of a single-quadrant regulator, the motor will have a specified rotation which is assumed to be counterclockwise, and the converter will operate in quadrant I. A negative reference causes clockwise rotation and converter operation in quadrant III. Whenever the speed feedback exceeds the reference voltage, regeneration occurs and power is pumped back into the ac line through inversion. Regulator logic performs this function by turning off the converter, which is on, and then turning on the idle converter. The switching interval typically is approximately 20 μs and results in a "dead band" of operation. During this period, the motor coasts and is neither accelerating nor decelerating (see Art. 11).

The preceding method describes regulator operation of a reversing drive where the thyristor gates of the forward and reverse armature converters are switched on and off to establish direction and acceleration or regeneration. Similar logic is performed to switch field-current polarities along with a single-armature rectifier for field-reversing drives.

Another method of control is used to achieve armature reversal where the thyristor gates are not switched on and off. Instead, the gates are modulated within limits. This type is referred to as a circulating-current method (Ref. 7) where current is allowed to circulate between the forward and reverse rectifiers. The magnitude of circulating current is kept low (a few percent) by using dc reactors and by proper forward-reverse converter gating relationships. Advantages and disadvantages of each method are as follows:

Method	Response	Dc reactors	Field circuit	Economics
Armature rectifiers . . .	Moderately fast	No	Standard	Moderate
Field switching	Relatively slow	No	Special	Reasonable, especially large drives
Circulating current . . .	Extremely fast	Yes	Standard	Expensive, especially large drives

PROTECTION

18. Classes of Protection. Various classes of overcurrent protection are used for source load and converter faults. Generally, it is desirable to protect against them to the fullest extent. However, protection of the converter is unique because of the small thermal capacity of a semiconductor. Typically, a semiconductor will reach thermal equilibrium in less than 10 s (exclusive of heat sink). This lack of thermal capacity results in a considerable thermal mismatch between the converter and the source or load equipment. Obviously, the same protection technique is not economically feasible for converters and other equipment such as transformers, reactors, cables, breakers, contactors, and motors. In recognition of this difference, classes of protection have been established which range from a very expensive, highly reliable, completely protected system to an economical, less reliable, marginally protected system. The proper choice of protection is based on equipment size, application reliability, shutdowns, probability of faults, and economics (Ref. 6).

a. Class I applications require the most reliable converter to minimize downtime, such as large main drives for metal-rolling mills. External faults, such as dc short circuits, dips, and loss of ac power leading to "conduction through" while inverting, must be cleared by gate protection and/or breakers. Internal faults, such as defective gate control or faulty regulator, must be similarly protected. Thyristor fuses and thyristors must not fail under these conditions. The thyristor fuses are used to isolate defective thyristors from the circuits to prevent service interruptions. They also provide backup protection for failure of this protective means.

b. Class II applications are similar to Class I, except utilizing smaller converters which cannot justify the overall economics of Class I. Externally induced faults are cleared the same as Class I, but internal faults allow fuse protection to prevent thyristor damage. This class of converter would be used for large metal-mill auxiliary drives, and especially for those used for four-quadrant operation.

c. Class III applications require good continuity of service but are less critical than Class II. External faults resulting in less than five times rated current caused by motor flashover, excessive impact loading, regulator failure, etc., are cleared by gate suppression and a dc contactor without fuse melting. External faults in excess of five times rated current or internal faults allow fuse protection of the semiconductors, but semiconductor failures must not occur. Typically, noninverting (first-quadrant)'drives, such as small-mill auxiliary types, fit into this category.

d. Class IV applications are for those converters (usually small) in which coordinated protection between thyristors and fuses cannot be economically achieved. Severe faults may lead to both fuse interruption and thyristor failures. Drives of this type are usually applied where the converter duty is relatively light, the probability of failures is low, and downtime for replacement of thyristors and fuses can be tolerated. These applications could be considered typical general-purpose drives.

The first class of protection is the most difficult to achieve and requires a fast gate-suppression scheme and, frequently, a high-speed current-limiting dc breaker. A high-speed breaker of this type generally current-limits in approximately 5 ms which is well before the anticipated 8-ms crest of fault current (Fig. 8). This protection minimizes the need for additional semiconductor capacity to ride through the crest fault current. Alternate methods of providing Class I protection can be achieved by using one or more economic combinations of the following:

1. Additional semiconductor capacity to withstand the crest value of fault current and the follow-through current until a semi-high-speed breaker can interrupt
2. Additional source reactance to limit fault current
3. Additional dc reactors to reduce the rate of rise of fault current to utilize slower breakers and to ensure positive gate-suppression protection.

As backup protection (Ref. 8), high-speed current-limiting fuses are required which limit the peak let-through current within the thyristor's peak-current capability, i.e., have an I^2t rating for subcycle values of time which is less than that of the thyristor for the same time t.

19. Overvoltage Protection. Protection against overvoltage is important and must be considered for converter designs. A thyristor is susceptible to fault operation due to excessive forward voltage (anode to cathode) and to excessive reverse voltage (cathode to anode). If the cell is subjected to excessive reverse voltage, permanent damage will occur and replacement is required. If excessive forward voltage is applied or an excessive rate of rise dv/dt occurs (Ref. 12) while the converter is operating at a fraction of its maximum output, the thyristor will have an unscheduled turnon and deliver an excessive output voltage. The application of this voltage to a motor is somewhat equivalent to an across-the-line start of a dc machine which results in excessive currents. The magnitude of this current, which is a function of the motor CEMF and the degree of premature thyristor turnon (output voltage delivered) may destroy a thyristor and cause fuses to interrupt [Eq. (2)].

Excessive voltage may be generated by transients on the ac line or by interruption of a thyristor fuse (Refs. 10 and 11). The latter cause is a result of the fuse arcing voltage, which usually rises to twice the nominal voltage rating for semiconductor current-limiting fuses. For example, a 500-V fuse will have an arcing voltage of approximately 1000

V. Usually, this voltage will be applied directly across one or more of the thyristors in the circuit. Since this value of voltage is an anticipated protective occurrence, thyristors must be capable of withstanding this voltage. In general, to ride through this peak voltage and source-voltage transients, thyristors with a minimum PRV rating of two times the peak ac line voltage are applied. This tolerance allows sufficient voltage margin for voltage protectors such as

1. Selenium devices
2. Variable resistors ("Varistors" or "Thyrites")
3. Capacitor-resistor networks

Source transient voltages (Refs. 12 and 13) or excessive dv/dt caused by turnon or turnoff of thyristors within a bridge must be prevented from appearing at the thyristor (anode-cathode). This protection is generally accomplished by connecting a series RC circuit across each thyristor and by inserting reactance in the ac line and/or in series with each thyristor. The theory is to have the reactor stressed with the voltage transient while a relatively slow buildup of voltage, typically less than 100 V/μs (refer to semiconductor manufacturer's published data sheets), occurs across the RC circuit in parallel with the thyristor. The use of an LRC circuit and fuses which are voltage-coordinated with the source and the semiconductor voltage rating will provide sufficient converter-voltage protection.

REFERENCES

1. Chester L. Dawes, "Electrical Engineering," vol. I, "Direct Currents," 3d ed., chap. 13, McGraw-Hill Book Company, New York.
2. NEMA Standards, MG 1-23.4, part 23, September 1966.
3. A. Schmidt, Jr., and W. P. Smith, Operation of Large Dc Motors from Controlled Rectifiers, *AIEE Proc.*, 1948, vol. 67.
4. H. Winograd and J. B. Rice, Conversion of Electric Power, in D. G. Fink and J. M. Carroll, "Standard Handbook for Electrical Engineers," 10th ed., McGraw-Hill Book Company, New York, 1968.
5. L. E. Nickels, "Power Control and Conversion," IEEE Education and Lecture Series, 1964.
6. "IEEE Standard Practices and Requirements for Thyristor Converters for Motor Drives," vol. I, 1971.
7. D. L. Duff and A. Ludbrook, "Reversing Thyristor Armature Dual Converter with Logic Crossover Control," IEEE-IGA, New York, March 1965.
8. F. B. Golden, Take the Guesswork Out of Fuse Selection, *Electron. Eng.*, July 1969.
9. "General Electric SCR Handbook," 5th ed., 1972.
10. J. Feenan, Protection of Electrical Apparatus, *Elec. J.*, 1961.
11. Carbone Ferraz, *Tech. Bull.* T1, TB1-7006.
12. J. B. Rice and L. E. Nickels, Dv/Dt Effects of Commutation in Thyristor Three-Phase Bridge Converters, *IEEE-IGA Trans.*, vol. IGA-4, no. 6, 1968.
13. J. B. Rice, Design of Snubber Circuits for Thyristors Converters, *IEEE Conf. Record*, 1969.
14. "Industrial Controls and Systems," NEMA Standards Publication 3-106A, part ICS, October 1971.

27

Regulation Feedback Control and Stability

RONALD A. HEDIN*

*Manager, Engineering Analysis and Simulation, Power Systems Technology, Allis-Chalmers Corporation; Member, IEEE.

FOREWORD

This section provides some fundamental concepts relative to stability and automatic-control-system design. These fundamental concepts are illustrated through simple examples of the application of system analysis, using the transfer function, and the frequency response to common control-system problems.

GENERAL

The proper design of an automatic control requires satisfying a number of criteria dictated by the particular system. A fundamental requirement of any feedback control is a specified steady-state accuracy that it must provide over a range of system operating conditions. These data will dictate both the power requirements and the sensitivity or gain of the system components at steady state (zero frequency).

The second major requirement deals with the dynamic response of the control system which can be specified in a large number of ways. The time required to make a given correction or change in operating level may be specified. The relative stability in terms of allowable overshoot or frequency-response phase margin may also be specified.

Another important consideration in the performance requirements is any periodic type of disturbance which can affect system operation. The control design must consider these effects in terms of whether the feedback should ignore the disturbance or make corrections required to reduce the oscillation. In terms of control-system frequency response, the frequency of the disturbance should be specified, relative to the design crossover frequency (the frequency at which the gain is unity).

1. **Criteria for good control design** include the following:

1. Satisfy steady-state accuracy requirements over the entire range of operation.

2. Be dynamically stable over the entire range of operation.

3. Reduce the effect of periodic and transient disturbances to specified levels over the entire range of operation.

4. Performance should be independent of small variations in control-system component and process parameters.

2. **Systems Analysis.** The analysis of a proposed or existing control system requires that the performance of individual system components be determined. Although most systems are nonlinear to the extent that they cannot be rigorously described by linear differential equations, linear-analysis methods are sufficiently accurate for the normal range of operation. In all but the simplest cases, the solution of system equations can best be accomplished by the use of operational methods such as the Laplace transform. Through the development of this transform, transfer functions are developed which are particularly useful in analyzing the performance of control systems.

LAPLACE TRANSFORMS

The Laplace transform of a function of time is defined as follows:

$$f(t) = \text{function of time}$$

$$F(s) \equiv \mathscr{L}\left[f(t)\right] \equiv {}_0\!\int^\infty e^{-st} f(t)\,dt$$

Variable s is the complex frequency having real and imaginary parts.

3. **Properties of Laplace Transforms.** Through the use of the Laplace transform a set of linear differential equations can be converted into a set of algebraic equations involving the complex variables. These algebraic equations can be used to define transfer functions, inverted to calculate the solution in the time domain or determine the sinusoidal frequency response of the system.

4. **Example of Laplace Transform**

Let $f(t) = e^{-at}$

$$F(s) = {}_0\!\int^\infty e^{-at} e^{-at} dt = {}_0\!\int^\infty e^{-(s+a)t} dt$$

$$= \frac{1}{s+a} e^{-(s+a)t} \Big|_0^\infty = \frac{1}{s+a}$$

5. Typical Transforms and Theorems

$f(t)$	$F(s)$
$Af(t)$	$AF(s)$
$f_1(t) \pm f_2(t)$	$F_1(s) \pm F_2(s)$
$f(t/a)$	$a\,F(as)$
$e^{-at}f(t)$	$F(s+a)$
$d/dt\,f(t)$	$s\,F(s) - f(0+)$
$d^n f(t)/dt^n$	$s^n F(s) - s^{n-1}f(0+) \cdots df/dt(0+)$
$u(t)$, unit step	$1/s$
$\delta_0(t)$, impulse	1
t	$1/s^2$
e^{-at}	$1/(s+a)$
$\sin \omega t$	$\omega/(s^2 + \omega^2)$
$\cos \omega t$	$s/(s^2 + \omega^2)$

TRANSFER FUNCTIONS

In a linear system, the transfer function $H(s)$ is the relationship which relates the transform of the input $F_1(s)$ to the output $F_2(s)$ (Fig. 1).

$F_2(s) = F_1(s)\,H(s)$

Fig. 1 Simple linear system.

Another characteristic of a transfer function is that the inverse transformation of $H(s)$ is the impulse response of the system defined by $H(s)$. This characteristic provides, in a mathematical sense, one definition of stability. A stable system defined by $H(s)$ is characterized by an impulse response which approaches a constant value after a large period of time has elapsed.

6. Properties of Transfer Functions. A transfer function of a complex system may be characterized by a high-order polynomial plotted in the complex plane where $s = \sigma + j\omega$ characterizes the stability of the system. Stable systems will have roots in the left-hand side where σ is negative (Fig. 2).

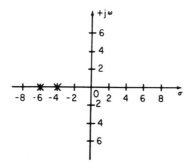

Fig. 2 Characteristic of a typical stable system.

7. Development of Transfer Functions. The following example demonstrates the procedure for determining transfer functions of simple systems (Fig. 3).

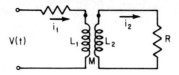

Fig. 3 Transfer-function example of linear transformer.

$$\text{Determine } H(s) = \frac{I_2(s)}{V(s)}$$

$$\text{Differential equations } V(t) = i_1 r + L_1 \frac{di_1}{dt} - M \frac{di_2}{dt}$$

$$0 = i_2 R + L_2 \frac{di_2}{dt} - M \frac{di_1}{dt}$$

Laplace transformed equations

$$V(s) = r\, I_1(s) + sL_1\, I_1(s) - sMI_2(s)$$

$$0 = R\, I_2(s) + sL_2 I_2(s) - sMI_1(s)$$

Eliminating $I_1(s)$,

$$\frac{I_2(s)}{V(s)} = \frac{SM}{s^2(L_1 L_2 - M^2) + s(rL_2 + RL_1) + Rr}$$

8. Applications of Transfer Functions. The transfer function derived in the previous section could be used to determine the time response for any given input $V(t)$ for which $V(s)$ can be determined. For $V(t) = u(t)$ unit step,

$$V(s)\frac{1}{s}$$

$$I_2(s) = \frac{M}{s^2(L_1 L_2 - M^2) + s(rL_2 + RL_1) + Rr}$$

For $V_1(s) = t$,

$$V_1(s) = \frac{1}{s^2}$$

$$I_2(s) = \frac{M}{s[s^2(L_1 L_2 - M^2) + s(rL_2 + RL_1) + Rr]}$$

$i_2(t) =$ inverse transforms of the above

A second application is the steady-state frequency-response calculation obtained by substituting $s = j\omega$.

$$\frac{I_2(j\omega)}{V(j\omega)} = \frac{j\omega M}{-\omega^2(L_1 L_2 - M^2) + j\omega(rL_2 + RL_1) + Rr}$$

9. Block Diagrams. A linear system can be described by a block diagram which shows all the relationships between system variables. It consists of interconnected and interrelated transfer functions of the individual transfer functions of the system components (Fig. 4).

10. Block-Diagram Reduction. A block diagram with several closed loops can be reduced to an equivalent block diagram having fewer or no closed loops. The following illustrates how the block diagram in Fig. 4 can be successively reduced to its simplest form.

The relationship between point b and point c is that

$$[b(s) - H_3(s)\, c(s)]\, G_3(s) = c(s)$$

$$c(s) = b(s)\, \frac{G_3(s)}{1 + G_3(s)\, H_3(s)}$$

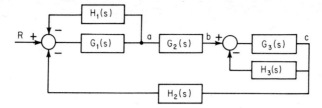

Fig. 4 Block diagram of a multilooped feedback system.

Applying this relationship in an identical manner to the $H_1(s)$ feedback loop, the resulting diagram is given in Fig. 5.

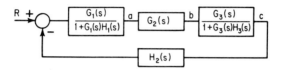

Fig. 5 Diagram of Fig. 4 simplified.

The general form of the reduction of a loop with forward gain $G(s)$ and negative feedback $H(s)$ is given in Fig. 6.

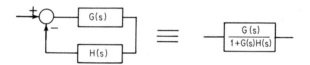

Fig. 6 General form of closed-loop function.

The final relationship between R and c is therefore

$$\frac{c}{R} = \frac{\dfrac{G_1(s)\,G_2(s)\,G_3(s)}{[1+G_1(s)\,H_1(s)]\,[1+G_3(s)\,H_3(s)]}}{1+\dfrac{G_1(s)\,G_2(s)\,G_3(s)\,H_2(s)}{[1+G_1(s)\,H_1(s)]\,[1+G_3(s)\,H_3(s)]}}$$

$$= \frac{G_1(s)\,G_2(s)\,G_3(s)}{[1+G_1(s)\,H_1(s)]\,[1+G_3(s)\,H_3(s)]+G_1(s)\,G_2(s)\,G_3(s)\,H_2(s)}$$

SYSTEM STABILITY

In any real control system, there are any number of time delays which prevent instantaneous or ideal corrective action from occurring. In a system which can be described by linear differential equations, the closed-loop response is stable when the roots of the characteristic equation contain no positive real roots or complex roots with positive real parts. The characteristic equation is the denominator of the closed-loop system transfer function $(1+GH)$ (Fig. 7).

$$\frac{O}{R} = \frac{G}{1+GH} = \frac{G}{(s-r_1)\,(s-r_2)\,(s-r_3)\cdots(s-r_n)}$$

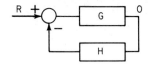

Fig. 7   Closed-loop system.

The transient response of the system will be of the form

$$O(t) = \sum_{n=1}^{m} A_n \, e^r n^t$$

since $r_1, r_2, \cdots, r_n$ are the roots of the equation.

Several examples are given for the sake of illustration.

Example 1

$$1 + GH = s^2 + s\,7 + 10$$
$$s^2 + s\,7 + 10 = 0 \qquad \text{when } s = -5 \text{ and } s = -2$$
$$O(t) = A_1 \, e^{-5t} + B_1 \, e^{-2t}$$

The system is stable because all roots of characteristic equations are negative.

Example 2

$$1 + GH = s^2 + s\,8 + 116$$
$$s^2 + s\,7 + 100 = 0 \qquad \text{when } s = -4 \pm j10$$
$$O(t) = A_1 \, e^{-t} + A_2 e^{+4t} e^{+j10t} + e^{-j10t}$$

The system is stable but oscillatory.

Example 3

$$1 + GH = s^3 - 7s^2 + s108 + 116$$
$$s^3 - 7s^2 + s108 + 116 = 0 \qquad \text{when } s = 4 \pm j10 \text{ and } s = -1$$
$$O(t) = A_1 e^{-t} + A_2 e^{+4t} e^{+j10t} + e^{-j10t}$$

The system is unstable with oscillatory buildup at a frequency of 10 rad/s.

Finding the roots of the characteristic equation will always determine stability, but in complex systems it may be very difficult. The method is not convenient for design purposes, since the determination of which system parameters critically affect stability can easily become lost in the algebra. Numerous methods of analysis and criteria for stability have been developed. The following paragraphs provide a description of several techniques particularly emphasizing those using frequency-response analysis.

11. Routh's Criterion—Stability.   This is an algebraic technique for determining the number of unstable roots in a given characteristic. Its disadvantage is that it gives no indication of relative stability or margin of stability. Details of the technique are not given here but are readily available (Ref. 5).

12. Root-Locus Analysis.   A method of analysis called root-locus can be used to determine the roots of the characteristic equation as a function of loop gain. The procedure is initially to plot the roots of the open-loop equation (gain equal to zero) in the complex plane. As the gain is increased from zero, the closed-loop roots move away from the open-loop roots. The trace of these roots is called the root loci.

Example.   If

$$1 + GH = 1 + \frac{K}{(s+5)(s+3)}$$

Characteristic equation is

$$1 + GH = (s+5)(s+3) + K = 0$$
$$s^2 + s8 + 15 + K = 0$$
$$s = -4 \pm \sqrt{16 - (15 + K)}$$
$$K = 0 \qquad\qquad s = -3, -5$$

$$K = 1 \qquad s = -4 \pm j0$$
$$K = 2 \qquad s = -4 \pm \sqrt{-1} = -4 \pm j1$$
$$K = 10 \qquad s = -4 \pm \sqrt{-3} = -4 \pm j3$$
$$K = 101 \qquad s = -4 \pm \sqrt{-100} = -4 \pm j10$$

As gain is increased, the real part remains constant with increasing imaginary part. The system is always stable but becoming more oscillatory as gain is increased (Fig. 8).

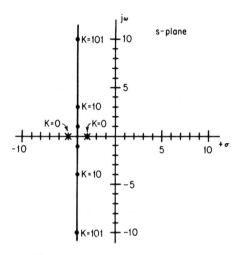

Fig. 8   Root-loci plot of stable system.

The preceding example of a simple second-order system showed that it was always stable. If the system contained an additional time constant, the system would become unstable at some finite value of system gain (Fig. 9).

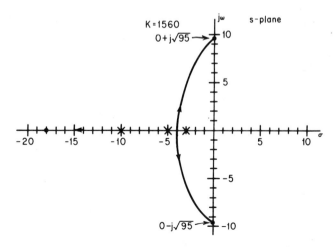

Fig. 9   Additional time constant causes loss of stability.

Example. If

$$1 + GH = (s + 5)(s + 3)(s + 10) + K = 0$$
$$s^3 + s^2 18 + s95 + 150 + K = 0$$

$K = 0$	roots $-3, -5, -10$
$K = 1560$	roots $-18, \pm j\sqrt{95}$
$K > 1560$	roots move to right-hand side (unstable)

For complex systems considerable effort is required in making root-locus plots. References 1 and 2 describe in detail graphical procedures which greatly simplify the effort required.

13. Frequency Response. The previous analysis dealt with the roots of the characteristic equation of closed-loop systems. These roots are in the time domain, the natural frequencies, or exponential time constants. There are techniques which deal with the open-loop function and develop criteria on the basis of system sinusoidal frequency response. These techniques may be illustrated by the following example of a system which is barely stable (Fig. 10).

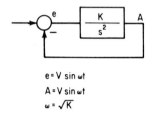

e = V sin ωt
A = V sin ωt
ω = √K

Fig. 10 Characteristics of an oscillatory open-loop system.

If disturbed, this system would oscillate continuously at its natural frequency with no further external excitation required. This behavior can be explained in terms of its steady-state sinusoidal frequency response.

$$\text{Point } A = V_m \sin \omega t$$
$$\text{Point } e = V_m \sin \omega t$$

The system function K/s^2 will amplify the error signal e by $k/(j\omega)^2$ or $-K/\omega^2$.

At $\omega^2 = K$ the gain is -1. The oscillation can sustain itself indefinitely. The open-loop function $GH = K/s^2 = -1$, unity gain with 180° phase shift at $\omega = \sqrt{K}$. This particular function at different gains will simply oscillate at different frequencies. In general, system gain will influence both the frequency and the relative stability of a system. Frequency analysis of the example given in Art. 12 demonstrates this point (Fig. 11).

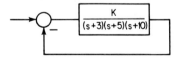

Fig. 11 Block diagram of example system.

To determine the open-loop frequency response, the gain and phase shift are calculated over the entire range of zero to infinity by substituting $s = j\omega$.

$$\frac{K}{(j\omega + 3)(j\omega + 5)(j\omega + 10)}$$

If $K = 1560$, the loop gain at $\omega = 9.78$.

$$GH = \frac{1560}{(j9.75 + 3)(j9.75 + 5)(j9.75 + 10)}$$

$$= \frac{1560}{(10.2 \ \llcorner 72.9°)(10.95 \ \llcorner 62.80)(14.0 \ \llcorner 44.40)}$$

$$= 1 \ \llcorner -180° = -1$$

With lower K, the system will reach unity gain with less than 180° of phase shift and be stable. With higher K, the system will have greater than 180° phase shift at unity gain, and the system will be unstable.

14. Nyquist Diagrams. The polar plot of the open-loop transfer functions frequency-response gain and phase shift, for values of $s = j\omega$ from $j0$ to $j\infty$ is a Nyquist diagram. Formal techniques for determining stability are given in references. For most common control systems which are stable, the plot will cross the unit circle between -90 and $-180°$ (Fig. 12).

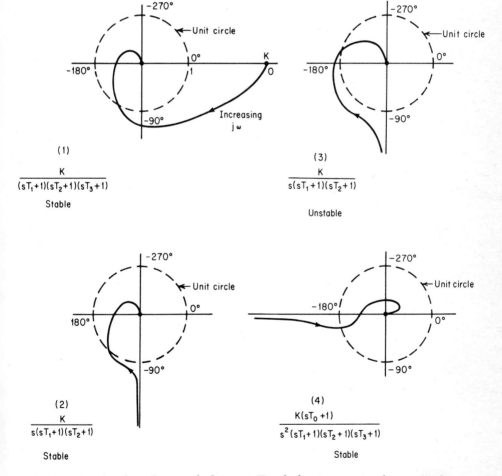

Fig. 12 Examples of open-loop transfer functions. Transfer functions are given for corresponding Nyquist diagrams.

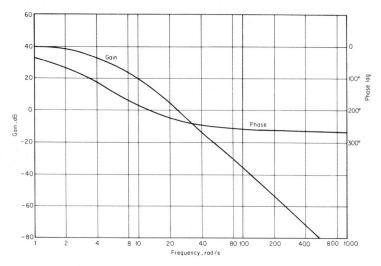

Fig. 13   Example of a Bode diagram where $GH = 15600/(s + 3)(s + 5)(s + 10) = 104/(s\,0.33 + 1)$ $(s\,0.2 + 1)(s\,0.1 + 1)$.

BODE DIAGRAMS

Bode diagrams plot the same information as the Nyquist plots but in a different form. Bode diagrams plot the logarithm of gain and the phase shift of open-loop transfer function as a function of frequency ω. The frequency ω axis is usually a log scale. In many practical cases the Bode diagram has advantages over the Nyquist diagrams. They are much easier to construct, and unless the gain function crosses unity gain or zero on the log GH plot numerous times, the system stability can be determined from the phase shift at crossover. In stable systems which cross over only once, the phase shift will be between 90 and 180°. The phase margin, defined as the difference from 180° at cross-over, is a measure of relative stability. As phase margin approaches zero, the damping ratio approaches zero and an unstable zone (Fig. 13).

Example

$$GH = \frac{15\,600}{(s + 3)\,(s + 5)\,(s + 10)} = \frac{104}{(s0.33 + 1)\,(s0.2 + 1)\,(s0.1 + 1)}$$

15. Plotting Techniques—Bode Diagrams.   With the appropriate choice of scales, plotting functions with simple lead and lag characteristics can easily be plotted. With the frequency axis on a log scale and the amplitude scale plotted in decibels, time constants can be approximated by straight-line segments, which have slopes equal to zero, plus 20 dB/decade, or minus 20 dB/decade (Fig. 14).

Higher-order functions such as a second-order polynomial will have zero slope at low frequencies and a slope of -40 dB/decade at high frequencies. At the point of the transfer function natural frequency corrections can easily be made (Fig. 15).

16. Phase-Angle Determination.   The phase angle of the total function at any frequency may be determined from the addition of all the individual transfer functions. At $j\omega = 1/T_1$, the phase shift of $1/(1 + sT_1)$ would be $-45°$. At a frequency equal to $j\omega = 2/T_1$, the phase shift would be 64°. A special scale can easily be constructed for a given log scale used for the frequency axis. By placing this scale on the log-magnitude diagram, one can read phase shift of each component in a relatively direct manner (Fig. 16).

17. Methods of Combining Functions.   The product of two functions is accomplished by adding the individual log-magnitude and phase angles of the two func-

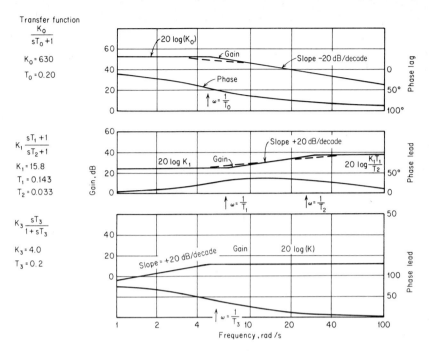

Fig. 14   Bode plot of typical first-order functions.

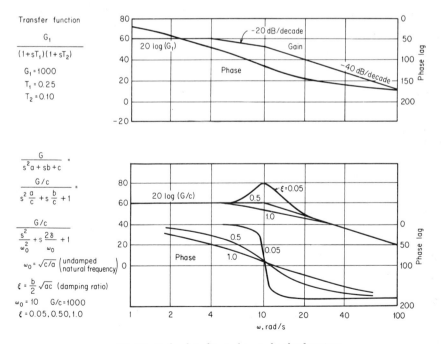

Fig. 15   Bode plot of typical second-order functions.

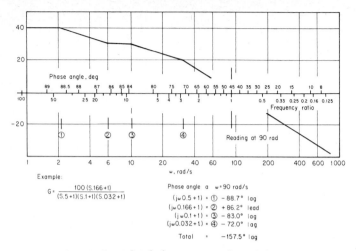

Fig. 16 Special scale for measuring phase shift.

tions. In evaluating compensation networks and in combining functions in multiloop systems, it is necessary to add more than one transfer function (Fig. 17).

DESIGN OF PRACTICAL SYSTEMS

The successful design of practical systems involves the proper selection and adjustment of components which, when combined, meet both the steady-state regulating accuracy and the dynamic-response and stability requirements. Analysis techniques are used to predict if given components will meet all requirements or conversely determine the specifications of alternate components which will meet them. Complex systems

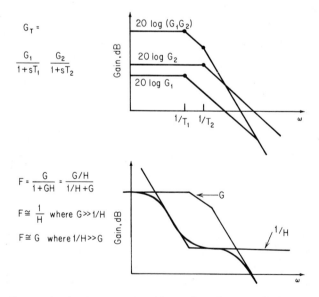

Fig. 17 Graphical construction of the product of two transfer functions.

having numerous feedback loops can be considered as an interrelated number of subregulators. Each subregulator must be designed for stable operation. Its closed-loop response then becomes a conponent of other control loops which make up the total system. The stability requirements of minor or subregulator loops in complex systems will be greater than that required in simpler systems. This difference occurs because a marginally stable minor loop will generally produce even more marginal or unstable total system behavior (Fig. 18).

18. Guidelines to design include the following items.

1. An ideal regulator would have one time constant T in its open-loop gain function GH. The closed-loop response would be exponential, with a time constant T divided by loop gain $(1 + K)$.

2. In a practical system which can have any number of delays, the open-loop function should cross the 1-dB (zero-gain) axis with a slope of -20 dB/decade. It would therefore appear to have only one large time constant and a number of relatively small time constants.

3. A system having an open-loop function with more than one large time constant will be inherently oscillatory and can easily be unstable. Low gain or suitable compensation must be incorporated in such systems to improve stability.

4. Pure transport time delay within a control loop is especially likely to produce stability problems. This occurs since it produces pure phase-angle lag directly proportional to frequency at all frequencies. Delay angle is equal to ωT_0, where T_0 is the delay period.

19. Simulation Design Study. The major purpose of a simulation study performed prior to final-component selection is to determine the specific requirements of all system components necessary to meet the overall system requirements. In instances where opposing requirements and system constraints require compromising certain control objectives, alternatives can be evaluated. The simulation model can also be used to determine which components most critically affect system performance. Working with the computer simulation, the designer should obtain an insight or feel of ex-

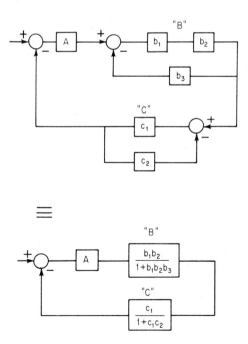

Fig. 18 System having three feedback loops.

actly how the system is supposed to operate when all components work in accordance with their design specifications. This insight can be the most significant benefit of the study, since it should enable the designer to maximize the performance of the real system and troubleshoot problems due to component malfunctions.

20. Computer Simulation.   Systems having many subregulators or nonlinear effects can more readily be analyzed through computer simulations. The accuracy of computer simulations is limited only by the fundamental knowledge and understanding of the individual system components. With a simulation it is possible to predict the behavior of systems before they are built or of existing systems under conditions too dangerous to test. The simulation models may consist of combinations of linear transfer functions, linear or nonlinear equations, or even empirically derived functional relationships between certain system variables.

21. Simulation Testing.   In simulation testing, real-system control-hardware components are tested in conjunction with a computer simulation which represents the remaining part of the system. In cases where simulation testing is used to check out control prior to shipment, the computer is usually used to represent the process and high-power components. The computer simulation must operate in real time and be connected to real hardware through suitable power-conversion equipment.

22. Experimental Modeling.   In certain instances, a portion of a system, such as a complex process already in operation, can best be modeled through experimental testing. This procedure requires that certain variables can readily be measured and key input variables can be suitably controlled. Under these conditions, step-response and frequency-response tests conducted over the entire range of interest will provide the data necessary to describe that portion of the system in a simulation study.

Simple systems may readily be characterized by simple transfer functions which match the empirically determined frequency response of the real system. Nonlinear systems will have characteristics which vary with operating point and/or amplitude of input. The linearity of a system should be tested by varying both the level and the amplitude of the disturbing function during the experiment.

TROUBLESHOOTING UNSTABLE SYSTEMS

The initial step taken to correct an existing problem is the identification of the exact cause of the problem. The following steps are suggested to distinguish if the problem is the result of an unstable control loop or of periodic disturbances. Tests should determine if the problem is poor design or a malfunctioning component.

1. Does the adjustment of loop gain directly change the frequency of system oscillation? Normally, in the case of system stability, it would change. If the problem is due to periodic disturbances, the frequency will remain the same.

2. Measurements using a multichannel oscillograph-type recorder should be taken, recording as many key variables as possible.

3. The data taken during oscillatory conditions from carefully calibrated recordings can be used to determine the gain and phase shift of the individual sections of the system at the exact frequency of concern.

4. These results can be compared directly with those individual gain and phase-shift values calculated using fundamental transfer-function considerations.

5. If all parts of the system correspond to the calculated values, the system is not malfunctioning but has, through poor design, produced poor response.

6. Areas where considerable discrepancies occur identify the source of the problem.

7. The isolated section should be thoroughly analyzed and tested further to identify if it is malfunctioning or possibly simply responding to external periodic disturbances.

8. For example (Fig. 19), the measured gain and phase shift may vary considerably from the expected design performance. Careful analysis may show that the measured phase shift cannot be attributed to the inherent phase shift unless that stage of the system has broken down. Where the measured phase shift simply exceeds the calculated value by a considerable amount (Fig. 20), the problem may prove to be a simple malfunction. Where a nearly full inversion (180°) shift occurs (Fig. 21), the problem may be an external periodic disturbance.

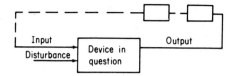

Fig. 19 System being analyzed.

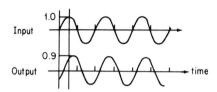

Measured: 45° lag — 0.90 gain
Calculated: 10° lag — 1.50 gain

Fig. 20 Case of excessive phase shift.

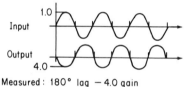

Measured : 180° lag — 4.0 gain
Calculated: 10° lag — 1.5 gain

Fig. 21 Case of external periodic disturbance.

23. Periodic Disturbances.   In instances where the process involves a mechanical or hydraulic drive system, periodic disturbances of a frequency near the crossover frequency of the control can produce sustained oscillations which appear to be control instability. Listed below are typical types of such systems and disturbances:

1. Mechanical-drive systems, i.e., metal-rolling mills
2. Gear backlash
3. Torsional natural frequency within the bandpass of control response
4. Rotational-component eccentricity

In such systems the frequency will be independant of regulator gain and often will vary in proportion to the rotational speed of the system.

REFERENCES

The following references are suggested for in-depth study and in practical applications where detailed analysis is needed.

1. Gordon J. Murphy, "Basic Automatic Control Theory," D. Van Nostrand Company, Inc., Princeton, N.J.
2. John G. Truxal, "Control Engineer's Handbook," McGraw-Hill Book Company, New York, 1958.
3. M.F. Gardner and J. L. Barnes, "Transients in Linear Systems," John Wiley & Sons, Inc., New York, 1942.
4. J. G. Truxal, "Automatic Feedback Control System Synthesis," McGraw-Hill Book Company, New York, 1955.

5. E. J. Routh, "Advanced Rigid Dynamics," The Macmillan Company, New York, 1882.
6. H. Nyquist, Regeneration Theory, *Bell Sys. Tech. J.*, vol. 2, pp. 126–147, January 1932.
7. H. W. Bode, "Network Analysis and Feedback Amplifier Design," D. Van Nostrand Company, Inc., Princeton, N. J., 1945.
8. G. A. Korn and T. M. Korn, "Electronic Analog and Hybrid Computers," 2d ed., McGraw-Hill Book Company, New York, 1957.

28

Automatic Control Systems

A. C. HALTER*

*Supervisor, Technical Services, East European Operations, Processing Group (Formerly Senior Staff Engineer, Electrical Systems Projects), Allis-Chalmers Corp.; Registered Professional Engineer (Wis.); Member, IEEE and AISE.

FOREWORD

Automatic control systems are now being recognized by all progressive industrial organizations for their contribution to improved plant efficiency. Control-system applications are extremely diversified and are rapidly expanding. Advantages for use of automatic control systems include:

Reduction in manpower
Greater production capacity
Increased production flexibility
Lower production costs
Improved quality control
Improved safety
Elimination of monotonous human operations
Elimination of operation under adverse conditions
Increased equipment utilization
Easier production control

In some applications there is no alternative to the use of automatic devices if the required performance is to be achieved. Control and guidance of high-speed missiles is an example of such an application. Human operators cannot perform the multiplicity of operations or provide the rapid response required. In other cases automatic control is used because it is the most effective way in which to achieve a given result.

This section provides a background on automatic control systems and their application.

GENERAL

1. Control Systems. Control systems are used in a wide variety of fields. Some representative applications of process-control systems are chemical plants, petroleum refineries, distilleries, and nuclear plants. Typical machines using automatic control systems include self-regulating rolling mills, numerically controlled milling machines, program-controlled lathes, automatic electronic assembly lines, automatic inspection and quality-control devices, packaging and bottling machines, and materials handling.

Before various system types are described, some definitions are given to assist in providing a common vocabulary.

Process. A process is an operation or series of operations which lead to the end result, by the addition or removal of energy.

Plant. The facilities or equipment in which the process is carried out.

Process Variables. Those variable conditions which can affect the maintenance of balance within the process. The variable which has the greatest effect upon the conditions to be controlled is generally the one selected for manipulation by the controller.

System. A system is a combination of components which act together and perform a certain function.

Disturbances. A disturbance is a signal which tends to affect the value of the output of the system. These disturbances may be either internal or external depending upon whether they are generated from within the system or from outside the system.

Feedback Control. Feedback control is an operation in which the output of the system is compared with a reference input or an arbitrarily varied desired state, and this difference is used to change the output to reduce the effect of the disturbances. In general, where possible, compensation is made for any known or predictable disturbances within the system, so that only the unpredictable disturbances will adversely affect the output.

Feedback Control System. A feedback control system is one which tends to maintain a prescribed relationship between reference input and system output, by comparing these two signals and using the difference as a means of control.

Servomechanism.   A servomechanism is one type of feedback control system in which the output is some mechanical position, velocity, or acceleration. Servomechanisms are used extensively in modern industry. For example, the positioning of rolls on a rolling mill or the completely automatic operation of machine tools from a set of programmed instructions may be accomplished through servomechanisms.

Automatic Regulating Systems.   An automatic regulating system is a feedback control system in which the reference input or desired output is either constant or changing very slowly with time, and in which it is desired to maintain the output at some desired value in the presence of disturbances.

Process-Control System.   A process-control system is an automatic regulating system in which the output is a variable such as temperature, pressure, flow, or liquid level. Programmed controls in which an output variable is controlled according to a preset program are often used in such systems.

The control system may vary in complexity from one in which it may be desired to hold one certain parameter at a fixed level, i.e., the speed of a drive motor, up to one that requires many functions to be performed in a unique sequence with each function controlled to a specific value or position, such as in an automatic machine tool for contouring special shapes. In a system required to hold or control a process to some desired value, there are two changes to be considered: the changes due to disturbances, and the change resulting from the regulating or corrective devices.

OPEN-LOOP AND CLOSED-LOOP SYSTEMS

Before basic control actions in automatic systems are considered, the closed-loop vs. the open-loop control system should be examined.

2. Open-Loop Control System.   In an open-loop control system, the output has no effect upon the control action; that is, the output is not compared against the input. A control system which operates strictly on an input-command vs. elapsed-time basis is an open-loop system. For example, an automatic washing machine in which the cycles are changed by a motor (clock)-driven selector switch is an example of an open-loop system.

The diagram in Fig. 1 shows that the controlled output variable c varies with the input reference r of the control system and the gain G of the system. An open-loop control system can be used to give good output control only if the relationship between input and output is well known and there are no large disturbances to the system.

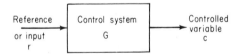

Fig. 1   Open-loop block diagram.

3. Closed-Loop Control System.   A closed-loop control system is one in which the change in output signal is used to produce a direct effect upon the control action.

A closed-loop system is also called a feedback control system.

Figure 2 shows that the input e to the forward circuit G is the difference between the

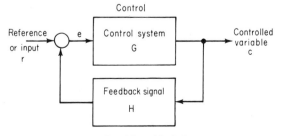

Fig. 2   Closed-loop block diagram.

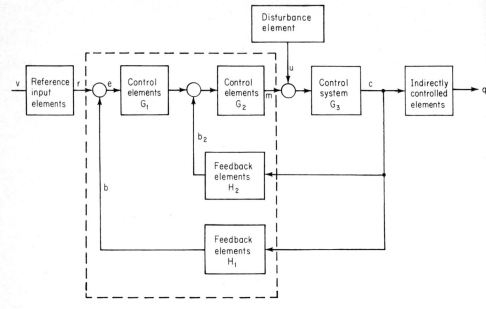

Fig. 3 Feedback-system block diagram.

reference input signal and the feedback signal. This self-correcting action tends to reduce the error and bring the output level of the system closer to a desired value.

4. Definition of Terms. The descriptive terms for the variable quantities in the block diagram (Fig. 3) are as follows:

v is the *desired value* and is the value that is desired from the control system. This value may or may not differ from the reference value depending upon the reference input elements.

r is the *reference* and is the actual reference input to the control system.

e is the *error* and is the difference between the reference r and the primary feedback signal b. It is the controller-actuating signal.

m, the *manipulated variable*, is the output of the control elements and manipulates the controlled system.

u, the *disturbance*, is the unwanted input or upset to the system that affects the controlled variable. The disturbance may enter the control system at any of a number of locations.

c, the *controlled variable*, is the output of the controlled system and is the basis of comparison with the referenced input.

q, the *indirectly controlled quantity*, is the resultant quantity that the control system actually produces. This value may or may not differ from the controlled variable, depending upon the characteristic of the indirectly controlled system elements.

b, the *feedback*, is derived from the controlled variable c but may differ from it because of the feedback-element characteristic. Occasionally additional feedback signals b_2 from the controlled variable or some other intermediate variable are used to improve accuracy or speed of response of the system.

Those parts of the system which are in effect the controller are shown enclosed by the dashed line in Fig. 3. Generally the conversion of the measured quantities is assumed, with the input reference and the feedback signals being expressed in the same units. The control system can then be represented by a simplified block diagram (Fig. 4).

5. Comparison of Open-Loop and Closed-Loop Regulation. The transfer function for an open-loop system (Fig. 1) is

$$c = rG \qquad (1)$$

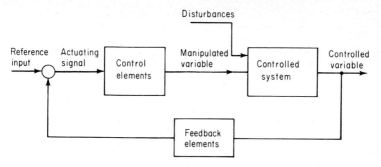

Fig. 4 Simplified feedback-control-system block diagram.

or

$$r = \frac{c}{G} \qquad (2)$$

Since the transfer function of the control elements G may be a function of the frequency of input as well as the parameters of the system including the gain, it is evident that there could be a change in controlled variable even if there is no change of the magnitude of the reference r. The effect of a change in gain ΔG can be expressed by the equation

$$c + \Delta c = r\,(G + \Delta G) \qquad (3)$$

or

$$c + \Delta c = rG + r\,\Delta G \qquad (4)$$

When the original conditions in Eq. (1) are subtracted from Eq. (4), the remainder is

$$\Delta c = r\,\Delta G \qquad (5)$$

and by substituting c/G for r from Eq. (2),

$$\Delta c = \frac{c}{G}\,\Delta G \qquad (6)$$

which represents the change in output due to a change of controller gain ΔG with no change of reference r. Thus the change in the controlled variable is directly proportional to the change in the transfer function. Therefore, to hold the controlled quantity within some accuracy band, the variation of the transfer function alone cannot exceed that allowable accuracy, even if the drift of reference and disturbances to the system are negligible.

In a closed-loop transfer function in which the input is the difference between the reference r and the feedback Hc (Fig. 2) the following equations apply:

$$c = G(r - Hc) \qquad (7)$$

or

$$c + GHc = Gr \qquad (8)$$

or

$$c = \frac{Gr}{1 + GH} \qquad (9)$$

To determine the effect of a change in gain G, this quantity will be added to Eq. (9) as was done for the open-loop functions.

$$c + \Delta c = \frac{(G + \Delta G)r}{1 + (G + \Delta G)\,H} \qquad (10)$$

$$c + cGH + c\,\Delta G\,H + \Delta c + \Delta c\,GH + \Delta c\,\Delta G\,H = Gr + \Delta Gr \qquad (11)$$

Subtracting Eq. (8), $c + cGH = Gr$ gives

$$c\,\Delta GH + \Delta c + \Delta c\,GH + \Delta c\,\Delta G\,H = \Delta Gr \qquad (12)$$

the term $\Delta c \, \Delta G \, H$, being the product of two extremely small quantities, can be dropped, and combining the rest of the terms gives

$$\Delta c \, (1 + GH) = \Delta G \, (r - Hc) \tag{13}$$

but from Eq. (7), $r - Hc = c/G$. Therefore,

$$\Delta c = \left(\frac{c}{G} \, \Delta G\right) \frac{1}{1 + GH} \tag{14}$$

Comparing this expression with that of the open loop, closed loop Δc = open loop $\Delta c \times 1/(1 + GH)$, which indicates that with feedback, the effect of changes of the forward transfer functions is greatly reduced from that without feedback. Assume, for example, that the gain of the feedback loop is unity ($H = 1$) and the controller forward gain is a nominal 20 ($G = 20$), a change of G with no change in reference value will result in a $1/(1 + 20)$ or just under a 5% change of the value of c from that value which it would be in the case of the open-loop control. In other words, for the above example instead of a 10% change of controlled variable (open loop) with a 10% forward-loop-gain change, the closed-loop output would change less than 0.5% for the same change in forward-loop gain.

However, it should be noted that a change in the transfer function of H gives an almost proportional change in the output variable [Eq. (13)]. In other words, the feedback system is more sensitive to variation of the feedback circuit than it is to variation of the forward circuit.

BASIC CONTROL ACTIONS

Most industrial controllers are electrically, pneumatically, or hydraulically driven. This section is limited primarily to those controllers which use electricity as the power source.

An automatic controller operates by comparing the output with the desired value, determining the magnitude and sign of the deviation, and then producing a control signal which is in a direction tending to reduce this deviation to zero. The manner in which the control signal is produced is called the control action.

Industrial automatic controllers can be classified according to their control action as follows:

1. Two-position or on-off controller
2. Proportional controller
3. Integral controller
4. Proportional plus integral controller
5. Proportional plus derivative controller
6. Proportional plus derivative plus integral controller

In an automatic controller there must be a means of measuring or detecting the difference between the desired or reference level and the actual controlled variable. This difference is the actuating error signal and is usually of low power level. Since the actuating device or manipulated variable generally requires considerable power, an amplifier is required. Figure 4 shows the essential elements along with a measuring element of the controlled variable. This measuring element is a device which converts the output variable such as speed, displacement, pressure, or electrical signals into another suitable variable which can then be compared with the input reference.

The actuator is an element or elements in the system which uses the amplifier output so as to cause the feedback signal to be brought into correspondence with the reference input signal.

6. Two-Position Control Action. A two-position control system usually has only two fixed positions, simply on and off. It does not recognize rates or magnitudes of deviation beyond the switching point but reacts only at fixed magnitudes of the controlled variable. This type of control is relatively simple and inexpensive, and is widely used for industrial and domestic control. The simple thermostat is an example of this type of control. When the temperature is below a certain point, a contact is closed to operate a contactor or open a valve. and when the temperature is above the desired point, the contact is opened (Fig. 5a). A differential is often used in conjunction with the two-position

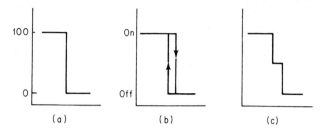

Fig. 5 On-off controller action.

controller to lessen the wear on the control mechanism. The differential may be adjustable, and is generally a small percent of full-scale operation (Fig. 5*b*).

The three-position control is a refinement of the two-position control in that it provides an additional step. For example, there may be a neutral point where the input power is held at some intermediate level, or a control valve will be set to a partially open position, when the controlled variable is within a band between maximum and minimum operating points. When the controlled variable moves out of the middle position, the control action causes an increase or decrease in input power (Fig. 5*c*).

The "on-off" controllers are also called discontinuous controllers and are used mainly because they are simple, inexpensive, light, and compact. They do, however, have a number of disadvantages, one of which is that they seldom give smooth control performances because of the abrupt action of the manipulated variable. Switching of large blocks of energy from the power source may also be a disadvantage.

7. Proportional Controller. The proportional, also called a throttling or modulating mode, is a type of control action in which the output of the controller has a linear relationship with the actuating error signal. It may be that for large error signals, saturation of the amplifier will occur, but at and around the regulating or balance point, the output will be proportional to the error input. The relationship for the unsaturated level can be expressed by equation as $m(t) = Kp\ e(t)$, where m is the output of the controller, e is the error signal, and Kp is called the proportional sensitivity, or the gain. For flexibility, an adjustment of gain is usually provided. A block diagram of a proportional controller is shown in Fig. 6.

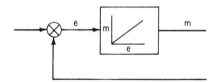

Fig. 6 Proportional-controller action.

In the above expression, the values are true only for static (steady-state) conditions. Since practically all amplifiers have time constants, the dynamic performance must be considered. This procedure is essential especially when analyzing performance of a large, multielement control system. When the dynamic response is considered, the above equation can be written in Laplace-transform quantities as $M(s) = Kp\ E(s)$. Both the steady-state gain and the time response to a step input are sometimes shown in the block diagram to indicate the kind of amplifier used. For example, an amplifier with a linear output vs. input and having a single time-constant lag can be shown in block-diagram form (Fig. 7).

8. Integral Controller. In an integral controller, the output value changes at a rate proportional to the value of the actuating signal. When the error signal is zero, the output level will remain stationary at some level. A negative error will cause output to move in one direction while a positive error will drive the output in the opposite direc-

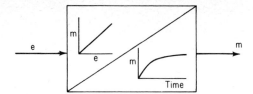

Fig. 7 Proportional gain and step-input response.

tion. This type of control action is also known as reset control. It is used where the large disturbances with relatively slow time rate are present. Without integrating action in the controller of such systems, large gains, with resulting stability problems, would be necessary to hold controlled variables within the desired tolerances.

The relationship for this controller can be expressed by the equation $m(t) = Kp \int_a^b e(t)dt$, or the transfer function $M(s) = (Kp/s) E(s)$, where Kp is an adjustable constant or gain and is equal to the inverse of the integral time $Kp = 1/Ti$. The relationship of e and m with time is shown in Fig. 8.

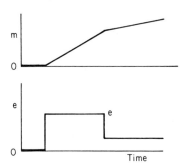

Fig. 8 Integral-controller action.

9. Proportional plus Integral. If a fast accurate control for small rapidly changing perturbations is desired while the entire system has a large slowly changing drift, a proportional plus integral controller is used. The equation for this type of controller is $m(t) = Kp\ e(t) + (Kp/T_i) \int_u^t e(t)dt$ or the transfer function $M(s) = Kp\ (1 + 1/T_i s)\ E(s)$, where Kp represents the gain and T_i represents the integral time in minutes. Changing of T_i changes only the integrating time, whereas a change of Kp changes both the gain and the time of integration. The inverse of T_i is the number of times per minute that the proportional part is duplicated, and is called the reset rate. Figure 9 shows e and m as a function of time.

10. Proportional plus Derivative. In a system which requires a control action to follow accurately a changing reference, as in position control, a controller in which the magnitude of the output is also proportional to the rate of change of error signal is desirable. Such a control action is provided by a proportional plus derivative controller and is defined by the following equation:

$$m(t) = Kp\ e\ (t) + Kp\ T_d\ \frac{de(t)}{dt}$$

or transfer function $M(s) = Kp\ (1 + T_d\ s)\ E\ (s)$, where Kp represents the gain (proportional sensitivity) and T_d represents the derivative time.

The relationship of e and m with time is shown in Fig. 10.

Derivative control action has an anticipatory character, but it can never be used alone because this action is effective only during transient periods. One of the disadvantages

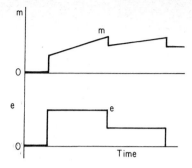

Fig. 9   Proportional plus integral controller action.

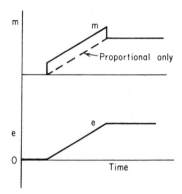

Fig. 10   Proportional plus derivative controller action.

of derivative control is that it amplifies noise signals and may cause saturation of control elements and/or actuator.

11. Proportional plus Derivative plus Integral.   It is possible to combine proportional control action, derivative control action, and integral control action into one control system.   The equation for this combination is

$$M(t) = Kp\ e(t) + KpT_d s\ \frac{de(t)}{dt} + \frac{Kp}{T_i} \int_a^b e(t)dt$$

or   Transfer function $M(s) = Kp \left(1 + T_d s + \dfrac{1}{T_i s} \right) E(s)$

The relationship of e and m with time is shown in Fig. 11.

CONTROL-SYSTEM RESPONSE AND APPLICATION

The fundamental rules which govern the action of a controlled system can be determined from the nature of the lags in the system.   These lags are process lag, measuring lag, and controller lag.   One of the previously mentioned controller modes is generally more useful than another in producing the desired kind of control.   The choice of the mode selected should be based upon the capabilities and limitations of each mode, and in general the choice of the least complicated type which will perform the function should be selected to keep the entire system as simple as possible.   Following are some advantages and disadvantages of each mode of control.

12. Two-Position Controller Response.   Because of its simplicity, the two-position controller is very popular.   Its control action is essentially cyclic, and in most cases a

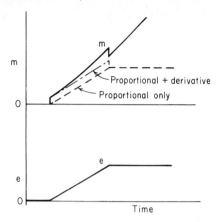

Fig. 11   Proportional plus derivative plus integral controller action.

controller with a differential gap is used. The period of cycling is inversely proportional to the process reaction rate and directly proportional to the magnitude of the controller differential gap.

A slow process rate with its large period is generally desirable because it causes less wear on the controlling mechanism. System dead time, however, causes overshoot beyond the controller differential gap. To minimize cyclic amplitude, the dead time should be small and the amount of input should not be much larger than that necessary to handle the greatest load. Also large-process capacities allow smaller amplitudes to be obtained.

In general, two-position control is most satisfactory when:
1. Transfer lag and dead time are negligible.
2. Process reaction rate is slow.
3. Measuring and controller lags are small.

13. Proportional-Controller Response.   Whenever a load change occurs in a proportional control system, the variable takes a permanent deviation from the control point. This deviation is called offset and the lower the gain of the controller the greater the offset becomes. Offset arises from the nature of proportional controllers, because when a load change occurs which requires a greater supply of energy to the system, an increased deviation or input is necessary to produce it. Higher gains in the controller result in faster corrective action and closer limits of control, but increase stability problems. Large-process transfer lags and dead times require lower controller gains with resulting increased offset for load changes.

Proportional control may be applied to a system when:
1. Large, rapid load changes are not present.
2. The transfer lag and dead time are not too great.
3. The process reaction rate is slow.

14. Integral-Controller Response.   In the integral controller the output of the final control element is changed at a rate dependent upon the magnitude of deviation. Actually this type of controller recognizes not only magnitude of deviation as in the proportional controller, but also the elapsed time of the deviation. An integrating rate too fast for the system results in excessive cycling before stabilizing, while a rate that is too slow produces overdamped or sluggish control action. Self-regulation is required in the process system to achieve stability. A characteristic of integral control is that regardless of the steady-state load, the output variable will, given sufficient time at steady state, always return to the control point, which is in effect zero offset. However, this type of controller does not immediately correct for suddenly changing process loads. In general, as the process reaction rate increases, large gain (faster rate of integration) is more effective. Any appreciable transfer lag or dead time produces an increasing period and requires a lower integrating controller gain. Although this mode may be used by itself, it is most often combined with the proportional mode where it is known as reset response.

The integrating controller provides the best results when:
1. Transfer and dead time are small.
2. Process range is large.
3. The process has self-regulation.
4. Measuring lag is small.
5. Controller lag is small.

15. Proportional plus Integral Response.   The proportional plus integral (also called proportional-reset) controller combines two of the modes already described to obtain the advantages of each. They are the proportional mode with its inherent stability, and the integral mode with its stabilization at zero offset from the control point. This type of control is the most generally useful of all types of control. The relative amount of each control response is generally made separately adjustable.

This mode of control may usually be applied to any system having characteristics suitable for the application of either mode separately.   But, in addition, there is no requirement for self-regulation as is the case with integral control action alone.   The proportional response is enough to provide the necessary stability.   Transfer lag is not a limitation as long as the measuring means is sensitive to small changes.   The limitation of applying proportional integral control lies in the large period of the controlled system and the slow response, when dead time is present.

If the proportional mode provides the major part of the controller action, the reset response is used to accomplish a slow movement of output level to eliminate offset of the controlled variable.   The effect of proportional response is generally more stabilizing than the effect of reset response.

If the control process has small capacity, but no transfer lag or dead time, the principal action is usually provided by the reset response.   Improvement in control of such a system is generally realized even with low-gain proportional response owing to its stabilizing effect.

16. Proportional plus Derivative and Proportional plus Integral plus Derivative Response.   Processes having large transfer lag, or large dead time, may prove difficult to control at times.   Large lags of these kinds practically preclude the choice of any mode of control except proportional-integral.   The proportional gain must be low and the reset rate slow to avoid continuous cycling.   Then, when load changes occur, excessive deviation results and a longer time is required to recover.

The application of derivative response to the proportional-integral controller substantially reduces the maximum deviation.   The most notable improvement is produced in the period of cyling, which is greatly reduced in time.   The proportional-integral adjustments of a controller generally require substantial change of settings when derivative response is added.

Derivative response is effective when transfer lag exists, but is more generally applicable for combating dead time, since it is the only means for doing so.   The adjustment of rate time for derivative response must be made carefully, since a large rate time can cause excessive oscillations.   When the rate time is adjusted to a low value, cycling may also result, but it is due to the integral response which had previously been satisfactory with higher derivative response.   The period of cycling is a means of recognizing the cause; i.e., a cycle caused by integral response is long, by proportional response is short, and by derivative response is even shorter.

Listed below are the basic control actions and the principal characteristics of a controlled system for which each is most suitable.   It should be noted that this comparison is useful only as a broad classification of controlled systems.

1. Two-position control is best suited for systems having no more than two energy-storage elements and very small dead time.
2. Proportional control is good in systems where changes in load variables are small, or where all the system time constants except one are small.
3. Integral control by itself is suitable only for systems with very few energy-storage elements in the process, and no dead time.   Even small dead times make this type uncontrollable.
4. Proportional plus integral has limitations in that the stabilizing time is excessive

when the process has many energy-storage elements. Dead time in the system has adverse affects.

5. Proportional plus integral plus derivative is the most effective control for systems with many energy-storage elements and/or dead time. This type could be used for every system; however, for economy and for ease of installation and maintenance the simplest mode which provides adequate control is the most desirable.

CONTROL-SYSTEM PERFORMANCE

17. Steady-State Performance of Feedback Control Systems. To evaluate how well a feedback control system performs or can be expected to perform under steady-state conditions, two criteria are used, namely, the operating deviation band and the service deviation band. These values are normally expressed in percentages of the maximum rated value of the controlled or specified variable, except where the system has no readily definable base, in which case it is expressed in absolute numbers.

The operating deviation band is the total excursion of the directly controlled variable as a result of specified operating variables under steady-state conditions and within a specified range of these operating variables (Fig. 12). The operating variable is a specified variable, other than those arising from service conditions and drift, for which the feedback control system must correct in attempting to maintain the ideal value of the directly controlled variable. The application of or the change of load is an example of an operating variable.

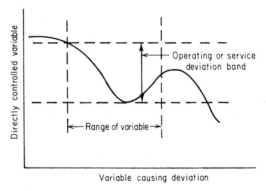

Fig. 12   Controller deviation vs. disturbance.

The service deviation is the total excursion of the directly controlled variable as a result of drift and changes in service conditions within specified limits. NEMA ICS 3-106A.11 defines the limits as those occurring any time during a 1-h interval following a warm-up period of at least 30 min. Variation of the ambient temperature within the limits of applicable service conditions shall be limited to a range of 15°C, and variations of supply voltage shall be limited to a range of 10% of rated voltage. The operating variables shall be held constant for determining service deviation band.

The above two performance factors are exclusive of changes in reference. When required by the application, the variations due to reference deviations should also be considered. The expected performance must take into account all three factors described above. However, it is not necessarily the sum total of all these factors unless they all happen to produce errors in the same direction and at the same time, which is unlikely.

18. Transient Performance of Feedback Systems. For many systems the static performance may be within limits, but the system is completely unacceptable because of its poor transient response. Figure 13 shows, for example, a typical effect of a step change in load, on the regulated variable. Of particular importance are the maximum transient deviation, the response time to recover to a specified value, and the settling

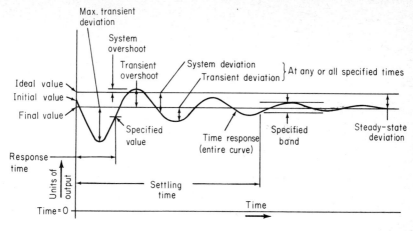

Fig. 13 Response following step increase in load.

time to recover to within a specified band. A typical response curve to a step change of reference is shown in Fig. 14. Here, the important parameters are the response time, the system overshoot, and again the settling time.

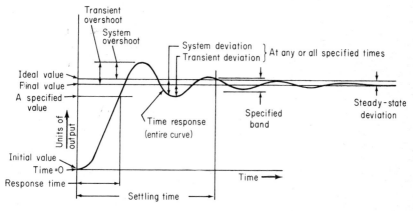

Fig. 14 Response following step change of reference input.

BASIC TYPES OF FEEDBACK SYSTEMS

Feedback control systems are classified into three general types according to their response characteristics. These groups are called Type O, Type I, or Type II systems.

19. Type O System. A system in which the controlled variable requires a constant error signal under steady-state conditions is generally referred to as a Type O system. A system of this type is used to maintain the controlled variable at some desired level despite disturbance conditions. It is not designed for or applied where it is desired to have the controlled variable follow a changing input with small amount of error. A Type O system is generally a regulator system (Fig. 15a). The open-loop transfer function of a Type O system has the general form

$$G(s) = \frac{K(1 + a_1 s + a_2 s^2 + \cdots)}{1 + b_1 s + b_2 s^2 + \cdots}$$

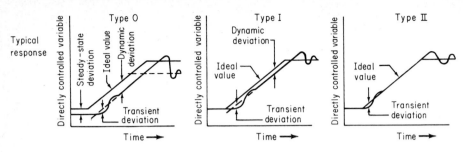

Fig. 15 Typical response characteristics of Types O, I, and II systems.

20. Type I System. A system in which a constant error signal produces a constant rate of change of the controlled variable is also known as a Type I system. This type of system is generally referred to as a servomechanism system, and because of its ability to maintain any constant value of the controlled variable with no steady-state error it is sometimes referred to as a velocity-type servomechanism system. Its use, however, is not restricted to servomechanisms. A characteristic of this type of system is that it has an integrator in its forward loop. An electric motor driving a device or load to some preselected position is an example of a Type I system. As long as the feedback and reference are not balanced, an error will exist, driving the motor until the error is reduced to zero. This system is sometimes referred to as a "zero displacement–error system." The transfer function of a Type I system has the same general form of the Type O, with the exception of an additional 1/s factor. The general open-loop transfer function is

$$G(s) = \frac{K(1 + a_1 s + a_2 s^2 + \cdots)}{s \, (1 + b_1 s + b_2 s^2 + \cdots)}$$

Figure 15b shows a typical performance curve of a Type I system.

21. Type II System. A system in which a constant error signal produces a constant acceleration of the controlled variable is called a Type II system. These systems are generally referred to as a servomechanism system. A system of this type is capable of maintaining any value of constant variable with no error as well as maintaining a constant controlled speed with no error. Systems of this type are sometimes referred to as "zero-velocity error" systems. An example of a system of this type is shearing specified lengths from a moving bar. In such a system, the knife blade must match the bar speed, as well as position itself correctly over the target point of the bar.

To accomplish this operation, two integrators are required in the forward loop, one to set the velocity or rate of change of the controlled variable and the second to indicate the position of the controlled variable. Since with zero error input, the velocity of the controlled variable will be maintained as set by output of the first integrator, it is possible for the load speed to duplicate the input speed without error. A system of this type, however, has an inherently difficult stability problem. The transfer function of a Type II system is again the same as a Type I except for the additional 1/s factor. The general form is

$$G(s) = \frac{K(1 + a_1 s + a_2 s^2 + \cdots)}{s^2 (1 + b_1 s + b_2 s^2 + \cdots)}$$

22. Feedforward Systems. The nature of feedback is such that there must be a measurable error to generate a restoring force; hence perfect control is unobtainable. A feedback controller solves the problem by trial and error, as it does not know what the output should be for any given set of conditions. If disturbances occur at intervals of less than three natural time periods of the system, steady state will never be reached.

For such systems there is a way to solve this problem by using "feedforward control." Here the principal factors affecting the process are used to compute the correct controller output to meet current conditions. Then, when a disturbance occurs, corrective action is begun immediately to cancel it out before it affects the controlled variable (Fig.

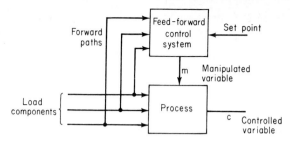

Fig. 16 Feedforward-loop block diagram.

16). With this type of control it is theoretically possible to achieve perfect control despite the difficulty of the process. Its performance is limited only by the accuracy of the measurements and computations.

In the figure three forward loops are shown to suggest that all the components of the load which may affect the controlled variable are used in solving for the manipulated variable. To be complete, the feedforward control system must be programmed to maintain the process balance in the steady state and also in transient intervals. It must consist of both steady-state and dynamic components like the process. In effect it is a model of the process.

DIGITAL SYSTEMS VS. ANALOG SYSTEMS

Systems in which the sensing elements, amplifiers, and output devices process signals of a continuously varying value over their entire range of operation are called analog control systems. These elements may or may not be linear throughout their range. Some elements may saturate during transient periods, but to provide good regulation and control they must not be in or near saturation level in the normal operating range under static conditions.

A variable-voltage dc motor drive system with a dc tachometer generator as the feedback element, an operational amplifier as the error sensor and preamplifier, and a dc generator or thyristor controller as a power amplifier is an example of an analog control system.

A digital system as contrasted to an analog system is one in which an interconnected set of elements is used to process discrete, finite-valued signals. Relay or static switching circuits, wired digital controls, and digital computers are examples of digital systems.

23. Digital Controls. There are two basic types of digital circuits, namely, (1) the combinational circuit where all the outputs at any particular time t are determined by the primary and feedback inputs at that same time t, or (2) the sequential circuit in which the output values at any given time depend not only upon the present inputs but also upon inputs applied previously. For example, the position of a rotating device can be determined by reading a position encoder (combinational circuitry) or by counting and storing pulse deviations from a zero or known position (sequential circuitry).

In addition, either of the above circuits may be synchronous or asynchronous, depending upon whether an external clock is or is not used respectively. For an asynchronous circuit to function properly, the following conditions must be met:

1. Sufficient time between input changes must be allowed to permit the circuit to settle down.

2. Successive inputs of sequential circuits must be adjacent; i.e., they must not differ by more than exactly one bit.

3. The application of any legitimate input to the circuit at any present stable level state will cause it to enter some new stable state.

4. If two or more unstable feedback lines change at a given time, the stable state eventually reached must be the same independent of the order in which the feedback lines became stable.

Because analog devices are generally subject to drift or gain changes due to aging, environmental noise, or other factors, high-accuracy systems are difficult to obtain. Digital devices, because of their on-off characteristics, do not drift and will maintain constant gain. They can therefore be designed into systems to give much higher resolution and accuracy than analog systems. However, the feature of the infinitely variable range of output of the analog device is desirable as an output controller. By combining the two systems, using the digital for measuring and comparison and the analog for power amplification and output power, the desirable qualities of each can be realized in a system.

24. Analog-to-Digital Converters (Encoders). To combine digital and analog elements into the same system requires the use of analog-to-digital and digital-to-analog converters. For example, a pulse tachometer or a coded disc attached to the output device can be used to convert analog position into digital values. When starting from a known position or when it is desired to know deviation from some arbitrary position, a pulse generator and counters can be used to track position. For bidirectional travel dual-channel pulse tachometers are used to determine the direction of count. For absolute-position indication, encoders are required.

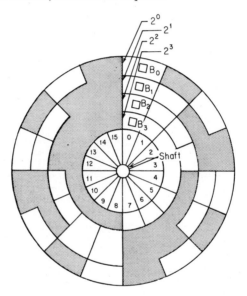

Fig. 17   Binary-coding disc.

Figure 17 is a diagram of a binary-coding disc which converts one shaft revolution into 16 parts. For this coding disc, four brushes along a fixed radius are required. The shaded areas are electrically conducting and are all connected to a common point, in this case the shaft, while the white areas are made of nonconducting material. As the shaft and disc rotate in the counterclockwise direction, the brushes will make contact in the order shown in Table 1.

Examination of this table shows that many times the transition between consecutive position requires that two or more brushes must transfer simultaneously. For example, between positions 7 and 8 all four brushes must change their status. If B_2 opened before B_0 and B_1, and before B_3 closed, a signal of position 3 between positions 7 and 8 would be obtained. This condition can occur if the brushes are lined up imperfectly or if the code pattern of the disc is not exact. Two techniques have been devised to avoid these errors.

In one technique, two or three brushes are used for each channel or ring of disc. The brushes are arranged such that when one brush is located over a transition point on the

TABLE 1. Binary-Coding-Disc Contact Order

Disc-position sector No.	Brushes making contact				Binary No.
0	—	—	—	—	0000
1				B_0	0001
2			B_1		0010
3			B_1,	B_0	0011
4		B_2			0100
5		B_2,		B_0	0101
6		B_2,	B_1		0110
7		B_2,	B_1,	B_0	0111
8	B_3				1000
9	B_3,			B_0	1001
10	B_3,		B_1		1010
11	B_3,		B_1,	B_0	1011
12	B_3,	B_2			1100
13	B_3,	B_2,		B_0	1101
14	B_3,	B_2,	B_1		1110
15	B_3,	B_2,	B_1,	B_0	1111

disc between conducting and nonconducting regions, a mating brush is located completely within the conducting or nonconducting area. This arrangement, in addition to requiring more brushes, also requires additional intelligence and circuitry to select the brush which is not directly over a transition.

The other technique for preventing transition errors is to use a code or disc in which only one digit channel is changing mode at each successive position. The Gray code (also called reflected or cyclic code) will accomplish this feature. This code eliminates reading errors even if brushes are slightly misaligned, or if the code pattern is not perfectly accurate.

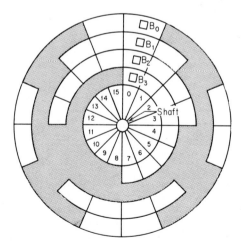

Fig. 18   Cyclic-coding disc.

Figure 18 shows a four-channel cyclic-coded disc, and Table 2 shows the equivalent cyclic code for decimal numbers of a five-channel code. In the cyclic code the sign of each of the successive 1s alternates with the most significant 1 having a positive value.

TABLE 2. Cyclic Code and Its Decimal and Binary Equivalents

Decimal No.	Binary No.	Cyclic code	Decimal No.	Binary No.	Cyclic code
	16 8 4 2 1	31 15 7 3 1		16 8 4 2 1	31 15 7 3 1
0	00000	00000	16	10000	11000
1	00001	00001	17	10001	11001
2	00010	00011	18	10010	11011
3	00011	00010	19	10011	11010
4	00100	00110	20	10100	11110
5	00101	00111	21	10101	11111
6	00110	00101	22	10110	11101
7	00111	00100	23	10111	11100
8	01000	01100	24	11000	10100
9	01001	01101	25	11001	10101
10	01010	01111	26	11010	10111
11	01011	01110	27	11011	10110
12	01100	01010	28	11100	10010
13	01101	01011	29	11101	10011
14	01110	01001	30	11110	10001
15	01111	01000	31	11111	10000

Example.

$$\overset{\text{Decimal}}{\overbrace{9}} = \overset{\text{Binary}}{\overbrace{8+1}} = \overset{\text{Cyclic}}{+\overbrace{15-7+1}}$$

Since most digital equipment uses the binary form, a converter must be used to change the cyclic code to binary code.

Coding discs using brushes are available in reasonable diameter with as many as 1024 sectors (10 binary digits). Greater resolution using standard coded discs can be obtained by gearing two encoders together through suitable gear ratio, or by using optically read discs which have much greater resolution potential. For example, if two 4-digit discs are geared together through a gear-reduction unit of 16/1, the total count covered can be $2^4 \times 2^4$ or 2^8 (256 bits). Any number of additional discs can be added through appropriate gearing to increase the total count. The resolution, however, is always determined by the amount of physical motion represented by one count of the fast disc. For the optical disc the code pattern is represented by opaque and transparent areas. A light source confined to a narrow radial line on one side of the disc matches up with photoelectric devices along the same radial line on the other side of the disc. Those photocells which receive light conduct while the others do not. Photoelectric discs of reasonable diameter have been made with an accuracy of 1 part in 65 536 (16 binary digits).

If the desired quantity to be sensed is in the form of a voltage and it is desired to express this quantity in digital form, voltage coders can be used. In general, voltage coders are not simple transducers and frequently consist of several pieces of equipment. One example in use is based upon measuring the time interval it takes to bring a linear sweep voltage to the same level as the voltage to be digitized. The time interval is measured by counting a fixed-frequency pulse source. Another type of voltage coder selects new digital values and then decodes them and compares them against a desired value. The next digital value is then increased or decreased by steps depending upon the results of the last comparison. When the comparator indicates equal voltage values, the digital signal is transmitted. The voltage comparisons are done very rapidly, and an entire conversion can be completed in a few microseconds. A typical accuracy figure for a practical system is 0.1%.

25. Digital-to-Analog Converters (Decoders). To convert numbers into a corresponding voltage amplitude, a decoding device or network is required. A method of decoding in common use is to switch electrical sources into and out of a network in accordance with the number to be decoded. Generally speaking, the accuracy characteristics of all networks are similar and are governed by the precision of the resistors used,

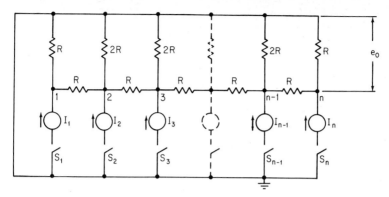

Fig. 19   Current-source decoder.

their behavior with environmental changes, and the precision of the voltage or current sources used. All the networks require storage of the number to be decoded. The major differences between the various network decoders are in the type of hardware best suited for each, and the applicability of some to a variety of digital codes. Decoders can be made using current sources, voltage sources, or weighted resistors which are switched in and out of the networks.

Figure 19 shows the configuration for decoding binary numbers with current sources. For this type of decoder to produce accurate output voltage, the current sources should not interact one with another. Since ideal infinite internal impedance is not available, and effects of aging must be considered, compensation is required.

A typical voltage-source decoder for binary numbers is shown in Fig. 20.

Each voltage source E_1 to E_n (with output value E) sees a total load resistance of $3R/2$, E_n having $^3/_4 R + ^1/_4 R + ^1/_2 R$ while the rest have a resistance R in series with two parallel circuits each having an equivalent total resistance of R also. The voltage at any node n, $n-1$, etc., due to its voltage source alone is $E/3$. Then at each interior node to the right of this point, the voltage is attenuated by 2 so that the voltage at $n-1$ node due to the kth voltage source is $(E/3) \times (1/2^{n-1-k})$. At the nth node (between the $^1/_4 R$ and $^3/_4 R$) the output voltage is further reduced to $^3/_4$ of $n-1$ node by the ratio of these resistors, so that the output voltage to an infinite-impedance load, from a single kth source, is $(E/4) \times 1/2^{n-1-k}$. For example, if there are 5 voltage sources ($n=5$) of 120 V each, the output voltage due to the first source (E_1 of Fig. 20) will be $(120/4) \times (1/2^{5-1-1}) = 30 \times (1/2^3) = 3.75$ V while for the fifth or nth source it is $(120/4) \times (1/2^{5-1-5}) = 30 \times (1/2^{-1}) = 60$ V. When more than one source is energized, the total output is the summation of the individual "energized" sources, i.e.,

$$e_0 = {}^1\!/_4\left(\frac{E_1}{2^{n-2}} + \frac{E_2}{2^{n-3}} + \cdots + \frac{E_n}{2^{-1}}\right)$$

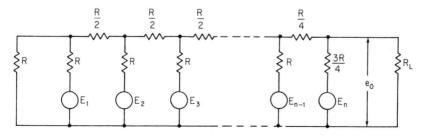

Fig. 20   Voltage-source decoder.

For a finite load resistance R_L the output voltage is

$$e_0 = \frac{R_L}{R_L + R_0} \times \frac{1}{4} \left(\frac{E_1}{2^{n-2}} + \frac{E_2}{2^{n-3}} + \cdots + \frac{E_n}{2^{-1}} \right)$$

where the factor R_0 is the internal impedance of the decoder as seen by the load and is $R_0 = 3/8\ R$.

Figure 21 shows a network used to decode a number by means of weighted resistors. In this scheme the relay K is operated when the corresponding digit is 1.

The equivalent circuit of this network is shown in Fig. 22.

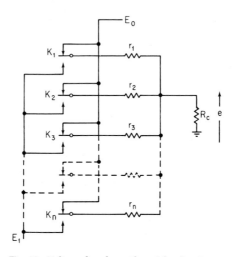

Fig. 21 Voltage decoder with weighted resistors.

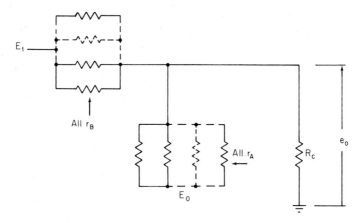

Fig. 22 Equivalent circuit of voltage decoder with weighted resistors.

E_0 and E_1 may both be separate voltage sources, in which case the output voltage would be made up of two parts, one from E_0 and one from E_1. If, however, E_0 is at ground potential, all the resistors which are connected by the deenergized relays can be lumped together with the load resistor R_c. If B represents all parallel resistors of the 1's digits and A represents all parallel resistors from zero digits and the load resistance, $e_0 = E_1[A/(A + B)]$.

For example, if $R_c = 10\ 000\ \Omega$, $r_1 = 1000\ \Omega$, $r_2 = 2000\ \Omega$, $r_3 = 4000\ \Omega$, and $r_4\ (r_n) = 8000\ \Omega$, then with relays K_1 and K_3 energized, $r_A = 10\ 000$, 8000, and $2000\ \Omega$ in parallel or $1379.3\ \Omega$ and $r_B = 1000$ and $4000\ \Omega$ in parallel or $800\ \Omega$. The output then is $e_0 = (1379.3/2179.3)\ E_1$ or $0.633\ E_1$.

Another type of decoder using weighted resistance is shown in Fig. 23. The resistors in this arrangement are connected in a voltage-divider circuit with a fixed voltage impressed across the input to the network.

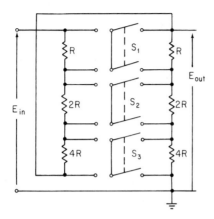

Fig. 23   Three-digit decoder in voltage-divider configuration.

If S_2 is closed to E_1 and S_1 and S_3 are closed to E_0, the voltage will be divided by the ratio $2R/(2R + 5R)$ or $E_0 = {}^2/_7\ E_1$.

A third decoder network using weighted resistors is shown in Fig. 24. Again, the relay K_n for a corresponding digit is energized by a 1 and opens its contact to insert its associated resistor in the amplifier feedback circuit. Since the output of an amplifier with infinite impedance is $E_{out} = -E_{in}R_F/R_{in}$, the output will vary directly as the resistance in the feedback circuit.

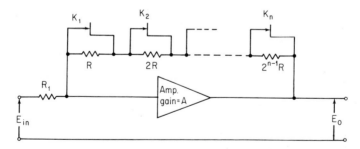

Fig. 24   Operation-amplifier decoder with weighted resistors.

26. Accuracy, Linearity, and Resolution.   Three important parameters of "digital-analog" converters are accuracy, linearity, and resolution. Absolute accuracy cannot really be stated as any given number because it depends on many factors. It is rarely mentioned over a component's entire operating range and even more rarely with time. Where a converter does not have internal references, an absolute-accuracy specification is meaningless, since the user supplies the external reference.

Relative accuracy and/or linearity is also known as quantization uncertainty. This factor is the uncertainty which is inherent because of the representation of an analog func-

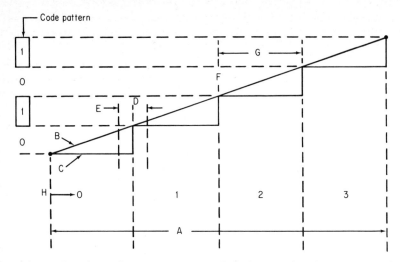

Fig. 25 Zero reference at code transition. *A*, range of an infinite number of analog points to be encoded; *B*, analog function; *C*, digital representation of the analog function; *D*, theoretical code transition point (transition error); *E*, deviation range of count transition (transition deviation); *F*, theoretical analog point represented by a digital 2; *G*, quantizing deviation of 1 quantum; *H*, numerical values of quanta.

tion by a digital value. Maximum uncertainty is either 1 quantum or $\pm \frac{1}{2}$ quantum depending upon the zero reference (Figs. 25 and 26).

Resolution in an analog-to-digital converter is the smallest input analog level for which a digital output code is produced. In a digital-to-analog converter resolution is the smallest input digital code for which an analog output level is produced. For example, an 8-bit converter has $1/2^8$ resolution or 1 part in 256 or about 0.4%.

27. Computer-Controlled System. Because of the increasing demands for greater output, lower costs, and improved quality of products, industry has expanded in both

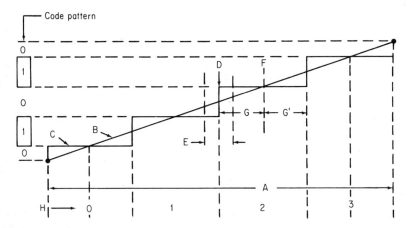

Fig. 26 Zero reference at center of code transition. *A*, range of an infinite number of analog points to be encoded; *B*, analog function; *C*, digital representation of the analog function; *D*, theoretical code transition point (transition error); *E*, deviation range of count transition (transition deviation); *F*, theoretical analog point represented by a digital 2; *G* and *G'*, quantizing deviations of plus and minus $\frac{1}{2}$ quantum; *H*, numerical values of quanta.

size and complexity and adopted faster and more intricate processes. As the scale of operations becomes faster and more complex, the human operator becomes an increasing limitation to the proper coordination of the necessary interrelationships. To overcome this limitation, industry is incorporating within the control loop modern computers capable of handling high-speed computations and complex mathematical relationships. Some of the ways in which a computer can be utilized to improve system controllability and performance are in the following applications.

1. Feedforward compensation for load and disturbance variations

2. Feedback control of performance indexes computed from measurements of process variables

3. Adaptive control which can compensate for large changes in system parameters

4. Optimizing control which maximizes a specific performance criterion

5. Learning system whereby the computer uses its past learned experiences to perform optimizing control

The first two types of control are similar to the open-loop and closed-loop control, respectively, which has already been described, except that a computer is used as the basic regulator.

An adaptive control system can be defined as one which continuously and automatically measures the dynamic characteristics (transfer functions) of the plant or process, compares them with the desired dynamic characteristics, and used the difference to vary adjustable system parameters or to generate an actuating signal so that the optimal performance can be maintained regardless of the environmental changes. To be a truly adaptive system, self-organizing features must exist. If the environmental changes are directly measured and used to make adjustments of the system, it is not an adaptive control system. Figure 27 shows a block-diagram representation of an adaptive control system. The performance index may be measured continuously or periodically, and then is compared with the optimal. Based upon this comparison, a decision as to how to modify the actuating signal is made. This control is in effect a closed-loop operation.

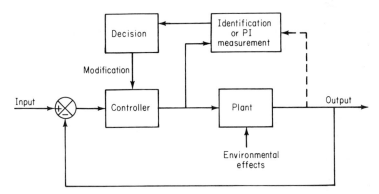

Fig. 27   Adaptive-control-system block diagram.

Identification of the dynamic characteristics of a plant may be made from normal operating data of the plant or by the use of test signals of small amplitude. The identification by the use of normal inputs is possible only when the signal has adequate bandwidth and amplitude characteristics. When test signals are used with the plant in normal operation, the signals used should not unduly disturb normal outputs, and furthermore the normal inputs and noise should not disturb and confuse the test.

The major contribution of computer control lies in the area of optimization. The objective here is to maximize (or minimize) a specified performance criterion for the control system subject to both disturbing and constraining influences. Basically, trial-and-error procedures are used to find the optimal operating point. In the steepest-descent approach (also called the derivative-sensing method) the performance is evaluated by noting the effect of small changes in the variable parameters and then moving this parameter in the direction of maximum slope. The amount of movement may be either a fixed amount or an amount as determined by the gradient of the performance-index surface.

For slowly changing parameters, the gradient may be calculated relatively infrequently, whereas for rapid parameter changes a procedure known as alternate biasing is preferred to the derivative-sensing method. When using alternate biasing the system is alternately operated at a fixed distance on either side of the calculated optimum.

The optimization approach is useful for the physical process whose performance surface has a single optimum, and where the variations are slow enough for the control system to handle them. Its greatest advantage is that it puts no restrictions on the plant as to linearity, multiple-input-output, time-varying, etc.

A control system which relies on past learned experience in behaving in an optimal manner is called a learning system. Such a system when subjected to new situations learns how to behave by adaptive approach, but if the system should experience the same situations again, it will recognize them and behave optimally without going through the adaptive approach. Once the system has learned the optimal control correction for each possible situation, it may operate near the optimal condition regardless of the disturbance.

REFERENCES

1. Donald P. Eckman, "Automatic Process Control," John Wiley & Sons, Inc., New York, 1958.
2. Harold Chestnut and Robert Mayer, "Servomechanism and Regulating System Design," John Wiley & Sons, Inc., New York, 1951.
3. W. Ahrendt and John Taplin, "Automatic Feedback Control," McGraw-Hill Book Company, New York, 1951.
4. Grabbe, Ramo, and Wooldridge, "Handbook of Automatic Computer and Control," John Wiley & Sons, Inc., New York, 1961.
5. A. J. Young, "An Introduction to Process Control System Design," Spottiswoode, Ballantine and Co., Ltd., London, 1955.
6. F. G. Shinskey, "Process Control Systems," McGraw-Hill Book Company, New York, 1967.
7. A. K. Susskind, "Notes on Analog-Digital Conversion Techniques," The Technology Press of the Massachusetts Institute of Technology, Cambridge, Mass., and John Wiley & Sons, Inc., New York, 1957.
8. "Ordnance Engineering Design Handbook," sec. 2, ORDP 20-137 Ordnance Corps Department of Army.
9. "Industrial Control and Systems," NEMA Standard Publication ICS-1970.

29

Small-Specialty-Motor Controls

RICHARD R. ANNIS*

FOREWORD

Small specialty motors—universal, shaded pole, permanent magnet, small synchronous, impulse—are in the fractional and subfractional horsepower sizes. In most cases they are designed for a specific business machine, military unit, toy, appliance, portable

*Assistant Vice President—Engineering, Oster Corporation; Registered Professional Engineer (Wis.); Member, WSPE, NSPE, SAM, IEEE.

29-1

tool, or automation component of a larger machine. These motors are commonly designed into the product. In most applications the duty cycle is intermittent, and large amounts of power may be required for short periods of time. Consumer-product use of these motors necessitates a relatively low cost motor. They are designed for mass production.

Controls for these small motors are also produced in large quantities at low cost and must be reliable and provide long trouble-free service.

In some cases control devices may be available as off-the-shelf packages, but in many cases the control is incorporated into the overall design of the product.

This section covers the control of various types of small specialty motors, and the characteristics of the various control types and devices.

GENERAL

Small specialty motors are of the fractional- and subfractional-horsepower class such as the universal, shaded-pole, synchronous, magnetic-vibrating, permanent-magnet, stepping, and impulse types (Ref. 1). In many cases, these motors are designed into the product. Generally the product housing provides the motor-bearing and brush-holder supports for commutator-type motors. Typical examples are the portable home appliances—can openers, blenders, ice crushers, knives, mixers, hair clippers, meat grinders, knife sharpeners, portable electric tools, drills, saws, sanders, garden appliances, hedge trimmers, grass trimmers, lawn mowers, and business machines—typewriters, time-card-imprinting clocks, bookkeeping machines, copiers, collators, desktop and stand-up computers, vending machines, major-appliance auxiliary drives, and windshield-wiper, heater, window-lift, and seat-control motors for automobiles.

The controls for this motor type, such as speed controls, timers, switches, and overload protectors, are also generally incorporated into the product design. The various speed-control techniques will be covered for each motor type. The various switches and motor protectors used on these motors will also be covered in this section.

CONTROLS FOR UNIVERSAL MOTORS

A review of the general equation for motor speed and the equivalent circuit (Fig. 1) shows that the speed of a universal motor can be varied by changing the impedance of the field or by inserting additional resistance or inductance in series with the armature and field.

1. Series-Resistor Type. An example of a series-resistance-type speed control is a foot-pedal control used to vary the speed of a sewing machine. The disadvantage of this method of controlling motor speed is the watts loss resulting from the line current flowing through the series resistance. Figure 2 shows typical speed-torque curves for a universal motor using series-resistance control.

2. Tapped Field Winding. Referring to the expression for motor speed in Fig. 1, it can be seen that motor speed can be varied by varying the impedance of the motor field by tapping the windings. Figure 3 shows a schematic of a typical tapped universal-motor field. Connecting the motor armature to fewer field turns shown by connecting to point 2 decreases the circuit impedance because resistance and inductance of the field are decreased. The line current increases but not as much as the decrease in impedance. The motor flux decreases because of the reduction in ampere turns. The increase in armature-generated voltage and the decrease in motor field result in increased motor speed. If the field is tapped such that more turns are added to the circuit, the reverse effects result and the motor speed decreases.

Figure 4 shows a schematic of a tapped multispeed salient-pole universal motor with a split-wound field with maximum distribution of the coils between both poles for optimum motor cooling. Figure 5 shows typical speed-torque curves for a universal motor utilizing a tapped field.

3. Diode and Variable Resistor across Motor Brushes. Figure 6 shows a schematic diagram of a universal motor with rectifier and variable series resistor placed across the motor brushes. By shunting the motor armature, additional line current flows through the field winding and increases the motor field strength, and lowers the armature-

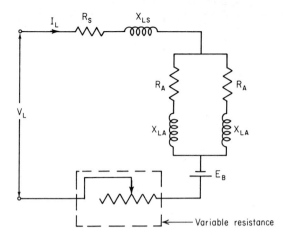

Fig. 1 Equivalent circuit for a universal motor.

$$N = K\frac{E_B}{BA} = K\frac{V_L - I_{LZ}}{BA}$$

where N = motor speed, r/min, for a design where armature diameter, length, number of parallel armature circuits, and flux distribution factor are constant

K = constant
E_B = armature back EMF
B = motor field strength, lines/in²
A = total number of armature inductors
V_L = line voltage, V
I_L = line current, A
Z = motor impedance, $\Omega = (R_S + R_{AT}) + j(X_{LS} + X_{LAT})$
R_S = resistance of stator, Ω
R_{AT} = total armature resistance, Ω
R_A = resistance in half of the armature, Ω
j = operator = $\sqrt{-1}$
X_{LS} = inductive reactance of stator = $2\pi F \times LS$, Ω
LS = stator inductance, H
X_{LA} = inductive reactance in half of the armature, Ω
X_{LAT} = total armature inductance
F = frequency, Hz

generated voltage, resulting in a reduction in motor speed. By varying the resistance, the speed is increased or decreased. This method of speed control results in greater incremental speed changes per unit change in resistance than a series resistance only. The disadvantage of this arrangement is the requirement for two additional components, a resistor and a rectifier. The arrangement also produces additional losses due to the I^2R in the resistor. Figure 7 shows the speed-torque curves for a universal motor where a variable resistor and rectifier placed across the armature, as shown in Fig. 6, control the speed.

4. Diode with Tapped Field. Figure 8 shows a diode that can be placed in series with a tapped field winding. By inserting the diode in series with the motor, the ac is rectified and half-wave current results. It has the same effect as reducing the line voltage by 30%. Switching the diode into the circuit adds eight more motor speeds. Figure 9 shows speed-torque curves for a tapped field in a universal motor using a series diode to obtain more speed-control points.

5. Shifting Motor Brushes. Shifting the brushes on a universal motor away from the designed neutral position results in a change in motor speed. Two factors affect the motor speed. As the brushes are shifted, the field produced by the armature conductors strengthens the main field produced by the field winding when shifted in the direction of rotation and weakens the main field when rotated opposite to the direction of rotation. The general equation (Fig. 1) shows that strengthening the field will lower the

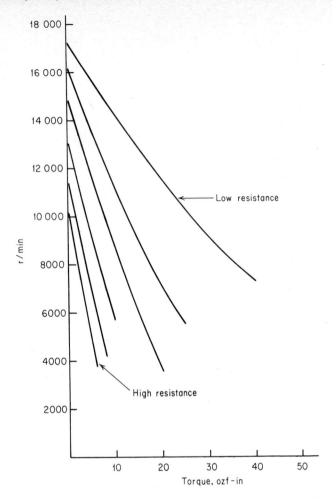

Fig. 2   Speed-torque curves for universal motor using series-resistance speed control.

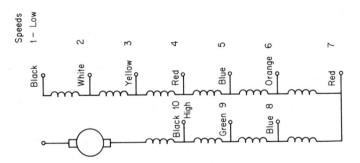

Fig. 3   Schematic for tapped universal-motor field.

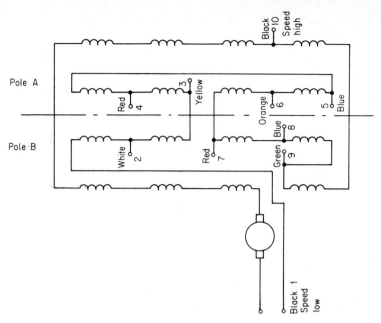

Fig. 4 Schematic of universal-motor field using balanced windings.

speed and weakening the field will increase the speed. Figure 10 shows the various fields produced in a salient-pole uncompensated universal motor and the vectorial summation of the stator and armature fields with different armature brush shift. The other factor that affects the motor speed is the negative torque produced in the armature conductors as the brushes are shifted. Figure 11 shows the torque developed by the armature conductors with different brush-shift positions. After the brushes have been shifted approximately 90°, the resultant torque produced is zero and the armature does not rotate. A further shift of the brushes produces a reversal in the motor torque, and the motor runs in the opposite direction and will increase in speed until the brushes are rotated approximately 180°. Shifting the brushes in a counterclockwise direction results in an increase in motor speed because the decrease in motor field is a larger factor than the decrease in torque owing to the armature current flowing in the wrong direction in some of the conductors. The change in speed per degree of brush shift will be less when the brushes are shifted opposite to the direction of rotation because the negative torque developed reduces the increase in speed caused by the weakening of the field. The line current increases more when the brushes are shifted opposite to the direction of rotation because of the reduction in armature inductive reactance.

Figure 12 shows a plot of motor speed and line current vs. brush shift for a universal motor running under a light load. Shifting the brushes results in increased generated voltages in the coils undergoing commutation and produces poor commutation (excessive sparking). The commutation is poorer when the brushes are shifted in the direction of rotation because the stronger field increases the generated voltage.

Because of the problems of commutation and the difficulty of adapting a design to shift the brushes by means of push buttons, this method of controlling universal-motor speed has only limited use.

6. Flyball Governor. Figure 13 shows a typical flyball-governor speed control. In a universal motor, speed can also be varied by use of an electromechanical governor. With low-speed operation the governor is not in the circuit in this motor design. A centrifugal governor is used to cause the opening and closing of contacts which place a series resistance in the motor circuit. The addition of the series resistance to the motor

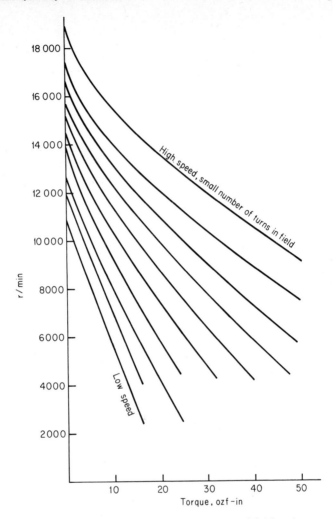

Fig. 5 Speed-torque curves for universal motor using a tapped field as shown in Fig. 3.

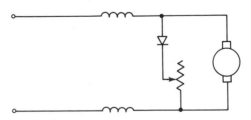

Fig. 6 Controlling universal-motor speed by use of a rectifier and armature-shunting resistor.

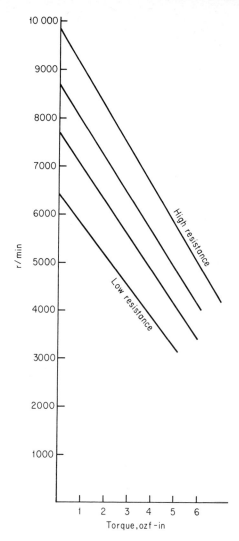

Fig. 7   Speed-torque curve of universal motor using a rectifier and variable resistor across the armature.

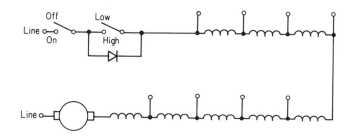

Fig. 8   Controlling universal-motor speed with diode in series with the field.

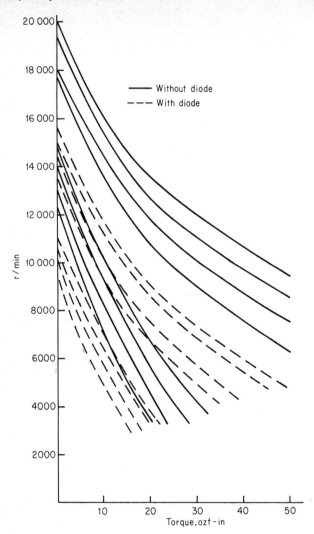

Fig. 9 Speed-torque curves for Fig. 8 using a diode to obtain more speeds.

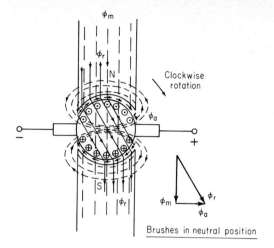

Brushes in neutral position

Brushes shifted counterclockwise

Results in a weaker field and
negative torque

ϕ_m = flux due to current in field winding

ϕ_a = flux due to current in armature winding

ϕ_r = resultant flux

Brushes shifted clockwise

Results in a stronger field
and negative torque

Fig. 10   Magnetic fields in a salient-pole uncompensated universal motor with the brushes shifted clockwise and counterclockwise.

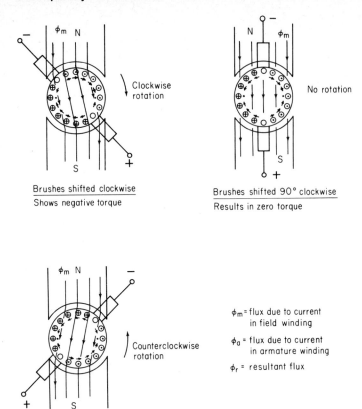

Brushes shifted clockwise
Shows negative torque

Brushes shifted 90° clockwise
Results in zero torque

Brushes shifted beyond 90° clockwise

Results in reverse direction of
rotation and increase in motor speed

ϕ_m = flux due to current
in field winding

ϕ_a = flux due to current
in armature winding

ϕ_r = resultant flux

Fig. 11   Motor torque developed with the brushes shifted in the direction of rotation.

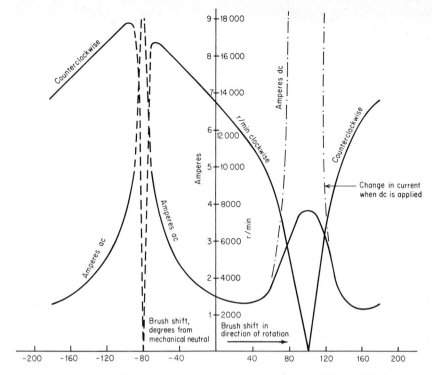

Fig. 12 Motor speed and line current in a universal motor as a function of brush shift when run on alternating current.

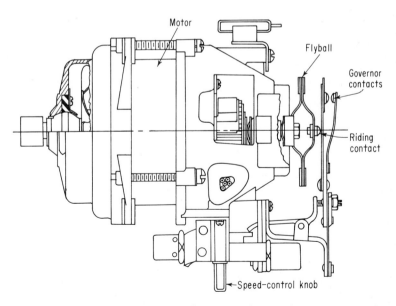

Fig. 13 Flyball-governor speed control for a universal motor. (*Oster Corporation.*)

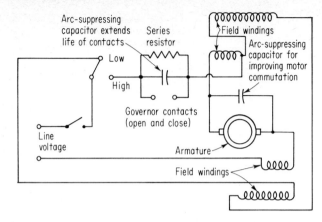

Fig. 14 Schematic for flyball-governor-controlled universal motor.

impedance creates a drop in voltage and results in a decrease in motor speed (Fig. 14). As the motor speed drops, the governor contacts close, taking the resistance out of the circuit and causing the motor to speed up again. This method results in a slight motor hunting at a very high rate.

The speed is varied in a governor-controlled universal motor by varying the contact pressure between the two governor contacts. In the design in Fig. 13, maximum speed is obtained by adjusting the control knob in a maximum clockwise position. The cammed interlock pushes the contacts together and requires maximum centrifugal force by the flyballs to cause the riding contact to move inward and the governor contacts to open. Figure 15 shows the curves of a universal motor utilizing an electromechanical type of governor. The speed can be varied with this type of control and also can be held constant over a range of load.

7. Solid-State Controls. With the advent of the thyristor silicon controlled rectifier (SCR), a wide variety of speed controls for universal motors have evolved. Silicon controlled rectifiers have the ability to control large amounts of power. These rectifiers can be triggered to conduct current through different phases of the ac cycle.

a. Half-Wave Speed Control without Counter EMF Feedback. The simplest speed control for a universal motor consists of a SCR, triggering device and potentiometer, capacitor, and diode for adjusting the conducting point of the SCR. Figure 16 shows a schematic of a typical half-wave solid-state speed-control circuit without feedback for a universal motor.

The resistor R varies the time for capacitor C to build up to a voltage level to fire the SCR. The length of time the SCR fires can be increased by decreasing the potentiometer resistance setting. Figure 17 shows a plot of the line voltage, capacitor voltage, SCR gate voltage, SCR voltage anode to cathode, and motor voltage with the control resistor R set for minimum current conduction resulting in low-speed operation. Current flows when the SCR fires and voltage appears across the motor.

Figure 18 shows plots of the same voltages with the control resistor R set for a lower resistance, allowing the capacitor to charge faster and trigger the SCR sooner, providing 90° conduction and a higher speed.

Figure 19 shows a high-speed setting with about 160° conduction.

Figure 20 shows a plot of typical speed-torque curves for a half-wave control without feedback for use with a universal motor with three different control resistor settings.

b. Half-Wave Speed Control with Counter EMF Feedback. Figure 21 shows a half-wave solid-state universal-motor control circuit with counter EMF feedback from the armature. The trigger voltage for the SCR is equal to the difference between the voltage across the armature and the voltage across the voltage-divider network. Initially the generated voltage will be low as the motor starts and the SCR can be triggered early in the cycle. As the motor gains speed, the back EMF increases, decreasing the

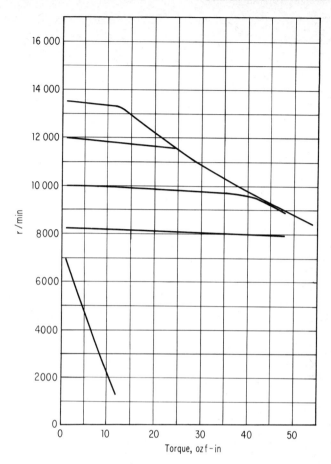

Fig. 15 Speed-torque curve for a universal motor using a flyball electromechanical-type governor.

gate voltage, and thus the SCR will fire later, decreasing the motor current. The net effect of putting the motor in the gate circuit, resulting in feedback, is to flatten the motor speed-torque curves.

Figure 22 shows a plot of the speed-torque curves with feedback. The curves for medium- and low-speed settings are much flatter than those of Fig. 20.

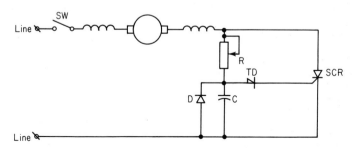

Fig. 16 Schematic of a half-wave solid-state speed control without feedback for a universal motor.

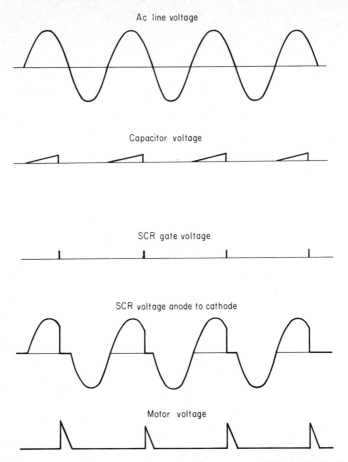

Fig. 17 Half-wave universal-motor speed control without counter EMF feedback—plot of line, capacitor, SCR gate, and anode-to-cathode and motor voltages with resistor *R* set for minimum current conduction.

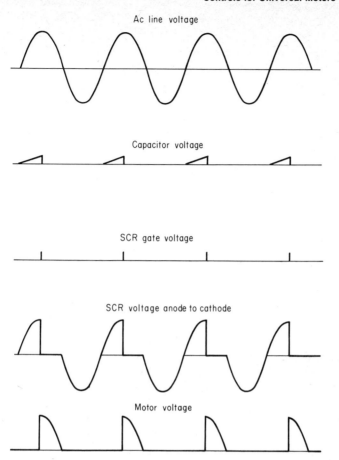

Fig. 18   Plots of voltages with R for a lower resistance and providing 90° conduction and a higher speed.

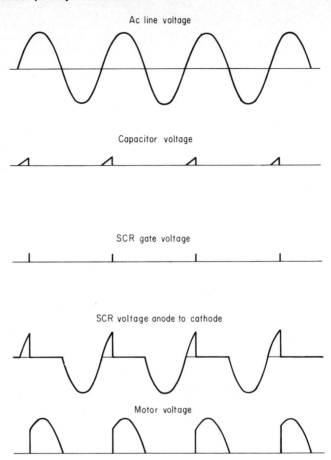

Fig. 19   Plot of voltages for a high-speed setting with 160° conduction.

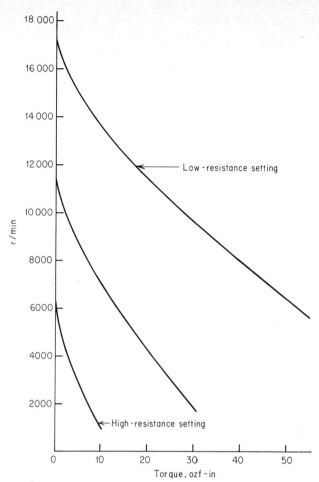

Fig. 20 Speed-torque-motor curves for a half-wave control without feedback for use with a universal motor with three different control resistor settings.

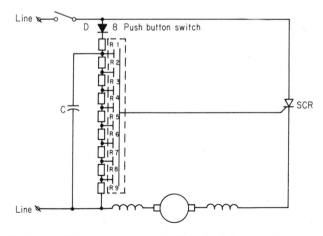

Fig. 21 Half-wave solid-state control circuit with feedback for use with a universal motor.

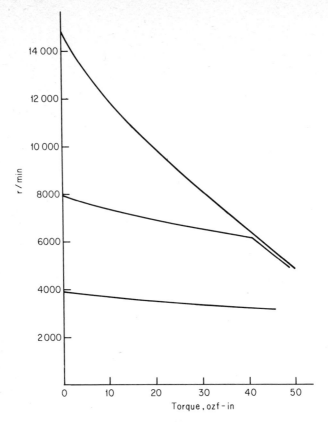

Fig. 22 Speed-torque curves for half-wave control circuit with feedback for use with a universal motor.

c. Full-Wave Universal-Motor Control without Feedback. Figure 23 shows a full-wave control. Capacitor *C* is charged. When the triggering voltage for the trigger diode *TD* is reached, the capacitor discharges, firing the Triac (two SCRs in parallel allowing conduction in both halves of the cycle). Figure 24 shows a plot of line voltage, capacitor voltage, Triac gate voltage, Triac voltage anode to cathode, and motor voltage for a low-speed setting.

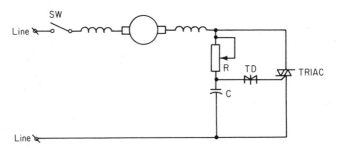

Fig. 23 Full-wave control circuit without feedback for a universal motor.

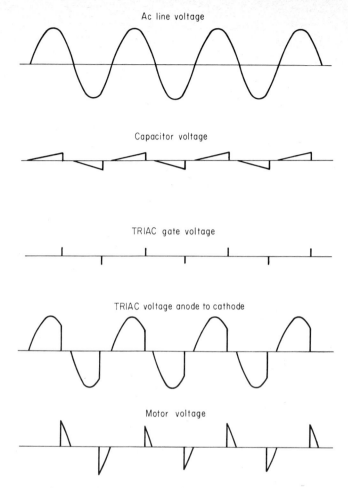

Ac line voltage

Capacitor voltage

TRIAC gate voltage

TRIAC voltage anode to cathode

Motor voltage

Fig. 24   Plot of voltages for full-wave control without feedback for a universal motor for a low-speed setting.

Figure 25 shows the plot of voltages for medium-speed and Fig. 26 for high-speed operation.

Figure 27 shows the speed-torque curves for the full-wave control. A full-wave control with feedback can be designed using current feedback or tachometer-generator feedback.

Many companies manufacture solid-state speed-control units where the appliance or tool can be plugged into the control package and infinite speed control is supplied.

8. Mechanical Timers.   Mechanical timers find application in appliances for controlling the running time or cycle times.

Mechanical timers are powered by winding up a mechanical power spring and releasing the energy through a series of gears and escapements. The output from these timers can be designed to operate switches for controlling motors, heaters, etc. Timer switches are available for timers from seconds to hours and can control a variety of switches for ratings up to 15 A, 125-250 V ac.

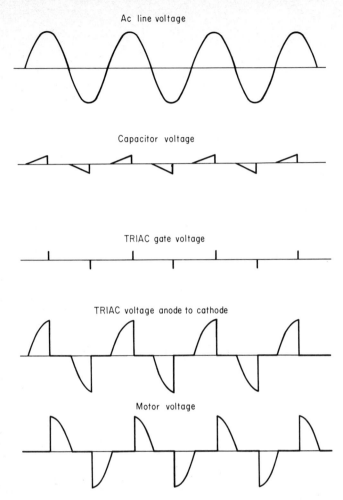

Fig. 25 Plot of voltages for full-wave control without feedback for a universal motor for medium speed.

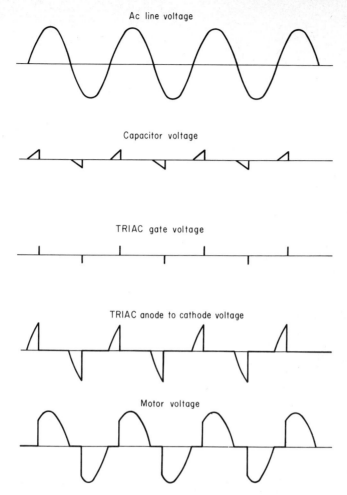

Fig. 26 Plot of voltages for full-wave control without feedback for a universal motor for high-speed operation.

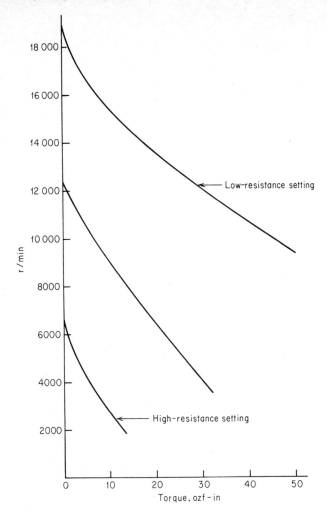

Fig. 27   Speed-torque curves for a full-wave control without feedback for a universal motor.

9. Solid-State Timers.   Solid-state circuitry for motor controls has resulted in combination speed control and timing circuits. Figure 28 shows the schematic of a solid-state timer and speed control for use in a blender. This circuit provides half-wave control without feedback. The solid-state timing circuit is designed to run the blender over a range of zero to 60 s. The circuit is designed to operate in a manual or timed mode.

In the manual mode of operation the timer, constant-current, and switch sections of the control circuitry are out of the circuit. Capacitor C_1 stays charged, keeping the switch portion of the circuit open and allowing operation of the speed-control and power portion of the circuit. Resistor R_{12} is the speed-control pot; R_{11}, R_{14}, R_{10}, and capacitor C_2 and transistor T_6 determine the speed settings.

In the timed mode, capacitor C_1 is charged up when the momentary switch time-start switch is pushed. With the capacitor voltage higher than the voltage midway between R_5 and R_6, the electronic-switch portion of the control made up of T_2, T_3, T_4, T_5, and R_7 and R_8 is open, allowing the function of the speed-control portion of this circuit.

Capacitor C_1 is discharged by the constant-current regulator at a rate determined by the setting of R_3. When the voltage of C_1 becomes lower than the voltage between R_5

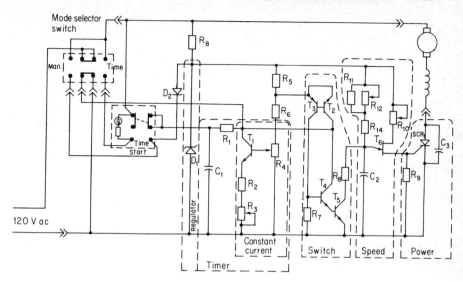

Fig. 28 Solid-state timer and half-wave speed control without feedback for use with universal motor.

and R_6, the electronic switch closes, allowing conduction, and transistor T_5 shorts out the speed-control portion, stopping current conduction through the motor.

OTHER TYPES OF SMALL-MOTOR CONTROL

10. Shaded-Pole-Motor Control. A shaded-pole motor is a particular type of single-phase induction motor that is made up of a stator winding, which in many cases is a simple bobbin-wound coil with shading coils around a portion of each pole face, and a squirrel-cage rotor which may have cast-aluminum or copper rotor bars and end rings (Ref. 1) (Fig. 29).

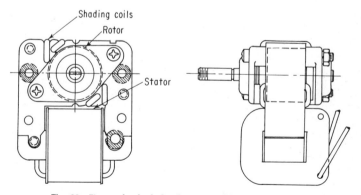

Fig. 29 Two-pole shaded-pole motor. (*Oster Corporation.*)

a. Varying Line Voltage. A common method of controlling the speed of this motor is to vary the applied voltage to the motor by the use of a variable autotransformer or series inductance (Fig. 30). The speed can also be changed by tapping the stator winding which varies the stator volts per turn and results in an increase or decrease in output torque and motor speed.

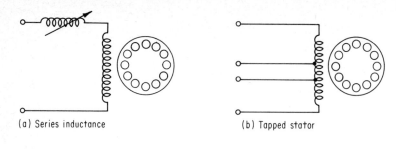

(a) Series inductance   (b) Tapped stator

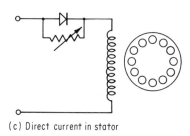

(c) Direct current in stator

Fig. 30 Three methods of controlling the speed of a shaded-pole motor.

$$T \approx E^2$$

where T = output torque
E = line voltage

b. Switching Poles. The speed of the shaded-pole motor can also be varied by winding the motor with two or more pairs of poles and switching from two poles to four or more poles, which will result in lower speeds. Switching from two to four poles would result in half speed.

$$\text{Synchronous-motor speed (r/min)} = \frac{120\ F}{P}$$

where F = line frequency, Hz
P = number of poles

Figure 31 shows a motor and switching circuit of this type of control.

c. Varying Line Frequency. Since the motor speed is dependent upon the line frequency, the motor speed can be varied by varying the power-supply frequency. This method requires a rather expensive variable-frequency power supply and thus is not practical in many cases.

d. Direct Current in the Stator. Figure 30 shows a rectifier and a current shunting resistor combination in series with the stator. The rectifier results in a dc flowing in the stator winding which produces a braking action to slow down the motor. Increasing the shunting resistance increases the dc and causes a larger reduction in motor speed.

e. Reversing Direction of Rotation. The direction of rotation of the shaded-pole motor can be changed by winding the motor with two sets of shading coils for each pole and switching the shading coils. The motor will run in the direction of shading.

f. Solid-State Controls. Phase-control solid-state packages, similar to the Triac control in Art. 7, are also available for controlling shaded-pole motors. By phase controlling the Triac the voltage applied to the motor can be reduced, lowering the speed of the motor.

11. Magnetic-Vibrating-Motor Control (Ref. 1). Since this motor type runs at a

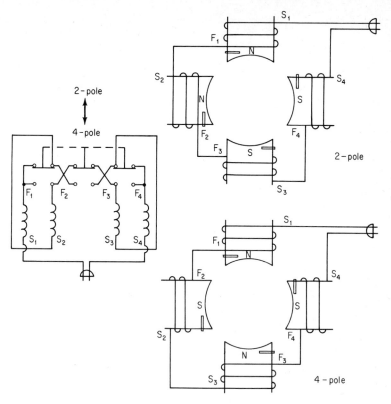

Fig. 31   Controlling shaded-pole-motor speed by switching poles.

speed of twice the line frequency for a motor without a permanent-magnet pole and at line frequency in a motor that contains a permanent-magnet pole, the motor speed can be varied by varying the line frequency. Since this control requires an expensive variable-frequency power supply, it is rarely used.

For a magnetic motor without a permanent magnet, a diode can be used to cut the vibrational frequency in half. Figure 32 shows a massage device utilizing a diode to produce a lower rate of vibration.

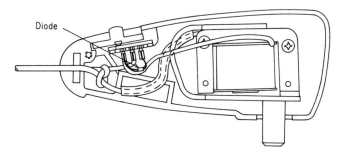

Fig. 32   Diode used to control vibration of magnetic vibrating motor. (*Oster Corporation.*)

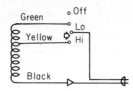

Fig. 33 Controlling amplitude of vibration in a magnetic vibrating motor by tapping the coil.

The amplitude of vibration in a magnetic motor can be varied by tapping the coil (Fig. 33). Increasing the turns in the circuit will result in a decrease in line current, and a resultant reduction in ampere turns and pole flux will produce less magnetic pull and thus reduce the amplitude of vibration. The amplitude of vibration and output force and power output can also be varied by adjusting the air gap of the motor (Fig. 34). By closing the air gap, a stronger magnetic field results, producing more force and amplitude of vibration.

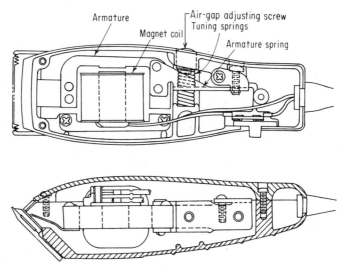

Fig. 34 Controlling amplitude of vibrator in magnetic vibrating motor by adjusting the air gap. (*Oster Corporation.*)

12. Permanent-Magnet-Motor Control. Many of the methods used to control universal motors can be used to control permanent-magnet motors: series resistor, shifting motor brushes, flyball governor, and solid-state controls.

a. Variation of Field Reluctance. The speed of a magnet motor can also be varied by varying the reluctance of the magnetic circuit by decreasing and increasing the magnetic flux.

b. Solid-State Control of Permanent-Magnet Motors. Solid-state controls similar to the circuits described in Art. 7 for universal motors can also be used effectively to control permanent-magnet motors. Some examples of these controls are as follows:

1. Half-wave permanent-magnet motor control without feedback. The SCR controls the amount of current flowing through the armature during the positive half cycle only (Fig. 35). Capacitor C is charged, and when the firing voltage of trigger TD is reached, the SCR conducts. The conduction angle can be varied by adjusting potentiometer R. The plot of line voltage, capacitor voltage, SCR gate voltage, SCR voltage anode to cathode, and motor voltage is similar to Fig. 17.

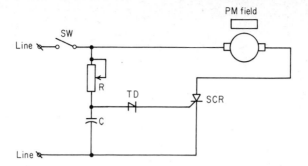

Fig. 35 Half-wave permanent-magnet motor control without feedback as used in a massage machine.

Figure 36 shows the motor-performance curves.

2. **Full-wave permanent-magnet motor control without feedback.** A Triac is used to obtain full-wave phase control (Fig. 37). The operation of this control is similar to the full-wave control described for universal motors (Art. 7) except for the addition of a bridge and an additional capacitor C_1 which improves the function of the control at low

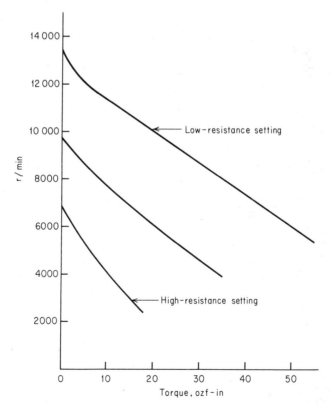

Fig. 36 Motor-performance curves for half-wave permanent-magnet motor control without feedback.

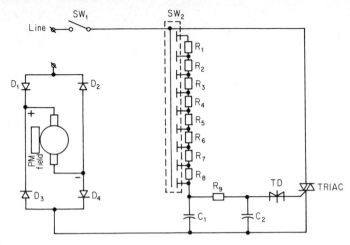

Fig. 37   Schematic of full-wave permanent-magnet motor control without feedback.

speeds by providing a steeper voltage across C_2 for firing the trigger diode. The pulsed output from the phase control is rectified by a full-wave bridge to obtain direct current required to power a permanent-magnet motor.

Figure 38 shows a plot of voltage across the line, capacitor voltage, Triac gate voltage, anode-to-cathode voltage, and motor voltage for a medium-speed setting.

3. **Full-wave tachometer-generator feedback control for permanent-magnet motors.** Figure 39 shows the same circuit, but a tachometer-generator feedback and comparator circuit has been added. This arrangement results in much flatter speed-torque characteristic curves (Fig. 40).

13. Impulse-Motor Control.   An impulse motor is a very simple ac-dc type of motor (Fig. 41). When the contacts are closed, a field is established in the stator, and the rotor moves. This motor is started by providing an external mechanical impulse to the rotor.

The speed of this motor type can be changed by modifying the shape of the rotor poles and by tapping the stator winding or inserting a variable inductance or resistance in series with the stator winding.

14. Stepping-Motor Control.   Stepping motors consist of the electromechanical types and the pulsed electromagnetic steppers (Ref. 1).

Free-running stepper motors operate from the line voltage, and the motor speed depends upon the line frequency. The input speed for a given design can be changed by altering the number of stator poles.

The large majority of stepper motors are electrically powered from digital dc pulses. The output speed of these motors can be varied by changing the pulse rate. The maximum rated output speeds for stepping motors range from $2\frac{1}{2}$ to 6000 steps per second, and standard stepping increments range from less than 1 to 90° without gear reduction (Ref. 2). The direction of rotation of many stepper motors can be reversed by reversing the pulse sequence to the stator coils.

15. Synchronous-Motor Control.   Synchronous motors are constant-speed motors with

$$\text{Synchronous speed (r/min)} = \frac{120\,F}{P}$$

where F = line frequency, Hz
P = number of poles

The motor speed can be varied by using a variable-frequency power supply and by changing the motor poles.

16. Switches for Small-Specialty-Motor Control.   A large variety of switches are used in consumer products to control the motors: on-off, momentary-on, slide, rotary,

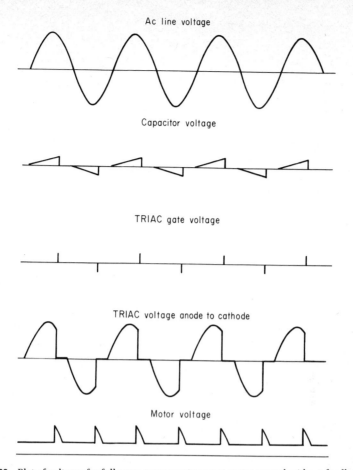

Ac line voltage

Capacitor voltage

TRIAC gate voltage

TRIAC voltage anode to cathode

Motor voltage

Fig. 38 Plot of voltages for full-wave permanent-magnet motor control without feedback.

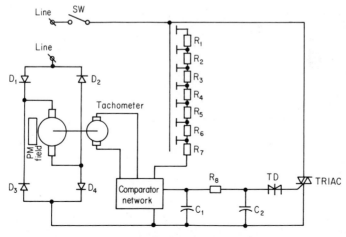

Fig. 39 Full-wave tachometer feedback control for permanent-magnet motors.

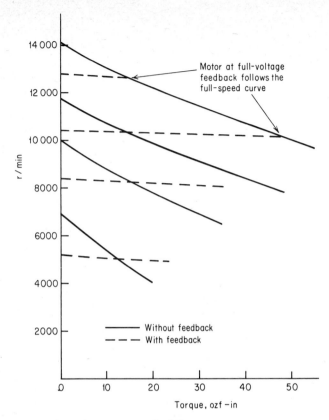

Fig. 40 Comparison of speed-torque curves for full-wave permanent-magnet motor control with and without feedback.

push-button, etc. On-off switches come in a variety of types: bat-type snap action, trigger types used in portable electric tools, slide action, push-button, on and off, and rocker type.

Typical motor-control switches include push-button, rotary, slide-type, lighted-pushbutton, momentary on-off, thumb-wheel rotary.

An experimental touch-control switch designed for a blender included a number of good features. Lightly touching the surface of the switch panel provided capacitance coupling for an oscillator circuit that fired an SCR, completing the circuit. This switch has many advantages such as a very light touch, smooth quiet operation, extremely long life due to the elimination of all mechanical parts and contacts, complete sealing from

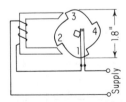

Fig. 41   Impulse motor.

dust, dirt, and liquids, and ease of cleaning. The present high cost of the required solid-state components, however, has prohibited its use in small electrical appliances.

PROTECTION

17. Motor Protectors.   Various types of bimetallic thermal-switch protectors are used to protect small specialty motors from heavy overloads or stalled armature or rotor conditions.   These thermal switches can be current-sensitive, temperature-sensitive, or both.   The current-sensitive thermal protectors have resistance-element heaters that respond to changes in load current.   The resulting heat causes the bimetallic switch to open, interrupting the load on the motor.   Many of these protectors have a manual-reset button which must be pushed after the bimetal has cooled sufficiently to close to reactivate the electric circuit.   Figure 42 shows several of these motor protectors.   Figure 43

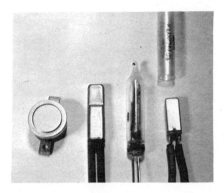

Fig. 42   Motor protectors.

Fig. 43   Motor protector with manual reset.

shows a protector with a manual reset.   Figure 44 shows typical response curves of a current-sensitive motor protector.

The temperature-sensitive type of thermal protector must be mounted close to the motor winding or in the hot-air discharge so the bimetal will sense the motor temperature.

18. Fuses.   Fuses can be used to protect motors against stalled-rotor conditions.   In most designs the fuse cannot be replaced by the customer but can be replaced easily by the appliance serviceman.   Figure 45 shows examples of these fuses.   In most applica-

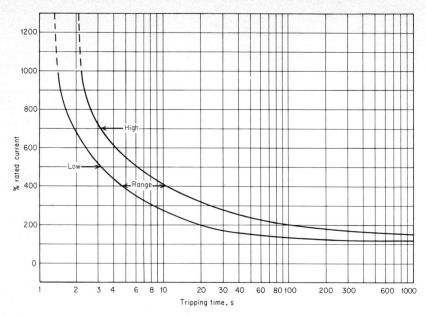

Fig. 44 Response curves for a current-sensitive motor protector.

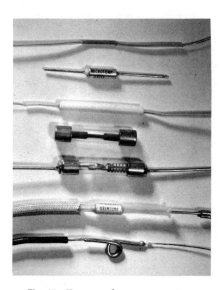

Fig. 45 Fuses used to protect motors.

tions the fuse is placed close to the motor stator windings and responds to motor temperature and current, usually under stalled conditions.

REFERENCES

1. Robert W. Smeaton, "Motor Application and Maintenance Handbook," McGraw-Hill Book Company, New York, 1969.
2. Stepping Motors Move Products, *Prod. Eng.*, Feb. 4, 1963.

30

Safety

ROBERT F. SCHOOF*

*Corporate Director, Safety and Security, Jos. Schlitz Brewing Company, Milwaukee, Wis.; Registered Professional Engineer (Wis.); Member, NSPE, ASSE, CSP.

FOREWORD

Safety as it relates to electrical equipment and to industrial switchgear and control specifically has made tremendous strides as an inextricable part of the development of better equipment. Progress has been made each decade in designing out hazards and providing built-in safeguards for operating and maintenance personnel. It is the intent of this section to assist those who engineer new electrical installations and those who maintain such systems to fulfill their safety responsibilities.

BASIC SAFETY PHILOSOPHY

While safety of personnel and equipment can never be guaranteed and can seem at times an elusive goal, it is generally true that we can be as safe as we want to be. The new aerospace concepts of systems safety engineering with strong emphasis on design safety and analysis of the overall system operation almost eliminate the consequences of human or mechanical failure. The prime objective in this approach is to look at the overall system rather than just an isolated problem, accident, or incident. For instance, in an investigation of an accident or a near-miss incident, questions similar to the following should arise:

1. Why was there an unsafe condition?
2. Was there failure of a maintenance or repair system or was there a system?
3. Was unusual skill or attention to detail required, and how can we assure such performance?
4. Why do we do it this way?
5. Are we expecting physical capabilities possessed by only a select group and only when they are feeling well?
6. How can we make the operation "fail-safe" mechanically and "human-error-proof" procedurally?

It does not take very long for an investigator of accidents to reach two conclusions.

First, the human being, consciously or unconsciously, is capable of and actually does some most unusual acts which eventually result in accident or injury ("If it can happen, it will").

Second, the only reasonably certain control acting to reduce accidents is the engineering approach.

Basic equipment-design safety as it affects function or basic reliability will not be covered in great detail. Attention is directed toward those equipment features or options which materially affect the safety of the operator or maintainer. Obviously some gaps remain as increasing probabilities for continuity of service almost automatically provide safety benefits to personnel.

1. **Safety engineering** has been defined as the art and science of dealing with the prevention of personal injury and property loss. There are safety-engineering specialists who engineer high-pressure systems, those who direct the safety of complex and highly dangerous chemical-plant operations, electrical-safety engineers, etc. Every engineer, however, is a part-time safety engineer and has as one of his basic responsibilities the safety of equipment and the people who operate it or are exposed to its hazards.

Fundamentally, safety engineering utilizes the following principles:

1. Eliminating (engineering out) the hazard, the possibility of harmful contact, the use of the toxic material, etc.
2. If the hazard cannot be eliminated, activities such as guarding, enclosing, ventilating, and controlling come into consideration. (Actually, both of the above are true engineering controls. If these controls are not feasible, the next best alternatives follow.)
3. The use and enforcement of use of personal protective equipment, safety glasses, safety respirators, gloves, safety shoes, hard hats, etc.

4. "Personal controls." These controls include defining procedures, training, the instruction manuals, caution signs, the exercise of mind and will, motivation programs, etc.

It must be emphasized that employment of the two latter alternatives does not always produce the result, the accident-free performance, that is the ultimate objective.

There has been much progress since the three Es—engineering, education, and enforcement—were accorded equal status. The following was expressed in 1947, and the progress at that time was also substantial: "The practice of safety has passed the stage in which it was compounded almost equally of statistics and evangelical exhortation. Although these two activities remain important, accomplishments will be small unless they are backed by a true engineering approach to all the problems of safety."[*]

The principles noted above can be stated more simply by: "You don't warn if you can guard, you don't guard if you can eliminate the exposure."

Many safety features that are now standard were once a part of local and state codes—emergency stops, lockouts, protected stairways, working platforms, and the like. Codes, however, should be considered a minimum standard. It is generally possible to engineer into an installation many additional specific safety guards, interlocks, and safety procedures that would not normally be apparent to the code inspector.

2. Total Safety Programs. It was noted earlier that safety is not composed of isolated segments such as statistics, promotions, or even safety engineering. Safety must be an objective and activity which pervades all functions and all the personnel within an organization. An executive of one large company listed what he considered the four basic safety fundamentals:

1. Management must really want a safe plant. They must be absolutely committed.

2. They must believe that safety can be managed, that accidents do not just happen.

3. They must organize their efforts in a systematic program—the whole works. Eliminating hazards, training workers, guarding machines, preventing unsafe work habits—establishing a climate for safety. They must never let up.

4. They must really promote safety with contests and publicity to keep everyone safety-minded.

With the objective stated and the desire instilled, the success of a program may well depend on the breadth of its activities. One major company developed a comprehensive outline for determining the overall components of a plant's safety effort. The following headings from the outline indicate the major considerations that should be included in a total safety program:

1. Safety policy and responsibility
2. Management participation and identification
3. Safety staff planning and coordination
4. Program evaluations and reports
5. Control of incoming materials
6. Special safety engineering
7. Fire safety controls
8. Safety and maintenance
9. Accident investigation, analysis, and reporting system
10. General safety inspections and surveys
11. Protective equipment
12. Job methods
13. Supervision communication and development
14. Employee motivation
15. Special employee training
16. Off-job safety
17. Medical controls
18. Product safety programs
19. Safety of visitors, tour groups, etc.

Also included were statistical exhibits detailing accident, fire, and loss experience both on a rate basis and on a dollar basis comparing the prior 3 years with "last year" and "current year to date."

---

[*]Col. Crosby Field, "The Study of Missiles Resulting from Accidental Exposures," *U.S. Atomic Energy Commission Safety and Fire Protection Bulletin* 10.

This large multiplant company employed in most locations half- to full-time safety supervisors. It must be noted that even with this level of staffing and with an extremely creditable safety experience there were many specific safety activities that were not being practiced.

Small specialized contractors, small plant operations, etc., obviously do not require formal programs for safety. Success is dependent upon development of a simple set of safety rules and procedures training with the major emphasis on the human factor. An example is the Richardson Paint Company, high-line and substation painting contractors, whose daily exposure to serious injury is almost beyond measurement. Their safety rule book states "Your job is just as safe as you make it," and follows this with five major rules of safe conduct:

1. Never take anything for granted.
2. Know where you are every minute.
3. Know where you are going before you make a move.
4. Plan every job before you start it.
5. Do not change your plan while the job is underway.

SAFETY MEASUREMENT

3. Frequency and Severity.   Safety is measured throughout the United States and many foreign countries by application of the rules and requirements of ANSI Z16.1 (1967), "Method of Recording and Measuring Work Injury Experience." The two basic measurements involve the rate of disabling injuries (frequency rate and severity rate).

The "frequency rate" is defined as the number of disabling injuries per million man-hours of exposure. Similarly, the "severity rate" is defined as the number of days lost (or charged) per million man-hours of exposure.

The National Safety Council publishes each year comprehensive statistics listing frequency and severity rates for major occupational groups.[*] The average rates published by the National Safety Council are considerably better than those published by the Bureau of Labor Statistics, which includes all industry and is not restricted to the members of the National Safety Council. The Z16.1 standard prescribes various rules covering situations such as, "Did the injury arise out of the course of employment?" "Are there special hernia and back injury qualifications?"

Definition or qualification of what is good safety performance is most difficult. It varies markedly from industry to industry and also among the large and small companies of a particular industry. The most recent (1975 edition) rates show an average of all occupational groups who are members of the Council of 10.2 for frequency and 614 for severity.

An appreciation of what these numbers actually mean in terms of an operating department is most important to grasp. If we consider a frequency rate of 10 and a severity rate of 200 as fairly representative of a reasonable safety performance, we will, by definition, have had 10 disabling injuries and 200 days of time lost or charged for a million man-hours of exposure. In terms of people the million man-hours is equivalent to 500 man-years at an average of 2000 h/year/man.

Carrying the arithmetic a little further, we see that there is one disabling injury and 20 days lost for every 50 man-years, slightly more than one worker's average work life-time. This performance is obviously not outstanding, and many companies are operating with frequency rates of less than 1 and severity rates of less than 100. Some further perspective is gained when one considers that for a frequency rate of 10 and a department staff of 10, one disabling injury will occur every 5 years (for even only average performance).

A most important rate not used universally or recorded in NSC statistics is the rate of "doctor's cases," those cases referred to an outside doctor or specialist for treatment which require more than routine first aid. It is enlightening to show this doctor's case rate as the number per 100 employees per year. Rates of 15, 20, and 25 are quite common in many industries. The rate may reach 100 per 100 employees per year. Safety controls and efforts are more quickly, and often more accurately, reflected by a rate of

[*]"Accident Facts," National Safety Council, 1970 ed.

this type as compared with frequency and severity rates. Disabling injuries on which the latter are based occur much more infrequently—on the order of a ratio of 1/10. Elementary statistics indicates that rates based on relatively rare occurrences cannot compare in reliability with rates developed by substantially greater amounts of data.

HAZARDS OF ELECTRICITY

Electric current kills and injures directly and also often serves as the trigger that initiates a chain reaction with injury or death from other causes.

4. Sources of injury arise from the following:

a. Electrical Shock or Current Flow. Current flow through chest or head area can cause cardiac arrest. A second possibility is stoppage of breathing, which is generally a temporary paralysis situation. A third major hazard especially from low voltage (low-amperage shocks) is that of "ventricular fibrillation." This subject and the "let-go" problem are treated in more detail in Art. 7. Also attributable to shock hazards are falls due to involuntary reactions while working at heights. Injuries also occur because of severe involuntary muscle contractions and also from contact with fixed objects due to jerking or falling-backward reactions.

b. Electrical Arcs. Electrical arcs, if not interrupted properly and confined, create enormous volumes of ionized hot gases which blow up apparatus, throw molten metal, blow open doors, etc. The hazards of serious burns and setting clothing afire are considerable. There is also the obvious hazard that arc generation in an explosive atmosphere will create havoc. Uncontrolled and controlled arcs such as those generated in arc welding cause flash burn to the eyes. Electrical arcs generate ultraviolet rays, which have a somewhat delayed effect upon the skin and particularly on the tissues of the eyes. It is still not too widely recognized that standard industrial safety glasses (3 mm of hardened crown glass) will provide adequate flash burn (ultraviolet) protection. Filter-type lenses are not necessary for flash protection during normal arc welding but are provided to prevent excessive pupil fatigue and to permit observation of the welding process. Often overlooked is hand and arm protection against flash burn.

c. Direct Burns. In addition to the hazards of burns from being in close proximity to arcs, a sizable number of injuries occur from direct contact with overheated electrical equipment. In many instances, overheating goes undetected and results in substantial fire loss.

d. Mechanical Hazards. Electrical equipment presents a host of mechanical hazards ranging from inadvertent start-up with "hands in the machinery" to housekeeping hazards, falling over leads, etc. Exposed couplings and overspeed rupture are becoming less common causes.

5. Injury Summaries. The great variety of injuries which occur in the electrical field are illustrated by a $2^1/_2$-year summary of electrical accidents occurring in the state of Wisconsin. Covering the period of May 1958 through September 1960, excluding the crane boom-line-contact class of which there were 13 cases (8 fatal and 5 nonfatal), there were 22 cases of other contact with power lines as indicated below:

1. Fatal cases—11
Painter's brush contacted 26 400-V line on tower—burns.
Lineman touched line on bus structure he was covering—electrocuted.
Lineman's climber spur broke out and he swung into line—electrocuted.
Primary line broke and fell on injured—electrocuted.
Lineman accidentally contacted transformer power loop—electrocuted.
Tree trimmer touched power line with tree pruner—electrocuted.
Two were electrocuted when a well pipe was pulled into a 2000-V line. Injured held guy wire which fell across 200-V line—electrocuted.
Grounding conductor accidentally connected to live line—electrocuted.
Telephone line being pulled in whipped into power line—electrocuted.

2. Nonfatal cases—11
Lineman accidentally contacted jumper wire on pole—burns.
Lineman burned by telephone line whipping into power line.
Lineman thoughtlessly touched live wire—burns.
Lineman accidentally swung hoist chain into live line—burns.

Well pipe contacted 7200-V line—burns.

Two were burned when a telephone line being pulled in whipped into power line. (Another employee was electrocuted in the same accident.)

Lineman contacted 4000-V line—burns and shock.

Injured on top of digger—contacted power line with head—shock and burns.

Line rubbed transformer case, shocking lineman—injured jumped to ground—fractures.

Injured carrying pipe on fire escape—contacted overhead power line—burns and fall.

The other cases, numbering a total of 35, are listed below:

3. Fatal cases—5

Faulty cable insulation at switch knockout—operator electrocuted.

Operator fell into power cubicle—electrocuted.

Improperly wired high-voltage test plug—electrocuted.

Ungrounded sump-pump motor frame was energized—electrocuted.

Inserted plug from truck in wrong receptacle—frame of truck was energized at 230 V—another truck was grounded—worker walked between the two trucks—electrocuted.

4. Nonfatal cases—30

Faulty switch exploded, burning operator.

Noncode trouble lamp in crane cab burned operator.

Electrician pushed fish tape onto live switch—burns.

Faulty switch connection burned electrician.

Hands burned replacing fuse with screwdriver.

Ground wire freed by electrician contacted live switch clips—burns.

Dripping water short-circuited switch—flash burns.

Metal dust in motor and switch caused flash burns.

Wrench slipped and injured contacted live bus on switch structure—burns.

Injured moved furnace lead with screwdriver—burns.

Injured opened switch under heavy load—arc and flash burns.

Injured placed 110-V lamp in 220-V indicator socket—burns.

Inspector's light broke, igniting gas fumes in tank—burns.

Lineman accidentally touched transformer jumper—burns.

Switch short-circuited when closed—burns.

Faulty insulation in cable to plug caused short circuit—burns.

Injured inserted plug with frayed cord strand in receptacle—shock.

Injured on top of bin—head touched hoist trolley wire—shock.

Faulty cord to conveyor—injured was on wet ground, touching conveyor—shock.

Injured received shock removing test clips from switch.

Injured came too close to electrostatic sprayer—shock.

Circuit ground to loosely grounded motor frame—shock.

Bare wires in handle of ungrounded machine—shock.

Injured inserted short-circuited plug in receptacle—flash and shock.

Bridge-type crane operator received shock in cab—cause unknown—shock and fall.

Arc welder touched electrode to shoulder—shock and burns.

Open ground circuit and bared conductors to hoist switch—shock and burns.

Injured was replacing hoist at factory assembly—shock and injury.

Improperly connected cord and plug—when used, injured received shock and burns.

Motor controller and fuse flash—injured fainted—vague complaints.

6. Injury Case Histories.   Some detailed case histories illustrate the complex nature of many electrical accidents. The circumstances are unusual but provide some insights on the fundamentally simple procedures which can prevent most, if not all, such occurrences.

1. A senior electrician was completing a rewiring job on a power press. As he inserted the fuses an arc occurred, causing second- and third-degree burns on the face, ears, and neck. It was found that the circuit leading to the oil-pump compartment had been disconnected and a short existed. The injured employee had tested the circuits with a test light and found no problem. The circuit had not been meggered.

2. Two electricians were servicing an induction heater with 440-V supply. The side and rear panels were removed, and the unit was being checked "hot." Unit tested out

satisfactorily. As one man attempted to go between the unit and an adjacent cabinet (a narrow 18-in space) to turn off the power, he tripped and fell, striking exposed 440-V terminals. Fatality.

3. Man contacted live terminal of ungrounded 440-V, 1600-A, 3-phase circuit with 12-in crescent wrench. Disconnect switch was inoperative and one phase was already grounded. Fatality.

4. A crew working on disconnected high-voltage leads when control circuit was inadvertently energized for turn-test-ratio (TTR) testing. Fatality.

5. Lineman contacted 7620 V (high side of potential transformer) and fell to ground, fatality resulting from broken neck. PT disconnects, open the day previous, had been reinstalled.

6. When installing a motor control, an operator closed disconnect on 2300-V cubicle to energize and check out control circuit. An improperly field-connected "start-stop" button having reversed connections energized a contactor coil which failed to fully close because of interlock. Operator, thinking he had closed on a fault, attempted to *open* disconnect carrying inrush current, which then arced to ground. Major flashover and fire resulted, operator vacating building on hands and knees without injury.

7. An electrician pushed a steel snake into an electric conduit which terminated in 480-V switchgear for refrigeration equipment. He was trying to find where the conduit went. In the resulting arc, a foot of the steel snake was burned off, a main breaker opened up, and equipment was shut down. Fortunately there was no injury.

8. In another steel-snake incident, a man was at the receiving end holding a piece of masonite to keep the end of the snake from hitting main buses while another man pushed the snake through the conduit. The snake got away, struck the live buses, and created a powerful arc that caused severe burns.

9. While disconnecting leads to a voltage regulator under test, the tester sustained burns and shock from 400 V ac. The accident happened because of failure of operator or helper to open their switches to kill the power to the regulator. There was no explanation of why this action was not taken. There was no highly visible "power on" indicator at the immediate site.

7. Special Hazards of 115 V Ac. Since 115 V ac is low voltage, it follows automatically in many minds that it is therefore safe, particularly for the experienced professional, the electrical tester, or the senior electrician. It is relatively safe if:

1. Skin resistance is high.
2. The path of the current is not through the heart, head, or respiratory area.
3. The duration of the contact is very brief.

A fatality may occur only once for every 10 000 to 40 000 contacts. The contacts unfortunately are frequent, and most statistical studies estimate that the majority of the 1000 electrical-shock deaths annually are from 115 V or less—550 from 115 V, 150 from 230–440 V and 300 from voltages over 440 V. A study of 22 fatalities on board navy ships over a period of 6 years showed 12 were caused by circuits of 115 V or less.

The major hazard of the low-voltage shock (inherently low current flow through the body) is from "ventricular fibrillation," a condition in which the heart rhythm is stopped and the muscle fibers contract individually (fibrillate or quiver) instead of acting to pump blood. If circulation is not restored within approximately 4 min, irreversible brain damage begins to occur and death follows shortly. First-aid procedures (Art. 8) can sustain life for up to 10 to 15 min until a defibrillator can be obtained or the victim transported to such equipment.

The factors which determine the magnitude of the shock and its effect include:

1. The condition of the skin and the nature of the contact. The resistance is lowered by moist or wet skin, cuts or breaks, and by increasing the amount of area contacted and by the pressure of contact. The resistance of wet skin can be $1/100$ of that of dry skin.

2. The duration of the contact. The outer layer of the skin is destroyed rapidly by the flow of even minute levels of current, and resistance drops rapidly as the current penetrates to the inner layers.

The duration is also significant in its effect upon the heart action. Currents of 25 to 80 mA lasting more than 25 to 30 s may cause fibrillation. Between 80 and 200 mA, even a very brief shock is likely to cause fibrillation. The effect is dependent upon the nature of the heart action when the shock occurs. The danger is least when the heart is pumping or after it has recovered from a beat.

The maximum level of reasonably safe current is largely determined by that current which will permit a person to "let go" of a live conductor. The following description of the "let-go" phenomenon is taken from Ref. 1:

> With increasing alternating current the sensations of tingling give way to contractions of the muscles. The muscular contractions and accompanying sensations of heat increase as the current is increased. Sensations of pain develop, and voluntary control of the muscles that lie in the current pathway becomes increasingly difficult. Finally a value of current is reached for which the subject cannot release his grasp of the conductor. At this point he is said to "freeze" to the circuit. The maximum current a person can tolerate when holding a conductor in one hand and still let go of the conductor by using the muscles directly stimulated by that current is called his "let-go" current. Let-go currents are important, as experience has shown that an individual can withstand, with no ill aftereffects, except possibly sore muscles following repeated exposure to his "let-go" current for at least the time required for him to release the conductor.

> Let-go currents [were] determined for 134 men and 28 women. In these tests the subjects held and then released a test electrode consisting of a No. 6 copper wire. The circuit was completed by placing the other hand or foot on a flat brass plate, or by clamping a conductive band lined with saline-soaked gauze on the upper arm. After one or two preliminary trials to accustom the subject to the sensations and muscular contractions produced by the current, the current was increased to a certain value and the subject was commanded to let go of the wire. If he succeeded, the test was repeated at a current of a slightly higher value. If he failed, a lower current was used, and the values were again increased until the subject could no longer release the test electrode. The end point was checked by several trials, and the highest value was taken as the individual's "let-go" value to eliminate the effects of fatigue.

> Other tests were made with dry hands, hands moist from perspiration and hands dripping wet from weak acid solutions. The effect of the size of the electrodes was also investigated. It was found that the location of the electrode, the moisture conditions at the point of contact and the size of the electrodes had no appreciable effect on the individual's let-go current. It is believed that the results obtained from tests in which hands wet with saline solution grasp and then release the small copper wire may be used to predict let-go currents of a specified degree of safety with an accuracy sufficient for most practical purposes.

> Sixty-cycle let-go currents were measured on 28 women. The women ranged in age from the late teens to the early twenties. They were light in stature and obviously not accustomed to hard physical work, and their forearm muscles were not particularly well developed. Although the women volunteered freely for the tests, it proved impossible to develop enthusiasm or any degree of competitive spirit at the higher currents. The results are probably representative for the sedentary type. However, from observation of the reactions of the subjects having the greatest muscular development, it is possible that values were considerably lower than those which would have been obtained had a group of mature, healthy women accustomed to physical labor been used. Results based on these data, therefore, should be conservative and on the side of safety.

> From these and other similar tests it is concluded that reasonably safe 60 cycle let-go voltages hand to hand are about 21 V, and hand to both feet ankle deep in salt water, 10 V. Reasonably safe dc release voltages are 104 V hand to hand, and 51 V hand to both feet ankle deep in salt water.

> The higher 60 cycle let-go currents were frequently sufficient to stop breathing during the period the current was flowing across the chest, and the reactions at the instant of current interruption during the dc release tests occasionally threw the subject a considerable distance. The muscular reactions during accidents frequently cause fractures, and the contractions resulting when a victim grasps bare overhead wires may be sufficient to freeze him suspended to the circuit in spite of his struggle to drop free. In many accidents a victim frees himself by breaking the conductor, or his body weight may assist him in interrupting the circuit; however, fortuitous circumstances must not be relied upon to provide safety to human life.

> Currents only slightly in excess of one's let-go current value are very painful, frightening and hard to endure for even a short time (Table 1). Failure to interrupt the current promptly is accompanied by a rapid decrease in muscular strength due to the pain and fatigue associated with the accompanying severe involuntary muscular contractions, and it would be expected that the let-go ability would decrease rapidly with the duration of contact. Prolonged exposure to currents only slightly in excess of a person's let-go limit may produce exhaustion, asphyxiation, collapse and unconsciousness followed by death.

Also noted are statements that currents that do not produce fibrillation may produce respiratory inhibition. An approved method of artificial respiration must be applied

TABLE 1. Quantitative Effects of Electric Current, mA, on Human Beings (Ref. 1)

			Ac rms values			
	Dc		60 Hz		10 000 Hz	
Effect	Men	Women	Men	Women	Men	Women
No sensation on hand.	1	0.6	0.4	0.3	7	5
Slight tingling. Perception threshold.	5.2	3.5	1.1	0.7	12	8
Shock—not painful and muscular control not lost	9	6	1.8	1.2	17	11
Painful shock—painful but muscular control not lost	62	41	9	6	55	37
Painful shock—let go threshold	76	51	16.0	10.5	75	50
Painful and severe shock— muscular contractions, breathing difficult	90	60	23	15	94	63
Possible ventricular fibril- lation from short shocks:						
Shock duration 0.03 s	1300	1300	1000	1000	1100	1100
Shock duration 3.0 s	500	500	100	100	500	500
Ventricular fibrillation— certain death	Multiply values immediately above by $2^3/_4$. To be lethal, short shocks must occur during susceptible phase of heart cycle					
Possible ventricular fibril- lation from impulse shocks:						
Dc short shocks and surge discharges.	27.0 W-s					
Power-frequency short shocks and low-frequency oscillatory discharges	13.5 W-s					

promptly to prevent suffocation. The mouth-to-mouth technique is now the approved method. One other point of interest is the unusual phenomenon of delayed death. Persons who have received a shock severe enough to cause unconsciousness and who have been revived sometimes die suddenly without cause minutes, hours, or days later. Serious shocks should never be treated lightly, and hospitalization for observation is highly recommended by many medical authorities.

The basic defenses against low-voltage shock are three:

1. Effective grounds
2. Stay dry
3. Insulate/isolate

Grounding continuity is difficult to maintain on portable tools, extension cords, receptacles, etc. One inexpensive portable ground continuity tester which is effective is the Daniel Woodhead No. 1750.

8. First Aid for Electric Shock. Not many years ago the only recourse for a heart in ventricular fibrillation was to open the chest and manually massage the heart. In some instances, an electrical countershock was applied directly to the heart itself. These massive surgical techniques were obviously limited in their application to those victims who were within or in close proximity to a major medical facility. Defibrillators which operate through electrodes placed on the surface of a closed chest are now available in almost all major medical centers. This, combined with the first-aid techniques of rescue breathing and manual heart massage, is saving many lives.

Manual heart massage is performed as follows (*caution*—It is imperative that this technique not be performed unless there is a strong indication that there is no heart action):

1. The victim should be on his back or a firm flat surface if possible.
2. Place one hand on top of the other with the heel of the lower hand on the flexible breastbone of the patient's chest.
3. Press down approximately one inch.
4. Release to let the chest expand and permit blood to circulate.

This action should be repeated about once a second until heart action is restored or until defibrillation equipment is available. This technique may maintain adequate heart function for periods of 10 to 15 min or longer. If there is heart function (even though weak), the above procedure should not be practiced but emergency rescue breathing should be employed. The preferred "mouth-to-mouth technique" is performed as follows:

Start immediately—seconds count.
Place victim—on his back.
Clear mouth—turn head to side. Remove all foreign matter.
Tilt head back—chin pointing upward to open air passage.
Jaw forward—pull or push jaw into jutting-out position.
Pinch nostrils—to prevent air leakage when you blow.
Blow—seal patient's mouth with yours, blow until you see chest rise. Then remove mouth.
Listen—for exchange of air or signs of throat obstruction.
Repeat—10 to 20 times per minute.
Continue rescue breathing until patient breathes independently.

It is critical to start "rescue breathing" immediately. The chances for restoration of breathing action have been reported as follows:

Time elapsed after shock, min	Chances for recovery
1	98 out of 100
2	92
3	72
4	50
5	25
6	11
7	8
8	5
9	2
10	1
11	1 out of 1000
12	1 out of 10 000

ENGINEERING/MAINTENANCE MANAGERS' SAFETY RESPONSIBILITIES

9. **Safe Environment.** With safe electrical equipment, safe portable tools and hand tools, and properly guarded mechanical exposures, etc., there are still many major safety factors in control of the basic environment.

There is the problem of safe storage of tools, supplies, and equipment. This kind of facility sometimes grows by inches while the work load doubles and the basic misarrangement that is all too common breeds contempt for safety rules because of its obvious hazards. It is necessary to plan suitable equipment of this type, racks, shelving, lighting, fire-protective equipment, etc. Housekeeping is obviously important. With proper safety considerations incorporated into major supply rooms and satellite storage areas, the problem of storing materials in, on, or near switch boxes and switchboards is eliminated.

Illumination in storage and electrical-equipment areas should be above 50 fc. Emergency lighting is definitely indicated in many of these areas. The use of flashlights is marginally safe, and efforts should be made to reduce the need for such equipment. Built-in work lights, connected to permit use with equipment deenergized, in large electrical cabinets is a major step forward.

Floor areas within major-equipment rooms or in front of test equipment should be provided with rubber insulating coverings or mats. Flooding potentials from groundwater, sprinkler failure, or other source should be reduced to near zero.

Portable-ladder misuse is often a source of injury. Safety feet of the appropriate type for the service are mandatory. Many accidents continue to happen with properly equipped ladders, however, because of workman error. Ladders are not placed properly at the top, men overreach, vehicles and doors strike the bottom, etc. A requirement should be made for ladders to be tied at the top if possible and if not possible at the bottom. Through experience, it was found necessary to provide tie ropes secured to the ladder at both the top and the bottom to assure their use. The available rope (not in the tool crib or buried in a tool cart) and a concerted effort to require their use has proved highly successful in eliminating ladder accidents.

A further consideration relative to safe environment is indicated. Safe for whom? Many codes permit location of hazardous open conductors for equipment in locked enclosures, open conductors 8 ft above a work area, guard-railed enclosures, etc. These practices are reasonably safe for the frequenter or general factory employee but not very safe for the electrician. There is still much to do in this area in "engineering out" the hazard for everyone. Locked enclosures have their place to prevent misoperation and intentional work interruption, but in time they should not be depended upon to substitute for insulation. Standard machine-guarding practices such as fixed enclosures with interlocks for doors and access points to deactivate units can minimize many electrical-hazard potentials now considered routine.

10. Safe Clothing and Hand Tools.   Unsafe personnel attire and small hand-held equipment have caused significant injuries. Many have learned through experience or by accident and follow to the letter all the rules of personal safety. There is an appreciation that skills, special alertness, and awareness do not afford a sufficient guarantee or compensation for an inherent hazard. It is recognized that in many instances the hazards are slight, but insistence on following the rules listed below will eliminate the potential of injury and also serve to emphasize to the maintenance electrician, installer, or contractor management safety objectives:

1. All metallic objects should be prohibited from being carried on the person. These include such items as wristwatches, rings, metal badges, belt buckles, metal pens, rules, flashlights, key or watch chains, and metal-rimmed or -framed eyeglasses.

2. Basic clothing, whether intended for use in warm or cold climates, should be tight-fitting without loose sleeves, cuffs, etc., to contact or catch upon equipment while working or climbing.

3. Shoes should be of insulating type, rubber-soled or worn with rubber boots. It is highly recommended that shoes be of safety-toe type. Subsidies or programs to encourage their use are common.

4. Miscellaneous small hand tools such as pliers, cutters, nut drivers, screwdrivers, and test leads should be insulated to the fullest extent possible through the use of rubber sleeves, taping, etc. Metal-handled brooms and shovels are to be avoided.

5. Portable powered hand tools should be of 3-wire grounded design or of double-insulated type. The use of the latter is growing rapidly and represents a somewhat increased factor of safety when properly maintained over that of the 3-wire grounding type. Grounded tools are obviously safe if the total circuit to ground is intact. Frequent testing and constant supervision are a necessity.

6. Aluminum ladders are usually prohibited for all electrical maintenance and installation work. They can be used with a reasonable degree of safety in low- and medium-voltage areas if they are equipped with a dielectric barrier. This barrier is commonly a phenolic or hardwood block inserted in the side stringers between the base and the first step. Their safety is compromised somewhat by the potential of completing a circuit between the upper part and an intermediate part of the ladder. The barrier is quite effective, however, against completion of a circuit through the base of the ladder.

11. Safety Protective Equipment.   Safety protective equipment provides a number of vital safety advantages, if conveniently available, if not used to permit unnecessary risk, and if use is properly supervised. Generally this type of equipment is intended to isolate one from a hazard. There are obviously other applications of equipment that go beyond the isolating function, such as the use of protection for working at altitudes, or breathing apparatus.

A minimum set of standard protective equipment for the average electrical maintenance operation includes:

Rubber gloves with leather protectors
Rubber sleeves
Rubber mats and blankets
Rubber line hose
Insulating platforms
Dielectric hard hats or caps (EEI specification°)
Safety glasses
Fuse tongs or pullers
Switch hooks
Safety belts and life lines

Note that rubber goods should be carefully inspected and tested electrically at least once a month, weekly if gloves are in active use. All such pieces of equipment should be visually inspected before each use and checked by trapping air in them.

12. Attitude Creation. Woven throughout the operating nucleus of any effective safety program is the concept of attitude. The approach can range from a very elementary "We have a good safety attitude" to a comprehensive psychological approach on many fronts. Suffice it to say that without the desire for safety and a good safety attitude at all levels of an organization, safety will not happen.

Examining the many personal causes for accidents, a listing can be developed:

Lack of alertness
Preoccupation
Lack of training or inadequate techniques
Lack of experience
Personal factors such as poor coordination and physical disabilities
Reckless or careless attitude

Putting aside the many other contributing factors which often are the real key to the accident (and prevention) such as basically poor human-factors design, including lack of standardization, identification inadequacies, and unnecessary complexity, the approach to many of these problems involves training. Training develops technical competence, the establishment of safe work practices, and ultimately desirable safety attitudes. However, the training should not be a one-time affair but should be followed up by the necessary supervisory controls and motivation efforts of varied styles. The problem of attitude creation is incredibly complex, involving basic human needs and desires. At the risk of oversimplification, safety attitude is built upon a foundation of safe equipment, safe procedures, and proper employee placement and training. Given these fundamentals, attitude becomes dependent upon management desire and a positive program to communicate these desires. The task is not easy.

Communicating a safety attitude can and should include special bulletins of many types. There should be frequent reminders such as the common safety signs and posters. Supplementing these, particularly in larger organizations, is the need for some further motivational techniques involving contests, awards, special recognition, etc. These kinds of efforts provide the reminders that are so necessary for a truly excellent safety effort.

13. Procedures and Training. Procedures covering basic safety requirements should be reduced to writing and distributed to all personnel. Most major heavy industrial concerns have developed specific sets of safety rules governing electrical departments or electrical maintenance departments. U.S. Steel, Bethlehem Steel, Allis-Chalmers, and Westinghouse are a few of many. Examples of the content of these safety-practice manuals follow. From Bethlehem Steel Electrical Department Safety

°It is highly recommended that the wearing of hard hats or caps be made a condition of employment.

Code 18170: "When opening or closing electrical safety switches use your left hand, stand to right side of the switch box not directly in front of it and turn head and eyes away from switch." They consider 440-V equipment extremely dangerous as evidenced by:

1. Never work on a 440-V circuit that is hot unless it is not feasible to make circuit dead.

2. Use rubber blankets and if necessary erect wooden barriers to isolate place of work.

3. Use rubber gloves with protector gloves and stand on rubber matting or a dry board.

An Allis-Chalmers Corporation safety manual includes in a section titled "Safety Rules for Electrical Maintenance" references to "live work, safe entrance procedures for pits, manholes, tunnels, etc."

In a following paragraph relating to "identification," it reads, "All electrical equipment junction boxes and wiring shall, at the time of installation, be properly marked with permanent tags, signs or strips to preserve the identity of the equipment and wiring, and to facilitate determination of the supply source."

Employee training generally should follow the following basic outline:

1. The hazards of electricity and the general environment.

2. First aid and mouth-to-mouth resuscitation. In some instances, closed-heart-massage training might be indicated.

3. Safe clothing and hand tools.

4. Safe protective equipment.

5. Lockout procedures.

6. Other safety control procedures, company safety handbooks, etc.

It is important and nearly mandatory that even small maintenance operations have basic safety procedures in writing, even though bound handbooks are not in order.

a. Start-up Procedures.   Proper start-up and operating procedures should be posted at all devices. These procedures should be attached to devices and enclosed in a sealed clear plastic envelope for ready reference.

Unless it is absolutely necessary to use two hands, only one hand should be used when closing or opening a disconnect, making connections, changing taps, or performing other operations on high- or low-voltage equipment.

b. Barricading Test Areas.   All areas in which there is present an electrical exposure to personnel should be adequately barricaded and signs posted warning of exposure. Yellow and black polyethylene rope is generally preferred for roping off danger areas. When exposure is terminated, the area should be cleared immediately.

c. Operating of Switches and Disconnects

1. Closing. Manual switches and disconnects should always be closed by a single unhesitating motion with one hand if possible. Do not reverse action under any circumstances.

2. Opening. Always reduce load wherever possible before opening circuit. Do not attempt to interrupt full-load, locked-rotor, or short-circuit currents unless switch, disconnect, or breaker is rated for such duty. Other backup protection should clear any irregularity.

d. Altering of Equipment.   Standard test equipment shall not be altered or modified unless urgently required and then only by express approval of supervisor. Such equipment shall be immediately tagged and note of change made thereon. Equipment shall be restored to its original condition at completion of test.

e. Grounding Procedure.   All fixed and portable test equipment shall be effectively grounded prior to being energized. Grounds should have adequate current-carrying capacity as fixed by code, not less than No. 8 copper except for instrument grounding. Grounds should generally be made to water-piping system of the building and should not exceed 3 Ω resistance. All ground connections should be checked at least annually.

f. Fuse Pulling and Replacements.   Fuse pulling should be done only with use of approved-type nonconducting fuse pullers and with circuit deenergized. Replacement fuses should be of the same rating and style unless it is determined by careful examination that the size and style being installed are proper for the circuit.

g. Prohibition of Work on Live Circuits Over 110 V.   Work shall not be performed on "live" circuits unless absolutely necessary. "Live" work shall not be permitted on circuits of more than 110 V except on instrument circuits and then only by qualified personnel.

h. Requirement for Test of Circuit to Determine Power Has Been Shut Off. Before working on any circuit, a check should be made with an approved-type voltage tester (neon or incandescent lights, voltmeter, or similar device) to determine definitely that the circuit is deenergized.

i. Lockout of Switches. Whenever service is performed on a machine or device, the main disconnect should be opened and locked in the open position. Pulling of fuses is also recommended as additional protection.

j. Requirement for Two Workers for over 750 V. On test setups involving voltages in excess of 750 V, work by one tester is prohibited. Where test is routine and no changes in circuit are required, the presence of another worker in the adjacent area shall satisfy the requirement for two workers.

k. Defective Equipment. Equipment which appears or becomes defective during test should be immediately removed from service and tagged with nature of defect.

l. Secondary Circuit of Current Transformers. The secondary circuit of a current transformer must be kept closed. Opening of this circuit will result in high induced voltages which can injure personnel or damage windings.

m. Assumption vs. Verification. Never take anything for granted when substantial amounts of voltage or power are being controlled. Examine circuits personally and verify step by step all test procedures.

n. Floor Protection. Main switching, control, and test areas which involve voltage of over 220 V should be equipped with rubber matting.

14. Accident Investigation. After an accident involving electrical shock or major property damage involving electrical equipment, the investigation commences. Because of the technical factors involved, the complexity of the equipment, etc., the investigation is generally thorough and detailed, and it usually goes far beyond the relatively standard one-page accident-investigation form used by most major industries (Fig. 1).

The investigator of electrical accidents will find a much higher proportion of mechanical (electrical) failures as opposed to what is termed loosely "human failure." The purpose of the form in Fig. 1 was to direct and focus attention upon the management failures involving equipment provisions, maintenance of such equipment, and improper or inadequate systems or procedures. The emphasis is obviously away from what human failures occurred. When a "human failure" is identified, often, unfortunately, the productive efforts to prevent recurrence stop at that point. It is most important to keep foremost in mind that when an accident occurs, "management" is responsible. Whatever the human failure, somewhere some member of management or management as a whole has failed in procedures control, training, placement, motivation, etc. It serves no useful purpose to pinpoint this responsibility in most instances. The emphasis must be on what can be done to "positively prevent recurrence." If this objective is thoroughly understood and demonstrated, witnesses will be factual and will assist with both the investigation and the action to prevent recurrence. Using this approach, progress is made in accident prevention and new safety records are established.

GENERAL-EQUIPMENT SAFETY CONSIDERATIONS

15. Electrical Interlocks. Safety interlocks can be arranged for almost any machine and any operating condition.

One of the briefest definitions of the term "interlock" is that given in the National Electrical Manufacturers Association's standards: "An interlock is a device actuated by the operation of some other device with which it is directly associated to govern succeeding operations of the same or allied devices."

Interlocks have three general functions: to assure personnel safety, to protect equipment, and to coordinate complex operations.

Adequate machinery guarding is, of course, basic to any organized safety program. Human habits and practices in the interest of safety are difficult to establish and main-

SUPERVISOR'S ACCIDENT REPORT

PLANT CITY	DATE OF ACCIDENT

NAME OF INJURED EMPLOYEE (GIVE LAST NAME FIRST)

CLOCK NO.	SHIFT	TIME	☐ A.M. ☐ P.M.

ARTMENT (CHECK ONE)

☐ BREWING - 1   ☐ QUALITY CONTROL - 4

☐ PACKAGING - 2   ☐ OFFICE - 5

☐ MAINTENANCE - 3   ☐ DISTRIBUTION - 6

SOCIAL SECURITY NO.	HIRING DATE	AGE	SEX ☐ Male ☐ Female
EXACT LOCATION			
SUPERVISOR	MANAGER OR SUPT.		

HOW DID THIS INJURY OCCUR:

NATURE OF INJURY (INDICATE PART OF BODY AFFECTED)

WAS THIS INJURY REPORTED TO THE SUPERVISOR OR FIRST AID BEFORE THE EMPLOYEE COMPLETED HIS WORK SHIFT: ☐ YES ☐ NO IF NO, WHEN:	FOR MED. OR SAFETY DIV. USE
	INDICATE IF: ☐ DOCTOR'S CASE ☐ DISABLING INJURY / EST. NO. OF DAYS OF DISABILITY:

WERE THERE WITNESSES: ☐ YES ☐ NO	IF YES, GIVE NAMES:

INDICATE CAUSE(S) AND EXPLAIN FULLY

☐ LACK OF PROPER EQUIPMENT   ☐ NORMAL JOB PROCEDURE NOT FOLLOWED   ☐ OTHER (SPECIFY)
☐ FAILURE OF EQUIPMENT   ☐ LACK OF PROTECTIVE EQUIPMENT (SPECIFY)
☐ INADEQUATE EQUIPMENT DESIGN OR GUARDING   ☐ PERSONAL FACTORS

WHAT ACTION DO YOU RECOMMEND BE TAKEN TO PREVENT RECURRENCE OF THIS TYPE OF INJURY:

SUPERVISOR	DATE

WAS THIS ACTION TAKEN: ☐ YES ☐ NO	WAS WORK REQUEST ISSUED: ☐ YES ☐ NO	WHAT TYPE OF DISCIPLINARY ACTION WAS TAKEN: ☐ NONE ☐ WRITTEN ☐ ORAL ☐ OTHER:

SUPERINTENDENT'S COMMENTS

SUPERINTENDENT	DATE

DEPARTMENT HEAD COMMENTS

DEPARTMENT HEAD	DATE

INDUSTRIAL RELATIONS/SAFETY COMMENTS

INDUSTRIAL RELATIONS/SAFETY	DATE

PLANT MANAGER COMMENTS

PLANT MANAGER	DATE

DISTRIBUTION NOTE- AFTER DEPT. HEAD'S APPROVAL ALL COPIES (3) TO BE FORWARDED TO SAFETY DIVISION NOT LATER THAN DAY FOLLOWING INJURY. SAFETY DIVISION TO FORWARD (2) COPIES TO PLANT MGR. FOR REVIEW. 1 APPROVED COPY RETURNED TO DEPT. MGR. 1 APPROVED COPY RETURNED TO SAFETY DIVISION.

Form 1638 - (2-70)

Fig. 1 Typical accident report involves all levels of plant management.

tain, but mechanical gains are permanent. Interlocking of equipment, either by the manufacturer or by the user, removes hazards and is a critical part of safe design and installation.

As a general rule, the starting point in determining the need for interlocking is to examine the accident history of the machine or equipment. If there is any history of injury or major material damage, the question of whether the use of an interlocking device would have prevented the injury should be carefully considered. It should be remembered that interlocking devices and their application go beyond protecting the point of operation during the normal work process. They can be used to restrict access areas such as gate-operated controls or through a very simple device such as a cord across an entryway which plugs into a receptacle and provides test power. Anyone entering the area automatically disconnects power. Interlocks can initiate visual or audible warnings or stop an operation on a malfunction. Key-type interlocks are often employed for access and sequence control.

If a visual warning is desirable, flashing red lights are often considered. Immediately, there is the problem of a "burned-out" light and the system is not "fail-safe." Two lights in parallel offer redundancy and are generally acceptable. An alternative solution, which is better in some instances, is to provide a red and a green light showing both "danger" and "safe" conditions with an appropriate sign indicating that one light must always be "on" or shutdown is required.

If adjustments are required in dangerous areas of the machines, interlocks can be provided which will require stopping the machine to make the adjustment. This arrangement may be in the form of an interlocking guard or the interlocking of adjustment tools with the starter. Limit switches used in this manner must be wired into the motor-start circuit and not into the control circuit.

In a series of process operations, interlocks can be provided which will afford the necessary safety for operator and equipment in the event of failure of sequence timers or controls.

Machine-control systems, in many cases, combine various electrical, mechanical, hydraulic, and pneumatic interlocked elements. Of the four types, electrical systems are predominant.

Electrical interlocking has three advantages: lower cost, greater flexibility, and ease of servicing. However, problems can arise in electrical interlocks, and precautions should be taken against them. These problems may result from the following:

1. Breaking of circuits
2. Short circuits
3. Grounds
4. Residual magnetism in relays and solenoids
5. Direct failure of the limit switch or other actuator

The most common electrical interlocks are used to prevent trouble despite an error by the operator. For this reason, enclosures for controllers, switchgear, and other apparatus often have interlocks to deenergize high-voltage compartments when access doors are opened and to prevent the opening of non-load-break switches under load.

Assuring a proper control sequence is another important use of electrical interlocks. Reduced-voltage open-transition motor starters use both electrical and mechanical interlocks for this purpose. The electric control circuit includes the elements shown in Fig. 2. The normally closed "run" contact shown is an auxiliary on the "run" contactor, preventing operation of the "start" solenoid when the controller has progressed to "run" position. Similarly, the normally closed "start" contact in the "run" circuit prevents energizing the "run" circuit when the controller is in the "start" portion of the cycle.

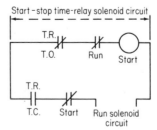

Start: "Start" solenoid circuit
Run: auxiliary contactor on "Run" solenoid
Start: auxiliary contactor on "Start" solenoid

Fig. 2 Simple relay-contact circuit prevents improper sequence in reduced-voltage motor control.

A mechanical-sequence interlock for this purpose can be a simple rod and latch, a latch keyed to the "start" contactor shaft, or a rod pivoted to the "run" contact shaft and passing through the guide bushing. A typical latch configuration is shown in Fig. 3.

Mechanical interlocking systems used on electrical equipment are of two basic designs: interference or latch type, and the teeter-totter type. The teeter-totter type has a good record of performance, but the weight and size of the devices required for strength are primary disadvantages.

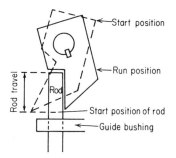

Fig. 3 Mechanical latch prevents accidental closing of reduced-voltage starter "run" contactor while start contactor is energized.

Interference systems using jam bars, latches, and other frictional devices are usually small and are more common in modern designs. Their one major difficulty is an inherent tendency to jam the entire mechanism because of friction, wear, or misalignment.

The design and application of interlocks usually affect a critical safety function. It follows that they must be extensively tested and proved, be convenient to use, have "fail-safe" provisions, and if applicable have detailed procedures to verify proper function. Often, the safety device that is on standby service, such as a crane-limit switch or a safety dog on a man lift, does not function when the need occurs owing to misalignment, wear, corrosion, etc. Checkout procedures for electrical-equipment interlocks must be considered a primary part of the safety-design responsibility.

16. Product safety has always been a major factor in design considerations. The Underwriters' Laboratories was originated as an outgrowth of product-safety problems. The current NEMA standards, while considering the factors of standardization, specification, etc., are very much oriented toward the overall safety of the product and the user. Medium-voltage electrical equipment has included many specific interlocking provisions, fail-safe devices, improved shielding of hot-circuit hazards, etc.

Within the past 7 years increasing emphasis has been focused on product safety because of changes in the basic law affecting recovery when injury or damage occurs and the upward spiral in the cost of settlements. These pressures have required many companies to formalize their procedures for assuring product safety and reliability.

The basic responsibility for a safe product lies with the product-design section. However, the functions within a company that have a direct responsibility almost include every function within a company. Table 2 illustrates the broad concepts that must be considered in building, selling, and maintaining a truly safe product. When the total-responsibility concept is established and the proper communications for coordination are developed, there are still the basic responsibilities of the design of a safe product within the product-engineering section. Various checklists have been developed to assist in a thorough analysis and critique (Ref. 2).

In addition to the problem of attempting to foresee all possible hazards, there is the necessity for legal purposes to document what safety considerations have been made and any recommendations for improvement. Table 3 is applicable to many different products and permits a senior engineer, a safety engineer, or a chief engineer to review for both quality and completeness the overall safety activity that was involved in the product evaluation. Areas of inadequate safety control are usually readily apparent.

TABLE 2.   Recommended Product Safety Responsibilities

OBJECTIVE: *To provide customers with safety-engineered products by appropriately integrating safety into design, manufacturing and marketing functions.*

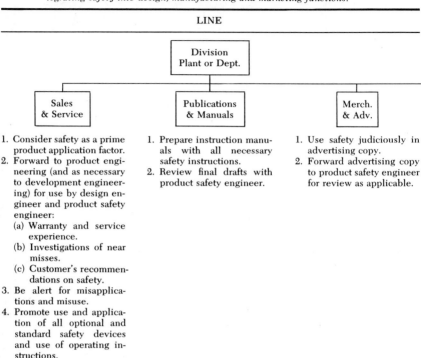

LINE

Sales & Service	Publications & Manuals	Merch. & Adv.
1. Consider safety as a prime product application factor. 2. Forward to product engineering (and as necessary to development engineering) for use by design engineer and product safety engineer: (a) Warranty and service experience. (b) Investigations of near misses. (c) Customer's recommendations on safety. 3. Be alert for misapplications and misuse. 4. Promote use and application of all optional and standard safety devices and use of operating instructions.	1. Prepare instruction manuals with all necessary safety instructions. 2. Review final drafts with product safety engineer.	1. Use safety judiciously in advertising copy. 2. Forward advertising copy to product safety engineer for review as applicable.

Another example of a safety-evaluation checklisting is given in Table 4. It represents only a starting point in most thorough safety evaluations.

There have been several good papers dealing with designing safety into the product (Refs. 3 to 5).

17. Danger, Warning, and Caution Signs.   After the basic safety-engineering principles of eliminating and guarding against an injury exposure are applied, it is still necessary in many cases to supplement these efforts with appropriate precautionary signs.

Signs and their use are guided by two national standards. ANSI Z35.1-1968, "Industrial Accident Prevention Signs," lists major categories of signs and includes "danger" and "caution" signs. The Manufacturing Chemists' Association, Inc. (MCA) recommends three categories as in the title of this guide.

DANGER denotes highest degree of hazard (ANSI Z35.1 defines use for "immediate" hazard) (Fig. 4).

WARNING is used for an intermediate degree of hazard (Fig. 5).

CAUTION is recommended for the least serious hazards (ANSI Z35.1 indicates use to warn against potential hazards or to caution against unsafe practices) (Fig. 6).

The use of the three hazard categories is recommended. Evaluations of degree of hazard and selection of the appropriate sign should consider the following factors:

1. Potential severity of injury or property damage.
2. Most probable severity of injury or property damage.
3. Duration of exposure. Is the hazard immediate and continuous, intermittent, or most infrequent under anticipated operating conditions?
4. Source of exposure. Are prior failures, human or mechanical, necessary which would warrant the use of "warning" instead of "danger"?

```
                    ┌─────────────┐
                    │ Engineering │
                    └─────────────┘
```

Development	Product Design	Product Safety Eng.-Coord.

Development

1. Analyze devices for injury potential.
2. Develop fail-safe and inherently safe concepts.
3. Review advanced prototypes with corporate safety services.

Product Design

1. Incorporate safety into product design by analyzing for
   (a) Operator injury potential.
   (b) Service life.
   (c) Misuse or abuse potential.
   (d) Compliance with state, ind. and trade standards.
   (e) Simplicity and safety of servicing.
2. Consult with product safety engineer on problem exposures.
3. Conduct tests to validate safety design.

Product Safety Eng.-Coord.

1. Review:
   (a) Accident reports.
   (b) Service reports.
   (c) Warranty experience.
   (d) New devices and components for inherent safety and standards compliance.
   (e) Operating and service instructions, aid in rewrite where necessary.
   (f) Proposed test procedures and recommend changes to assure adequate testing for operational safety.
   (g) Product advertising before release.
2. Analyze overall experience for trends and report quarterly to management.
3. Coordinate activities with corporate safety services
4. Guide design engineers in product safety standards. Communicate conceptual or long range safety problems to development engineering.
5. Recommend and coordinate component redesign.
6. Cooperate with Legal Div. in investigations, cases in suit, witness appearances, etc.
7. Represent Company on trade committees dealing with product safety.
8. Coordinate application of "Warning" and "Caution" labels for language and placement.
9. Coordinate and document product safety reviews.

Plant Safety°	Mfg.

Plant Safety°

1. To observe and analyze mfg. and test operations for user hazards.
2. Cooperate with product safety engineer in product reviews.
3. Act as advisor to product safety engineers.

Mfg.

1. Maintain close control over quality. and mfg. techniques.
2. Review with product safety engineer product hazards uncovered in assembly and final test
3. Maintain adequate final test techniques and documentation.

STAFF

Safety Advisory
Committee

1. Provide overall line/staff direction and review of product safety activities.
2. Establish ad-hoc committees for specific problem solution.
3. Provide administration and operating reviews of plans, programs, standards, etc.
4. Facilitate policy, program and activities communications.

Safety Services

1. Overall product safety administration, coordination, consultation and review.
2. Preparation of standards and guides.
3. Communications—distribution of case histories.
4. Review and analysis of Product Safety Review reports.
5. Review of significant accident and near miss reports.
6. Direct safety consultation for division development engineering.
7. Review corporate advertising before release.
8. Organize a library of standards and regulations, and analyze for overall Company effect.
9. Maintain liaison with insurance carrier on product safety.

Legal Division

1. Consultation—standards, language.
2. Preparation of defense in suits.
3. Alert management, Safety Services and Safety Advisory Committee to changes in laws affecting product safety.

° Position should be full or part time dependent on variety and complexity of plant products. Either the plant safety engineer or a design engineer with necessary experience and qualifications (graduate engineer) could serve.

CARBON TETRACHLORIDE

DANGER! HAZARDOUS VAPOR AND LIQUID
MAY BE FATAL IF INHALED OR SWALLOWED

Use only with adequate ventilation.
Do not breathe vapor
Avoid prolonged or repeated contact with skin.
Do not take internally.

 POISON

Call A Physician

MCA Chemical Safety Data Sheet SD-3 available

Fig. 4 Typical danger sign.

TABLE 3. Product Safety Evaluation

Function or component	Exposure(s) evaluated	Design spec. or limitation(s)	Accident contributing elements			Accident prevention or control provisions	Recommendations or comments
			Equipment failure or malfunction	Environment factors	Misapplications, operator or human factors		

TABLE 4. Safety Evaluation Checklist

A. What standards or safety regulations apply?
B. Have known safety problems with similar products been considered?
C. Will the product perform safely under conditions of unintended but foreseeable misuse?
D. Will the product safely withstand unexpected serious abuse?
E. What are the results if the product is misapplied?
F. Does the design present any risk to an operator?
G. Does the design minimize installation problems?
H. Who will use, maintain or service the product?
 I. What risks are involved in servicing the product?
J. How much maintenance or service does the product require?
K. Does the design minimize maintenance problems?
L. Has product packaging been reviewed for potential safety problems?
M. Has advertising, sales promotion literature, labels and instruction material been reviewed?

HYDROCHLORIC ACID
WARNING! CAUSES BURNS

Avoid contact with skin and eyes.
Avoid breathing vapor.
In case of contact, immediately flush skin or eyes with plenty of water
    for at least 15 minutes; for eyes, get medical attention.

MCA Chemical Safety Data Sheet SD-39 available

Fig. 5 Typical warning sign.

ACETIC ACID, 28%
CAUTION! MAY CAUSE BURNS

Avoid contact with skin, eyes, and clothing.
In case of contact, immediately flush skin or eyes with plenty of water
    for at least 15 minutes; for eyes, get medical attention.

MCA Chemical Safety Data Sheet SD-41 available

Fig. 6 Typical caution sign.

5. Purpose of sign. Is it intended primarily to outline safe procedures ("caution"), to warn against unsafe practices ("warning"), or to sound an alert about the significant danger of a specific hazard ("danger")?

Recommendations
1. Use "danger" for immediate or near-continuous hazards usually involving high risk of severe injury or fatality. Hazards arising solely from unsafe practices should be included in this category only when there is a significant exposure to severe injury or fatality.

Examples
Unguarded electrical exposures higher than 115 V ac or 125 V dc
Vehicle-overturn potentials—misuse of drawbars, high hitches

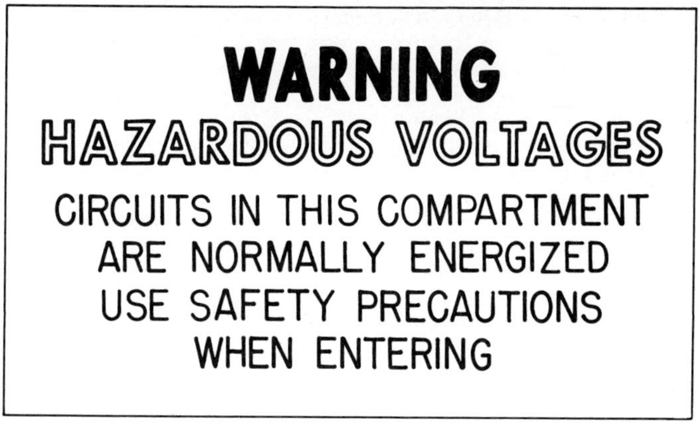

WARNING
HAZARDOUS VOLTAGES
CIRCUITS IN THIS COMPARTMENT ARE NORMALLY ENERGIZED USE SAFETY PRECAUTIONS WHEN ENTERING

Fig. 7   Warning sign used when conditions involve moderate risk of severe injury.

Use of tractors for carrying passengers

Exposed points of operation—rotary mowers, shredders, etc.

2. Use "warning" for hazards which involve moderate risk of severe injury or which are present only intermittently but have a severe risk, fatality, etc. (Fig. 7). This classification also can be used for hazards involving significant damage to equipment or property.

Examples

Instructions to shift transmission into neutral and disengage power takeoff on tractor when dismounting

Immediately ceasing operation on signal of low engine oil pressure

Use of safety belts on automotive equipment

Procedures on guard removal and replacement

Grounding portable electrical tools

3. Use "caution" normally for instructional purposes for the least serious, possibly intermittent, hazards. They are primarily procedure-oriented (Fig. 8). The risk of severe injury or major property damage should be fairly remote. Multiple or sequential factors or failures should be necessary to cause injury or loss.

Examples

Most safe-driving rules

Hazards which exist only during certain weather conditions for a certain group of operators, human-factor problems, etc.

CAUTION
ENTRY MAY BE HAZARDOUS

1. HAVE ATMOSPHERE CHECKED FOR:
   TOXIC OR FLAMMABLE GASES
   LACK OF OXYGEN
2. USE SAFETY BELT AND LIFE LINE
3. ONE MAN MUST STAY OUTSIDE TO KEEP WATCH

Fig. 8   Caution sign used for least serious or possibly intermittent hazards.

Tire loadings, tire-wear criteria (On high-speed or specialty equipment, this could require a "warning" sign)

Procedures to follow upon overheating of vehicle components or systems

Maintenance procedures

Protective-equipment requirements

18. Lockout Procedures. The failure to lock out properly and render a piece of equipment safe is the primary cause of electrical injuries. Not too many years ago a safety switch placed ahead of the primary controller was considered adequate. The location of this switch, the ability to lock it out properly, and whether it could interrupt the maximum circuit amperage expected were given little thought. Some codes still consider location within 50 ft satisfactory. Instead of readily accessible as required by code, disconnects should be immediately accessible. Equipment that is controlled and driven from one floor and that requires maintenance on another floor encourages maintenance activity without lockout. This thinking is in keeping with the basic safety axiom that the employee will not consistently follow even a very reasonable safety procedure (lockout) if it is not easy, convenient, and efficient to do so.

One of the major problems still present even with a good lockout program is exposure from movement of mechanical parts due to stored energy. The energy can be due to simple gravity, arising from parts on an inclined conveyor, upper rams of presses, stored energy from oil, water, pneumatic reservoirs and surge tanks, etc. If it is not reasonably convenient to deactivate and render these systems safe, injuries will occur. It is critical, therefore, that the designer of an overall system incorporate devices to bleed off or safely contain all stored-energy potentials as part of a complete lockout shutdown. An example is the discharge-resistor provision which automatically bleeds energy from a capacitor bank. Having a general lockout and tag-out procedure has proved successful, and is recommended (Fig. 9).

Fig. 9 Typical lockout tag.

a. Scope of Equipment Lockout and Hold-off Tag Procedure. A mandatory procedure to safeguard maintenance employees against injuries or property damage from accidental operation of equipment or processes during installation, maintenance, repair, adjustment, and lubrication. Under certain applications, this procedure may be extended to production employees.

For automated equipment, a standard design practice details specific additional safe practices supplementing the requirements of this procedure.

b. Responsibility. Each employee is responsible for taking personal positive measures in accordance with this procedure to assure against inadvertent start-up. These measures must be sufficient to protect against injury or equipment damage despite misoperation of equipment or human error.

c. General Procedures. Before the employee to be protected starts his work, the following shall be done:

1. Notify production and maintenance supervisors.

2. The main switch, valve, or operating lever shall be placed in the "off," "closed," or "safe" position.

3. Check or test to make certain that the proper controls have been identified and deactivated.

4. A lock shall be placed to secure the disconnection whenever possible. If a lock cannot be used on electrical equipment, an electrician or otherwise qualified individual shall remove the fuses from the circuit.

5. A hold-off tag shall be attached to the switch, valve, or lever. This tag shall bear the name, department, and telephone extension of the employee (or department) performing the work.

6. When auxiliary equipment or machine controls are powered by separate supply sources, such equipment or controls shall also be locked and tagged to prevent any hazard that may be caused by operating the equipment or by exposure to live circuits.

7. When equipment uses pneumatic or hydraulic power, pressure in the lines or accumulators shall be relieved. If pressure-relief valves have not been provided, the equipment shall be cycled until the pressure is dissipated or the pressure lines shall be opened or disconnected at appropriate connections.

8. When stored energy is a factor as a result of position, spring tension, or counterweighting, the equipment shall be placed in the bottom or closed position or it shall be blocked to prevent movement. The latter action is appropriate to prevent movement of motor or generator armatures while being serviced even though an electrical disconnect (lockout) has been accomplished.

9. When the work involves more than one person, additional employees shall attach their locks and tags as they report.

10. When outside contractors are involved, the equipment will be locked out and tagged in accordance with this procedure by the project engineer supervising the work. Only in emergency cases is equipment to be shut down by other than a company representative.

d. Equipment. Each employee who is exposed to accidental start-up hazards shall be provided with an individually keyed lock stamped or marked with his name and a sufficient number of hold-off tags.

Special lock attachments which will accommodate more than one lock shall be made available at tool cribs for use as required.

Additional locks identified by number shall be available at tool cribs for immediate issuance as required.

e. Lock control and usage requires the following procedures:

1. Master keys to fit all locks issued should be made available only to operating and maintenance superintendents.

2. Locks may be removed by the use of a master key only in the immediate presence of two management representatives (one each from maintenance and operations), one of whom must have the authority to possess a key.

This procedure must be used with extreme caution and good judgment. There is danger that the worker involved will return and proceed on the assumption that the machine is still locked out.

3. Any individual leaving a job with his lock attached for more than a few minutes shall notify a production or maintenance supervisor. In every instance after leaving a

job, he must recheck upon returning to make certain the machine is still locked out. If the shift ends before the job is completed and the machine is unsafe to start, the supervisor should be notified. He will apply his own lock.

4. Safety locks shall not be used for any purpose other than to safeguard personnel and equipment.

f. Use of Equipment with Hold-off Tags Affixed. Operation with hold-off tags in place (lockout not feasible) is permissible only for necessary testing or adjustment and only upon express direction of a supervisor.

g. Enforcement. All employees who fail to follow the requirements of this procedure will be subject to disciplinary action.

19. Disconnects. The problem of unsafe or improper application of knife switches should be noted specifically, as many of these installations are still in use. The hazard is normally the blowing up of the knife switch while attempting to interrupt inrush, stalled-rotor, or short-circuit currents in excess of its capacity. Many older switches do not have a spring or snap action, interlocked cover, or arc-interruption devices. Many are severely underrated for the circuit. It has been common practice in many progressive concerns to review all such installations, catalog their hazard potentials, and then begin a concerted replacement program.

Some additional criteria for safety switches include:

1. Blades dead when circuit is open.

2. Operating handle with lockout provisions for a minimum of three locks.

3. Position of handle and circuit condition easily seen with large legible fail-safe indicators. Red handles sometimes used.

4. Heavy-duty portable-equipment switches with interlocks to prevent connecting or interrupting circuit to the equipment with circuit energized.

5. Where not of load-break design, must be interlocked. Examples are tap changers on transformers, and regulators.

6. Double-throw switches should be mounted in a horizontal plane to be unaffected by gravity. If necessary to mount vertically, latching devices are indicated.

20. Basic Grounding Procedures and Requirements. Article 250 of the National Electrical Code, "Grounding," details with the specifics of grounding requirements for electrical systems in general and refers in Art. 250-2 for additional requirements for specific locations and equipment. The hazards to life from inadequate or improper grounding are obviously most significant, and even minor deviations are therefore completely unacceptable.

Problems with grounding circuits usually are not a fault of the original installation (particularly when reviewed by a good inspector) but arise through general wear, corrosion, and quite often burnouts due to stray welding return currents (Art. 27). One sometimes overlooked item is the failure to ground wire-link fencing.

A good starting point, supplemented by the NEC specifics, is to ground all exposed metallic non-current-carrying parts of electrical supply equipment. These items include such equipment as frames of motors and generators, controllers, switch boxes, cabinets, conduit, transformer cases, operating handles, and machine tools. As noted previously, problems generally arise through the failure over a period of time of safety grounding circuits. The solution is to institute periodical ground-inspection procedures. These procedures are followed on varying time schedules and should include visual examinations most frequently, followed up periodically by mechanical inspections to verify tightness and bondings of joints, etc., and less frequently, electrical-continuity checks. It is considered standard good practice to test grounds of portable electrical tools at least once every month and machine tools and other fixed equipment at least annually. Welder grounds should be checked often, perhaps weekly. A megger ground tester is often employed to test the impedance levels of safety grounds. This device is not the common megger insulation tester.

There is a continuing divergence of opinion as to the safety advantages of grounded-neutral vs. ungrounded-neutral control systems. While there are safety advantages of each, the major advantage of the ungrounded-neutral system is that it minimizes interruption of service on vital production equipment. The advantages of ungrounded systems are considerably negated if action to correct ground problems immediately is not given high priority. A minimum in such systems is the use of ground-fault indicators or lights. Special ground-fault locator devices are available and can make ground-fault location a straightforward maintenance problem.

SPECIFIC-EQUIPMENT SAFETY CONSIDERATIONS

21. Control Systems. The basic standard covering design safety requirements for industrial-control systems is the NEMA Standard IC-1, "Industrial Control." These standards in many instances supplement the NEC and UL requirements, even though the specific type of equipment may be separately listed in the NEC. These specifications, however, are not sufficient if one is interested in the highest degree of safety possible.

The first concerted effort to standardize and require additional specific safety requirements was begun by the National Machine Tool Builders Association in 1941 and revisions continued through 1960. Paralleling this move, the Joint Industry Conference issued its first standards in 1948. These revisions continued through 1957. Since that time two new electrical standards involving industrial equipment have been developed. They are "Electrical Standards for Mass Production Equipment," EMP-1-1967, and "Electrical Standards for General Purpose Machine Tools," EMT-1-1967. Both apply to systems of 600 V or less.

Both of the standards are now handled administratively through the National Machine Tool Builders Association.

These standards go into considerable and specific detail regarding such important safety considerations as equipment nameplates, supply-circuit disconnects, circuit design and interlocking, mounting of controls and switches, and work lights and grounding. Full familiarity with these standards for the designer or purchaser of industrial controls is important. It is impossible to detail the many requirements they contain. Many provide major increases in safety factors over the minimums required by the NEC.

a. Human Error—Voltages. Simplicity should be a prime consideration in the design of operating and alarm panels. Flow charts between indicating lights and overall layouts of the system enable new operators to assume their duties without long, intensive training.

An important concept is that control systems not only need to meet all codes and standards but must compensate for and be safe in the event of human error and machine malfunctions. This concept requires a thorough analysis.

The question, "What would happen if...?," must be asked at every point of a production-cycle operation or multimachine start-up procedure. When the machine is in trouble, the injury hazard to the operator or maintainer either makes its first appearance or, if already present, increases manyfold.

During the 1950s many safety-progressive companies standardized all control voltages at 115 V ac for ac-controlled equipment. There were some difficulties with additional transformers and, in some instances, problems in achieving function with this voltage. A voltage of 115 V ac is dangerous and 230, 440, and 550 V ac are more so. For these higher voltages, the substantial degree of protection afforded by dry skin is very much reduced. Good practice is to make 115 V ac control standard and rarely deviate. In special situations 125 and 250 V dc are reasonably safe.

b. Limit Switches. The misoperation or failure of limit switches can result in severe equipment damage and personal injury. A major concern of an otherwise properly applied limit switch is its location with respect to accidental actuation. This condition can occur either while in normal operation or during repair or troubleshooting.

Often the type of switch such as a recessed type with push-rod actuator can provide inherent protection. In other instances a guard or shield is indicated. Unless designed for such service, different sources of power or sources of opposite polarity should not be connected to one switch (Sec. 3).

Another elementary consideration, although often overlooked, is placing the switch in an adverse environment, subject to accumulation of debris, chips, oils, etc., where such positioning is not really necessary. At times the design of the conduit connections are such that oils, water, etc., are led to the switch instead of having drip-free provisions. Sealing conduits to prevent such an occurrence often lasts only until the first breakdown. The potential problems of service and replacement should also be considered and the limit switches placed in such a location as to minimize or eliminate the need for someone to climb on or enter into equipment—a positive safety contribution.

c. Push Buttons. Specifications and application of standard push buttons are detailed in NEMA Standard IC-I and EMP and EMT Standards. The position and identification of associated start and stop buttons is detailed. Also noted is a requirement for mushroom-type stop buttons.

A nonstandard requirement is a full shroud for all buttons which initiate a motor start or machine movement (Ref. 7). The standard push button for emergency stops should, in many cases, not be limited to a single control-station location but should also be provided at the front, back, or sides of a machine. Many times their nearness to others besides the operator has averted major damage or serious injury. Conveyor lines often are inadequately equipped with emergency stops. They should be provided at all transfer points and at frequent intervals along the line. A chain or cable operator is an alternative. Warning buzzers or horns combined with delayed starts are also desirable for large equipment.

It is important that emergency stop buttons be equipped with reset protection, either mechanically or electrically, operating at the point of shutdown to prevent restart at another point of a conveyor line or at the control panel of a large machine. On circuits where operation is by repetitive cycling or start controls, the circuitry should be "anti-repeat" with "antiholddown" provisions. This arrangement will minimize the potentials of repeat cycling and assures that each push button must be released and reactivated for each cycle. As noted previously, these types of controls require shrouding or equivalent protection to prevent actuation by elbows, foreheads, etc.

22. Medium-Voltage Motor Control. Some basic safety features in medium-voltage motor controllers include:

1. Primary disconnect interlocked to prevent opening under load, incorporates lockout provisions.

2. Horizontal rack-out design, movement prevented (in or out) when contactor is closed.

3. Automatic mechanically driven shutters for line-terminal shielding.

4. Isolated, separate-compartment buses. Compartment door interlocked against opening unless carriage is racked out.

23. Switchgear safety considerations include:

1. Horizontal drawout breakers interlocked to prevent opening under load.

2. Disconnect device employing multiple fingers and springs, inherently more fail-safe.

a. Arc chutes may be a source of safety problems. Lift-out designs have potential problems of cracking and failure to replace. Hinged-type units are much more satisfactory and inherently safer. There have been innumerable instances of failing to replace arc chutes in contactors and switchgear circuit breakers with obvious consequences.

It should be standard procedure to note either on the lockout tag or on a special tag whenever chutes are removed:

DANGER—Arc Chutes Removed—REPLACE before operating

Some of the current designs incorporate safety interlocks or other provisions (hinged as noted above) which protect against operation with the chutes removed.

b. Other safety features include:

1. Antipumping and trip-free circuitry

2. Stored-energy mechanisms or provisions

3. Automatic shielding and isolating shutters, mechanically driven, nonmagnetic

4. Key interlocks among functions

5. Auxiliary-equipment control system, etc., positioned for safe servicing

6. Position indicators fail-safe and distinctively marked

7. Load-break switches with fuses on load side

24. Overhead Handling. At some point during the design stage or the installation of a major piece of electrical equipment, there is the consideration of overhead handling. There are the problems for the designer of "what handle to pick it up with?" which will be compatible with plant facilities. The purchaser often has a similar problem unloading the equipment from the railroad car or truck and moving it to the point of final utilization. By reference to any table of safe-load capacities, the hardware for the lift is usually quickly and safely specified or selected. Other than gross error, the two major difficulties in this process of providing "handles" and transporting arise from two sources. These difficulties involve the use of eye bolts and the problem of load stability.

  a. Eye Bolts. Eye bolts, because of their extremely wide variation of allowable load to the angle or pull, are continual problems and must be carefully selected and specified. Nonshoulder-type eye bolts are obviously completely unsuitable for side loading and should not be used except for vertical lifts. Often eye bolts are required to have a safety factor of 10 in straight tension. Safe loads for high-quality drop-forged-steel eye bolts with relieved shanks at the shoulder to permit full engagement are given in Table 5.

TABLE 5. Eyebolt Safe Loading vs. Pull Angle

Shank dia.	Vertical	60° Top pull	45° Top pull	90° or less side pull
Inches	Pounds/Bolt	Pounds/Bolt	Pounds/Bolt	Pounds/Bolt
¼	300	50	30	40
½	1300	200	140	150
¾	3000	400	250	300
1	6000	800	500	600
1 ¼	9000	1300	800	900
1 ½	13000	1800	1200	1300
2	23000	3300	2100	2300
2 ½	37000	6000	3800	4300

  A much more satisfactory device for side loads is a welded lug properly sized and positioned so that the anticipated sling pull is in line with the axis of the lug.

  Standards and specifications for welded lugs were described in Ref. 8. It indicated that welded lift lugs when properly dimensioned and applied can provide safe and efficient overhead material handling. The use of the tables in Ref. 8 makes their application a straightforward design situation. Full utilization will provide maximum safety in handling procedures with a minimum of construction cost and handling-time requirements. A major consideration in control and switchgear is the minimizing or complete elimination of sling and product damage which frequently results from excessive slippage or bearing pressures between the sling and the product.

  b. Instability Problems. Whenever overhead-handling equipment is used to move a load, there is the risk of having the load tip if the slings or rigging are attached below the load's center of gravity. It is not uncommon, because of the unitized construction of switchgear and control assemblies, to attach them to the main frame or undercarriage for overhead lifting. This hitch obviously creates the danger of load instability, and there have been numerous cases of lost loads, damage to equipment, and personal injury (Ref. 9).

  The five following questions should be explored:
  1. What are the situations in which instability is likely to go unrecognized by riggers?
  2. What causes vertical instability?
  3. Why do riggers fail to detect the unstable conditions?
  4. What are the best ways to prevent vertical instability?
  5. How can we get an adequate margin of safety in load stability?
  Specific guides are established in Ref. 7 to setting safe margins of stability using

graphical procedures. They generally revolve about reducing the base angle or the distance of the center of gravity above the attachment point with respect to the distance of the hitch point above the spreader bar (Figs. 10 and 11). Two special types of rigging, the "umbrella-tent" and the "bird-cage," should be checked carefully (Figs. 12 and 13). The "bird-cage" type of rigging should be avoided if at all possible. Such potential stability problems are positively avoided by providing load-attachment points which are above the center of gravity of the device. The riggers' difficulties in using these hitches may be further complicated by

1. Lack of headroom at some point along the line of load movement
2. Need to protect upper surfaces of the load from the inward, crushing, or chafing force of the slings or load lines

Fig. 10   Pyramidal rigging is best for good vertical stability if base is wide and center of gravity G is at any point on line AF.

Fig. 11   Parallel-load rigging is stable if base angles b, c are greater than base angles d, e.

Fig. 12   Umbrella-tent rigging is made stable by greater spread of attachment points.

Fig. 13   Bird-cage rigging is inherently unstable and should be avoided if possible.

Figure 14 illustrates a situation in which loads look stable but may be dangerously unstable.

25. Floor-controlled Hoisting Equipment.   There are a number of significant electrical safety considerations that should be included in the specification of hoisting equipment regarding disconnect, limit, and safety switches.

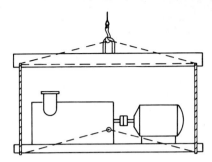

Fig. 14 Beam spreader may be equivalent to a parallel-load line rigging with very shallow base angles on the hook-spreader triangle (dotted lines).

a. Main Disconnect Switch. All crane and hoist installations shall incorporate a wall-mounted main disconnect switch easily accessible to the operator.

For units operated on a 3-phase electric circuit, the main disconnect must be arranged to interrupt all three phases. On units operated by direct current, both leads must be disconnected.

b. Secondary Disconnect Switch. All cranes and hoists which may travel more than 25 ft from their main disconnect switch shall be equipped with a secondary disconnect switch which will completely isolate the unit from the main power supply. This secondary disconnect shall be mounted on the bridge or hoist and shall have a pull cord suspended near the operator. As an acceptable alternative, this secondary disconnect may be electrically operated through a stop button mounted on the pendant.

c. Safety Switches. Units of over 3 tons capacity shall incorporate a secondary load-break switch which will be actuated in the event an upper-limit switch should fail.

26. Power Presses. Electrical controls for power presses have undergone a major revolution, beginning in the early 1950s. With the advent of reliable air-friction clutch systems a speed of response and versatility was available that lent itself to process electrical controls. A part of this near revolution was a major improvement in the inherent safety to the operator. Controls developed from the two-hand control permitting tie-down to anti-tie-down with "time-out" features. For a number of years the complex safety oriented electrical controls for presses had significant reliability problems. They were safe but they did not run too regularly, and it took a high-grade electrician to service them. In more recent years these problems have largely disappeared. Listed below is a recommended safety checklist for purchase of a new power press. Many of these should be more carefully defined. This checklist provides the starting point for negotiation with the press manufacturer.

1. Air-friction clutch-brake system.

2. Dual clutch valves or equivalent with electrical monitoring. If not monitored, valves shall be pneumatically in parallel.

3. Dual control relays.

4. Two independently driven rotary limit switches (if single with loss-of-motion detection system).

5. Low-air-pressure interlock.

6. Motor drive and control circuit interlocked to prevent normal "run" operation if motor is not running in forward direction.

7. Circuitry providing (*a*) anti-tie-down protection, (*b*) single-stroke protection, (*c*) key-lock-selector switches, (*d*) long-hold timing (120 to 160° recommended).

8. Safety blocks interlocked with control circuit.

9. Guarded two-hand controls (one set for every operator).

10. Multiple-station controls with lockout or dummy plug provisions to incorporate closed-circuit interlock protection.

11. Two-"inch"-button operation preferred or guarded single-"inch" button where necessary.

12. Flywheel brake (normally 20 to 30 s stopping time) interlocked with drive motor.

13. Mufflers on valve exhaust ports.

14. Main-control safety-lockout provisions.

27. Welder Grounding. The proper and safest way of grounding welding equipment is a matter of continuing controversy. Many welders are provided with isolated secondaries. Operation in this manner is attractive because it eliminates the possibility of welder return currents from passing through various items of mechanical equipment, antifriction bearings, crane cables, etc., and burnout of the welder main power-supply cables. There is, however, personnel hazard of the large metal case or work terminals becoming energized to supply voltage, which is, in most cases, 440 V, 3-phase, ac.

If the secondary output of the welder is connected to case ground and from there to building ground, a high degree of personnel protection is provided if the system is maintained. There is, however, the problem of return welding current which can occur without proper secondary connection. This current returns through building equipment and equipment-supply cables. Frequent burnout of supply and grounding cables is the result (Fig. 15). Most equipment is provided with an isolated secondary by the manufacturer. It is imperative that secondary ground and case connections be made external to the case and easily visible. A basic program for maintaining safe welder operation is as follows:

1. Maintain an adequate building ground on all welding fixtures, preferably strap type. The connection point for the ground cable or strap should be painted white or bright yellow. A 10- or 12-in circle is suggested.

2. Maintain an external case ground on all welders. In some instances, this case ground would supplement the ground lead contained in the 4-wire primary-service cord.

3. Provide a welding secondary (output) ground connected externally to the case ground. By code this ground lead must have a wire size at least two AWG size numbers smaller than the protective ground as per item 2.

4. The grounds should be checked at least once weekly by the supervisor of the welding operation.

5. Training in the grounding provisions and purposes and the shock-exposure potential should be conducted regularly and should include both welder operators and electrical-maintenance personnel.

28. Field Servicing and Installation

1. Available instruction books and warning tags should be read carefully to obtain complete familiarity with the equipment and safety information pertaining to personnel.

2. All electrical equipment, especially where exposed terminals are present, should be considered as being energized or hot until the isolating switch or breaker has been personally checked and locked open and then appropriately tagged per a standard lockout procedure. Removal of low-voltage line fuses is also recommended, particularly when using special test equipment.

3. Wherever possible, grounding chains or bars should be connected across incoming power connections. Precautions should be taken to assure proper removal when units are to be reenergized. It is standard good practice to check high-voltage circuits with a megohmmeter (megger) before release of equipment.

4. The operation of all safety devices and interlocks should be questioned and checked as a standard procedure.

5. All voltage, regardless of classification as high-voltage or low-voltage, should be considered dangerous.

6. Safety devices should be defeated only where absolutely necessary. Prominent tags should be immediately affixed detailing changes. When equipment must be checked out while energized, the removal of main-line fuses may eliminate unnecessary exposure to high voltages without interfering with checkout procedure. Where equipment operation has not yet been verified on energized equipment, appropriate rubber gloves are recommended. Rubber blankets, shields, etc., may be indicated.

7. Use safety tools as provided or as required such as hook sticks, fuse pullers, manual operating tools, gloves, and handles.

8. Make certain adequate equipment grounding has been provided.

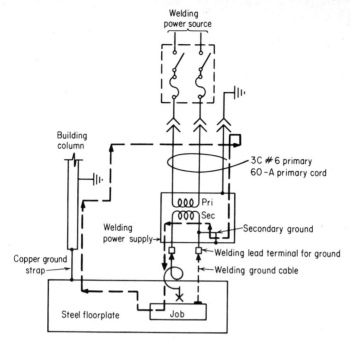

Fig. 15 High degree of personnel protection is provided by proper grounding of welding equipment using this method if system is maintained.

9. If at all possible, expose only one hand in live-current areas, keeping one hand in the pocket or behind the back locked in the belt.

10. When working on live equipment, have another person close by who is familiar with treatment of electrical shocks.

REFERENCES

1. Charles F. Dalziel, "The Effects of Electric Shock on Man," *U.S. Atomic Energy Commission Safety and Fire Protection Bulletin,* July 1959.
2. Dean H. Robb, "Safety Is Not Just Common Sense . . . A Trial Lawyer's View," *J. Am. Soc. Safety Eng.,* December 1965.
3. M. F. Biancardi, "Designing for Product Safety," SAE Paper 700679, September 1970.
4. Clare Wise, Product Liability, *Machine Design,* Mar. 28, 1968.
5. Willie Hammer, Systems Design . . . Designing a Safe System, *Machine Design,* Sept. 3, 1970.
6. "Industrial Accident Prevention Signs," ANSI Z35.1-1959.
7. R. F. Schoof and T. F. Bellinger, How to Check Electrical Hazards, *Rock Prod.,* October 1960.
8. R. F. Schoof and J. A. Churchill, Safer Lifting Built In, *Allis-Chalmers Elec. Rev.,* 2d Quarter, 1962.
9. J. A. Churchill and R. F. Schoof, How to Prevent Suspended Loads from Tipping, *Material Handling Eng.,* July 1963.

31

Maintenance

ROBERT C. BLAKEY*

FOREWORD

When starting up a new plant or modernizing an older plant, particular care and attention must be given to power-distribution equipment because it not only supplies power to the various areas of the plant but also provides fault protection for the power system serving the plant.

While this section covers maintenance for industrial substations, it also provides practical information on power-distribution equipment which may not be a part of the substation.

Knowledge in handling distribution substation equipment will aid in maintaining industrial-control equipment also operated on the distribution system.

Because this range of equipment controls and protects vital distribution voltage systems, emphasis is given to safety for personnel and equipment.

*Manager, Apparatus Repair, Westinghouse Electric Corporation, Phoenix, Ariz. (Retired).

GENERAL

Preventive maintenance first and foremost requires equipment knowledge, equipment records, and careful planning.

Equipment knowledge is obtained from instruction leaflets and descriptive bulletins on maintenance and adjustment techniques, all readily supplied by reputable manufacturers of such equipment. This valuable instruction material should be divided into three sets of identical material. The first set of materials should be circulated among maintenance personnel to make them aware of the general nature, problems, procedures, and necessary adjustments. The second set of materials should be available for study and procedure when actual maintenance or a breakdown is experienced. The third set of materials should remain in permanent file and be removed only for duplicating the first or second set of materials.

Many plants have old or obsolete equipment, working satisfactorily, but the manufacturer no longer has such material in print. Many companies in this age of "conglomerates" have sold sections of their manufacturing lines to other companies, and during the transfers many valuable instruction leaflets have disappeared.

Equipment records, properly maintained, give a history of past performance which serves as an excellent guide for future performance and planning. Experience records implanted only in an operator's mind disappear when the operator is transferred or otherwise leaves the department.

Insurance companies frequently base their premium and insurability upon past and present records. For example, they may feel more confident of an older transformer which has maintained a relative constant megger reading throughout the years than of a much newer transformer whose megger readings fluctuate wildly throughout its short history.

Some utilities will allow customers to connect any power-consuming load to their lines; others, however, will insist upon reviewing customer maintenance schedules and corresponding records before and during the time such loads are connected.

For this reason, no attempts have been made to set limits on various test results. Utilities, insurance companies, and experienced maintentenance personnel frequently disagree. It is suggested that the manufacturer's recommendation be the base of limits, tempered by utility, insurance recommendation, and later by seasoned personal experience.

Finally, the total preventive program will be a success or failure in direct proportion to the planning which is expended in such programs. Failure nearly always costs more than maintenence, and the downtime failures may mount to astronomical costs.

INDUSTRIAL SUBSTATION-EQUIPMENT CARE

Figure 1 shows an electrical-distribution system and various parts of associated equipment. Consideration based on this arrangement may be given to maintenance planning, setting file records, renewal parts, instruction leaflets, descriptive bulletins, etc.

1. Disconnect Switches. These air-break, high-voltage, hand-operated switches are used for insulating high-voltage equipment, so that this equipment may be worked on safely. There are three general types, according to their usage.

1. A no-load disconnect switch is not designed to interrupt any current, but physically to disconnect equipment after all current has been interrupted by other means. These knife switches are usually opened by a hook stick and are frequently single-pole switches.

2. A no-load air-break switch is designed to break transformer magnetizing current, but not any load current. These switches have arcing horns so the sparking will not damage the main contacts and all arcing during closing or opening will occur at the arc horn or auxiliary contacts.

3. A partial-load air-break switch is designed to break a specified amount of load current. In addition to the arcing horns, they have an arc chute or arc box to extinguish quickly the arcing produced by load current.

Both no-load air-break and partial-load air-break switches are gang-operated, as opening of single-pole switches cause momentary single phasing of equipment which can severely upset a system.

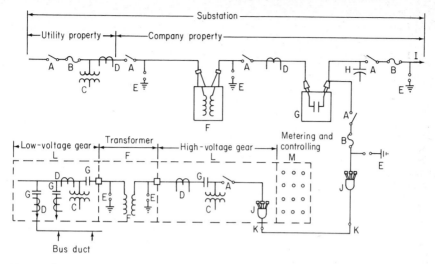

Fig. 1 Electrical-distribution equipment as commonly found in industrial substations. (A) Disconnect switch. (B) High-voltage fuse. (C) Potential transformer. (D) Current transformer. (E) Arrester. (F) Transformer. (G) Breaker. (H) Capacitor. (I) Overhead line. (J) Pothead. (K) Underground cable. (L) High-voltage–low-voltage gear. (M) Metering and controlling. (N) Plant start-up.

If there is any possibility of a feedback in transformer circuits, both primary and secondary sides should have disconnect switches. Before any work is performed on the disconnected high-voltage equipment, all buses or windings should be grounded and shorted by a grounding chain. Switches should be padlocked in the open or closed position to prevent unauthorized operation. Usually only one key is issued to the lead operator.

It is not considered safe practice to work on high-voltage equipment when disconnected by means of an oil circuit breaker only. Carbonized oil may allow sufficient "sneak current" to pass to produce a serious shock.

Maintenance of the following parts may be involved:

a. Switch Insulators. These porcelain or epoxy insulators are "stand-off insulators," as live parts are insulated from ground and between phases and no current flow passes in a lead within the insulator.

Inspection of porcelain consists of examining the insulator glazing for chipping, cracks, or possible "crazing" (small cracks), and the insulator surface for cleanliness. In some localities, atmospheric contamination causes a conducting film to build up on the insulators which reduces the flashover value to ground, or phase to phase, particularly when wet. Soft rags, water, and nonabrasive soap is usually a satisfactory cleaning agent. In stubborn cases, the local utility can recommend the best cleaning agent. Glazed porcelain surfaces should not be scratched while cleaning.

b. Connections. All connections should be examined for signs of heating. All connections must be kept clean and tight. Corrosion may set in on copper-to-aluminum connections in salt air or contaminated air. If such problems develop, the local utility will have the problems worked out and will be glad to provide information.

c. Mechanical Linkage. The linkage should be operated and examined for indications of friction, misalignment, or corrosion. Gang-operated switch linkage should be adjusted so all switches close and open simultaneously. The linkage should be padlocked in the open or closed position to prevent unauthorized operations. (Note that there are also insulators between the steel mechanical linkage and the live parts of switches.)

d. Arcing Tips or Auxiliary Contacts. These devices must be adjusted so they close before main contacts and open after main contacts. The factory maintenance manual should be consulted for such adjustments, including contact pressure, which is usually specified when the main contacts are fully closed. Badly pitted or burned contacts are usually replaced.

e. Main Contacts. Oxidizing of silver-tipped contacts should be ignored, but copper contacts should be polished when oxidized. Moving contacts must be carefully aligned to stationary contact. Contact pressure, which is usually a function of travel distance after tips touch, should be checked. All contacts should be adjusted for simultaneous closing of three-pole switches (or the simultaneous closing of a double-break, single-pole switch as frequently used on oil circuit breakers). All springs and shunts should be examined for signs of heating. Frayed shunts or weak springs should be replaced. The factory maintenance manual should be checked for adjustments.

f. Arc Chutes or Arcing Boxes. These devices are designed to contain the arc (from main contacts) within a certain confined area so that it will not flash phase to phase or to ground, and to extinguish the arc quickly. Some devices operate magnetically to push the arc upward, while others allow the arc to rise (because of heat) into a series of insulated steel plates, thus breaking the arc into a series of smaller arcs for ease of extinguishing. Excessively burned or pitted arc boxes should be replaced or repaired. If repairable, the factory maintenance manual should be referred to. Arc boxes must be in proper position for effectiveness.

2. High-Voltage Fuses. These fuses are not designed to protect individual equipment, but to protect lines. When fuses "blow" and interrupt the circuit, they drop a flag or stringer so observations will show they have "blown."

Replacement fuses or replacement links must be of the proper voltage and current rating. All connections must be tight, and insulators clean (Art. 1*a* and *b*).

These fuses are individual fuses and are not interconnected. Since one fuse can blow without affecting the other two, a single-phase condition can occur which can be harmful to motor operation. If two fuses blow, one leg of the circuit will still be alive, which might be a safety hazard for the unsuspecting maintenance operator. Under this condition power current can still flow on a grounded wye-wye circuit or a midpoint grounded delta.

When a high-voltage fuse or fuses blow, the best procedure is to drop all load, open no-load disconnect switches, replace all blown fuses, close no-load disconnectors, and apply individual loads by their breakers or switches.

If the system fault is still present, the fuses will again blow, and individual circuits must be examined for faults.

3. Potential Transformers. A potential transformer (PT) is a small rated transformer, used to reduce high voltages to a low-voltage source, usually 120 V, for metering or controlling.

Potential transformers used as control source have primary voltage and secondary voltage given on the nameplate. Frequently taps are brought out in the low-voltage windings so that a selected low voltage can be obtained. Unused taps must be insulated.

Potential transformers used for metering do not have taps, as a fixed ratio is required between the primary and secondary voltages. The nameplate will state primary and secondary voltages and the ratio which is primary voltage/secondary voltage.

Metering transformers are more expensive than control transformers, as metering transformers have carefully designed constants to reduce the primary-secondary phase angle to a minimum.

Potential transformers are constructed for indoor or outdoor use. For safety, the case or mounting structure should be solidly grounded.

Potential transformers have a polarity marker to indicate relative instantaneous polarity between primary and secondary. This polarity marker may be a white dot or button on one primary and one secondary lead, or the marking may be H1 on the primary lead and X1 on the secondary lead. When replacing a potential transformer, the polarity markers must be observed.

Primary and secondary fuses, if used, are usually nonrefillable cartridge fuses of the proper voltage rating. The fuses are not used to protect the transformer, but to take it off the line in case of internal or secondary shorts.

Outdoor transformers have tanks which are designed to be used in outdoor conditions. These units may be compound- or oil-filled.

Indoor units must be kept clean and dry, as they must not be subjected to weather. Porcelain bushings must be kept clean (Art. 1a). Secondary connections must be kept clean and tight (Art. 1b).

Units above 12 kV are usually liquid-filled and should have maintenance as outlined in Art. 6.

4. Current Transformers.   A current transformer (CT) is a small transformer used to produce a secondary current which is in a given ratio to the primary current, and to supply this secondary current at a low-voltage value. Current transformers are used for metering or controlling.

Current transformers used for metering will not have taps, as a fixed primary-secondary current relationship is always required. When full-load current is flowing in the primary, the secondary current is usually 5 A rated. The nameplate will specify primary and secondary currents and the ratio, which is primary current/secondary current.

Current transformers used for control will usually have taps in the secondary windings, so that the secondary current can be a varying function of the primary current. The taps might be from 3.5 to 7.5 A, when full load is flowing in the primary circuit.

Another important rating is the primary insulation level. This is the insulation level for which the primary winding is insulated. A current transformer may be used on lower voltages than its rating, but never on a circuit higher than its insulation level.

Metering transformers are more expensive than control transformers as metering transformers have carefully designed constants to reduce the primary-secondary phase angle to a minimum.

Caution. The secondary circuits of current transformers must never be opened and are never fused, as dangerously high voltage will appear across the secondary leads of an open-circuited current transformer when load current is flowing in the primary.

Current transformers are constructed for outdoor or indoor use. For safety, the case or mounting structure should be solidly grounded, and usually one side of the secondary connection is grounded.

Polarity Markers.   Primary and secondary leads are marked to show the relative instantaneous polarities of primary and secondary windings. A white dot or button may be used on one primary and one secondary lead, or one primary lead may be lettered H1 and the corresponding secondary lead lettered X1. When replacing current transformers, polarity markers should be observed.

Outdoor Units.   These units may be compound- or oil-filled. All units above 12 kV are usually oil-filled and should receive maintenance as outlined in Art. 6. Primary bushings must be kept clean and all connections tight (Art. 1a and b).

Indoor units must be kept clean and dry and all connections tight.

5. Arresters.   Arresters are used to drain power lines of steep-wave, high-voltage surges, and to divert such surges to ground to prevent them from entering windings which may produce a winding failure. These steep-wave surges are produced by direct or indirect lightning strokes or by switching. The voltage magnitude of these surges may be many times normal voltage. The duration is usually less than 50 μs, with the steep-wave portion in the order of $1\frac{1}{2}$ to 5 μs.

Underground cables and their associated equipment are also subject to steep-wave surges produced by switching or induced strokes on the overhead section of the lines. Underground cables must also be protected against surges by the application of arresters.

Arresters are connected from lines to a good solid ground, and as near the apparatus to be protected as possible. Long lines frequently have arresters scattered along their length to drain the line of surges, without having all line insulators subjected to the surge.

Years ago choke coils and a gap were used to block surges and direct them to ground through the gap. Power current would frequently follow the surge current, and the lines would become momentarily grounded until the arc finally broke after rising in the gap horns. Modern arresters will allow the high-frequency steep-wave surges to pass to ground but will interrupt any low-frequency power current. Some arresters have built-in gaps, permanently set at the factory, and others have external gaps which must be adjusted for factory recommendations.

The only maintenance required is to keep the insulators clean and connections tight and a good solid ground on the grounded side of the arresters. Inspection consists of checking porcelains for cracks or damage to the porcelain glazing due to flashover.

Special testing techniques are used for checking the adequacy of the grounds. Such testing is usually beyond the scope of the average plant maintenance service.

Oil-filled transformers, dry-type transformers, and rotating equipment have different basic-impulse-level (BIL) ratings for a given voltage rating. Careful engineering should be applied in selecting and replacing arresters for the above equipment.

6. Transformers. In the original plant-planning process, the question should always be raised as to the purchase of primary power with company-owned transformers and associated equipment, or the purchase of power at secondary voltages with the corresponding higher energy rate.

Transformer selection and maintenance is considered on the basis of common transformer items, liquid-filled transformers, and dry-type transformers.

a. Common Transformer Items

1. Single-phase vs. 3-phase transformers. Some plants purchase one 3-phase transformer and hope they never have a failure; or if they do, they try to borrow a similar unit from the local utility. If a 3-phase transformer fails, it is out of service until it is repaired or a replacement is secured.

Other plants purchase four single-phase units, place three in a bank, and have the fourth for a spare. Such a system gives the greatest reliability factor. If one phase of a three-single-phase bank fails, a spare single-phase unit can quickly be connected. If no spare is available, and if the bank is connected delta-delta, the failed unit can be removed and the bank can operate in open delta at a reduction in rating of 58% of the bank rating. Or a plant might have a number of substations, and should one bank fail, switching equipment would isolate the failed bank and all load could be carried by the remaining substations.

If transformers are suitably supplied, properly protected, and loaded within manufacturer's recommendations, they are probably the most trouble-free of any electrical equipment. However, it is still a maintenance problem, and maintenance people should be prepared to make recommendations in case of a transformer failure.

2. Bushings (Art. 1a). The bushings must be sealed to prevent oil from escaping the tank, or to prevent air or moisture from entering the tank in the breathing process.

3. No-load tap-changer switch. Taps may be provided in the primary or secondary windings. These taps must never be changed unless the transformer is deenergized. The taps change the primary-to-secondary winding ratio, and thus select another voltage within its range. On dry-type units, the tap changer can be examined for signs of heating. Usually no such examination is possible on liquid-filled units.

All units in the bank must have the same tap setting when energized. The packing gland around the tap-changer shaft must be sealed against oil leaks, or against entrance of moisture on liquid-filled transformers.

4. Arresters. All transformers, both primary and secondary, should have arresters applied as close to the terminals as possible (Art. 5).

5. Connections. All connections should be examined occasionally for signs of heating due to corrosion (Art. 1b).

6. Grounding. All tanks and cases should be solidly grounded.

7. Parallel operation. Transformers will operate in parallel successfully, and share the load in proportion to their size, provided they have the

Same percent impedance ($\pm$ 7^1/$_2$% of nameplate impedance)

Same primary and secondary voltage

Same polarity (single-phase) or same vector rotation (3-phase)

Same tap position

b. Liquid-filled Units

1. These units are usually filled with a high grade of petroleum product, commonly called transformer oil. This oil serves the dual purpose of conducting the coil heat to the tank surface for dissipation, and also it is an excellent dielectric, and its dielectric strength is incorporated into the design of internal electrical clearances. For proper cooling, the level should never be allowed to drop, and for proper insulation, the oil must never be allowed to become contaminated, particularly with water, as a few parts per million of water can ruin its insulation value. Further, if the oil level drops, certain live parts might be out of the oil and flash to tank or other parts.

Transformer oil will burn if subjected to proper conditions. Such units used indoors must usually be placed in special vaults to meet various codes. A nonflammable liquid, "askarel," has been developed for transformer cooling and insulating. Each manufacturer of askarel transformers has applied a special name for this material.

Transformer rooms or vaults for either oil or askarel units should be ventilated to prevent buildup of dangerous gases in the room due to normal operation or possible arcing. Transformer oil is lighter than water; so any free water will sink to the bottom if allowed to settle. Askarel, however, is heavier than water, so that any free water will rise to the top if allowed to settle. Transformer-oil samples are therefore taken from the bottom, and askarel samples are taken from the top. Minute water droplets may stay for years in suspension in either oil or askarel.

2. Cooling. Liquid-filled units may be cooled in a number of ways. A "self-cooled" unit is one which can dissipate sufficient heat by convection by its tank and radiators. By the addition of fans, its cooling and its consequent rating can be increased and is called "forced air." Its rating can be further increased by forcing cooled oil through the unit. Such a unit is called "forced-oil-cooled." The oil may be cooled by a water-cooled heat exchanger. Some units have two or all three methods of cooling and usually have that many ratings. Obviously, the fans must come on at certain temperature ratings, and the forced-oil pump must be energized at higher values. The cooling water must have a predetermined flow through the heat exchanger, and the incoming cooling water must not exceed a maximum temperature.

A number of transformers were made "water-cooled," with internal piping through which water was allowed to flow or be pumped. Specifications state the flow and temperature of the cooling water. Precautions must be taken in colder climates to prevent the water from freezing, with the possibility of breaking tubes.

Any unit except the self-cooled must have suitable controls to prevent the transformer from reaching excessive termperature, in the event of water, fan, or pump failure.

In emergencies, water has been sprayed on the sides of a transformer for additional cooling in case of emergency overloads or certain failures of cooling equipment.

3. Liquid gages. Such gages are usually installed on liquid-filled transformers. They are calibrated at cold temperatures, as all cooling mediums expand at elevated temperatures. The liquid gages may have contacts to sound an alarm or to deenergize the unit in case of low level.

4. Temperature gages. Such gages are usually installed on liquid-filled transformers. In case of several ratings, several different calibrated gages bring on additional cooling as required. A top-rated temperature gage may sound alarms or shut down the unit.

5. Breathing. Between maximum cold and maximum hot, liquid volume may change over 5%. This volume change varies the airspace over the liquid, and the air, if allowed to move in and out, is called breathing. If breathing is not allowed to occur, the internal pressure will change, varying from positive to possibly negative pressure.

Many original transformers were sealed with felt gaskets which readily allowed breathing. Hot oil, in contact with air, however, caused sludging, producing a mild acid which would attack winding insulation and shorten its life. It was usually necessary to filter the oil or replace it every few years. In addition, when breathing in, moist air would be drawn into the unit, which would condense on the walls and inside cover and eventually would drop into the oil, thus lowering its dielectric strength.

Later models were sealed but would breathe through a conservator tank. This conservator tank was divided into two sections, but piped together. One side was allowed contact with the internal transformer oil, and the other side was allowed contact with the outside air. In this manner only half of the conservator-tank oil would become contaminated with moisture and sludge. The contaminated oil could be scrapped without too much effort or expense.

Later a breathing device was placed on sealed units. The unit readily expelled air, but the intake was filtered through a moisture-absorbing unit with silica gel or similar material. When the moisture-absorbing material reached moisture saturation, it showed a color change through an observation glass. The main problem was that most maintenance departments forgot about it and failed to dry or replace the breathing material. Eventually the unit would become a regular free breather.

Still later, the tanks were designed to stand more pressure, and the unit would breathe out at pressures, perhaps above 6 to 8 lb/in². When cooling, and the pressure dropped to

2 to 3 lb/in², a dry-nitrogen bottle would allow nitrogen to enter the tank, always keeping the tank pressure above atmospheric. This system works beautifully if the inlet-outlet gages are correctly set and if the nitrogen bottle is not allowed to go empty. The nitrogen gage frequently has a pressure switch on the primary-pressure side which would sound an alarm when the primary pressure dropped to about 100 lb/in².

The latest design in certain size units is a tank built to withstand rather high pressure, perhaps 10 to 12 lb/in², and a large vacuum. The unit is completely sealed and allowed to fluctuate between these values. If the tank bushings and all tank openings are sealed, there is absolutely no breathing and no chance for oxidation or moisture absorption.

6. Liquid tests. There are perhaps 15 to 20 items in most factory specifications dealing with the quality of transformer oil. However, most utilities and industrial users consider only two tests on routine oil testing necessary. These tests are for dielectric strength and acidity tests. Dielectric testing consists of placing approximately 0.5 pt of liquid in a special cup, having two electrodes 1 in. in diameter with a gap spacing of 0.1. Voltage is raised across the gap by a variable-voltage transformer at a uniform rate of 3 kV/s, until breakdown occurs, at which time the voltage is read. Five consecutive readings are averaged for the breakdown voltage.

Acidity tests consist of neutralizing acids in the oil to obtain a certain color. Readings are expressed in milligrams per liter of neutralization. Color and smell are usually functions of acid in the oil. The frequency of such tests depends upon the breathing method of the transformer. A free breather should have its oil tested yearly, a completely sealed unit perhaps every 5 years. Utilities are not in close agreement as to maximum limits of acidity or minimum limits of dielectric. Personal judgment and unit importance usually set the criteria.

7. Tests. Routine inspections should be performed on a monthly basis as far as observations are concerned. Routine tests involve liquid tests and winding megger readings. The frequency depends upon the type of breathing.

c. Dry-Type Units

1. Cooling. Dry-type units are usually self-cooled. They usually have a much higher temperature rise than liquid-filled units. The main problem experienced in cooling is when later construction closes off case ducts and restricts free circulation of air.

Some smaller special units are water-cooled, with the tubes embedded in secondary windings. Care must be exercised to turn off the cooling water when not in use; otherwise condensation will take place on the windings and insulation failure is apt to occur.

Some older units used forced-air cooling which incorporates hot-air temperature switches and cooling fans. With the advent of higher-temperature insulation, forced-air cooling of dry-type transformers has almost disappeared.

Another fairly recent modification of the dry-type transformer is the gas-filled dry type. The unit is placed in a sealed steel tank, all air is evacuated, and then it is filled with an inert gas. The gas conducts the heat from the coils to the tank surface, and the gas (or lack of air) serves as an insulating medium. This construction completely eliminates the contamination-moisture problem as found in the usual dry-type unit. A pressure gage shows the gas pressure on the unit, which must be maintained for successful operation.

2. Inspection. Routine inspections of dry-type units should involve inspection for cleanliness and megger readings. If dirt is allowed to build up on insulations, it will impair cooling and insulation levels. The longer the accumulation of dirt is allowed to remain on the windings, the more it bakes on and becomes more difficult to clean. The dirt also has an affinity for surface moisture, which is a detriment to all dry-type transformers.

The rule for dry-type transformers is to keep them dry and clean and the connections tight. If these simple operations are performed, routine megger readings will usually be high and satisfactory.

7. **High-Voltage Oil or Air Circuit Breakers.** High-voltage circuit breakers are series switches used primarily to disconnect the load circuit under predetermined faults or conditions. They may be applied to a single load or to a line or a transformer bank. They may be in a 3-phase bank, or three single-phase units, mechanically tied together to operate as a gang switch. They may be of outdoor construction (watertight) or of in-

door construction (not watertight). Transformer oil is used for insulation and to help quench the arc on oil circuit breakers. Arc chutes or arc boxes are usually used on all air circuit breakers (ACB) and higher-voltage oil units. Cooling is seldom a problem.

The contacts may be closed magnetically, with pneumatic motor operation, or by hand. They are usually tripped electrically by a tripping circuit. The tripping circuit can be controlled by overcurrent, low or no voltage, single phasing, overvoltage, manually, or by any other fault or undesirable operating condition.

The tripping circuit may be supplied by the load circuit, by a battery bus, or by capacitor-stored energy. The battery bus system is considered the most reliable.

High-speed breakers for industrial use may trip within a few cycles of operation. The longest delay may be the time of closing of the tripping contacts.

Oil-circuit-breakers (OCB) maintenance involves the following:

1. Bushings (Art. 1*a*).
2. Mechanical linkage (Art. 1*c*).
3. Auxiliary contacts (Art. 1*d*).
4. Main contacts (Art. 1*e*)
5. Tank liner. This insulating liner gives additional internal electrical clearances to ground. Liners must be carbon-free, clean, and dry. Drying (in case of moisture) is usually beyond the scope of usual plant maintenance.
6. Transformer oil (Art. 6*b*1).
7. Closing and tripping mechanism (see factory instruction manual). Checking, testing, and adjusting are usually beyond the scope of plant maintenance owing to the special test equipment required.
8. Maintenance schedule. This is usually a yearly schedule, depending upon the severity of service. After a severe fault a special inspection schedule should be made.

8. Capacitors. Capacitors are used for power-factor correction, or to help regulate voltage on power lines. When used for power-factor correction, they are usually tied permanently across the load to be power-factor-corrected. When used to help regulate line voltage on power lines, they are usually connected by an oil switch, controlled by a voltage-sensitive relay. When the capacitors are connected, they absorb magnetizing current, keeping the magnetizing current from flowing in the lines, and thus reduce line drop.

Modern capacitors are completely sealed and are applicable for outdoor use.

Capacitor banks should be fused, and frequently have individual fuses per tank. Maintenance consists of:

1. Bushings (Art. 1*a*).
2. Connections. Check all line connections for tightness (Art. 1*b*).
3. Grounds. Cases are usually grounded, but this connection depends upon the wiring diagram. Some capacitors use the case as a terminal connection. The wiring diagram should be carefully followed.
4. Capacitor faults. Capacitors are subject to opens, grounds, and shorts. Fuses will clear grounds and faults from the line. Grounded or shorted units can usually be detected by bulged tanks.
5. Open circuits. Opens or partial opens are checked by current measurements, comparing one bank or one tank with other banks or tanks. Calculations can also be made to determine current values. Line voltages may be checked and compared.

Open capacitors (individual units or banks) will cause a voltage unbalance because of their failure to absorb lagging power-factor currents. This unbalance voltage can severely upset the operation of 3-phase motors, and has caused winding burnouts.

6. Case inspections. Cases and bushing seals should be carefully inspected for signs of leaking. Leaking units should be removed, repaired, or replaced.

While modern capacitors usually have an internal bleeding resistance to drain stored energy when disconnected from its power source, no capacitor or its line should be handled unless it has been shorted and grounded by grounding chains.

9. Overhead Power Lines. Probably the greatest hazard that high-voltage overhead lines in industrial premises are subject to is mechanical damage. Some industrial plants have painted the lower section of poles and guys with an "alert orange or yellow." While this may not aesthetically beautify surroundings, it seems to keep trucks and other equipment from knocking down poles and loosening guys. Large signs should call attention to overhead lines crossing roadways, especially with low-overhead-clearance lines.

Apart from the mechanical maintenance, insulators should be kept clean (Art. 1*a*) and connections tight (Art. 1*b*). All too frequently, plant maintenance overlooks line maintenance, until the line flashes to ground and becomes inoperative. If lines have arresters, they should be checked as well as the arrester grounding (Art. 5). Sectionalizing no-load disconnect switches should be examined (Art. 1).

While utilities frequently work on their lines while "hot," their personnel are highly trained and have special equipment. Industrial-plant maintenance departments usually are not so equipped.

Overhead lines should be disconnected at both ends, disconnect switches locked open, and both ends of the lines be grounded and shorted before any line maintenance is started. Overhead-line capacitance will store considerable energy between lines even after disconnect switches are opened, hence the shorting and grounding procedure.

10. Potheads.   Potheads make the connection between underground cables or conduit and open lines. Potheads consist of the case, bushings, cable-entrance seal, internal connection between the cables and bushing studs, and some form of insulation around the internal connection. This insulation may be sealed dry air, petrolatum, or oil. If gasket seals, bushing gaskets, and solder seals at the cable entrance remain airtight, little deterioration will occur in this insulating medium (Art. 1*a* and *b*).

The insulation within the pothead is checked along with the cables at their regular maintenance schedule (Art. 11). If this insulation is deteriorating, it will show up in the cable (phase-to-phase) testing.

11. Underground Cables.   Underground cables are used to improve the aesthetic appearance of power distribution, and in certain cases under buildings where overhead lines would prove to be a hazard.

Underground cables may or may not have a ground sheath. The cables may be laid in steel or tile conduit, or laid directly in a trench. Both ends (internal and external) usually terminate in potheads.

Underground cables, owing to their high dielectric value and high capacitance, may store considerable energy even after they are disconnected. Safety precautions should be followed as outlined in Art. 9.

Great care must be exercised while excavating, because of the extreme danger of puncturing a live cable and to prevent damage to underground cables. Telephone company practice of clearly marking underground-cable routes should be followed.

Except for potheads, no maintenance is usually applied to underground cables unless a cable failure occurs. Cable testing should be performed on a scheduled program. The testing consists of applying a high-voltage dc between each cable and ground sheath, or between each cable and all other cables in the bundle, and measuring the small leakage current. Analyzing the test is a matter of cable manufacturer's recommendations, test-equipment manufacturer's recommendations, certain electrical standards, and personal experience. Wide latitude is frequently experienced.

Should certain line-to-line or line-to-ground readings be extreme, the cable potheads should be examined. Special test equipment is required for cable checking, and is usually beyond the scope of the usual industrial maintenance departments.

Careful records should be maintained from each cable test, recording temperature. Previous records should be compared at each new test.

12. High-Voltage and Low-Voltage Switchgear.   Years ago all high- or low-voltage switching and controlling was done in enclosed buildings and suitable wire screens kept personnel away from live parts. Modern designs call for such equipment to be enclosed in steel cubicles, usually referred to as gear. The gear may be weathertight and installed outside, or dustproofed and installed inside. Low-voltage gear is 600 V or lower, and high-voltage gear is 2400 V or higher.

High-voltage gear is frequently connected to the high-voltage side of a transformer, and the low-voltage side is connected directly to low-voltage gear. This arrangement is sometimes referred to as a unit substation when all protection control and transformation is self-contained. High-voltage entrance is usually from underground cables and low-voltage feed is to bus duct. The transformer in the center is termed "throat-connected," to the high- and low-voltage gear.

The high-voltage gear usually contains disconnect switches, oil, air, or vacuum circuit breakers, arresters, potential and current transformers for control or metering, and the primary bushings of the transformer and potheads.

The low-voltage gear usually contains arresters, air circuit breaker, current and potential transformer for control or metering, and the secondary bushings of the transformer. The secondary load may be fed by conduit or by bus duct.

Indoor switchgear is not watertight and free moisture from condensation, drips, leaks, or other source of moisture must not be allowed to settle on the roof.

Fans are sometimes used to move air for cooling or to prevent internal condensation. The air is frequently moved through filters to remove as much contamination as possible. Heaters are sometimes installed to help remove moisture condensation.

The gear, or the vault in which the gear is installed, should always be locked to prevent unauthorized personnel from harming themselves or equipment.

Equipment and corresponding maintenance procedures are as follows:

1. Disconnect switches (Art. 1).
2. Oil circuit breakers (Art. 7).
3. Air circuit breakers (Art. 7). Air circuit breakers use the medium of air as insulation instead of oil. Contact opening is made extremely fast to interrupt power current. In addition, arc chutes or arc boxes contain the arc, break it into many small segments, and quickly extinguish it. High-speed air blasts are sometimes directed at the arc to distort it further for quenching.
4. Oil or air circuit breakers, as used in high-voltage switchgear, can usually be lowered by hand-crank arrangement and/or pulled out on rails for inspection and adjustment. Interlocks require that the breaker must be in the open position before it can be removed.

Maintenance for air circuit breakers is similar to that for oil circuit breakers except for the arc chutes. The arc chutes must be inspected for signs of burning or charring. Some arc chutes are repairable and others are replaceable only. The arc chutes must be in proper place over the main contacts.

5. Vacuum breakers. The use of this type of breaker is fairly recent in the field of circuit closing or interrupting. Consequently, even experienced maintenance men may need additional information concerning these breakers.

At present, electrical ratings range up to the 13 kV class, with full load current values of perhaps 1000 A. The breaker is of the typical draw-out type, mounted on steel frames with wheels and floor channels. Primary disconnects are of the typical cluster contact spring assembly. The closing mechanism is a high-speed device, consisting of a toggle spring actuated to close the contacts quickly, but without high impact to the contacts to prevent wear damage.

The load contacts open the circuit in a vacuum container and current will arc across the gaps until the zero current cycle point is reached. The arc will not be re-established across the gap due to the vacuum surrounding the opening gaps.

Since visual contact inspection is not possible within the vacuum chamber, contact checking is performed by applying a high potential across the open contacts. This high voltage may be ac or dc, and must be raised from zero, or a low voltage, at a uniform rate up to rated test voltage, and held for a specified time.

Inspection also consists of measuring contact erosion, or the distance the contacts move from open to closed position, by means of a spacing gap external to the vacuum chamber. When the contacts have eroded to the maximum amount permissible, the entire gap assembly and vacuum chambers are replaced as a unit. A step-by-step outline of the procedure and checks is contained in the manufacturer's instruction leaflet.

Another periodic inspection consists of measuring the closing and opening times and comparing these to previous tests. This determines any friction or wear or required adjustments in the mechanical closing system. Breaker mechanisms must be kept clean and properly lubricated according to manufacturer's instruction leaflets.

Since the vacuum breakers are relatively new, manufacturer's instructions dealing with installation, inspection, and checking are quite thoroughly detailed, and should be carefully followed.

Special care should be exercised in following the instruction leaflets as to cautions dealing with the safety of personnel and equipment while installing, making inspection checks, and maintenance operations.

6. Bus duct (Sec. 19) consists of bus bar insulated in a steel-enclosed duct. Removable panels allow a circuit breaker to be connected to the bus by bus stabs, which are spring-loaded contacts straddling each bus. The individual circuit breaker can easily be moved to new locations as required. Little maintenance is required or given to bus ducts provided that moisture is not allowed to drip on the bus duct and run inside. The bus stabs have proved to be maintenance-free if installed and rated properly. Occasional maintenance includes meggering of bus insulation and inspection for signs of heating at the bus stabs.

13. **Metering and Controlling Equipment.** It is here, regardless of whether all controls are located on one panel or on a large number scattered throughout the plant, that some of the largest frustrations will occur. It is here a diabolical red glowing lamp indicates that the most important machine has been shut down because of one of a dozen possibilities. It is here the electronic signal is received and converted into another form of action; the tachometer generator output is received, analyzed, and magnified; the zero-speed switch locks out a relay; and where the hydraulic or pneumatic pressure or flow switches remove a machine from the line.

It is here where the usual plant maintenance people have their greatest headaches, if they are not knowledgeable in the theory of operation, adjustment, and troubleshooting of a thousand and one different gadgets, each with the built-in capability of analyzing, changing, adjusting, controlling, recording, and if need be, shutting down the plant or a section of it.

Generating utilities have their own section of relay experts, trained in the technology of testing, calibrating, adjusting, and repairing. Industrial plants usually do not have such personnel available and are frequently dependent upon outside service personnel to furnish this talent.

If this sounds like a hopeless task, it is not. There are a number of things a well-organized maintenance department should do, and if they do, probably 95% of their problems will never occur in the first place. The following rules should be observed:

1. Keep it clean. Watches seldom wear out. They get so dirty they will not run. It is imperative that dirt and dust be kept out of sealed relay cases. When the case cover is removed for adjusting, be sure it is properly resealed. Are cover screws missing? Is the gasket hanging loose? Is the glass case cracked? Many relays are as delicate as watches, but cost considerably more.

2. Keep it dry. Condensing pipes, leaking roofs, broken windows, all contribute to accumulation of moisture. Unless built for outdoor use, controls are usually not moisturetight. Moisture causes coils to fail, corrosion on contacts, springs, and pivots.

3. Eliminate vibration. An unlatched cabinet door may vibrate; a panel with screws missing may pick up a resonant vibration. Swinging cabinet doors should not be slammed. A newly installed machine close by may set up vibrations. The passing of an overhead crane may give the panel a real workout. Industrial relays are not designed to cope with vibration. They are not built to the same standards as navy relays.

4. Keep connections clean and tight. Atmospheric contamination will cause corrosion, especially on loose and moving copper strands. Vibration tends to move or loosen connections. Many electrical impulses are too weak to break down a corroded current path.

5. Do not tinker. Many of us have torn down clocks, but would they run better—or run at all—when reassembled? Usually a number of extra parts were available after reassembly. Many relays are as delicate as a clock mechanism. The testing, repairing, calibrating, and adjusting should be left to the meter-repair technicians.

6. Schedule inspections. On a planned basis, a specialist should test, repair, calibrate, and adjust. If the above suggestions are followed, every 3 years should be adequate. Good judgment should always override, as plant conditions vary tremendously.

7. Maintain a file of instruction leaflets. These leaflets are invaluable, even to a skilled relay technician, as test procedures and values differ from manufacturer to manufacturer and even from model to model of a given manufacturer. Relays are sometimes replaced, simply because calibration data could not be obtained.

Finally, it is well to consider relays and control devices as much like a watch. If it is properly handled, it will give years of service, but under improper handling, inaccuracy and failure will result.

PLANT START-UP

Plant start-ups are usually frustrating for maintenance departments, as all the mistakes made by designers, manufacturers, and installers are brought together and focused upon a new maintenance crew, usually with the result that the original start-up date assigned is missed.

14. Planning. Careful planning can eliminate much of this frustration. After the maintenance crew is trained, and installation problems solved, start-up work can be initiated. The following procedures will greatly reduce maintenance problems:

1. Keep equipment clean during construction periods (Art. 13-1) Dust and dirt must be kept out of most equipment. Tarps or plastic wrappers should be used for this purpose.

2. Keep it dry (Art. 13-2). Do not allow indoor equipment to become moisture-laden during construction-period storage.

3. Maintain a file of instruction leaflets (Art. 13-7), which are usually tied to the equipment by manufacturers, or which are available and used by factory representatives for installing, testing, or calibrating equipment.

4. Keep up with changes in circuits and services which the maintenance department will be required to maintain in the future. Pay particular attention to changes in concealed or underground conduits. Be certain drawings clearly show all changes.

5. As each piece of equipment is installed and tested, make and keep a record of test data for future use.

6. As each piece of equipment is installed and tested, try to determine at that time whether future inspection and maintenance will be performed by the maintenance group or by an outside technical group because of special skills or test or calibrating equipment required.

7. Review each building structure and the equipment it contains to ensure specification requirements are met in regards to enclosure (weatherproof, dusttight, etc.).

8. Before test runs are made, be sure protective equipment is properly wired and ready for use, such as:

Contacts of protective relays are usually blocked to prevent shipping damage.

If current transformers are not mounted and wired in at the factory, the terminals of the terminal board and current transformer are usually jumpered. Be sure the jumpers are removed after proper wiring of the secondaries of CTs.

Check for correct polarity of potential and current transformers.

Time-delay apparatus must be properly set; oil dashpots are usually shipped dry.

Breaker contacts are frequently blocked closed to prevent shipping damage.

Index

1